Practical Plastic Surgery

Zol B. Kryger, M.D.
Division of Plastic Surgery
Northwestern University Feinberg School of Medicine
Chicago, Illinois, U.S.A.

Mark Sisco, M.D.
Division of Plastic Surgery
Northwestern University Feinberg School of Medicine
Chicago, Illinois, U.S.A.

Austin, Texas
U.S.A.

VADEMECUM
Practical Plastic Surgery
LANDES BIOSCIENCE
Austin, Texas U.S.A.

Please address all inquiries to the Publisher:
Landes Bioscience, 1002 West Avenue, 2nd Floor, Austin, Texas 78701, U.S.A.
Phone: 512/ 637 6050; FAX: 512/ 637 6079

ISBN: 978-1-57059-696-4

Library of Congress Cataloging-in-Publication Data

A C.I.P. Catalogue record for this book is available from the Library of Congress.

Dedication

This book is dedicated to the students and house-staff whose interest and commitment to learning plastic surgery inspired us to create a practical guide to help them along the way.

Z.B.K. and M.S.

Contents

Foreword xix

Preface xxi

Section I: General Principles

1. **Wound Healing and Principles of Wound Care** 1
   Leonard Lu and Robert D. Galiano

2. **Basic Concepts in Wound Repair** 4
   Zol B. Kryger and Michael A. Howard

3. **Dressings** 12
   Anandev Gurjala and Michael A. Howard

4. **Pharmacologic Wound Care** 20
   Peter Kim and Thomas A. Mustoe

5. **Negative Pressure Wound Therapy** 24
   Peter Kim and Gregory A. Dumanian

6. **Leeches** 27
   Mark Sisco

7. **Local Anesthetics** 29
   Zol B. Kryger and Ted Yagmour

8. **Basic Anesthetic Blocks** 33
   Zol B. Kryger

9. **Surgery under Conscious Sedation** 38
   Zol B. Kryger and Neil A. Fine

10. **Principles of Reconstructive Surgery** 43
    Constance M. Chen and Robert J. Allen

11. **Principles of Surgical Flaps** 49
    Constance M. Chen and Babak J. Mehrara

12. **Microvascular Surgical Technique and Methods of Flap Monitoring** 56
    Sean Boutros and Robert D. Galiano

13. **Tissue Expansion** 61
    Zol B. Kryger and Bruce S. Bauer

14. **Alloplastic Materials** 65
    Jason H. Ko and Julius Few

Section II: The Problematic Wound

15. The Chronic Infected Wound and Surgical Site Infections 71
Kevin J. Cross and Philip S. Barie

16. Diabetic Wounds 80
Roberto L. Flores and Michael S. Margiotta

17. Wounds Due to Vascular Causes 87
Kevin J. Cross and Robert T. Grant

18. Radiated Wound and Radiation-Induced Enteric Fistulae 95
Russell R. Reid and Gregory A. Dumanian

19. Pressure Ulcers 100
Zol B. Kryger and Victor L. Lewis

20. Infected and Exposed Vascular Grafts 111
Mark Sisco and Gregory A. Dumanian

21. Management of Exposed and Infected Orthopedic Prostheses 114
Mark Sisco and Michael A. Howard

Section III: Integument

22. Hypertrophic Scars and Keloids 117
Zol B. Kryger

23. Benign Skin Lesions 121
Zol B. Kryger

24. Basal Cell and Squamous Cell Carcinoma 126
Darrin M. Hubert and Benjamin Chang

25. Melanoma 131
Gil S. Kryger and David Bentrem

26. Vascular Anomalies 138
Robert D. Galiano and Geoffrey C. Gurtner

27. Skin Grafting and Skin Substitutes 145
Constance M. Chen and Jana Cole

28. Burns: Initial Management and Resuscitation 154
Baubak Safa

Section IV: Head and Neck

29. Head and Neck Cancer 163
Zol B. Kryger

30. Ear Reconstruction 168
Amir H. Taghinia, Theodore C. Marentis, Ankit I. Mehta, Paul Gigante and Bernard T. Lee

31. Eyelid Reconstruction 178
Amir H. Taghinia and Bernard T. Lee

32. Nasal Reconstruction 188
Clark F. Schierle and Victor L. Lewis

33. Lip Reconstruction 193
Amir H. Taghinia, Edgar S. Macias, Dzifa S. Kpodzo and Bohdan Pomahac

34. Mandible Reconstruction 211
Patrick Cole, Jeffrey A. Hammoudeh and Arnulf Baumann

35. The Facial Nerve and Facial Reanimation 216
Zol B. Kryger

36. Frontal Sinus Fractures 225
Joseph Raviv and Daniel Danahey

37. Orbital Fractures 231
John Nigriny

38. Fractures of the Zygoma and Maxilla 237
Zol B. Kryger

39. Nasal and NOE Fractures 243
Clark F. Schierle and Victor L Lewis

40. Mandible Fractures 248
Jeffrey A. Hammoudeh, Nirm Nathan and Seth Thaller

Section V: Trunk and Lower Extremity

41. Breast Disease and Its Implications for Reconstruction 258
Kristina D. Kotseos and Neil A. Fine

42. TRAM Flap Breast Reconstruction 263
Amir H. Taghinia, Margaret L. McNairy and Bohdan Pomahac

43. Latissimus Flap Breast Reconstruction 274
Roberto L. Flores and Jamie P. Levine

44. Tissue Expander Breast Reconstruction 278
Timothy W. King and Jamie P. Levine

45. Nipple Reconstruction and Tattooing 283
Kristina D. Kotseos and Neil A. Fine

46. Reduction Mammaplasty 288
Timothy W. King and Jamie P. Levine

47. Sternal Wounds 294
Jonathan L. Le and William Y. Hoffman

48. Chest Wall Defects 299
Jason Pomerantz and William Hoffman

49. Coverage of Spinal Wounds 304
Jason Pomerantz and William Hoffman

50. Abdominal Wall Defects 309
Mark Sisco and Gregory A. Dumanian

51. Pelvic, Genital and Perineal Reconstruction 314
Mark Sisco and Gregory A. Dumanian

52. Lower Extremity Reconstruction 318
Mark Sisco and Michael A. Howard

Section VI: Craniofacial Surgery

53. Basic Dental Concepts 324
Mark Sisco and Jeffrey A. Hammoudeh

54. Cephalometrics 327
Matthew Jacobsen and Jeffrey A. Hammoudeh

55. Craniofacial Syndromes and Craniosynostosis 334
Zol B. Kryger and Pravin K. Patel

56. Craniofacial Microsomia 340
Zol B. Kryger

57. Microtia Repair 343
Zol B. Kryger

58. Cleft Lip 348
Alex Margulis

59. Cleft Palate 356
Alex Margulis

Section VII: Aesthetic Surgery

60. Rhytidectomy 368
Stephen M. Warren and James W. May, Jr.

61. Browlift 377
Clark F. Schierle and John Y.S. Kim

62. Otoplasty 382
Clark F. Schierle and Victor L. Lewis

63. Blepharoplasty 386
Robert T. Lancaster, Stephen M. Warren and Elof Eriksson

64. Rhinoplasty 394
Ziv M. Peled, Stephen M. Warren and Michael J. Yaremchu

65. Genioplasty, Chin and Malar Augmentation 400
Jeffrey A. Hammoudeh, Christopher Low and Arnulf Baumann

66. Augmentation Mammaplasty 406
Richard J. Brown and John Y.S. Kim

67. Gynecomastia Reduction 413
Richard J. Brown and John Y.S. Kim

68. Mastopexy 417
Richard J. Brown and John Y.S. Kim

69. Abdominoplasty 423
Amir H. Taghinia and Bohdan Pomahac

70. Liposuction 430
Zol B. Kryger

71. Laser Resurfacing 433
Keren Horn and Jerome Garden

72. Chemical Rejuvenation of the Face 440
Keren Horn and David Wrone

73. Fat Injection and Injectable Fillers 446
Darrin M. Hubert and Louis P. Bucky

74. Cosmetic Uses of Botulinum Toxin 451
Leonard Lu and Julius Few

75. Dermabrasion 454
Zol B. Kryger

76. Hair Restoration 458
Anandev Gurjala

Section VIII: The Hand and Upper Extremity

77. Anatomy of the Hand 462
Zol B. Kryger

78. Radiographic Findings 468
Zol B. Kryger and Avanti Ambekar

79. Examination of the Hand and Wrist 476
Zol B. Kryger

80. Soft Tissue Infections 481
Zol B. Kryger and Hongshik Han

81. Compartment Syndrome of the Upper Extremity 486
Zol B. Kryger

82. Replantation 492
Zol B. Kryger

83. Fractures of the Distal Radius and Ulna 497
Craig Birgfeld and Benjamin Chang

84. Wrist Fractures 502
Gil Kryger

85. Finger and Metacarpal Fratures 512
Oliver Kloeters and John Y.S. Kim

86. Brachial Plexus Injuries 518
Mark Sisco and John Y.S. Kim

87. Nerve Injuries 523
Zol B. Kryger

88. Vascular Trauma 529
Zol B. Kryger

89. Extensor Tendon Injuries 533
Zol B. Kryger

90. Flexor Tendon Injuries 539
Zol B. Kryger

91. Injuries of the Finger 547
Millicent Odunze and Gregory A. Dumanian

92. Soft Tissue Coverage 561
Hongshik Han

93. Carpal Tunnel Syndrome 566
David S. Rosenberg and Gregory A. Dumanian

94. Cubital Tunnel Syndrome **574**
David Rosenberg and Gregory A. Dumanian

95. Trigger Finger Release **580**
Hakim Said and Gregory A. Dumanian

96. Ganglion Cysts **583**
Hakim Said and Thomas Wiedrich

97. Stenosing Tenosynovitis **586**
Zol B. Kryger

98. Radial Artery Harvest **589**
Zol B. Kryger and Gregory A. Dumanian

99. Common Anomalies of the Hand and Digits **592**
Zol B. Kryger

100. Dupuytren's Disease **597**
Oliver Kloeters and John Y.S. Kim

101. Reflex Sympathetic Dystrophy **601**
Zol B. Kryger and Gregory A. Dumanian

Appendix I—

Part A: Important Flaps and Their Harvest **607**
Zol B. Kryger and Mark Sisco

Groin Flap 607
Rectus Abdominis Flap 611
Fibula Composite Flap 612
Pectoralis Major Flap 614
Latissimus Dorsi Flap 615
Serratus Flap 616
Omental Flap 617
Gracilis Flap 618
Radial Forearm Flap 619
Gluteus Flap 620
Anterolateral Thigh (ALT) Flap 621

Part B: Radial Forearm Free Flap **622**
Peter Kim and John Y.S. Kim

Appendix II—Surgical Instruments **625**
Zol B. Kryger and Mark Sisco

Index **633**

Editors

Zol B. Kryger, M.D., Resident
Division of Plastic Surgery
Northwestern University Feinberg School of Medicine
Chicago, Illinois, U.S.A.

Mark Sisco, M.D., Resident
Division of Plastic Surgery
Northwestern University Feinberg School of Medicine
Chicago, Illinois, U.S.A.

Contributors

Robert J. Allen, M.D.,
Professor and Chief
Division of Plastic Surgery
Louisiana State University
New Orleans, Louisiana, U.S.A.
and
The Center for Microsurgical Breast
Reconstruction
Mount Pleasant, South Carolina, U.S.A.
Chapter 10

Avanti Ambekar, M.D., Fellow
in Musculoskeletal Imaging
Department of Radiology
UCSF School of Medicine
San Francisco, California, U.S.A.
Chapter 78

Philip S. Barie, M.D.,
Professor of Surgery and Chief
Division of Critical Care and Trauma
Weill Medical College
Cornell University
New York, New York, U.S.A.
Chapter 15

Bruce S. Bauer, M.D., Professor
Division of Plastic Surgery
Northwestern University
Feinberg School of Medicine
and
Chief, Division of Plastic Surgery
Children's Memorial Hospital
Chicago, Illinois, U.S.A.
Chapter 13

Arnulf Baumann, M.D., D.D.S., Ph.D.,
Resident
Division of Plastic Surgery
University of Miami
Miami, Florida, U.S.A.
Chapters 34, 65

David Bentrem, M.D.,
Assistant Professor
Department of Surgery
Division of Surgical Oncoloty
Northwestern University
Feinberg School of Medicine
Chicago, Illinois, U.S.A.
Chapter 25

Craig Birgfeld, M.D., Resident
Division of Plastic Surgery
University of Pennsylvania
Medical School
Philadelphia, Pennsylvania, U.S.A.
Chapter 83

Sean Boutros, M.D.
Houston Plastic and Cranofacial Surgery
Houston, Texas, U.S.A.
Chapter 12

Richard J. Brown, M.D., Resident
Department of Surgery
Northwestern University
Feinberg School of Medicine
Chicago, Illinois, U.S.A.
Chapters 66-68

Louis P. Bucky, M.D.,
Assistant Professor
Division of Plastic Surgery
University of Pennsylvania
Medical School
Philadelphia, Pennsylvania, U.S.A.
Chapter 73

Benjamin Chang, M.D.,
Associate Professor
Division of Plastic Surgery
University of Pennsylvania
Medical School
Philadelphia, Pennsylvania, U.S.A.
Chapters 24, 83

Constance M. Chen, M.D., M.P.H.,
Resident
Division of Plastic Surgery
New York-Presbyterian Hospital
New York, New York, U.S.A.
Chapters 10, 11, 27

Jana Cole, M.D., Assistant Professor
Division of Plastic Surgery
University of Washington
Seattle, Washington, U.S.A.
Chapter 27

Patrick Cole, M.D., Resident
General Surgery
University of Miami/JMH
Miami, Florida, U.S.A.
Chapter 34

Kevin J. Cross, M.D., Resident
Division of Plastic Surgery
Weill Medical College
Cornell University
New York, New York, U.S.A.
Chapters 15, 17

Daniel Danahey, M.D. Ph.D.,
Assistant Professor
Department of Otolaryngology
Northwestern University
Feinberg School of Medicine
Chicago, Illinois, U.S.A.
Chapter 36

Gregory A. Dumanian, M.D.,
Associate Professor
Division of Plastic Surgery
Northwestern University
Feinberg School of Medicine
Chicago, Illinois, U.S.A.
Chapters 5, 18, 20, 50, 51, 91, 93-95, 98, 101

Elof Eriksson, M.D. Ph.D.,
Professor and Chief
Division of Plastic Surgery
Brigham and Women's Hospital
Harvard Medical School
Boston, Massachusetts, U.S.A.
Chapter 63

Julius Few, M.D., Assistant Professor
Division of Plastic Surgery
Northwestern University
Feinberg School of Medicine
Chicago, Illinois, U.S.A.
Chapter 14, 74

Neil A. Fine, M.D., Associate Professor
Division of Plastic Surgery
Northwestern University
Feinberg School of Medicine
Chicago, Illinois, U.S.A.
Chapters 9, 41, 45

Roberto L. Flores, M.D., Resident
NYU Medical Center
Institute of Reconstructive
Plastic Surgery
New York, New York, U.S.A.
Chapters 16, 43

Robert D. Galiano, M.D.,
Assistant Professor
Division of Plastic Surgery
Northwestern University
Feinberg School of Medicine
Chicago, Illinois, U.S.A.
Chapters 1, 12, 26

Jerome Garden, M.D., Professor
Department of Dermatology
Northwestern University
Feinberg School of Medicine
Chicago, Illinois, U.S.A.
Chapter 71

Paul Gigante, B.S., Medical Student
Harvard Medical School
Boston, Massachusetts, U.S.A.
Chapter 30

Robert T. Grant, M.D.,
Associate Professor and Chief
Division of Plastic Surgery
Weill Medical College
Cornell University
New York, New York, U.S.A.
Chapter 17

Anandev Gurjala, M.D., Resident
Division of Plastic Surgery
Northwestern University
Feinberg School of Medicine
Chicago, Illinois, U.S.A.
Chapters 3, 76

Geoffrey C. Gurtner, M.D.,
Associate Professor
Department of Plastic Surgery
Stanford University School of Medicine
Stanford, California, U.S.A.
Chapter 26

Jeffrey A. Hammoudeh, M.D. D.D.S.,
Resident
Division of Plastic Surgery
University of Miami/JMH
Miami, Florida, U.S.A.
Chapters 34, 40, 53, 54, 65

Hongshik Han, M.D.
Private Practice
Fresno, California, U.S.A.
Chapters 80, 92

Peter E. Hoepfner, M.D.,
Assistant Professor
Departmnet of Orthopedic Surgery
Northwestern University
Feinberg School of Medicine
Chicago, Illinois, U.S.A.
Chapters 84, 87, 89, 90

William Y. Hoffman, M.D.,
Professor and Chief
Division of Plastic
and Reconstructive Surgery
UCSF School of Medicine
San Francisco, California, U.S.A.
Chapters 47-49

Keren Horn, M.D., Resident
Department of Dermatology
Northwestern University
Feinberg School of Medicine
Chicago, Illinois, U.S.A.
Chapters 71, 72

Michael A. Howard, M.D.,
Assistant Professor
Division of Plastic Surgery
Northwestern University
Feinberg School of Medicine
Chicago, Illinois, U.S.A.
Chapters 2, 3, 6, 21, 52

Darrin M. Hubert, M.D., Resident
Division of Plastic Surgery
University of Pennsylvania
Medical School
Philadelphia, Pennsylvania, U.S.A.
Chapters 24, 73

Matthew Jacobsen, D.M.D., M.D.
Division of OMFS
Massachusetts General Hospital
Boston, Massachusetts, U.S.A.
Chapter 54

John Y.S. Kim, M.D., Assistant Professor
Division of Plastic Surgery
Northwestern University
Feinberg School of Medicine
Chicago, Illinois, U.S.A.
Chapters 61, 66-68, 81, 82, 85, 86, 88, 97, 99, 100, Appendix IB

Peter Kim, M.D., Resident
Division of Plastic Surgery
Northwestern University
Feinberg School of Medicine
Chicago, Illinois, U.S.A.
Chapters 4, 5, Appendix IB

Timothy W. King, M.D. Ph.D., Resident
NYU Medical Center
Institute of Reconstructive
Plastic Surgery
New York, New York, U.S.A.
Chapters 44, 46

Oliver Kloeters, M.D., Resident
Department of Hand, Plastic
and Reconstructive Surgery
BG-Burn and Trauma Center
University of Heidelberg
Ludwigshafen, Germany
Chapters 85, 100

Jason H. Ko, M.D., Resident
Division of Plastic Surgery
Northwestern University
Feinberg School of Medicine
Chicago, Illinois, U.S.A.
Chapter 14

Kristina D. Kotseos, M.D., Resident
Division of Plastic Surgery
Northwestern University
Feinberg School of Medicine
Chicago, Illinois, U.S.A.
Chapters 41, 45

Dzifa S. Kpodzo, M.D., Resident
Division of Plastic Surgery
Harvard Medical School
Boston, Massachusetts, U.S.A.
Chapter 33

Gil S. Kryger, M.D., Resident
Department of Plastic Surgery
Stanford University School of Medicine
Stanford, California, U.S.A.
Chapters 25, 84

Zol B. Kryger, M.D., Resident
Division of Plastic Surgery
Northwestern University
Feinberg School of Medicine
Chicago, Illinois, U.S.A.
Chapters 2, 7-9, 13, 19, 22, 23, 29, 35, 38, 55-57, 70, 75, 77-82, 87-90, 92, 97-99, 101, Appendix IA, Appendix II

Robert T. Lancaster, M.D., Resident
Department of Surgery
Massachusetts General Hospital
Boston, Massachusetts, U.S.A.
Chapter 63

Jonathan L. Le, M.D., Resident
Division of Plastic Surgery
UCSF School of Medicine
San Francisco, California, U.S.A.
Chapter 47

Bernard T. Lee, M.D.,
Instructor in Surgery
Division of Plastic Surgery
Beth Israel Deaconess Medical Center
Harvard Medical School
Boston, Massachusetts, U.S.A.
Chapters 30, 31

Jamie P. Levine, M.D., Assistant Professor
NYU Medical Center
Institute of Reconstructive
Plastic Surgery
New York, New York, U.S.A.
Chapters 43, 44, 46

Victor L. Lewis, M.D.,
Professor of Surgery
Division of Plastic Surgery
Northwestern University
Feinberg School of Medicine
Chicago, Illinois, U.S.A.
Chapters 19, 32, 39, 62

Christopher Low, M.D., Resident
General Surgery
University of Miami/JMH
Miami, Florida, U.S.A.
Chapter 65

Leonard Lu, M.D., Resident
Division of Plastic Surgery
Northwestern University
Feinberg School of Medicine
Chicago, Illinois, U.S.A.
Chapters 1, 74

Edgar S. Macias, B.S., Medical Student
Harvard Medical School
Boston, Massachusetts, U.S.A.
Chapter 33

Theodore C. Marentis, M.S.E.E.,
Medical Student
Harvard Medical School
Boston, Massachusetts, U.S.A.
Chapter 30

Michael S. Margiotta, M.D.,
Assistant Professor
NYU Medical Center
Institute of Reconstructive
Plastic Surgery
New York, New York, U.S.A.
Chapter 16

Alex Margulis, M.D., Associate Professor
Department of Plastic Surgery
Haddassah Hospital
Jerusalem, Israel
Chapters 58, 59

James W. May, Jr., M.D.,
Professor and Chief
Division of Plastic Surgery
Massachusetts General Hospital
Harvard Medical School
Boston, Massachusetts, U.S.A.
Chapter 60

Margaret L. McNairy, B.S.,
Medical Student
Harvard Medical School
Boston, Massachusetts, U.S.A.
Chapter 42

Babak J. Mehrara, M.D.,
Assistant Professor
Division of Plastic Surgery
Memorial Sloan Kettering
New York, New York, U.S.A.
Chapter 11

Ankit I. Mehta, B.S., Medical Student
Harvard Medical School
Boston, Massachusetts, U.S.A.
Chapter 30

Thomas A. Mustoe, M.D.,
Professor and Chief
Division of Plastic Surgery
Northwestern University
Feinberg School of Medicine
Chicago, Illinois, U.S.A.
Chapter 4, Foreword

Nirm Nathan, B.S., Medical Student
University of Miami
Miami, Florida, U.S.A.
Chapter 40

John Nigriny, M.D., D.D.S., Resident
Department of Plastic Surgery
Stanford University School of Medicine
Stanford, California, U.S.A.
Chapter 37

Millicent Odunze, M.D., Resident
Division of Plastic Surgery
Northwestern University
Feinberg School of Medicine
Chicago, Illinois, U.S.A.
Chapter 91

Pravin K. Patel, M.D.,
Assistant Professor
Division of Plastic Surgery
Northwestern University
Feinberg School of Medicine
and
Chief, Division of Plastic Surgery
Shriner's Hospital for Sick Children
Chicago, Illinois, U.S.A.
Chapter 55

Ziv M. Peled, M.D., Resident
Division of Plastic Surgery
Harvard Medical School
Boston, Massachusetts, U.S.A.
Chapter 64

Bohdan Pomahac, M.D.,
Instructor in Surgery
Division of Plastic Surgery
Harvard Medical School
Brigham and Women's Hospital
Boston, Massachusetts, U.S.A.
Chapters 33, 42, 69

Jason Pomerantz, M.D., Resident
Division of Plastic Surgery
UCSF School of Medicine
San Francisco, California, U.S.A.
Chapters 48, 49

Joseph Raviv, M.D., Resident
Department of Otolaryngology
Northwestern University
Feinberg School of Medicine
Chicago, Illinois, U.S.A.
Chapter 36

Russell R. Reid, M.D. Ph.D., Assistant Professor and Bernard Sarnat Scholar
Section of Plastic Surgery
University of Chicago
Chicago, Illinois, U.S.A.
Chapter 18

David S. Rosenberg, M.D., Resident
Division of Plastic Surgery
Northwestern University
Feinberg School of Medicine
Chicago, Illinois, U.S.A.
Chapters 93, 94

Bauback Safa, M.D., Resident
Department of Plastic Surgery
Stanford University School of Medicine
Stanford, California, U.S.A.
Chapter 28

Hakim Said, M.D., Resident
Division of Plastic Surgery
Northwestern University
Feinberg School of Medicine
Chicago, Illinois, U.S.A.
Chapters 95, 96

Clark F. Schierle, M.D., Ph.D., Resident
Division of Plastic Surgery
Northwestern University
Feinberg School of Medicine
Chicago, Illinois, U.S.A.
Chapters 32, 39, 61, 62

Mark Sisco, M.D., Resident
Division of Plastic Surgery
Northwestern University
Feinberg School of Medicine
Chicago, Illinois, U.S.A.
Chapters 6, 20, 21, 28, 50-53, 86, Appendix IA, Appendix II

Amir H. Taghinia, M.D., Resident
Harvard Plastic Surgery Residency Program
Harvard Medical School
Boston, Massachusetts, U.S.A.
Chapters 30, 31, 33, 42, 69

Seth Thaller, M.D., D.M.D., Professor and Chief
Division of Plastic Surgery
University of Miami/JMH
Miami, Florida, U.S.A.
Chapter 40

Stephen M. Warren, M.D., Associate Professor
Institute of Reconstructive Plastic Surgery
NYU Medical Center
New York, New York, U.S.A.
Chapters 60, 63, 64

Thomas Wiedrich, M.D., Associate Professor
Division of Plastic Surgery
Northwestern University
Feinberg School of Medicine
Chicago, Illinois, U.S.A.
Chapter 96

David Wrone, M.D., Assistant Professor
Department of Dermatology
Northwestern University
Feinberg School of Medicine
Chicago, Illinois, U.S.A.
Chapter 72

Ted Yagmour, M.D., Associate Professor
Department of Anesthesia
Northwestern University
Feinberg School of Medicine
Chicago, Illinois, U.S.A.
Chapter 7

Michael J. Yaremchuk, M.D., Professor
Division of Plastic Surgery
Massachusetts General Hospital
Harvard Medical School
Boston, Massachusetts, U.S.A.
Chapter 64

Foreword

At first glance, it would seem that the field of plastic and reconstructive surgery is so inundated with texts that there is hardly a need for one more. First impressions, however, can be deceiving. There is a book that has been lacking—until now. *Practical Plastic Surgery* was conceived to address the need for a comprehensive, compact, and concise handbook. One that is as useful in the "trenches" as it is for the In-Service exam preparation.

One hundred chapters and two extremely useful appendices have been authored by most of the residents and faculty in the Division of Plastic Surgery at Northwestern University, as well as by many of our colleagues around the country. Over the course of several years this collaborative effort has evolved into an outstanding text for fellows, residents, students, and other physicians interested in the practical aspects of our field. I predict that this book will find a place close to every plastic surgery resident's fingertips, and that its usefulness will apply to other areas of surgery as well.

Thomas A. Mustoe, M,D.
Professor and Chief, Division of Plastic Surgery
Northwestern University Feinberg School of Medicine
Chicago, Illinois, U.S.A.

Preface

What attracted us to plastic surgery are its tremendous scope and the multitude of approaches for every clinical situation. Yet precisely these aspects of plastic surgery make its study and teaching especially challenging. Although there are several excellent atlases and texts, we have come across few references that are compact, affordable, timely, and focus on the practical, day-to-day practice of plastic surgery.

The purpose of *Practical Plastic Surgery* is to provide a guide to plastic surgery as it is practiced in academic medical centers. As such, it is written with the resident and fellow in mind. It is also our intention that this book be useful to general surgeons and other healthcare providers. Many of the chapters, such as *Basic Concepts in Wound Repair* and *Dressings* provide information relevant to all surgical specialties.

The book contains over a hundred chapters organized into eight sections that cover the breadth of plastic surgery, starting with *General Principles*, *The Problematic Wound* and *Integument*. The next five sections address the principle disciplines, and include *Head and Neck*, *Trunk and Lower Extremity*, *Craniofacial Surgery*, *Aesthetic Surgery*, concluding with *Hand and Upper Extremity*. The book concludes with two large appendices and a comprehensive index. Appendix I lists most of the commonly used flaps and their harvest and has many illustrations of these flaps. Appendix II is comprised of illustrations and the names of the common surgical instruments used by most plastic surgeons.

The text is written by over 75 authors, many of whom are considered among the leaders in their respective fields. Each chapter is concise and focused on the practical aspects of the topic. Historical and out-dated procedures are largely ignored. Every chapter concludes with a section titled "pearls and pitfalls," as well as a handful of important references.

Tremendous efforts have been made to ensure the accuracy and verify the information provided in this text. However, the art and science of plastic surgery are dynamic and constantly evolving. Procedures that are standard of care today may fall out of favor tomorrow. Therefore, this book should serve as a guide and not as an authoritative text.

We hope you enjoy reading and using this text as much as we enjoyed editing it. We welcome any comments that may help us improve future editions.

Zol Kryger and Mark Sisco

Chapter 1

Wound Healing and Principles of Wound Care

Leonard Lu and Robert D. Galiano

Introduction

Wound healing involves a broad range of overlapping cellular and metabolic processes that are orchestrated as a fundamental homeostatic response to injury. An understanding of these concepts is essential to care for wounds in all disciplines of surgery. Plastic surgeons are often consulted by other practitioners to deal with difficult, nonhealing, compromised wounds. Therefore, an understanding of the basic science of wound healing allows one to identify the variables involved in a given wound, and ultimately modulate the process to restore the structure and function of the injured tissue.

Classically, wound healing is divided into three distinct phases: **inflammatory**, **proliferative** and **remodeling** (Table 1.1). Even though each phase is described as a separate event, there is a large degree of temporal overlap and variability in these phases. Factors that influence the timing and length of these events include ischemia, age of the host, nutrition, radiation, smoking, systemic diseases such as diabetes, contamination or infection, desiccation, and the amount of devitalized or necrotic tissue in the wound. This chapter outlines the cellular, vascular and physiologic events underlying wound healing, focusing on the clinically relevant aspects.

Inflammatory Phase

Immediately after injury, bleeding occurs as a result of disruption of the blood vessels. Hemostasis is obtained by initial transient vasoconstriction and subsequent platelet plug and clot formation. Platelet degranulation of alpha and dense granules releases various substances, including platelet-derived growth factor (PDGF) and transforming growth factor-β (TGF-β), which ignite the chemotaxis and proliferation of inflammatory cells that characterize this phase of wound healing. Following the period of vasoconstriction, the migration of cells to the site of injury is aided by vasodilation and increased endothelial permeability (mediated by histamine, prostacyclin and other substances).

The first cells to arrive are the polymorphonuclear leukocytes (PMNs), which increase in numbers over the first 24 hours. These cells aid in the process of clearing the wound of debris and bacteria. Over the next 2-3 days, macrophages replace the PMNs as the predominant cell type. Macrophages have several critical roles in healing wound, including phagocytosis, release of multiple growth factors and cytokines, and recruitment of additional inflammatory cells. The importance of macrophages is exemplified by studies that have shown that wound healing is significantly impaired without their participation. In contrast, blocking or destroying PMNs during the inflammatory phase still results in a normally healing wound in the absence of bacteria. Finally, lymphocytes populate the wound, although their direct role in wound healing requires further investigation.

Practical Plastic Surgery, edited by Zol B. Kryger and Mark Sisco. ©2007 Landes Bioscience.

Table 1.1. The phases of wound healing

Phase	Cellular Response	Vascular Response	Time Course
Inflammatory	PMNs, macrophages, lymphocytes	Vasconstriction, followed by vasodilation	Injury to 7 days
Proliferative	Fibroblasts, endothelium	Angiogenesis, collagen deposition	5 days to 3 weeks
Remodeling	Fibroblasts	Collagen crosslinking and increasing tensile strength	3 weeks to 1 year

Note: these are overlapping processes and the time course varies depending on local and systemic factors.

Proliferative Phase

The clot formed during the inflammatory phase provides the provisional matrix and scaffolding for the proliferation of the dominant cell type during this phase—the fibroblast. In addition, growth factors stimulate angiogenesis and capillary in-growth by endothelial cells. The capillaries and fibroblasts form a substrate recognized clinically and histologically as granulation tissue. Fibroblasts produce collagen, which is the principal structural molecule in the final scar. Initially, type III collagen is produced in relative abundance in the healing wound; the normal adult 4:1 ratio of type I to type III collagen is gradually restored during the remodeling phase. The formation of collagen is a multi-step, dynamic process with both intracellular and extracellular components. Procollagen is synthesized and arranges as a triple-helix. After the secretion of procollagen from the intracellular space, peptidases trim residues from the terminal ends, allowing the collagen molecule to associate with other secreted fibrils. Ultimately, hydroxylation and cross-linking of collagen is required for the strength and stability of this protein.

Remodeling Phase

Approximately 2-3 weeks after the initial injury, collagen accumulation reaches a steady-state, where there is no change in total collagen content. During this time, there is replacement of the random collagen fibrils with organized, cross-linked fibrils. This process of remodeling persists for up to a year. Scars continue to gain strength over this phase; however, the tensile strength of scars never reaches that found in unwounded skin, approaching approximately 70% of normal strength.

Epithelialization

The skin is composed of the epidermis and dermis. Among the many important functions of the epidermis is to provide a barrier against bacteria and other pathogens and to maintain an aqueous body environment. When the skin is wounded, epithelialization begins to reconstitute the surface of the wound soon after the initial injury. In partial-thickness wounds, the epithelium derives from dermal appendages, hair follicles and sweat glands. In contrast, in full-thickness wounds, the epithelium migrates from the edges of the wound at a rate of 1 to 2 millimeters per day. A delay of epithelialization leads to a prolonged inflammatory phase, compromising the body's ability to restore the structure and function of the skin.

Wound Contraction

Myofibroblasts are fibroblasts that contain actin microfilaments, and allow wound contraction to occur. Under certain circumstances, wound contraction is advantageous, because it creates a smaller wound area. However, wound contraction that occurs across a joint, such as the elbow, knee or neck, may create functional limitations.

Pearls and Pitfalls

Understanding the basic science of wound healing has important clinical implications. Hemostasis, adequate debridement of dirty or contaminated wounds, and gentle handling of tissues reduces the inflammatory phase of wound healing. Allowing patients to cleanse their wounds with nonirritating solutions such as water further decreases inflammation. In addition, minimizing tension and dead space during wound closure increases the chance for creating an acceptable scar. Moist wound healing is superior to the healing in a desiccated wound; therefore, dressings should be tailored to create a moist local environment. Finally, an often overlooked facet of wound healing is to optimize nutrition. Patients with chronic or poorly healing wounds often require supplementation to provide the substrates necessary for collagen formation and epithelialization.

Suggested Reading

1. Fine NA, Mustoe TA. Wound healing. In: Greenfield LJ et al, eds. Surgery: Scientific Principles and Practice. 3rd ed. 2001:69.
2. Winter GD, Scales JT. Effect of air drying and dressings on the surface of a wound. Nature 1963; 197:91.
3. Mustoe TA, Pierce GF, Thomason A et al. Accelerated healing of incisional wounds in rats induced by transforming growth factor-β. Science 1987; 237:1333.
4. Burns JL, Mancoll JS, Philips LG. Impairments to wound healing. Clin Plast Surg 2003; 30(1):47-56.
5. Leibovich SJ, Ross R. The role of the macrophage in wound repair: A study with hydrocortisone and antimacrophage serum. Am J Pathology 1975; 78:71.

Chapter 2

Basic Concepts in Wound Repair

Zol B. Kryger and Michael A. Howard

Definitions

- **Primary closure** is defined as the surgical closure of a wound in one or more layers, within hours of its occurrence. Most surgical incisions and traumatic lacerations are closed primarily.
- **Delayed primary closure** is the surgical closure of a wound, days to weeks later. The granulation tissue is excised, the edges of the wound are freshened and the wound is closed. An example of this technique is the closure of a fasciotomy incision.
- **Skin grafting** is indicated when a defect is too large to close primarily, and creation of flaps is not desirable or feasible. It can be performed immediately following the injury or in a delayed manner. The indications and principles of skin grafting are discussed elsewhere in this book.
- **Surgical flaps** allow the recruitment of local or distant tissue for wound coverage. They are discussed in detail in an upcoming chapter.
- **Healing by secondary intention** is the choice a surgeon is left with when a wound cannot be surgically repaired. This doesn't mean that the surgeon can leave the wound to heal on its own; daily care and a long-term commitment by the patient and the care-givers are required. The wound must be kept clean and bacterial colonization should be minimized by daily washing, debridement of necrotic tissue and antibiotics when indicated. Healing by secondary intention involves the wound's progression through granulation tissue formation, epithelialization and contraction.

Suturing Techniques

The commonly used suturing techniques are illustrated in Figure 2.1 and described below. Some important points are applicable to all the techniques. The tissue should be entered as close to 90° as possible. The path of the needle should follow its curve. The suture should be pulled forward through the tissue as gently as possible. These steps will help minimize trauma to the tissues.

- **Simple interrupted sutures** are used to achieve optimal wound edge alignment. This technique is quick and easy to master. It is ideal for most traumatic lacerations. Nylon sutures are commonly used. Knots should never be tied tightly since the tissue can swell and undergo pressure necrosis under the suture.
- **Continuous running** (over and over) closure is the most rapid suturing technique; however it is difficult to achieve precise edge alignment when tension is present. In tension-free regions it can be used with a good cosmetic result. It is useful for achieving hemostasis (e.g., in scalp lacerations). If additional hemostasis is required, the stitch can be locked.

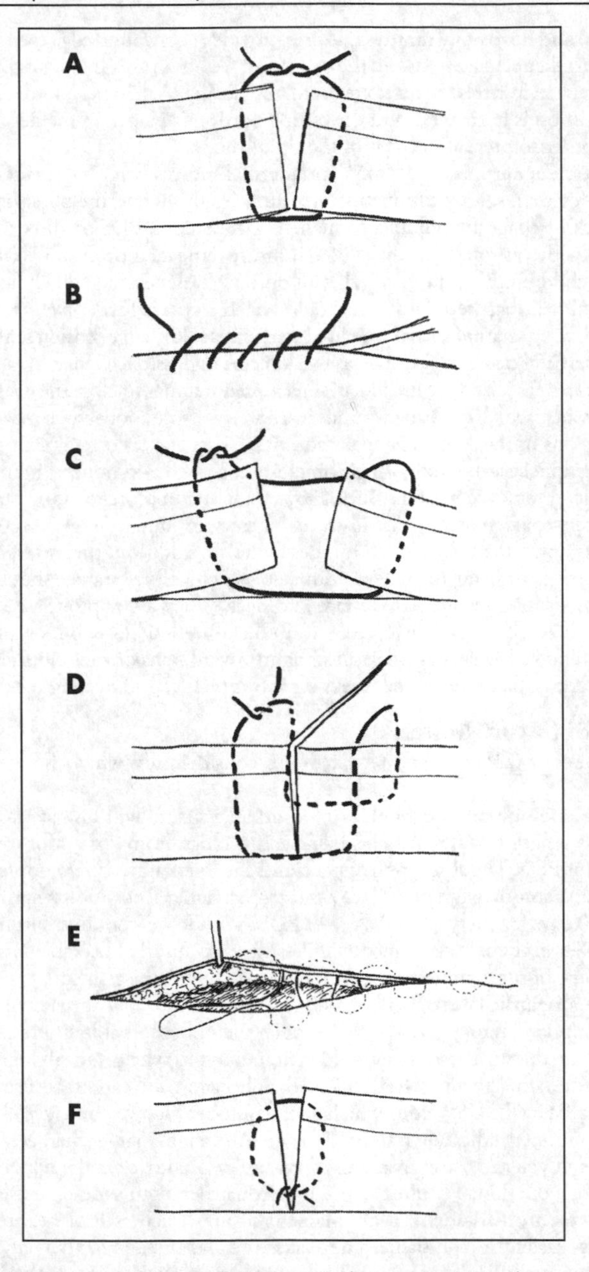

Figure 2.1. The commonly used suturing techniques. A) Simple interrupted. B) Continuous running. C) Vertical mattress. D) Horizontal mattress. E) Subcuticular. F) Buried dermal.

- **Vertical and horizontal mattress sutures** provide good wound edge eversion. They are an excellent choice for use in the hands and feet, or in areas of high skin tension.
- **Half-buried mattress sutures** are useful for closing V-shaped wounds. The mattress portion is horizontal, and the buried portion is placed in the dermis of the tip in order to prevent necrosis of the tip of the V.
- **Subcuticular sutures** are running, intradermal sutures that can provide an excellent cosmetic result by eliminating any surface sutures and the potential epithelial tracking that can result in a permanent suture mark. PDS or other absorbable sutures with low reactivity can be used if suture removal is problematic, such as in young children. If suture removal is an option, Prolene is a good choice since it has minimal tissue reactivity and should be left in place for 2-4 weeks.
- **Buried, deep dermal sutures** are used to decrease skin-edge tension and to allow the superficial closure to be done as tension-free as possible. Generally, absorbable sutures such as Vicryl are used in an interrupted manner to close the deep dermis.
- **Staples** are useful for closing wounds in a variety of situations, such as lacerations or incisions of the scalp. The main advantage that staples offer is that they provide the quickest method of incision closure, and they produce minimal tissue reactivity if removed within a week. However, if left in place too long, staples will produce a characteristic "railroad-track appearance" due to migration of epithelial cells down the tract created by the staples. In addition, precise wound edge alignment is difficult to achieve with staples. Therefore, staples should not be used on visible sites such as the face and neck. They are appropriate for use in reconstructive cases in which precise wound closure is of lesser importance. They can be removed as early as 7 days in straightforward, tension-free closures, or they can be left in place for several weeks if suboptimal wound healing is expected.

Choice of Suture Material

A number of factors should be taken into consideration when choosing suture material:

- **Absorbable or nonabsorbable.** An absorbable suture will lose at least half its tensile strength by 60 days. This half-life can range from 7 days for catgut to 4 weeks for PDS. The absorption of plain and chromic catgut is very unpredictable. Synthetic, absorbable sutures have a more predictable absorption length, ranging from 80 days for Vicryl to 180 days for PDS. With few exceptions, sutures should not be left in the skin permanently unless they are absorbable. Table 2.1 summarizes some of the commonly used sutures and their characteristics.
- **Tensile strength.** The strength of a suture is determined by the material of which it is comprised and by its diameter. Among the nonabsorbable sutures, polyester sutures are the strongest, followed by nylon, polypropylene and silk. For absorbable sutures, the order is polyglycolic acid, polyglactin and catgut. Suture diameter is indicated by the USP rating which gives a number followed by a "zero," with the higher number indicating a thinner suture. Although a larger diameter suture is stronger, it will also cause greater tissue reactivity and leave a more noticeable scar. Therefore, the thinnest suture that is of adequate strength should be used.
- **Mono- or multifilament.** Monofilament sutures, such as Prolene, are smooth and pass easily through tissue. They cause the least tissue reactivity and trauma and are more difficult for bacterial adhesion. The drawback is that they are difficult to handle compared to multifilament sutures such as silk. In addition, knot security, which is proportional to the coefficient of friction of the suture, is usu-

Table 2.1. Commonly used absorbable and nonabsorbable sutures

Nonabsorbable Sutures

Brand	Material	Strength	Knot Security	Reactivity
Silk	Silk braided	Low	Very good	High
Prolene, Surgilene	Polypropylene monofilament	Average	Very low	Low
Ethilon, Dermalon	Nylon monofilament	High	Poor	Low
Nurolon	Nylon braided	High	Average	Average
Ethibond, Tevdek	Polyester braided-coated	Very high	Average	Average
Mersilene, Dacron	Polyester braided-uncoated	Very high	Average	Average

Absorbable Sutures

Brand	Material	Duration	Knot Security	Reactivity
Plain catgut	Sheep/cattle intestine	1 week	Poor	Very high
Chromic catgut	Treated intestine	2 weeks	Average	High
Vicryl	Polyglactin	2-3 weeks	Average	Average
Monocryl	Poliglecaprone 25	2-3 weeks	Good	Low
Dexon	Polyglycolic acid	2-3 weeks	Good	Low
Maxon	Polyglyconate	4 weeks	Average	Average
PDS	Polydiaxanone	4 weeks	Poor	Low

ally greater in multifilament sutures, especially those that are braided. The lower the knot security, the more throws are required to create a secure knot.

- **Needle types**. There is no uniform nomenclature that describes the characteristics of the needles. A simplified approach is to classify needles as tapered, cutting or reverse-cutting. Tapered needles minimize trauma to the tissue. They are used to suture tissue that is fragile and can tear easily. Examples include cartilage and bowel wall. Cutting and reverse-cutting needles are typically used in closing dermis, with the latter being more commonly used due to the creation of a tract that is less likely to tear through the skin.
- **Suture removal**. The optimal timing for removal of sutures varies widely from surgeon to surgeon. Sutures that are left in place too long can lead to epithelial tracking down through the skin along the length of the suture. This may result in punctate scars left from the sutures themselves. In cases in which impaired wound healing is expected and cosmesis is of secondary importance, sutures can be left in place for weeks or even months. The following is a guideline for the timing of suture removal:

  Eyelids → 3-5 days
  Face → 5-7 days
  Breast → 7-10 days
  Trunk → 7-10 days
  Hands → 10-14 days
  Feet → 10-14 days

2

Considerations in Wound Healing and Scar Formation

Important factors that contribute to a worsened scar outcome include:

- Tension on the closure
- Infection
- Delayed epithelialization
- Imprecise alignment of the wound edges
- Impaired blood flow to the healing scar
- Genetic factors beyond control

By minimizing these factors, an incision will heal more rapidly and the resulting scar will be more cosmetically acceptable.

- **Tension** on the closure should always be minimized. Closure of the deeper dermis with absorbable sutures will help reduce tension. Whenever possible, incisions should be placed in lines of election. These are the natural creases of minimal skin tension corresponding to wrinkle lines. They are also known as relaxed skin tension lines (RSTL). In the face they usually lie perpendicular to the direction of pull of the muscles of facial expression. If the edges of a wound cannot be brought together without undue tension, undermining or creation of a flap is required. Undermining of the wound edges should be performed with extreme care in order to avoid compromising blood supply. Techniques for creating surgical flaps are discussed in detail elsewhere in this book.
- **Infection** is of greatest concern in areas of poor vascularity such as the extremities. The face and scalp, in contrast, rarely become infected due to their robust blood supply. In general, wounds older than 12 hours should not be closed. This rule can often be violated when dealing with uncontaminated facial lacerations. Grossly contaminated wounds, such as human bites, are at high risk of developing an infection and are not usually closed primarily. Devitalized tissue should always be debrided from all wounds since it will become a nidus for infection. Pulse lavage of wounds is probably the single most effective method for decreasing bacterial count. Either normal saline or an antibiotic solution can be used.

Systemic antibiotics should be used with care. A single dose of preoperative antibiotics is usually indicated. In routine clean cases, there is little evidence to support the use of antibiotics beyond the first 24 hours postoperatively. A patient who presents to the emergency department with a wound requiring surgical repair should probably receive a dose of intravenous antibiotics and his tetanus status should be determined.

- **Delayed epithelialization** has been shown in many studies to delay overall wound healing and to worsen scar outcome. The presence of a foreign body will interfere with epithelialization; therefore all wounds should be explored carefully prior to closure. Infection will also delay epithelial migration. Finally, there is mounting evidence that moist wounds epithelialize faster and heal better. A moist healing environment is achieved by occlusion of the incision with a semi-permeable, occlusive dressing such as a Steri-strip®. Such a dressing should be used for the first week postoperatively. Under optimal circumstances, an incision will epithelialize within the first 24 hours.
- **Improper wound edge alignment** occurs during primary closure. It can be minimized by ensuring that the suture traverses the dermis on each side of the incision at the same depth. Once the wound is completely closed, the edges should appear tightly apposed and maximally everted. In irregular wounds, such as stellate-shaped lacerations, one must take care to properly match the

two sides. Initial closure of the apex of the laceration can help the pieces properly fall into place.

- **Impaired blood flow** will prevent the wound from receiving adequate oxygen, nutrients, growth factors and the essential cells involved in the wound healing and scarring process. Little can be done to improve blood flow during primary closure; however a number of factors will worsen it. Smoking has been shown to worsen ischemia in healing wounds by vasoconstriction. External pressure on the wound greater than capillary perfusion pressure (>35 mm Hg) must be avoided. Care should be taken whenever placing circumferential bandages, compression dressings or casts. In addition, sutures that are placed too close to one another can also create areas of ischemia.

Other factors that have an unclear effect on scarring, but will impair wound healing, include elevated blood glucose levels, poor nutritional status, venous and lymphatic insufficiency, chronic corticosteroid use, and a variety of comorbid conditions. Finally, genetics play a definite role in scarring as illustrated by the fact that certain ethnic groups and families have a predisposition to hypertrophic scarring and keloid formation.

The "Dog-Ear"

In certain instances, misaligned closure of a wound can result in a bunching or outpouching of skin termed a "dog-ear." This will commonly occur when closing oval or circular defects. The "dog-ear" can be excised at its base; however this will result in a scar that is longer than the length of the original defect (Fig. 2.2). In somc cases a "dog-ear" will settle with time or can be treated at a later time if it becomes bothersome to the patient.

Z-Plasty

The Z-plasty is a technique that can be used to help prevent scar contracture, or more commonly, as a method of treating scar contracture. Essentially, two interdigitating triangular flaps are transposed resulting in: (1) a change in the orientation of the common limb of the Z; and (2) a lengthening of the common limb of the Z (Fig. 2.3). The change in orientation can be used for managing wounds, in which direct closure may result in undue tension and distortion of nearby structures, such as in the face. The gain in length can be used for treating contracted scars.

Both the length of the transverse limbs and their angle with the common limb can be varied. First, the greater the angle, the greater the amount of lengthening that will occur. A 45° angle will lengthen the common limb up to 50%, and a 60° angle

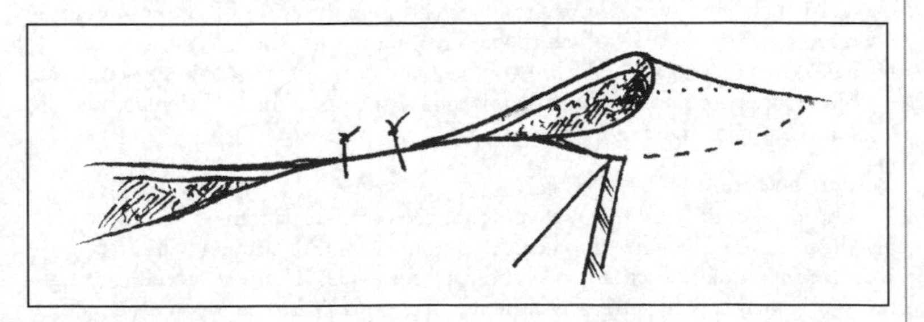

Figure 2.2. "Dog-ear" excision.

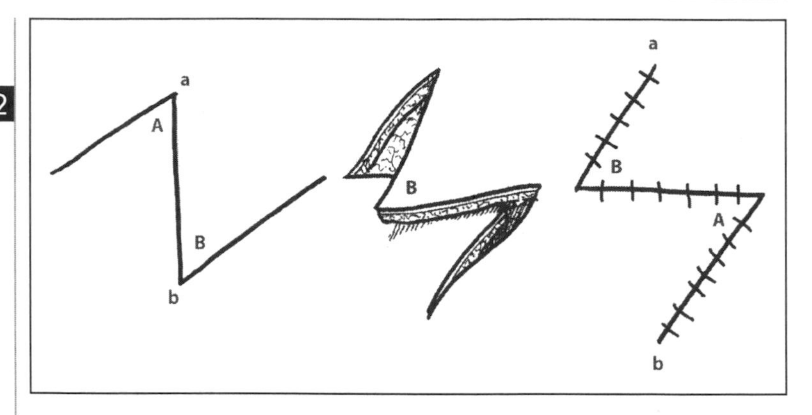

Figure 2.3. The Z-plasty technique.

up to 75%. The angles should generally not exceed 60° since excessive transverse shortening and tension will occur. Second, the limb length is determined by how much tissue is available on either side: the more tissue is available, the longer the limbs can be.

Planning the Z-Plasty

When releasing scar contracture, the Z-plasty is created as follows:

1. The common limb of the Z is drawn along the length of the scar. The parallel, transverse limbs are drawn at 60° to the common limb.
2. The skin is incised along the Z shape, and any contracted scar is also incised.
3. Vascularity to the tips of the triangles must be maintained, since they are at the highest risk of necrosis. This is achieved by maintaining a broad base to the triangles, keeping the flaps as thick as possible, avoiding undue transverse tension and handling the tissue with care.
4. The triangles are transposed, resulting in a reorientation of the transverse limbs and a lengthening of the common limb.

When reorienting the direction of a facial scar, the Z-plasty is created as follows:

1. The common limb of the Z is drawn along the length of the scar. The new direction of the common limb is planned so that it will lie in a natural skin crease such as the nasolabial fold.
2. The parallel, transverse limbs should extend from the ends of the common limb up to the skin crease in which the new common limb will lie.
3. The skin is incised along the lines of the Z and the triangles are transposed. If the blood supply to the tips of the flaps is robust, such as in the face, tip necrosis will not occur and angles more acute than 60° can be used.

Patient Selection

The ideal candidate is one with a pronounced wrinkle pattern. In such individuals, the scar can be reoriented to lie in a pronounced line of election. Children, with their lack of wrinkles are not good candidates for Z-plasties on the face. If the original scar is markedly hypertrophic, the use of a Z-plasty is questionable. Scars that cross a hollow (bridle scars), such as the angle of the jaw, are also amenable to Z-plasty.

Multiple Z-Plasties

A single Z-plasty is limited by the transverse shortening resulting from reorientation of the transverse limbs of the Z. This creates lateral tension that is concentrated most heavily at the apices of each triangle. The use of multiple Z-plasties can provide the same degree of scar lengthening while significantly limiting the amount of transverse shortening. In addition, when a scar is very long and would require enormous transverse limbs, multiple Z-plasties with shorter limbs may be more appropriate. In practice, multiple Z-plasties are usually performed with their common limbs as one continuous unit.

Pearls and Pitfalls

The choice of wound closure technique and suture material vary widely. The goal remains the same: achieving a tension-free closure with clean skin edges that are well approximated. The following questions are useful to address before attempting to repair most types of lacerations, especially in the acute-care setting:

1. Are there other potential life-threatening injuries that must be dealt with first?
2. When did the wound occur and what was the mechanism? Is there gross contamination, or is the risk of infection too great to allow primary closure?
3. Has the patient received prophylactic antibiotics and a tetanus shot (when indicated)?
4. Has all devitalized tissue been excised and have all foreign bodies been removed?
5. Can the wound be closed primarily without excess tension? Is there a role for undermining or creation of a flap?
6. What suture material should be used and which suturing technique should be chosen?
7. Which points match up in order to recreate the pre-injury anatomy?
8. Have the wound edges been adequately approximated?
9. Will the dressing provide adequate occlusion? Is it too tight or will it be too tight if postoperative swelling occurs?
10. Has the patient received proper postoperative counseling (how to keep the wound clean, when to get it wet, which activities to avoid, when to follow up, and what the signs of a wound infection are)?
11. When should the sutures be removed?
12. Long-term: Is the final outcome acceptable? Is scar revision necessary?

Suggested Reading

1. Furnas DW, Fisher GW. The Z-plasty: Biomechanics and mathematics. Br J Plast Surg 1971; 24:144.
2. Karounis H, Gouin S, Eisman H et al. A randomized, controlled trial comparing long-term cosmetic outcomes of traumatic pediatric lacerations repaired with absorbable plain gut versus nonabsorbable nylon sutures. Acad Emerg Med 2004; 11(7):730-5.
3. McGregor AD, McGregor IA, eds. Fundamental Techniques of Plastic Surgery, and Their Surgical Applications, 10th ed. New York: Churchill Livinstone, 2000.

Chapter 3

Dressings

Anandev Gurjala and Michael A. Howard

Introduction

A surgeon's goal to achieve successful wound healing rests on two basic principles: (1) **optimizing** conditions for the body's natural wound healing mechanisms to occur, and (2) **minimizing** the manifold detriments which interfere with this process. A few, if any methods currently exist to actually **enhance** wound healing, the classic tenets of "keep the wound moist, clean, free of edema and free of bacteria" are based on these two fundamentals, which in turn are rooted in basic principles of wound healing biology. Although wound healing products available today are increasingly varied and sophisticated, their primary function remains to support the intrinsic wound healing process.

Wound Healing

Wound healing occurs as an orchestrated series of four overlapping phases: coagulation (immediate), inflammation (0-7 days), proliferation (4-21 days) and remodeling (14 days-2 years). Actual physical closure of the wound occurs during the proliferative phase by granulation (fibroblasts, endothelial cells), contraction (myofibroblasts) and epithelialization (keratinocytes). An uncompromised wound will progress normally through these phases, however a compromised wound can arrest in the inflammatory or proliferative phases; resulting in delayed wound healing. If a wound is able to overcome its compromising factors and reach the remodeling phase within one month, it is termed an acute wound. If a wound is overwhelmed by its compromising factors and fails to heal within three months it is called a chronic wound.

Table 3.1 lists the factors that impair wound healing. They can be classified as **intrinsic** (underlying conditions that inhibit normal wound healing), **extrinsic** (external factors imposed on the body that inhibit wound healing) and **local** (factors which influence the condition of the wound bed). These impediments must be eliminated or limited in order to optimize wound healing.

Goals of Wound Dressings

The purpose of wound dressings is to control the local factors and create an environment that will optimize the wound bed for healing. The ideal dressing would achieve this goal by having the following properties:

1. Maintain a moist wound healing environment
2. Absorb exudate
3. Provide a barrier against bacteria
4. Debride—both macroscopic and microscopic material

Practical Plastic Surgery, edited by Zol B. Kryger and Mark Sisco. ©2007 Landes Bioscience.

Table 3.1. Factors that impair wound healing

Intrinsic	Extrinsic	Local
Wound:	- Smoking	- Desiccation
- Hypoperfusion	- Radiation	- Inflammation
- Hypoxia	- Chemotherapy	Infection
	- Drugs (e.g. steroids)	Bacterial burden
Systemic:	- Temperature (cold)	Hematoma
- Age	- Mechanical	Foreign Body
- Obesity	- Pressure	Ischemia
- Malnutrition	- Sheer	- Necrotic burden
- Hormones	- Trauma	Dead cells, exudate
- Disease (e.g., diabetes, cancer, uremia, EtOH)		- Cellular burden
- Venous insufficiency		Senescent and nonviable cells
- Extremity edema (e.g., CHF)		- Edema
- Arterial insufficiency/PVD		

5. Reduce edema
6. Eliminate dead space
7. Protect against further injury from trauma, pressure and sheer
8. Keep the wound warm
9. Promote skin integrity of the surrounding tissue and to do no harm to the wound

Types of Dressings

There are hundreds of commercially available wound dressings. It is beyond the scope of this chapter to cover all of them. A practical approach is to classify dressings by the material of which they are composed. The commonly used dressings are summarized in Table 3.2.

Gauze

Gauze is composed of natural or synthetic materials and may be woven (lower absorptive capacity, higher tendency to adhere to the wound bed, and high amount of lint) or nonwoven (superior absorbency, less adherence, low lint). This dressing is commonly used to cover fresh postoperative incisions. Other popular uses of gauze have remained the wet-to-dry (WTD) dressing and wet-to-moist (WTM) dressings. WTD dressings should be avoided as a method for mechanical microdebridement, since this debridement is nonselective and will harm viable tissue during dressing removal. As WTD dressings dry out, they also lead to wound desiccation, violating one of the central wound healing principles. WTM dressings—used to maintain a moist environment—are also less than ideal because they are labor intensive requiring many dressing changes, and in the process tend to dry out anyway, achieving the opposite of their intended purpose.

Other uses of gauze exist. It is indicated for wounds with exudate so heavy that other more sophisticated dressing types would not be cost-effective. Examples include drainage from a seroma or a fistula requiring many daily dressing changes. Gauze is also indicated as a primary dressing over ointments and as a secondary dressing over wound fillers and hydrogels.

Table 3.2. A list of common dressing materials, their characteristics and the appropriate situations for their use

Dressing Type	Function(s)	Use On	Do Not Use On	Composition	Characteristics
Dry gauze	*To absorb bleeding in first 24 hrs following wound closure/ debridement*	Closed surgical wounds	Partial- or full-thickness wounds (require moist environment)	Gauze	Of limited use
Wet-to-dry gauze	*To debride devitalized tissue from the wound bed* To fill dead space To absorb exudate and wick drainage	Full-thickness and open wounds	Partial-thickness wounds (require moist environment, WTD would debride epithelium)	Gauze and saline	Has been suggested that WTD dressings are grossly overused, not necessarily the least expensive alternative, and have higher infection rates than moistureretentive dressings; labor intensive
Wet-to-moist gauze	*To maintain a moist environment* To debride moist necrotic wounds To fill dead space To absorb exudate and wick drainage	Partial-thickness, open and infected wounds	Highly exudative wounds Severe maceration of surrounding tissue (WTM would overhydrate)	Gauze and saline	Possibly overused like WTD; difficult to use correctly since WTM often becomes WTD, achieving opposite of desired effect; labor intensive
Transparent films (Tegaderm, Op-site)	*To maintain a moist environment* To provide a barrier to bacteria and contaminants	Partial-thickness/ superficial wounds with minimal exudate Nondraining, primarily closed surgical wounds	Moderate to heavily exudative wounds Friable surrounding skin Wounds with sinus tracts Full-thickness wounds	Semipermeable polyurethane or copolyester	Mimics skin: permeable to water vapor and impermeable to water and bacteria; transparent and adhesive

continued on next page

Table 3.2. Continued

Dressing Type	Function(s)	Use On	Do Not Use On	Composition	Characteristics
Nonadherent dressings (Adaptec, Xeroform)	*To provide a surface that won't stick to the wound bed* To help reduce bacterial proliferation	Skin grafts and donor sites with minimal to moderate exudate Abrasions and lacerations	Heavily exudating wounds	Nonimpregnated: nylon or polyurethane covering; Impregnated: gauze with petrolatum, antibacterial or bactericidal compund	Used as primary dressings; require a secondary cover or wrap to secure them
Hydrogels (Curasol, Aquaphor, Elastogel)	*To maintain a moist environment* To minimize pain To hydrate and promote autolytic debridement (water is donated from gel to rehydrate and soften necrotic tissue, aiding with debridement)	Shallow, noninfected pressure ulcers Abrasions, minor burns and other partial-thickness wounds Radiation injuries Donor sites Superficial and partial-thickness burns	Moderate to heavily exudating wounds Full-thickness burns Cavity wounds	80-99% water and a cross-linked polymer (polyethylene oxide, acrylamide, polyvinyl pyrrolidone); amorphous and sheet forms	Does not adhere to wound, low absorbency, requires secondary dressing (Telfa or Tegaderm)
Hydrocolloids (Duoderm, Cutinova Hydro)	*To facilitate autolysis and debride soft necrotic tissue* To provide a moist environment To mechanically protect wound To encourage granulation To promote reepithelialization	Partial- or full-thickness wounds with minimal exudate Wounds with slough or granulation tissue and minimal to moderate exudate Pressure ulcers (stages I-IV)	Infected wounds (do not want an occlusive dressing) Wounds with sinus tracts Deep cavity wounds Heavily exudating wounds Wounds with friable surrounding skin Full-thickness burns	Hydrophilic colloidal particles (e.g., quar, karaya, gelatic, carboxymethyl cellulose) within an adhesive mass (polyisobutylene)	Good adhesiveness without sticking to wound bed

continued on next page

Table 3.2. Continued

Dressing Type	Function(s)	Use On	Do Not Use On	Composition	Characteristics
Foams (Lyofoam, Cutinova Plus)	*To absorb exudate* To debride	Partial- or full-thickness wounds which are heavily exudative and require mecahnical debriding	Dry wounds Partial-thickness wounds with minimal exudate Exposed muscle, tendon, bone Arterial ischemic lesions	Hydrophilic or phobic nonocclusive foam coated with polyurethane or gel	Water vapor permeable
Collagen dressings (Fibracol, hyCURE)	*To absorb exudate* To fill dead space To provide a moist environment	Deep, tunneled or cavity wounds with drainage Partial- and full-thickness wounds Partial-thickness burns	Wounds with dry eschar	Bovine avian or porcine collagen available in pads, gels and particles	Promotes deposition of newly formed collagen in wound bed and acts as a hemostatic agent; requires a secondary dressing
Alginates (Sorbsan, Carrasorb)	*To absorb exudate* *To fill dead space* To promote debridement autolytic	Deep wounds with slough and heavy exudate	Full-thickness burns Heavily bleeding wounds Dry wounds Demonstrated sensitivity to alginates	Composed of naturally occuring mannuronic or glucuronic acid polymers from brown seaweed, available as pads or ropes	Highly absorbent becoming a gel which facilitates moist wound healing; requires a secondary dressing; severe adherence if the dressing dries out
Negative pressure dressing (wound VAC)	*To absorb exudate and minimize edema* *To stimulate production of granulation tissue* *To aid in wound closure/contraction* *To increase blood flow* To fill dead space To provide a moist healing environment	Small to large deep, open or dehisced wounds Skin grafts Chronic wounds: pressure ulcers (stage II-IV), diabetic ulcers, venous ulcers, neuropathic ulcers	Partial-thickness wounds Necrotic tissue with eschar Untreated osteomyelitis Cancer in wound Fistula in vicinity of wound	Reticulated polyurethane sponge covered with transparent plastic self-adhesive sheets; an embedded evacuation tube leading to a suction device	Versatile, cost-effective in the long run

Transparent Films

Films are composed of a thin, clear polyurethane with an adhesive side which sticks only to dry, intact skin and not moist wound bed. They are designed to **regulate the correct amount of moisture** beneath the dressing and are water vapor and gas permeable, yet impermeable to bacteria and liquids. These dressings are indicated for partial-thickness or superficial wounds with minimal exudates (e.g., split-thickness skin graft donor site). They are also ideal for primarily-closed, nondraining surgical wounds. Films may also be used as secondary dressings over absorptive fillers on full-thickness wounds. Because they are impermeable to liquids, they are contraindicated for infected wounds and heavily exudative wounds. Bacteria will be retained rather than debrided, and collection of fluid can lead to maceration of the wound and the surrounding healthy skin.

The moisture regulation of transparent films provides an ideal environment for softening eschar through autolytic debridement. Hydrogels can even be added under the film to speed up this process. Use of films reduces pain and provides protection from friction and sheer forces. Transparency of the films also facilitates easy observation and postoperative monitoring of the wound. When applying films, at least 1 inch of surrounding skin should be utilized for good adherence, which can be aided by use of Benzoin™ tincture or Mastasol™. Films can be left in place for several days, but must be changed earlier if exudate leaks onto intact skin.

Examples of transparent films are Tegaderm™ and Op-site™.

Nonadherent Dressings

These dressings can be nonimpregnated (nylon or polyurethane) or impregnated (gauze with petrolatum or an antibacterial compound). Their purpose is to **provide an interface which will not stick to the wound bed** and maintain some degree of moisture in the wound bed. Indications for nonadherent dressings are skin grafts and donor sites with minimal to moderate exudate, and abrasions or lacerations. Contraindications are heavily exudative wounds.

Examples include Xeroform™, Adaptec™ and Vaseline-impregnated gauze.

Hydrogels

Hydrogels are composed of 80-99% water suspended in a cross-linked polymer, and are available in amorphous (dispensed from a tube) and sheet forms. Hydrogels **preserve wound hydration** and actually donate water to the wound, but are not at all absorptive of exudate. They are thus indicated for dry to minimally exudative wounds, such as shallow pressure ulcers, abrasions and other partial-thickness wounds, skin graft donor sites and superficial partial-thickness burns. Sheet gels should be used on superficial wounds <5 mm in depth, and amorphous gels should be used on deeper full-thickness wounds. Hydrogels should not be used for heavily exudative wounds and should not be applied to intact skin.

The moisture provided by gels promotes autolytic debridement, however overuse of hydrogels will cause wound maceration. Sheet hydrogels can be changed every 4-7 days and amorphous gels added as needed to titrate wound hydration. Either gauze or a transparent film is used as secondary dressing depending on the tendency of the wound to dry out.

Examples of hydrogels are Curasol™, Aquaphor™ and Elastogel™.

Hydrocolloids

Hydrocolloids are composed of hydrophilic colloidal particles (such as gelatin or cellulose) within an adhesive mass (polysobutylene). They come as adhesive wafer dressings designed to interact with the wound bed by forming a gel over it as exudate is absorbed by the dressing, thus forming a protective layer over the wound and creating a moist wound healing environment. The absorptive layer of the dressing is covered by a completely impermeable film.

Hydrocolloids are indicated for wounds with low to moderate exudate, partial- or full-thickness wounds, and granulating or necrotic wounds. A frequent use of these dressings is for venous stasis ulcers. Hydrocolloids may be also used over absorptive wound fillers. They should not be used on infected wounds, heavily exudative wounds or on wounds with fragile surrounding skin. In addition to offering protection from sheer force, these dressings protect against exogenous bacterial contamination, and are relatively painless. They are ideal at providing an environment for autolytic debridement of fibrinous slough. When applied, at least one inch of surrounding skin should be covered to ensure adherence. These dressings should be changed when exudate is within an inch of the dressing edge, which may be daily until exudate slows down, at which time hydrocolloids can be left on for up to seven days.

Examples of hydrocolloids are DuoDERM™ and Cutinova Hydro™.

Foams

Foams are composed most commonly of polyurethane polymers whose primary function is to absorb wound exudate. They come in thin and thick foams, adhesive and nonadhesive foams, foams used to pack wounds and sheet foams. Foams are indicated for wounds with moderate to high exudate, partial- or full-thickness wounds, and granulating or necrotic wounds. They can be used on infected wounds if changed daily, and they can be used over creams or ointments. Foams are not recommended for dry wounds. Foams protect wounds well (although do not reduce force on pressure ulcers), facilitate autolytic debridement, and minimize granulation tissue. As with other dressings, at least one inch of surrounding skin should be covered, and foams can be left on the wound surface for up to seven days.

Examples of foams are Lyofoam™ and Allevyn™.

Alginates

Alginates are composed of naturally occurring mannuronic or glucuronic acid polymers from brown seaweed. They come available as pastes, granules, powders, pads or ropes which soften, gel and conform to the wound, thereby functioning to absorb, exudate and fill dead space. These dressings are indicated for wounds with moderate to high exudate, and may be used on partial- or full-thickness, and granulating or necrotic wounds. They may also be used on infected wounds if changed daily. Foams are well-suited to be used under other dressings such as hydrocolloids to increase dressing wear time. Sheets are generally used on shallow wounds, and ropes, pastes and strands are used to fill deep wounds. Alginates should not be used on minimally exudative wounds because they will adhere to the wound and cause damage when removed. They should not be packed into very deep or tunneling wounds as they may easily be left behind and become a nidus for infection (iodoform gauze™ is the only appropriate packing for this type of wound).

When applied, alginates should only fill one-third to one-half of the wound since they will expand with time. They require a secondary dressing based on the tendency of the wound to dry out, for example gauze coverage for a highly exudative wound. Alginates should be changed when exudate reaches the secondary dressing.

Examples of alginates are Sorbsan™ and Carrasorb™.

Pearls and Pitfalls

Prior to selection and application of wound dressings, there are several key components to wound management that must be addressed:

- One must establish the etiology of the wound and address its causative factors (see Table 3.1).
- A clean, viable and well-vascularized wound is the goal. If a wound contains any more than a minimal amount of devitalized (shades of white, brown or black) or infected tissue, dressing changes alone will not be enough to achieve adequate debridement. Surgical debridement and revascularization may be required. A form of debriding dressing can then be used.

The choice of dressing material in clinical practice is often arbitrary and based on the clinician's personal experience. Many different dressings can achieve the same goal. The key to optimizing wound healing to adhere to these basic principles:

1. Perform dressing changes with sufficient frequency so that the dressing provides a moist wound environment, but not an overly saturated one.
2. If the wound is pink, healthy and free of infection, it requires only constant moisture.
3. Wounds must heal from the inside without the overlying skin sealing over unhealed deeper tissue. Therefore, all dead space must be eliminated by packing the wound. If packing a wound requires more than a single 4x4 gauze dressing, a Kerlix™ role should be used instead to eliminate the risk of a 4x4 gauze being retained.
4. Given that the pre-dressing conditions above are met, almost all wounds can be treated using only saline gauze or hydrogel and gauze.
5. Most importantly, frequent wound monitoring by the clinician is paramount—an ignored wound will not just "go away." Evaluation of a therapy's effectiveness is required. Failure of a wound to progress indicates that the patient's medical condition, the wound environment or the choice of wound dressing must be revisited.

Suggested Reading

1. Falanga V, Grinnell F, Gilchrest B et al. Workshop on the pathogenesis of chronic wounds. J Invest Dermatol 1994; 102:125.
2. Falanga V. The chronic wound: Impaired healing and solutions in the context of wound bed preparation. Blood Cell Molec Dis 2004; 32:88.
3. Harding KG, Morris HL, Patel GK. Science, medicine and the future: Healing chronic wounds. BMJ 2002; 324:160.
4. Hunt TK, Hopf HW. Wound healing and wound infection. What surgeons and anesthesiologists can do. Surg Clin North Am 1997; 77:587.
5. Ovington LG. Wound care products: How to choose. Adv Skin Wound Care 2001; 14:259.
6. Seaman S. Dressing selection in chronic wound management. J Am Podiatr Med Assoc 2002; 92:24.
7. Singer AJ, Clark RA. Cutaneous wound healing. N Eng J Med 1999; 341:738.

Chapter 4

Pharmacologic Wound Care

Peter Kim and Thomas A. Mustoe

Patient Assessment

As the wound healing process has been better elucidated, the practice of wound care has evolved. However, the basic principles of good wound care have not changed significantly over time. The process begins with an assessment of the entire patient. Any underlying medical conditions that impair wound healing must be treated. These include systemic infection, hyperglycemia, inadequate nutritional status, poor circulation, a deficient immune system, and the absence of someone dedicated to caring for the wound. Once patient factors have been adequately addressed, attention should be turned to the wound itself.

The majority of chronic wounds encountered in hospitalized patients can be categorized into four types of ulcers: pressure, diabetic, venous and ischemic. In addition, chronically infected wounds can occur in the setting of underlying osteomyelitis or a foreign body (such as orthopedic hardware). With respect to local wound care, the principles are always the same: eradicate infection, debride necrotic tissue, remove any nonessential foreign material, maximize arterial inflow and venous outflow, and keep the wound **moist** and clean.

Topical pharmacologic wound care can help achieve these goals to a certain extent. In broad terms, agents commonly used in local wound care can be grouped into three categories: **antimicrobial agents**, **enzymatic agents** and **growth factors**.

Antimicrobial Agents

Topical antimicrobials are moderately effective in treating infected wounds. They provide the benefit of delivering a high therapeutic dose of the drug to a local area with minimal systemic side effects, particularly in wounds with relatively underperfused tissues. They are not, however, a substitute for systemic antimicrobial therapy when indicated (e.g., surrounding cellulites of the wound). The goal of topical antimicrobial therapy is to diminish the burden of bacteria to a level that is manageable by the host immune cells. In fact, sub-infection levels of bacteria have been shown to accelerate wound healing and granulation by promoting the infiltration of neutrophils, monocytes and increased collagen deposition.

There are a variety of commercially available topical antimicrobials (Table 4.1). Silver sulfadiazine (Silvadene®) is used in superficial soft tissue infections. Its effectiveness against Pseudomonas makes it a favored choice in burn treatment. Two other commonly used antibiotic agents are Bacitracin® and Neosporin®. These petroleum-based ointments are useful more for superficial infections. In addition, they can be used on surgical incisions, particularly of the face, to minimize bacterial load and provide a moist wound environment to promote epithelialization.

Table 4.1. Commonly used topical antimicrobial agents

Agent	Mechanism	Spectrum	Resistance	Comments
Silver Sulfadiazine (Silvadene®)	Bactericidal; interference with cell wall	Gram negatives and positives, including Pseudomonas and Staph sp.	Pseudomonas (rare)	Contraindicated in sulfa-allergic patients. Rarely may cause neutropenia. Forms a thick scum layer that must be wiped off.
Bacitracin®	Bactericidal; inhibits cell wall synthesis	Gram negatives and positives	Staph sp. (rare)	1-6% incidence of contact dermatitis. Higher incidence of systemic toxicity.
Neomycin/Polymyxin B Sulfate/Bacitracin zinc (Neosporin®)	Bactericidal; multimodal	*S. pneumo*, *S. aureus*, *E. coli*, Pseudomonas	Rare	Zinc thought to promote wound healing
Povidone-iodine (Betadine®)	Bactericidal; delivers iodophor through cell wall which inactivates cytoplasmic substrates	Gram negatives and positives, MRSA, fungi, some viral	Very rarely	Potential hypersensitivity and systemic toxicity. Possibly diminishes fibroblast and keratinocyte proliferation. Desiccates tendon and bone.
Iodosorb® gel/ Iodoflex® pads	Microbeads soak up wound exudate and release iodine	Gram negatives and positives, MRSA, fungus, some viral	Rare	Does not need to be changed as frequently.

4

Table 4.2. Commonly used topical enzymatic agents

Agent	Mechanism	Indication	Comments
Papain-urea (Accuzyme®)	Nonspecific proteinase extracted from papaya, activated in moist environment. Potency increased when combined with urea.	Wounds with necrotic tissue that have had any thick eschar removed or cross-hatched	Painful when applied to surrounding healthy tissue
Papain urea-chlorophyllin-coppper (Panafil®)	Added effect of chlorophyllin and copper promotes epithelialization	Wounds with mild to moderate burden necrotic	May be used in tandem with other modalities, such as wet-to-dry dressings and vacuum assisted closure
Collagenase ointment (Novuxol®)	Obtained from *Clostridium histolyticum*, cleaves native and denatured collagen	Burns, wounds with minimal necrosis	No difference between Novuxol and Fibrolan have been demonstrated
Firbrinolysin/DNAse (Fibrolan®)	Fibrinolysin from bovine plasma breaks down fibrin in clots and subsequently stimulates debridement by macrophage. DNAse from bovine pancreas cleaves nuclear proteins in wounds.	Most ulcers with fibrinous debris	

The role of povidone-iodine (Betadine®) in topical wound care is somewhat controversial. Several animal studies demonstrate no adverse affect on wound tensile strength or reepithelialization rates. On the other hand, several human in vitro and in vivo studies have shown that Betadine inhibits fibroblast proliferation, kerotinocyte growth and migration, and hampers the phagocytic effect of monocytes and granulocytes. In addition, any admixture of blood, pus or fat has been proven to diminish the antimicrobial effect of Betadine. Given this data, many plastic surgeons do not use Betadine as a topical antimicrobial, although it is still commonly used in the operating room as a prepping agent. Other specialties that treat wounds still use Betadine due to the lack of convincing clinical trials and a long history of its use.

Enzymatic Agents

In addition to appropriate antimicrobial therapy, the wound must be properly debrided of any devitalized tissue. Necrotic tissue can serve as a culture medium for further bacterial proliferation, and its presence will impede the healing process. Sharp debridement is the simplest, most effective means of eliminating nonviable tissue. Enzymatic debriding agents are an adjunct to surgical debridement. As with antimicrobial agents, there is a spectrum of enzymatic agents that is commercially available (Table 4.2).

Growth Factors

Perhaps the realm with the greatest therapeutic potential in the pharmacologic treatment of wounds is the use of growth factors. A variety of growth factors and chemotactic agents have been discovered since the 1970s, and many have been probed for possible clinical applications. Platelet-derived growth factor (PDGF) is present in acute surgical wounds, however *not* in chronic, nonhealing wounds. In several randomized controlled trials, topical application of PDGF increased wound tensile strength and accelerated the healing process overall. Recombinant PDGF is currently the only cytokine approved for use in chronic wounds, specifically in neuropathic diabetic foot ulcers. It is available commercially as beclapermin (Regranex®); however the extremely high cost makes its use prohibitive in many centers.

Pearls and Pitfalls

1. Choosing a topical wound care agent must be done in conjunction with choosing the appropriate dressing.
2. If the wound is complicated by underlying osteomyelitis, topical antibiotics will not suffice in eradicating the infection.
3. Close observation is warranted when instituting a new therapy for the wound since many of these agents have side effects and can even induce an allergic reaction.
4. Enzymatic agents cannot penetrate thick eschar. Necrotic tissue should be sharply debrided before applying a topical enzymatic substance.
5. Some topical wound care agents, such as papain-containing enzymatic agents, are painful when they come in contact with surrounding healthy skin.

Suggested Reading

1. Ladin D. Becaplermin gel (PDGF-BB) as topical wound therapy. Plast Reconstr Surg 2000; 105(3):1230.
2. Mustoe TA. Understanding chronic wounds: A unifying hypothesis on their pathogenesis and implications for therapy. Am J Surg 2004; 187(5A):65S.
3. Steed DL. Debridement. Am J Surg 2004; 187(5A):71S.

Chapter 5

Negative Pressure Wound Therapy

Peter Kim and Gregory A. Dumanian

Introduction

Management of difficult acute and chronic wounds poses a significant challenge to the patient and the caregiver. The application of negative pressure wound therapy (NPWT) has proven to ease some of this burden by promoting a favorable wound-healing environment, decreasing the need for frequent dressing changes, improving patient comfort, and reducing associated costs. In NPWT, a pliable foam dressing is cut to shape, placed into a wound, and covered with an occlusive dressing. Controlled sub-atmospheric pressure is then applied to the wound by evacuating air and liquids from the foam dressing. The most commonly used device for applying NPWT is the wound VAC®.

Mechanism

Initial research on pigs demonstrated the superiority of NPWT when compared with moist saline dressings. Although few randomized controlled trials exist in humans, one review suggests that NPWT improves granulation, augments wound contraction, and reduces the need for systemic antibiotics. Several mechanisms may be responsible for these observations. NPWT has been shown to improve local tissue perfusion and reduce the bacterial load on wounds. NPWT may also improve granulation tissue formation by reducing proteolytic enzymes found in wound exudates, by promoting a moist wound, and by applying shear forces that induce cellular hyperplasia.

Indications

Since NPWT has become commercially available, the list of indications has continued to grow (Table 5.1). NPWT is indicated for almost any open wound where surgical closure is not feasible or desirable. While it may be used as a sole treatment toward achieving wound closure, NPWT is often used as a bridge toward definitive surgical management. Much of its utility is in creating favorable conditions for subsequent wound reconstruction.

With the success seen in treating a variety of wounds, many authors have tried to extend the application to improve graft take and flap survival. When flaps are used to cover wounds, some studies suggest that additional use of the NPWT may promote improved flap survival and overall wound healing. In several case series, skin graft take was shown to be 90% or greater when the VAC was employed in lieu of a traditional bolster dressing. Recipient sites with irregular contours, susceptibility to shear forces, and excess drainage were thought to be particularly amenable to VAC dressings. Nevertheless, these results have yet to be confirmed in randomized control trials.

Practical Plastic Surgery, edited by Zol B. Kryger and Mark Sisco. ©2007 Landes Bioscience.

Table 5.1. Indications and contraindications for use of NPWT

Indication	Notes
Chronic open wounds	
Diabetic ulcer	
Pressure sore	Debridement must be performed prior to application of NPWT
Traumatic wounds	
Extirpative defects	Brachytherapy and external-beam irradiation can be performed through the dressing
Spinal and orthopedic wounds	Dressing can be placed directly over hardware after debridement has been performed
Sternotomy defects	
Open abdomen	Excellent for the temporary management of bowel edema and gross peritoneal contamination
Burns	May be applied over allograft of skin substitute
Skin graft bolster	NPWT improves graft take by limiting shear forces and evacuating fluid collections
Contraindications	
Malignancy in the wound	
Untreated underlying osteomyelitis	
Nonenteric or unexplored fistulas	
Undebrided necrotic tissue	
Untreated active soft tissue infection	
Exposed internal organs	
Exposed blood vessels or vascular prosthetic grafts	
Coagulopathic patients (relative)	

Technique

Application of NPWT can be performed by anyone with the appropriate training, provided that the wound is hospitable. Prior to application, the wound should be debrided of any necrotic or fibrinous debris and adequate hemostasis achieved. The surrounding skin is then cleansed and dried. The sponge is cut to be slightly smaller than the volume of the wound. The adhesive dressing is then applied over the sponge such that there is at least a 6 cm overlap on adjacent skin; it is imperative that a hermetic seal be achieved. Once the adhesive dressing has been applied, it is pierced and the adhesive suction tube is applied over this opening. The device is then turned on and continuous suction is applied. When placed properly, the dressing will create a closed suction environment. Depending on the nature of the wound, the NPWT dressing can be changed every 48 to 72 hours. The dressing should be taken down sooner should the patient show signs of infection or if the seal on the dressing becomes compromised.

The VAC® device comes with two types of foam available for use. The original foam is black, and it is made of polyurethane. It is hydrophobic which enhances exudate removal. It has reticulated pores and is considered to be the most effective at stimulating granulation tissue while aiding in wound contraction. The second, newer available foam is white. It is a denser foam with a higher tensile strength. It is hydrophilic and possesses overall nonadherent properties. The white foam does not require the use of a nonadherent layer. It is generally recommended for situations in which slower growth of granulation tissue into the foam is desired or when the patient cannot tolerate the black foam due to pain. Due to the fact that it has a higher density than the black foam, higher pressures must be utilized in order to provide adequate negative pressure distribution throughout the wound. Newer foams are constantly emerging, such as silver-impregnated foams.

Pearls and Pitfalls

- Though the NPWT dressing is applied less frequently than the comparable saline dressings, some patients find it painful and require appropriate premedication with analgesics.
- The skin surrounding the wound should be completely dry prior to placement of the adhesive. Shaving of hair and application of Benzoin™ may facilitate adhesion.
- The foam should be cut down to the proper size so that it fits within the borders of the wound, otherwise it will compress the surrounding healthy skin.
- After placing the dressing and applying the vacuum, ensure an adequate seal by clamping the tubing leading to the dressing and then disconnecting the tubing from the machine. If the seal is adequate, the sponge should **slowly** return to its original shape.
- If poor hemostasis is presumed or if the wound is particularly "weepy", close monitoring of the patient's hemodynamic and fluid status is warranted.
- The foam should not encroach on normal surrounding skin. However, two separate wounds can be hooked up to a single suction tubing by bridging the two sponges with a thin piece of foam that traverses the normal interfering skin.
- A useful trick for dressing change analgesia is to clamp the tubing and inject 1% lidocaine with epinephrine (10-30 ml) into the tubing distal to the clamp. The vacuum will suck the local anesthetic into the wound.

Suggested Reading

1. Argenta LC, Morykwas MJ. Vacuum-assisted closure: A new method for wound control and treatment: Clinical experience. Ann Plast Surg 1997; 38:563.
2. Morykwas MJ, Faler BJ, Pearce DJ, Argenta LC. Effects of varying levels of subatmospheric pressure on the rate of granulation tissue formation in experimental wounds in swine. Ann Plast Surg 2001; 47(5):547-51.
3. Evans DL, Land LL. Topical negative pressure for treating chronic wounds: A systematic review. Br J Plast Surg 2001; 54:238.

Chapter 6

Leeches

Mark Sisco and Michael A. Howard

Leeches have been used for medicinal purposes for 2,500 years. Their contemporary use in plastic surgery, first described in 1836, is for the relief of soft tissue venous congestion, most commonly in compromised flaps and in avulsed or replanted appendages such as the ear and finger. Leeches have proven especially useful in microsurgery, in which venous anastamoses may prove difficult. The success rate of salvaging tissue with medicinal leech therapy has been reported to be up to 70-80%. In 2004, the U.S. Food and Drug Administration approved the commercial marketing of leeches for medicinal purposes.

Medicinal leeches, typically *Hirudo medicinalis*, are unique in their ability to effect prolonged venous bleeding, because they inject salivary substances that have anticoagulant, antiplatelet and vasodilatory effects. These components cause bleeding for up to 24 hours, long after the leech has been removed. Leeches also release a local anesthetic, rendering bites painless.

The indication for the use of leeches is venous congestion. This diagnosis can be made by observing the following signs: cyanosis, edema and brisk capillary refill. Pricking the affected area with a needle results in dark bleeding. Intraoperative issues, such as difficulty with a venous anastomosis or undue pedicle tension, also suggest the diagnosis. When flaps are congested, other mechanical means to improve venous outflow should be considered first, including removing tight sutures, decompressing tunneled pedicles, and evacuating hematomas.

While the initial leech bite causes about 5-15 ml of blood loss, each wound can ooze an additional 50-150 ml of blood over a period of up to 24 hours. As such, the number and timing of leeches to be applied should be tailored to the area involved. Venous ingrowth can be anticipated in 3-5 days. Treatment should be continued until signs of venous congestion subside. This may take up to 10 days.

Leeches are commercially available from several sources. After receipt, leeches can be stored in the pharmacy or on the patient floor. They must be refrigerated and kept in a feeding medium (either dissolved in distilled water or a gel) that arrives with them.

A general approach is as follows:

1. Clean the skin thoroughly with soap and water. It is especially important to remove old antiseptic or other noisome substances, as they may affect the leech's appetite.
2. Cut a 1 cm hole in the middle of a saline-moistened gauze sponge. Place this sponge so that the hole overlies the area to which the leech is to be applied.
3. Place the leech on the gauze pad such that its head (the end that tends to move the most) is against the skin. It may be helpful to place the leech in the barrel of a 5 ml syringe (after removing the plunger) and inverting the syringe against the skin so that the leech can be specifically applied.

4. Leeches will usually attach immediately. If not, prick the skin with a needle before reapplying the leech.
5. Leeches will typically remain in the same place until they are completely distended, at which point they will fall off. This usually takes 30-45 minutes. Instruct the patient's nurse to check on the patient often so that leeches are not lost after detachment.
6. Wounds can be encouraged to bleed after detachment by occasionally scraping the eschar off.

7. Used leeches can be discarded by anesthetizing and then euthanizing them in 8% and 70% alcohol, respectively. They should be considered biohazardous and disposed of as such.

If several leeches are used concurrently, it may be necessary to check the patient's hemoglobin/hematocrit at regular intervals. All patients should be started on an oral antibiotic while on leech therapy. Suggested antibiotics include a fluoroquinolone or amoxicillin/clavulanic acid. Patients with HIV or taking immunosuppressive medications should not undergo leech therapy because of the risk of bacterial sepsis.

Pearls and Pitfalls

Leeches should be used as a treatment of last resort when all other means of venous outflow establishment are exhausted. It is imperative to relieve a mechanical or iatrogenic cause of venous compromise.

It is critical to rule out arterial insufficiency as the cause of flap necrosis or pallor, since leeches will not work in this situation.

Flaps demonstrate significantly decreased survival after 3 hours if venous congestion is not relieved. As opposed to arterial ischemia, venous stasis tends to cause irreversible damage. Since leeches must be flown in, it is wise to anticipate their need as early as possible. We have ordered them intraoperatively in some cases.

Although leeches can be reused on the same patient, they tend not to work as well. Used leeches should not be stored with unused ones to prevent cross-contamination. Used leeches should never be applied to another patient.

The importance of an appropriate bedside manner in ensuring acceptance of and compliance with this regimen cannot be understated. Most patients are willing to accept treatment when it is explained in a thorough and confident manner. It is also critical to include nursing and ancillary staff in the discussion of leeches, as many will not have seen them used before. We have found that by observing the first application of a leech, most nurses are willing to apply subsequent leeches without supervision.

Leech Suppliers

Carolina Biological Supply Co.—(800) 262-2922

Leeches U.S.A.—(800) 645-3569, after hours: (800) 488-4400 Ext. #2475

Suggested Reading

1. de Chalain TM. Exploring the use of the medicinal leech: A clinical risk-benefit analysis. J Reconstr Microsurg 1996; 12(3):165-172.
2. Haycox C, Odland PB, Coltrera MD et al. Indications and complications of medicinal leech therapy. J Am Acad Dermatol 1995; 33(6):1053-1055.
3. Utley DS, Koch RJ, Goode RL. The failing flap in facial plastic and reconstructive surgery: Role of the medicinal leech. Laryngoscope 1998; 108(8 Pt 1):1129-1135.
4. Whitaker IS, Izadi D, Oliver DW et al. Hirudo medicinalis and the plastic surgeon. Br J Plast Surg 2004; 57(4):348-53.

Chapter 7

Local Anesthetics

Zol B. Kryger and Ted Yagmour

Introduction

As the number of plastic surgical procedures performed under local anesthesia continues to grow, a thorough understanding of local anesthetic techniques has become essential. Furthermore, emergency care of lacerations, avulsions and other acute injuries also necessitates an adequate grasp of local anesthesia. It is important to obtain informed consent prior to using local anesthesia. Discussion of the risks and benefits of the surgery alone is not sufficient. Anesthetic-related issues such as adverse reactions, systemic toxicity, nerve damage, hematoma and pain both during and after the injection should be addressed.

Mechanism of Action

Local anesthetics exert their effect by temporarily blocking nerve conduction. This is achieved by interference with influx of sodium ions through the sodium channel. This leads to a slowing of the rate of membrane depolarization, a lowering of the threshold potential, and the inhibition of propagation of the action potential down the length of the axon. The smallest unmyelinated sensory nerves (C fibers) are affected first. The motor nerves are usually larger and myelinated, and are unaffected or only mildly affected by the actions of local anesthetics at the doses commonly used.

Pharmacodynamics

Local anesthetics can be classified based on their molecular structure as either amides or esters (Table 7.1). The amides, such as lidocaine, are metabolized in the liver by microsomal enzymes and excreted in the urine. The esters, such as cocaine, are quickly metabolized by plasma pseudocholinesterase into PABA and excreted in the urine.

Table 7.1. Commonly used local anesthetic agents and their duration of action

Anesthetic Agent	Class	Duration of Action
Lidocaine (Xylocaine)	Amide	1.5-2 hours
- Lidocaine with epinephrine		up to 3 hours
Bupivicaine (Marcaine)	Amide	3-6 hours
- Bupivicaine with epinephrine		up to 10 hours
Mepivicaine (Carbocaine)	Amide	2.5-3 hours
Cocaine (Cocaine)	Ester	0.5-3 hours
Tetracaine (Pontocaine)	Ester	1-3 hours

Local anesthetics are acidic, in the pH range of 5-7 . Their pH further decreases with the addition of epinephrine to the anesthetic solution. Once they enter the tissue, the body's bicarbonate buffer system converts the acidic solution to a more basic form. This is the active, uncharged form of the drug that can diffuse through the plasma membrane of the neurons. Bupivicaine, with its higher pKa, has a slower onset of action than lidocaine, which has a lower pKa. Acidic tissue, such as a hypoxic or infected wound, increases the fraction of ionized drug, thus delaying the onset and decreasing the efficacy of local anesthetics.

The Addition of Epinephrine

A vasoconstricting agent such as epinephrine, is often added to local anesthetic solutions. This provides the following benefits:

- Decreases the rate of systemic absorption
- Reduces the risk of systemic side effects
- Prolongs the duration of action of the anesthetic
- Improved hemostasis due to its vasoconstrictive effects

Premixed solutions containing epinephrine are acidified even further than plain local anesthetics. This increased acidity delays the onset of action and is more painful on injection. There is no utility in using greater than 1:100,000 epinephrine solutions. No additional vasoconstrictive benefit is offered, whereas the risk of toxicity increases in a dose-dependent manner. Adequate hemostasis relies greatly on allowing adequate time for the vasoconstrictive effects to occur. This usually takes 7-10 minutes.

Contraindications to the use of epinephrine-containing solutions include patients with unstable angina, cardiac dysrhythmias, severe uncontrolled hypertension, or pregnant patients with placental insufficiency. Relative contraindications include hyperthyroidism and concurrent use of MAOI or tricyclic antidepressants. When contraindicated, phenylephrine (1:20,000) can be substituted, however it is not as effective as epinephrine.

The Addition of Bicarbonate

Sodium bicarbonate can be added to local anesthetics in order to alkalinize the solution. This neutralization of the low pH creates a solution that is less irritating to the tissues and less painful on administration. The limiting factor in the addition of bicarbonate is the tendency for the lipid soluble agents, such as bupivicaine, to precipitate at the more neutral pH values. Therefore, bicarbonate can be added to lidocaine but should generally not be used with bupivicaine.

Lidocaine

Lidocaine is the most widely used local anesthetic. It is prepared as a 1% (10 mg/ml) or 2% (20 mg/ml) solution with or without epinephrine. Its duration of action is about 1.5 hours without epinephrine, and this is doubled to 3 hours with the addition of epinephrine to the solution (1:100,000). Lidocaine can also be used as a dilute solution (0.2%–0.5%) for certain procedures such as a rhytidectomy. This solution is adequately anesthetizing and vasoconstrictive. A commonly used dilute solution, the modified Klein solution, can be prepared as follows: 20 ml of 2% lidocaine, 5 ml of sodium bicarbonate, and 1 ml of 1:1,000 epinephrine all mixed in 500 ml of lactated Ringer's solution. The maximum safe dose for plain lidocaine is reported as 3-4 mg/kg. With the addition of epinephrine, this increases to 7 mg/kg.

Recent literature, however, refutes this figure, providing evidence for a much higher maximal safe dose-up to 35 mg/kg when combined with epinephrine.

Bupivicaine

Bupivicaine is widely used in plastic surgery because of its long duration of action. It is effective for 3-6 hours, significantly longer than lidocaine. The addition of epinephrine can increase this duration to 10 hours. It comes as a 0.25% or 0.5% solution, with or without epinephrine. It is somewhat more painful than lidocaine on administration. It should not be used for large volume infiltration because of its high toxicity profile. It can, however, be combined with lidocaine for lengthy facial procedures such as a rhytidectomy. This combination has a rapid onset of action due to the lidocaine, and a long duration of action due to the bupivicaine. The maximum safe dose of bupivicaine is 2.5 mg/kg, and this increases to 3 mg/kg with the addition of epinephrine.

Mepivicaine

Mepivicaine is similar to lidocaine except for its slightly longer duration of action. Its anesthetic effects can last up to 3 hours. It is prepared as a 0.5% or 1% mixture. It is much less commonly used than lidocaine due to its higher cost and lesser availability. It also has a slightly increased risk of toxicity compared to that of lidocaine.

Eutectic Mixture of Local Anesthetics (EMLA)

EMLA is typically a cream composed of 2.5% lidocaine and 2.5% prilocaine. It provides dense topical anesthesia 45-60 min after application. It must be covered with an occlusive dressing for this period in order for the cream to be effective. Within 2 hours, the maximal depth of penetration is reached. EMLA cream is not widely used because of the long latency until onset of action and the need for the occlusive dressing. It is effective in children who will not tolerate a needle stick, as long as it is applied sufficiently in advance.

Cocaine

Cocaine is used primarily as a topical agent for septo-rhinoplasty procedures. It comes in 4% or 10% solutions. As opposed to other local anesthetics, cocaine produces significant local vasoconstriction without the addition of epinephrine. Its onset is extremely rapid (1-2 minutes), but it takes an additional 5 minutes for its vasoconstrictive effects to begin. Its duration of action is up to 3 hours. Cocaine can be highly toxic by sensitizing the heart to circulating catecholamines. This can lead to tachycardia, hypertension, coronary vasospasm and dysrhythmias. Its CNS effects are stimulatory before leading to confusion, dysphoria and seizures. The maximum safe dose is about 3 mg/kg.

Tetracaine

Tetracaine, similar to cocaine, is used as a topical agent in nasal surgery. It can also be combined with EMLA as a topical agent for anesthesia for closed nasal reduction. It comes as a 0.05% to 4% solution. It has a rapid onset and is effective for 1-3 hours. Tetracaine is several times more potent than cocaine. It is extremely toxic due to its slow rate of metabolism, and the maximum safe dose is 1 mg/kg.

Toxicity

The risk for adverse reactions with local anesthetics is low, but it is important to be familiar with the signs and symptoms of toxicity. Some sites on the body are at greater risk for toxicity due to their robust blood supply. The face and scalp are rich in vascularity, and the systemic absorption of the drug from these sites is higher. In addition, patients with pseudocholinesterase deficiency, myasthenia gravis and those taking cholinesterase inhibitors are at a higher risk for overdose. Certain local anesthetics pose a higher risk of toxicity due to their lipid solubility. For example, bupivicaine is more lipid soluble than lidocaine and has a higher risk of toxicity.

The **cardiovascular** and **CNS** are the two systems most commonly affected by local anesthetic toxicity. **CNS manifestations occur before cardiac signs**, and the early signs and symptoms include restlessness, headache, disorientation, dizziness, blurred vision, tinnitus, slurred speech, nystagmus and twitching. Late signs of toxicity include generalized seizures, apnea and death. Treatment of seizures is by administration of a benzodiazepine such as diazepam or midazolam. Cardiovascular manifestations appear after those in the CNS and include myocardial depression, hypotension or shock, and dysrhythmias such as prolonged P-R interval and widening of the QRS complex. Of the commonly used local anesthetics, bupivicaine is the most cardiotoxic due to its strong affinity for the cardiac calcium channels.

Allergic Reactions

Allergies to local anesthetics are extremely rare, and account for less than one percent of adverse drug reactions during anesthesia. Reactions can range from a subtle rash to a full-blown anaphylactic response. The amides, such as lidocaine, rarely cause allergic reactions. The esters, however, such as cocaine, are metabolized by plasma pseudocholinesterase into PABA, and allergic reactions to these anesthetics are more common. If an allergic reaction does occur following administration of a local anesthetic, the culprit is usually one of the preservatives or additives in the solution rather than the anesthetic agent itself.

Pearls and Pitfalls

It has become increasingly clear that the maximum safe dose of lidocaine is much higher than previously thought. The traditional value of 7 mg/kg as the maximal dose of lidocaine with epinephrine is probably much too low. Recent literature supports a value closer to 35 mg/kg. Furthermore, the common use of dilute solutions of lidocaine with epinephrine has demonstrated that concentrations above 1% are not required. In the vast majority of cases, dilute solutions of lidocaine will provide adequate anesthesia, and the addition of epinephrine will greatly increase the maximal dose that can be safely used, while decreasing blood loss. One should wait at least 7-10 minutes for the vasoconstrictive effects of the epinephrine to take effect.

Suggested Reading

1. Ahlstrom KK, Frodel JL. Local anesthetics for facial plastic procedures. Otolaryng Clin N Am 2002; 35(1):29.
2. Baker IIIrd JD, Blackmon Jr BB. Local anesthesia. Clin Plast Surg 1985; 12(1):25.
3. Klein JA. Tumescent technique for regional anesthesia permits lidocaine doses of 35 mg/kg for liposuction. J Dermatol Surg Oncol 1990; 16(3):248.
4. Zilinsky I, Bar-Meir E, Zaslansky R et al. Ten commandments for minimal pain during administration of local anesthetics. J Drugs Dermatol 2005; 4(2):212.

Chapter 8

Basic Anesthetic Blocks

Zol B. Kryger

Introduction

Regional anesthetic blocks can be a valuable supplement or even replacement to the more common field block used in plastic surgery. The principle behind a regional nerve block is to anesthetize a sensory nerve that supplies innervation to the area of injury at a single more proximal site. The advantages of this technique over a field block are that it is usually much faster, it requires a smaller volume of local anesthetic, and it avoids distortion of the surgical site, as well the bleeding that often ensues after multiple needle sticks. It does, however, require a thorough knowledge of the anatomy of the nerve, and it does not always provide complete anesthesia to the desired site secondary to collateral innervation.

This chapter will focus on regional nerve blocks in two key anatomic regions: the face and hands.

Choice of Anesthetic Agent

The previous chapter discussed the various anesthetic agents in detail. Briefly, most blocks can be achieved using 1% lidocaine with epinephrine (1:100,000). The addition of epinephrine prolongs the duration of action of the anesthetic, as well as providing vasoconstriction of the site. Epinephrine can be used anywhere in the face; however it should not be used in the fingers or penis. The addition of bupivicaine to the lidocaine solution can prolong the duration of anesthesia for several hours, providing additional post-procedure pain relief. Furthermore, sodium bicarbonate can be added to the lidocaine solution to cut back on the burning sensation from the injection.

Choice of Syringe and Needle

A 5 ml syringe is usually sufficient for most blocks since rarely is more than this amount required. The smaller syringe is also easier to maneuver. The needle should be a 25 or 27 gauge. The length of the needle must be sufficient to reach the target. For example, the infraorbital foramen is usually reached through the oral cavity, requiring at least a 1 inch needle.

Regional Block of the Scalp

Indication

Anesthesia of the scalp down to the periosteum.

Technique

The scalp is innervated by branches of the trigeminal and cervical nerves. These nerves can be anesthetized as they penetrate the scalp. They become subfascial along

Practical Plastic Surgery, edited by Zol B. Kryger and Mark Sisco. ©2007 Landes Bioscience.

a line that encircles the head (like a skull-cap). This line passes just above the tragus and through the glabella and occiput. A wheal should be raised in the subdermal plane along this line. About 10 ml of lidocaine is required every few centimeters.

Supraorbital Nerve Block

Indication

Anesthesia of the upper eyelid and medial forehead.

Technique

Palpate the supraorbital notch/foramen at the junction of the medial and middle thirds of the orbital ridge (about 2.5 cm off the midline of the face). Raise a wheal using a 25 gauge 1 inch needle. Advance the needle until the tip meets the foramen, and inject 1-2 ml while withdrawing.

Infraorbital Nerve Block

Indication

Anesthesia of the lower eyelid, medial cheek region or upper lip.

Extraoral Technique

Place the index finger in the canine fossa pointing caudal towards the infraorbital foramen. Raise a wheal using a 25 gauge 1 inch needle about 1 cm lateral to the ala of the nose. Advance the needle towards the tip of the finger until the tip meets the foramen on the maxilla. Inject 1-2 ml into the foramen and while withdrawing. The infraorbital canal runs in a superolateral direction.

Intraoral Technique

Retract the cheek with the thumb and introduce the needle into the upper gingival sulcus above the second bicuspid. Rest the syringe on the lower lip of the patient. Aim slightly laterally away from the midline along the maxilla until the infraorbital foramen is encountered. Inject 1-2 ml into the foramen and inject while withdrawing. The infraorbital canal runs in a superolateral direction.

Mental Nerve Block

Indications

Anesthesia of the lower lip, anterior portion of the lower jaw (including the anterior lower teeth).

Extraoral Technique

The mental foramen is located directly below the root of the second lower bicuspid at the midpoint between the lower and upper margins of the mandible. The needle is inserted into the skin and a wheal is raised. It is aimed inferolaterally towards the mental foramen, and anesthetic is injected while the needle advances until bone is met. After instilling 1 ml of anesthetic, the needle is used to palpate the mental foramen after which an additional 1 ml is injected into the foramen.

Intraoral Technique

With the mouth closed, the cheek is retracted and needle inserted into the gingivobuccal sulcus below the bicuspids. A wheal is raised, and the needle is aimed towards the root of the second bicuspid and advanced at 45° until bone is reached. After instilling 1 ml of anesthetic, the needle is used to palpate the mental foramen after which an additional 1 ml is injected into the foramen.

Regional Block of the External Nose

Indications

Anesthesia of the skin of the nose.

Technique

The two sides of the nose should be anesthetized separately. The needle is introduced into the skin about 1 cm lateral to the alar base. A wheal is raised, and the needle is advanced towards the radix; 2-3 ml is injected along this line. The needle is withdrawn almost completely and then directed downward towards the oral commissure. An additional 1-2 ml is injected along this course. The entire procedure is repeated for the other side of the nose.

Regional Block of the External Ear

Indication

Anesthesia of the ear.

Technique

The anterior ear is supplied by the auriculotemporal nerve and the posterior ear by the greater auricular nerve and occipital nerve (including its mastoid branch). These nerves all reach the ear from the superior, posterior and inferior directions only. A needle is inserted 2 cm above the helix and advanced anteroinferiorily and posteroinferiorily. The needle is removed and inserted 3 cm posterior to the ear and advanced anterosuperiorily and anteroinferiorily. The needle is removed and inserted 1 cm below the ear, advancing it posterosuperiorily and anterosuperiorly. When these three injections are completed, a continuous infiltration around the entire ear (excluding the anterior portion) has been achieved.

Radial Nerve Block

Indication

Anesthesia of the radial dorsum of the hand and proximal thumb, index and middle finger. The ring finger should also be blocked with an ulnar nerve block.

Technique

1. Identify extensor pollicus longus (dorsal tendon of the anatomical snuffbox).
2. Insert the needle over the tendon at the base of the first metacarpal.
3. Inject superficial to the tendon (about 2 ml) and over the snuffbox (1 ml).

Median Nerve Block

Indication

Anesthesia of the palmar side of the thumb index finger and middle finger, and radial side of the ring finger. Also, the nailbeds of the above fingers can be blocked with this technique. The thenar region (palmar cutaneous branch of the median nerve) can also be blocked.

Technique

1. Identify flexor carpi radialis and palmaris longus by having the patient make a clenched fist and slight wrist flexion.
2. Insert the needle 2 cm proximal to the proximal wrist crease.
3. As the needle passes through the flexor retinaculum, 3 ml of anesthetic is injected.
4. Injection of an additional 1 ml above the retinaculum will anesthetize the palmar cutaneous branch supplying the thenar eminence.

Ulnar Nerve Block

Indications

Anesthesia of the little finger and ulnar side of the ring finger.

Technique

1. Identify flexor carpi ulnaris by having the patient forcefully ulnar deviate the wrist slightly with the fingers fully extended.
2. The ulnar nerve lies radial to the flexor carpi ulnaris tendon.
3. Insert the needle 2 cm proximal to the wrist on the radial side of the tendon directed towards the midline.
4. After parasthesias are felt, inject 4 ml of anesthetic in a fanwise fashion along the course of the nerve.

Digital Nerve (Ring) Block

Indications

Anesthesia of the digit.

Technique

1. With the dorsum of the hand facing upward, insert the needle into the dorsal skin at the midpoint between the digits (the apex of the "V" of the web space) and raise a wheal.
2. Advance the needle towards the palm perpendicular to the skin and infiltrate along this course about 2 ml of anesthetic.
3. Withdraw the needle almost completely and then begin advancing the needle towards the middle of the digit, infiltrating the skin on the dorsum of the finger base.
4. The digital nerves on either side of the finger should be anesthetized in this manner.

Pearls and Pitfalls

1. The supraorbital, infraorbital and mental foramena all lay along a vertical line that also includes the pupil in the midgaze position. Therefore, if any two of the foramena have been located, the third can be easily found.
2. Epinephrine requires about 10 minutes until full effect, and the same is true for lidocaine used in a regional block. Therefore, one should administer the block in advance.
3. Several studies reviewing thousands of cases of digital anesthesia have found that using epinephrine in the digits is entirely safe, with almost no cases of digital ischemia secondary to the epinephrine. However, until a prospective trial demonstrates the absolute safety of this practice, epinephrine should not be used in the digits.

4. An adequate block is not always 100% successful at eliminating pain from the site of injury. Often a supplemental field block is required after the initial regional block has taken effect.

Suggested Reading

1. Stromberg BV. Anesthesia. In: McCarthy JG, ed. Plastic Surgery. 1st ed. Philadelphia: WB Saunders Company, 1990.
2. Wedel DJ. Anesthesia in hand and upper extremity surgery. In: Berger RA, Weiss AC, eds. Hand Surgery. Philadelphia: Lippincott Williams and Wilkins, 2004.
3. Zide BM, Swift R. How to block and tackle the face. Plast Reconstr Surg 1998; 101(3):840-51.

Chapter 9

Surgery under Conscious Sedation

Zol B. Kryger and Neil A. Fine

Introduction

Conscious sedation is a technique that combines the use of local anesthesia and intravenous sedation. It is defined as a depressed level of consciousness to the point that the patient is in a state of relaxation, but maintains respiratory drive and the ability to protect the airway. The patient is also capable of purposefully responding to physical and verbal stimulation. This is in contrast to deep sedation, in which the patient is unable to respond to verbal stimuli, will only respond to painful stimulation with withdrawal and has potential compromise of airway protection and respiratory drive. As opposed to monitored anesthesia care (MAC), in which an anesthesiologist or nurse anesthetist are required, conscious sedation can be performed by a nurse under the supervision of the operating surgeon.

Conscious sedation is rapidly gaining acceptance and popularity among plastic surgeons. It has been utilized for many years by other specialties, and now with the growth in office-based procedures and surgicenters, there has been a corresponding increase in the role of conscious sedation. Currently, almost all aesthetic procedures can be performed using a local anesthetic combined with some form of intravenous sedation. These include breast augmentation, breast reduction, mastopexy, abdominoplasty, rhytidectomy, rhinoplasty, blepharoplasty and liposuction.

Benefits and Disadvantages of Conscious Sedation

There are a number of benefits to the use of conscious sedation instead of general anesthesia or deep sedation. First, the complications associated directly with the administration of a general anesthetic are avoided. These are not negligible, and include adverse cardiopulmonary effects, airway injury and positional nerve injuries. Such complications occur in roughly 1-2% of aesthetic procedures performed under general anesthesia. The incidence of postoperative nausea and vomiting, which account for most unintended admissions after outpatient surgery, is much less than that associated with general anesthesia. Secondly, the risk of developing deep vein thrombosis (DVT) as a result of blood pooling in the lower extremities during general anesthesia is greatly reduced due to the continued contraction of leg muscles and the spontaneous shifting of the patient during the procedure. Third, as a result of the relatively large dose of an amnestic medication that is used, most patients have no memory of the procedure, no recollection of experiencing pain, and many choose to undergo conscious sedation at subsequent procedures. Finally, because it can be performed safely without the presence of an anesthesiologist, there is a considerable saving in cost to the patient.

Conscious sedation is not suited to all patients. Furthermore, the use of conscious sedation requires a surgeon who can "multi-task," focusing on the operation

Practical Plastic Surgery, edited by Zol B. Kryger and Mark Sisco. ©2007 Landes Bioscience.

as well as on the vital signs and level of arousal of the patient. The fact that the patient is conscious and can shift position or move freely, necessitates that the surgeon be prepared to stop working at any moment. Nevertheless, many patients are well-suited for conscious sedation.

Preoperative Considerations

Prior to using conscious sedation for the first time, the surgeon must familiarize herself with the medications she will be using, as well as their side effects and reversal agents. She must also be familiar with ACLS protocol, airway management and have readily available resuscitation equipment. Immediate access to an anesthesiologist in case of emergency is strongly recommended.

Proper patient selection is an important preoperative decision. Those with moderate to significant cardiopulmonary disease are poor candidates. Patients should meet the criteria of the American Society of Anesthesiologists status I or II. This means that candidates for conscious sedation should be healthy or have only a mild systemic disease that results in no functional limitation (e.g., obesity, diabetes, hypertension and extremes of age). All other patients should receive monitored anesthesia care by an anesthesiologist or general anesthesia. Furthermore, individuals with anxiety disorders and extreme fear of the operating room may not be suited for conscious sedation.

Prior to the procedure, patients may benefit from premedication with intravenous diazepam (Valium), administered in increments of 5-10 mg. The dose administered usually ranges from 10 to 50 mg, with the goal being adequate preoperative subjective relaxation of the patient with the desired endpoint being of slurred speech. **Oral** diazepam is also an option; however, it has to be given almost an hour prior to the procedure in order to be effective. A second medication that should be administered preoperatively is an antiemetic. Ondansetron (Zofran), given as a single 4 mg intravenous injection is used routinely at our institution. Recently, we have found that clonidine (0.1-0.3 mg PO) given 30 minutes prior to the procedure is not only effective in lowering blood pressure during surgery, it also contributes significantly to patient relaxation during the procedure. It does, however, cause post-procedure orthostatic hypotension.

Intraoperative Considerations

Tumescent Anesthesia

As stated previously, conscious sedation—as it pertains to plastic surgery, involves the administration of local anesthesia in addition to the intravenous sedation. In fact, it is the methodical use of tumescent anesthesia that ensures a smooth, relatively pain free procedure. Tumescence, or wetting solution as it is more appropriately termed, should be infiltrated into the surgical field. Two goals should be kept in mind: anesthesia of the sensory nerves and vasoconstriction of the blood vessels in the region. Achieving these goals requires at least 10 minutes for the wetting solution to exert its effects. Two solutions are commonly used at our institution:

Liposuction solution	1 liter bag of Lactated Ringer's solution
	50 ml of 1% plain lidocaine
	1 ml of epinephrine (1:100,000)
Face/breast solution	250 ml bag of normal saline
	100 ml of 1% lidocaine + epinephrine (1:100,000)
	10 ml of sodium bicarbonate

Table 9.1. Simple medication regimen that can be used for conscious sedation

Medication	Dosage Range	Purpose	Reversal Agent
Preoperative			
Diazepam	5-10 mg (up to 50 mg)	Preoperative sedation	Flumazenil (0.2 mg/min; up to 5 doses; reversal in 1-2 min)
Ondansetron	2-4 mg	Prevention of postop nausea and vomiting	None
Intraoperative			
Midazolam	0.5-2 mg	Anxiolytic, sedative	Flumazenil (0.2 mg/min; up to 5 doses; reversal in 1-2 min)
Fentanyl	12.5-50 mcg	Analgesia	Naloxone (0.1-0.2 mg/ 2 minutes) reversal in 2-3 minutes

Note: all medications listed are given intravenously.

9

Intravenous Sedation Regimens

Although there are a number of intravenous sedation regimens available, an excellent choice is the combined use of midazolam (Versed) and fentanyl (see Table 9.1). The advantage of using this combination is that midazolam has both anxiolytic and amnestic effects, whereas fentanyl is a potent, short-acting analgesic. The combination of fentanyl and midazolam is superior to midazolam alone in decreasing patients' subjective report of pain and anxiety. The main drawback of fentanyl is respiratory depression; however unlike other commonly used intravenous opiates such as morphine, it does have a very short half life. Midazolam, in contrast, has minimal effects on the respiratory system except in the elderly, in which lower doses should be utilized. Both of these medications have antagonists. Flumazenil (Mazicon) and naloxone (Narcan), the antagonists of midazolam and fentanyl respectively, should be readily available in the operating room.

Another method of intravenous sedation involves the use of propofol in combination with an opiate and benzodiazepine. The fact that a deeper level of sedation can be maintained makes this technique preferable for selected patients who are very anxious. Nevertheless, the disadvantage of this combination is the higher risk of respiratory depression, and the lack of a reversal agent for propofol. This technique necessitates a higher degree of experience and training in anesthetic technique including the ability to intubate the patient if needed. The use of propofol is not discussed in this chapter.

In the operating room, one nurse should be responsible for continuously monitoring patient status using pulse oximetry, blood pressure and cardiac monitoring. This should be performed by a nurse with appropriate experience and background in continuous patient monitoring; however specialized anesthesia training is usually not needed. It is important to emphasize that this nurse should have no other duties to perform during the procedure. The patient's oxygen saturation, blood pressure, heart rate, level of arousal and respiratory status should be monitored every 5 minutes. Changes in vital signs, level of arousal and the oxygen saturation are communicated to the surgeon. In addition, the surgeon should make his own assessment of arousal based on response to verbal stimulation, as well as the patient's degree of discomfort.

Based on the patient's condition, 0.5 to 2 mg of midazolam should be administered at the 5 minute intervals. In addition, fentanyl should be given in increments of 12.5 to 50 mcg. After local anesthetic is infiltrated, fentanyl administration is infrequently required, except in preparation for subsequent local anesthetic administration to a new surgical site. The total dose of fentanyl should rarely exceeded 200 mcg over the course of the procedure. Toward the end of the case, the amount of sedation should be decreased to allow the patient to slowly return to a normal state of arousal and awareness.

During conscious sedation, supplemental oxygen is usually not necessary. The ability of the patient to maintain an oxygen saturation over 95% without supplemental oxygen is a useful guideline to avoid oversedation (crossing from conscious to deep sedation). Occasional periods of deep sedation may occur, usually lasting for a few minutes at most. Brief stimulation and *rarely* jaw thrust may be required to maintain adequate ventilation. The use of small incremental doses of midazolam, limited use of narcotics and effective local anesthesia help to limit episodes of deep sedation. Nevertheless, as a safety measure, the capability to convert to general anesthesia or immediate assistance from an anesthesiologist should always be available. Foley catheters and sequential compression devices are generally not required due to the relatively short length of procedures utilizing conscious sedation, and the fact that venous stasis is minimal due to spontaneous patient movement and leg muscle contractions. For cases involving large volume liposuction or those that are longer than a few hours, a Foley catheter should be used to monitor fluid status and to allow greater flexibility in intraoperative fluid resuscitation.

Postoperative Considerations

Following the procedure, many hospitals will allow patients to bypass the recovery room and proceed directly to the outpatient day surgery area. This saves the patient the extra costs of recovery room care. Patients are monitored postoperatively in a standard manner. Those who choose to go home the day of surgery must meet criteria for discharge (ability to ambulate to a chair and the bathroom, bladder control, tolerate oral intake without emesis). Patients who received preoperative clonidine must be monitored for orthostatic hypotension.

Inpatient stay in an observation unit is appropriate for longer cases that involve multiple procedures, as well as for older patients who live alone. Postoperative nausea and vomiting is the major factor contributing to unintentional hospital admission after outpatient surgery. It begins shortly after arrival in the recovery room and usually lasts no longer than 12-24 hours postoperatively. A number of studies support the administration of a preoperative antiemetic (see preoperative considerations).

Pearls and Pitfalls

One of the risks of conscious sedation is crossing over into deep sedation. The responsible surgeon and monitoring nurse should be able to identify and handle patients who briefly slip into deep sedation. In very rare instances, a patient may require jaw thrust, mask ventilation or narcotic reversal. It is critical that the surgeon be comfortable performing these steps if necessary. A common pitfall leading to over sedation is to administer excessive amounts of fentanyl, instead of maximizing the use of the local anesthetic. Towards the end of the procedure, there is nothing wrong with cutting back on the amount of sedation, and allowing the patient to become more awake. In addition, excessive administration of versed can result in the opposite effect: an overly anxious, and occasionally claustrophobic patient. When this occurs, it is best to withhold sedation, reassure the patient and allow her to reorient herself.

Suggested Reading

1. American Society of Anesthesiologists task force on sedation and analgesia by nonanesthesiologists. Practice guidelines for sedation and analgesia by nonanesthesiologists. Anesthesiology 1996; 84:459.
2. Byun MY, Fine NA, Lee JY et al. The clinical outcome of abdominoplasty peformed under conscious sedation: Increased use of fentanyl correlated with longer stay in outpatient unit. Plast Reconstr Surg 1999; 103:1260.
3. Dionne RA, Yagiela JA, Moore PA et al. Comparing efficacy and safety of four intravenous sedation regimens in dental outpatients. J Am Dent Assoc 2001; 132:740.
4. Iverson RE. Sedation and analgesia in ambulatory settings. American society of plastic and reconstructive surgeons. Task force on sedation and analgesia in ambulatory settings. Plast Reconstr Surg 1999; 104:1559.
5. Finder RL, Moore PA. Benzodiazepines for intravenous conscious sedation: Agonists and antagonists. Compendium 1993; 14:972.
6. Kallar S. Conscious sedation in ambulatory surgery. Anesth Rev 1991; 18:9.
7. Klein JA. Tumescent technique for regional anesthesia permits lidocaine doses of 35 mg/kg for liposuction. J Dermatol Surg Oncol 1990; 16(3):248.
8. Marcus JR, Few JW, Chao JD et al. The prevention of emesis in plastic surgery: A randomized, prospective study. Plast Reconst Surg 2002; 109:2487.
9. Marcus JR, Tyrone JW, Few JW et al. Optimization of conscious sedation in plastic surgery. Plast Reconst Surg 1999; 104:1338.

Chapter 10

Principles of Reconstructive Surgery

Constance M. Chen and Robert J. Allen

Introduction

Plastic and reconstructive surgery is a field that relies upon basic principles to restore form and function to the human body. Whether it is a gunshot wound to the face, a congenital hand deformity, or a malformed breast, plastic surgeons must be adept at adapting a fundamental knowledge of human anatomy and physiology to create ingenious solutions to ever-changing challenges. Unlike techniques which must be modified with each new advance in medical technology, the use of principles makes it possible for the plastic surgeon to address problems as varied as the infinite diversity of the human species. Rote memorization of operative steps and mathematical formulas are insufficient. The reconstruction of the human body depends upon the ability to devise creative solutions based on core principles. Over the years, numerous efforts have been made to categorize these principles. Despite changes in technique, the fundamental principles of plastic and reconstructive surgery have withstood the test of time.

Ambrose Paré

The earliest principles of reconstructive surgery may be attributed to the French surgeon, Ambrose Paré, who in 1564 published five basic principles of plastic surgery. The first principle was "to take away what is superfluous." Whether applied to the excision of redundant tissue or the complete amputation of a surplus structure such as a digit or a supernumerary nipple, this first principle emphasized the need to eliminate that which served no purpose. The second principle was "to restore to their places things which are displaced." Whether applied to a congenital deformity, such as a cleft lip, or an acquired deformity, as in trauma, this principle required recognition of normal parts and diagnosis of the abnormal position. Likewise, the third and fourth principles, "to separate tissues which are joined together," and "to join those tissues which are separate," also required the ability to conceptualize a hypothetical norm. Indeed, a given defect could often be determined accurately only after distorted tissue was returned to its normal shape. This was true whether applied to a congenital defect, such as syndactyly, or an acquired defect, such as a burn contracture. Finally, the fifth principle, "to supply the defects of nature," also required the ability to visualize restoration to a normal state.

Modern Plastic Surgery: Gillies and Millard

Building upon these early ideas, Sir Harold Gillies and D. Ralph Millard took the principles of Paré to the next level. Recognizing that the remodeling of human tissue was different from clay, Gillies and Millard took as their founding principle

Practical Plastic Surgery, edited by Zol B. Kryger and Mark Sisco. ©2007 Landes Bioscience.

Table 10.1. Gillies' ten commandments of plastic surgery

1. Thou shalt make a plan.
2. Thou shalt have a style.
3. Honor that which is normal and return it to normal position.
4. Thou shalt not throw away a living thing.
5. Thou shalt not bear false witness against thy defect.
6. Thou shalt treat thy primary defect before worrying about the secondary one.
7. Thou shalt provide thyself with a lifeboat.
8. Thou shalt not do today what thou canst put off until tomorrow.
9. Thou shalt not have a routine.
10. Thou shalt not covet thy neighbor's plastic unit, handmaidens, forehead flaps, Thiersch grafts, cartilage nor anything that is thy neighbor's.

that "plastic surgery is a constant battle between blood supply and beauty." That is to say, the reshaping of human structures demanded that its vitality as living tissue be respected. Drawing upon the wisdom of his mentor, Sir Harold Gillies, Millard produced one of the most widely recognized efforts to outline the principles of reconstructive surgery. In 1950, Millard codified rules learned from Gillies and published them as the "ten commandments" of plastic surgery (Table 10.1). Shortly thereafter, the pair expanded these ideas to 16 principles that would apply not only to plastic surgery problems but also to a philosophy of life in general. Millard went on to develop the concept of principles still further in his classic tome, *Principalization of Plastic Surgery*. Divided into four broad sections, this work offered 33 commonsense rules to help plastic surgeons fashion answers to a variety of surgical problems.

Preparational Principles

Millard's first 12 principles fell under the framework of "Preparational Principles"-that is, principles to keep in mind before making the opening incision. The first principle was to "correct the order of priorities." Applied broadly, this could mean emphasizing integrity and ethics; it could mean prioritizing function over form; and it could also mean performing a blepharoplasty before a facelift since the latter could affect the former but not vice versa. The bottom line was that whether in life or in a specific procedure, each part needed to be considered in the context of the whole.

The second principle was that "aptitude should determine specialization," meaning that the plastic surgeon should play to strengths when deciding whether to focus on reconstructive surgery, cosmetic surgery, microvascular surgery, craniofacial surgery, head and neck oncology, hand surgery, burn physiology or laboratory research. Millard emphasized that a person who initially appeared inept in one area could later progress to excel above all others in the same area. Using himself as an example, Millard revealed that he took an aptitude test early in his career that determined that he would be well-suited to writing and possibly medicine, but completely unsuited for surgery due to a perceived inability to visualize objects in three dimensions. Despite this, he went on to become one of the most accomplished plastic surgeons in history, known especially for the three-dimensional rotation-advancement flap that is the standard of care for cleft lip repair today.

The third principle was to "mobilize auxiliary capabilities." That is to say, the plastic surgeon should incorporate individual talents to develop a "personal style with individual flair." Advised to develop one primary capability and several secondary talents such as sculpture, music, writing or painting, the ideal plastic surgeon would be multi-talented for maximal depth and versatility in the operating room. The fourth principle was to "acknowledge your limitations so as to do no harm," a self-evident principle that spoke to the temptation to persevere on a case with endless complications. Instead, the successful surgeon should know when to stop. The flip side of this was the fifth principle, which was to "extend your abilities to do the most good." This spoke to the moral obligation to use plastic surgical training to alleviate human suffering, that is, to reconstruct mutilated or severely deformed patients instead of limiting one's practice to purely aesthetic procedures. The sixth principle was to "seek insight into the patient's true desires." Delving into the psyche, this principle directed the plastic surgeon to decipher a patient's actual problems instead of merely taking the stated problem at face value to preempt patient disappointment, improve public relations and prevent postoperative legal complications.

The seventh principle was to "have a goal and a dream." In plastic surgery, this principle shifted depending on whether a procedure was primarily cosmetic, in which the goal would be to surpass normal, or primarily reconstructive, in which the goal would be to attain normal. Either way, the plastic surgeon should have a target in mind before beginning an operation. The eighth principle was to "know the ideal beautiful normal." While this ideal beautiful normal could vary among different ethnic backgrounds, it was important for the plastic surgeon to be able to define it in order to attain pleasing aesthetic proportions and visual harmony.

The ninth principle was to "be familiar with the literature." Knowing what had already been described assisted a surgeon in discriminating between procedures that would and would not be successful; it also gave the surgeon access to a collective bank of experience that allowed extension beyond what one person could accrue in a lifetime. The tenth principle, to "keep an accurate record," was like the sixth principle in that its underlying purpose was both to further patient care and provide legal protection for the surgeon. In addition, since memory was inherently unreliable, accurate written and photographic records provided baseline references that allowed the plastic surgeon to coordinate multi-staged procedures to achieve a successful final result.

The eleventh principle was to "attend to physical condition and comfort of position." Often overlooked by single-minded surgeons, the basis of this principle was the belief that the optimal surgical performance depended upon good physical condition and a comfortable working position for the surgeon. Finally, the twelfth principle, "do not underestimate the enemy," acknowledged that peril lay behind every procedure. Thus, whether the enemy was hypertrophic scar formation or inadequate vascular supply, it was never possible to be overly vigilant in preventing surgical complications.

Executional Principles

The second category of principles addressed the wielding of the blade. The thirteenth principle, "diagnose before treating," emphasized that observation was the basis of surgical diagnosis. The plastic surgeon must use all senses—particularly visual and tactile cues—to accurately determine a problem before proceeding with an operation. The fourteenth principle was reminiscent of Paré, in that it advised the plastic surgeon to "return what is normal to normal position and retain it there." As previously mentioned, displacement of structures could be due to failure in normal embryonic development or as a direct result of trauma, ablation, scar contraction, or even the aging process, but correction required the ability to recognize the norm in order to restore displaced parts to their correct place.

The fifteenth principle stated that "tissue losses should be replaced in kind." More specifically, when attempting reconstruction of lost body parts, bone should be replaced with bone, muscle with muscle and glabrous skin with glabrous skin. If exact replacement was impossible, then a similar substitute should be made, such as a beard with scalp, thin skin for an eyelid, thick skin for the sole of a foot, and a prosthesis for an eye. The idea was that replacing like with like would give the most natural outcome. The sixteenth principle advised the plastic surgeon to "reconstruct by units." By basing reconstruction on unit borders demarcated by creases, margins, angles and hairlines, surgical scars could often be concealed by the meeting of light and shadow.

The seventeenth principle was to "make a plan, a pattern and a second plan (lifeboat)." By visualizing an entire operation from beginning to end, the plastic surgeon could anticipate possible difficulties and then proceed to devise a secondary plan for use should the primary plan fail. The eighteenth principle was to "invoke a Scot's economy." This involved thrift in surgery, in which no tissue was ever discarded until it was certain that it was no longer needed. A corollary of this was to discard the useless, as once a piece of tissue was determined to have no further value it should be removed—but refrigerated storage was advised even then in case the tissue could be used later.

The nineteenth principle was to "use Robin Hood's tissue apportionment." That is, Robin Hood would steal from the rich to give to the poor. Likewise, this principle advised using excess tissue to make up for areas with tissue deficits by rotating, transposing, or transplanting expendable tissue flaps to areas in need. The corollary to this was the twentieth principle, to "consider the secondary donor site." That is, while reconstructing deficient areas with tissue taken from areas that were more ample, the resulting secondary defect must also be considered to make sure that its sacrifice was not too deforming. The twenty-first principle was to "learn to control tension." In opening, tension usually facilitated a clean cut with the scalpel; in closure, tension could lead to tissue necrosis or excess scarring; in flap design, skin tension lines could be identified and used to camouflage scars. The twenty-second principle was to "perfect your craftsmanship." For the plastic surgeon, "good" suggested mediocrity, and nothing short of perfection was acceptable. The twenty-third and final executional principle was "when in doubt, don't!" Doubt should function as a deterrent, and if a solution to a problem left seeds of doubt, it was better to develop a better idea.

Innovational Principles

The third category of principles governed the generation of new concepts in plastic surgery. The twenty-fourth principle was to "follow up with a critical eye." That is, it was important to follow patients postoperatively over time to critically evaluate results, as regular review of one's handiwork was the best way to spur advancement and improvement of surgical procedures. Likewise, the twenty-fifth principle, to "avoid the rut of routine," exhorted surgeons to shun mindless and tenacious clinging to unthinking rituals. Again, by thinking outside the box, the plastic surgeon could make the advance to the next level of innovation and development. The twenty-sixth principle, "imagination sparks innovation," was the "breakthrough" or problem-solving principle that encouraged free-spirited thinking and creativity.

The twenty-seventh principle, "think while down and turn a setback into a victory," was labeled the "prince of principles" by Millard. It admonished the surgeon not to panic or despair, or compound error when faced with possible defeat. Instead, the surgeon should keep cool while determining the cause of loss, expend no energy in worrying about a compromised position, and make certain not to repeat the same mistake while thinking one's way to recovery. Finally, the twenty-eighth principle was to "research basic truths by laboratory experimentation." By testing even minor theories in the laboratory, the surgeon could discover answers to plastic surgical questions in a controlled setting.

Contributional Principles

The fourth set of principles governed ways to contribute to the field of plastic surgery. The twenty-ninth principle was to "gain access to other specialties' problems." By consulting with physicians or surgeons from other specialties, it could be possible to learn management of common complications that would both benefit patients as well as broaden the base of plastic surgery. The thirtieth principle was that "teaching our specialty is its best legacy." The implication was that the best way to extend plastic surgery was to transmit knowledge via lectures, books, symposiums and personal experiences to ensuing generations. The thirty-first principle was to "participate in reconstructive missions." Moreover, the ideal method to conduct such missions was to lend specialists not just to operate, but to teach people in underdeveloped countries how to perform the operations and manage all the postoperative care themselves.

Inspirational Principles

The final set of principles attempted to prod the plastic surgeon to strive for perfection. Toward this end, the thirty-second principle was to "go for broke!" That is, the plastic surgeon should use every means possible to overcome obstacles, strive for the very best, and seek perfection. The thirty-third and last principle was to "think principles until they become instinctively automatic in your modus operandi." By incorporating principles constantly and consistently into plastic surgical practice, it would become second nature to avoid rote memorization of techniques and instead stimulate the imagination to engage in innovative problem solving.

The Reconstructive Ladder

The traditional approach to the reconstruction of a variety of defects is based on the concept of the reconstructive ladder (Fig. 10.1). The basic notion is that one should use the simplest approach to solving a reconstructive problem, before advancing up the ladder to a more complex technique. Consequently, if the procedure fails, one can climb to the next level of complexity. For example, a lower extremity venous stasis ulcer should be treated by dressing changes alone or by a split-thickness skin graft if these are applicable. A more complex reconstruction with a free flap should be reserved as a last resort if all simpler options have been ruled out or have failed. More recently, however, experienced reconstructive surgeons are beginning to realize that certain problems are not amenable to simple solutions. In select cases, bypassing the lower rungs of the reconstructive ladder and proceeding directly to microvascular free tissue transfer is the optimal approach. A good example of this is post-mastectomy breast reconstruction. For many surgeons, the free TRAM or DIEP flaps have become the standard of care.

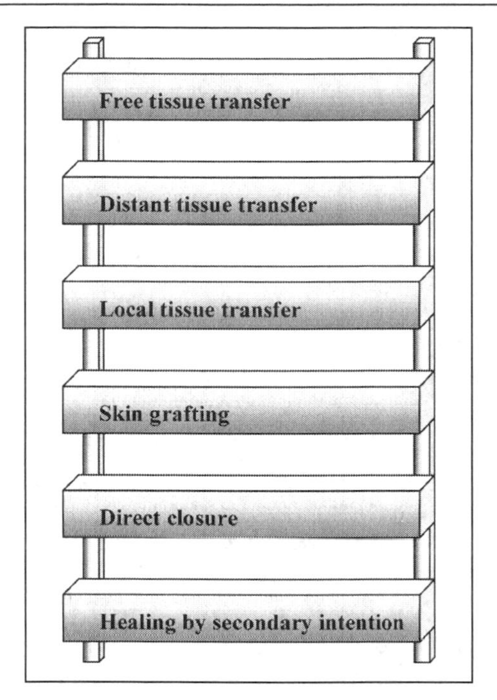

Figure 10.1. The reconstructive ladder.

Pearls and Pitfalls

Plastic surgery takes passion, determination and sacrifice. As plastic surgeons, we would like to create perfection. Yet techniques and procedures are always evolving, so the operative process must be based upon principles. Without a commitment to perfection, a concept of what beauty is, and what the end result will be ahead of time, the surgeon is lost. Poets can be our role models, because poets are creative and can help show us how to get going with the creative process. Ultimately, however, plastic surgery involves sacrifice, focus, determination and, above all, will power. When these qualities are combined, the plastic surgeon is able to elevate the work that is performed. A person who is able to go to work and create something close to perfection, striving for perfection, will lead a very satisfying life. By its very nature, then, plastic surgery gives us the opportunity to enjoy that ideal life.

Suggested Reading

1. Chase RA. Belabouring a principle, milestones in modern plastic surgery. Ann Plast Surg 1983; 11:255-60.
2. Gillies HD, Millard Jr DR. The Principles and art of plastic surgery. 1st ed. Boston: Little, Brown and Co., 1957.
3. Millard Jr DR. Plastic Peregrinations. Plast Reconstr Surg 1950; 5:26-53.
4. Millard Jr DR. Principlization of plastic surgery. 1st ed. Boston, Toronto: Little Brown and Co., 1986.
5. Rana RE, Puri VA, Baliarsing AS. Principles of plastic surgery revisited. Indian J Plast Surg 2004; 37:124-125.

Chapter 11

Principles of Surgical Flaps

Constance M. Chen and Babak J. Mehrara

Introduction

The underlying principle of all surgical flaps is the ability to maintain a viable blood supply upon transfer of flap tissue from a donor site to a recipient site. Given this fundamental capacity to retain vascular circulation, surgical flaps may be classified in many ways. One approach is by composition, as a flap may be made up of many different kinds of tissue. Another is by vascularity, and several different schemata have been developed to categorize flaps by the type of vascular supply. A third manner of categorizing flaps is by method of movement, and it is important to understand the basic techniques of flap transfer. Unlike a graft, which is wholly dependent upon the recipient bed to provide blood supply, a flap by definition is able to preserve its own vascular supply for survival. Thus, whether classifying a flap by composition, vascularity or method of movement, the core principle essential to all flaps is how to maintain blood supply so that the flap tissue will remain robust after transfer to its new site.

Composition

The most basic way to think about a flap is to consider what tissues are contained within it. A flap may contain skin, fascia, muscle, bone or various combinations of these tissues. As the underlying principle of any flap is its ability to retain its own blood supply, the amount of tissue that may be carried within it is dictated by the minimum or maximum amount of tissue that can be transferred with intact vascularity. When more than one type of tissue is contained within a flap, it is called a "composite flap."

The simplest type of flap is the skin flap. The blood supply of the skin is contained largely in the dermal and subdermal plexus and derives from two main sources: a musculocutaneous vascular system and a direct cutaneous vascular system. When the blood supply to the skin is via a named artery, the skin flap is called an "axial flap." When the blood supply to the skin lacks a significant pattern in its vascular design, the skin flap is called a "random flap." Either way, the survival of a cutaneous flap depends on the number and type of blood vessels at the base of the flap. For an axial flap, the survival pattern of the flap is based on the length of the underlying feeding artery. For a random pattern flap, the length and width should be designed in a 2:1 ratio, as a wider base width increases the chance that a large vessel will be incorporated to provide an adequate blood supply to the enclosed dermal-subdermal plexus. Even in an axial flap, the distal borders of the flap are also random pattern with distal perfusion from the dermal-subdermal plexus (Fig. 11.1).

Skin flaps may also be transferred based on the vascular plexus of the deep fascia, in which case they are termed "fasciocutaneous flaps." The blood supply of the deep fascia is derived from perforating vessels of regional arteries that pass along the fibrous septa of muscle bellies or muscle compartments. Including the deep fascia

Practical Plastic Surgery, edited by Zol B. Kryger and Mark Sisco. ©2007 Landes Bioscience.

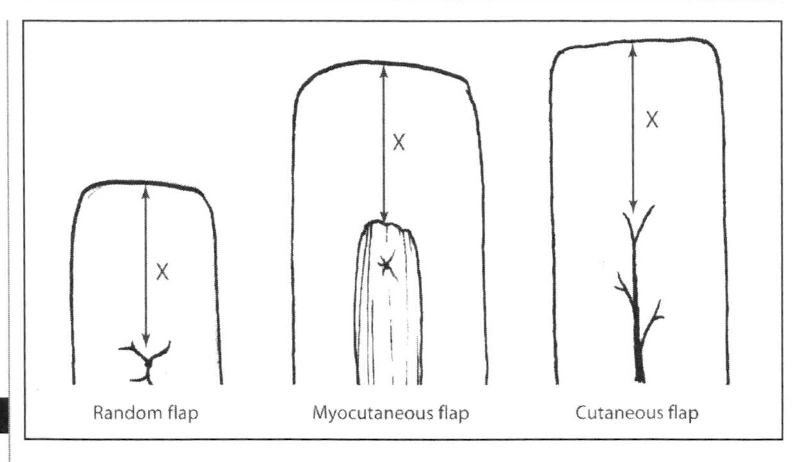

Figure 11.1. Survival pattern of skin flaps. X = subdermal plexus. The distal end of axial flaps (cutaneous and myocutaneous) also have a random pattern.

along with the skin avoids tedious dissection and may also preserve adjacent subfascial arteries. Among the advantages of fasciocutaneous flaps in reconstructive surgery are ease of elevation and transfer, decreased bulk, good reliability, and decreased functional morbidity at the donor site. Depending on the size of the skin paddle, however, the secondary defect at the donor site may require coverage with a split-thickness skin graft.

Progressing one layer deeper still, another common flap in reconstructive surgery is the "myocutaneous" or "musculocutaneous" flap, which combines muscle, skin, and the intervening fascia and subcutaneous tissue. Supplied by one or more dominant vascular pedicle within the muscle instead of a direct cutaneous arterial source, the essential feature of a myocutaneous flap is that the underlying muscle "carries" the blood supply for the overlying skin. Myocutaneous flaps have two key advantages. First, the increased bulk better allows it to fill dead space. Secondly, myocutaneous flaps are also more resistant to bacterial infection than fasciocutaneous flaps by a factor of 100. This makes them very reliable and useful, particularly when increased bulk is needed with a robust arterial supply to fill a defect that has been subjected to chronic infection. If a skin paddle is not needed, muscle can also be transferred alone, without the overlying fascial and cutaneous tissue.

A final type of tissue commonly incorporated into a flap is bone. When taken with the overlying skin, this is called an "osseocutaneous flap." A dominant vascular pedicle with perforating branches supplies the skin and periosteum. Usually taken as a free flap, the bone is harvested with a cuff of muscle and/or skin to reconstruct a skeletal framework with soft tissue. The long bones of the extremities, such as the fibula, are often used as they provide more length for shaping according to the required need.

Type of Blood Supply

Once the composition has been determined, flaps can be further categorized according to their blood supply. As mentioned earlier, random flaps are based primarily on the cutaneous blood supply from the dermal-subdermal plexus. Pedicled or axial flaps are based on anatomically mapped or named blood vessels.

Fasciocutaneous flaps have been classified into three categories based on their vascular patterns.

Type A: Direct cutaneous pedicle
Type B: Septocutaneous pedicle
Type C: Musculocutaneous pedicle

Muscle flaps may be classified in two different ways. First, Mathes and Nahai developed a system of muscle classification based on circulatory patterns.

Type I: Single pedicle (e.g., tensor fascia lata)
Type II: Dominant pedicle(s) with minor pedicle(s) (e.g., gracilis)
Type III: Dual dominant pedicles (e.g., gluteus maximus)
Type IV: Segmental pedicle(s) (e.g., sartorius)
Type V: Dominant pedicle, with secondary segmental pedicle(s) (e.g., latissimus dorsi)

Second, Taylor developed a system of muscle classification based on mode of innervation.

Type I: Single, unbranched nerve enters muscle (e.g., latissimus dorsi)
Type II: Single nerve, branches prior to entering muscle (e.g., vastus lateralis)
Type III: Multiple branches from the same nerve trunk (e.g., sartorius)
Type IV: Multiple branches from different nerve trunks (e.g., rectus abdominis)

Finally, the body can be further segregated anatomically into three-dimensional vascular territories called "angiosomes." The angiosome is a composite unit of skin and underlying deep tissue that is supplied by a source artery. Each angiosome defines an anatomic unit of tissue from skin to bone that may be safely transferred as a composite flap. The angiosomes are interconnected by either true anastomotic arteries, in which there is no change in caliber between the vessels of adjacent angiosomes, or reduced-caliber, choke anastomotic vessels. The junctional zone between adjacent angiosomes usually occurs within the muscles of the deep tissues rather than between them, so that most muscles span across two or more angiosomes. Thus, when designing musculocutaneous flaps it is possible to capture the skin island from one angiosome by using muscle supplied from the adjacent angiosome.

Flap delay is defined as the surgical interruption of a portion of the blood supply in a preliminary stage prior to tissue transfer. The purpose of delay is to augment the surviving portion of the flap. There are two schools of thought regarding the pathophysiology of the delay phenomenon. The first holds that delay conditions tissue to ischemic conditions so that it is able to survive with less vascular inflow. The second believes that delay actually increases vascularity by dilating reduced-caliber choke anastomotic vessels and stimulating additional vascular ingrowth.

Another way to increase survival of a myocutaneous flap is by **supercharging** the blood supply. This method involves augmenting arterial inflow by using microsurgical techniques to bring in an additional vascular pedicle. Classically described for use in a pedicled TRAM flap, the supercharging technique may be performed in one of two ways. First, in the pedicled TRAM flap, the contralateral deep inferior epigastric vessels may be retained in a cuff of inferior rectus muscle in a planned vascular augmentation to a single-pedicle flap. Alternatively, the inferior epigastric vessels on the pedicled side may be used to save a flap during the immediate postoperative period in an emergency "supercharged" TRAM flap.

Techniques of Flap Transfer

The final way to categorize flaps is by the technique of flap transfer. Broadly speaking, flaps can either be **pedicled flaps** or **free flaps**. Pedicled flaps remain attached to the underlying blood supply, while the tissue connected to it is transferred to another site. Free flaps are temporarily disconnected from their blood supply, and then the feeding vessels are surgically anastomosed to the blood supply at the recipient site. Flaps can be further categorized by the distance between the donor site and recipient site. **Local flaps** are used to close defects adjacent to the donor site. **Distant flaps** imply that the donor site and the recipient site are not in close proximity so that closure cannot be facilitated by a local method.

There are several different types of local flaps. An **advancement flap** moves along an axis in the same direction as the base to close the defect simply by stretching the skin. Examples of an advancement flap are the V-Y flap, Y-V flap, and the bipedicled flap (Fig. 11.2). A **rotation flap** has a curvilinear design and rotates about a pivot point to close a wound defect. The donor site is closed primarily by reapproximating the skin edges or with a skin graft. A back cut in the direction of the pivot point can be made to facilitate closure, but this can also compromise the blood supply to the flap by decreasing the base width. A Burow's triangle can also be made external to the incision to decrease tension and facilitate primary closure of the donor site (Fig. 11.3). Finally, a **transposition flap** is a rectangular flap that is rotated laterally about a pivot point into an adjacent defect to be closed. The farther the flap rotates, the shorter the effective length of the flap, so that the flap must be designed longer than the defect in

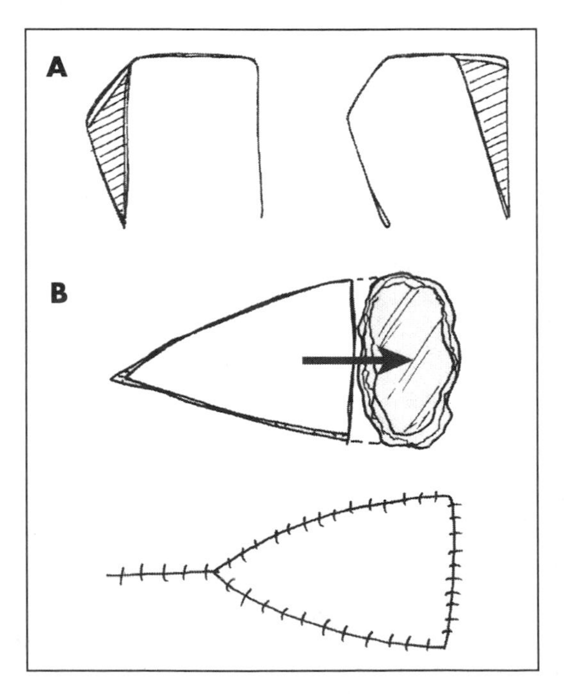

Figure 11.2. A) The rectangular transposition flap. B) The V-Y advancement flap.

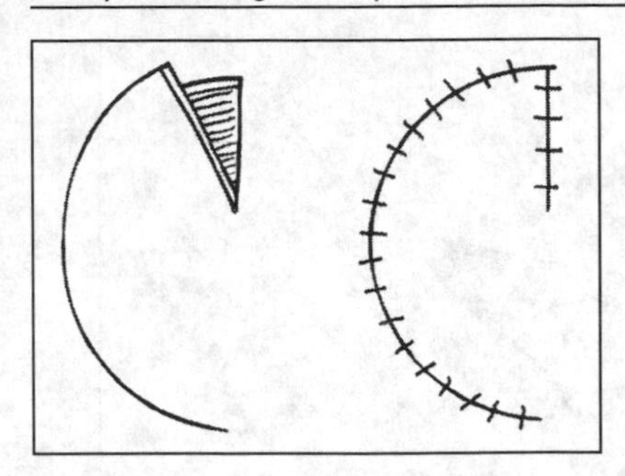

Figure 11.3. The rotation flap.

order to close the donor site. Otherwise, the donor site may be closed primarily with a skin graft or with an additional transposition flap, as in a bilobed flap (Fig. 11.4).

There are several important types of transposition flaps. The first is the **Z-plasty**, in which adjacent triangular flaps are interchanged to exchange the width and length between them. The three limbs of the Z must be equal in length, and the amount of length obtained depends upon the intervening angles, with 60° being the classic angle to obtain optimal increase in length while preserving blood supply to the triangular flaps (Fig. 11.5). The **rhomboid** or **Limberg flap** is another type of transposition flap that can be used to close a skin defect. Four different flaps can be designed at angles of 60°, with the longitudinal axis paralleling the line of minimal skin tension (Fig. 11.6). The **Dufourmentel flap** is like the rhomboid flap, except the angles are at 90°. Finally, the **double opposing semicircular flap** can be used to close circular skin defects (Fig. 11.7).

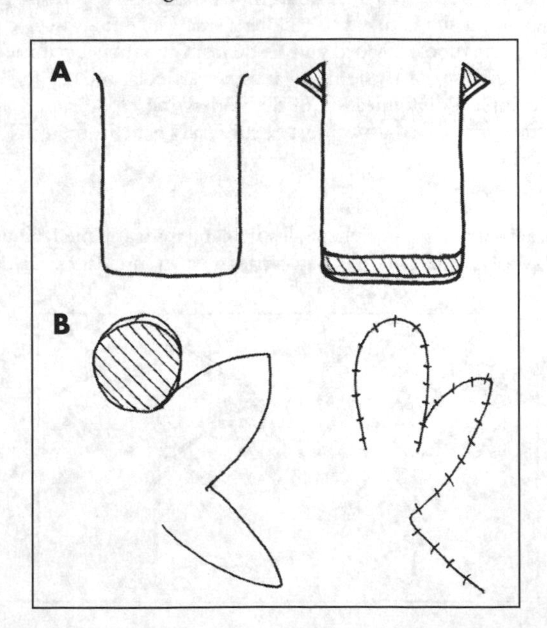

Figure 11.4. A) The rectangular advancement flap. B) The bilobed flap.

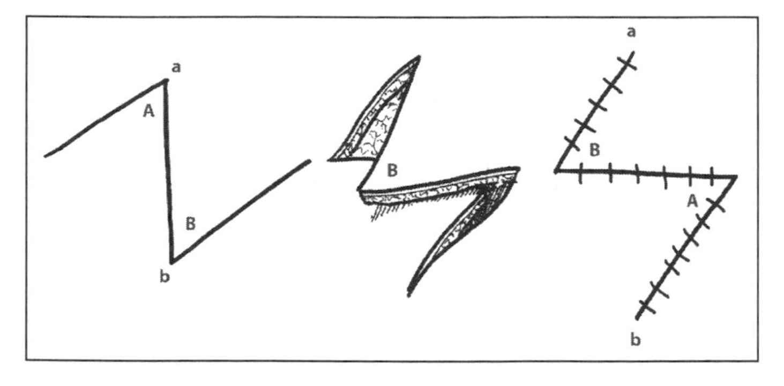

Figure 11.5. The Z-plasty technique.

Interpolation flaps also rotate about a pivot point, but they are either tunneled under or passed over intervening tissue to close a defect that is not immediately adjacent to the donor site. Examples include the Littler neurovascular island flap and the pedicled TRAM flap.

Distant flaps involve tissue transfer from one part of the body to another in which the donor site and the recipient site are not in close proximity to each other. There are three types of distant flaps: direct flaps, tubed flaps and free flaps. The **direct flap** involves the direct transfer of tissue from a donor site to a distant recipient site. Examples of direct flaps include the thenar flap, cross-leg flap and groin flap. **Tubed flaps** are used when tissue cannot be directly approximated, so that tissue from the donor site is tubed to recipient site. Once the vascular supply has been established, the tube is divided and tissue from the tube is returned to donor site. Examples of this are the forehead flap and the clavicular tubed flap. Finally, **free flaps** involve complete disconnection of the underlying blood supply, so that the blood vessels from transferred tissue must be surgically reanastomosed to reestablish vascular circulation.

Summary

In sum, the underlying principle of all surgical flaps is the meticulous preservation of blood supply. Unlike grafts, a flap carries its own vascular circulation with it.

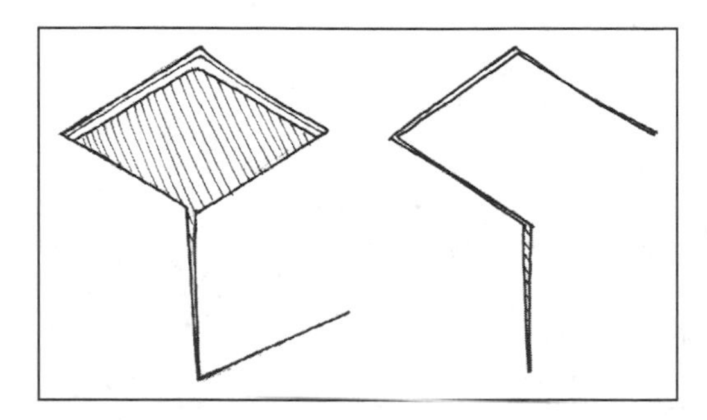

Figure 11.6. The rhomboid (Limberg) flap.

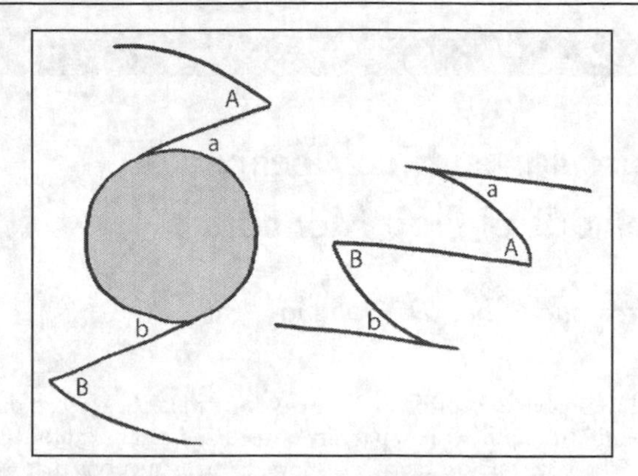

Figure 11.7. The double opposing semicircular flap.

The amount and type of tissue that a flap can contain is wholly dependent on the maintenance of adequate blood supply. Knowledge of vascular anatomy is crucial to flap design. Techniques of flap transfer must take care to safeguard the vascular circulation of the flap. With the careful protection of blood supply, it is possible to successfully plan and implement any surgical flap.

Pearls and Pitfalls

The success or failure of a flap is dependent upon blood supply. The ingrowth of new blood vessels from the surrounding tissue occurs over several weeks. As a general rule, the tissue that is most distant from the arterial inflow is at the highest risk of necrosis. Efforts to reduce this risk include the following: (1) preferentially discarding excess tissue from the distant tip; (2) for skin flaps, designing a flap with as broad a base as possible, away from any previous incisions sites; (3) minimizing tension; (4) maximizing inflow.

When designing a flap for covering or filling a defect, it is prudent to follow the carpenter's rule of "measure twice, cut once." Defects must be examined and measured three-dimensionally, since the width, depth and length will not always conform to a two-dimensional plane. The final desired contour should also be considered (e.g., if a convex contour is desired, the length of the flap should be greater than the direct length of the defect). Furthermore, it should be determined whether or not moving adjacent structures (such as the arms or legs) will change the dimensions of the defect. For instance, a supraclavicular skin defect will significantly increase in size when the patient's head is turned away from the defect.

Suggested Reading

1. Jackson IT. Local Flaps in Head and Neck Reconstruction. St Louis, Mosby: 1985.
2. Kayser MR. Surgical flaps. Selected Readings in Plastic Surgery 1999.
3. Mathes SJ, Nahai F. Clinical atlas of muscle and musculocutaneous flaps. St Louis, Mosby: 1982.
4. McGregor AD, McGregor IA. Fundamental techniques of plastic surgery. London: Churchill Livingston, 2000.
5. Taylor GI, Palmer JH. The Vascular territories (Angiosomes) of the body: Experimental study and clinical applications. Br J Plast Surg 1987; 40:113.

Chapter 12

Microvascular Surgical Technique and Methods of Flap Monitoring

Sean Boutros and Robert D. Galiano

Introduction

The hand is capable of coordinated activity finer than the eye can direct. With the aid of magnification, the true capability of the hand can be exploited. As a tool for the plastic surgeon, microsurgery has allowed reconstructions that were simply not possible before. However, microvascular free tissue transfer is not a technique for the occasional microsurgeon. The catastrophic complication of flap failure looms over every microsurgical case; therefore, expertise in the execution of a free flap as well as its postoperative surveillance is key to a successful outcome.

Experience has shown that flap loss is a preventable complication and that elective microsurgery should have a failure rate of less than 2%. Most cases of flap loss are technical in nature. The fault may lie in the choice of flap, the harvest of the flap, preparation of donor vessels, insetting of pedicle or microsurgical technique. In general, it is best to think of all possible errors as additive in the process of thrombosis. Failure will occur if the procoagulatory factors outweigh the intrinsic ability of the vessels, in particular intact and uninjured intima, to prevent clot formation.

Flap Choice

The first step for success in microsurgery is flap choice. The specifics of different flaps are discussed in subsequent chapters. The most important determining factors for flap choice should be the surgeon's experience and the goals of reconstruction.

In general, each surgeon should identify at least four flaps they feel comfortable with. These flaps should include a bulky muscle flap, a bulky fasciocutaneous flap, a thin fasciocutaneous flap, and a bone flap. With this armamentarium, the reconstructive surgeon will have tools that can be applied to most situations. By limiting himself to a small number of flaps, more experience can be obtained with each one. This increased experience translates to increased success. It is not advantageous to explore every novel flap that is reported, as this dilutes the experience and increases the chance of failure. With increasing experience with each flap comes increasing success and a lower failure rate.

This does not imply that specific flaps may not be beneficial over others in certain situations. There is no doubt that the donor properties of a latissimus dorsi flap differ from those of the gracilis flap and that each may be a better choice for a specific patient. However, the patient is best served with successful reconstruction. If there is significant benefit in a flap where the surgeon has no experience, the surgeon should consider referral or should seek additional training in order to add that flap to his or her armamentarium. This may include time in a cadaver lab or observing a surgeon with a particular skill.

Practical Plastic Surgery, edited by Zol B. Kryger and Mark Sisco. ©2007 Landes Bioscience.

Having mastered the tools of reconstruction, the surgeon should judiciously consider the requirements for reconstruction. Bulky muscle flaps are best for contaminated defects and bony injuries with high risk for infection. Thick fasciocutaneous flaps are useful for contour and shape reconstruction. Thin fasciocutaneous flaps provide stable, noncontracting coverage. Bone flaps provide structural integrity.

Flap Harvest

Specific aspects of each flap harvest are discussed elsewhere in this book. Certain principles, however, hold true despite the flap chosen. While harvesting a flap, the pedicle should be carefully dissected with as much length as possible. It is important not to limit the pedicle length to the anticipated need, but to harvest the maximum that can safely be obtained. It is much more advantageous to discard unneeded length than to find oneself requiring more pedicle length. Vein grafts should be avoided unless absolutely necessary.

While harvesting the flap and dissecting the pedicle, the most common mistake is damaging the vessels. Forceps should only touch the adventia and never purchase the vessel as the intimal layer is extremely fragile and easily fractured or crushed by manipulation. Any grasping of the vessels will cause damage to the intima which increases the likelihood of clot formation. This intimal injury leads to platelet deposition and thrombosis as the injured endothelial cell layer loses its natural thrombolytic properties.

Division of the pedicle should be reserved until the last possible moment. Prior to division, the donor vessels should be dissected, isolated, prepared and positioned for the anastomosis. It is helpful to mark the vessels in their natural state to assure that they are not twisted when transferred to the recipient site. Prior to division, the artery should be occluded first, followed by the vein. This will avoid excess blood pooling in the flap. Immediately after the flap is removed, one can consider cooling the flap with chilled saline as this decreases the metabolic activity of the tissue and allows the luxury of a longer ischemic time.

There is seldom a need to separate the artery and vein within the pedicle for anything more than a minimal distance. The only exception is the case where the recipient vessels are not paired. The vessels should not be skeletonized until they are brought to the recipient site and carefully prepared under the microscope. Any branches within 2 mm of the anastomosis are best sutured closed with microtechnique to avoid blood pooling near the anastomosis.

Preparation of Recipient Site

Preparing the recipient site mirrors the harvest of the flap. Vessels should be chosen that are simple to use and of the largest caliber available. They should be expendable when possible and have sufficient length. Again, care should be taken in the preparation of the vessels. They should not be extensively manipulated or injured. They should only be skeletonized for 2-3 mm around the anastomotic site, and this should be done under the microscope.

Microsurgical Technique

The anastomosis can be done in several fashions. These include end-to-end or end-to-side. They can be performed by multiple suture techniques or with coupling devices. The general philosophy is to gain experience with two or three techniques and apply those techniques to different situations. With careful planning, preparation, and mobilization of both the pedicle and recipient vessels, this is generally possible.

General principles of proper microsurgical technique are:

1. Pass sutures perpendicularly through the adventitia into the intima.
2. Avoid grasping or manipulating the intima.
3. Avoid multiple suture passes.
4. Avoid torquing the needle in the vessel; grasp and regrasp the needle in order to pass it through the vessel following the curve of the needle perfectly.
5. Dilate and visualize the inside of the vessels with heparinized saline irrigation on an ocular anterior chamber needle.
6. Use polished vessel dilating forceps to gently open spasmodic vessels or for vessel expansion.
7. Leave long tails on the sutures for manipulation and visualization.
8. Perform both anastomoses prior to reperfusion.
9. Release clamps on the vein first.
10. Inspect the anastomosis using the long suture tails as handles.
11. Place additional sutures in gaps with pulsatile or pressurized bleeding.
12. Avoid the temptation to place excess sutures in cases of mild oozing of blood from the anastomosis.
13. Apply warm saline to the flap and papaverine to the anastomosis after reperfusion to dilate the vessels and relax spasm.

Anastomotic Techniques

End-to-End

The end-to-end anastomosis is the simplest and the most reliable method. There are several techniques of suture placement including the 180°-180° and triangulation methods. The easiest is probably the 180°-180° technique. This can be applied to any situation and is probably the best technique for size-mismatched vessels. Important points to remember are:

1. The vessels must not be twisted prior to placement in the double clamp holder. This can be ensured by inking one surface of the pedicle and recipient vessels prior to their division or dissection.
2. The first two sutures are placed at opposite poles of the vessels.
3. The third suture is placed midway between the poles.
4. In most cases, the next sutures bisect the gap though on occasion, two sutures will be needed in the gap.
5. Once the anterior wall is complete, twist the entire double clamp to show the backwall.
6. Visually inspect every suture of the anterior wall from the posterior view to assure that they are evenly spaced and have not purchased the back wall of the vessel.
7. Place another bisecting suture midway between the poles on the back wall.
8. All remaining sutures can be placed and left long (not tied).
9. Dilate the vessel with saline when tying the back wall to assure that there is no purchase of the anterior wall.

End-to-Side

The end-to-side technique is occasionally necessary. For example, it is used in the leg when there is only one vessel available or for an anastomosis in the head and neck (for example, to the internal jugular vein). Principles are:

1. The pedicle vessels should enter the recipient vessel at a gentle angle.
2. Perform a limited arterioectomy, removing a small window of vessel.
3. Place heel and toe sutures first.

4. Initially close the heel.
5. Follow with closure of the toe.

Coupling Devices

Coupling devices are useful for veins or thin-walled arteries. They save some time in the anastomosis. They, however, are not a panacea. The major time consumption in a microsurgical case is not the anastomosis, but the set up and preparation. If coupling devices are used, the set up and preparation time remain the same. Principles of gentle handling of vessels are still required as is avoidance of damage to the intima. Overall, the devices appear to have a place in the venous anastomosis, where they can also act as a stent, or in cases with significant size mismatch. Points to consider are:

1. Use the largest size coupler that will comfortably fit (range 2-3.5 mm).
2. Draping of the vessel over the spikes is performed by one surgeon while the other maintains the engagement of the spike as the vessel is seated.
3. Seat the vessel 180° apart to assure even spacing on the coupler.
4. Avoid grasping the intima of the vessel as it is draped over the spikes.
5. Assure that the coupling device is closed and guide it off of the coupling applier.

Draping of the Pedicle

After the anastomosis is complete and the flap is successfully revascularized, it is not uncommon for significant problems to arise. Kinking or unnatural curvature of the pedicle will certainly cause thrombosis. In fact, any turbulent, nonlaminar flow increases the likelihood of thrombosis and flap loss. The pedicle should be carefully draped. Gelfoam sponge or Alloderm can be used to help maintain the proper position of the pedicle.

Closure

A sound closure technique is again crucial for success. Both the flap and pedicle can be compressed by a tight closure. Anticipation of this is critical, as well planned incisions will allow closure after the edema of these long cases has set in. If there is any question, the liberal use of skin grafts to allow tensionless closure is recommended. The anastomosis should never be situated immediately under a suture line.

Monitoring

There is no "perfect" monitoring technique. Despite numerous ingenious techniques and improvements in technology, the ideal monitoring technique should be the one that surgeons and ancillary staff at a particular hospital are most familiar with and meet the restraints (budgetary or manpower) of the institution. What is ideal at one institution may not be practical at another. What is clear over many years of clinical experience, although this remains to be formally proven, is that the presence of dedicated staff in a dedicated unit stands the best chance of picking up problems earlier. The impetus to closely monitor a flap comes from the enormous investment undertaken on the part of the patient as well as the surgeon regarding microvascular free tissue transfer. The utility of postoperative flap surveillance has been proven, with an increase in the salvage rate of the failing flap from 33% to about 70% in some series.

The clinical exam is useful when performed by the experienced clinician. The transition of a healthy, plump flap or vibrant replanted digit to cold, flat, lifeless tissue can proceed via either arterial occlusion or venous congestion. These characteristics are useful in deciding whether to explore a flap or perhaps treat with leech therapy. Although it is the least technologically-based, much information can be gleaned from a thorough physical exam. Turgor can indicate the state of arterial inflow or venous outflow. Like a balloon, the flap or digit will inevitably declare itself if it has arterial insufficiency or venous congestion. Bleeding can be useful, as the qualitative and quantitative flow in response to pinpricks or rubbing of wound edges can declare the state of circulatory flow to the flap. In particular, a congested flap may bleed briskly, but the blood will appear dark and unoxygenated. The blood flow of a flap with compromised arterial inflow will be weak or absent. A caution regarding the pinprick test is that it is useful for evaluating a flap, but will occasionally cause trauma leading to vasospasm or hematoma in the confined space of a finger.

It is possible to monitor free flaps with a temperature probe. This method consists of placing surface temperature probes on the skin of the free flap and comparing them to probes placed on neighboring native skin. The probes are attached to a temperature monitor that will give off an alarm if there is a difference in temperature between the two sites greater than the specified amount (typically, 2-3°C). Although appealing, there are limitations to the use of temperature probes, as the readings may be affected by regional changes in blood flow that are not secondary to flap flow disturbances.

Doppler ultrasonography is perhaps the most widely used monitoring tool. Two permutations exist. The first is the external Doppler. A recent innovation is the implantable internal Doppler. This tool permits monitoring of the segment of artery and vein a short distance downstream of the anastomosis. Its use has obviated the need for an external sentinel skin segment, and is ideally suited for buried anastomosis (e.g., jejunal free flaps in the head and neck, or vascularized bone transfers). These techniques are extremely useful; however, complications such as probe dislodgement and the occasional monitoring of an adjacent vessel that is not the pedicle can result.

In replants, the pulse oximeter is extremely useful. Some centers have reported success with fluorescein infusion and fluorescent lamp observation. This technique is not as useful in pigmented skin. Other techniques that at this time must be considered experimental include pH monitoring, duplex ultrasound, photoplethysmography, reflection photometry and radioisotope studies. None of these are currently widely used.

Pearls and Pitfalls

Although the microsurgical trainee may be eager to execute a large variety of occasionally exotic flaps, it is much more important to master a limited number of flaps and apply these flaps to different defects throughout the body. The principles outlined in this chapter serve as the basis to successfully execute any type of microsurgical transfer the plastic surgeon will encounter, even unusual flaps. In summary, it is essential to:

1. Sharpen microsurgical skills in the lab.
2. Handle the vessels gently.
3. Place significant attention on closure and pedicle position.
4. Familiarize oneself with one or two monitoring techniques. This will maximize salvage of the inevitable free flap failure.

The most important indicator of a problem with the free-flap is a change in the clinical exam. This necessitates that the flap be seen as soon as possible by a surgeon who has been actively managing the patient.

Suggested Reading

1. Hidalgo DA, Disa JJ, Cordeiro PG et al. A review of 716 consecutive free flaps for oncologic surgical defects: Refinement in donor-site selection and technique. Plast Reconstr Surg 1998; 102(3):722.
2. Disa JJ, Cordeiro PG, Hidalgo DA. Efficacy of conventional monitoring techniques in free tissue transfer: An 11-year experience in 750 consecutive cases. Plast Reconstr Surg 1999; 104(1):97.
3. Khouri RK. Avoiding free flap failure. Clin Plast Surg 1992; 19(4):773.

Chapter 13

Tissue Expansion

Zol B. Kryger and Bruce S. Bauer

Introduction

Tissue expansion relies on the ability of skin and soft tissues to generate in response to tension. In plastic surgery, tension is generated by implanting a subcutaneous balloon (expander) that is inflated over a period of weeks; new tissue is generated in response to the constant stretch caused by the progressive inflation of this expander. This tissue can be used to reconstruct extirpative or traumatic defects such as those encountered after mastectomy, burn excision, or removal of giant nevi.

Biological Basis of Tissue Expansion

A number of studies support the concept that the increase in skin surface area after expansion is due to the generation of new tissue rather than the stretching of existing skin. In culture, mechanical stress induces fibroblast and epidermal hyperplasia. These cells preserve their phenotype without malignant degeneration. This observation is supported by the fact that there has never been a reported case of skin malignancy secondary to tissue expansion.

From a histological standpoint, adult and pediatric skin responds to expansion in the same manner. Within 1 week of expansion, the epidermis begins to thicken and the dermis thins. The skin appendages do not change. The subcutaneous fat and muscle atrophy. Cellular proliferation reduces the resting tension of the skin over time, enabling another round of expansion to take place. Once the process is complete, the expanded skin eventually returns to its baseline thickness. The vessels of the skin and subcutaneous tissue also resume their pre-expanded size and number; however anecdotally, some flaps demonstrate increased vascularity.

Indications

In general terms, expansion of tissue is used to improve rotation, transposition or advancement of local or regional flaps, or to increase the harvest of full-thickness skin grafts. Recently, tissue expansion has been successfully applied to myocutaneous and free flaps. In adults, aside from their use in breast reconstruction, tissue expanders are used primarily for secondary burn and trauma reconstruction in the head and neck region. In the pediatric population, expanders have been used in a multitude of reconstructive procedures. The most common indication in children is to reconstruct defects left by excision of giant congenital nevi. Tissue expansion is contraindicated in infected skin. Although expansion is possible in radiated or scarred tissue, it is associated with a much higher complication rate and should be avoided whenever possible.

Technique

Expanders come in a variety of shapes and sizes, and there is no absolute ideal expander for a given site or condition. Expanders can have either internal or external

Practical Plastic Surgery, edited by Zol B. Kryger and Mark Sisco. ©2007 Landes Bioscience.

filling ports. Most experienced surgeons recommend using remote ports. These should be placed away from the expander. Internal ports have both a higher failure rate and a greater incidence of accidental expander rupture. In children, the use of internal ports is associated with a higher rate of exposure of the expander due to the pressure exerted on the skin by the port. Whenever possible, the incision should be placed within tissue destined to be excised, as in the case of congenital nevi. Straight incisions along the border of the defect should be avoided because this will enlarge the defect and may interfere with flap coverage. An alternative is to use a U- or V-shaped incision that is hidden and remote from the defect. Such incisions should be perpendicular to the direction of expansion in order to maximize skin blood supply. When doing serial expansion, longitudinal blood supply must be preserved. This holds true especially in the trunk and extremities.

The expander should be placed on top of the deep fascia (or subgaleal in the scalp), unless the plan is to incorporate muscle into the expanded flap. The pocket should always be larger than the base diameter of the expander. Blunt dissection in a single fascial plane is safest for preserving blood supply. Most surgeons overinflate tissue expanders beyond the manufacturer's recommended maximum capacity. Studies have demonstrated that significant overinflation is possible before weakening or rupturing. The rate of inflation is variable and largely based on surgeon preference. Patient comfort and signs of tissue perfusion, such as tension, color, and capillary refill, guide the filling rate. Filling is usually initiated one week after surgery.

Tissue expansion should continue until the expanded area is larger than the defect, because of the length that is lost upon advancement and inset of the flap. The use of rotation and transposition flaps enables the transfer of tension from the tip of the flap more proximally to its base. A single or double back-cut can be performed prior to inset in order to gain extra length. Lastly, the donor site should be closed in layers after the implant capsule is excised. Pre-expansion of distant pedicle- or free-flaps facilitates closure of otherwise tight donor sites.

Intraoperative Expansion

Most surgeons fill the expanders intraoperatively with sufficient saline to eliminate dead space and tamponade raw surfaces to help prevent postoperative bleeding. There is, however, an alternative to traditional prolonged expansion. Immediate intraoperative expansion combined with broad undermining of the defect can help reduce the tension that occurs on the distal parts of a local flap. In rapid expansion, the skin initially expands due to its elasticity and the displacement of interstitial fluid. Within minutes, the alignment of the collagen fibers changes due to the stretch. This process yields up to 20% more tissue for flap coverage. Intraoperative expansion is indicated for relatively small defects, such as in coverage of defects of the ear.

Scalp

Although tissue expansion does not increase the number of hair follicles, the size of the hair-bearing region can be doubled without a noticeable decrease in hair density. As such, tissue expansion may be used a means of treating male pattern baldness in addition to reconstructing the scalp. Expanders are most commonly placed in the occipital or posterior parietal regions. They should be placed under the galea, superficial to the periosteum. It usually requires 6-8 weeks to complete the expansion in adults, and up to 12 weeks in children. Radial scoring of the galea at the time of surgery can speed the process. Once the expansion is complete, flaps are advanced or transposed, ideally based on named arteries of the scalp. It is important to orient flaps so that the correct direction of hair growth is maintained. Although galeal scoring or capuslotomy incisions can be useful, wide undermining is a safer method of recruiting tissue.

Forehead

The brow position is the most important structure to preserve during forehead expansion. When possible, two or more expanders are used with incisions hidden within the hairline. For mid-forehead lesions, bilateral, temporal expanders are used, and the skin is advanced medially based on the superficial temporal arteries. Expanders should be placed deep to the frontalis muscle. Expansion can usually begin 7-10 days postoperatively. When a large forehead flap is required for nasal reconstruction, the forehead skin can be pre-expanded prior to flap transfer.

Face and Neck

The skin of the neck and face is relatively thin. Therefore, multiple expanders with smaller volumes are preferable to a single large expander. In general, however, a single larger expander is preferable to several smaller expanders. Careful planning is essential in determining where to place the expanders, and where incisions should be located. Considerations such as preserving aesthetic units, matching skin color, avoiding distortion of the eyelids and oral commissure, and facial symmetry are all essential. The expander is usually placed above the platysma muscle in order to avoid risk of facial nerve injury and to keep the flap from being excessively bulky. The expanded flaps can be advanced, rotated, or transposed. Incisions should be placed in skin creases such as the nasolabial fold. Expanding the hairless skin adjacent to the mastoid region can increase the available tissue for reconstructive procedures of the ear. The skin above the clavicle can be expanded to provide full-thickness skin grafts to the face.

Trunk

Unlike the head and neck, there are very few critical landmarks on the trunk that must be preserved. Aside from the breast and nipple-areola complex, distortion of the skin and soft tissues of the trunk is well-tolerated. For defects requiring excision, multiple expanders surrounding the defect are often employed. Many myocutaneous flaps of the trunk, such as the latissimus dorsi, TRAM and pectoralis flaps, can be pre-expanded in order to increase their size and facilitate donor site closure. Expanders can also be used to expand the skin of the abdomen for use as a donor site of full-thickness skin grafts.

Extremities

Tissue expansion in the extremities has been reported to have a higher complication rate in comparison to other regions and therefore should not be a first choice among the reconstructive options. The blood supply and drainage of the extremities is inferior to that of the trunk and head. This predisposes limbs, especially below the knee, to an increased rate of infection and wound complications. Multiple expanders are usually required in the extremites.

Complications

Proper placement and filling of tissue expanders has a steep learning curve. With experience, the complication rate drops dramatically. Among all patients, the major complication rate is about 10% and includes implant exposure, deflation, and wound dehiscence. Minor complications also occur in about 10% of patients. These include filling port problems, seroma, hematoma, infection and delayed healing.

Patients under the age of 7 have the highest risk of complications. One explanation for this is that young children are more prone to expander rupture due to external pressure on the expanded skin. Expansion in the extremities caries twice the risk of complication compared to other regions. The use of tissue expansion in burn reconstruction and soft tissue loss has a 15-20% major complication rate, whereas for congenital nevi it is 5-7%. Finally, tissue that has undergone serial expansion (two or more prior expansions) is at a higher risk for a major complication.

Pearls and Pitfalls

Tissue expansion should be avoided in infected fields, in close proximity to a malignancy, in skin-grafted regions, and in skin that has been previously radiated.

Every effort should be made to place the incision as far as possible from the region to be expanded, unless the incision can be incorporated into the tissue that is destined to be excised. If the incision is subject to the tension of expansion, it becomes at risk for dehiscence and hypertrophic scarring.

A key point in tissue expansion is the development of an adequately sized pocket. If the pocket is too small, expansion will likely fail. If the pocket is overly large, the expander can shift positions, resulting in expansion of the wrong tissue. Textured expanders are less likely to shift after placement.

The rate of expansion is variable and depends both on the body site as well as patient factors. Some skin is more amenable to expansion, and some patients can tolerate the discomfort better than others. It is possible to be overly aggressive with the rate of expansion, resulting in overlying skin ischemia, necrosis and ultimately implant extrusion.

As a general rule, the diameter of the expanded flap should be 2-3 times the diameter of the skin that is to be excised.

Suggested Reading

1. Argenta LC. Reconstruction of the paralyzed face. Grabb and Smith's Plastic Surgery. 5th ed. Philadelphia: Lippincott-Raven, 1997:91.
2. Bauer BS, Johnson PE, Lovato G. Applications of soft tissue expansion in children. Clin Plast Surg 1987; 14:549.
3. Bauer BS, Few JW, Chavez CD et al. The role of tissue expansion in the management of large congenital pigmented nevi of the forehead in the pediatric patient. Plast Reconst Surg 2001; 107:668.
4. Bauer BS, Margulis A. The expanded transposition flap: Shifting paradigms based on experience gained from two decades of pediatric tissue expansion. Plast Reconst Surg 2004; 114:98.
5. Chun JT, Rohrich RJ. Versatility of tissue expansion in head and neck burn reconstruction. Ann Plast Surg 1998; 41(1):11.
6. De Filippo RE, Atala A. Stretch and growth: The molecular and physiologic influences of tissue expansion. Plast Reconst Surg 2002; 109:2450.
7. Friedman RM, Ingram Jr AE, Rohrich RJ et al. Risk factors for complications in pediatric tissue expansion. Plast Reconst Surg 1996; 98(7):1242.
8. LoGiudice J, Gosain AK. Pediatric tissue expansion: Indications and complications. J Craniofac Surg 2003; 14(6):866.
9. Sasaki GH. Intraoperative sustained limited expansion (ISLE) as an immediate reconstructive technique. Clin Plast Surg 1987; 14(3):563.

Chapter 14

Alloplastic Materials

Jason H. Ko and Julius Few

Introduction

Advances in medical technology have allowed plastic surgeons to utilize synthetic materials as an alternative to autologous tissues when performing many of today's aesthetic and reconstructive surgeries. Although autologous materials are generally preferred, synthetic materials provide several advantages over tissues obtained from the patient:

- Not resorbed over time (unless they are designed to do so)
- Do not require a second surgical donor site
- Provide more material than can often be obtained from the patient
- Can be custom-tailored to the individual patient
- Reduce operating time since graft harvesting is not performed

Because of the many benefits to using alloplastic materials, there is currently a strong interest in developing the ideal implant material which would possess the following characteristics: it should (1) be chemically inert; (2) be incapable of producing hypersensitivity or a foreign body reaction; (3) be easily contoured; (4) retain stable shape over time (except when desired); (5) be noncarcinogenic; (6) become ingrown or replaced by living tissue; (7) be easy to remove and sterilize; and (8) not interfere with radiographic imaging. Despite much effort and ingenuity, creation of the ideal implant material has yet to be accomplished. However, various alloplastic materials are being used today in plastic and reconstructive procedures, and many of them have proven quite promising.

Pre- and Intraoperative Considerations

The vascularity of the recipient site and the ability to provide sufficient soft tissue coverage of the implant must be assessed preoperatively. Decreased vascularity secondary to scar tissue (from previous surgeries) or radiation impairs the establishment of normal fibrovascular tissue encapsulation and may interfere with the normal inflammatory response if the implant were to become infected. In order to prevent implant exposure or extrusion, soft tissue coverage over an implant should be as thick as possible. The size of the implant should be comparable to that of the tissue pocket or wound cavity in order to avoid tension of the overlying soft tissue, and the implant should be fixated to a stable adjacent structure to prevent migration of the implant. All patients should receive perioperative intravenous antibiotics followed by a postoperative oral course, although the optimal antibiotic choice and duration have yet to be determined for most implants. What is clear is that intraoperative handling of the implant should be minimized in order to prevent bacterial transmission, and strict adherence to sterile technique is essential.

Practical Plastic Surgery, edited by Zol B. Kryger and Mark Sisco.

Table 14.1. Classification of synthetic materials used in plastic and reconstructive surgery

Silicone-based materials:	BioPlastique Injectable silicone Silastic sheets Silicone Silicone gel
Polytetrafluoroethylene:	Gore-Tex Proplast I and II Teflon
High density polyethylene:	Medpor
Polymer mesh:	Dacron (Mersilene) Dexon Prolene Supramid Vicryl
Biological glasses:	Bioactive glasses (Bioglass) Glass ionomer
Tissue adhesives:	Cyanoacrylate
Acrylics:	HTR Polymer Methylmethacrylate

Choice of Alloplastic Material

The type of procedure as well as the size and character of the defect being augmented often dictate the type of implant material. In the preantibiotic era, inert materials such as gold, silver, platinum and paraffin were used with little success and were quickly abandoned. Currently, there are numerous implantable materials being used today (Table 14.1). These materials are used in a wide range of procedures, such as aesthetic procedures, craniofacial surgery, maxillofacial trauma, breast reconstruction and hand surgery. Table 14.2 lists the common uses for the various allopastic implants.

Silicone

Silicone-based prosthetics have been used as medical implants since the 1950s due to their chemically inert nature, resistance to degradation, and lack of significant allergic reactions. Silicone is useful for a variety of aesthetic surgeries, complex contouring and reconstructive procedures. Silicone comes in the form of silicone gels, silicone rubber or solid silicone implants. Silicone gels can provide a more natural feel, as seen with breast implants, but the risk of rupture requiring capsulectomy is a distinct disadvantage. The use of silicone gel has been surrounded by controversy related to concerns about migration, toxicity and an unproven association with systemic disease, leading to restriction of the use of silicone gel implants by the FDA in 1992. This ban was recently lifted after an extensive unbiased review by the Institutes of Medicine.

Silicone rubber is used for tissue expanders, the outer shell of both saline-filled and silicone gel-filled breast implants, and as an onlay material for the augmentation of the bony skeleton and soft tissues. However, silicone rubbers are relatively weak and tend to tear, leading to implant failure. Solid silicone implants are commonly used for chin and malar augmentation, and have been used in nasal, chest and calf augmentation, as well as in joint replacement and tendon reconstruction.

Table 14.2. A list of the procedures that commonly employ allopastic materials

Procedures	Materials Used
Cranioplasty and forehead augmentation	Glass ionomer and bioactive glass Hard-Tissue-Replacement (HTR) polymer Methylmethacrylate Medpor Poly(L-lactide) and polyglycolic acid plates and screws Silicone
Anterior mandibular augmentation	Medpor Polyamide mesh Silicone
Mandibular body and angle augmentation	Glass ionomer and bioactive glass Medpor Methylmethacrylate Poly(L-lactide) and polyglycolic acid plates and screws
Malar and maxillary reconstruction	Glass ionomers Gore-Tex Medpor Methylmethacrylate Polyamide mesh Silicone Teflon
Zygomatic reconstruction	Glass ionomers Medpor Gore-Tex Poly(L-lactide) and polyglycolic acid plates and screws Silicone
Orbital reconstruction	Gore-Tex HTR Polymer Medpor Poly(L-lactide) Silicone Teflon
Ear reconstruction	Medpor Silicone
Tendon repair	Gore-Tex Cyanoacrylate
Soft tissue augmentation	BioPlastique Gore-Tex
Breast augmentation and tissue expansion	Silicone (saline or silicone gel filled)
Wound repair and scar revision	Cyanoacrylate Silastic sheets
Chest and abdominal wall reconstructions	Dacron mesh Gore-Tex Prolene mesh Vicryl mesh
Nasal augmentation	Gore-Tex Polyamide mesh Silicone

Because silicone is not porous, tissue ingrowth does not occur. A fibrous capsule forms around the implant that is relatively avascular and can contract which may lead to implant migration. This avascular capsule is a potential space for infection and in the setting of infection may require removal of the implant.

BioPlastique

BioPlastique® is a nonbiodegradable, relatively inert injectable liquid used for soft tissue augmentation. The textured surface of the particles allows for tissue ingrowth, and the particle size is large enough to prevent engulfment by macrophages but small enough to become encapsulated within 3 to 6 weeks. Studies on the use of BioPlastique demonstrate good-to-excellent results in augmenting small defects on the dorsal nose, malar area, cheeks and chin with no adverse immunologic reactions. Although the clinical results with Bioplastique have been encouraging, it is not FDA approved at this time.

Polymethylmethacrylate

Polymethylmethacrylate (PMMA) is an acrylic polymer used as a bone substitute in plastic surgery and neurosurgery. PMMA is radiolucent, extremely durable and completely biocompatible, making it a widely used material for cranial bone reconstruction alone or in combination with wire or mesh reinforcement. When powdered granules of methylmethacrylate polymer are mixed with methylmethacrylate liquid monomer, a moldable dough forms as the monomer polymerizes and hardens in about ten minutes. Near the end of the polymerization process, an exothermic reaction occurs that can potentially damage the local tissues, the major complication associated with the use of PMMA. This can be avoided by continually irrigating the implant bed with cool saline during the polymerization. A rare, but serious complication is the inadvertent entry of the PMMA into the venous or arterial systems. If this occurs it can cause complete heart block, cardiac arrest and other arrhythmias. This complication is most often seen during orthopedic procedures where PMMA is used for joint replacements or fracture repair.

Hard-tissue-replacement (HTR) polymer is a porous form of PMMA that allows for fibrous ingrowth, leading to an implant that is nonresorbable and very stable. Applications for HTR include chin and malar augmentation, with potential for additional uses in craniofacial reconstruction.

Polyester (Dacron®, Mersilene®)

Polyethylene terephthalate (Dacron) is a biocompatible, flexible, nonabsorbable polymer that is used as a suture material, as a prosthetic material for arterial replacement, and as a mesh (Mersilene) in abdominal and chest wall reconstruction. Its use has also been described for chin and nasal augmentation.

Biodegradable Polyester (Polyglycolic Acid, Poly-L-lactic Acid)

Polyglycolic acid (PGA) and Poly(L-lactide) (PLLA) are polymers that are degraded in the body at physiologic pH over the course of weeks to months. These resorbable polymers are available as mesh sheets for body wall reconstruction and as rods for the internal fixation of fractures and osteotomies. They have also been fashioned into resorbable miniplates and screws for the fixation of bones of the craniofacial skeleton. Although they do not appear to have any cytotoxic effects, they do provoke an inflammatory or foreign body response after implantation.

Polyamide Mesh (Supramid®, Nylamid®)

Polyamide mesh is a woven, polymer mesh implant that is biocompatible, can be easily shaped and sutured, allows for fibrous tissue ingrowth, and has been used for the repair of orbital floor defects. It seems to be well tolerated and has a low rate of extrusion, even in areas of thin skin such as the nasal dorsum. However, polyamides do undergo resorption and induce an inflammatory response, making their use in facial augmentation and reconstruction somewhat limited.

Porous Polyethylene (Medpor®)

Medpor is a high-density, porous polyethylene implant used frequently in facial surgery because it is nonantigenic, nonallergenic, nonresorbable, highly stable and easy to fixate. In addition, Medpor is available in a wide variety of preformed shapes for its use as a malar, chin, nasal, orbital rim, orbital floor and cranial implant, as well as an auricular framework in postburn ear reconstruction. Overall, complications of Medpor, such as exposure or infection, are rare.

Polytetrafluoroethylene (Teflon®, Gore-Tex®, Proplast®)

Polytetrafluoroethylene (PTFE) is an inert and highly biocompatible polymer that is extremely useful in soft tissue augmentation but has limited use in bony repair due to its low tensile and compressive strength. Teflon, the first PTFE graft to be used in plastic surgery, is a chemically inert polymer used for soft tissue augmentation in the past, but the main application for Teflon has been orbital floor reconstruction.

Gore-Tex is a pliable, durable, inert, biocompatible material that has some tissue ingrowth, little inflammatory reaction and almost no encapsulation. In addition to being used in abdominal fascial reconstruction, chest wall reconstruction and soft tissue reconstruction, Gore-Tex has also been utilized for lip, nasal, chin and malar augmentation. It has also been utilized for the treatment of nasolabial and glabellar creases.

Proplast I is a highly porous, black graphite/PTFE composite with a spongy consistency. Because Proplast I led to discoloration of the surrounding soft tissues when implanted, Proplast II-a more rigid, white PTFE/alumina compound-was developed as an alternative. Proplast, with a wide variety of applications including the reconstruction of the chin, zygoma, orbital rim, maxilla, mandible, skull and rib cage, was originally regarded favorably. However, reports of biomechanical failure, intense inflammation, infection and extrusion related to the Proplast temporomandibular joint implant, led to the removal of all Proplast implants from American markets by the FDA in 1990.

Calcium Phosphate Ceramics

Calcium phosphate implants have been available as bone replacement/augmentation materials for 20 years. The primary calcium phosphates in clinical use are hydroxyapatite and tricalcium phosphate. These materials are osteoconductive (providing a scaffold for bone ingrowth) thus allowing for integration into the recipient site after placement. As a result, calcium phosphates are very well tolerated with essentially no inflammatory response, minimal fibrous encapsulation, and no adverse effects on local bone mineralization.

Metals

Metals have been used for the past 35 years for skull reconstruction and repair, in addition to reconstruction of craniofacial and upper extremity skeletal injuries. Stainless steel, cobalt-chromium (vitallium), pure titanium and titanium alloys are the principal metals currently available. Characteristics of a desirable metal implant include biocompatibility, strength, resistance to corrosion and imaging transparency.

Postoperative Considerations

Although numerous potential complications may occur with any implant-related procedure (e.g., migration, extrusion, palpability), the one common denominator shared by all alloplastic implants is their inherent risk of infection. The majority of postoperative infections appear within weeks to months after the initial surgery. Low-grade infections manifested only by fevers and signs of mild cellulitis are managed by intravenous antibiotics. More serious infections involving wound breakdown, implant exposure, gross purulence or systemic spread of the infection require prompt removal of the implant as antibiotics and drainage alone are usually insufficient. Reimplantation should not be performed for at least 3 to 6 months to allow for complete treatment and resolution of both the infection and the inflammation in the surrounding tissues. Several studies suggest that smooth, nonporous, nonresorbable implants have lower rates of infection, but it remains to be seen whether any true infectious risk differences exist among the various alloplastic implant materials available today.

Pearls and Pitfalls

- Incisions should be placed as far as possible from the final position of the implant. This will decrease the risk of implant exposure or extrusion in the setting of a minor wound infection.
- The implant should be covered with as much soft tissue as possible. The pocket should be of adequate size; too large and the implant will shift position, too small and the implant will be at risk for extrusion due to tension on the closure.
- Whenever possible, always try and close a second layer of tissue between the skin and implant. This is critical if the implant lies directly beneath the incision.
- Implants with sharp corners should be smoothed down, since sharp edges can erode through the skin with time.
- The implant should be touched as little as possible. Clean, powder-free gloves should be worn and instruments should be used to handle the implant whenever feasible. The risk of infection and abnormal capsule formation is increased by the presence of any bacteria or foreign material on the implant.
- Do not use an implant composed of a rigid material to replace soft, pliable tissue.
- Keep an organized registry of all alloplastic implants in the event that the device fails or has to be removed. Give the patient a copy of the device name, model, manufacturer and serial number, in case failure occurs in the care of another physician.

Suggested Reading

1. Ousterhout DK, Stelnicki EJ. Plastic surgery's plastics. Clin Plast Surg 1996; 23(1):183.
2. Eppley BL. Alloplastic implantation. Plast Reconstr Surg 1999; 104(6):1761.
3. Park JB, Lakes RS. Polymeric implant materials. In: Park JB, Lakes RS, eds. Biomaterials: An Introduction. New York: Plenum Press, 1994:164.
4. Mladick RA. Twelve months of experience with BioPlastique. Aesthetic Plast Surg 1992; 16:69.
5. Manson PN, Crawley WA, Hoopes JE. Frontal cranioplasty: Risk factors and choice of cranial vault reconstruction material. Plast Reconstr Surg 1986; 77:888.
6. Kulkarni RK. Polylactic acid for surgical implants. Arch Surg 1966; 93:839.
7. Romano JJ, Iliff NT, Manson PN. The use of Medpor porous polyethylene implants in 140 patients with facial fractures. J Craniofac Surg 1993; 4:142.
8. Mole B. The use of Gore-Tex implants in aesthetic surgery of the face. Plast Reconstr Surg 1992; 90:200.
9. Bucholz RW, Carlton A, Holmes RE. Hydroxyapatite and tricalcium phosphate bone graft substitutes. Orthop Clin North Am 1987; 18:323.
10. Ellerbe DM, Frodel JL. Comparison of implant materials used in maxillofacial rigid internal fixation. Otolaryngol Clin North Am 1995; 28:365.

Chapter 15

The Chronic Infected Wound and Surgical Site Infections

Kevin J. Cross and Philip S. Barie

Introduction

The skin is the organ of the body that bears the responsibility of protecting it from environmental pathogens. It is important to understand, however, that the skin surface is not germ-free. In fact, healthy skin can be colonized with multiple species of virulent organisms. Therefore, the mere presence of bacteria in a wound does not in itself define infection. Rather, it is the amount and type of bacteria that determines whether a wound is truly infected. There is a continuum spanning from the sterile wound to the grossly infected wound, and determining where a wound falls on this continuum is a challenge for the bedside clinician. It is important for the plastic surgeon to become proficient in this skill, because it will help predict the likelihood and rate of wound closure versus the potential for the wound to become a systemic threat. Understanding the severity of bacterial involvement will also help to guide therapy.

Chronic Wounds

Normal Skin Flora

The skin is covered with microorganisms. These can be either resident organisms, those that can typically be found on the subject's skin, or transients that are often seen on the skin surface but are quickly shed during normal body hygiene or by skin sloughing. While these organisms are usually bacteria (Table 15.1), the yeast *Pityrosporum* and skin mite *Demodex* are also commonly found. These colonizing microbes take residence in the crypts and crevices that favor bacterial growth, and prevent pathologic species from gaining access to these areas.

Human beings are protected from bacterial overgrowth and invasion at the surface by a number of defense mechanisms. A layer of dead, keratinous epithelial cells known as the stratum corneum is the outermost layer of skin. As the keratin sloughs, it removes attached organisms with it. Sebaceous glands secrete an oily, lipid-rich, acidic substance, (pH range of 4.2 to 5.6) that acts to retard bacterial growth. Bacteria become more active on the skin surface as the pH rises above 6.5, as is seen with the use of many cleansing and moisturizing agents. Should foreign organisms get past these defenses, the antigen-presenting Langerhans cells found in the epidermis and the phagocytosing macrophages and immune-stimulating mast cells present in the dermis rapidly mobilize the body's cellular and humoral immune responses.

Contamination vs. Infection

Cutaneous wounds, by definition, have lost their protective barrier and are subject to invasion by not only foreign bacteria introduced through the environment,

Practical Plastic Surgery, edited by Zol B. Kryger and Mark Sisco. ©2007 Landes Bioscience.

Table 15.1. Normal bacterial skin flora

Bacteria
Staphylococcus
Micrococcus
Peptococcus
Corynebacterium
Brevibacterium
Propionibacterium
Streptococcus
Neisseria
Acinetobacter

but also the local bacterial flora that is present on intact skin. These wounds occur in the setting of various pathologies and are usually chronic in nature before being brought to the attention of a plastic surgeon. Unlike most surgical incisions, these wounds heal by secondary intention and are always colonized by bacteria. They require extensive granulation tissue formation and keratinocyte migration for closure, involving endothelial cells and fibroblasts for the purposes of neovascularization and matrix production, respectively. For this to occur, **macrophages** and a varying milieu of growth factors must be present. Along with neutrophils, macrophages also act to disinfect the wound, killing foreign organisms by the generation of peroxide and superoxide radicals. The clinical spectrum of bacterial invasion exists on a continuum from least to most severe: contamination, colonization, local infection or critical contamination, invasive infection and sepsis.

15

- **Contaminated wounds** have nonreplicating organisms within their borders. These wounds will go on to heal normally.
- **Colonized wounds** have replicating bacteria, but these bacteria are nondestructive and contained within the wound. A hallmark of colonization is that it does not delay the wound healing process.
- **Local infection** or critical contamination is an intermediate level of bacterial invasion characterized by granulation tissue that has an unhealthy appearance, and wound healing that may be delayed. In this type of wound, however, tissue invasion is not present. This stage is notable for the absence of other signs of infection such as cellulitis or pus formation.
- **Invasive infection** occurs once bacteria have invaded through the wound bed, tissue destruction has begun, and an aggressive immune response is present. Signs and symptoms of invasive infection include pain, edema, erythema and fever. The finding of a chronic, nonhealing wound, often with pus formation and tissue necrosis is often evident.
- **Sepsis** occurs when the infection spread systemically, and cardiovascular instability and organ-system dysfunction develop.

The Molecular Biology of Bacterial Infection

Low levels of bacteria in wounds actually help to promote wound healing by stimulating brisk monocyte and macrophage activity. However, as their number or virulence increases, the tissue response to their presence disrupts and prolongs the inflammatory phase of wound healing, depletes the components of the complement cascade, interferes with normal clotting mechanisms, and alters leukocyte function. The level of pro-inflammatory cytokines, including interleukin-1 and tumor necrosis factor-alpha, rises and stays elevated. Elevated levels of matrix metalloproteinases and a lack of their inhibitors lead to tissue breakdown and growth factor inhibition. Bacteria also compete with local cells for oxygen, reducing its availability to these cells and stimulating an angiogenic response, leading to friable granulation tissue that is prone to bleeding.

Bacteria in Wounds

Classic teaching is that wounds with greater than 10^5 organisms/gram should be considered infected whereas those with a lower bacterial count should not. Although studies do show that wounds with bacterial counts higher than this heal more slowly and have a higher rate of infection, a more practical approach to diagnosing the infected wound is encouraged. As wounds mature, not only do the species of organisms present in the wound change, the wounds begin to carry a higher level of bioburden, meaning a higher baseline number of colonies without being infected. Conversely, the more virulent bacteria, such as beta-hemolytic streptococcus and some rare Clostridium species, can easily cause infection at lower quantitative levels than the more commonly occurring species (Table 15.2). Finally, the status of the patient's immune response has a role in the patient's likelihood of developing an infected wound. Therefore, the surgeon is encouraged to study the appearance of the wound and the overall clinical picture when deciding whether a wound is infected.

Although it is important to note the classic signs and symptoms of infection including erythema, edema, fever and an elevated white blood cell count, recent studies attempting to establish evidence-based criteria for the determination of a chronic wound infection have shown that increasing pain, friable granulation tissue, foul odor and wound breakdown are the most sensitive indicators.

Bacteria in chronic wounds often establish a **biofilm**. This is an extracellular, polysaccharide-rich matrix in which the organisms are embedded. Within this glycocalyx is a system of channels, like a primordial circulatory system, that allows the bacteria to remain viable with less direct dependence on the host tissue. Cells in this environment become more sessile and less metabolically active. As a result, they are resistant to host immune responses and antibiotic therapy. Biofilms often coat foreign and implanted material, making infections in this setting more difficult to treat, and certain bacteria such as *Pseudomonas aeruginosa* have a predilection to biofilm production.

Table 15.2. Bacteria commonly infecting chronic wounds

Bacteria	Occurrence (%)
Staphylococcus aureus	20
Coagulase-negative staphylococci	14
Enterococci	12
Escherichia coli	8
Pseudomonas auruginosa	8
Enterobacter species	7
Proteus mirabilis	3
Klebsiella pneumonia	3
Other streptococci	3
Candida albicans	3
Group D streptococci	2

Reproduced with permission from: Barie PS. Surgical Infections 2002; 3:S-9.

Clinical Evaluation

A thorough history includes information related to the chronicity of the wound, any changes to the wound appearance, and details that should make the clinician suspicious of a more invasive bacterial involvement (e.g., pain, fever). Mitigating factors such as comorbid conditions that could lead to immunosuppression, the use of any immunosuppressive medications, previous radiation in the wound area and the overall functional status of the patient are important to explore. In addition to a white blood cell count and blood cultures, laboratory tests can include the erythrocyte sedimentation rate and C-reactive protein. Although not specific, in a patient with no recent history of surgery or acute illness, their value is in helping to determine the level of systemic response to a wound and in helping to determine the presence of a deep wound infection.

When examining a wound, its depth and width should be measured and a careful inspection and probing should be done. Attention to findings such as erythema at least 5 mm beyond the wound edges, expressed pus, necrotic debris or granulation tissue that is dark, friable or heaped above the wound edges can help to determine the extent of infection. Foreign bodies such as old strands of gauze should be removed and the presence of underlying foreign material such as sutures or mesh should be ruled out. Care must be taken to ensure that wounds overlying osseous structures do not have any exposed bone at their base that would suggest the presence of osteomyelitis.

As stated earlier, bacterial cultures can help to make a diagnosis and guide appropriate therapy. In a wound that has been appropriately cleaned and prepared, a swab of the deeper tissue can give a qualitative notion of which bacteria are present. It does not, however, allow the clinician to quantitate the amount of bacteria within the wound. For this, the gold standard is a biopsy culture. A punch biopsy is taken and ground into a liquid state from which serial dilutions are cultured. A measure of colonies per milligram can then be reported.

Treatment

Antibiotics are ineffective in penetrating chronic, nonhealing wounds. **Debridement** is the best option for clearing bacterial loads and removing nonviable tissue. If not performed, necrotic material can release endotoxins that inhibit keratinocyte migration and matrix production, and can prolong the inflammatory response, promoting matrix-destroying proteases.

Methods of debridement include sharp, mechanical, chemical and biodebridement.

Sharp, or surgical debridement affords the luxury of speed, since it can be performed at the bedside with nothing more than scissors and a pair of forceps. More extensive debridement may require anesthesia, and should be performed in the operating room.

Mechanical debridement is the eradication of dead tissue by the sequential changes of dressings that are inserted moist into the wound and removed after they are allowed to dry. Exudative and necrotic tissues adhere to the drying gauze and are pulled out of the wound with the gauze. This technique has only a limited ability to remove structurally intact or strongly adherent devitalized tissue. The classic "wet-to-dry" dressing is a mechanically debriding dressing.

Chemical debriding agents are enzymatic compounds that break down tissue. They are most effective in moderately sized areas of necrosis or in those patients that will not tolerate an operation. In order to gain maximum benefit, larger eschars should be cross-hatched or excised to allow for better penetration of the agent. The papain-containing cream, Accuzyme®, is commonly used for this purpose.

Biodebridement involves the application of sterile maggots into a wound for periods of 48-72 hours. The maggots feed on, and thus remove dead tissue before being irrigated out. This process can be repeated as necessary. Needless to say, it is not a commonly used technique.

Antibiotics do have a role in the treatment of chronic, infected wounds once debridement has achieved healthy wound borders. Empiric antibiotics should be selected based on the bacteria that are likely to be involved. For example, empiric antibiotics for wounds near the oropharynx and diabetic foot wounds should include coverage for anaerobic species. The Gram stain can give a general idea of whether Gram-positive, Gram-negative or a combination of bacteria is present. Once culture results return over the subsequent 2-3 days, antibiotic coverage should be tailored to the involved organisms. Topical antibiotic preparations can help to reduce bacterial load and can be used with some success in an adjuvant setting in the select wound population. Prudence should be taken with their use, however, because many of these preparations also impair the function of the superficial cells necessary for wound healing. They should never be used in wounds related to venous disease, as these wounds are more prone to sensitivity reactions. Examples include: iodine or iodophor paint, sodium hypochlorite solution, hydrogen peroxide, acetic acid, antibiotic creams or the newer cadhexomer iodine and nanocrystalline silver.

For wounds that arise in the setting of underlying pathology, treating the disease process can increase the speed and likelihood of wound healing. For venous stasis ulcers, reducing edema fluid with Unna boot compression, elevation and diuresis can improve oxygen delivery and thus cellular function. Patients with diabetic foot ulcers should have their blood sugar strictly controlled given the deleterious effects of hyperglycemia on neutrophil and monocyte function. If ischemia is believed to be contributing to the etiology or chronicity of a wound, smoking cessation and elimination of dehydration and anemia should all be considered in the treatment plan. Ultimately arterial revascularization with or without surgical reconstruction using local or microvascular flaps may be necessary.

Hidradenitis Suppurativa

This condition is due to infection of the apocrine sweat glands, most commonly in the axillary, perineal and groin regions. It results in recurrent, draining abscesses and sinus tracts that can lead to severe pain and debilitation. Lesions in the axilla that heal may scar and secondarily cause contracture limiting arm motion. Active infection should be treated with a 1-2 week course of oral antibiotics and is usually due to Gam positive cocci. Cultures should always be taken since other bacterial infections can occur, and the antibiotic should be appropriately selected.

Surgical treatment consists of full-thickness excision of the infected dermis and any involved subcutaneous fat. Primary closure can be obtained in small- to moderate-sized wounds without active infection. Larger defects, or those that are grossly infected, should not be closed primarily. They should be allowed to granulate with dressing changes, followed by split-thickness skin grafting or healing by secondary intention. Incomplete excision of the involved tissue is common due to retained sinus tracts or deep infected glands. When this occurs, there is a high likelihood of recurrence and skin graft failure.

Surgical Site Infections

Definitions

Overall, surgical site infections (SSIs) are the leading cause of nosocomial infections, accounting for 38% of these complications. By definition, to be an SSI, an infection must occur within 30 days of the operation. SSIs can be broken down into three general categories. Superficial incisional SSIs involve only the skin or subcutaneous tissue of the incision. Signs and symptoms of this type of infection may include pain, swelling, redness, warmth and tenderness. Deep incisional SSIs demonstrate either purulent drainage from deeper tissue, a deep incisional dehiscence, or an abscess in the depth of the incision. Lastly, organ or deep space SSIs involve infections in manipulated regions other than the skin and subcutaneous tissue that was opened during the procedure. By definition, these infections must contain purulent drainage, positive cultures with fluid aspiration or documentation of the presence of an abscess. If a foreign body such as mesh or titanium was left in the wound an SSI can occur up to one year postoperatively.

Risk Factors

Generally speaking, the overall well being and the severity of any comorbid conditions determine how susceptible a patient is to wound infections (Table 15.3). The American Society of Anesthesiology rates patients' operative risk according to their level of illness and comorbidities, termed the ASA class. There is a close correlation between the severity of the preoperative risk and the risk of wound infection. Furthermore, greater operative time is also associated with an increased risk of developing an SSI.

When planning an operation, the surgeon must consider the level of expected contamination. Clean surgical procedures are those that involve only skin and the musculoskeletal soft tissue and carry approximately a 2% chance of developing an SSI (although it must be noted that wound infection rates are probably underreported). Clean-contaminated procedures are those that involve the planned opening of a hollow viscus (e.g., the respiratory, biliary or gastrointestinal tracts) and have a 7-15% risk of becoming infected. Contaminated procedures are those that introduce nonsterile, bacteria-rich contents into the wound for a short period of time (e.g., penetrating abdominal trauma, unplanned enterotomies) and lead to SSIs in 20% of cases. Dirty procedures take place in an infected setting (e.g., bowel resection for an abscess related to Crohn's disease, removal of infected prosthesis). Approximately 20-40% of these wounds will become infected if closed primarily.

Bacteria and Prophylaxis

Whereas most SSI are caused by skin derived Gram-positive cocci, including *Staphylococcus aureus*, coagulase-negative staphylococci such as *Staphylococcus epidermidis* and *Enterococcus* species, site-specific pathogens, may infect wounds. Consideration for Gram-negative bacilli should be given to any wound that is located near the site of bowel injury or repair, and when either bowel or tracheopharyngeal structures are violated, both enteric aerobic bacteria such as *Escherichia coli* and anaerobic bacteria such as *Bacteroides fragilis* may be of concern.

Prophylaxis for clean surgery is controversial. It is generally accepted that when bone is violated (e.g., during cranial vault reconstruction) or when a prosthesis is inserted, preoperative antibiotics are indicated. Less convincing data exists for straightforward soft tissue surgery (e.g., scar revisions).

Table 15.3. Risk factors for the development of a surgical site infection (SSI)

Patient factors	Anemia (postoperative)
	Ascites
	Chronic inflammation
	Corticosteroid therapy (controversial)
	Obesity
	Diabetes
	Extremes of age
	History of irradiation
	Hypocholesterolemia
	Hypoxemia
	Malnutrition
	Peripheral vascular disease
	Recent operation
	Remote infection
	Skin carriage of staphylococci
	Skin disease in the area of infection (e.g., psoriasis)
Environmental factors	Contaminated medications
	Inadequate disinfection/sterilization
	Inadequate skin antisepsis
	Inadequate tissue oxygenation
Treatment factor	Drains
	Emergency procedure
	Hypothermia
	Inadequate antibiotic prophylaxis
	Prolonged preoperative hospitalization
	Prolonged operative time

When choosing an antibiotic agent, the following factors should be considered:

- It should have minimal side-effects and be safe for the patient.
- It should have a narrow spectrum of coverage for the expected organisms.
- It should not be overused (making it less likely that bacteria have developed resistance).
- It should cover typical infections that are specific for the institution.
- It can be used for a brief period of time (less than 24 hours).

Long prophylactic courses have been associated with an increased risk of nosocomial infections and multi-drug resistance. For clean and most clean-contaminated cases, a first-generation cephalosporin should be used. If a patient has a documented penicillin allergy, clindamycin is an alternative. Only in the setting of a hospitalized patient in an institution that carries a high rate of methicillin-resistant *S. aureus* (MRSA), should vancomycin be considered for prophylaxis.

It is important to recall that the timing of the antibiotic dose determines its effectiveness. Preoperative prophylaxis should be closed within two hours of incision time. Given too early, the antibiotic can be cleared before the case is started. Some benefit can be gained from intraoperative dosing if antibiotics are not given before the case begins, but no benefit has been shown when the first dose is given after the case ends. This loss of benefit after skin closure is related to the fact that sutured wounds exist in a low blood flow state owing to vasoconstriction, the use of electrocauterization for hemostasis, and the constrictive effects of the suture closure. Therefore, antibiotics will not reach the surgical site. In extremely lengthy cases, redosing intraoperatively is recommended.

Prevention and Treatment

In the weeks to months before a planned operation, much can be done to maximize the immune state and wound healing capabilities of the patient. Smokers should be encouraged to stop at least one month prior to their surgery. Smoking is a known vasoconstrictor that can reduce oxygen delivery to wounded tissue, and its effects have been found to last weeks beyond the point of smoking cessation.

The nutritional status of the patient should be taken into consideration as well. Obese patients should be encouraged to lose as much weight as possible while maintaining a healthy, protein-rich diet, and in the malnourished hospitalized patient, even a short 5-7 day course of parenteral or enteral nutrition has been shown to significantly reduce the risk of SSIs.

Studies show that having a patient take a preoperative shower with an antiseptic soap (e.g., hexachlorophene) can reduce skin bacterial load. However, shaving the planned surgical site with a razor either the night before surgery or immediately preoperatively should be discouraged due to the transient bacterial infestation that it promotes. Studies report greater than a 3-fold increase in infection rates with shaving versus hair clipping (5.6 vs. 1.7%). Finally, known *S. aureus* carriers should have their nasal orifices treated with topical 2% mupirocin.

Intraoperatively, care should be taken to keep the patient warm and well hydrated. This will improve blood flow to the wound and maximize oxygen delivery. Even 30 minutes of preoperative warming can reduce patient risk for SSI by two-thirds in some cases. Adequate oxygenation is important for cellular function and bacterial destruction via superoxide and peroxide formation. Case length should be kept to a minimum, given the fact that infection rates almost double for each hour an operation lasts. Tissues should be handled gently and electrocautery for hemostasis should be kept to a minimum. During the case, wounds should be kept moist and retractors should be released periodically to restore blood flow. The smallest possible suture diameter should be used to minimize foreign material in the wound (studies show that on average, surgeons use sutures one size larger than needed), and the prudent use of drains should be encouraged. By acting as a conduit for bacterial invasion and preventing epithelial closure of wounds, drains probably cause more SSIs than they prevent and they should be removed as soon as possible. Antibiotic prophylaxis of an indwelling drain is never indicated. High pressure pulse irrigation and topical antiseptic washes have been proven to be of some benefit in the contaminated or dirty wound. Both during the case and postoperatively, blood glucose concentration should be kept under tight control (80-110 mg/dl). And finally, postoperative nutrition should be optimized.

Controversy exists on whether it is appropriate to close contaminated wounds primarily. Studies in adults show that this practice can lead to a higher rate of wound failure and a greater cost of care. It is recommended that a delayed primary closure of the incision be used. This involves either placing untied sutures during the case that can later be cinched down, or using adhesive strips for closure when the wound is ready. Until the time when the wound appears to have minimal debris and no apparent progressing erythema, wet-to-dry, twice daily packing should be used (usually for 4-5 days).

Pearls and Pitfalls

Antibiotic prophylaxis of clean surgical procedures (e.g., elective operations on skin and soft tissue) is controversial based on a single randomized trial that showed benefit in breast and groin hernia surgery. The controversy persists because the incidence of superficial surgical site infection was so high (4%, versus an expected incidence of about 1%) in the placebo group. Evidence that antibiotic prophylaxis is indicated for soft tissue procedures of other types is lacking entirely, and prophylaxis cannot be recommended. If administered, antibiotic prophylaxis should be given before the skin incision is made, and only as a single dose. Additional doses are not beneficial because surgical hemostasis renders wound edges ischemic by definition until neovascularization occurs, and antibiotics cannot reach the edges of the incision for at least the first 24 hours. Not only is there lack of benefit, prolonged antibiotic prophylaxis actually increases the risk of postoperative infection.

Increasingly in the practice of plastic surgery, there is a tendency to leave closed-suction drains in place for prolonged periods in the erroneous belief that the incidence of wound complications is reduced by prolonged drainage. Nothing could be further from the truth. Data indicate that the presence of a drain for more than 24 hours increases the risk of postoperative surgical site infection with MRSA. Closed suction drains must be removed as soon as possible, ideally within 24 hours. Prolonged antibiotic prophylaxis is often administered to "cover" a drain left in place for a prolonged period. This is a prime example of error compounding error, and is a practice that must be abandoned.

Suggested Reading

1. Barie PS. Surgical site infections: Epidemiology and prevention. Surg Infect 2002; 3:S-9.
2. Bratzler DW. Antimicrobial prophylaxis for surgery: An advisory statement from the National Surgical Infection Prevention Project. Clin Infect Dis 2004; 38:1706.
3. Classen DC, Evans RS, Pestotnik SL et al. The timing of prophylactic administration of antibiotics and the risk of surgical-wound infection. N Engl J Med 1992; 326:281.
4. Edwards R, Harding KG. Bacteria and wound healing. Curr Opinion Infect Dis 2004; 17:91.
5. Hunt TK, Hopf HW. Wound healing and wound infection: What surgeons and anesthesiologists can do. Surg Clin N Am 1997; 77:587.
6. Nichols RL. Surgical wound infection. Am J Med 1991; 91:3B.
7. Platt R, Zaleznik DF, Hopkins CC et al. Perioperative antibiotic prophylaxis for herniorrhaphy and breast surgery. N Engl J Med 1990; 322:153.
8. Schultz GS, Sibbald RG, Falanga V et al. Wound bed preparation: Systematic approach to wound management. Wound Rep Regen 2003; 11:S1.
9. Wysocki AB. Evaluating and managing open skin wounds: Colonization versus infection. AACN Clinical Issues 2002; 13:382.

Chapter 16

Diabetic Wounds

Roberto L. Flores and Michael S. Margiotta

Epidemiology

Diabetes is currently one of the most common diseases in the United States with approximately 800,000 new cases diagnosed annually. This number is expected to increase with the growing incidence of obesity in this country, particularly among the young. Diabetes is the seventh leading cause of death in the United States with an annual health care cost at several billion dollars. For patients with diabetes, foot disease is the most common cause of hospitalization and up to one in five are expected to develop foot ulceration within their lifetime. Additionally, the relative risk of lower extremity amputation is 40 times greater in diabetics. The strongest risk factor for lower limb loss, diabetes accounts for half of all lower extremity amputations in the United States. Of diabetics undergoing amputation, 50% are expected to have another amputation within five years.

The reconstructive plastic surgeon should be well versed in the pathophysiology, evaluation and treatment of the diabetic foot. Our role in treatment is to protect the limb by optimizing wound healing and providing soft tissue coverage to exposed bone, tendon, muscle and nerve, with the goal of delaying or averting amputation.

Pathophysiology

The pathogenesis of diabetic foot ulcers is due to the combined effects of ischemia, neuropathy and infection.

Ischemia

Both macrovascular and microvascular circulation are impaired by diabetes. Macrovascular occlusive disease secondary to atherosclerosis is well characterized among diabetics. The relative risk of developing peripheral arterial disease is 2 to 3 times higher in diabetic versus nondiabetic patients. Additionally, diabetics tend to have a faster progression of disease from intermittent claudication to limb-threatening ischemia compared with their nondiabetic counterparts. The cause of the accelerated atherosclerosis observed in diabetics is likely a result of dyslipidemia, enhanced platelet adhesiveness and endothelial injury/dysfunction. The spatial pattern of large vessel disease also differs between the two groups. In the lower extremity, nondiabetics tend to have proximal disease, which affects the superficial femoral and popliteal arteries. Diabetics tend to have distal disease affecting the tibial and peroneal arteries, with the superficial femoral and proximal popliteal vessels spared from advanced disease. The arteries of the foot, especially the dorsalis pedis, are typically affected to a lesser degree, allowing for extreme distal bypass reconstruction in select patients.

Microvascular disease involves a nonocclusive microcirculatory impairment, characteristically affecting the capillaries and arterioles of the kidneys, retina and peripheral nerves. The sequelae of this process results in the characteristic complications of

Practical Plastic Surgery, edited by Zol B. Kryger and Mark Sisco. ©2007 Landes Bioscience.

nephropathy, retinopathy and neuropathy commonly seen in diabetic patients. Microcirculatory disease may also impair wound healing and increase susceptibility to wound infection. The physiologic mechanisms underlying microcirculatory disease have not been fully elucidated. Nonenzymatic glycosylation of the capillary basement membrane, seen in diabetics, may increase vascular permeability, leading to protein transudation and local edema. The well-characterized thickening of the capillary basement membrane is thought to impair migration of leukocytes to diseased areas.

Neuropathy

Peripheral neuropathy, a common complication in diabetics, poses a lifetime risk approaching 60%, and is seen in up to 80% of patients presenting with foot disease. The disease process includes sensory, motor and autonomic neural dysfunction. The pathogenesis remains to be fully elucidated, but the mechanism is likely due to both metabolic and microvascular defects. Diabetic neuropathy affects all nerve fibers with a predilection to the longest and finest fibers, leaving the nerves innervating the distal lower extremity most susceptible. Neuropathic disease characteristically affects the lower legs symmetrically, with a distal to proximal progression.

Sensory nerve disease initially involves vibratory fibers, followed by pain and temperature fibers, and ultimately, all sensory nerves leading to the insensate foot. Poorly fitting shoes, small stones and particulate matter can lead to undetected focal trauma, which can later develop into full-thickness ulceration. In addition to diminished sensibility, patients with sensory neuropathy may experience burning, tingling and hyperesthesia in affected areas. These sensory changes, fortunately, diminish with time.

Motor dysfunction leads to atrophy and weakness of the intrinsic muscles of the foot, resulting in an imbalance between the flexor and extensor mechanisms, clawing of the toes and increased prominence of the metatarsal heads. The resultant claw-foot deformity creates abnormal pressure points along the plantar metatarsal heads, tips of the toes and the dorsal surface of the PIP joints. Combined with an insensate foot, the pressure points become susceptible to undetected focal trauma, skin breakdown and foot ulceration.

Autonomic neuropathy leads to loss of sympathetic tone and increased arteriovenous shunting in the foot, resulting in decreased nutrient flow. Deinnervation of sweat glands results in dry skin, predisposing to skin cracking and local infection. In addition, loss of sympathetic innervation to the bone may lead to increased bone blood flow, resulting in osteopenia and "bone washout."

The combination of advanced sensory, motor and autonomic dysfunction contributes to the development of the **Charcot foot**, a deformity now seen almost exclusively in diabetics. The combination of foot deformity, walking on an insensate foot, and bone resorption results in joint instability, leading to bone and joint destruction. Breakdown of the tarsometatarsal joints and eversion of the plantar arch create the characteristic "rocker-bottom" foot. The deformity creates new pressure points on the plantar surface, which can lead to ulceration and represents a total breakdown of foot architecture.

Infection

Diabetic foot infections are the end result of the combined effects ischemia and neuropathy. The lesion typically starts as a neglected or unrecognized infection in the insensate foot. A break in skin integrity, such as a puncture wound, a neuropathic ulcer, an ingrown nail or a crack in the skin allows bacteria to locally disseminate. Poor tissue perfusion, diminished sensibility and an impaired inflammatory response lead to a rapid progression of what would otherwise be a local infection. The bacterial flora present in the wound is variable. In outpatients with localized and superficial infections, the culprit is usually a Gram-positive cocci such as *Staphylococcus* or *Streptococcus*. Deep ulcers and more advanced limb infections are usually polymicrobial. Organisms include Gram-positive cocci, as well as Gram-negative bacilli such as *Escherichia coli*, *Klebsiella*, *Enterobacter*, *Proteus* and *Pseudomonas*. Anaerobic flora such as *Bacteroides* is also common.

Clinical Assessment

Assessment of the diabetic foot starts with a thorough history and physical examination as well as directed laboratory studies to assess the metabolic status of the patient. The foot exam is directed toward assessing the primary pathophysiological mechanisms of disease, namely vascular and neurological compromise, and a search for occult or advanced infection.

Vascular Exam

Assessment of vascular compromise is critical, as it defines the degree of ischemic change to the foot. Evaluation of the femoral, popliteal and pedal pulses may suggest the site of large vessel arterial compromise if present. Additionally, prolonged capillary refill time, dependent rubor, pallor on elevation and loss of foot hair are all signs of arterial compromise. Although sensibility and motor function should be a part of any vascular exam, impairment may be due to ischemia as well as neuropathic changes.

When distal pulses are absent or impaired, the ankle-brachial index (ABI) can assess the degree of large vessel perfusion to the extremity. These values can be helpful in determining the potential for wound healing. An ABI of 0.4-0.6 correlates with significant claudication and a 10%-40% chance of progression of disease to amputation or revascularization to prevent amputation. Most diabetics need an ankle pressure of at least 80 to 90 mm Hg to heal a digit or metatarsal amputation. However, the ABI must be interpreted within the context of the disease. Medial arterial calcinosis, frequently seen in diabetics, especially those with end-stage renal disease results in artificially inflated indices. Although calcification appears to spare the vessels of the toes, disease in this area can often limit the utility of toe pressures.

Segmental Doppler analysis can warn the clinician of arterial compromise in the background of an elevated ABI. An attenuated waveform indicates a proximal occlusion, whereas a normal waveform is suggestive of insignificant proximal disease. Although segmental Doppler waveforms and pulsed Doppler recordings are not affected by calcification, these are qualitative, not quantitative, measures. Also, the quality of the waveforms is affected by local edema, and cuff placement may be affected by ulceration. The hindrances the diabetic foot presents to these noninvasive tests can significantly limit the usefulness of these modalities. The clinician is encouraged to assess these values in the context of the disease state and physical exam.

When foot pulses are absent and ulceration present, large vessel disease can be assumed to be a significant component of the disease process. Ischemic ulcers are usually painful, superficial lesions with a rim of vascularized tissue and a necrotic center. Arteriography, or magnetic resonance angiography (MRA) of the pelvis and entire lower extremity will provide a road map for a planned bypass procedure. As the proximal leg and foot vessels are frequently spared in diabetics, extreme distal bypass is not an uncommon procedure. Judicious hydration and renal protective agents prior to infusion of IV contrast are suggested for diabetics with compromised renal function. Although arteriography is considered the gold standard in vascular imaging, MRA is a useful alternative, particularly in patients with compromised renal function.

Neurological Exam

The foot should be thoroughly inspected for signs of neuropathic disease. Claw-foot deformity or other changes representative of motor disease should be noted. High-risk areas such as the plantar surface of the metatarsal heads, the dorsal PIP joints and the tips of the toes are inspected for ulceration. The toes are also inspected for ingrown nails; the web spaces and the plantar surface of the foot are inspected for dry, cracked skin. Sensibility, which is usually impaired in a bilateral, symmetrical pattern is assessed. Vibratory sensation, typically lost first, may be tested with a 128-cycle tuning fork. Light touch, temperature and pain can be assessed with a cotton swab, warm and cool tubes, and a pinprick, respectively. Neuropathic ulcers, known as mal perforans ulcers, present as nontender ulcers found on the plantar contact areas, such as the metatarsal heads or under the heel. They have a deep, punched-out appearance with a hypertrophic callous formation at the edges.

Searching for Infection

A thorough search for ulceration, purulent drainage, crepitus, erythema and sinus formation should be part of the initial evaluation. Heavily calloused areas are unroofed and cultures taken from the base of all ulcers. All wounds are probed to determine if the lesion exclusively affects the superficial tissue or if the deeper planes are involved. The tissue surrounding the ulcer is compressed to express occult purulence. **Osteomyelitis** is a common complication of the diabetic foot disease and is seen in up to 70% of diabetic foot ulcers. The presence of exposed bone in an ulcer is associated with bony infection. This observation can be made visually or with a sterile probe. In addition, plain X-rays, 3-phase bone scans, labeled leukocyte scans, CT and MRI may be used to assess osteomyelitis. Bone biopsy is the gold standard of diagnosis. It should be noted that the hallmarks of infection such as erythema and pain may be absent in the diabetic due to the sequelae of microvascular and neuropathic dysfunction. Even patients with deep abscesses may not present with expected fever, chills and leukocytosis. Uncontrolled hyperglycemia may be the only harbinger of an active infection.

16

Necrotizing fasciitis, an infection of the subcutaneous tissue and fascia, is a frequently missed condition affecting diabetics. Although not a common infection, it is a devastating disease that can rapidly result in large tissue loss, sepsis and death. Clinicians may mistake this disease with cellulitis, and the diagnosis of necrotizing fasciitis is not usually considered until the patient is floridly septic. The diagnosis is frequently missed because the skin often shows no evidence of deep tissue infection. Crepitus, cyanosis or bronzing of the skin should raise suspicion. Easy and painless introduction of a probing instrument into the necrotic subcutaneous space is highly suggestive of the diagnosis. A X-ray CT or MRI of the affected limb may show subcutaneous air. Fascial thickening and stranding may also be appreciated on CT. Affected patients should be taken emergently to the operating room for aggressive debridement of all infected tissue. Patients may require repeat debridements, which may ultimately lead to amputation.

Management of Diabetic Ulcers

Eradication of Infection

Foot infections in diabetic patients can rapidly produce a septic state or spread widely through the deeper planes of the foot, and therefore demand immediate intervention. Not all foot infections, however, require hospitalization. Superficial infections may be treated with a first-generation cephalosporin, nonweight-bearing status on the involved extremity, and close follow-up. More commonly, patients present with ulceration or gangrene involving the deeper planes of the foot, including tendon and bone. These patients require immediate hospitalization, foot elevation, debridement and initial broad-spectrum antibiotics, which can be narrowed when wound culture results are complete. The duration of antibiotic therapy depends on clinical resolution of the infection. In the face of osteomyelitis, prolonged intravenous antibiotic therapy is required, usually for four to six weeks. The course may be curtailed if the infected bone is thoroughly debrided.

Abscesses and deep-space infections are promptly incised and drained. All necrotic tissue must be surgically debrided. In advanced infections, amputation may be necessary to allow for complete drainage and excision of all devitalized tissue. Although all necrotic tissue should be removed, care is taken to conserve as much viable tissue as possible. An overly aggressive approach to debridement can lead to higher amputations and problems during closure.

After debridement and drainage, the wound is followed closely to ensure appropriate healing and eradication of the infection. Wounds are kept moist and weight-bearing on the affected limb should be kept to a minimum. During convalescence, nutrition should be optimized and serum glucose levels strictly controlled. Progressive foot necrosis, in the face of optimal medical management and wound care, may signify underlying ischemia or may be indicative of an undrained abscess.

Local Wound Care

There are several adjuncts to wound care that can maximize healing in these difficult wounds. Vacuum assisted closure (VAC) has revolutionized the management of a variety of wounds. Animal and human studies have shown an accelerated rate of granulation tissue formation and increased nutrient blood flow to the wound compared to saline moistened gauze. Early data on diabetic wounds report an accelerated rate of healing and decreased wound surface area, compared with saline-gauze dressing. The greatest strength of the VAC device is its ability to contract the wound, thereby decreasing its depth. The patient may then be spared a major reconstruction, with all its associated risks and morbidities. Disadvantages of the device include added cost and patient discomfort. We stress that VAC therapy does not replace debridement. This therapy should commence after adequate debridement has occurred. A more detailed discussion of this modality can be found in the VAC chapter.

Wet-to-dry gauze dressing, in conjunction with surgical debridement has been the standard wound treatment by which all others have been measured. There is some evidence in the literature that debriding agents such as hydrogels and collagen-alginate preparations may be more efficacious in the treatment of diabetic foot ulcers. Further research is required to define the optimal applications of these this emerging modalities.

Of the numerous clinical studies that have tested the efficacy of various growth factors for the treatment of chronic foot ulceration, only topical platelet-derived growth factor (PDGF) has shown significant improvement in healing the diabetic foot. Becaplermin gel, also known as Regranex, is the recombinant human PDGF isoform BB and the only growth factor licensed for the treatment of chronic, full-thickness diabetic foot ulcers. Due to its significant cost, ($300-$400 for a 30-gram tube), becaplermin is recommended only for well-perfused, chronic diabetic foot ulcers that have failed standard wound therapy.

Peripheral Vascular Surgery

After the infection is controlled and signs of systemic toxicity have resolved, attention should be focused on maximizing pedal perfusion. Macrocirculatory dysfunction, as well and an impaired inflammatory response, predisposes the diabetic foot to ulceration in the face of even moderate ischemia. Restoration of impaired inflow is essential for healing and limb salvage. Options include endovascular techniques (i.e., angioplasty and stenting), bypass grafting or a combination of the two. The procedure of choice should be tailored to each individual patient with regards to anatomy, comorbidities and operative risk. Details of the different interventions are beyond the scope of this chapter. The vascular surgery team will lead in this endeavor. Communication between the vascular and plastic surgeons is critical at this point if a reconstructive procedure is planned. For example, the vascular surgeon may not consider the fact that his target vessels in a bypass procedure may also serve as the site of anastomosis for a future free flap.

Offloading

Changes in the bony architecture of the foot create abnormal pressure points resulting in characteristic patterns of ulceration. By relieving the local stress created at these points, recurrence of the ulcer can be prevented and the reconstruction of the affected area may be protected. Redistribution of pressure can be achieved surgically (e.g., metatarsal head resection, arthrodesis, partial calcanectomy, or Achilles tendon lengthening) or by custom orthotic footwear, tailored to evenly distribute pressure to the dorsum of the foot. These custom-made orthotic devices made of plaster or fiberglass allow for distribution of pressure off the wound so that the patient can remain partially active while the wound heals. These devices may be worn at home. Some patients complain of discomfort from the device, but it is an effective means of protecting the foot from further damage while it heals.

Reconstruction

The reader is referred to the lower extremity reconstruction chapter for the various strategies used to cover foot defects. Reconstructive issues particular to the diabetic foot are discussed here. Prior to surgery, systemic and local infection, including osteomyelitis, should be eradicated. All necrotic tissue is debrided and local perfusion is optimized by peripheral bypass surgery and/or local wound care. The patient should be nutritionally optimized, and blood sugar tightly controlled. Lastly, the patient must **understand** the type of reconstruction planned and the measures required for complete wound healing. Patient compliance with postoperative care (e.g., leg elevation, no weight bearing, glucose control etc.) cannot be overstated. The convalescence phase may take several weeks and the patient should be aware of their role in protecting the wound.

Most foot ulcers, due to the inelasticity of the surrounding skin, are not amenable to primary closure. Skin grafts are not recommended to cover wounds over primary weight-bearing areas. Small, well-perfused, noninfected neuropathic ulcers can be repaired in a single-stage procedure that includes debridement, bony reconstruction (for offloading of the affected area) and primary closure using a random flap. Scarring around the wound secondary to chronic inflammation limits the reach, and therefore the use, of random flaps.

Wounds less than 3x6 cm with exposed tendons, joints or bone may be repaired with a **local muscle flap**. These flaps have the advantage of straightforward

Table 16.1. Commonly used local muscle flaps for various diabetic ulcers

Local Flap	Location of Ulcer
Abductor digiti minimi flap	lateral foot defects
Abductor hallucis flap	medial mid-foot and heel defects
Flexor digitorum brevis flap	plantar heel defects
Extensor digitorum brevis	anterior ankle defects

dissection, with minimal donor defects and primarily closure. The procedure can be performed with regional anesthesia, minimizing anesthetic risk. Some of the commonly used local flaps for various diabetic ulcers are listed in Table 16.1.

Again, it should be noted that such procedures require a commitment from both patient and surgeon. Although a long-term salvage rate of 89% has been reported, an average hospital stay is 27 days, and the average time to wound healing is 125 days.

Larger defects or defects not within reach of local muscle flaps may be repaired with free flaps in selected patients. Free flap success rates are equivalent in diabetic and nondiabetic patients. However, the diabetic patient is usually more debilitated and has significant comorbidities, which may preclude free flap reconstruction.

Pearls and Pitfalls

Prevention of diabetic foot ulcers through proper education is one area in which primary care providers often fall short. The plastic surgeon should take an active role in educating diabetic patients on this topic. Patients should appreciate the importance of strict glucose control, smoking cessation and routine foot surveillance. Foot protection, meticulous foot care, with particular attention to the skin and nails, regular podiatry visits and warning signs of an at-risk foot should all be reviewed. Proper shoe selection and sizing is critical in prevention of disease and may require a custom-made device. Patients should know that this is a devastating but preventable problem and they are at the helm of influencing the course of their disease.

Previous ulceration puts the patient at risk for future ulceration. Every effort should be made to reduce abnormal pressure loading. Immobile patients should have appropriate cushioning and ulcer precautions. Mobile patients will benefit from custom-fitted footwear to minimize mechanical trauma.

Suggested Reading

1. Akbari CM, Macsata R, Smith BM et al. Overview of the diabetic foot. Sem Vasc Surg 2003; 16:3.
2. Attinger CE, Ducic I, Cooper P et al. The role of intrinsic muscle flaps of the foot for bone coverage in foot and ankle defects in diabetic and nondiabic patients. Plast Reconstr Surg 2002; 110:1047.
3. Bennett SP, Griffiths GD, Schor AM et al. Growth factors in the treatment of diabetic foot ulcers. Br J Surg 2003; 90:133.
4. Blume PA, Paragas LK, Sumpio BE et al. Single-stage treatment of noninfected diabetic foot ulcers. Plast Reconstr Surg 2002; 109:601.
5. Evans D, Land L. Topical negative pressure for treating chronic wounds: A systemic review. Br J Plast Surg 2001; 54:238.
6. Jeffcoate WJ, Harding KG. Diabetic foot ulcers. Lancet 2003; 361:1545.
7. Sumpio BE. Foot ulcers. New Eng J Med 2000; 343:787.

Chapter 17

Wounds Due to Vascular Causes

Kevin J. Cross and Robert T. Grant

Introduction

Vascular diseases leading to wounds and chronic ulcers affect greater than 1% of the adult population and nearly 5% of the population over 65 years of age. The low blood flow state that occurs in many vascular conditions and its sequelae, ischemia and venous stasis, cause maladaptive changes in cell function and maintenance. These changes include a decreased ability for cells to use oxidative radicals to fight bacteria and a propensity towards cellular apoptosis, or cell death. The combined effect of these changes leads to tissue breakdown and high levels of bacterial invasion, establishing chronic, nonhealing ulcers.

Though wounds related to vascular diseases are often multifactorial, they largely fall into four main categories. Over 50% of these wounds are caused by venous insufficiency. Arterial insufficiency, diabetes and a group of rare underlying disorders comprise the remainder of vascular wounds. An incorrect diagnosis can lead to incorrect treatment; therefore, a careful assessment and detailed knowledge of the clinical picture, pathogenesis, likely causes and treatment options are essential.

Etiologies

Venous Insufficiency

Wounds of venous origin are caused by increased pressure in the venous system. This pressure results from a malfunction of the valvular system in the deep veins of the leg. Normally, the retrograde flow of blood with each contraction of the calf muscle ("the muscle pump") is prevented by a valvular system present in the deep veins. This leads to an obligatory anterograde flow as pressure builds in the veins as a result of this muscular contraction. In patients with venous insufficiency, the prevention of retrograde flow is lost, or the promotion of forward flow through contraction does not occur. Causes can range from congenital weakness of the valves or loss of normal valvular function, to conditions that weaken muscle contraction such as immobility or paralysis (a condition known as dependency syndrome). Valvular function is often compromised after anatomical changes to the valves resulting from either venous thrombosis (post-thrombotic syndrome), a history of phlebitis in the deep veins, or from trauma (fractures, burns or crush injuries to the limb). Furthermore, valvular dysfunction is also more likely to occur in patients with a family history of varicose veins, suggesting a genetic component.

Regardless of the initiating event, valvular dysfunction leads to the pooling of blood in the capillary system, vascular congestion and resultant dilatation and congestion of the lymphatic system. Poor blood flow can lead to ischemia as a result of slowed oxygen delivery, the formation of capillary microthrombi and leakage of plasma and erythrocytes into surrounding tissue. With this nonphysiologic ischemic

Practical Plastic Surgery, edited by Zol B. Kryger and Mark Sisco. ©2007 Landes Bioscience.

environment, the slightest trauma puts too great a demand on the cells in the region and a chronic, nonhealing ulcer is established. Fibroblasts migrate to the area and lay down fibrin, resulting in the characteristic "fibrin cuff" seen in many ulcers of venous origin.

Classically, wounds are found in the "gaiter" distribution around the lower extremities and around the medial malleoli. Although not usually very painful, they can cause discomfort and even moderate pain in certain patients. Trophic changes including hair loss on shiny, indurated, scaly, erythematous skin. Deeper induration can present as panniculitis, followed by the fibrinous, exudative ulcers of venous stasis.

Arterial Disease

The traditional form of peripheral vascular disease, arterial insufficiency, results from the occlusion of vessels and prevention of forward flow of blood, leading to ischemia and tissue death. Typically this is a result of atherosclerosis of peripheral vessels and presents in the lower extremities. Embolic events and vasculititis (mentioned in the rare causes section of this chapter) are other less common causes of arterial disease and can also occur in the upper extremities as well. Risk factors for arterial insufficiency include: diabetes, smoking, hyperlipidemia, hypertension, obesity and age. Any artery along the course of the femoropopliteal distribution can be affected. If larger vessels such as the popliteal, peroneal or either the tibialis anterior or posterior are involved, severe tissue injury can be seen. Limited tissue damage will occur with more distal arterial involvement. Usually, ulcers in arterial disease occur in the distal regions of the feet and toes up to the malleoli or on the shin.

Typically presenting in men ages 50 to 70, patients will report symptoms such as intermittent claudication, rest pain, pain with extremity elevation and disuse atrophy. Physical findings include asymmetric or absent distal pulses and doppler signals, temperature discrepancies between limbs, delayed capillary refill, audible bruits, skin color changes (mottled, dusky appearance) and an ankle-brachial index (ABI) less than 0.8.

Embolic phenomena are seen in patients with risk factors such as atrial fibrillation, atrial or ventricular thrombus (either left-sided, or right-sided with a patent foramen ovale), valvular vegetation, recent surgery or angiography, or more proximal vascular disease. Other causes include amniotic fluid, fat or bile emboli and intra-arterial injections of air or medication; however, these are much less common than emboli from a cardiac source or a more proximal plaque in the extremity. Embolic events and in situ thrombosis should be considered in hypercoagulable states such as in patients with a known or suspected malignancy or disorders of the coagulation system (e.g., lupus, protein c/s or antithrombin deficiencies). Embolic occlusion presents with acute and often painful distal symptomatology. When smaller vessels are affected, it is not uncommon to see distal finger or toe color changes and necrosis in the setting of palpable pulses. Finally, any patient who has a distal arterial embolism should undergo an echocardiogram to evaluate the heart and should be ruled out for an aneurysm (aortic, iliac or popliteal) as the sources of the embolus.

Diabetes

Though not a disease of the vascular system primarily, it is important to have an understanding of the role of diabetes in the lower extremity wound. At least 15% of diabetics will develop a chronic, nonhealing wound during their lifetime, and approximately 40% of these will have a component of peripheral vascular disease as a contributing factor. Furthermore, diabetic patients are at risk of arteriosclerosis of peripheral vessels at a younger age.

Many diabetic ulcers result from peripheral neuropathies leading to pressure ulcers. When the usual sensory signaling that control distribution of weight and pressure across a large surface of the skin is lost, micro-ischemic environments occur at the sights of prolonged pressure, leading to tissue death and bacterial invasion. Furthermore, it has been shown that there is impairment to neovascularization, prolonged inflammation and decreased wound repair in diabetic wounds. Combining these pathophysiologic changes with the tendency towards arterial occlusive disease in diabetics, a wound with multiple contributing factors can occur.

Rare Vascular Diseases

A subset of patients will present with a picture that does not fit one of the classic causes for vascular wounds or will describe a history that suggests one of the rare causes for wounds of vascular origin. The more common of these will be described, with a more extensive list present in Table 17.1. Due to the fact that the vasculitides comprise a broad group of different diseases, no one disease process explains the common final pathway of vessel wall inflammation. Immune complex disease, antibody-dependent cellular cytotoxicity (ADCC), endothelial activation and coagulopathy have been invoked in models of inflammatory disease of the vasculature.

The vasospastic condition known as Raynaud's disease classically presents in a young woman after exposure to cold or after cigarette use. The patient reports paresthesias, and a typical triphasic pattern of color change is seen in the affected acral body part, progressing from blue to white to red. The disorder is thought to result from increased serotonin release or an imbalance in vasoactive prostaglandins, and is known as Raynaud's phenomenon when associated with one of a number of underlying diseases, including lupus erythematosus, scleroderma or drug use. Treatment involves avoidance of the instigating stimuli, the use of vasodilators such as calcium-channel blockers or serotonin antagonists and surgical debridement of any necrotic tissue.

Inflammatory vasculititis such as thromboangiitis obliterans (Buerger's disease) or Wegener's disease often present in younger patients. These conditions are characterized by the size of the vessels involved (large, medium or small vessel disease) and cause distinct cutaneous findings. Small-vessel vasculitis often presents with urticarial-like lesions and palpable purpura, while medium- and large-vessel vasculitis commonly results in livedo reticularis, purpura, necrosis and ulceration. End-stage renal disease can lead to a phenomenon known as calciphylaxis syndrome. The findings of hyperparathyroidism, hypocalcemia, hyperphosphatemia, hypertriglyceridemia and hypomagnesemia make up the metabolic derangement of this disorder. Patients

Table 17.1. Rare vasculitic diseases that may lead to lower extremity wounds

Small vessels	Small vessel-leukocytoclastic vasculitis, microscopic polyangiitis, allergic granulomatosis (Churg-Strauss), Henoch-Schonlein purpura, essential cryoglobulinemic vasculitis, erythema induratum Bazum, livedo reticularis, livedo vasculitis and Sneddon syndrome
Medium vessels	Polyarteritis nodosa, Kawasaki disease
Large vessels	Giant cell arteritis (polymyalgia rheumatica, Takayasu arteritis)

Reproduced from: Mekkes JR et al. Causes, investigation and treatment of leg ulceration. Br J Dermatol 148:390; ©2003 with permission from Blackwell Publishing.

develop cutaneous calcium deposition typically in a trunk and limb girdle distribution, and go on to have painful full-thickness necrosis in these areas. Treatment involves correction of the metabolic changes and debridement and skin grafting once disease progression has been controlled.

Initial Workup

Following an interview aimed at understanding the history of the disease and the presence of risk factors for such things as embolic events or venous thrombosis, examination of the affected area follows (Table 17.2). The location of the wound may give the first clue as to the origin of disease. Venous stasis ulcers can often be diagnosed by the presence of edema, lack of hair, pigmentation changes from hemosiderin deposition and ulcers in a distribution around shin and calf regions. These findings may present so classically as to require no further work-up. A wound with an irregular border and black necrosis or with purple or reddish discoloration of surrounding tissue should raise the clinician's suspicion for a vasculitis. A biopsy as well as lab test such as erythrocyte sedimentation rate, antinuclear antibody test, rheumatoid factor, antineutrophil cytoplasmic antibodies, and compliment and immune complex levels should be sent.

If arterial disease is suspected or if a compression dressing is to be applied, ABIs should be performed with the lower limit permissive for the use of compression dressings being above 0.8. This cutoff is set to avoid causing ischemic conditions in marginally viable tissue with the use of the compression dressing. If pressures appear abnormally high during this test, calcification of vessels preventing vessel occlusion during the test must be considered. This finding invalidates the ABI as a useful measure of arterial disease. Toe pressure measurement in the setting of inaccurate ABIs or in a patient with painful ulcers in the ankle region is another option. Further evaluation includes transcutaneous oxygen pressure measurement to evaluate the level of tissue ischemia, lower extremity color Doppler exam (as well as duplex scanning), and angiography (or MRA) for arterial occlusive disease.

In patients who present with evidence of peripheral neuropathies or ulcers over the sights of pressure points (especially the distal metatarsal joints) vibration perception (using a biothesiometer) and light touch sensation (using Semmes-Weinstein monofilaments) should be tested. Further workup for undiagnosed diabetes should include routine serum glucose and, if elevated, HbA1c levels should be checked.

Treatment

Wounds of Venous Etiology

Treatment for venous stasis ulcers is directed toward controlling lower extremity edema. Compression dressings create an external pressure which acts against the intravascular pressure to push plasma back into the vascular system and to prevent blood from pooling in the lower extremity. Options include a multilayer, zinc paste dressing bandage (the **Unna boot**), short stretch compression stockings and elastic bandages. There is no proven benefit of one over another; however the Unna boot should only be used in patients with good hygiene and minimal wound exudate. Compression stockings are often difficult to get on, especially for an elderly, arthritic patient. When dressing an extremity with the Unna boot, it is important to remember to wrap from distal to proximal and to avoid wrapping so tightly that the arterial flow is occluded. If using compression stockings, options include stockings graded from 20 mm Hg to greater than 60 mm Hg depending on the severity of disease.

Table 17.2. Morphologic findings in various wounds of vascular etiology

Wound Type	Morphology	Location	Surrounding Skin	Pain
Venous	Exudative, irregular margins	Gaiter area	Indurated, dermatitis weeping, edema, varicose veins, hair loss hyperpigmentation	Mild to moderate, better with elevation
Arterial	Dry, necrotic, pale, fibrotic, punched-out	Toes, ankles, anterior tibia	Cool, cyanotic, delayed capillary refill	Moderate to severe, better with dependency, worsens with exercise
Embolic	Punched out	Toes, malleoli	Cyanotic, pustules	Severe
Diabetic	Hyperkeratotic, bulla formation	Pressure sites, metatarsal heads, heels	Charcot's deformity	Less severe to absent due to neuropathy
Vasculitic	Hemorrhagic, palpable purpura, punched-out, jagged	Variable	Retiform purpura, livedo reticularis, atrophie blanche	Moderate to severe, not positional

Reproduced from: Choucair MM, Fivenson DP. Leg ulcer diagnosis and management. Dermatologic Clinics 19:659-678; ©2001 with permission from Elsevier, Inc.

Due to the likelihood of multifactorial disease, it is important to rule out an arterial contribution to the chronic wound before applying a compression dressing. The patient should also be encouraged to keep their affected extremities out of the dependent position while resting to further encourage forward flow of blood and edema fluid. Keeping vascular wounds moist and clean is extremely important. To this end, saline dressing changes are the most cost effective choice but must be done twice a day to maintain the moist environment. A hydrogel can be used with dressings that will be changed less often. Vacuum-assisted therapy (VAC) is now being used for vascular wounds with greater success, especially in larger wounds. Whirlpool treatment may also be combined with dressing changes. Many patients find the water soothing, especially in painful venous ulcers.

The surgical approach to venous disease has been met with little success and is often limited to debridement of necrotic wounds and skin grafting. There has been little proven benefit to the restoration or replacement of deep veins or their valves and ligation of veins to prevent retrograde flow of blood has been equally unsuccessful. If a healthy wound bed can be established, autologous split-thickness skin grafts are likely to take. Often, in a healthy wound bed, skin grafting followed by tight wrapping is used in the hopes that there is enough oxygen perfusing the surface level to allow the skin graft to take. Allogenic skin grafts such as Apligraf (Novartis, East Hanover, NJ) are eventually rejected, but provide a "biologic dressing" to prevent bacterial growth and help promote cell migration and speed healing rates. Autologous skin equivalents, cultured from a donor sight, are expensive and time consuming to grow, but are another option. Should more definitive closure of these wounds be necessary, with adequate arterial blood flow, a host of rotational and free flap surgical techniques are available. These range from rotational skin flaps, to local myocutaneous flaps (gastrocnemius flaps), to free tissue transfers (radial forearm flap).

Pharmacologic options include the use of pentoxifylline, a drug thought to decrease excessive white blood cell activity and to increase oxygen delivery to tissue. Diuretics should be considered to help alleviate edema.

Generally speaking, antibiotics for vascular wounds should be reserved for cases where systemic involvement is suspected. Wound cultures are of little value in that they only sample superficial bacteria which are often simply skin contaminates. The decision to use antibiotics should be guided by culture and sensitivity results from deep tissue biopsies. In the setting of suspected osteomyelitis, a radiographic workup can be pursued with plane films followed by bone scan if necessary. When ruling out osteomyelitis, it is important to remember that X-rays often lag behind disease progression by two weeks. More recently, MRI, and specifically those using gadolinium, have been used in the workup for osteomyelitis. The definitive procedure is a bone biopsy and culture with a six-week course of antibiotics following positive culture results.

Wounds of Arterial Etiology

Contrary to wounds resulting from venous disease, those with an arterial origin are often managed surgically. In addition to debridement of the ulcer, a revascularization bypass procedure should be performed. In cases of more proximal, focal disease in a larger artery of the leg, intravascular approaches such as balloon dilatation (percutaneous transluminal angioplasty) and stenting can be considered. Anticoagulation shows little benefit, but often aspirin with or without an antiplatelet agent is instituted especially in the setting of stent placement. Drugs often used include pentoxifylline, clopidogrel and cilostazol. Hyperbaric oxygen has proven to be of benefit in chronic ischemic wounds and is often used as a limb salvage technique with arterial disease, or in the setting of osseomyelitis.

In patients with nonhealing wounds and arterial insufficiency that cannot be treated surgically (poor distal target vessels for bypass options, poor surgical candidate, etc.), recent data suggest that pneumatic compression stockings that deliver retrograde sequential pressure at 120 mm Hg can improve popliteal and distal arterial flow and improve blood delivery to distal tissue.

For embolic disease, rapid institution of therapy is essential. The patient should be kept warm and the affected limb made dependent. Full heparin anticoagulation should be started immediately and early consideration for thrombolytic agents or embolectomy is appropriate.

Diabetic Wounds

Initial treatment for diabetic wounds should be aimed at eliminating pressure at the wound site. Total contact casting for a diabetic foot ulcer is very effective. This ensures that when the extremity meets a hard surface like the floor, the pressure is distributed across the entire foot. Another option is the orthopedic shoe. It serves a similar purpose but is easily removed for wound care and daily cleaning. Finally, the patient should be encouraged to avoid using the foot as much as possible, preferably relying on the use of crutches, or wheelchair for as brief a time as is necessary to allow wound healing.

Along with foot care, aggressive surgical wound debridement is a vital part of the healing process of diabetic wounds. Devitalized tissue that can act as a site for bacterial growth and as a barrier to the migration of new granulation tissue should be excised. Although they are expensive and little benefit has been shown in clinical trials, enzymatic debridement dressings are often used as part of institutional wound care protocols.

Diabetic wounds hold the distinction of being the first class of wounds shown to benefit from growth factor therapy. Topical recombinant platelet derived growth factor BB (becaplermin) and granulocyte-colony stimulating factor have both been shown to be beneficial in randomized control trials. Synthetic skin substitutes using neonatal dermal fibroblasts and Apligraf (Novartis) are often used and, as is the case in venous wounds, may help promote cellular infiltration. Lastly, strict glucose control is of utmost importance in promoting a more effective healing process.

Pearls and Pitfalls

Definitive reconstructive procedures such as flaps and grafts are doomed to failure as treatments for chronic wounds unless the primary 'wound diathesis' has been identified and optimally treated. Flap failure, graft loss or wound recurrence, not cure, are the more likely outcomes if the patient's arterial, venous or metabolic issues have not been corrected. Most chronic wounds can be addressed in the ambulatory setting, through medical management, optimization of comorbidities, local debridement and proper wound care. Once tissue oxygen delivery has been assured, converting the chronic wound back into an acute wound by serial debridements leads to improved outcomes with less chronic inflammation and improved wound repair.

As is the case with antibiotics, there will likely be no single topically applied growth factor that will promote healing and be truly effective in the absence of satisfactory blood flow and local wound care. The promise of improved angiogenesis with gene therapy-enhanced wound healing agents is also under active investigation.

Suggested Reading

1. Choucair MM, Fivenson DP. Leg ulcer diagnosis and management. Dermatol Clin 2001; 19:659.
2. Lautenschlager S, Eichmann A. Differential diagnosis of leg ulcers. Curr Prob Dermatol 1999; 27:259.
3. Mekkes JR, Loots MAM, Van Der Wal AC et al. Causes, investigation and treatment of leg ulceration. Brit J Dermatol 2003; 148:388.
4. Miller IIIrd OF, Phillips TJ. Leg ulcers. J Am Acad Dermatol 2000; 43:91.
5. Simon DA, Dix FP, McCollum CN. Management of venous leg ulcers. BMJ 2004; 328:1358.
6. Weingarten MS. State-of-the-art treatment of chronic venous disease. Clin Infect Dis 2001; 32:949.
7. Weitz JI, Byrne J, Clagett GP et al. Diagnosis and treatment of chronic arterial insufficiency of the lower extremities: A critical review. Circulation 1996; 94:3026.

Chapter 18

Radiated Wound and Radiation-Induced Enteric Fistulae

Russell R. Reid and Gregory A. Dumanian

Introduction

No other clinical dilemma is more challenging, and no other substrate is more difficult to work with than irradiated tissue. This chapter concisely reviews the etiology, presentation and management of chronically irradiated tissue. In the new millennium, greater than 50% of cancer patients receive some form of radiotherapy. A working knowledge of the tissue effects of radiation and current approaches to the rising epidemic of radiation injury is essential for every plastic surgeon.

Types of Radiation

Ionizing radiation exerts its effects by the energy transference to biologic material, which results in excitation of tissue electrons. Radiation exists principally in three forms: **X-rays** (short wavelength rays produced by an electrical device), **gamma rays** (short wavelength rays emitted from unstable isotopes) and **particulate radiation** (rays produced by electrons, protons, α-particles, neutrons and p-mesons). Those clinically relevant are of the X-ray and gamma varieties. These rays have both **direct** (alteration of intracellular DNA/RNA) and **indirect** (generation of oxygen free radicals) mechanisms of cell toxicity. The former effect confers its therapeutic benefit on rapidly dividing cancer cells, but indirect mechanisms are a detriment to rapidly dividing, normal tissue such as skin and the lining of the gastrointestinal tract.

Therapeutic radiation is typically delivered locally at a low-dose rate (brachytherapy) or high-dose rate through megavoltage devices (external beam or tele-therapy). In either case, radiation administered is measured as the radiation absorbed dose (rad) or more recently, the Gray (Gy) unit. **1 Gray equals 100 rads**. Each type of radiation has a characteristic depth of penetration that is used to establish its effect on a lesion. The following treatment parameters therefore have influence on the overall radiation-induced damage: (1) total dose; (2) dose fraction size; (3) total volume of tissue treated; (4) elapsed time during irradiation. Dose fractionation regimens have been created by radiation therapists to minimize injury to normal tissue. Standard dose regimens in practice today occur at a rate of 100-200 cGy per minute.

Biology of Radiation: Effects on Skin Cells

Overall, the extent of radiation damage has been categorized as **lethal** (irreversible), **sub-lethal** (reversible/correctable by cellular repair mechanisms) and **potentially lethal** (modifiable by cellular environment). Toxic effects can manifest as an acute injury (<6 months) or chronic injury (>6 months). On a cellular level, the toxic manifestations are multiple. Keratinocytes, being the most superficial and proliferative cell type in the skin, are the most radiosensitive. Erythematous reactions, which signal epidermal damage, are said to be trimodal. The first, often not clinically evident, occurs just 1-24 hours post radiation and is most likely due to activation of proteolytic enzymes and increased local capillary permeability. This is typically followed by a more intense reaction appearing approximately one week after therapy; this phase is caused by injury to the basal layer of the epidermis. Inflammatory and immune response to this injury lead to a third phase of erythema, which may occur 6-7 weeks after radiation. Complete destruction of the epidermal layer results in the characteristic **moist desquamation** seen in early radiation damage. Owing to the turnover time of the epidermis (59-72 days), the daily rate of epidermal loss in irradiated tissue occurs at 2.6% ± 0.2%. Other epidermal residents are also affected by radiotherapy. Exposure of melanocytes to ionizing radiation results in increased melanin transfer to keratinocytes and thus hyperpigmentation of the skin. This is seen early in treatment, whereas at a later time, melanocyte death leads to patches of hypopigmentation characteristic of chronic cases.

Of the cells in the dermis, the fibroblast is the primary target in radiation injury. Fibrotic response to multiple radiation insults ("reactive" fibrosis) has been shown to be mainly due to alterations in the physiology of this cell type. Overexpression of collagen, differentiation of fibroblast progenitors into myofibroblasts and increased production of TGF-β1, a profibrotic cytokine, all appear to contribute to clinically evident fibrosis. Paradoxically, long-term radiation leads to fibroblast depletion and thus the poor wound healing potential and compromised tensile strength inherent in chronic wounds. Reconstitution of chronic wounds with nonirradiated fibroblasts or with platelet-derived growth factor-BB (PDGF-BB), which indirectly stimulates fibroblasts via activated macrophages, restores normal wound breaking strength and time to healing.

Clinical Presentation

Clinical manifestations of radiation injury can be divided into **acute** and **chronic**. Acute effects include erythema, dry desquamation (which occurs at moderate radiation doses) and moist desquamation (which occurs with ablation of most skin cancers). As the skin cancer is treated, the basal epidermal layer becomes denuded, resulting in serous oozing characteristic of the last condition. It follows that as radiation injury becomes chronic, dermal and adnexal structures are affected. Hypo- and hyperpigmentation, thickening of the dermis and loss of sebaceous and sweat gland function result in dessicated, poorly vascularized tissue that is difficult to handle. These changes are irreversible. Finally, necrosis and even cancer can arise from a chronic damage of a radiated target. Considering these changes, any nonhealing ulcer that arises within a radiated field must always be biopsied to rule out neoplasia.

Local Wound Care

Given these characteristics of the chronic radiation wound, its propensity to dessicate and its inability to generate the normal inflammatory response, meticulous local wound care is essential. Maintaining a moist wound bed to prevent bacterial intrusion is important. In early phases of injury, patient education is critical: one must be informed of avoidance of sun exposure, alcohol-based emollients, cosmetic-based agents and trauma to the radiated area. Gentle cleansing with normal saline or mild soap solutions is recommended. Moist desquamation requires copious irrigation with dilute hydrogen peroxide or normal saline, followed by light application of Silvadene. When dealing with dry desquamation, one must compensate for the loss of moisture secondary to sebaceous/sweat gland destruction. Several hydrophilic preparations (e.g., hydrogels) and antipruritic agents have been used with good effect.

Beyond topical therapy, an irradiated wound must be kept clean at all times. Manual irrigation with pressurized water helps to clean off surface exudates. A helpful suggestion to patients is to use multiple forceful showers for large wounds, or the irrigating pulse of a dental cleaner for small wounds. In wounds where surface cleansing does not suffice, meticulous sharp debridement, with the overall goal of reducing or eliminating bacterial counts by clearance of devascularized tissue, is critical for wound closure. Debridement should not only be aggressive, but frequent; the presence of small amounts of devascularized tissue can result in progressive bacterial overgrowth and subsequent necrosis. The "poorly" vascularized tissues of tendon, bone and cartilage are the most troublesome to treat, due to the difficulty in maintaining a moist environment and the prevention of new tissue dessication.

Surgical Considerations

Radiation Enteric Fistula

The incidence of chronic radiation injury to the intestine, occurring in 2-5% of patients who receive abdominal or pelvic radiation, is on the rise. Its manifestations, including abdominal strictures, hemorrhage, perforation and fistulae result in complex abdominal wall defects. Most frequently, small bowel adherent to and fixed by scar into the pelvis receives an inordinately high local dose of radiation. This happens after treatment for rectal, bladder and gynecologic malignancies. Surgical goals in these cases consist of damage control, return of intestinal continuity and reconstruction of the abdominal wall.

Generally a multidisciplinary approach, involving nutritional, colorectal, oncological, and reconstructive specialists, is mandatory. When perforation or fistulization is present, the first objective is to decrease the amount of fluid flowing across the fistula point. This can be achieved with simple measures such as bowel rest, or require a more aggressive approach, such as proximal diversion.

A second objective is to decrease local inflammation of the soft tissues with adequate drainage of feculent material, gentle surface cleansing and antibiotics. A surgical repair can be contemplated when the wound is less acute and inflamed, when nutrition has been optimized, and when the anatomy is thoroughly understood. Even then, the complication rate can be formidable. Anastomosis of diseased and even normal bowel segments within the irradiated field is fraught with morbidity and potential mortality. Along these lines, studies have demonstrated a leak rate of 36% and mortality as high as 21% in patients who underwent resection and anastomosis in cases of radiation enteritis. The difficulty inherent in such scenarios is pinpointing exactly what segment of bowel is diseased and what is normal. Neither intraoperative frozen section nor Doppler survey of bowel segments have been beneficial in reducing complications. From this experience, it makes the most clinical sense to widely resect the bowel in the area of the fistula and to locate the new anastomosis far away from the radiation field.

At our institution, several principles are employed to guide treatment of the abdominal wall. The debridement of all inflamed tissue is critical, so that in no case should the success of the procedure depend on the healing of fibrotic, scarred tissue without pulsatile blood flow. After debridement, two independent decisions must be made to guide intraoperative planning: first, what is the quality of the abdominal wall, and second, what is the quality of the soft tissue (skin) cover? Integrity of the abdominal wall is obtained using the separation of parts procedure for midline defects. For nonmidline defects, our tissue of choice is a sheet of autogenous fascia lata. There is interest in the use of new biologic agents such as Alloderm® and Surgisys® in these situations as well to avoid the donor site morbidity of fascia lata harvest. These sheets of tissue are sewn to the undersurface of the intact abdominal wall using horizontal mattress sutures with as much overlap to good tissue as possible. When the skin can be mobilized and closed (either in the midline or over the fascia lata laterally), the procedure is finished. In certain instances, myocutaneous TFL and rectus abdominis flaps are used to provide full-thickness vascularized coverage to the abdominal wall reconstruction.

The radiated pelvis is its own subject, as radiated bowel loops often can become adherent and fistulize out the perineum. The abdominal wall reconstruction is usually not as important as keeping bowel out of the pelvis after surgery. After a wide bowel resection and placement of the new anastomosis away from the radiated field, a flap is chosen to separate the intraabdominal contents from the pelvis inflammation which can not be easily debrided. In these cases, a rectus flap with a skin paddle based obliquely from the periumbilical perforators and angled toward the tip of the scapula is raised and dropped into the pelvis. This oblique rectus abdominis myocutaneous flap (ORAM) is preferable to the standard VRAM flap, as this technique significantly decreases the amount of muscle harvested, while still adequately filling the pelvic dead space. The subcutaneous fat does not atrophy over time, and so the bowel loops do not have the opportunity to slowly reenter the pelvis.

Summary

Management of the irradiated wound requires a multidisciplinary approach. Knowledge of the pathophysiology of radiation damage will lead to successful care. Aggressive local wound measures, and keeping the wound clean and moist with a low threshold for surgical debridement, are critical steps towards healing. In terms of enteric damage, control of fistula in conjunction with staged bowel repair and definitive abdominal wall reconstruction with **autogenous tissue**, minimize local wound morbidity and recurrence.

Pearls and Pifalls

- Discard all stiff, nonpliable, inflamed radiated tissue. The best repairs discard the most tissue. En bloc resections of bowel, fistula and skin are preferred.
- Avoid the use of alloplastic materials as much as possible.
- For midline defects, use the separation of parts repair.
- Use fascia lata (or Alloderm) for nonmidline defects with good skin cover.
- Use flaps such as the TFL and/or myocutaneous rectus abdominis flaps for full-thickness abdominal wall defects.
- Use the oblique rectus abdominis myocutaneous flap to obliterate radiated pelvic defects.

Suggested Reading

1. Bernstein EF, Sullivan FJ, Mitchell JB et al. Biology of chronic radiation effect on tissues and wound healing. Clin Plast Surg 1993; 20(3):435.
2. Dumanian GA, Llull R, Ramasastry SS et al. Postoperative abdominal wall defects with enterocutaneous fistulae. Am J Surg 1996; 172:332.
3. Hall EJ. Radiobiology for the Radiologist. 3rd ed. Philadelphia: JB Lippincott, 1988:108-136.
4. Hopewell JW. The skin: Its structure and response to ionizing radiation. Int J Radiat Biol 1990; 57:751.
5. Mendelson FA, Divino CM, Reis ED et al. Wound care after radiation therapy. Adv Skin and Wound Care 2002; 15(5):216.
6. Mustoe TA, Porras-Reyes BH. Modulation of wound healing response in chronic irradiated tissues. Clin Plast Surg 1993; 20(3):465.
7. Nussbaum ML, Campana TJ, Weese JL. Radiation-induced intestinal injury. Clin Plast Surg 1993; 20(3):573.
8. Sukkar SM, Dumanian GA, Szcerba SM et al. Challenging abdominal wall defects. Am J Surg 2001; 181(2):115.

Chapter 19

Pressure Ulcers

Zol B. Kryger and Victor L. Lewis

Epidemiology and Risk Factors

Pressure ulcers are a major health care problem costing billions of dollars annually. Patients can be classified into two groups: insensate (denervated ulcers) or sensate (innervated ulcers). The major risk factor for the patients in the first group is inadequate reposition and turning. A history of a previous pressure sore is also a significant risk factor. Roughly 60% of spinal cord injury patients will develop a pressure ulcer during their lifetime. Sensate patients also develop pressure sores due to prolonged pressure. The patients at the highest risk are elderly with femoral neck fractures, of which 65% will develop a pressure ulcer. Other high risk groups include ICU patients, burn patients and elderly residents of long-term care facilities, of which about 20-30% will develop a pressure sore. Among the sensate patients, risk factors include a low ratio of staff to patients, infrequent turning, prolonged immobility, poor nutritional status, significant weight loss, use of catheters and the required use of positioning devices.

Pathophysiology

Pressure ulcers are caused by continual pressure that exceeds the normal capillary pressure (20-32 mm Hg) for a length of time that is sufficient to cause tissue death. Relieving pressure for a few minutes every hour will save the tissue from dying. The duration required for cell death to occur is variable, ranging from 4 hours for muscle to 12 hours for skin. Nevertheless, most authorities agree that continual pressure for greater than 2 hours will result in severe tissue damage, especially if it has already been subjected to prolonged ischemia. Shearing forces and other factors also play a role in pressure sore formation.

Pressure sores occur characteristically over bony prominences, where the overlying soft tissue is compressed between the bone and a firm surface. Normally, people reposition themselves with sufficient frequency; however when patients are insensate or incapable of repositioning, there is a risk of developing a pressure sore. The early stages of a pressure ulcer begin with erythema and then ulceration of the skin. If proper wound care and pressure relieving measures are not taken, the ulcer will progress.

Pressure Ulcer Description

When describing a pressure ulcer, the following points should be addressed:

- The anatomic location of the pressure ulcer
- The depth of the ulcer (Stage I-V described below)
- The dimensions of the ulcer and the presence of sinus tracts
- Whether there is evidence of infection
- Whether it is clean (pink-red) or necrotic (white or black)
- If there is evidence of scars from prior flaps or from healing by secondary intention

Common Sites of Occurrence

Pressure ulcers can develop over any bony prominence. There are four areas that encompass most of the pressure sores seen by plastic surgeons:

- Ischium—due to prolonged pressure in the sitting position
- Trochanter—due to prolonged pressure in the lateral decubitus position
- Sacrum—due to prolonged pressure in the supine position
- Calcaneus and occiput—due to prolonged pressure in the supine position

Preoperative Considerations

Infection and Antibiotics

The majority of pressure ulcers do not develop invasive, soft tissue infection because they drain freely. They are all, however, colonized with bacteria. It is rare for a pressure sore to be the cause of a fever in a paraplegic patient who presents to the emergency department. For the noninfected pressure sore, antibiotics are not required. It is important, however, to rule out soft tissue infection with a thorough evaluation of the ulcer because a pressure ulcer must be free of infection prior to flap coverage. Any undrained abscess cavity should be incised, drained and packed. Grossly necrotic tissue should be debrided as it serves as a nidus for infection. Tetanus prophylaxis should be administered when necessary.

Underlying osteomyelitis is critical to diagnose prior to coverage. If a flap is placed over infected bone, there is a significantly higher risk of flap failure and other complications such as deep abscess or sinus tract formation. The diagnosis of osteomyelitis is accomplished with the measurement of an ESR level and a core needle biopsy. The combination of an ESR greater than 120 and a positive bone biopsy has the highest combined sensitivity and specificity. A simple method of biopsy is the Jamshidi core needle bone biopsy. Culturing the specimen is useful for identifying the specific organism, but since almost all bone will be colonized with bacteria, culture alone is not sufficient to make the diagnosis of osteomyelitis. When the diagnosis of osteomyelitis is made, the patient should receive a 4-6 week course of intravenous antibiotics prior to surgery.

The use of MRI and other imaging modalities is time consuming and not cost effective. Furthermore, with the exception of MRI, most of these tests have a relatively low specificity. In academic centers in which MRI is readily available, it remains an option for diagnosing osteomyelitis.

Patient Compliance

There is a difference between a pressure sore that is appropriate for surgery, and a patient that is a good surgical candidate. Due to the high risk of recurrence and complications, a compliant patient is essential to offer any chance of treatment success. For those patients who will not be compliant with the postoperative instructions, nonsurgical management should be considered. Many individuals with pressure ulcers suffer from depression and often feel socially isolated. This is especially true for paralyzed patients. A psychiatric evaluation can be of benefit in certain cases.

Nutrition Optimization

Most insensate patients are young and can achieve an adequate nutritional status. In contrast, bed-bound, debilitated patients, especially the elderly, are often malnourished. It is critical to optimize nutritional status prior to surgery. Many surgeons use a serum albumin level of 2.0 as their minimum cut-off for surgery.

However, albumin has a very short half life and is not a precise reflection of nutritional status. Serum prealbumin and transferrin levels are more accurate. Patients unable to obtain adequate caloric intake on their own should receive supplemental enteral feeding with TPN being a second choice. In addition to adequate protein levels, wound healing requires zinc, vitamin C, and other vitamins and minerals. Nutritionally depleted patients should receive a daily multivitamin.

Comorbid Conditions

A number of medical conditions, if not treated, will have a detrimental effect on the healing of any surgical flap. Diabetes must be managed aggressively, and blood sugars should be kept below 150. It is well known that uncontrolled blood sugars have a significant negative impact on healing tissue.

Active infections such as urinary tract or pulmonary infections must be completely treated prior to surgery. It is ill advised to operate on an actively infected patient. Any individual with bacteremia should demonstrate a negative blood culture prior to surgery.

For patients with severe peripheral vascular disease, a preoperative angiogram or magnetic resonance angiography should be considered. Since a number of flaps survive based on blood supply from the internal iliac vessels, it is important to rule out significant disease of these vessels. If indicated, a vascular bypass or stent/angioplasty should precede flap surgery.

Psychiatric conditions, especially depression, should be addressed. These are common in many pressure sore patients, especially the elderly. Since patient compliance is essential to the successful healing of a pressure ulcer, it is vital to ensure that a psychiatric evaluation be performed when appropriate. In addition, patients who abuse alcohol or illicit drugs should undergo drug rehabilitation prior to surgery. There is a very high risk of pressure sore recurrence among illicit drug users with spinal cord injuries.

Finally, patients taking steroids should receive vitamin A, (10,000 IU daily) to counteract the detrimental effects of steroids on the wound healing process. Vitamin A may stimulate macrophages that have been inhibited by the steroids. Furthermore, some evidence shows that vitamin A reverses the inhibitory effects that steroids have on TGF-beta.

Nonsurgical Management

Pressure Sore Staging System

It is useful to classify pressure sores according to the depth of the ulcer. There are several classification schemes of pressure ulcers that are commonly used. The National Pressure Ulcer Advisory Panel uses a four stage system. The following staging system is used at our institution and is based on Shea's modification of this system:

Stage I — intact skin with nonblanchable erythema
Stage II — ulceration through the epidermis and most of the dermis
Stage III — ulceration through the dermis into deeper subcutaneous tissue
Stage IV — ulceration down to underlying bone
Stage V — closed large cavities with a small sinus opening

This staging scheme is useful for treatment purposes since most pressure sores which are either stage I or II usually can be healed with nonsurgical, conservative treatment. Stage III and IV ulcers will often require surgical intervention to heal.

Stage IV ulcers should be evaluated for osteomyelitis as described above. It is important to note that most stage III pressure ulcers may have a component of stage IV, i.e., ulceration down to bone. Stage V ulcers should be unrooted.

Pressure Relieving Strategies

Aside from adequate wound care, relieving pressure is the most important measure to stop the progression of a pressure ulcer and to allow healing to begin. Frequent turning (every 1-2 hours) is mandatory. If possible, the patient should perform this himself. If not, it is imperative that the care-givers be instructed on the importance of this.

A number of hospital beds have been designed to relieve pressure on any one point of the body below 30 mm Hg. Low-air-loss beds or air-fluidized beds are the most common beds used. These are partially effective in reducing pressure on the ulcer, but they are not substitutes for frequent turning. For heel ulcers, there are a number of boots designed to relieve pressure on the calcaneus. Patients confined to wheelchairs can use cushions or foam padding under pressure points (i.e., the ischial tuberosities).

Dressings and Debriding Agents

Simple gauze dressings have withstood the test of time as a first-line treatment of pressure ulcers. They are the cheapest, least invasive means of treating pressure sores, and their use requires minimal training. Gauze dressings should only be applied to ulcers after manual debridement of any thick, superficial eschar. Dressings can be used both as a moist barrier to facilitate wound granulation, and as a method of mechanical debridement.

For superficial stage I ulcers, keeping the sore moist and clean is sufficient. This can be achieved with a transparent thin film adherent dressing such as a Tegaderm or Opsite. These will heal quickly as long as external pressure is relieved. For stage II ulcers, a moist dressing should be applied. It should not be allowed to dry out. This can be achieved by saturating gauze with normal saline and changing it before it has dried (wet-to-moist dressing). Another method is to coat the gauze with a water based gel such as Hydrogel or Aquaphor. Alternatively, gauze impregnated with a petroleum-based substance (Vaseline gauze, Xeroform, Xeroflow) will also prevent the ulcer from drying out.

Stage III and IV ulcers usually require some amount of mechanical debridement. The classic wet-to-dry dressing can achieve this by drying out within the ulcer and in the process sticking to any necrotic cellular debris. This dressing should be changed 3-4 times a day. Once the ulcer demonstrates pink, healthy granulation tissue, a switch should be made to a continuously moist dressing. Grossly contaminated, foul smelling ulcers can be treated with a modified wet-to-dry dressing using 0.25% Dakins bleach instead of saline. Dakins dressings should not be used for extended periods of time because they cause wound desiccation. It is important to emphasize to anyone involved in dressing changes not to pack large ulcers with individual, smaller pads of gauze (i.e., 2 x 2 or 4 x 4 gauze), but rather to use one continuous long piece of gauze (e.g., Kerlix gauze). There have been many cases of gauze pads being unintentionally left in ulcers for weeks.

An alternative form of debridement is with the use of enzymatic debriding agents. Most of these contain a papain derivative (an enzyme found in papaya) that enzymatically digests necrotic tissue. Common agents include Accuzyme® and Panafil®. Another commonly used substance is collagenase. They are applied 1-2 times a day directly onto the ulcer and covered by dry gauze. The efficacy of these agents is limited; they cannot be expected to debride a grossly contaminated wound or a thick, tough eschar. Both dressings and debriding agents are discussed in greater detail in Section 1 of this book.

Many pressure ulcers will heal with time by secondary intention as long as there is not excessive pressure, no underlying infection and adequate wound care is performed. For large ulcers this can take weeks to months, and therefore flap closure should be considered for all large stage III and IV pressure sores.

Negative Pressure Therapy (wound VAC)

Vacuum-assisted closure has become the nonsurgical method of choice for treating stage III and IV pressure ulcers at many centers. Studies have demonstrated that small- to moderate-sized pressure ulcers being treated with VAC therapy heal faster and with fewer complications than with conventional dressing changes alone. The VAC is also an excellent temporizing measure to help prepare the ulcer for flap coverage. Surgical debridement should be performed prior to initiation of the VAC. The wound VAC is described in detail in the first section of this book.

Hydrotherapy

Also termed whirlpool, hydrotherapy is sometimes used for mechanical debridement of pressure ulcers, especially those on the lower extremities. It's more common use is for venous stasis ulcers. There are no randomized prospective trials that demonstrate the efficacy of hydrotherapy. Benefits of this modality include less pain than many other forms of treatment for sensate patients, easier removal of dressings stuck to the wound, and high patient satisfaction. Pulsed lavage has largely replaced whirlpool as the preferred hydrotherapy treatment for pressure sores. Studies have shown it to increase the rate of granulation tissue formation compared to whirlpool. Its primary use is to decrease bacterial count in the ulcer. Optimal lavage pressures are in the range of 10-15 psi. Local wound care and dressing changes should be performed between hydrotherapy treatments.

Surgical Management

Debridement

Surgical debridement is required for all necrotic ulcers, in which case, the reconstructive procedure should be delayed until the ulcer is clean and granulating. Some surgeons, however, will perform both the debridement and coverage as a single-stage procedure. Debridement can be performed at the bedside or in the operating room. Most denervated patients will tolerate bedside debridement; however only superficial debridement of necrotic tissue should be done at the bedside. Patients with coagulopathies or those with severe thrombocytopenia should be debrided in the operating room. Operative debridement should include all necrotic soft tissue. This will speed up the granulation process, as well as decrease the bacterial load in the ulcer. Bone should be debrided conservatively, especially in the ischial region. Total ischiectomy should be reserved for cases in which there is extensive destruction of the ischium due to the high rate of complications from this procedure. Bone specimens should be sent for pathology and culture to rule out osteomyelitis. Tissue may be sent for culture; however it will usually be colonized with polymicrobial flora. Diagnosing soft tissue infection is therefore determined clinically rather than by the microbiology laboratory.

Femoral Head Excision (Girdlestone arthroplasty)

Excision of the femoral head is appropriate in the permanently nonambulatory patient when there is evidence of extension of a pressure ulcer (usually trochanteric or ischial) into the femoral joint space. This diagnosis can be made clinically. The patients usually presents febrile, with a septic picture. There is communication of the joint space with the ulcer, and this can be demonstrated with a cotton-tip applicator. Femoral head excision can be performed as part of the debridement procedure or in a single stage along with the flap coverage. When a Girdlestone arthroplasty is performed, the resulting dead space must be obliterated with a bulky, muscle or musculocutaneous flap.

Flap Selection

Flap selection is based on the location of the pressure sore, the local tissue that is available, and whether muscle is required in addition to subcutaneous fat and fascia. As a general guideline, large dead-spaces should be filled with muscle. In addition, ulcers with infected bone or those that are at a high risk of reinfection, should be covered by a muscle flap. There is no overriding rule that dictates whether an ulcer should be covered with a fasciocutaneous perforator flap or a myocutaneous flap. This is based largely on surgeon preference. There are studies that strongly advocate each type of flap as a first-line choice. Another consideration in patients with distal spinal cord injuries is whether to use a sensory flap. This involves using sensate tissue from above the level of injury to cover the ulcer in hopes that sensation will result in behavior modification by the patient to avoid pressure on ulcer-prone areas and prevent recurrent ulceration.

The following flaps are commonly used for pressure ulcer reconstruction:

- For ischial pressure ulcers, gluteal perforator flap or posterior thigh flaps
- For sacral pressure ulcers, the V-Y advancement type gluteal flaps
- For trochanteric pressure ulcers, the TFL or posterior thigh flaps

Prior to elevation of the flap, the pressure ulcer should be prepared by excising any nonhealthy tissue, removing the pseudobursa and smoothing out the underlying bone. The skin edges should be trimmed back until healthy bleeding is seen. Attention is then turned to the flap. After marking the skin paddle, the site of the main pedicle or perforating vessels should be located with a hand-held Doppler.

An outline of the more commonly used flaps for pressure ulcer repair is listed below:

Gluteal Perforator Flap

A fasciocutaneous flap based on perforators from the gluteal vessels. It is commonly used for sacral ulcers (Fig. 19.1) but can also be used for ischial and trochanteric defects.

1. The skin paddle is marked: either a transposition or V-Y advancement pattern is used, keeping in mind the ability to close the donor site directly.
2. The distal skin is incised, down to the gluteal muscles.
3. The flap is elevated along with the muscle fascia. Care is taken to locate the perforators.
4. The proximal perforators to be saved are skeletonized to allow greater flap mobility. Those that hinder flap mobility are ligated and divided.
5. Once the perforators are isolated, the proximal skin is incised, creating an island flap.

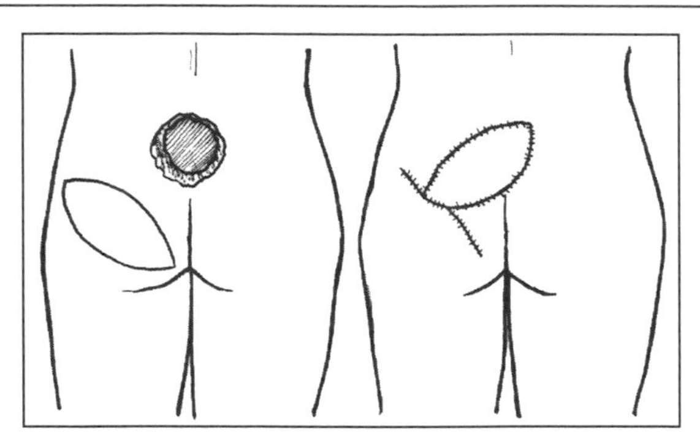

Figure 19.1. The superior gluteal artery perforator (SGAP) flap for coverage of a sacral wound.

6. The flap is transposed or advanced into the defect and sewn into place with two layers.
7. The donor site is closed directly over a suction drain, and an additional drain is placed under the flap.

Gluteus Myocutaneous Flap

A Type III myocutaneous flap based on either the inferior or superior gluteal artery or both. It is a versatile flap commonly used for both ischial and sacral pressure sores. Bilateral flaps can be advanced medially to close a large midline sacral defect.

1. Either a superior or inferior skin paddle can be utilized (Fig. 19.2). The superior skin paddle is based on the superior gluteal artery and the inferior paddles on the inferior gluteal artery (The entire muscle and buttock skin can be based on the inferior artery).
2. The muscle flap can be advanced or rotated (see Fig. 19.1).
3. The skin and subcutaneous tissue are divided. For rotational flaps, the muscle insertion (greater trochanter and IT band) is also divided.
4. The inferior and lateral borders of the muscle are divided. The muscle is detached from its origin. For ambulatory patients, the inferior portion of the muscle with its insertion and origin should be preserved.

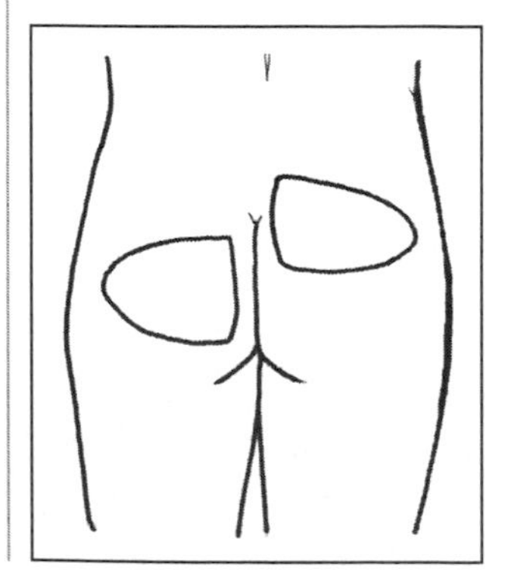

Figure 19.2. The gluteus myocutaneous flap. Examples of a superior and inferior skin paddle.

5. The pedicle and the sciatic nerve are located using the piriformis muscle as a landmark (the sciatic nerve emerges from beneath this muscle).
6. The flap is inset into the defect, and the muscle fascia is sewn to the contralateral gluteus maximus fascia. The subcutaneous tissue is closed in a second layer followed by skin.
7. The donor site is closed directly over a suction drain and additional drains placed under and over the flap. If direct closure is not possible, skin grafting the donor site is an option.

TFL Fasciocutaneous Flap

A fasciocutaneous flap based on cutaneous branches off the lateral descending branch of the lateral circumflex femoral artery. It is commonly used for trochanteric pressure ulcers.

1. The skin paddle is marked: A classic V-Y advancement pattern is used with the central axis running down the mid-lateral thigh.
2. The skin is incised beginning at the apex of the V.
3. The flap is elevated incorporating the TFL.
4. The dissection should be slow at two points: the midway point between the lateral femoral condyle and the greater trochanter, and 12 cm below the greater trochanter. The pedicles are located in these areas.
5. Once the perforators are isolated, the proximal skin is incised, creating an island flap.
6. The flap is advanced into the defect in a V-Y fashion and sewn into place in two layers.
7. After undermining, the donor site is closed directly over a suction drain from distal to proximal, and an additional drain is placed under the flap.

TFL Myocutaneous Flap

A Type I musculocutaneous flap based on the ascending branch of the lateral circumflex femoral artery. It is used for trochanteric pressure ulcers.

1. The skin paddle is marked: A classic V-Y advancement pattern is used with the central axis running down the mid-lateral thigh (Fig. 19.3).

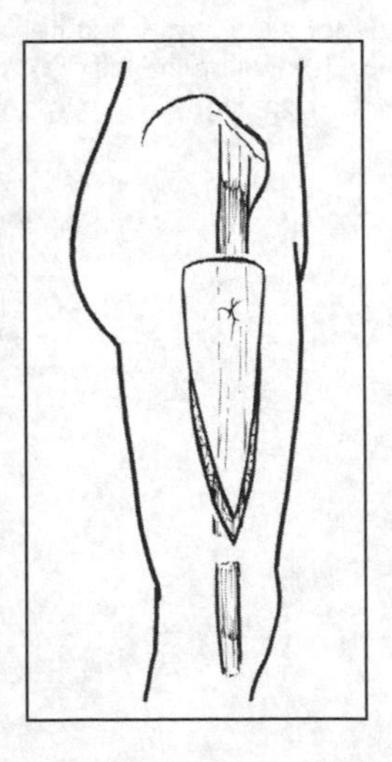

Figure 19.3. The Y-Y advancement TFL myocutaneous flap.

2. The skin and subcutaneous tissue are incised beginning at the apex of the V. The muscle insertion into the lateral aspect of the knee is divided.
3. The flap is elevated along with TFL muscle. Dissection progresses superiorly in the plane deep to the TFL and superficial to the vastus lateralis.
4. The pedicle is encountered roughly 10 cm below the anterior superior iliac spine.
5. Once sufficient mobility of the flap is obtained, it is advanced into the trochanteric defect and sewn into place in two layers.
7. After undermining, the donor site is closed directly over a suction drain from distal to proximal, and an additional drain is placed under the flap.

Posterior Thigh Flap

A Type I myocutaneous flap based on the descending branch of inferior gluteal artery. It is commonly used for ischial and trochanteric pressure ulcers. This flap can be solely based on the biceps femoris muscle in ambulatory patients or on all of the hamstring muscles in spinal cord injury patients.

1. The skin paddle is marked either as a rectangle for a medially based rotation or as a triangle for V-Y advancement (Fig. 19.4). Maximum skin paddle dimensions are 12 cm wide by 34 cm in length (never closer than 4 cm to the popliteal fossa).
2. The flap can be advanced, rotated or a combination of both. The point of rotation is 5 cm above the ischial tuberosity.
3. The skin and deep fascia are incised beginning at the most inferior point.
4. The correct plane of dissection is determined by locating the posterior femoral cutaneous nerve and inferior gluteal vessels between the semitendinosus and biceps femoris muscle.
5. The nerve and vessels are divided where necessary and elevated with the flap as the dissection progresses superiorly until the inferior border of the gluteus maximus is reached. If more reach is needed, fibers of the gluteus maximus can be included.
6. The medial and lateral skin borders are incised creating a skin island on the flap. If very minimal rotation/advancement is needed, the medial skin border can be maintained.

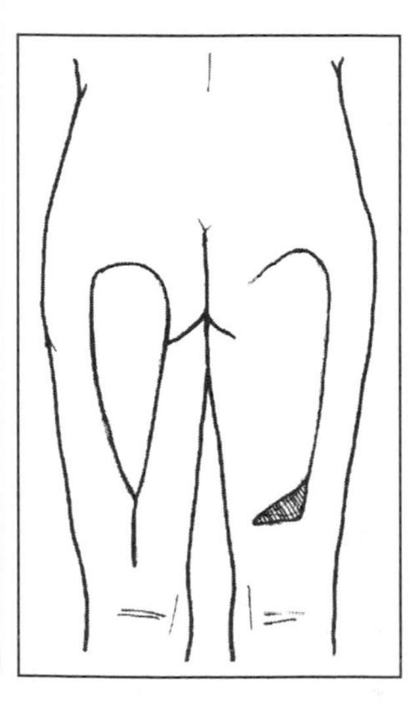

Figure 19.4. The posterior thigh flap. It can be advanced as a V-Y flap (shown on the left leg) or rotated (shown on the right leg).

7. The flap can be advanced or rotated into the defect, or even turned over for a more distal defect.
8. After undermining, the donor site is closed directly over a suction drain from distal to proximal, and an additional drain or two is placed under the flap.

Note: when the descending branch of the inferior gluteal artery is absent—a relatively common anatomical variant—the flap can be elevated as a superiorly based, random, fasciocutaneous flap based on proximal perforators from the cruciate anastomosis of the fascial plexus.

Postoperative Care and Complications

The greatest foes of a fresh flap are pressure or shearing forces. Strict use of an air-fluidized bed for a minimum of 2 weeks postoperatively is a common regimen. Some surgeons will only insist on 10 days of immobilization. Over the next 1-2 weeks, progressive sitting is done starting at 15 minutes twice a day and then increasing the frequency and duration from there. Pressure release maneuvers are performed at least every 15 minutes.

A long course of postoperative antibiotics is not routinely necessary. In cases of known osteomyelitis, or in which the intraoperative bone specimen is positive, 10-14 days of postoperative antibiotics is indicated. Infected ulcers that have been debrided and covered in a single stage procedure should also receive 1-2 weeks of postoperative antibiotics. Suction drains (e.g., Jackson-Pratt bulb drains) are mandatory for pressure ulcer repairs and should be left in place for a minimum of 10 days and often longer if their output remains elevated. Drains help eliminate dead space and reduce the risk of seroma formation. They do not, however, prevent infection or hematoma, the two common early postoperative complications of pressure ulcer repair. After 1-2 weeks, a breakdown in part of the suture line is not uncommon. It should heal with local wound care within 7 days. If not, it may indicate a deeper wound dehiscence.

Postoperative nutrition must be optimized in order for healing to occur. Adequate protein intake must be maintained by keeping the patient in positive nitrogen balance.

Debilitated patients unable to consume sufficient protein should receive supplemental nutrition. The enteral route is always the first choice. In addition to protein, vitamin C, zinc and other minerals can be supplemented with vitamins or nutritionally rich shakes such as Boost or Ensure. When available, a nutritionist's help can be valuable in designing a diet that will promote wound healing and a catabolic state.

Later complications include pressure ulcer recurrence (discussed below) and malignant degeneration. "Marjolin's ulcer" is the term used to describe malignant degeneration of a pressure ulcer and other chronic wounds. The latency is about 20 years. All chronically draining ulcers, especially those that have a change in the character of their drainage or appearance, should be biopsied to rule out malignancy.

Outcomes

Long term follow of patients who have undergone pressure ulcer repair reveal a wide range of recurrence, ranging from 25-80%. Pressure ulcers recur in up to 50% of patients who undergo flap coverage. This high rate of recurrence is independent of the original site of the pressure sore. There is also minimal difference in the recurrence rates between fasciocutaneous and myocutaneous flaps. In some studies, fasciocutaneous flaps have a lower rate of recurrence; however pressure sore recurrence appears to be more dependent on patient factors rather than on the type of flap that was used. As a group, young spinal cord injury patients have the highest risk of recurrence.

Pearls and Pitfalls

- When consulted for a fever in a patient with a pressure sore, remember that bacteremia due to a pressure sore is rare, especially when the clinical exam fails to demonstrate an abscess or obvious purulence. Urosepsis is at the top of the differential.
- A pressure ulcer that has a "fish bowl" appearance-a small opening in the skin with a large ulcer cavity beneath it, must have the opening widened so that it can be packed adequately.
- The diagnosis of osteomyelitis should be excluded in all stage IV pressure sores prior to flap coverage. This is easily done with a bone biopsy. Remember to always check for a coagulopathy first.
- Take the time to instruct the nursing staff or consulting service on proper dressing changes and wound care. They will appreciate your input and the patient will have a much better chance of healing.
- Position, prep and drape the patient in a manner that will keep all your options open, including using an alternative flap or harvesting a skin graft.
- When elevating the flap, work from distal to proximal. Dissect quickly where it is safe, but slow down near the pedicle. Once the pedicle is identified, inspect it frequently, especially when changing the direction of dissection.
- Never inset a flap that is under tension, especially the distal tip of the skin paddle. If it doesn't reach, don't force it. Careful dissection around the pedicle can provide the extra needed length.

Suggested Reading

1. Coskunfirat OK, Ozgentas HE. Gluteal perforator flaps for coverage of pressure sores at various locations. Plast Reconstr Surg 2004; 113:2012.
2. Cuddigan J, Berlowitz DR, Ayello EA et al. Pressure ulcers in America: Prevalence, incidence and implications for the future. An executive summary of the National Pressure Ulcer Advisory Panel monograph. Adv Skin Wound Care 2001; 14(14):208.
3. Evans GRD, Lewis Jr VL, Manson PN et al. Hip joint communication with pressure sore: The refractory wound and the role of Girdlestone arthroplasty. Plast Reconstr Surg 1993; 91:288.
4. Foster RD, Anthony JP, Mathes SJ et al. Ischial pressure sore coverage: A rationale for flap selection. Br J Plast Surg 1997; 50:374.
5. Goetz LL, Brown GS, Priebe MM. Interface pressure characteristics of alternating air cell mattresses in persons with spinal cord injury. J Spinal Cord Med 2002; 25:167.
6. Han H, Lewis Jr VL, Wiedrich TA et al. The value of Jamshidi core needle bone biopsy in predicting postoperative osteomyelitis in grade IV pressure ulcer patients. Plast Reconstr Surg 2002; 110(1):118.
7. Hess CL, Howard MA, Attinger CE. A review of mechanical adjuncts in wound healing: Hydrotherapy, ultrasound, negative pressure therapy, hyperbaric oxygen, and electrostimulation. Ann Plast Surg 2003; 51:210.
8. Horn SD, Bender SA, Ferguson ML et al. The national pressure ulcer long-term care study: Pressure ulcer development in long-term care residents. J Am Geriatrics Soc 2004; 52(3):359.
9. Kroll SS, Rosenfield L. Perforator-based flaps for low posterior midline defects. Plast Reconstr Surg 1988; 81:561.
10. Morykwas MJ, Argenta LC, Shelton-Brown EI et al. Vacuum-assisted closure: A new method for wound control and treatment: Animal studies and basic foundation. Ann Plast Surg 1997; 38(6):553.
11. Ramirez O, Hurwitz D, Futrell JW. The expansive gluteus maximus flap. Plast Reconstr Surg 1984; 74:757.
12. Shea DJ. Pressure sores: Classification and management. Clin Orthop 1975; 112:89.
13. Thomas DR. Improving outcome of pressure ulcers with nutritional interventions: A review of the evidence. Nutrition 2001; 17:121.
14. Yamamoto Y, Tsutsumida A, Murazumi M et al. Long-term outcome of pressure sores treated with flap coverage. Plast Reconstr Surg 1997; 100(5):1212.

Chapter 20

Infected and Exposed Vascular Grafts

Mark Sisco and Gregory A. Dumanian

Infected Arterial Bypass Grafts

Vascular graft infections remain challenging clinical problems for the vascular surgeon, especially when prosthetic material is involved. Traditional approaches, which until the 1960s consisted of removal of all prosthetic grafts with extra-anatomic bypass or amputation, has given way to more conservative approaches in patients whom graft removal is not feasible. These approaches have significantly reduced mortality and improved limb salvage. Contemporary management consists of an escalating algorithm of interventions that range from retention of the graft with healing by secondary intention to graft replacement with muscle flap reconstruction of the defect. When definitive extra-anatomic bypass is not possible, in situ replacement of infected prosthetic grafts with cadaveric homografts or autogenous tissue is preferred. The decision to replace the graft is typically made by the vascular surgeon. This section will focus on reconstructive options in the groin, since it is the most common site of graft exposure and infection requiring plastic surgical intervention.

Preoperative Considerations

Salvage of an arterial graft may be considered in patients who have patency of their reconstruction, an intact anastomosis and localized infection. Systemic antibiotics should be administered preoperatively. Superficial infections that do not extend to the graft itself may be managed with debridement alone, followed by local wound care.

Flap reconstruction is the management of choice for deep infections that involve the graft. This is due to the its well-established utility in lowering bacterial counts, improving antibiotic delivery, filling dead space and providing tension-free soft tissue coverage. Patency of donor vessels to the intended flap must be assessed preoperatively using MRA or angiography. The reconstructive surgeon must know the patency status of the graft before commencing the procedure. Staged debridements may be required prior to reconstruction. Wet-to-dry dressings, hydrogels, or closed suction drains may be used in the interim.

Operative Considerations

Flap options for coverage of groin wounds are listed in Table 20.1. The most commonly used flap in the groin is the sartorius. Harvest of the flap involves detaching from the anterior superior iliac spine, transposing it over the graft and suturing it to the groin musculature and inguinal ligament. Care must be taken not to interrupt more than three perforating vessels to this flap, since its segmental blood supply from the superficial femoral artery puts it at risk for necrosis.

Practical Plastic Surgery, edited by Zol B. Kryger and Mark Sisco. ©2007 Landes Bioscience.

Table 20.1. Options for flap coverage of the groin

Flap	Pedicle	Advantages	Disadvantages
Sartorius	Branches from the superficial femoral artery (inferior/ minor pedicle)	Easy harvest Minimal donor site	SFA occlusion common in vasculopathies Limited arc of rotation Proximity to infected tissue
Gracilis	Medial circumflex femoral, profunda femoris	Minimal donor site	More difficult harvest
Rectus femoris	Lateral circumflex femoral, profunda femoris	Easy harvest Bulky flap Skin may be transferred	May weaken extension at the knee
Rectus abdominis	Deep inferior epigastric, external iliac	Wide arc of rotation	May develop ventral hernia at the donor site DIEA occlusion is common

The senior author has described an extended approach to the gracilis, which can also be used successfully. It is approached through an overlying medial thigh incision. If a recent saphenectomy has been performed, this incision may be used. After medial reflection of the adductor longus, the gracilis is identified by the absence of nerves around the muscle, the lack of attachments deep and superficial to the muscle, and tapering of the muscle as it dissection proceeds down the thigh. The dominant pedicle is identified on the deep medial aspect of the muscle. The insertion of the gracilis to the femur is divided distally, and the minor segmental pedicles are ligated, as are small vessels from the dominant pedicle that supply the overlying adductor longus. The origin of the gracilis at the pubic symphysis is then divided and the muscle is tunneled beneath the adductor into the femoral triangle. Drains are placed in the donor site and beneath the flap if possible.

The use of the rectus femoris is another option. This flap is reached through a midline thigh incision. The tendon is divided 4 cm from the patella. The distal minor pedicle is ligated, and the flap can be folded up into the wound. The rectus abdominis has also been used successfully in this setting. Finally, the ipsilateral or contralateral rectus muscle can be used for coverage of the groin when the profunda femoris vascular pedicle is compromised. Intraoperatively, blood flow to the flap should be confirmed with a handheld Doppler, before it is elevated.

Postoperative Considerations

Intravenous antibiotics should be administered based on intraoperative culture results. Some authors have suggested that therapy be continued for six weeks in the case of autogenous grafts and up to one year for prosthetic grafts. Lifelong suppressive doses of oral antibiotics should be considered in the latter group.

Infected Prosthetic Hemodialysis Grafts

Although exposed dialysis grafts have traditionally been removed, the paucity of vascular access sites in long-term hemodialysis patients has led to several successful strategies to salvage them. In contrast to the bypass grafts described above, dialysis grafts are not acutely imperative to life or limb. As in arterial bypass grafts, hemorrhage or systemic infection mandates total graft excision. Since such grafts have a lifespan that averages 2-3 years, simple local flaps are typically used. Flaps such as the flexor carpi ulnaris and lateral arm flap allow coverage in the proximal forearm. The radial artery island fasciocutaneous flap, meanwhile, may provide coverage to the mid and distal forearm. Random pattern flaps should be used with caution, as they do not provide as reliable coverage. Vascular puncture can usually be continued during healing.

Pearls and Pitfalls

Patients with infected or exposed vascular grafts are in the highest risk group for subsequent wound complications. They have already demonstrated wound healing problems from their initial surgery. Their vascularity and wound healing is compromised due to their underlying peripheral vascular disease. Many of these patients are also renal failure patients and malnourished.

In considering the coverage procedure, one should expect donor site complications ahead of time. The donor site should be distant from the graft site and away from important structures. For example, harvesting a sartorius flap for coverage of a groin bypass graft may not be wise if the donor site is in close proximity to the infected wound. In regards to the exposed graft, muscle coverage should be placed transversely over a longitudinal graft, so that if the coverage breaks down, only a small area of graft will be exposed. In addition, the graft should be covered in staggered layers by closing the muscle and skin layers separately without the skin incision lying directly over the muscle incision.

Suggested Reading

1. Morasch MD, Sam IInd AD, Kibbe MR et al. Early results with use of gracilis muscle flap coverage of infected groin wounds after vascular surgery. J Vasc Surg 2004; 39(6):1277-83.
2. Alkon JD, Smith A, Losee JE et al. Management of complex groin wounds: Preferred use of the rectus femoris muscle flap. Plast Reconstr Surg 2005; 115(3):776-83.

Management of Exposed and Infected Orthopedic Prostheses

Mark Sisco and Michael A. Howard

Background

Wound dehiscence and periprosthetic infections complicating orthopedic implants are a significant source of postoperative morbidity that may be limb, or life-threatening. While the treatment of superficial wound complications is relatively straightforward, there is less agreement in the literature about the correct course of action when an implant itself is exposed or infected. Several authors advocate the traditional approach, which consists of removal of all exposed or infected implants with delayed flap closure. This approach tends to result in prolonged hospitalization and significant functional loss. Several recent reports suggest that early debridement with definitive muscle flap coverage may make it possible to salvage prostheses in select patients. The plastic surgeon is often consulted after the orthopedic surgeon has determined that the wound cannot be closed primarily or that the implant is infected. The cornerstones of prosthesis salvage include:

1. Early identification of infection
2. Aggressive debridement
3. Appropriate antibiotic therapy
4. Prompt soft-tissue coverage

All options in the reconstructive ladder may be needed depending on the size of the defect and the extent of the exposed structures.

Preoperative Considerations

Orthopedic patients are susceptible to wound complications for several reasons. Surgery may involve wide undermining of the soft tissues. Postoperative edema may be significant since joints have relatively poor lymphatic drainage, further compromised by surgery, leading to undue tension of the skin and incision. Prostheses may be positioned directly underneath the skin incision in an area that is poorly vascularized. Finally, many patients have had prior surgery, the scarring from which contributes to decreased tissue pliability and blood flow. In sum, the altered blood flow, increased edema and wound tension result in decreased oxygen delivery to the healing incision.

Persistent drainage or problematic wound healing has been described in up to 20% of total knee arthroplasty (TKA) patients, with an infection rate of 1-12%. The total incidence of infection following total hip arthroplasty (THA) is smaller: about 1%. When evaluating such problems, it is critical to distinguish wound infection from wound failure, since the treatment algorithms are different. Wound infection may cause wound failure. Wound failure, meanwhile, may cause implant

Practical Plastic Surgery, edited by Zol B. Kryger and Mark Sisco. ©2007 Landes Bioscience.

contamination and subsequent infection. The most common symptoms and signs of infection following TKA and THA are pain, erythema and purulent wound drainage. Additional laboratory tests such as a CBC, ESR, C-reactive protein and joint aspiration may also help to establish the diagnosis.

It is also important to consider the patient's comorbidities. Factors that predispose to failure of implant salvage include: previous surgeries, diabetes, adjuvant radiation therapy, connective tissue disease, peripheral vascular disease, tobacco use, prior steroid treatments and rheumatic disease. Factors that predict successful implant salvage in the setting of infection include: <2 week duration of symptoms; susceptible gram positive organism (especially Streptococcus); lack of radiologic evidence of infection or loosening of the prosthesis; and absence of a sinus tract.

Operative and Postoperative Considerations

Open Wounds without Evidence of Infection

In the absence of infection, prostheses underlying open wounds are often salvageable, even when exposed. Immediate closure may be considered, provided that there are no signs of infection and well-vascularized tissue exists. Primary closure may be especially difficult due to the lack of mobile soft tissue adjacent to the wound. Wounds that do not involve exposed tendon, bone or joint may be treated with local wound care, negative pressure wound therapy followed by skin grafts, or fasciocutaneous flaps-depending on the size of the defect. Wounds in which the tendon, bone or joint are exposed should be treated with debridement, irrigation and flap reconstruction. Some of the more commonly used flaps are listed in Table 21.1.

Infections

The keys to initial management of wound infections are: identification of the anatomic extent of infection, aggressive debridement of devascularized tissue, thorough irrigation, and culture-specific systemic antibiotic therapy. Thereafter, superficial infections may be managed with local wound care, skin grafts or fasciocutaneous flaps. Medium-depth infections, which extend to the joint capsule without involving the bone or joint structures, may be treated with skin grafts, fasciocutaneous flaps or muscle flaps depending on the size of the defect and, in the case of the knee, whether tendon is involved.

Deep infections that involve the bone or joint structures require more aggressive management. Acutely infected wounds should be thoroughly debrided, irrigated and

Table 21.1. A hierarchical list of flaps useful for treatment of exposed and infected artificial joints

Hip	Knee	Ankle
Vastus lateralis	Medial gastrocnemius	Free flap
Rectus femoris	Lateral gastrocnemius	Tibialis anterior
Rectus abdominis (inferior pedicle)	Fasciocutaneous flaps	Medial plantar
Gluteus medius	Free flap	
Tensor fascia lata		
Free flap		

treated with broad spectrum, anti-staphylococcal antibiotics. Definitive surgical closure of the wound, often with muscle flaps, may be undertaken when signs of infection have abated, preferably within 7 days. Chronic infections (greater than 4 weeks) often require implant removal and placement of an antibiotic-impregnated spacer, followed by second-stage reimplantation after long-term IV antibiotic therapy. Patients who require implant removal are at increased risk of wound failure following prosthesis reimplantation due to inadequate local tissues and patient compromise. As such, a low threshold should be used for muscle flap coverage at the second stage.

Pearls and Pitfalls

Successful salvage of a threatened or exposed implant requires a good working relationship with the orthopedic surgeon and prompt evaluation and treatment of the patient. Serial debridements may be needed until the site is clean and all tissues are deemed viable. Final closure should proceed in an expeditious manner when the wound is deemed ready. The major pitfalls to success lie in both delaying treatment and conversely, rushing closure. If one adopts the "wait and see" attitude, a worsening picture may develop due to bacterial contamination and subsequent infection. Hurrying the closure before the wound is clean may result in the subsequent development of osteomyelitis.

Suggested Reading

1. Adam RF, Watson SB, Jarratt JW et al. Outcome after flap cover for exposed total knee arthroplasties. A report of 25 cases. J Bone Joint Surg Br 1994; 76(5):750-3.
2. Attinger CE, Ducic I, Cooper P et al. The role of intrinsic muscle flaps of the foot for bone coverage in foot and ankle defects in diabetic and nondiabetic patients. Plast Reconstr Surg 2002; 110(4):1047-54.
3. Browne Jr EZ, Stulberg BN, Sood R. The use of muscle flaps for salvage of failed total knee arthroplasty. Br J Plast Surg 1994; 47(1):42-5.
4. Burger RR, Basch T, Hopson CN. Implant salvage in infected total knee arthroplasty. Clin Orthop Relat Res 1991; (273):105-12.
5. Eckardt JJ, Lesavoy MA, Dubrow TJ et al. Exposed endoprosthesis. Management protocol using muscle and myocutaneous flap coverage. Clin Orthop Relat Res 1990; (251):220-9.
6. Gusenoff JA, Hungerford DS, Orlando JC et al. Outcome and management of infected wounds after total hip arthroplasty. Ann Plast Surg 2002; 49(6):587-92.
7. Jones NF, Eadie P, Johnson PC et al. Treatment of chronic infected hip arthroplasty wounds by radical debridement and obliteration with pedicled and free muscle flaps. Plast Reconstr Surg 1991; 88(1):95-101.
8. Markovich GD, Dorr LD, Klein NE et al. Muscle flaps in total knee arthroplasty. Clin Orthop Relat Res 1995; (321):122-30.
9. Nahabedian MY, Mont MA, Orlando JC et al. Operative management and outcome of complex wounds following total knee arthroplasty. Plast Reconstr Surg 1999; 104(6):1688-97.
10. Nahabedian MY, Orlando JC, Delanois RE et al. Salvage procedures for complex soft tissue defects of the knee. Clin Orthop Relat Res 1998; (356):119-24.

Chapter 22

Hypertrophic Scars and Keloids

Zol B. Kryger

Scar Definitions

An **immature scar** is red, slightly elevated and may be pruritic or tender. With time, it will usually become mature.

A **mature scar** is flat and usually slightly paler, but occasionally darker than the surrounding skin.

A **linear hypertrophic scar** is red, raised and confined to the original borders of the incision. It usually occurs weeks after surgery and can continue to increase in size over the next few months. It will often become less raised with time.

A **widespread hypertrophic scar**, such as a burn scar, is red, raised and confined to the original borders of injury.

Minor keloids are raised and usually pruritic. They extend beyond the borders of original injury, over the normal skin. They can develop up to a year post injury. They do not regress spontaneously, and if excised, usually return.

Major keloids are over 5 mm in diameter. They can be painful, and often will continue to spread over years. Keloids have a familial predilection. They are much more common in blacks and Asians than in whites.

Pathogenesis

Hypertrophic scarring and keloid formation are the result of excess collagen accumulation in a healing wound. The collagen is largely Type III, the form found in normal immature scars. Causes include excessive skin tension, wound infection, delayed healing, abnormal fibroblast metabolism, and an array of hereditary abnormalities. The importance of excess tension cannot be overemphasized. Hypertrophic scars tend to form in areas of high tension, such as the anterior chest and upper back. All measures for reducing tension on a healing scar should be employed.

The molecular mechanisms for pathological scarring are under intense investigation. Many studies have shown that levels of various cytokines are elevated in hypertrophic scars and keloids. For example, the transforming growth factor-beta (TGF-β) superfamily has been implicated in hypertrophic scarring. Other cytokines, such as tumor necrosis factor and interleukin-1, show decreased levels in keloids. Research is ongoing, and several active clinical trials are evaluating agents that inhibit or increase key mediators in the process of excessive scarring.

Surgical Excision

Surgery alone is not recommended for keloids due to the very high rate of recurrence (50-100%). It should be combined with additional treatment modalities such as steroid injection or silicone sheeting. For hypertrophic scars, excision alone may be indicated if it is felt that the abnormal scarring was due to excessive tension or

Practical Plastic Surgery, edited by Zol B. Kryger and Mark Sisco. ©2007 Landes Bioscience.

wound complications (e.g., infection or delayed healing). In such cases, excision should be accompanied by a measure to decrease tension. Splinting of the incision is most effectively accomplished by intradermal sutures left in place for 6 weeks to 6 months, depending on the degree of tension. Z-plasties and other techniques of reorienting the direction of tension can be used as well.

Silicone Gel Sheeting

This modality has become standard of care for the treatment of hypertrophic scars. It should be used as a first-line agent for linear hypertrophic scars. Numerous randomized, double-blind studies have shown that it is efficacious for treating hypertrophic scars and small keloids. The benefits of silicone sheeting have not been demonstrated with other types of semi- or total occlusive dressings. Since this treatment is painless, it is an excellent option for children or adults unwilling to tolerate more painful options. For prophylaxis in those at risk for hypertrophic scarring, treatment should begin a few days postoperatively. Silicone sheeting should be worn a minimum of 12 hours a day, and preferably the entire day. It should be continued for several weeks postoperatively.

Corticosteroid Injections

Years of clinical experience and many randomized, prospective trials have shown that triamcinolone injected into the scar is efficacious at decreasing scarring. Response rates range from 50-100%, with a recurrence rate of 10-50%. It should be the first-line therapy for keloids and second-line for hypertrophic scars. In combination with other therapies such as surgery and cryotherapy, corticosteroid injections can be even more effective. Side effects are common and include skin atrophy, telengiectasias and pigment changes. The exact mechanism by which steroids diminish scarring is still largely unknown. What has been shown is that topical steroids are not effective in reducing scarring.

Pressure Therapy

Compression is the first-line treatment for post-burn, widespread hypertrophic scars. In order to be effective, pressure must be maintained between 24-30 mm Hg for at least 6 months duration. The longer the treatment, the more efficacious pressure therapy has been shown to be.

Steri-Strips®

Adhesive microporous paper tape applied to fresh incisions for several weeks postoperatively is moderately useful in preventing hypertrophic scaring. The mechanism is not entirely clear. It is likely a combination of occlusion and splinting of the incision.

Radiation Therapy

Radiotherapy should be reserved for adults with keloids resistant to other treatment modalities. Monotherapy is controversial, and most authors recommend using it following surgical excision. Response rates range from 10-90% for radiotherapy alone, and in combination with surgery it is more likely to be effective. Recurrence is very common, ranging from 50-100%.

Cryotherapy

This modality has shown benefit in acne-induced scarring. It should not be used for large scars. Side effects are common and include hypo- or hyperpigmentation, skin atrophy and pain.

Laser Treatment

Various lasers have been used in an attempt to treat hypertrophic scars and keloids, and the results have been largely disappointing. The flashlamp-pumped pulsed-dye laser appears promising, but more studies are needed. In combination with other modalities, such as corticosteroid injection, it has been shown to be effective. Its primary role is in reducing scar erythema and flattening mildly atrophic and hypertrophic scars. At this point, it is largely an emerging technology.

Other Emerging Therapies

A number of chemotherapeutic agents have demonstrated efficacy in treating hypertrophic scarring and keloids. These include intralesional injections of **interferon**, **5-fluorouracil**, and **bleomycin**, as well as topical administration with retinoic acid. Other emerging novel treatments focus on interfering with collagen synthesis and the cytokines involved in scarring, such as inhibitors of TGF-β.

Summary

Hypertrophic scars and keloids can be a formidable problem. Many approaches have been tried, but the recurrence rate is high for most treatments, especially for keloids. Surgical excision should be followed by an additional preventative measure, such as silicone gel sheeting. Ongoing research is attempting to pinpoint the mechanisms for these pathologic processes in hopes of uncovering more effective therapies.

Pearls and Pitfalls

The recommended treatment modality for the various types of pathologic scars is summarized in Table 22.1.

Table 22.1. Recommended treatment modality for various types of pathologic scars

Type of Scar	Recommended Management
Prophylaxis	Avoidance of infection and delayed healing; Steri-strips® for occlusion; silicone sheeting
Immature hypertrophic scar	Follow closely and treat as a linear scar if it progresses; pulsed-dye laser for redness
Linear hypertrophic scar	Silicone gel sheeting; pressure garments and intralesional corticosteroid if it fails. Surgical excision with silicone sheeting if 1 year of conservative management fails
Widespread hypertrophic scar	Pressure garments and silicone gel sheeting; massage and/or physical therapy can help
Minor keloids	Silicone sheeting combined with intralesional corticosteroids; add surgical excision if these fail (use epidermis as a STSG); add postoperative radiotherapy only for refractory cases
Major keloids	No consensus on treatment; radiation therapy or other emerging modalities (e.g., 5-fluorouracil) should be attempted

Suggested Reading

1. Chang P, Laubenthal KN, Lewis IInd RW et al. Prospective, randomized study of the efficacy of pressure garment therapy in patients with burns. J Burn Care Rehabil 1995; 16:473.
2. Gold MH. A controlled clinical trial of topical silicone gel sheeting in the treatment of hypertrophic scars and keloids. J Am Acad Dermatol 1994; 30:506.
3. Mustoe TA, Cooter RD, Gold MH et al. International clinical recommendations on scar management. Plast Reconstr Surg 2002; 110:560.
4. Poston J. The use of silicone gel sheeting in the management of hypertrophic and keloid scars. J Wound Care 2000; 9:10.
5. Rockwell WB, Cohen IK, Ehrlich HP. Keloids and hypertrophic scars: A comprehensive review. Plast Reconstr Surg 1989; 84:827.
6. Tang YW. Intra- and postoperative steroid injections for keloids and hypertrophic scars. Br J Plast Surg 1992; 45:371

Chapter 23

Benign Skin Lesions

Zol B. Kryger

Introduction

Although this chapter deals with benign skin lesions, a number of these conditions are premalignant and must be regularly evaluated and biopsied if they become suspicious. The benign lesions and disorders of the skin are tremendously diverse and extensive. This chapter focuses on the common lesions encountered by the plastic surgeon. Since many patients arrive with the question, "is this cancer?" an attempt has been made to classify every lesion as benign or premalignant. Some common terminology used in describing disorders of the skin is listed in Table 23.1.

Lesions of the Epidermis

Seborrheic keratosis is a common lesion, particularly in the elderly on sun-exposed areas. Multiple lesions are usually present. It demonstrates variable pigmentation and its borders have a sharp, "pasted on" appearance, allowing it to be scraped off with a scalpel. Clinically, it may be confused with melanoma; pathologically, it appears similar to squamous cell carcinoma. It is benign and has no malignant potential, so shave excision or freezing is adequate.

Actinic (solar) keratosis is a dysplastic, premalignant lesion. Like the seborrheic keratosis, multiple lesions may present in sun-exposed areas. It appears as a scaling, poorly demarcated plaque. Suspicious sites should undergo excisional biopsy. Multiple solar keratoses may be treated with topical 5-fluorouracil.

Keratoacanthoma is a rapidly growing papule with a round, smooth, pink rim encircling a keratinous plug. It is premalignant, and some consider it to be a variant of squamous cell carcinoma. Diagnosis is made by excisional biopsy; treatment consists of wide local excision or injection with 5-fluorouracil. Occasionally, these lesions regress spontaneously.

Epithelial Cysts

Epithelial cysts, previously termed sebaceous cysts, are epithelium-lined cysts filled with a keratinous and lipid core. They are benign and have several forms:

Dermoid cysts are congenital cysts that usually occur along the midline or at the lateral end of the eyebrow. They represent developmental inclusion of the embryonic epidermis. Like the teratoma, their core may contain material from all three germinal cell layers (e.g., glandular material, hair follicle, cartilage, bone). Patients with midline lesions of the face should undergo computed tomography to check for intracranial extension. Treatment consists of excision.

Epidermoid cysts are firm, fluctuant nodules, often with a central comedo that represents an epithelial opening. They are variable in size. They contain a cheesy,

Practical Plastic Surgery, edited by Zol B. Kryger and Mark Sisco. ©2007 Landes Bioscience.

Table 23.1. Terminology of various skin lesions and disorders

Term	Definition	Size
Papule	A palpable elevated skin lesion	< 1 cm
Plaque	A larger palpable elevated skin lesion	> 1 cm
Macule	A flat, colored, nonpalpable lesion	< 1 cm
Patch	A larger flat, colored, nonpalpable lesion	> 1 cm
Vesicle	A fluid-filled blister	< 0.5 cm
Bulla	A larger fluid-filled blister	> 0.5 cm
Hyperkeratosis	Increased thickness of the stratum corneum	variable
Parakeratosis	Hyperkeratosis with retained keratinocyte nuclei	variable
Acanthosis	Thickening of the epidermis	variable
Hypertrophic scar	Raised scar that is confined to the borders of the wound or incision	variable
Keloid	Raised scar that extends beyond the borders of the wound or incision	variable

lipid rich material and have a tendency to become infected. Treatment consists of excision during which the entire lining must be removed to avoid recurrence. Mildly infected cysts should be treated with antibiotics for one week prior to excision. Severely infected cysts should undergo incision and drainage.

The **pilar cyst**, or trichilemmal cyst, is the epidermoid cyst equivalent on the scalp. Multiple epidermoid cysts can be seen in Gardner's syndrome (familial polyposis).

Moles

The **nevocellular nevus** (common mole) is a benign congenital lesion derived from the melanocyte that occurs in clusters or nests. The **junctional nevus,** commonly seen in children and younger adults, consists of cells confined to the epidermal-dermal junction. It is a smooth and flat nevus with irregular pigmentation. The **compound nevus** has cells in both the epidermal-dermal junction and the dermis. It is usually raised and represents a progression from the junctional nevus. The **intradermal nevus** contains cells in clusters that are confined to the dermis.

The **Juvenile melanoma (Spitz nevus)**. usually occurs in children as a pale red papule on the face. Despite its name, it is a benign lesion. It is treated by excision with margins due to the risk of recurrence.

The **atypical mole,** formery referred to as dysplastic nevus, is premalignant. It is unevenly pigmented and has irregular borders. The lesion is usually solitary and is clinically indistinguishable from melanoma. Atypical moles should be biopsied. The familial form, B-K mole syndrome, presents with hundreds of atypical moles and confers a 100% risk of malignant transformation. Patients must have an annual photographic examination of their entire bodies. Any lesions that change must be excised.

The **nevus of Ota** is benign blue-grey macule that is found on the face in the V1 or V2 distribution of the trigeminal nerve. It is usually congenital and has a female predisposition. It is treated with laser therapy.

The **blue nevus** is similar to an intradermal nevus in that it is composed of intradermal melanocytes. It presents as a well-defined papule that can be distinguished from venous lesions by the fact that it does not blanch. Distinguishing it from Kaposi's sarcoma or melanoma can be difficult. It has a very small risk of malignant degeneration. It is treated by excision.

The **freckle** is a pigmented lesion that represents excessive melanin granule production without an increase in the number of melanocytes. Freckles are benign and have no malignant potential.

Lesions of the Epidermal Appendages

This group of miscellaneous benign lesions includes those derived from hair follicles and sebaceous, apocrine and eccrine glands. Most of the lesions requiring treatment occur in the scalp and face. They are summarized in Table 23.2.

Lesions of the Dermis and Subcutaneous Fat

A number of benign tumors occur in the dermis and subcutaneous tissue. Their behavior may range from completely benign to recurrent and locally aggressive.

Lipomas are fatty tumors of the subcutaneous fat. They are soft, mobile and usually painless. They can range in size from one to tens of centimeters. Larger lipomas are more likely to recur; they are often classified as low-grade liposarcomas. Small lipomas do not need to be biopsied. Patients with very large lipomas should undergo MRI to determine whether local invasion has occurred. Treatment consists of total excision as lipomas may recur if not completely removed.

Dermatofibromas are encapsulated intradermal masses that are painless, firm and mobile. They usually are less than 2 cm and occur on the extremities. Due to their accumulation of hemosiderin, they can display the range of colors seen in an evolving bruise. Treatment consists of excision, mostly for cosmetic purposes. They may recur if incompletely excised.

The **dermatofibrosarcoma protuberans** is a locally aggressive, indolent, nodular intradermal mass. It occurs on the trunk and thighs. They are often painful and can be complicated by ulceration or superinfection. The overlying epidermis may appear waxy. Dermatofibrosarcoma protuberans tends to grow extensions into the surrounding dermis and fascia, creating a gross "cartwheel" appearance. Treatment consists of wide excision to encompass the extensions of the tumor. Metastasis has been described in neglected or incompletely excised lesions.

Angiofibromas present as small, erythematous, telangiectatic papules, usually located on the cheeks, nose or around the lips. They are benign when solitary. Superficial excision is the standard treatment. When multiple, these papules may be associated with tuberous sclerosis (seizures, mental retardation, renal angiomyolipoma). Dermabrasion can offer some cosmetic improvement for patients with multiple angiofibromas.

Skin tags, also termed **cutaneous papillomas**, are soft and often pedunculated, arising from a central stalk. They may become infected or may necrose. Treatment consists of amputation at the stalk and electrodissection of the remaining base. They are benign and do not recur.

Glomus tumors present as a painful, firm, blue nodules on the hands and feet, especially subungually. They are benign vascular hamartomas derived from the glomus body. Excision is often required due to the pain caused by pressure from these tumors.

Table 23.2. Lesions of the epidermal appendages

Lesion	Tissue of Origin	Clinical Appearance	Location	Treatment
Apocrine cystadenoma	Apocrine gland	Translucent dome shaped papule	Face	Excision
Chondroid syringoma	Apocrine gland	Firm, fixed painless nodule	Face	Excision
Syringoma	Eccrine gland	Clear papule occuring in adults	Periocular	Laser, cautery or cryotherapy
Eccrine poroma	Eccrine gland	Soft red nodule at any age	Foot	Excision
Cylindroma	Eccrine gland	Smooth pink papule or plaque	Scalp, forehead	Excision
Nevus sebaceus	Sebaceous gland	Linear yellowish plaque, onset at birth	Forehead	Excision
Sebaceous adenoma	Sebaceous gland	Yellow nodule in adults	Head, neck	Excision
Sebaceous hyperplasia	Sebaceous gland	Ulcerated yellow-white papules	Face	Laser, cautery or cryotherapy
Pilomatricoma	Hair follicle	Firm nodule stretching the skin	Face, neck, arms	Excision
Trichoepithelioma	Hair follicle	Pink, shiny papules or a plaque	Face	Excision
Trichofolliculoma	Hair follicle	Skin colored papule with central hair	Face, scalp	Excision

Pearls and Pitfalls

Although most of the lesions described in this chapter are benign, many are difficult to clinically distinguish from malignant tumors of the skin. Furthermore, some of the lesions, such as atypical moles, may remain unchanged for years before undergoing dysplastic changes. Even the most experienced surgeon has been fooled on occasion by a lesion he was sure could not be malignant. These facts emphasize the importance of excisional biopsy of any suspicious skin lesion. Skin incisions should adhere to the principles of relaxed skin tension lines in the neck and face, and should be longitudinal in the extremities. The excision should be as complete as possible; any part of the lesion that appears to have been left behind should also be excised. All excisional biopsies should be sent to a pathologist. When applicable, specimens should be oriented with a marking stitch. Finally, it is the obligation of the surgeon to follow up on the pathology report and notify the patient of the results.

Suggested Reading

1. Demis DJ, ed. Clinical Dermatology. 22nd ed. Philadelphia: Lippincott-Raven, 1995.
2. Kumer et al, eds. Robbins Basic Pathology. 6th ed. Philadelphia: WB Saunders, 1997.
3. Zarem HA, Lowe NJ. Benign growths and generalized skin disorders. In: Grabb and Smith's Plastic Surgery. 5th ed. Philadelphia: Lippincott-Raven, 1997:141.
4. Slade J et al. Atypical mole syndrome: Risk factors for cutaneous malignant melanoma and implications for management. J Am Acad Dermatol 1995; 32:479.

Chapter 24

Basal Cell and Squamous Cell Carcinoma

Darrin M. Hubert and Benjamin Chang

Overview

Basal cell carcinoma (BCC) and squamous cell carcinoma (SCC) comprise the vast majority of nonmelanoma skin cancers. Over one million white Americans are affected by these two entities yearly. They predominantly affect fair-skinned individuals, and their incidence is rising rapidly. Etiology may be multifactorial, but sun exposure appears to play a critical role. When detected early, their prognosis is generally excellent. However, both are malignant cutaneous lesions with inherent metastatic potential. Thus appropriate diagnosis, treatment and surveillance are of utmost importance. Malignant melanoma was discussed in the previous chapter.

Premalignant Lesions

The most common precursor of cutaneous squamous cell carcinoma is the **actinic keratosis**, also known as the solar keratosis. It appears as a scaly, discrete, maculopapular lesion that arises primarily on sun-damaged skin. Palpation of these flat lesions may reveal roughness that is not apparent on visual inspection. Due to the potential for progression to SCC, actinic keratoses are commonly treated by curettage and electrodessication, liquid nitrogen or topical 5-FU (Efudex).

Bowen's disease is a type of squamous cell carcinoma-in-situ marked by a solitary, sharply demarcated, erythematous, scaly plaque of the skin or mucous membranes. A second form of squamous cell carcinoma-in-situ is **erythroplasia of Queyrat**, which appears as glistening red plaques on the uncircumcised penis. Both have the potential for progression to invasive carcinoma and should be resected completely with conservative surgery.

Leukoplakia is a condition found on the oral mucosa commonly in association with smokeless tobacco use. These white patches may undergo malignant transformation to SCC in 15% to 20% of cases if left untreated. Epidermodysplasia verruciformis is a rare autosomal recessive disorder in which the body is unable to control human papilloma viral infections. It manifests itself as flat wart-like lesions that frequently degenerate into SCC. A keratoacanthoma, on the other hand, grows rapidly to form a nodular, elevated lesion with a hyperkeratotic core. It may involute spontaneously or appear indistinguishable from a SCC, and early conservative excision is recommended.

Tumor Staging

All nonmelanoma skin cancers are staged by the TNM system established by the American Joint Committee on Cancer (AJCC). The staging system is shown in Table 24.1. Characteristics of the primary tumor (T), regional lymph node status (N) and distant metastasis (M) are considered. BCC rarely metastasizes, although it may be locally destructive. The malignant potential of SCC is real and is related to the size and location of the tumor, as well as the degree of anaplasia.

Practical Plastic Surgery, edited by Zol B. Kryger and Mark Sisco. ©2007 Landes Bioscience.

Table 24.1. Staging of nonmelanoma skin cancers according to the TNM system established by the American Joint Committee on Cancer (AJCC)

Primary Tumor (T)	
TX	Primary tumor cannot be assessed
T0	No evidence of primary tumor
Tis	Carcinoma in situ
T1	Tumor 2 cm or less in greatest dimension
T2	Tumor more than 2 cm, but not more than 5 cm in greatest dimension
T3	Tumor more than 5 cm in greatest dimension
T4	Tumor invades deep extradermal structures (e.g., cartilage, muscle or bone)
Regional Lymph Nodes (N)	
NX	Regional lymph nodes cannot be assessed
N0	No regional lymph node metastasis
N1	Regional lymph node metastasis
Distant Metastasis (M)	
MX	Distant metastasis cannot be assessed
M0	No distant metastasis
M1	Distant metastasis

Staging System			
Stage 0	Tis	N0	M0
Stage I	T1	N0	M0
Stage II	T2	N0	M0
	T3	N0	M0
Stage III	T4	N0	M0
	Any T	N1	M0
Stage IV	Any T	Any N	M1

Basal Cell Carcinoma

BCC is the most common skin cancer and, indeed, the most common malignancy in the United States and Australia. It outnumbers cutaneous squamous cell carcinoma by approximately four to one. Its origin lies in the basal layer of the epithelium or the external root sheath of the hair follicle. Classic teaching holds that BCC requires stromal participation for survival, not the malignant transformation of preexisting mature epithelial structures seen in SCC. Although its metastatic potential is very low, basal cell carcinomas exhibit oncogene and tumor suppressor gene characteristics that question this classic explanation. Basal cell carcinoma tends to follow the path of least resistance, spreading into adjacent tissues. It only rarely metastasizes to distant sites.

Classification

Multiple histologic classifications have been proposed for subtypes of BCC, however, only the most common are mentioned here. **Nodular BCC** is the most common (45-60%), found typically as single translucent papules on the face. It is firm, may ulcerate, and often exhibits telangiectasia. **Superficial BCC** (15-35%) occurs as multiple scaly lesions on the trunk. Lightly pigmented or erythematous, it may resemble psoriasis or eczema. The less common subtypes are usually more aggressive. These include **infiltrative BCC** (10-20%), **morpheic BCC** (9%), which is associated with the highest recurrence rate, **micronodular BCC** (15%) and **adenoid BCC** (precise incidence unknown).

Risk Factors

Exposure to ultraviolet radiation appears to play a major role in the development of BCC. A thorough history during the preoperative evaluation should investigate this, making special mention of any significant sunburns during childhood or adolescence. Other risk factors include exposure to radiation or chemical carcinogens such as arsenic, Fitzpatrick skin type 1 or 2 (fair skin), increasing age, male sex, xeroderma pigmentosum, albinism and immunosuppression. Patients with **basal cell nevus syndrome** may develop multiple basal cell carcinomas. This syndrome, known eponymously as Goltz-Gorlin syndrome, is characterized by odontogenic keratocysts, palmar or plantar pits, cleft lip or palate, rib anomalies and areas of ectopic calcification. **Nevus sebaceous** lesions also predispose to BCC. As hairless yellow plaques present at birth, these lesions are typically found in the head and neck region. They may undergo malignant transformation in 10% of cases.

Special mention should be made of the importance of the "**H-zone**" of the face. This designation, roughly in the shape of an "H," is defined by the preauricular regions and ear helices, nasolabial folds, columella and nose and lower eyelids. BCC lesions located in this area are associated with both a higher risk of recurrence and greater morbidity as a consequence of treatment.

Treatment

There are several modalities available for the treatment of BCC. For a given lesion, one must weigh the treatment in terms of effectiveness in eliminating the malignancy against the functional and cosmetic implications before choosing the appropriate route.

First, surgical excision involves the full-thickness removal of the lesion, down to subcutaneous fat, along with a rim of "normal" tissue. Current literature recommends margins of 3 mm for small (<10 mm) and 5 mm for larger (10-20 mm) BCC of the face. For lesions found in any other location, margins of 5 mm are recommended. These wounds are typically either closed primarily or allowed to heal by secondary intention. For lesions located in delicate areas of the face, such as the eyelids, where removal of a margin of normal tissue may have profound functional consequences, Mohs micrographic surgery may be indicated. This technique has demonstrated the highest cure rates of any treatment modality. Cure rates of 99% for primary BCC and 93-98% for recurrent BCC have been demonstrated with the use of Mohs surgery. The technique of Mohs surgery is discussed in detail below.

An additional accepted treatment is cryotherapy, which is typically followed by curettage and healing by secondary intention. Local anesthesia is used, and the lesion is rapidly frozen with liquid nitrogen. There is no histological control with this method, and the tissue typically becomes initially very edematous. Its use has been advocated particularly near underlying cartilage. Recurrence rates of 3.7-7.5% have been reported.

Curettage and electrodessication have been employed in the past, with recurrence rates of 3.3% for low risk lesions to 18.8% for high risk ones. Due to unacceptably high recurrence rates, poor cosmetic outcome and lack of histological control, it is generally not accepted as a first line therapy for BCC. Radiation therapy has also been used to treat BCC, but the risk of radiation dermatitis, increased risk for future skin malignancy, and lack of histological control have discouraged its current use.

Squamous Cell Carcinoma

Risk Factors

Similar to BCC, cutaneous SCC typically occurs in areas of skin receiving the greatest sun exposure. The etiology appears to be the result of UVB radiation (wavelength range of 290-320 nm), which produces thymidine dimers in the DNA of the p53 tumor-suppressor gene. Fair-skinned individuals, albinos and those with xeroderma pigmentosum seem to be at particularly increased risk. Other risk factors include infection with human papillomavirus, chronic immunosuppression such as that seen in the organ-transplant population, exposure to chemical carcinogens such as arsenic, and exposure to ionizing radiation. Chronically inflamed or damaged skin may predispose to carcinoma as well, termed **Marjolin's ulcer**. These are most commonly squamous cell carcinomas, and they can arise in long-standing, chronic wounds such as pressure sores, fistulae, venous ulcers, lymphedema and burn scars.

Recurrence and Metastasis

The metastatic potential of SCC is greater than that of BCC, and the relative risk of recurrence and metastasis can be assessed according to characteristics of the lesion. The most important predictor is **size** of the tumor, with lesions greater than 2 cm in diameter recurring at a rate of 15% and resulting in metastasis at a rate of 30%. Anatomic location predicts greater malignant potential, especially the lip and ear, but also the scalp, forehead, eyelid, nose, dorsum of the hands, penis and perineum. Other features associated with higher risk of recurrence and metastases are rapid tumor growth, host immunosuppression, prior local recurrence, depth of invasion greater than 4 mm or into the subcutaneous tissues and location in a Marjolin's ulcer. Perineural invasion denotes a particularly poor prognosis and is lethal in a majority of patients by five years.

Treatment

With the exception of cryotherapy, treatment options for SCC are similar to those for BCC. However, there have been no randomized controlled trials comparing the efficacy of the various techniques. Direct surgical excision has demonstrated a recurrence rate of approximately 8% and metastatic rate of 5% at five years. Some authors have advocated surgical margins of 4 mm for low risk lesions and 6 mm for high-risk lesions whenever feasible. Mohs surgery has demonstrated the highest cure rates, about 97% for primary SCC, and is especially recommended for high-risk lesions. Due to lack of histological control and unacceptable cure rates, curettage with cautery, cryosurgery and radiotherapy are not recommended.

Mohs Surgery

Mohs micrographic surgery (MMS) was developed by Dr. Frederic Mohs in the 1930s. Dermatologists with specialized fellowship training in Mohs surgery typically perform the technique today. Fresh-tissue horizontal frozen sections are the hallmark of the procedure. Another salient feature is that the surgeon who performs the excision also interprets the histological results. The process can be summarized briefly in the following steps:

1. Gross debulking of the tumor
2. Excision of a narrow (2-3 mm rim) of normal tissue, beveled at 45° at the edges
3. Color-coding of specimen to mark margins and orientation
4. Mapping of the specimen and division into sections
5. Frozen-section processing in on-site laboratory
6. Microscopic examination
7. Repeat cycle if any residual tumor is noted
8. Healing by secondary intention, primary closure, skin graft or flap closure

This approach combines the highest cure rates for nonmelanoma skin cancer with maximum preservation of normal tissue and function. It provides the theoretical benefit of being able to examine 100% of the surgical margin in a single sitting, compared to the standard "bread-loafing" histological examination of surgical specimens. The disadvantages are that is can be time-consuming, labor intensive and expensive. Indications for Mohs micrographic surgery over other treatments include location of the lesion in the "H-zone," tumors greater than 2 cm in diameter, aggressive tumors, recurrent tumors, tumors in previously irradiated areas, incompletely excised tumors, clinically ill-defined tumors, presence of perineural invasion and aggressive rare tumors such as dermatofibrosarcoma protuberans, microcystic adnexal carcinoma or sebaceous gland carcinoma.

Follow-Up Examinations

Follow-up after treatment of skin cancers consists of monthly whole-body skin self-examinations and twice yearly skin examinations by a dermatologist or other qualified medical provider for at least five years to check for local recurrence as well as the occurrence of a second primary. Patients with larger SCC should be examined every three months, including palpation of regional lymph nodes, for several years and then followed at six-month intervals for the rest of their lives.

Pearls and Pitfalls

The two main pitfalls in treating skin cancers are missed or delayed diagnosis and inadequate biopsy. Although some skin cancers can be diagnosed by visual inspection alone, histologic confirmation is essential. The adage "when in doubt, cut it out" should be applied when dealing with skin lesions. A corollary to that rule is "if the patient wants it out, cut it out." If one fails to biopsy a lesion, which later turns out to be a malignancy, the delay in treatment can lead to a higher risk of recurrence, metastases and malpractice suits. An excisional biopsy is preferable because it gives the pathologist the entire lesion to examine. If the lesion is large, a representative incisional or punch biopsy, taking the full thickness of the skin, is preferable to a shave biopsy, particularly if the lesion turns out to be a melanoma.

As in other areas of tumor surgery, reconstructive concerns should not compromise surgical margins. One way to avoid this "conflict of interest" is to have one surgeon excise the lesion and another reconstruct the defect. Mohs surgery is particularly well suited to this two-surgeon approach. For extensive tumors treated by standard excision, delayed reconstruction should be considered in order to wait for the final histological margins, unless the defect can be closed primarily. One should not proceed with a complex flap reconstruction solely on the basis of an intraoperative frozen section examination of the margin because of the potential for a false negative.

Suggested Reading

1. Greene FL et al, eds. AJCC Cancer Staging Manual, 6th ed. Philadelphia: Lippincott Raven Publishers, 2002:203.
2. Alam M, Ratner D. Cutaneous squamous-cell carcinoma. N Engl J Med 2001; 344(13):975.
3. Kuijpers DI, Thissen MR, Neumann MH. Basal cell carcinoma, treatment options and prognosis, a scientific approach to a common malignancy. Am J Clin Dermatol 2002; 3(4):247.
4. Motley R, Kersey P, Lawrence C. Multiprofessional guidelines for the management of the patient with primary cutaneous squamous cell carcinoma. Br J Plast Surg 2003; 56:85.
5. Nelson BR, Railan D, Cohen S. Mohs' micrographic surgery for nonmelanoma skin cancers. Clinics Plast Surg 1997; 24(4):705.
6. Snow SN, Madjar Jr DD. Mohs surgery in the management of cutaneous malignancies. Clinics Derm 2001; 19(3):339.

Chapter 25

Melanoma

Gil S. Kryger and David Bentrem

Introduction

Malignant melanoma is the leading cause of death from skin cancer, despite only accounting for a fraction of all cutaneous cancers. The incidence of melanoma has doubled over the past 35 years in the United States. In 2002, there were 53,000 new cases in the U.S. The mean age of diagnosis was 45 years and the lifetime risk (in the U.S.) is 1 in 58 for men and 1 in 82 for women. The lowest rates occur in China and Japan (0.3/100,000) and the highest rates are in Australia and New Zealand (36/100,000).

In addition to geographic risk factors, there are several patient-related factors that are associated with an increased incidence of disease. Individuals with fair skin have a higher risk of developing melanoma. The incidence is 10 times higher in whites than blacks. People with red hair have a 3.6 times higher incidence of melanoma than those with black hair. Other risk factors include a history of severe sunburns in childhood, freckling after sun exposure, and people with more than 20 nevi on their body. Of patients with melanoma, 5-11% will have a family member with melanoma.

Classification

Histologic level of invasion (Clark's level) and tumor thickness (Breslow thickness or depth) are important indicators of metastatic risk and outcome (Fig. 25.1). Lymph node status is the single most powerful predictor of survival. When lymph nodes are not involved, the most important factors for prognosis are Breslow depth

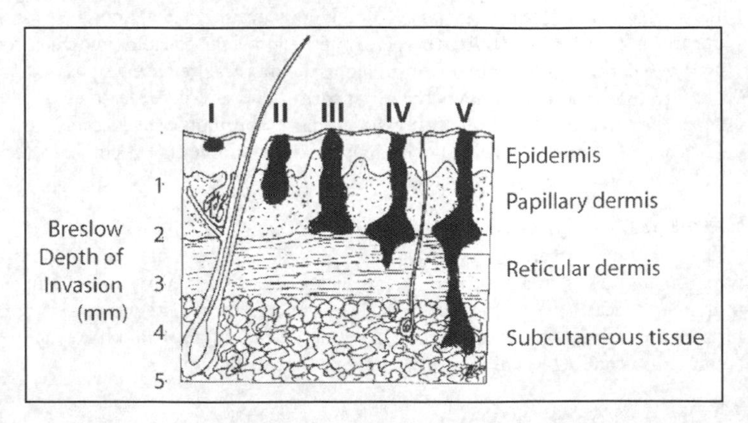

Figure 25.1. A schematic diagram of Breslow depth and Clark's levels of invasion.

and ulceration. Surgical margins and the need for sentinel node biopsy are based on tumor thickness.

Pathophysiology

The intermittent exposure hypothesis states that intermittent high energy exposure of melanocytes to sunlight is more damaging than the total cumulative dose. This is because continuous exposure can actually increase the amount of melanin (tanning) which protects the nucleus of the cell. People who have had three or more blistering sunburns before age 20 are at increased risk for disease.

Melanoma arises in epidermal melanocytes. Melanocytes produce melanin in response to sunlight and transport it to keratinocytes via dendrites. The keratinocytes regulate melanocyte growth, thus maintaining homeostasis. When these melanocytes escape from regulation, a dysplastic nevus arises. These lesions are premalignant and grow in a radial growth phase, parallel to the skin surface. A primary melanoma arises when a vertical growth phase begins with the capacity to invade deeper structures.

Clinical Characteristics

There are four subtypes of melanoma, each with a different appearance and clinical behavior.

- **Superficial spreading melanoma** is the most common subtype (70-75%). This type frequently arises from a pre-existing dysplastic nevus and always grows in a radial growth phase. These are characterized by the ABCD's described below.
- **Nodular melanoma** is the second most common type (15%). This subtype lacks a radial growth phase, making it very difficult to diagnose at an early stage as it does not arise from a pre-existing pigmented lesion. The appearance of these melanomas is shiny and smooth with a uniform color.
- **Acral lentiginous melanoma** is a rare subtype with a poor prognosis. These lesions are notoriously difficult to diagnose as they are commonly masked by thick stratum corneum of the palms and soles of the feet or on difficult to find mucosal surfaces. They are flat, dark brown or black and have irregular borders. Subungal melanomas belong to this category. These lesions occur equally among all races and arise in the nail bed or matrix. Hyperpigmentation of the nail fold (Hutchinson's sign) or a dark band (~3 mm in width) along the nail are presenting signs.
- **Lentigo maligna melanoma** is the rarest form (5%) and is the least aggressive type. They arise on chronically sun-exposed parts, commonly the face and neck. They are dark brown or black, larger than the other subtypes (1-3 cm diameter) and have highly irregular borders.

Diagnosis

The mainstay of treatment is early diagnosis and excision. A suspicious lesion is any pigmented lesion that is **a**symmetrical, has **b**order irregularity, **c**olor change, and **d**iameter greater than 6 mm (the ABCD's of malignancy). The presence of red, white or blue variegation in a brown or black lesion is highly suspicious. Any rapidly changing or ulcerating lesion is highly suspicious for melanoma.

Biopsy

Whenever possible, suspicious lesions should undergo complete excisional biopsy with a 1-2 mm rim of normal skin in an elliptical shape. If the lesion is large (over 1.5 cm), or in a location where skin removal is critical, an incisional biopsy

Table 25.1. Guidelines for surgical margins during excision of malignant melanomas

Type of Lesion	Recommended Margin
Dysplastic nevus	0.5 cm
Lentigo maligna/melanoma in situ	0.5-1 cm
Melanoma 0-1 mm thick	1 cm
Melanoma 1-2 mm thick	1-2 cm (2 cm preferable; 1 cm in aesthetically important areas)
Melanoma 2-4 mm thick	2 cm
Melanoma >4 mm thick	2 cm

(punch biopsy) should be taken from the most raised or irregular area. If incisional biopsy is used for diagnosis, the management should not be based on the Breslow depth because there may be thicker areas that were not sampled. The biopsy is carried down to normal subcutaneous fat, but not through the underlying muscle fascia. Biopsy can be done in the clinic with local anesthetic (a mixture of 1% lidocaine with 1:100,000 epinephrine, and an equal volume of 0.5% Bupivicaine). One must consider the relaxed skin tension lines and orientation as future wide local excision may be needed. The specimen is handled carefully and sent to the pathologist according to hospital protocol. The skin is closed with care, in layers if necessary, as frequently no more surgery will be required (if the lesion is benign). Subungual lesions are biopsied by removing the nail and performing an excisional biopsy down to, but not including periosteum.

Surgical Treatment

Once the diagnosis of melanoma is made, definitive surgical treatment should be scheduled as expeditiously as possible. The surgical plan is based on the histopathologic depth of the tumor (Breslow depth). For incisional biopsies, one must ensure that the plan is based on the deepest part of the tumor, which may not have been the part first incised. The lesion is staged based on the depth of invasion and the presence of palpable nodes on exam. Tumors can then be categorized into **thin tumors** (<1 mm thick) amenable to wide local excision only down to the muscle fascia, **intermediate tumors** (1-4 mm thick) that need sentinel lymph node biopsy and possibly total lymph node dissection, and **thick tumors** (>4 mm thick) that have likely metastasized at the time of diagnosis. The recommended guidelines for margins are shown in Table 25.1.

Wound closure is typically performed immediately after excision. Reconstruction should follow the reconstructive ladder: primary closure when possible, skin grafts and flaps if primary closure is not possible. Complex defects with limited reconstructive options are generally treated with temporary dressings or skin grafts until permanent pathology sections confirm negative margins.

Sentinel Node Biopsy

Indications

Patients with intermediate lesions 1-4 mm thick historically underwent elective lymph node dissection (ELND) with resulting morbidity. The role of sentinel node biopsy (SNB) is to evaluate the local nodal basin for metastases without the morbidity and expense of total lymph node dissection. Any patient with ulceration or regression, males with Clark level III or greater lesions, and patients with intermediate thickness tumors (1-4 mm) should undergo SNB.

Rationale for Sentinel Node Biopsy

The basis for SNB is that nodal metastases follow an orderly progression and that the histology of the sentinel node reflects the entire nodal basin. Therefore, if the sentinel node is free of disease, no further dissection is needed. If tumor is found, a therapeutic nodal dissection is performed removing the entire nodal basin. Palpable nodes (regardless of tumor depth) that are not suspicious can be evaluated with surveillance and fine needle aspiration (or open biopsy). Palpable nodes that are in the basin draining the primary site, or are otherwise suspicious, should undergo elective lymph node dissection without biopsy.

Description of the Procedure

For patients undergoing SNB, preoperative lymphoscintigraphy aids in localization of the tumor. The technique involves the intradermal injection of a radio-labeled colloid and dynamic imaging over a 2 hour period. This allows the surgeon to find the approximate area of the sentinel node and is especially important for regions with irregular drainage (such as the head and neck). The patient should then be informed of the risks of SNB, specifically the risk that the procedure may need to be converted to a complete lymphadnectomy if the sentinel node is positive. Intraoperative vital blue dye lymphatic mapping (1-3 ml of lymphazurin blue dye injected into the dermis) can be an adjunct to intraoperative lymphoscintigraphy detection. Histologic examination is performed on all excised lymph node using routine and immunoperoxidase S-100 and HMB-45 stains. Intraoperative frozen section is not routinely performed.

Intraoperative complications include anaphylaxis, retained blue hue and inaccurate oxygen saturation readings. Cooperation between the radiologist (lymphoscintigraphy), surgeon (biopsy), and pathologist (histology) is essential to the success of the procedure and has resulted in only a 4-11% false negative rate.

Additional Diagnostic Studies

Patients with thin lesions (<1 mm) should be evaluated with liver function studies (LFTs), lactate dehydrogenase (LDH), and chest X-ray. Patients with thicker lesions should undergo chest/abdomen/pelvis CT and possibly MRI of the brain (although prospective studies have shown no survival benefit for CT and MRI in asymptomatic patients). Patients with stage IV disease undergo PET scanning.

Staging and Adjuvant Therapy

Once the tumor size, node status and presence or absence of distant metastases hve been established, the TNM stage is determined. The American Joint Committee on Cancer has recently revised the stage groupings for melanoma (Table 25.2). Stages I and II denote localized disease, stage III indicates nodal involvement and stage IV disease implies distant metastases. The treatment for localized disease (stages I and II) has been described: wide local excision and lymph node dissection if indicated (sentinel node or complete lymph node dissection). High risk stage II and stage III patients, as well as all patients with stage IV disease, should be offered adjuvant treatment. For patients with metastases, surgery is palliative and systemic therapy is the mainstay of treatment.

Table 25.2. A simplified version of the TNM classification and staging scheme for malignant melanoma according to the American Joint Committee on Cancer (AJCC)

Stage	Definition
Stage I	T1-2 N0 M0*
Stage II	T3-4 N0 M0
Stage III	Any T with N1-3 M0
Sage IV	Any T with any N M1

T = Thickness of the Primary Tumor

T1	≤ 1 mm
T2	1-2 mm
T3	2-4 mm
T4	> 4 mm

-a: no ulceration or level II/III; -b: ulceration or level IV/V

N = Regional Lymph Node Status

N0	No nodal involvement
N1a	One microscopic positive node; N1b: one macroscopic positive node
N2a	2-3 microscopic positive nodes; N2b: 2-3 macroscopic positive nodes
N3	>4 nodes, matted nodes, or nodal involvement as well as in-transit mets

M = Distant Metastases

M0	No evidence of metastasis
M1a	Distant skin, subcutaneous, or nodal metastasis
M1b	Lung mets
M1c	All other visceral mets, or metastasis combined with elevated LDH

* T2b (ulcerated) is Stage II

A number of immunomodulators are currently being studied, but only interferon-alpha 2b (IFN alpha) has been shown to decrease the rate of disease recurrence and significantly impact relapse-free survival. High dose IFN alpha, when given within 56 days of surgery, decreases recurrence by 26%. Other agents, specifically interleukin-2 (IL-2), vaccines, and gene therapy are being investigated for their anti-tumor effects and show promise. Chemotherapy with dacarbazine (DTIC) has shown a response rate of 10-20% in certain studies, but its use is considered to be premature outside of clinical trials.

Isolated limb perfusion (hyper- and normothermic) with cytotoxic agents has been used for over 30 years in the treatment of melanoma. A prospective study from 1990 demonstrated increased five-year survival with hyperthermic limb perfusion with melphalan for extremity melanomas. In this study, even patients with intermediate thickness tumors (1.5-3 mm) benefitted from perfusion over conventional surgery alone. The value of limb perfusion is even more evident in patients with local recurrences and in-transit metastases. Severe adverse reactions include tissue breakdown and the need for fasciotomies due to post-perfusion leg compartment swelling.

Long-Term Care and Follow Up

All patients with a history of melanoma should undergo an annual skin exam for the the rest of their life. In addition, follow-up is based on the tumor depth at diagnosis; for patients with stage I disease (<1 mm thick, negative nodes) recommended follow-up is physical examination every 6 months with studies (LDH, LFTs and CXR) every other visit for five years. After 5 years, examine patients every 12 months and obtain studies every 2 years. For patients with intermediate lesions (1-4 mm thick) and negative nodes, physical exam is performed every 4 months and studies every other visit for the first 3 years. After 3 years, follow-up is the same as for thin (<1 mm) tumors. Patients with thick tumors (>4 mm) and negative nodes should be seen every 3-4 months, with studies every other visit, for the first three years. After 3 years, annual follow up with yearly studies is recommended. Patients with positive nodes (and any thickness tumor) should be seen every 3 months for the first three years and then every 6 months for 2 years (with studies every other visit), and yearly for the rest of their lives.

Pearls and Pitfalls

Certain anatomic locations deserve special mention:

- **Eyelid** lesions are rare and the importance of tumor thickness is uncertain. Recommended treatment is complete excision with a margin of normal skin. One series was unable to show any survival benefit with regional lymph node dissection.
- Tumors of the **ear** should be resected full-thickness (wedge resection) to the auditory canal. If a wedge resection is not performed, minimum treatment requires excision of the underlying perichondrium. Elective node dissection is of benefit in selected cases (the ear drains to the preauricular and postauricular, parotid, and jugulodigastric nodes).
- Patients with melanomas of the **neck and scalp** fare worse than those with tumors on the face or ears. Excision should include the underlying galea. The unpredictable lymphatic drainage dictates a complete neck dissection if indicated.
- **Subungual** melanomas and those of the distal digit are treated with amputation just proximal to the DIP joint (or IP joint of the thumb) regardless of tumor thickness (Fig. 25.2). Amputation at the metatarsal joints is indicated in the toes. Melanoma of the proximal fingers is treated by excision following the recommended soft tissue margins with flap or graft reconstruction.

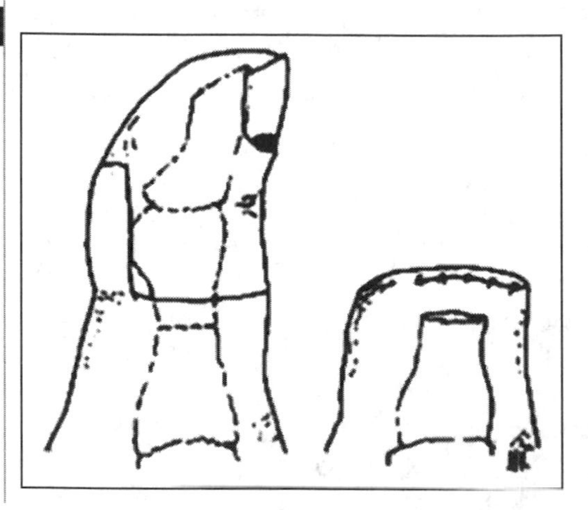

Figure 25.2. Amputation of the distal digit for subungal melanoma.

Suggested Reading

1. Balch CM et al. Efficacy of 2 cm surgical margins for intermediate thickness melanomas (1-4mm). Ann Surg 1993; 218:262.
2. Breslow A. Thickness, cross-sectional areas and depth of invasion in the prognosis of cutaneous melanoma. Ann Surg 1970; 172:902.
3. Clark Jr WH. A classification of malignant melanoma in man correlated with histiogenesis and biological behavior. In: Montagna W, Hu F, eds. Advances in Biology of the Skin, Vol VIII. Elmsford: Pergamon Press, 1967:621-647.
4. Evans GRD, Manson PN. Review and current perspectives of cutaneous malignant melanoma. J Am Col Surg 1994; 178:523.
5. Ho VC, Sober AJ, Balch CM. Biopsy techniques. In: Balch CM, Houghton AN, Sober AJ, Soong SJ, eds. Cutaneous Melanoma. 3rd ed. St. Louis: Quality Medical Publications, 1998, (Chapter 7).
6. Morton DL, Wen DR, Cochran AJ. Management of early stage melanoma by intra-operative lymphatic mapping and selective lymphadenectomy: An alternative to routine lymphadenectomy or "watch and wait." Surg Oncol Clin N Am 1992; 1:247.

Vascular Anomalies

Robert D. Galiano and Geoffrey C. Gurtner

Introduction

The discipline of vascular anomalies has matured with the acceptance of a common terminology that replaces the descriptive, flowery and imprecise jargon previously used to describe the vascular lesions seen by plastic surgeons. The publication by Mulliken and Glowacki of a rational classification system in 1982 and ultimately its incorporation within the International Society for the Study of Vascular Anomalies Classification System has permitted clinicians and researchers to apply scientific precision to the diagnosis, treatment and research on vascular anomalies. This chapter will describe the vascular lesions seen by plastic surgeons, with an emphasis on their diagnosis and treatment.

Vascular Anomalies: A Descriptive Classification

Vascular lesions can be broadly categorized as either **vascular tumors** (including hemangiomas, the emphasis of this chapter) or **vascular malformations** (see Table 26.1).

Hemangiomas are differentiated from vascular malformations by their characteristically rapid growth over several months followed by a prolonged involution. In contrast, vascular malformations never truly involute. At the core of all types of vascular lesions is the involvement of the endothelial cell in its pathogenesis. These lesions are differentiated from vascular malignancies by their benign behavior at the cellular level; they consist of mature, differentiated endothelial cells that behave aberrantly without evidence of dysplasia. Their dysregulation is more subtle than that found in a cancer, and this may be why, until recently, little progress was made on their etiology. Most are nonsyndromic, with a few important exceptions listed in Table 26.2.

Vascular malformations that concern plastic surgeons are always present, although not always apparent, at birth, whereas most of the vascular tumors present later in life. Congenital hemangiomas are an exception; their life cycle is such that they are present at birth, whereas the common infantile hemangioma grows in size after the neonatal period. Vascular malformations typically grow in pace with the patient, whereas the growth of the hemangioma outraces the development of the child. Despite such dissimilarities, the distinction between these two groups is not always so apparent (e.g., the violaceous hue of a deep hemangioma may be confused with a venous or lymphatic malformation). To further complicate matters, some vascular malformations may demonstrate periods of precipitous growth, often following trauma, sepsis, or hormonal fluctuations. Finally, there exists a cohort of vascular malignancies which can mimic benign vascular lesions. Therefore, it is best if these lesions are seen in the context of a multidisciplinary vascular anomalies clinic, or at

Table 26.1. Categorization of vascular lesions

Vascular Tumors
- Hemangioma of infancy
  - Superficial
  - Deep
  - Mixed
- Congenital hemangioma
  - Rapidly involuting congenital hemangioma (RICH)
  - Noninvoluting congenital hemangioma (NICH)
- Kaposiform hemangioendothelioma
- Tufted angioma
- Pyogenic granuloma
- Hemangiopericytoma

Vascular Malformations
- Simple malformation
  - Capillary (port-wine stain)
  - Venous
  - Lymphatic
  - Arteriovenous malformation (AVM)
- Combined malformation
  - Capillary-lymphatic-venous
  - Capillary-venous
  - Capillary-venous with AV shunting and/or fistulae
  - Cutis marmorata telangiectatic congenita

least by the practitioner who makes a dedicated investment in the care of patients with these lesions. Although benign in terms of cellular behavior, these lesions can have devastating sequelae on vital organs and may in fact be life-threatening.

Hemangiomas and Other Vascular Tumors

The common hemangioma of infancy (infantile hemangioma) is a unique lesion whose course of rapid growth and delayed involution has eluded a mechanistic explanation to date. It is unknown whether the origin is embryonic, maternal or placental, or whether it arises from a clonal endothelial or vascular progenitor precursor.

There are several statistics associated with these lesions which are worth mentioning:
- Occur in a 3:1 female: male ratio
- Present in 12% of Caucasian infants, with a lower incidence in other races
- Up to 30% incidence in premature infants
- 20% occur as multiple lesions
- 60% of these lesions are situated in the head and neck area
- 25% are found on the trunk
- 15% are present on the extremities

In addition, 5% of cases will be complicated by infection, bleeding or ulceration at some point during their proliferative phase. They may occasionally be associated with a low-grade consumptive coagulopathy but are not a cause of the Kasabach-Merritt syndrome. Although it is classically stated that 50% involute by age 5, true involution likely is complete before this. Most of the changes that occur

Table 26.2. Syndromes associated with vascular lesions

Syndrome	Description
PHACES	**P**osterior fossa brain anomalies, **H**emangiomas (usually facial), **A**rterial anomalies, **C**oarctation of the aorta and/or **C**ardiac defects, **E**ye abnormalities, **S**ternal/midline defects
Sturge-Weber syndrome	Capillary malformation in V1, occasionally V2 dermatome. Can have hypertrophy of underlying soft tissue and bone.
Klippel-Trenaunay syndrome	A capillary-lymphatico-venous malformation, associated with multiple port wine stains, hypertrophy of the underlying soft tissues and bone, often involving an extremity, and varicosities.
Parkes Weber syndrome	The triad of the Klippel-Trenaunay syndrome with the addition of vascular malformations.
Blue-rubber bleb nevus syndrome	Cutaneous venous malformations associated with venous malformations of the gastrointestinal tract.

past age 3-4 represent post-involutional changes and scar maturation, rather than true involution. Although in the past these lesions were described as either cavernous, strawberry or capillary, these ambiguous terms are now considered obsolete, and instead hemangiomas are described as either superficial, deep, or mixed, depending on their tissue depth.

The histologic appearance of the growing hemangioma of infancy will show rapidly proliferating endothelial cells, displaying numerous mitoses. This is distinct from vascular malformations, where the endothelial cells are flat and nonproliferating, forming ectatic vessels. Recently, the glucose transporter GLUT-1 has been shown to be a universal marker for hemangiomas of infancy; this immunohistochemical marker will in the future undoubtedly be utilized more frequently.

Another type of hemangioma is the congenital hemangioma. This lesion is invariably clinically apparent at birth and is divided into two subtypes: the rapidly involuting congenital hemangioma (RICH), which will involute by 10 months postpartum, and the noninvoluting congenital hemangioma (NICH) which persists. Other vascular tumors are occasionally seen by the plastic surgeon. A pyogenic granuloma is an angioma which appears at any age and is characterized by bleeding disproportionate to its size. Kaposiform hemangioendothelioma is responsible for most cases of Kasabach-Merritt syndrome. Notably, none of these hemangiomatous lesions express the histochemical marker GLUT-1.

Diagnosis

The diagnosis of hemangioma of infancy is made by clinical observation and a careful history. Pertinent points include the location of the lesion as well as other comorbid conditions. Occasionally, an imaging study may be used. Although computed tomography (CT) can be utilized, the most useful imaging study remains magnetic resonance imaging (MRI). A benefit of MRI is the delineation of flow characteristics, tissue involvement and penetration with excellent resolution of involved adjacent structures, without the risk of ionizing radiation. These advantages are most useful for imaging of hemangiomas of the head and neck, as well as visceral hemangiomas. The reader is referred to radiologic reviews for an overview of imaging characteristics specific to these lesions. A vessel density of greater than 5/cm^2

and a peak arterial shift greater than 2 kHz give a positive predictive value greater than 97% for a proliferative hemangioma. Other lesions that can be included in the differential diagnosis in an infant include sarcomas (rhabdomyosarcoma, fibrosarcoma, neurofibroma) as well as other vascular malformations.

Occasionally, the need for definitive distinction between a proliferating and an involuting hemangioma exists. Some markers of proliferating hemangiomas include VEGF, bFGF and PCNA. Involuting hemangiomas will have fewer levels of PCNA, more TIMP, endostatin, angiostatin and IL-12 along with an increased number of mast cells.

Treatment

There are several levels of urgency for treatment. At one end of the spectrum of surgical urgency lie those hemangiomas that mandate immediate intervention. These include airway-obstructing oro-tracheal-laryngeal hemangiomas, hemangiomas of the periorbit that can produce visual disturbances, including blindness, lesions that exhibit severe hemorrhage and ulceration, and the occasional liver hemangioma that can result in cardiac failure. Most hemangiomas, however, can be managed by observation, as they will eventually involute spontaneously. About half of hemangiomas of infancy will involute in a manner that leaves minimal or no disfigurement. However, the rest will leave a conspicuous scar and a residuum of fibro-fatty tissue that can be aesthetically disfiguring. These will require reconstructive interventions following involution to obtain a pleasing cosmetic result.

Recently, a fourth tier of lesions has been defined as those that are best prophylactically removed by surgery before resolution of the involution phase, most commonly to avoid the social-psychological trauma in affected children, or in order to prevent further distortion of structures and avert a more complex reconstruction. This includes hemangiomas of the nasal tip, where the delicate nasal cartilages are often involved. Although the management of hemangiomas requiring urgent treatment is clear, the controversy lies in management of the nonurgent lesions. Unfortunately, there have been few prospective studies to guide the management of these patients. A difficulty in managing these patients is the entrenched belief of many referring physicians, unfortunately communicated to the parents, that hemangiomas should not be treated until involution is complete. One recent useful tool that may assist in therapeutic decision making stratifies patients into groups that are high or low risk. Those hemangiomas with a significant dermal component will likely leave greater residual cutaneous deformations with post-involution changes and scarring. The location of hemangiomas is also a significant risk factor, with those of the face having high likelihood of surgical intervention due not only to the involvement of eyes, ears and nose, but also because those in a beard-like pattern have a likelihood of airway involvement. Future research should further define subgroups that will best respond to specific interventions, whether medical or surgical.

Those hemangiomas that need to be urgently treated are first treated with high doses of systemic steroids and intra-lesional steroid injections. Second-level therapeutics for nonresponders or poorly tolerant patients include interferon-α or occasionally vincristine. Interferon-α is less commonly used than in the past because of its side effects, the most serious of which is spastic diplegia. Orbital and airway lesions can be treated with laser cauterization and surgical debulking as a second choice.

The treatment of less catastrophic lesions begins with supportive or hygienic measures; this is particularly important for ulcerated lesions which may benefit from topical analgesics and dressings. Pharmacologic treatment is brought into play for the above-mentioned absolute and relative indications. Small lesions can be treated with direct intralesional infiltration of a corticosteroid. Multiple injections (4-7) are typically needed. For more extensive hemangiomas, prednisone given at 2-3 mg/kg daily is initiated; for life-threatening lesions, intravenous steroids are given, as stated above. Eighty percent of hemangiomas of infancy will respond to corticosteroid therapy. Responses should be seen within 2 weeks, and the steroids should then be tapered gradually over a period of several months. The clinician should be on guard for nonresponsiveness or rebound growth.

Laser therapy can occasionally prove useful for superficial hemangiomas and is perhaps most useful for telangiectasias remaining in the residual tissue following involution of the hemangioma. Mixed and deep lesions will have a greater tendency to result in distorting fibro-fatty sequelae, which may require excision or reconstruction. Surgery, as mentioned above, is definitively indicated for removal of a symptomatic hemangioma, as in the nasal tip. Uncomplicated hemangiomas are rarely resected; involuting hemangiomas may be resected for psychosocial reasons before the child reaches school age or, if in the opinion of the surgeon, the remaining scar would be persistent and would eventually require surgical removal anyway.

The other vascular tumors are treated differently. Pyogenic granulomas will typically require curettage and cauterization to stop bleeding. A NICH will require surgical excision, as it will not involute. Kasabach-Merrit syndrome is a consumptive coagulopathy in the setting of an enlarging soft tissue mass, usually a kaposiform hemangioendothelioma (KHE); the patient will display a thrombocytopenia and bleeding diathesis that may be unresponsive to platelet transfusions. These patients require admittance to a hospital, steroid infusions and antiplatelet/antifibrinolytic medications, with or without embolization of the lesion. Interferon-α is not as effective as other antineoplastics such as vincristine and cyclophosphamide in the treatment of a KHE involved in the Kasabach-Merritt syndrome. The KHE lesion should be surgically resected when possible; otherwise, these lesions are difficult to manage medically. As mentioned above, this syndrome does not occur in hemangioma of infancy; it is found only in tufted hemangiomas or kaposiform hemangioendotheliomas.

Vascular Malformations

The classification of the various types of vascular malformations was described above. These lesions are brought to the attention of the plastic surgeon for several reasons. The mainstay of treatment, except for small ones in the extremities, is a combined approach with radiologic-guided superselective embolization, and surgical debulking when necessary. Properly done, risks are minimized, although these can include ulceration, infection and occasionally catastrophic complications such as limb loss.

Unfortunately, due to their extensive intercalation and incorporation within tissues, dissection planes are nearly impossible to reliably discern. These tend to be space-occupying lesions, and recidivism is common. Like endothelial cells elsewhere in the body, the endothelial cells in these lesions are quiescent until stimulated to proliferate by injury, distressingly often by a surgical insult. This accounts for their recalcitrance following monotherapy such as surgery. Therefore, the treatment goals need to be clearly explained to the patient. They rarely include cure and instead are mostly palliative to control the complications of ulceration, bleeding, and for improved hygiene. Reoperation is unfortunately often the rule, rather than the exception.

Vascular malformations can be categorized as arteriovenous (AVM), venous, capillary, or lymphatic malformations. Unlike hemangiomas of infancy, these lesions do not display a proclivity for either sex.

Diagnosis

The history and physical can guide the diagnosis. Capillary malformations are typically noted at birth; they include the well-known port wine stain. Eighty percent of lymphatic lesions will become apparent either at birth or during the first year of life. Venous malformations will be noted at any time from birth to adulthood, a fact which is important when evaluating the young adult with prominent leg veins who presents for sclerotherapy. They are typically darker lesions seen on the skin or mucosal surfaces. Arteriovenous malformations can become evident during periods of hormonal fluctuations, such as puberty or pregnancy. Lymphatic malformations are composed of dilated lymphatic channels and vesicles; these lesions will typically swell with an infectious episode. They may be either macrocystic or microcystic.

Plain radiographs are only utilized when there exists a potential for bone or joint involvement. MRI and ultrasound can often play a complementary role in the treatment of these lesions. Ultrasound can be a confirmatory screening test to demonstrate the vascular nature of a soft tissue mass, and MRI will show the lesion's relationship to adjacent structures as well as feeding and draining vessels. Because of these benefits, MRI is clearly the first-line diagnostic tool. However, angiography plays a larger role in the diagnosis and treatment of vascular malformations than in hemangiomas. The differential diagnosis of an AVM includes malignant vascular tumors such as angiosarcoma and rhabdomyosarcoma. When the benign nature of these lesions cannot be definitively ascertained through an imaging approach, a biopsy of the lesion is warranted.

Treatment

The optimal treatment should be tailored to the type of lesion and its location. Because of the wide variety of presenting symptoms and multitude of lesions, this is a discipline that is continuing to evolve. There are some lesions that require immediate treatment. Although the plastic surgeon may not be directly involved in the treatment phase of these lesions, he should be aware of what constitutes a life-threatening lesion so that the patient can be promptly referred if necessary.

Venous malformations are treated mainly by sclerotherapy (occasionally by an interventional radiologist), typically when the lesion becomes symptomatic. Debulking surgery, when indicated, is performed 6-8 weeks following sclerotherapy. Occasionally, lesions in or near vital structures (such as the hand/digits) are treated by surgical excision alone.

Lymphatic malformations are notoriously difficult to treat. Most patients present for treatment of cellulitis and will benefit initially from elevation of the affected body part and intravenous antibiotics. Macrocystic lymphatic malformations are best treated with sclerotherapy. Most lesions will require surgical resections, a procedure which is tedious, bloody and fraught with recidivism. Because of the often bloody procedures, they are best staged.

Capillary malformations can often be treated with laser therapy. Up to 75% of patients will benefit from this therapy. As these lesions can be accompanied by underlying soft tissue and bony hypertrophy, surgical resections of the involved tissues may be indicated.

AVMs are the most treacherous of vascular malformations to treat. They are treated when they become symptomatic; unfortunately, the herald symptoms may include life-threatening bleeding or heart failure. There are some guidelines to treatment of the AVM which all members of a multidisciplinary vascular anomalies team should be aware of. In terms of embolization of these lesions, proximal feeder vessels should never be embolized, as the lesion will only become more vascular via reactive neovascularization. In addition, as these lesions may require multiple interventions, a poorly planned embolization will make it difficult or impossible to perform future embolizations. These lesions will require wide, deep, extensive resections, often necessitating a microvascular free flap to achieve wound closure.

Pearls and Pitfalls

The most important point in treating vascular anomalies is establishing the correct diagnosis. Incorrect treatment for a lesion can have dire, even life-threatening, consequences. Radiologic imaging by a specialist radiologist is essential to preventing the above. Furthermore, there exist a set of vascular malignancies which demand treatment and not a "watchful waiting" approach. Even within the more benign subset of patients with hemangiomas, a sizable proportion will require some surgical treatment even if the lesions involute, particularly on the face. This is generally not appreciated by the referring physicians. In addition, these lesions often take years to involute, and the parents need to be so informed.

The use of novel markers, with GLUT-1 being a paradigm for this, will in the future enable more precise identification of the type and life-stage of these lesions. Novel approaches using anti-angiogenic factors may ease the effects of steroids or other anti-neoplastic regimens currently used to treat these lesions, and may be particularly beneficial for patients with vascular malformations, many of which end up as medical and surgical orphans. Finally, it may be possible to block vascular stem cell incorporation within these lesions, or conversely, exploit the proclivity of these lesions to attract vascular progenitors by genetically modifying vascular stem cells to impair the growth of hemangiomas or accelerate their involution.

Suggested Reading

1. Hand JL, Frieden IJ. Vascular birthmarks of infancy: Resolving nosologic confusion. Am J Med Genetics 2002; 108:257.
2. Mulliken JB, Glowacki J. Hemangiomas and vascular malformations in infants and children: A classification based on endothelial characteristics. Plast Reconstr Surg 1982; 69:412.
3. Chang MW. Updated classification of hemangiomas and other vascular anomalies. Lymphatic Res Biol 2003; 1:259.

Chapter 27

Skin Grafting and Skin Substitutes

Constance M. Chen and Jana Cole

Introduction

Skin is the largest organ in the human body, measuring approximately 1.6-1.8 square meters in the adult. While its function is often taken for granted, any violation quickly reveals itself in pain and suffering for the owner, and extensive damage is life-threatening. Bacteria, viruses, fungi and harmful chemicals must penetrate the skin before causing harm to deeper tissues. By providing a barrier to the outside world, the integument protects internal organs against injuries and also prevents insensible fluid losses. In addition, skin helps to regulate body temperature through the activity of sweat glands and blood vessels; nerves in skin also receive stimuli that are interpreted by the brain as touch, heat, cold and vibration.

Relevant Anatomy

Skin is composed of three layers: **epidermis, dermis** and **subcutaneous tissue**. As the deepest layer of the skin, the subcutaneous tissue connects the dermis to the deeper structures of the body, insulates the body from cold and stores energy in the form of fat. The integument varies in thickness depending on anatomic location, sex and the age of the individual. On the back, buttocks, palms and soles of the feet, skin can be as thick as 4 mm or more. In marked contrast, the skin of the eyelids, postauricular and supraclavicular region may be as thin as 0.5 mm. In all anatomic locations, female skin is characteristically thinner than male skin. Both the young and the old also have thin skin; the thin integument of children thickens with age until it reaches a peak in the fourth or fifth decade of life, when it begins to thin again. In the elderly, thin skin primarily represents a dermal change, with a loss of elastic fibers, epithelial appendages and ground substance.

The epidermis has no blood vessels, and it can only receive nutrients by diffusion from the underlying dermis through the basement membrane. On the other hand, the dermis, which is composed of the superficial papillary dermis and the deeper reticular dermis, contains capillaries and larger blood vessels as well as connective tissue, elastic fibers, collagen, fibroblasts, mast cells, nerve endings, lymphatics, ground substance and epidermal appendages. The epidermal appendages include sebaceous (holocrine) glands, eccrine and apocrine sweat glands and hair follicles. The hair follicles are epithelial structures lined with epithelial cells that can divide and differentiate. Found deep within the dermis and in the subcutaneous fat deep to the dermis, they are responsible for the ability of the skin to resurface even very deep cutaneous wounds that are nearly full thickness.

At times, however, skin cannot regenerate sufficiently, aesthetically and functionally to cover an open wound. Examples include full-thickness or deep partial-thickness burns as well as large exposed surfaces from surgical or traumatic extirpations. Skin grafts and skin substitutes are commonly used to provide coverage over a broad spectrum of open soft tissue defects. Unlike surgical flaps, however, skin grafts and skin substitutes do not have their own blood supply and are limited by their thinness. Thus, the recipient bed must be well-vascularized in order to provide the transferred skin with nutrients to survive. Tendon may be grafted if the paratenon is intact; likewise, bone may be grafted if there is intact periosteum. On the other hand, irradiated tissue is not a good candidate for a skin graft, as radiation often leads to capillary depletion and inadequate nutrition to the transferred skin. Furthermore, open wounds with exposed bone often require more soft tissue padding than a skin graft is able to provide.

Types of Skin Grafts

Skin harvested from another species is called a **xenograft**. Skin harvested from another person from the same species is called an **allograft**. Skin that is harvested from one part of the body and used to cover another part of the same person's body is called an **autograft**. While xenografts and allografts may be used to provide temporary coverage over open wound defects, their MHC class II mismatch will cause rejection over time. They are very useful as biological dressings, however, to allow a recipient bed to improve before it is ready for an autograft. In particular, allografts are useful to test a questionable recipient bed. Ultimately, however, an autograft is necessary for permanent coverage of the wound.

Split-Thickness vs. Full-Thickness

There are two primary types of skin autografts: **split-thickness skin grafts (STSG)** and **full-thickness skin grafts (FTSG)**. A STSG includes the entire epidermis as well as varying portions of the underlying dermis. A FTSG includes both the epidermis and the entire dermis and thus retains more of the normal characteristics of skin including color, texture and thickness. Since the FTSG contains more dermis, it has greater primary contraction than a STSG; however, full-thickness dermis impairs secondary wound contraction to a greater degree than a split-thickness graft. Of course, thicker grafts demand greater vascularity in the recipient bed in order to maintain cellular respiration.

Whether to use a full- or a split-thickness skin graft depends upon the condition and location of the defect, its size, as well as aesthetic considerations. Areas of high cosmetic concern, such as visible parts of the face or the hands, may benefit from a FTSG. An example is using full-thickness postauricular skin to cover the thin skin of the eyelid. In addition, areas that tolerate little wound contraction, such as the interphalangeal joints or antecubital fossa, do better with a full-thickness skin graft.

STSG-Advantages

STSG are much more versatile. Since a skin graft does not have its own blood supply, any condition that impairs the means by which the graft is nourished and oxygenated (**imbibition, inosculation** and **neovascularization**-described below) may threaten graft take. Not only does this include infection and poor vascularity in the underlying recipient bed, but any condition that prevents firm adherence of the skin graft to the recipient bed. Such conditions include seroma, hematoma, foreign body, graft wrinkling or tenting, shear forces, necrotic material and bumpy irregularities in the underlying wound surface. STSG will do better than FTSG in these conditions. Furthermore, STSG can be used to cover a larger surface area. Donor sites for STSG are able to heal spontaneously due to the epidermal appendages that remain in the unharvested dermis. Additionally, once the donor site has healed, it may be used again to harvest more skin.

STSG-Disadvantages

While the STSG has broader use, it also has its drawbacks. A healed STSG does not accurately resemble uninjured skin. The lack of epidermal appendages gives it an abnormally shiny and smooth appearance, and the color is often noticeably lighter or darker than that of the surrounding skin. In addition, while the thinner graft may take more easily, it also fails to prevent wound contraction, which can lead to functional and aesthetic problems. Thinner grafts are more fragile from lack of bulk and do not tolerate subsequent radiation therapy well. Despite its aesthetic deficiencies, STSG are used for many purposes including replacing lost skin in burn injuries, resurfacing large wounds and muscle flaps, lining cavities and covering mucosal deficits.

Preparing the Recipient Bed

The key factor in the successful take of skin grafts is a healthy well-vascularized recipient bed. Necrotic tissue must be completely debrided, as decaying tissue contains no blood supply and produces toxins that retard wound healing. Likewise, the wound must be free of infection, defined as bacterial counts less than 100,000/cm^2, since bacterial overgrowth can infect and destroy the skin graft. Devascularized tissues, such as bone without periosteum, cartilage without perichondrium, tendon without paratenon, and nerve without perneurium, need an overlying layer of granulation tissue in order to be grafted. While skin grafts can survive on periosteum, perichondrium, peritenon, perineurium, dermis, fascia, muscle, fat and granulation tissue, irradiated wounds have a compromised blood supply and may not be able to nourish a skin graft. Finally, vasculopaths with arterial insufficiency or venous stasis ulcers need to correct underlying vascular problems to support a skin graft.

A contaminated or chronic wound often needs to undergo a prolonged course of wound care in preparation for skin grafting. Multiple surgical debridements may be required either at the bedside or in the operating suite. Enzymatic debriders may be used to dissolve necrotic tissue. Topical or systemic antibiotics may be necessary to reduce the bacterial count. Wet-to-dry dressing changes or vacuum-assisted closure therapy should be carried out until the underlying recipient bed appears clean, healthy and red with punctate bleeding. The importance of adequate recipient bed preparation cannot be overstated. Not only does graft failure prolong the time the primary wound stays open, but it creates a secondary wound at the donor site that will also need to heal. Indeed, the exposure of raw nerve endings in the donor site defect often makes it more painful to the patient than the original wound itself.

Operative Technique

Once a clean, well-vascularized recipient bed has been achieved, the patient may be taken to the operating room for skin grafting. The wound is prepped and draped in the usual sterile fashion. Any overlying layer of granulation tissue should be scraped off with a blunt instrument; the back end of a pair of forceps or the handle of a scalpel work nicely. The top layer of contaminated tissue should be eliminated and healthy bleeding tissue exposed. If the underlying tissue is covered with eschar or severely burned skin, then the overlying dead tissue needs to be excised sharply.

Pre-Grafting Excision

Surgical excision may be either fascial or tangential. Fascial excision removes both viable and nonviable tissue down to the muscle fascia and yields an easily defined plane that is well-vascularized. Blood loss is minimized with fascial excision because vessels are more easily identified, and the entire excision can be performed with electrocautery. The excision inevitably includes healthy, viable subcutaneous tissue, however, and a contour deformity often results. To address these issues, tangential excision is designed to limit excision only to nonviable tissue, leaving the healthy tissue intact, and it also minimizes the contour deformity. Instruments used for tangential excision include the Weck/Goulian blade and the Watson knife, which are handheld knives that tangentially excise slices of eschar until a layer of viable tissue is seen with diffuse punctate bleeding. Sharp excision of eschar makes it more difficult to control diffuse bleeding, however, and it can also be difficult to assess the suitability of the underlying fat for accepting a graft. Hemostasis can be obtained with epinephrine-soaked Telfa pads, topical pressure and, if necessary, electrocautery. Only after complete hemostasis has been obtained and the wound bed defined, can the skin be grafted.

FTSG Harvest

A full-thickness skin graft is harvested with a scalpel. The wound defect is measured and a pattern made, and then this is traced over the donor site. The FTSG is then harvested sharply by dissecting the dermis off of the underlying subcutaneous tissue. Any fat that remains on the graft must be removed sharply with scissors, as fat is poorly vascularized and will prevent firm adherence between the graft dermis and the recipient bed. All yellow fat must be trimmed until all that remains is the shiny white undersurface of the dermis. The donor site is usually closed primarily or, occasionally, with a STSG.

STSG Harvest

A split-thickness skin graft is most commonly harvested with a dermatome. Dermatomes may be air-powered, electric, or manually operated, and they facilitate skin harvest of uniform thickness. Frequently used dermatomes include the Padgett-Hood, Zimmer, Brown and Davol-Simon. Care must be taken to set the desired thickness of the skin graft prior to harvest, and then to double-check the setting again prior to operating the device. A 15-blade scalpel simulates a thickness of 0.015 inches and can be run along the blade to check a uniform depth setting. The blade must also be checked for proper orientation and the appropriate guard width chosen.

After the wound is measured, the surgeon calculates the area of coverage needed and marks the donor site. The most common donor site is the anterior or lateral thigh, as it is easily accessible and straightforward to camouflage, although any region with available skin may be harvested. The donor site is often lubricated with mineral oil or Hibiclens solution although other substances can be used according to surgeon preference. One or more assistants press along the edges of the donor site to create a firm, flat surface. The dermatome is then moved along the donor site at a 45-degree angle with equal pressure on all areas. An assistant picks up the harvested skin with forceps to prevent the graft from getting caught in the instrument. Multiple strips of skin graft should be harvested without any gaps between adjacent donor sites. When the skin has been harvested, epinephrine-soaked Telfa pads are placed on the donor site to obtain hemostasis.

To Mesh or not to Mesh?

STSG may be meshed or unmeshed. Meshed grafts are much more versatile, and are better able to survive under less than ideal conditions. The mesh allows serum or blood to drain, minimizing seroma or hematoma. Meshed grafts take more easily on bumpy surfaces and can be expanded by stretching the width. Meshed grafts will heal with an unsightly cobblestone appearance, however, as the meshed areas heal by reepithelialization. Unmeshed graft, known as sheet graft, maintains a smoother appearance. The primary advantage of sheet graft is cosmetic, however, and it requires better conditions to survive. The underlying wound bed must be smooth and flat, and sheet graft must be inspected on postoperative day one for hematoma and seroma. If any fluid collections are observed, they must be "deblebbed" by either gently rolling out the fluid or clot with a cotton-tip applicator or by cutting slits in the graft itself to release the fluid. Occasionally, sheet graft will be "pie-crusted" in the operating room by creating slits that allow fluid to drain. This results in an intermediate appearance between the perfectly smooth sheet graft and the rough, pebbly meshed graft.

Graft and Donor Site Dressings

Once the skin graft is placed on the wound bed, it may be secured by staples or by small dissolvable sutures. The intention is to maximize contact between the graft and the recipient bed and to prevent shearing. The graft may be dressed in a variety of ways. Traditionally, a bolster dressing is placed to keep the skin graft firmly adherent to the wound bed. This may be created by placing Xeroform gauze wrapped around cotton balls over the graft, which is then secured with nylon sutures tied firmly over the dressing. Another method is to place Xeroform and Reston self-adhering foam over the graft, and securing the dressing with staples.

The donor site of a STSG may be dressed in a variety of ways. The donor site may be left open to air, but this is associated with prolonged healing time, more pain and a higher risk of complications. Dressings that promote a moist wound environment are associated with faster healing. Op-Site, Tegaderm and Jelonet are all associated with rapid, relatively painless healing and low infection rates. Xeroform, Biobrane and DuoDERM have slightly longer healing times, more pain and more infection. The donor site dressing is removed once the skin reepithelializes. Prior to healing, the donor site contains exposed nerve endings that often make it more painful that the grafted wound bed itself. As healing progresses, the donor site becomes less painful and more itchy.

Skin Graft Healing

As the skin graft becomes incorporated into the recipient bed, it undergoes three predictable stages of graft "take." The first stage, **plasmatic imbibition**, consists of simple diffusion of nutrients from the recipient bed to the skin graft. Lasting 24-48 hours, imbibition prevents the graft from drying out and keeps the graft vessels patent so that the graft can survive the immediate postgraft ischemic period. Grafts appear plump during this time and can add as much as 40% to their pregraft weight through fluid movement from recipient bed to graft.

After 48 hours, the second phase, **inosculation**, and the third phase, **revascularization**, occur to restore blood flow to the graft. During inosculation, capillary buds from the recipient bed line up with graft vessels to form open channels. This establishes blood flow and allows the skin graft to become pink. The connection between graft and host vessels develops further as the graft revascularizes and newly formed vascular connections differentiate into afferent and efferent vessels between days four and seven. Lymphatic drainage is present by the fifth or sixth postgraft day, and the graft begins to lose weight until it reaches its pregraft weight by the ninth day.

Graft Contracture

Once the skin graft is harvested, it also begins to shrink. Primary contraction is passive and occurs immediately after harvest. A FTSG loses about 40% of its original area; a medium-thickness skin graft about 20%; and a thin STSG about 10%. Secondary contraction occurs when the skin graft is transferred to the recipient bed, and it is only seen in STSG. The amount of wound contraction depends upon the amount of dermis in the skin graft. The more dermis there is in a skin graft, the less the wound bed will contract. Secondary contraction is minimal in a FTSG, which is actually able to grow after it heals. Wound contraction can be useful in reducing wound size, but a contracted wound can also be tight and immobile, leading to distortion of surrounding normal tissue.

Return of Function

As the skin graft heals, it reinnervates and undergoes color changes. Nerves grow into skin grafts from the wound margins and recipient bed, and skin grafts may begin to show sensory recovery anywhere from 4-5 weeks to 5 months postgraft. Full return of two-point sensation is usually complete by 12-24 months, but temperature and pain sensation may never return. With regard to color, grafts from the abdomen, buttocks and thigh tend to darken over time, whereas grafts from the palm tend to lighten. Thin grafts are also usually darker than thick ones.

Depending on thickness, skin grafts may also regain function of transferred epithelial appendages, such as sweat glands, sebaceous glands and hair follicles. Usually, only FTSG are reliably capable of sweat production, oil secretion and hair growth. Sweat production depends on the number of sweat glands transferred during grafting and the extent of sympathetic reinnervation. The graft will sweat according to incoming sympathetic nerve fibers so that a graft on the abdomen sweats in response to physical activity while the same graft on the palm will sweat in response to emotional stimuli. With regard to oil secretion, sebaceous gland activity can be seen in both full- and split-thickness grafts, but they are usually not functional in thin STSG. Similarly, FTSG produce hair, while STSG produce little or no hair. In addition, inadequate revascularization or disruptions in graft take will damage graft hair follicles and result in sparse, random and unpigmented hair growth.

Skin Substitutes

When the amount of skin loss has been too great to allow adequate replacement with skin autografts, it may be necessary to use tissue-engineered skin substitutes (Table 27.1). Skin substitutes are designed to be left in place for long periods of time, and may be autologous, allogeneic, xenogeneic, or recombinant. They can either be used for wound coverage or wound closure. Materials intended for wound coverage provide a barrier against infection, control water loss, and create an environment suitable for epidermal regeneration. Materials used for wound closure restore the epidermal barrier and become incorporated into the healing wound.

Table 27.1. Skin substitutes

Product	Indications for Use	Clinical Considerations
Allograft	Temporary covering for excised burns or wounds; may be used as a protective covering over autografts	• If adherent to the wound bed, can remain intact for 2 to 3 weeks
Biobrane	Temporary covering for clean, debrided superficial and partial-thickness burns and donor sites; may also be used as a protective covering over meshed autografts	• Minimizes painful dressing changes • May be left open to air • Remains in place until wound is healed, then is trimmed away
TransCyte	Temporary covering for mid-dermal to indeterminate-depth burns that typically require debridement and may be expected to heal without surgical intervention; also indicated as a temporary covering for surgically excised full-thickness and deep partial-thickness burns prior to autografting	• Minimizes painful dressing changes • Remains in place until wound is healed • Contraindicated in patients with a known hypersensitivity to porcine dermal collagen or bovine serum albumin
Apligraf (Graftskin)	For use in conjunction with standard compression for treatment of noninfected partial- and full-thickness skin ulcers due to venous insufficiency of greater than 1 month's duration that have not adequately responded to conventional ulcer therapy; for use with conventional diabetic foot ulcer care in the management of diabetic foot ulcers of greater than 3 weeks duration	• Does not require additional autografting • Contradicted for use on clinically infected wounds, in patients with a known allergy to bovine collagen, and in patients with a known hypersensitivity to the contents of the agarose shipping medium
Dermagraft	For treatment of foot ulcers in diabetic patients (Canada and United Kingdom); clinical trials ongoing in the United States for treatment of diabetic foot ulcers	• Applied weekly for up to 8 weeks to promote healing • Does not require additional autografting
AlloDerm	For treatment of full-thickness burns and use in plastic and oral surgery	• Immunologically inert • Allows for immediate wound closure with thin epidermal autografting during same procedure • Postoperative dressing may remain in place for 14 days or longer

continued on next page

Wound Coverage

Skin substitutes used for wound coverage include Biobrane, TransCyte, cultured epidermal allogeneic keratinocytes, Dermagraft and Apligraf.

Table 27.1. Continued

Product	Indications for Use	Clinical Considerations
Integra	For treatment of life-threatening, full-thickness or deep partial-thickness skin burns where sufficient autograft is not available at the time of excision or not desirable due to the physiologic condition of the patient	•Provides immediate post-excisional physiologic wound closure •Allows the use of a thin epidermal autograft of 0.005 inch •Must remain in place for 21 days before epidermal autografting; must be protected against shearing forces and mechanical dislodgment •Contraindicated in patients with a known hypersensitivity to bovine collagen or chondroitin materials •Contraindicated in the presence of infection
Epicel (cultured epitheilial autografts)	For treatment of deep dermal or full-thickness wounds where sufficient donor sites are unavailable.	•Cultured keratinocytes can be grown in 3 weeks •Graft take varies from poor to fair •Process is expensive •Grafts extremely fragile and may remain so for months after grafting

Biobrane is a bilaminar material made of nylon mesh bonded to a thin, semipermeable silicone membrane and used as a temporary skin replacement for superficial partial-thickness burns or skin graft donor sites. It eliminates the need for dressing changes and reduces the length of inpatient treatment.

TransCyte is Biobrane with the addition of neonatal fibroblasts seeded to the collagen-coated nylon mesh. The benefits are similar to Biobrane, but TransCyte is considerably more expensive. Cultured allogeneic keratinocytes are obtained from neonatal foreskin or elective surgical specimens, and are used to cover burn wounds, chronic ulcers and skin graft donor sites. While they do not achieve wound closure, they can survive up to 30 months and produce growth factors that facilitate host dermal and epidermal cell proliferation and differentiation, but they are thin, fragile and require meticulous wound care to survive.

Apligraf and **Dermagraft** are multilaminar materials designed to overcome the fragility of cultured allogeneic keratinocytes. Apligraf is a type I bovine collagen gel with living neonatal allogeneic fibroblasts overlaid by a cornified epidermal layer of neonatal allogeneic keratinocytes, and it is used to treat chronic ulcers, pediatric burns, epidermolysis bullosa and full-thickness wounds from Mohs' surgery pending definitive repair. Dermagraft is a cryopreserved dermal material made up of neonatal allogeneic fibroblasts on a polymer scaffold, and it stimulates ingrowth of fibrovascular tissue from the wound bed and reepithelialization from the wound edges. It is used to promote healing of chronic lesiongs and to replace lost dermal tissue beneath meshed split-thickness skin grafts on full-thickness wounds.

Wound Closure

Skin substitutes used for wound closure include Alloderm, Integra and cultured epithelial autografts.

Alloderm is acellular deepithelialized human cadaver dermis, which is used as a dermal graft in full-thickness or deep partial-thickness wounds. A STSG must be placed over Alloderm in a one- or two-stage procedure.

Integra is a bilaminar skin substitute made up of a cross-linked bovine collagen-glycosaminoglycan matrix coated with silicone elastomer barrier on one side. Integra is used in a two-stage procedure, in which a thin split-thickness skin graft is applied in the second stage after the silicone "epidermis" is removed. It is very reliable, with good elasticity and cosmesis, and low risk of infection, but it requires two operations and is expensive.

Epicel, or **cultured epithelial autografts** were developed in the 1970s, and have been used for burns, chronic leg ulcers, giant pigmented nevi, epidermolysis bullosa and large areas of skin necrosis. A one square centimeter skin harvest is expected to grow to one square meter in 21 days. Cultured epithelial autografts must be applied on a wound bed with granulation tissue or muscle fascia for proper take. Sheets are fragile, however, and often result in friable, unstable epithelium that spontaneously blisters, breaks down, and contracts long after application. Cultured epithelial autografts are also very sensitive to infection and are only able to tolerate maximum bacterial counts of 100-1000/cm^2, compared to 10,000-100,000/cm^2 for standard split-thickness skin grafts. Finally, cultured epithelial autografts are extremely expensive.

Pearls and Pitfalls

Over time, the vacuum-assisted closure (VAC) device is gaining a reputation as the ultimate bolster dressing. The VAC bolster dressing consists of a sponge cut in the shape of the graft, and then sealed and placed to 75-125 mmHg continuous suction. Usually, an intervening layer such as Conformant or Adaptik may be placed between the skin graft and VAC to prevent the graft from lifting off the bed when the dressing is removed. We have found that the VAC promotes graft adherence to the recipient bed and removes any accumulating serous fluid or blood. When the VAC is used correctly, STSG take will approach 100%.

Suggested Reading

1. Branham GH, Thomas JR. Skin grafts. Otolaryngol Clin North Am 1990; 23(5):889-97.
2. Gallico IIIrd GG. Biologic skin substitutes. Clin Plast Surg 1990; 17(3):519-26.
3. Jones I, Currie L, Martin R. A guide to biological skin substitutes. Br J Plast Surg 2002; 55(3):185-93.
4. Hauben DJ, Baruchin A, Mahler D. On the history of the free skin graft. Ann Plast Surg 1982; 9(3):242-5.
5. Petruzzelli GJ, Johnson JT. Skin grafts. Otolaryngol Clin North Am 1994; 27(1):25-37.
6. Place MJ, Herber SC, Hardesty RA. Basic techniques and principles in plastic surgery (Skin grafting). Grabb and Smith's Plastic Surgery. 5th ed. Philadelphia: Lippincott-Raven, 1997:17-9.
7. Ratner D. Skin grafting. From here to there. Dermatol Clin 1998; 16(1):75-90.
8. Thornton JF. Skin grafts and skin substitutes. Selected Readings in Plastic Surgery. 2004; 10(1):1-24.

Chapter 28

Burns: Initial Management and Resuscitation

Baubak Safa and Mark Sisco

Epidemiology

Of the 2-3 million thermal injuries that occur in the United States each year, approximately 100,000 require hospital admission to a burn unit. Furthermore, 5,000-6,000 people die as a direct result of thermal injury in this country. Although thermally injured patients require 1-1.5 days in the hospital per percent total body surface area (TBSA) burned, this period only represents a small fraction of the total treatment for these patients which also includes rehabilitation and physical therapy, reconstruction and readaptation.

Burn victims tend to reflect four general populations: The very young, the very old, the very unlucky and the very careless. Indeed, National Burn Information Exchange (NBIE) data indicate that up to 75% of burn injuries result from the victim's own actions.

Scald burns represent the most common burns in the United States. Fifty percent of these burns occur in children in the kitchen, followed by burns from hot water in the bathroom. The burn depth is directly proportional to the time of exposure of the hot liquid and therefore, this aspect of the history is important in evaluating a burn patient. Many adult scald burns are also caused by automobile radiator injuries.

Heating unit failure is the most common cause of residential fires. The requirement of smoke detectors in new construction buildings has resulted in increased warning time and a decreased chance of death. A key part of the history in patients involved in residential fires is the presence or absence of ignition of their clothing. Full-thickness burns are six times more likely when clothing ignition is present. Furthermore, mortality increases approximately four times when clothing is ignited.

The Burn Wound

There are three zones of burn. The **zone of coagulation** is the central area and is composed of nonviable tissue. The **zone of stasis** surrounds the central zone of coagulation. The adequacy of the initial burn resuscitation will affect the extent and outcome of this zone. Typically, blood flow is initially present, but ischemia and hypoperfusion prevail in the subsequent 24 hours, especially with inadequate resuscitation. The **zone of hyperemia** surrounds the zone of stasis and contains viable tissue.

Burn depth, along with the extent of the burn (TBSA) and age of the patient, are primary determinants of mortality following thermal injury. The depth of injury is also a major determinant of a patient's long term function and appearance. The varying depth of a burn as well as the changing perfusion of the zone of stasis render the precise determination of burn depth difficult in the first 24 to 48 hours. Therefore, the most accurate method of determining the depth of burn is clinical assessment based on experience.

Practical Plastic Surgery, edited by Zol B. Kryger and Mark Sisco. ©2007 Landes Bioscience.

Burn Classification

Superficial Burns (First Degree)

Superficial burns are easily diagnosed. The typical superficial burn is a bad sunburn with erythema and mild edema. The area involved is tender and warm to the touch and there is rapid capillary refill. Topical antimicrobial therapy is unnecessary and all layers of the epidermis and dermis are intact. Healing should occur within five to seven days and some superficial epidermolysis may be seen. These burns are not included in the assessment of the TBSA of a burn victim.

Partial-Thickness Burns (Second Degree)

Partial-thickness burns may involve a wide spectrum of dermal injury and present a diagnostic dilemma. Superficial partial-thickness burns involving the uppermost layers of the dermis may only be slightly more serious than a superficial burn. A deep partial-thickness burn, however, may behave in a similar fashion to a full-thickness burn and require excision and grafting. The formation of blisters is the hallmark of partial-thickness burns and implies some integrity of deeper dermal layers. Other signs of dermal viability include blanching with pressure and capillary refill. These signs may be absent in deep partial-thickness burns, and there may be a red-and-white reticulated appearance after blister debridement. In general, if complete reepithelialization is not expected within three weeks, or if the resulting wound will lead to contractures or a less-than-ideal cosmetic appearance, excision and skin grafting is performed.

Full-Thickness Burns (Third Degree)

Full-thickness burns have an easily recognizable appearance and may extend into fat, fascia, muscle and even bone. There is complete destruction of all epidermal and dermal elements and the wounds are insensate. The patient, therefore, has little or no discomfort. The burn wound is leathery, waxy, or translucent and thrombosed vessels may be visible beneath the skin. All full-thickness burns need surgical excision and skin grafting.

Burn Triage

The American Burn Association has defined the criteria used in triaging burns to be admitted and treated in a specialized burn unit:

- Partial- and full-thickness burns >10% TBSA in patients under 10 or over 50 years of age
- Partial- and full-thickness burns > 20% TBSA in other age groups
- Full-thickness burns >5% TBSA in any age group
- Partial- and full-thickness burns involving the face, hands, feet, genitalia, perineum, or major joints
- Electric burns, including lightning injury
- Chemical burns with serious threat of functional or cosmetic impairment
- Inhalation injuries
- Any burn patient with concomitant trauma
- Lesser burns in patients with preexisting medical problems that could complicate management
- Combined mechanical and thermal injury in which the burn wound poses the greater risk
- Any case in which abuse or neglect is suspected

On admission and initial evaluation, the burn victim is treated as any trauma patient and is evaluated for the "ABCs" (Airway, Breathing and Circulation). Large-bore peripheral IVs are placed in unburned skin. In patients who will likely need invasive hemodynamic monitoring, a central venous catheter, pulmonary arterial or Swan-Ganz catheter, or a peripheral arterial catheter may be used.

One should have a low threshold for endotracheal intubation especially if inhalational injury is suspected. Facial burns, for example, can lead to severe edema rendering later intubation extremely difficult or impossible thereby necessitating a surgical airway. Once airway, ventilation and systemic perfusion have been established, the next priority is diagnosis and treatment of concomitant life-threatening injuries. The patient's tetanus status should also be obtained and updated if necessary.

If inhalational injury is suspected, early intubation is necessary to prevent respiratory distress, especially if the patient is being transferred to a burn center. The physician must maintain a high degree of suspicion for the presence of inhalational injury. If it is suspected, arterial blood gases and carboxyhemoglobin (CHgb) levels should be obtained. If CHgb levels are elevated (>10%), 100 percent oxygen must be administered.

Burn Resuscitation

Burn Shock

Burn shock develops from hypovolemia and cellular breakdown. This type of shock is characterized by decreased cardiac output and plasma volume, increased extracellular fluid and edema and oliguria.

Fluid Resuscitation

The volume of fluid needed for resuscitation depends on both the area and the depth of burn. The "rule of nines" is a simple and relatively accurate method of estimating TBSA burned in adults (Fig. 28.1). The goal of fluid replacement is to restore and maintain adequate tissue perfusion and oxygenation, prevent organ ischemia, save as much of the zone of stasis as possible, and minimize the iatrogenic contribution to edema. Various fluid replacement protocols have been described in caring for the burn patient. It is important to keep in mind that these formulas are merely a starting point and that precise monitoring of the patient's status should be used to fine-tune fluid replacement. Urine output remains an excellent guideline for the adequacy of fluid replacement. An output of 30-50 ml/hour for adults and greater than 1.0 ml/kg/hour for children is used as a guideline for adequate fluid resuscitation. Central venous pressure and Swan-Ganz monitoring can also be used to fine-tune resuscitation, especially in elderly patients with preexisting cardiopulmonary pathology.

In pediatric burn patients, the Lund and Bowder chart or the Berkow formula are used for establishing the extent of burn injury (Table 28.1).

The Parkland Formula is the most widely used method for calculating resuscitation volume. It is simple, safe and inexpensive. This formula calls for 4 mL/kg/% TBSA of Lactated Ringer's solution over the first 24 hours of injury. Half of this volume is given over the first eight hours and half over the next 16 hours after injury. It is important to keep in mind that the volume calculated is to be given from the time of injury and not from the time of initial evaluation of the patient.

Parkland Formula

$$[0.5 \cdot (4\text{ mL} \cdot \text{kg} \cdot \%\text{ TBSA})] / 8\text{ hr} = \text{mL/hr (first eight hours)}$$

$$[0.5 \cdot (4\text{ mL} \cdot \text{kg} \cdot \%\text{ TBSA})] / 16\text{ hr} = \text{mL/hr (next sixteen hours)}$$

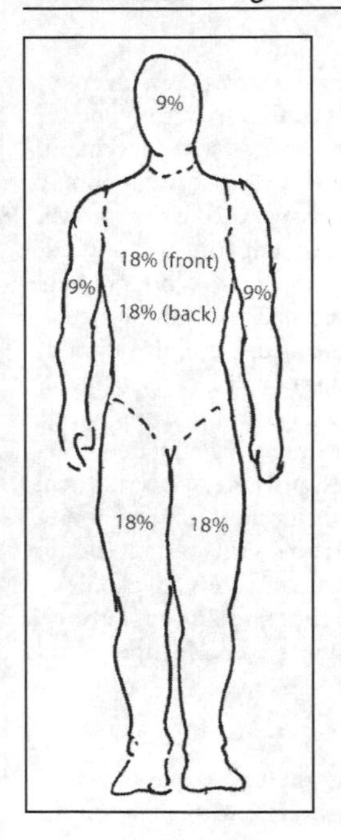

Figure 28.1. "Rule of Nines" for estimating percent TBSA in adults.

Table 28.1. The Berkow formula for calculating percent TBSA in children and adolescents

Site Age (years):	<1	1-4	5-9	10-14	15+
Head	19	17	13	11	9
Neck	2	2	2	2	2
Trunk (anterior)	13	13	13	13	13
Trunk (posterior)	13	13	13	13	13
Buttocks	5	5	5	5	5
Genitalia	1	1	1	1	1
Right upper arm	4	4	4	4	4
Right forearm	3	3	3	3	3
Right hand	2.5	2.5	2.5	2.5	2.5
Left upper arm	4	4	4	4	4
Left forearm	3	3	3	3	3
Left hand	2.5	2.5	2.5	2.5	2.5
Right thigh	5.5	6.5	8	8.5	9
Right leg	5	5	5.5	6	6.5
Right foot	3.5	3.5	3.5	3.5	3.5
Left thigh	5.5	6.5	8	8.5	9
Left leg	5	5	5.5	6	6.5
Left foot	3.5	3.5	3.5	3.5	3.5

Colloid Therapy

The timing of colloid therapy remains highly controversial. Plasma proteins counteract the outward hydrostatic force in the capillaries by generating an inward oncotic force. Massive interstitial edema occurs in many burn patients after the leaking of plasma fluid into and around the burned tissue. Albumin is the most commonly used colloid infusion. Although the precise amount of protein needed is not known, experience has shown that infusions at a constant rate seem superior to bolus administration. It is important to note that protein administration does not decrease burn edema but does limit edema in nonburned tissues and maintains intravascular volume better than crystalloid infusion. The exact protocol for administration of protein infusions varies widely between burn centers. In our burn unit, crystalloid is given in the first 8 hours, and protein infusions are initiated 8-12 hours after injury.

Other nonprotein colloid solutions available include dextran and hetastarch. Dextrans are colloids consisting of glucose molecules polymerized to form high-molecular-weight polysaccharides. Dextran has an osmotic effect and consequently increases urine output. When using dextran, therefore, urine output cannot be used to judge the adequacy of resuscitation. Hetastarch is an alternative to dextran and has volume expanding properties similar to a 6% protein solution. Furthermore, hetastarch has been used successfully as colloid administered during the second 24 hours of burn resuscitation.

Escharotomy

Thick, leathery eschar from a circumferential burn to the upper and lower extremities or trunk can be life- or limb-threatening. A circumferential eschar on the extremities can cause severe constriction resulting in compartment syndrome, ischemia and necrosis. This phenomenon is worsened by massive capillary leak and edema caused by a deep partial- or full-thickness burn. In a similar fashion, constricting eschar in the region of the trunk may cause decreased chest wall excursion and an inability to adequately ventilate the patient due to high peak inspiratory pressures.

In the extremities, peripheral perfusion can be easily assessed using a Doppler apparatus. Alternatively, transcutaneous oxygen saturation can be used to assess peripheral oxygenation. Typically, an oxygen saturation of less than 95 percent correlates with a need for emergent escharotomy. The need for chest escharotomies can be assessed by noting increased peak inspiratory pressures on a ventilator or more simply, limited chest excursion on physical examination.

If chest wall excursion is limited, escharotomies should be performed bilaterally in the anterior axillary lines using either an electrocautery device or a scalpel. These escharotomies may be joined with a chevron-shaped incision over the costal margin if needed. Extremity escharotomies are done in the mid-lateral lines of the affected limb. The location of the ulnar nerve in the upper extremity (posterior to the medial epicondyle) and the common peroneal nerve in the lower extremity (posterior to the fibular head) should be noted and extreme care should be taken to avoid injury to these structures. Finger escharotomies are done on the mid-lateral lines as well. Such a release is performed on the ulnar aspects of the index, middle and ring fingers and on the radial aspect of the thumb and small finger to avoid an incision over the "working surface" of these digits. Escharotomies may also be performed over the thenar and hypothenar muscles if necessary. In deep hand burns, escharotomies are performed over the dorsum of the hand, and the interosseous compartment fasciotomies are performed through the same incisions. Typically two longitudinal

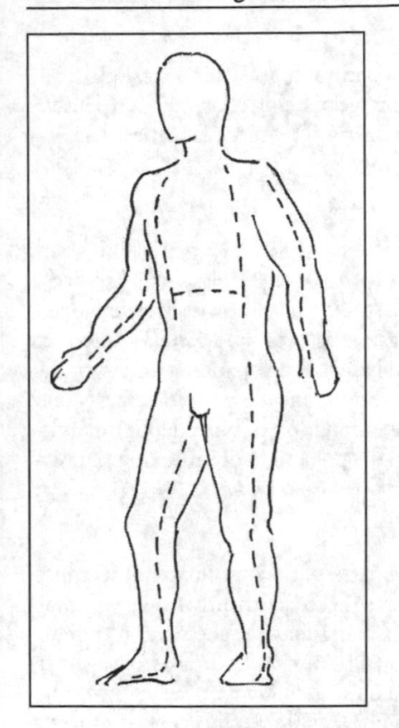

Figure 28.2. The commonly used escharotomy sites.

incisions (over the second and fourth metacarpals) are sufficient for release of all compartments. Figure 28.2 illustrates the commonly used escharotomy sites.

Escharotomies should be performed through the entire length of an eschar, from normal to normal skin since even a small area of remaining circumferential constriction can lead to peripheral ischemia. Typically, an obvious release of the underlying soft tissue indicates an adequate incision. Inadequate fluid resuscitation or the need for compartment fasciotomies should be suspected in patients in whom sufficient release of a constricting eschar fails to lead to a return of peripheral perfusion.

Rarely, aggressive fluid resuscitation may lead to a massive accumulation of intra- or retroperitoneal fluid. This phenomenon has been observed, for example, in patients with previous intra-abdominal pathology such as chronic pancreatitis. In such patients, one should have a high index of suspicion for the development of abdominal compartment syndrome (ACS). Elevated peak inspiratory, bladder and central venous pressures as well decreased urine output are suggestive and if diagnosed, immediate abdominal myofascial release should be performed. The development of abdominal compartment syndrome portends a very poor prognosis in the burn patient.

Burn Wound Care

After initial resuscitation of the burn patient, debridement and wound care is commenced in the burn unit. This is typically done in a heated hydrotherapy room with access to warm water to minimize heat loss. At this time, any loose skin and blisters are debrided and topical agents are applied. Effective debridement of burn eschar increases penetration of topical agents, improves time-to-healing for partial-thickness burns, and may allow faster skin graft coverage of full-thickness burns. Administration of appropriate analgesia, oral or intravenous, is extremely important during the debridement process. A variety of topical antibiotics are available for application to burn wounds.

Xeroform/Bacitracin

A combination of xeroform and bacitracin is commonly used for superficial partial-thickness burns. This application has the advantage of being relatively inexpensive and widely available and is applied on a daily basis until reepithelialization has occurred.

Silver Sulfadiazine

Silver sulfadiazine is the most widely used topical agent for deep partial-thickness burns or partial-thickness burns of indeterminate depth as well as full-thickness burns. It can be applied either daily or twice daily. In vitro studies have shown silver sulfadiazine to be active against gram-positive and gram-negative bacteria, and *Candida albicans*. Minimal pain is associated with its application and in fact, many patients find it soothing when applied to partial-thickness burns. The formation of a pseudoeschar is one detriment of silver sulfadiazine and may render determination of burn depth difficult. Other downsides of its use include leukopenia and the possibility of induction of resistant organisms such as *Pseudomonas aeruginosa* or *Enterobacteriaceae*.

Silver Nitrate

Silver nitrate has significant antimicrobial properties and is nontoxic in its most commonly used formulation. It is important to keep in mind, however, that gram-negative bacteria and some gram-positive bacteria may reduce silver nitrate to silver nitrite which can lead to methemoglobinemia in rare cases. It is active against gram positive and some gram negative bacteria, such as *Pseudomonas aeruginosa*, and is now commonly incorporated into commercially available wound care products.

Mafenide (Sulfamylon)

Mafenide has a broad antibacterial spectrum and has the best eschar penetration of any available agent. Due to its efficient penetration of cartilage, it is used over cartilaginous areas such as the nose or the ear. It is usually applied twice daily and due to its action as a carbonic anhydrase inhibitor may cause a metabolic acidosis if applied to large surface areas. Of the silver containing products, Sulfamylon is the most painful to the patient.

Inhalation Injury

Currently, inhalation injury is a more common **acute** cause of death from a burn injury than the surface burns themselves. The mechanisms of injury may involve carbon monoxide inhalation, thermal injury to the upper airway and digestive tract, and inhalation of the products of combustion.

Carbon Monoxide Poisoning

Carbon monoxide (CO) is a tasteless, odorless gas. It preferentially binds to hemoglobin and displaces oxygen from the hemoglobin molecule thereby impairing tissue oxygenation. The major deleterious effects of carbon monoxide result from its displacement of oxygen. It is estimated that it has more than 200 times the affinity for hemoglobin than oxygen.

Thermal Injury to the Upper Airway

The most common cause of direct thermal injury to the upper airway is inhalation of steam. This is because steam has 4000 times the heat-carrying capacity of air. Edema of the face and perioral tissues can result in narrowing of the airway and increased work of breathing. In addition to direct thermal injury to the airway, edema of the airway often parallels the generalized edema in a burn patient. Therefore, airway protection should always be kept in mind in a patient with severe generalized burn edema.

Inhalation of Products of Combustion

Inhalation of aldehydes, ketones and organic acids, all products of combustion, is the most significant component of inhalation injury. All of the above chemicals cause significant chemical injury to the respiratory tract and mimic aspiration of acidic gastric contents. Increased capillary permeability and alveolar injury may lead to pulmonary edema and ultimately adult respiratory distress syndrome (ARDS). Secondary pneumonias may develop after plugging of the lower airways with sloughed bronchial mucosa. Patients with secondary pneumonias after inhalation injury have a poor prognosis.

Diagnosis

A history of a flame burn in an enclosed space, singed nasal hairs and facial or oropharyngeal burns, and the presence of carbonaceous sputum should raise suspicion of an inhalation injury. CHgb levels should be obtained in all patients with such findings. Levels above 10% are significant and denote inhalation injury and levels above 50% are associated with a high likelihood of death. A patient with documented inhalation injury should be immediately placed on 100% oxygen therapy which reduces the washout time of carbon monoxide. Fiberoptic bronchoscopy of the upper respiratory tract is the gold standard for diagnosis and may show edema of the vocal cords and charring, sloughing, or edema of the hypopharyngeal and upper tracheal mucosa

Management

The management of inhalation injury is primarily supportive care. Endotracheal intubation should be performed liberally due to the potentially disastrous consequences of a delay in diagnosis. If carbon monoxide poisoning is present, 100% oxygen is administered. Furthermore, the airway can be cleared of mucosal plugs and airway debris by aggressive pulmonary toilet and bronchoscopy.

Nutrition

Proper nutrition in the burn patient is of the utmost importance. A major burn results in a hypermetabolic state that is even more profound than that found in the stress response to trauma and sepsis. Enteral nutrition should be used whenever possible, and therefore a feeding tube should be inserted immediately in patients with major burns. This is important in light of the fact that delayed placement of a post-pyloric feeding tube is often extremely difficult. It is also important to keep in mind that large burns are also associated with a transient ileus. The precise calculations in basal energy expenditures and caloric requirements are beyond the scope of this text, but the inclusion of a nutritionist as part of the multi-disciplinary approach to the management of the burn patient is extremely important. Finally, ulcer prophylaxis should also be considered in all patients with large burns.

Pearls and Pitfalls

- If not treated rapidly and adequately, the central area of the burn, the zone of coagulation, can progress and enlarge. This occurs because the surrounding zone of stasis converts to coagulated tissue. This underscores the importance of early debridement and wound care.

- The treatment of blisters in partial-thickness burns is controversial. Some advocate keeping the blisters intact since the fluid contained within is sterile and will aid in healing. Others prefer to debride the blisters and remove the separated skin layers since they can rapidly become colonized and act as a nidus for infection. Either approach is appropriate, as long the burns are cleaned and dressed daily and all non-viable tissue is debrided.
- The need for aggressive fluid resuscitation in extensive burns cannot be overemphasized. It is better to over- than under-resuscitate a burn patient. Following urine output is the most accurate means of tracking fluid status.
- Numerous burn dressing regimens exist. None of these are a replacement for adequate debridement and daily cleansing of the burn. Despite the many products that are commercially available, simple silver-containing gauze dressings have withstood the test of time.

Suggested Reading

1. Press B. Thermal, electrical, and chemical injuries. Grabb and Smith's Plastic Surgery. 5th ed. Philadelphia: Lippincott-Raven, 1997.
2. Heimbach DM, Engrav L. Surgical management of the burn wound. New York: Raven Press, 1985.
3. Hunt JL, Purdue GF, Zbar RIS. Burns: Acute burns, burn surgery and post-burn reconstruction. Selected readings in plastic surgery, Vol. 9. 2000; (No. 12).
4. Moylan Jr JA, Inge Jr WW, Pruitt Jr BA. Circulatory changes following circumferential extremity burns evaluated by the ultrasonic flowmeter. An analysis of 60 thermally injured limbs. J Trauma 1971; 11:763.
5. Carvajal HF, Parks DH. Burns in children: Pediatric burn management. Chicago: Year Book Medical Publishers, 1988.

Chapter 29

Head and Neck Cancer

Zol B. Kryger

Introduction

This chapter covers primary head and neck cancer, excluding neoplasms of the skin. Ninety percent of these tumors are squamous cell in origin and tend to affect elderly men. Most have a history of alcoholism and tobacco abuse. Tumors of the salivary, thyroid and parathyroid glands are not of the squamous cell type and can be found in a wider age distribution. Thyroid and parathyroid tumors are not usually treated by the plastic surgeon.

Anatomical Definitions

- The oral cavity extends from the vermilion border of the lip to the junction of the hard and soft palates.
- The pharynx is divided into three cavities:
    1. The **oropharynx** extends from the hard and soft palate junction anteriorly to the posterior pharyngeal wall. The lateral borders are the tonsils and tonsilar pillars. The ceiling is the soft palate, and the floor is the soft tissue between the base of the tongue and the epiglottis.
    2. The **nasopharynx** extends from the nasal septum to the posterior pharyngeal wall. Its ceiling is the skull base, and its floor is the soft and hard palate.
    3. The **hypopharynx** spans from the vallecula and aryepiglottic folds to the inferior aspect of the cricoid cartilage.
- The larynx is divided into the **supraglottis**, **glottis** and **infraglottis**, with the true vocal cords defining the glottic region. The anterior and posterior commissure are also considered part of the glottis. The epiglottis, ventricles, erytenoids, and false vocal cords are all supraglottic structures.

Intraoral Tumors

Etiologic Factors

A history of tobacco use is the number one risk factor for head and neck cancer in the United States. As the duration and quantity of tobacco use increases, so does the risk of developing intraoral cancer.

Alcohol is another major risk factor. Heavy consumption increases the risk of developing aerodigestive cancer by sixfold. Furthermore, the risk from concomitant tobacco and alcohol use is synergistic compared to either one alone. Other risk factors include dentures and poor oral hygiene. In countries with poor dental care, these factors may play a greater etiologic role.

Table 29.1. TNM classification

T = extent of the primary tumor	
Tis	In-situ tumor
T1	≤ 2 cm
T2	2-4 cm
T3	> 4 cm
T4	Invades adjacent structures (varies by site of tumor)
N = regional lymph node status	
N0	No nodal involvement
N1	Movable, ipsilateral nodes
N2	Movable, contralateral or bilateral nodes
N3	Fixed nodes
M = distant metastases	
M0	No mets
M1	Distant mets present
Stage I	T1N0M0
Stage II	T2N0M0
Stage III	T3N0M0 or T1-3N1M0
Sage IV	One or more of the following: T4, N2, N3 or M1

Pathology

Squamous cell carcinomas can present as white patches, termed leukoplakia, or as an erythematous patch, known as erythroplakia. As a general rule, erythematous lesions have a higher risk of malignancy than leukoplakic lesions. The need to biopsy every leukoplakic area is controversial. Most advanced squamous cell carcinomas are endophytic (ulcerated, deeply infiltrating). They may also be exophytic (projecting outward).

Staging and Treatment

Staging is based primarily on the TNM classification shown in Table 29.1. The treatment of intraoral cancers is summarized in Table 29.2.

Tongue

The tongue is the most common site of intraoral malignancy. In addition to alcohol and tobacco, Plummer Vinson syndrome is a risk factor. Most lesions are on the anterolateral two thirds of the tongue. These tumors are usually painless and thus are often neglected. The average stage at presentation is T2 (2-4 cm). T1 tumors are treated with either wedge resection or radiation. T2 lesions require partial glossectomy and T3 tumors require total or subtotal glossectomy. T2 and T3 lesions are resected in combination with an elective neck dissection even if there are no palpable neck nodes (N0) due to the high risk of occult nodal metastasis.

Table 29.2. Summary of the treatment of intraoral cancers

Stage	Treatment	Adjuvant Therapy
Stage I	Excision or radiation	None
Stage II and III	Surgical resection (see below)	Pre- or postop radiation
Stage IV	Chemo± palliative resection	Radiation

Floor of Mouth

The floor of the mouth is the second most common site of intraoral cancer. Lesions are usually anterior and often present with a palpable submandibular node. Fifty percent of patients present with stage III or IV disease due to the paucity of symptoms. The survival rate for stage I and II lesions is high (80-90%); advanced disease has a poorer prognosis (30-60%). An important consideration in any form of treatment is the risk of submandibular duct stenosis with subsequent enlargement of the gland. If this occurs, referral to a specialist in this condition is warranted.

Alveolar Gingiva and Buccal Mucosa

The third most common site of intraoral tumors is the lower alveolar gingiva. Eighty percent of these tumors occur on the lower alveolus. Lesions begin on the alveolar ridge and spread laterally. Nodal metastasis at time of presentation is common. Cancer of the buccal mucosa is found primarily in tobacco chewers in the U.S. There is a higher incidence in India due to the custom of betel leaf chewing.

Tonsil

The tonsil is the most frequent site of squamous carcinoma in the oropharynx. Most tumors present late, as stage III or IV lesions. Hence, the prognosis is poor.

Hypopharynx

As with cancer of the tongue, Plummer Vinson syndrome is a risk factor. Patients often present with advanced disease with dysphagia and clinically positive neck nodes. Extensive resections with free flap reconstruction are usually required.

Larynx

In cancer of the larynx, the vocal cords are involved in about 50% of the cases. This allows for relatively early detection due to hoarseness and respiratory symptoms. If the glottis is involved, radiotherapy is more successful at preserving speech than surgery. Advanced disease occurs more frequently with subglottic tumors and requires total laryngectomy with neck dissection.

Nasopharynx

Cancer of the nasopharynx is unique among head and neck tumors in its etiology. Chronic inflammation of the mucosa is the main risk factor; chronic sinusitis, human papilloma virus and Epstein-Barr virus infections have all been implicated. There is also in increased incidence among individuals from mainland China for unknown reasons. Tumors of the nasopharynx often present as locally advanced neck masses. The primary mode of treatment is radiation rather than surgery. Since most tumors present with nodal metastases, the neck should also be irradiated. In certain cases, chemotherapy is indicated.

Salivary Gland Tumors

Parotid Gland

The parotid glands are the largest of the salivary glands. They are located in the infra-auricular region. The parotid has a deep and a superficial lobe. The facial nerve traverses the deep lobe. The parotid has a fascial covering which is continuous with the SMAS. The parotid duct (Stenson's duct) passes superficial to the masseter muscle and pierces the buccinator muscle. It enters the oral cavity at the level of the upper second molar. All parotid masses should undergo fine needle aspiration for diagnosis. Imaging is usually not necessary, although an MRI may be useful if the tumor is large (over 3 cm).

The most common tumor of the parotid is the benign pleomorphic adenoma (benign mixed tumor). They are treated by superficial parotidectomy (unless the tumor is in the deep lobe which is rare). If incompletely resected, they can recur and become locally invasive. If left untreated, they can degenerate into a malignant mixed tumor that is prone to early metastasis.

Warthin's tumor is a benign cystic tumor. Ten percent are bilateral, making it the most common bilateral parotid tumor. It occurs primarily in male smokers. Superficial parotidectomy is sufficient treatment.

The most common malignancy in the parotid is the mucoepidermoid carcinoma. Such malignant tumors require excision of both the superficial and deep lobes of the parotid. The facial nerve is spared unless the tumor directly invades the nerve. Nerve contaminated with tumor that is left behind should receive adjuvant radiation postoperatively. High-grade (anaplastic) mucoepidermoid carcinoma warrants a simultaneous neck dissection.

Complications from parotidectomy are uncommon. Facial nerve injury is the most devastating; any recognized facial nerve injury should be repaired at the time of injury. Hematoma can be avoided by meticulous hemostatic technique. Frey's syndrome (auricular temporal syndrome) is gustatory sweating due to reinnervation of sweat fibers by severed salivomotor fibers. Botox injections have shown some promise in treating this condition. Development of a sialocele is treated by aspiration and compression.

Submandibular and Sublingual Glands

The paired submandibular and sublingual glands are located below the mandible and in the floor of the mouth, respectively. Wharton's duct is the submandibular duct that enters the floor of the mouth. The most common malignancy in these glands is the mucoepidermoid carcinoma. Complete excision of the affected gland is required.

Minor Salivary Glands

Minor salivary glands are mucus-secreting glands located below the oral submucosa. Most minor salivary gland tumors are malignant. The most common variety of these glands is the adenoid cystic carcinoma. Treatment consists of wide local excision.

Neck Dissection

Nodal metastases spread in a predictable fashion:

- Level I nodes are found in the submental/submandibular triangle
- Level II nodes are in the upper jugular region
- Level III nodes are in the middle jugular region
- Level IV nodes are in the lower jugular region
- Level V nodes are in the posterior triangle

Patients with a clinically positive node or a large primary lesion should undergo simultaneous neck dissection. Nodal metastasis is an indication for adjuvant radiation therapy. Bilateral neck dissections for midline lesions are staged. At least one of the two internal jugular veins should be preserved.

The modified radical neck dissection has become the procedure of choice for elective cases. It involves removal of all the nodes from level I-V as described above. The internal jugular vein, accessory nerve and sternoclidomastoid muscle are spared. This results in decreased facial edema, shoulder dysfunction and a less cosmetic defect. If possible, the neck dissection should be done in-continuity with the primary resection.

Reconstruction

The primary goals of reconstruction include the following: adequate deglutition, preservation of speech, avoiding salivary fistulas and drooling and achieving an acceptable cosmetic result. Small defects are often amenable to primary closure. Moderate-sized defects may require tongue flaps and split-thickness skin grafts (if no bone or buccal mucosa is exposed).

Large defects often necessitate pedicled skin and myocutaneous flaps including the deltopectoral, pectoralis major, forehead, latissimus and trapezius flaps. Many surgeons favor microvascular free flaps as a first choice. The radial forearm free flap may be used for mucosal defects alone. For defects involving both bone and mucosa, the fibula free flap is often used. For circumferential defects of the hypopharynx or esophagus, the jejunal interposition free flap is the first choice; however, the anterolateral thigh flap is rapidly becoming more popular than the jejunal flap for esophageal reconstruction. These free flaps are described in detail in the flap harvest section of this book.

Post maxillectomy—these defects can often be skin-grafted and a dental prosthesis placed once the tissue has healed.

Pearls and Pitfalls

In elderly patients, especially those with a history of tobacco or alcohol use, any neck mass is cancer until proven otherwise. Patients with a neck mass should undergo a thorough head and neck exam including examination of the auditory canal, nasopharynx, oral cavity and oropharynx. Furthermore, all suspicious palpable neck masses should undergo fine needle aspiration, and most plastic surgeons have limited experience in performing this procedure. Therefore, it would be prudent to refer patients suspected of having carcinoma of the head and neck to a specialist who routinely treat such cancers. Treatment is often multimodal and requires the collaboration of multiple disciplines. Experienced head and neck surgeons have an established network of such experts and are better suited to coordinate the patient's diagnosis and treatment. The role of the plastic surgeon should focus on the post-resection reconstruction.

Suggested Reading

1. Ballantyne AJ. Modified neck dissection. Recent Adv Plast Surg 1985; 3:169.
2. Cooper JS et al. Postoperative concurrent radiotherapy and chemotherapy for high risk sqaumous-cell carcinoma of the head and neck. New Eng J Med 2004; 350(19):1937.
3. Day TA et al. Salivary gland neoplasms. Curr Treat Options Onc 2004; 5(1):11.
4. In: Fleming ID et al, eds. AJCC Cancer Staging Manual. Philadelphia: Lippincott-Raven, 1997.
5. Hidalgo DA. Fibula free flap: A new method of mandible reconstruction. Plast Reconstr Surg 1989; 84:71.
6. Jackson IT. Intraoral tumors and cervical lymphadenectomy. Grabb and Smith's Plastic Surgery. 5th ed. Philadelphia: Lippincott-Raven, 1997:439.
7. Robinson DW, MacLeod A. Microvascular free jejunum transfer. Br J Plast Surg 1982; 35:258.
8. Shaha AR et al. Squamous carcinoma of the floor of the mouth. Am J Surg 1984; 148:455.
9. Soutar DS et al. The radial forearm flap: A versatile method of intraoral reconstruction. Br J Plast Surg 1983; 36:1.

Chapter 30

Ear Reconstruction

Amir H. Taghinia, Theodore C. Marentis, Ankit I. Mehta, Paul Gigante and Bernard T. Lee

Introduction

Acquired ear deformities are usually the result of trauma, burns, or ablative skin cancer operations. Reconstruction of these deformities is primarily an aesthetic endeavor. Nevertheless, seemingly minor deformities can cause significant patient anxiety and concern. Auricular deformities can be divided into defects of the helical rim, upper third, middle third and lower third of the ear. Defects of the upper ear complicate the use of eyewear, but they are more easily camouflaged by hair. Defects of the lower ear are harder to hide and thus, more aesthetically important. Unfortunately, these defects are also the hardest to reconstruct well. Despite many variations in ear anatomy, there are several consistent landmarks shown in Figure 30.1.

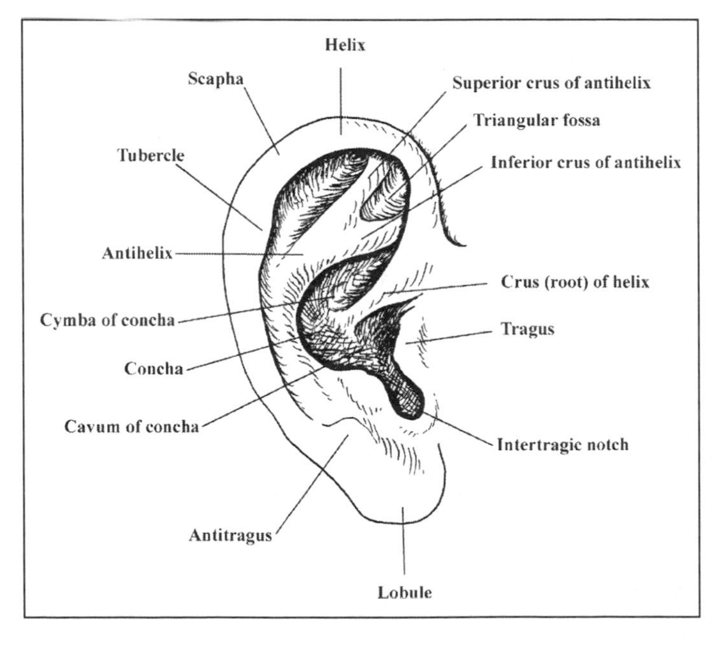

Figure 30.1. Topographic anatomy of the ear.

Acute Auricular Trauma

Otohematoma

An injury frequently associated with contact sports, otohematoma ('cauliflower ear') results from blunt trauma or excessive traction that causes hemorrhage between the perichondrium and the cartilage. Subperichondrial blood produces a clot that, if left untreated, leads to the formation of neocartilage and eventual deformity of the ear's convolutions. Treatment should be administered soon after injury. Needle aspiration drains the fluid but rarely removes the clots. Preferred treatment is incision and drainage followed by the placement of sutured bolsters or a thermoplastic splint for 7-10 days to maintain a broad area of pressure.

Burns

The ear is uniquely susceptible to thermal injury because of its exposed, unprotected position. Deep ear burns destabilize the skin and are likely to involve the cartilage. Chondritis is a serious infectious complication that occurs most commonly between the third and fifth weeks post-burn. General burn management should include liberal use of mafenide, frequent soap and water cleansing, and avoidance of pressure on the affected ear. Adequate healing time should be allowed, and a maximal amount of viable cartilage should be salvaged prior to reconstruction. Should chondritis occur, systemic antibiotics must be administered. Severe infections may require incision, drainage and debridement of skin and cartilage.

Lacerations

The ear is protected from traumatic forces by its resilient, pliable cartilaginous framework. Nevertheless, lacerations are the commonest form of auricular trauma. Preservation of tissue is critical to ensuring optimal aesthetic outcome in these injuries. The ear's rich blood supply allows for excellent tissue recovery in most cases. Compromised flaps of skin (with or without cartilage) usually survive, even if based on a thin pedicle. At the initial time of treatment, one should debride grossly necrotic tissues only. Animal or human bites require thorough irrigation and systemic antibiotics. In laceration repair, discernable landmarks must be approximated meticulously to avoid poor aesthetic outcome or exposed cartilage. As with auricular burns, major reconstructive intervention is usually delayed until adequate healing is complete.

Acquired Auricular Deformities

Traditionally, deformities of the ear have been classified based on the location, and methods have been developed to address each anatomic site (e.g., helix or lobule). However, many of these methods can be applied in multiple locations, thus confusing the novice. In this chapter, methods for reconstruction will be presented based on defect size and the affected part of the ear.

Wedge and Star-Wedge Excisions and Primary Closure

Application: Defects anywhere on the ear
Defect size: Small
Illustration: Figure 30.2.

Small defects of the ear can usually be repaired using wedge excision and primary closure. If the angle of the wedge is too obtuse, apposition of the edges creates a standing cone that makes the ear bulge outward. This problem is alleviated by additional triangular excisions at the wedge borders yielding a star-shaped

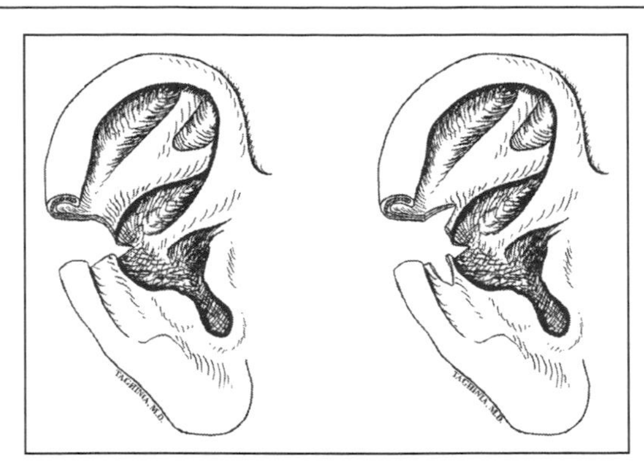

Figure 30.2. Wedge excision and star-wedge excision. Small defects of the ear can usually be closed primarily after wedge excision. Additional triangular excisions creating a star-shaped wedge facilitate closure of larger wounds.

excision. Closure of wedge-shaped earlobe defects requires an offset flap or Z-plasty to prevent a notch in the inferior contour (see earlobe reconstruction).

Chondrocutaneous Helical Advancement (Antia-Buch Procedure)

Application: Helical defects of the upper- and middle-third of the ear

Defect size: Small, moderate, or large

Illustration: Figure 30.3.

This method is commonly used because most ear lesions occur on the periphery. Advancing the helical stumps from both directions can repair helical defects up to 3 cm. To achieve adequate mobility, the entire helix must be freed from the scapha using an incision in the helical sulcus that extends through the anterior skin and cartilage but not through the posterior skin. The posterior skin is then undermined (above the perichondrium) until the entire superior and inferior helical remnants are hanging as composite flaps. The flaps are then brought together and sutured. Additional length can be gained by a V-Y advancement of the helical root.

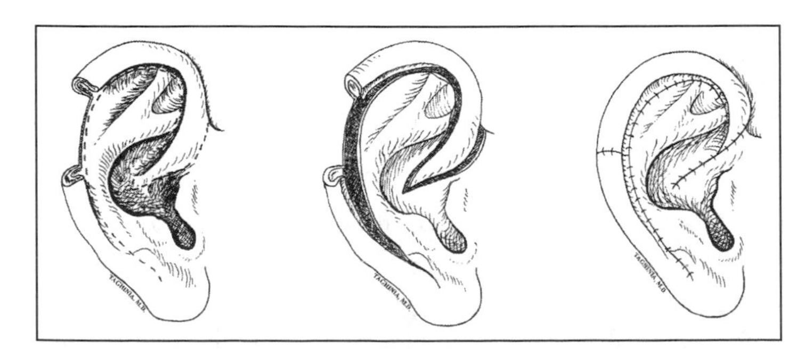

Figure 30.3. Antia-Buch helical advancement. A V-Y advancement closure at the root of the helix can be used for additional flap mobility.

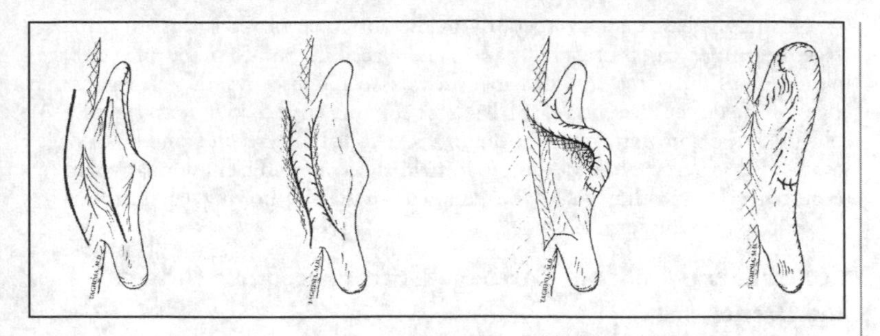

Figure 30.4. Tube skin flap. A tube skin flap can be fashioned in the retroauricular sulcus and transferred in stages to reconstruct the helical rim. The donor site can be closed primarily or with a skin graft if a large tube is made.

Tubed Skin Flaps

Application: Helical rim defects
Defect size: Moderate or large
Illustration: Figure 30.4.

Minor burns frequently destroy the helical rim but leave the posterior ear intact. Thin, tubed skin flaps can then be fashioned from skin in the auriculocephalic sulcus. These flaps require multiple delay and inset procedures to achieve a final result. Furthermore, it is technically difficult to obtain the required length and width using this technique; often the flaps are too bulky or too thin. If performed well, however, these flaps can give a pleasing aesthetic outcome.

Banner Flap (Crikelair)

Application: Upper-third defects of the ear
Defect size: Moderate
Illustration: Figure 30.5.

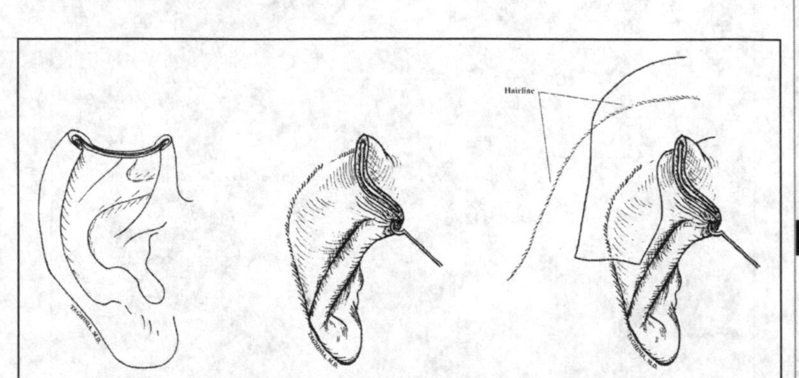

Figure 30.5. Banner flap (Crikelair flap). The ear is retracted anteriorly and a superiorly-based flap is designed in the retroauricular sulcus. The flap is raised and placed on an anchored cartilage graft in the defect (not shown).

This flap is based anterosuperiorly in the auriculocephalic sulcus. In the first stage, a cartilage graft is procured, carved and placed in the area of missing cartilage in the upper-third of the ear. The flap is then raised and inset over the cartilage at the edges of the defect. Presence of cartilage for support ensures long-term stability. A split-thickness skin graft covers the donor site defect. In the second (and final) stage several weeks later, the base of the flap is divided leaving sufficient nonhairy skin to reconstruct the ascending helix. The remainder of the hair-bearing flap is sutured to the scalp to reestablish the hairline.

Contralateral Conchal Cartilage Graft Procedures (Adams and Brent)

Application: Upper-third (Adams) and lower-third (Brent) defects of the ear

Defect size: Large

Illustration: Figures 30.6 (upper-third, Adams), 30.7 (lower-third, Brent)

Initially described for upper-third defects, this method was then used by Brent for reconstruction of lower-third losses. The success of these procedures relies on supple, intact temporal and mastoid skin. Initially, a conchal cartilage graft is removed from the other ear using an anterolateral or posterior approach. For reconstruction of the upper-third, the cartilage graft is implanted under mastoid skin in the upper ear and anchored to the remnant of the helical root. Anchoring of the

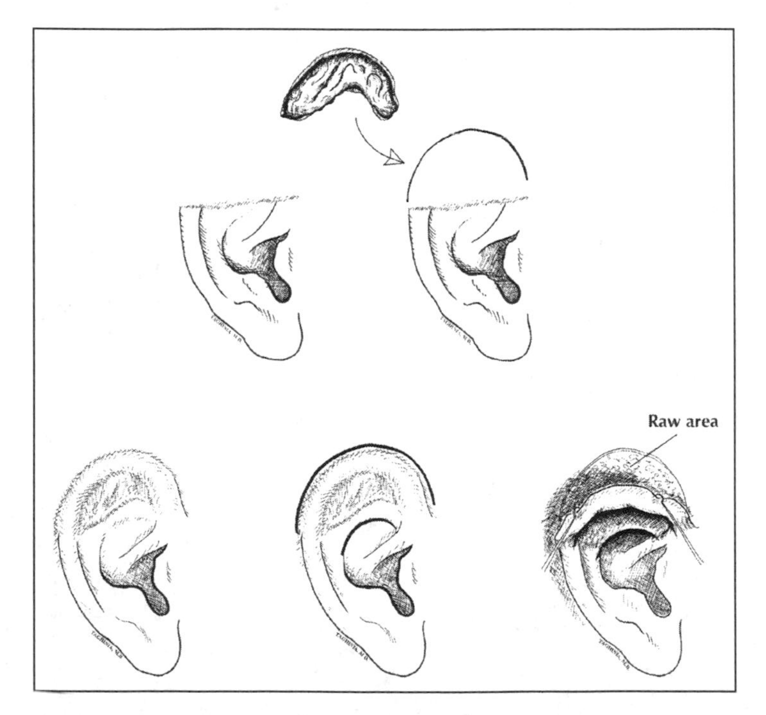

Figure 30.6. Contralateral conchal cartilage graft procedure for upper-third deformities (Adams flap).

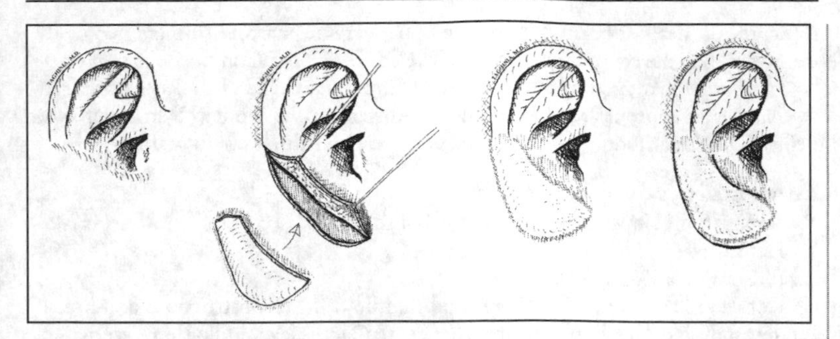

Figure 30.7. Contralateral conchal cartilage graft procedure for lower-third deformities described by Brent. A bipedicle flap is outlined in a second stage, and a 'valise handle' is raised. The raw surface underneath is skin grafted.

cartilage graft ensures stability and long-term helical continuity. At a second stage, an incision is made in the hair margin and the upper ear is undermined, lifted, and the raw area is skin grafted. Brent (1977) prefers to make another incision in the proposed posterior conchal wall and inferior crus so the entire upper ear bipedicled flap can be lifted as a 'valise handle'. This maneuver elevates and defines the inferior crus and gives depth to the superior concha.

A similar approach is used for large defects of the lower third of the ear (Brent). Creation of a bipedicle flap, lifted as a 'valise handle' gives definition to the posterior conchal wall. A tube skin flap can be used for reconstruction of the lower helix and earlobe.

Conchal Chondrocutaneous Transposition Flap (Davis)

Application: Upper-third defects of the ear
Defect size: Large
Illustration: Figure 30.8.

If the mastoid skin is unfavorable for flap reconstruction, a conchal transposition flap may be used for large defects of the upper ear. This flap is technically

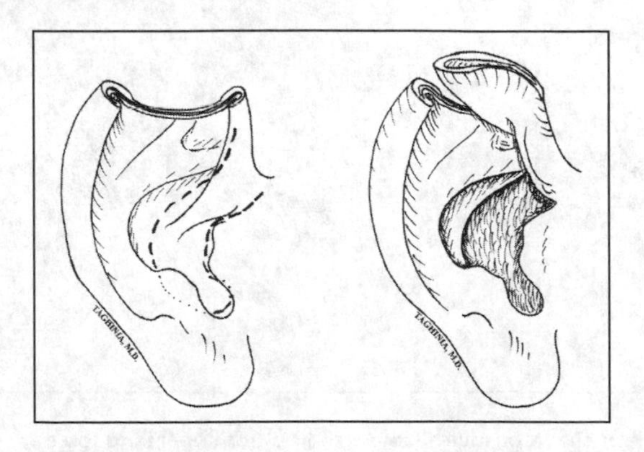

Figure 30.8. Conchal transposition flap (Davis technique).

difficult but ideal for burns because the upper ear and mastoid skin are frequently destroyed, but the central ear is spared. The entire concha (composite flap of cartilage and skin) is rotated upward on a small anterior pedicle of the crus helicis. Irregularities in contour are tailored and the lower part of the flap is inset into the defect. The conchal donor bed and the newly created rim are skin grafted.

Tunnel Procedure (Converse)

Application: Upper- and middle-third of the ear
Defect size: Moderate to large
Illustrations: Figure 30.9.

Converse's techniques for upper- and middle-third reconstructions are similar. Interestingly, most textbooks incorrectly refer to Converse's middle-third reconstructive method as the 'tunnel' procedure, presumably because a subcutaneous tunnel (for a cartilage graft) is created between the two edges of the middle-third defect. In fact, Converse called his technique for upper-third reconstruction the 'tunnel' procedure. A tunnel is created in the superior auricular sulcus as the upper ear is apposed to the edge of the mastoid skin pocket (see below). This tunnel is lined by native skin epithelium and ultimately gets opened at the second stage operation. It needs frequent cleansing by the patient to prevent infection. For simplicity, we will refer to both of these procedures as the 'tunnel' procedures.

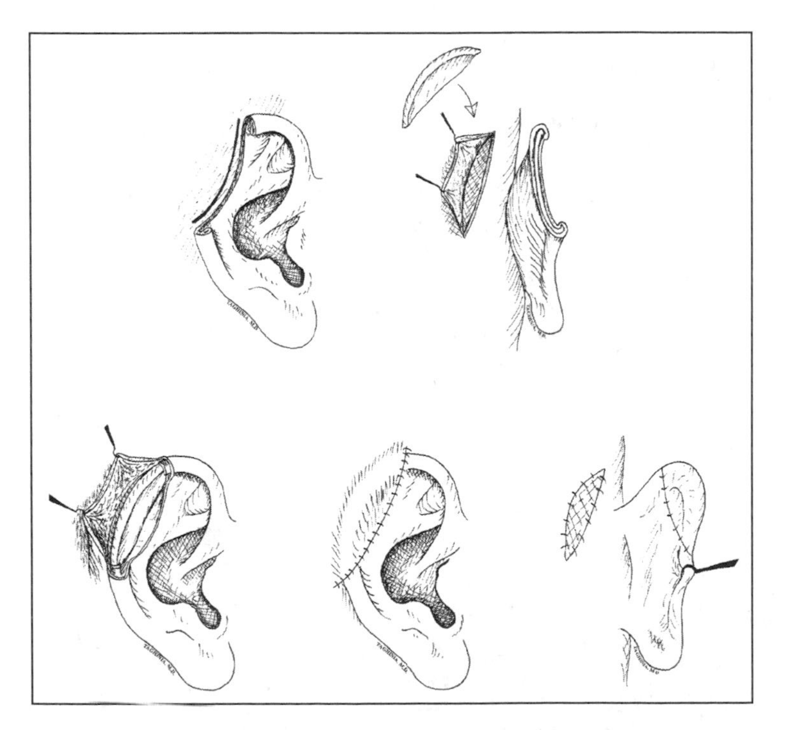

Figure 30.9. Tunnel procedure (Converse flap) for upper-third losses.

When applied for upper-third defects (Fig. 30.9), Converse's method has the advantage of preserving the superior auricular sulcus. The ear is pressed against the scalp, and incisions are planned and made along the edge of the ear defect and the mastoid skin. A mastoid skin flap is then raised posterosuperiorly. The medial edge of the ear incision is sutured to the anterior edge of the mastoid skin incision, thus creating a pocket where a carved costal cartilage graft can be placed and secured. The mastoid skin flap is then advanced to cover the cartilage graft. In a second stage, the ear is separated from the mastoid area and the raw areas (on the back of the ear and the mastoid donor site) are skin grafted.

For middle-third defects, a pocket is created in the mastoid skin flanking the edges of the defect and a cartilage graft is placed in the pocket. The edges of this pocket are approximated to the edges of the defect. A few weeks later, an incision posterior to the cartilage graft is made and the retroauricular sulcus is created. The raw areas on the mastoid and posterior ear are skin grafted. For large defects, tissue expansion may be needed to get additional nonhairy, thin skin.

Retroauricular Flap with Cartilage Graft (Dieffenbach)

Application: Middle-third defects of the ear
Defect size: Moderate to large
Illustration: Figure 30.10.

Similar to the Converse tunnel procedure, this method relies on a skin flap and a cartilage graft. The defect is drawn out and a posterior skin flap is planned in the mastoid skin. The flap is raised and a cartilage graft is placed and covered by the flap. At a second stage, the base of the flap is divided and the flap is folded around the cartilage graft. A skin graft then covers the mastoid donor site.

Chondrocutaneous Composite Graft

Application: Upper- and middle-third defects of the ear
Defect size: Small to moderate
Illustration: Figure 30.11.

Composite grafts from the contralateral ear may be used for reconstruction of small to moderate defects. A wedge-shaped graft of 1.5 cm width can be harvested from the helix/scapha of the unaffected ear. The success of these grafts is enhanced by converting part of the composite graft to a full-thickness skin graft (i.e., by excising some skin and cartilage). In this manner, more of the skin component is in direct contact with a vascular bed, thereby improving plasmatic imbibition. These grafts have a tendency to shrink in the long-term.

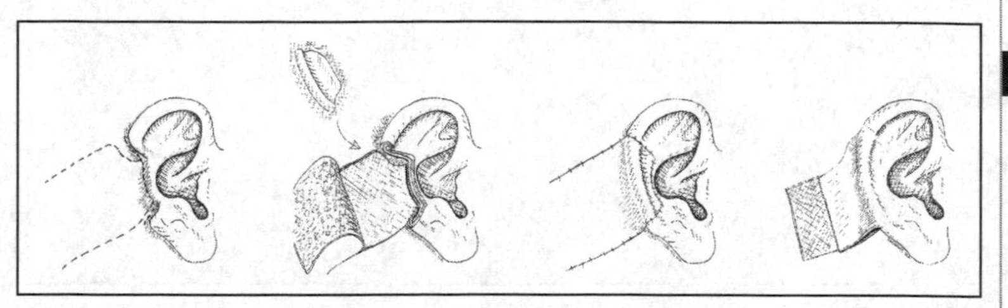

Figure 30.10. Retroauricular flap (Dieffenbach procedure).

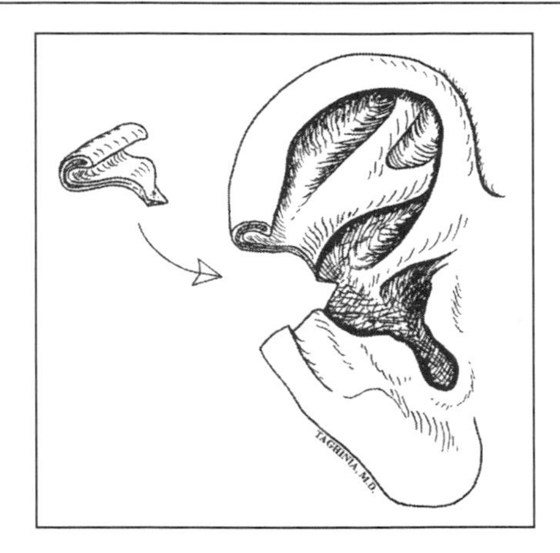

Figure 30.11. Contralateral composite graft for middle-third reconstruction. This method can also be applied to small to moderate defects of the upper-third. Brent recommends increasing chance of graft 'take' by creating more raw surface contact.

Earlobe Reconstruction

Application: Earlobe defects
Defect size: Small to large
Illustration: Figures 30.12, 30.13

The most common acquired earlobe deformities are keloids and traumatic clefts. Keloids are treated with steroid therapy and pressure-spring earrings. Radiation can sometimes be effective for recurrent or recalcitrant keloids. When a keloid is excised, the skin overlying the keloid can be harvested as a split-thickness skin graft and used to cover the remaining defect.

Earlobe clefts can be repaired using a variety of methods. Straight-line excision and closure of these defects is not recommended because the scar contracts upwards, creating a notch in the inferior contour. Most surgeons prefer to repair these clefts

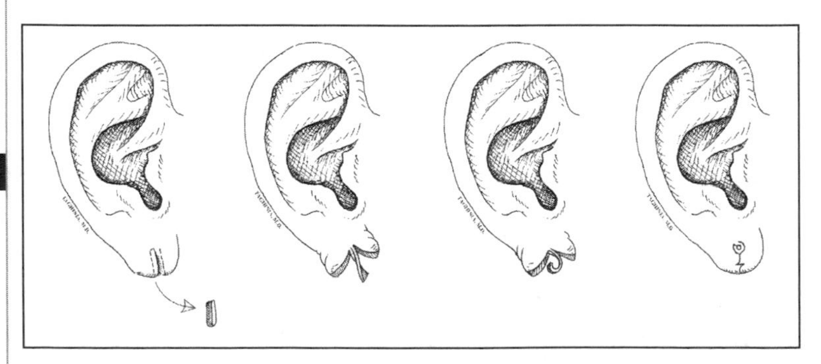

Figure 30.12. Earlobe cleft reconstruction (Pardue technique). A clever method using a rolled-up skin flap to allow continued wearing of earrings.

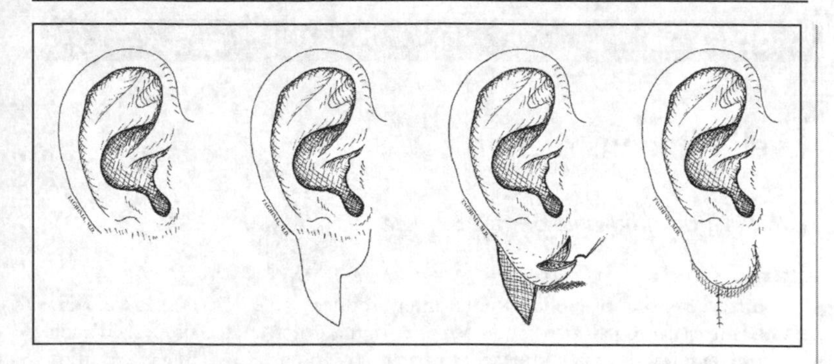

Figure 30.13. Earlobe reconstruction (Alanis technique).

by elongating the incision using Z-plasties. Pardue suggested a clever method whereby patients can continue wearing earrings (Fig. 30.12). Multiple methods have been devised to reconstruct missing earlobes. Most of these methods use mastoid skin flaps that roll or fold to create an earlobe (Fig. 30.13). These earlobe constructs are prone to shrinking, especially if the blood supply is tenuous.

Pearls and Pitfalls

Cartilage is the key aesthetic element of the ear; it creates the hills and valleys that make an ear look like an ear. During ear reconstruction, replacing missing cartilage is important for construct stability and aesthetic contour. Without cartilaginous support, skin and soft tissue constructs do not uphold in the long term.

Recreation of the helix is one of the more important tasks in aesthetic ear reconstruction. Our brain is trained to visualize a 'normal' ear as one with a smooth, uninterrupted outer contour. Any slight deviation from this outline is strikingly obvious. Symmetry is less important in ear reconstruction because both ears are rarely seen at the same time. As long as the ears look normal, great discrepancies in ear size can go unnoticed.

Suggested Reading

1. Alanis SZ. A new method for earlobe reconstruction. Plast Reconstr Surg 1970; 45(3):254.
2. Antia NH, Buch VI. Chondrocutaneous advancement flap for the marginal defect of the ear. Plast Reconstr Surg 1967; 39(5):472.
3. Adams WM. Construction of upper half of the auricle utilizing composite concha cartilage graft with perichondrium attached on both sides. Plast Reconstr Surg 1955; 16(2):88.
4. Brent B. The acquired auricular deformity. Plast Reconstr Surg 1977; 59(4):475.
5. Brent B. Reconstruction of the auricle. In: McCarthy JG, ed. Plastic Surgery. New York, NY: WB Saunders Co., 1990:2094.
6. Crikelair GF. A method of partial ear reconstruction for avulsion of the upper portion of the ear. Plast Reconstr Surg 1956; 17(6):438.
7. Converse JM. Reconstruction of the auricle, Part I. Plast Reconstr Surg 1958; 22(2):150.
8. Donelan MB. Conchal transposition flap for postburn ear deformities. Plast Reconstr Surg 1989; 83(4):641.
9. Dujon DG, Bowditch M. The thin tube pedicle: A valuable technique in auricular reconstruction after trauma. Br J Plast Surg 1995; 48(1):35.

Chapter 31

Eyelid Reconstruction

Amir H. Taghinia and Bernard T. Lee

Introduction

Surgical excision of eyelid tumors often produces large defects that can cause significant visual impairment and cosmetic stigma if not treated properly. Reconstruction requires thorough knowledge of anatomy and precise technical execution. Although there are many reconstructive options in the form of flaps and grafts, the basic principles are: 1) replace lining, support and skin cover; 2) restore corneal protection and lubrication; and 3) optimize the aesthetic outcome. This chapter briefly summarizes eyelid reconstruction while providing a useful, quick-reference algorithm (Table 31.1).

Relevant Anatomy

The cross-sectional anatomy of the upper and lower eyelids is shown in Figure 31.1. The eyelids consist of skin, orbicularis muscle, tarsal plate and conjunctiva. The upper lid also contains slips of the upper eyelid retractors (see below). The orbicularis oculi muscle consists of three separate divisions: pretarsal, preseptal and orbital. The pretarsal orbicularis is primarily responsible for involuntary blinking. The preseptal orbicularis pumps tear through the lacrimal system and aids in voluntary lid closure. The orbital orbicularis depresses the medial brow and performs protective forced eyelid closure.

Eyelid layers have been arbitrarily divided into the anterior and posterior lamellae. The anterior lamella contains the skin and orbicularis oculi muscle while the posterior lamella contains the tarsus, eyelid retractors and conjunctiva.

The tarsal plates contain vertically oriented Meibomian glands that exit on the lid margin. These glands secrete oils that mix with tears to provide lubrication for the conjunctiva. Medially and laterally, the tarsal plates narrow into fibrous bundles that ultimately converge to form the medial and lateral canthal tendons. The medial aspects of the upper and lower lids host the upper and lower punctae, respectively. Tears generated from the lacrimal gland drain via the punctae into the lacrimal canaliculi and ultimately into the lacrimal sac. The lacrimal sac empties into the inferior meatus via the nasolacrimal duct. Blinking enhances tear drainage by squeezing the lacrimal sac and forcing tears down the nasolacrimal duct.

Upper eyelid positioning is achieved through the action of the levator palpebrae superioris and Müller's muscles (the upper eyelid retractors). The levator muscle originates from the orbital cone and is innervated by the oculomotor nerve (CN III). As it approaches the upper lid, it broadens into the levator aponeurosis and becomes closely approximated with Müller's muscle before attaching to the upper tarsal plate. Müller's muscle is controlled by sympathetic nervous fibers.

Practical Plastic Surgery, edited by Zol B. Kryger and Mark Sisco. ©2007 Landes Bioscience.

Table 31.1. A basic algorithmic approach to eyelid reconstruction

		Defect Size		
		Up to 1/3	**1/3 to 2/3**	**>2/3**
Lower Eyelid	*Partial Thickness*	Primary Closure	FTSG or local flap	FTSG or local flap
	Full Thickness	Primary Closure	Lateral canthotomy and cantholysis with primary closure Semicircular (Tenzel) flap with primary closure Lateral advancement with Z-plasty (McGregor) and primary closure	Upper eyelid tarsoconjunctival flap (modified Hughes) covered by local flap or FTSG Chondromucosal graft covered by local flap or cheek advancement flap (Mustarde)
Upper Eyelid	*Partial Thickness*	Primary Closure	FTSG	FTSG
	Full Thickness	Primary Closure	Lateral canthotomy and cantholysis with primary closure Semicircular flap (Tenzel) with primary closure Lateral advancement with Z-plasty (McGregor) and primary closure Sliding upper eyelid tarsoconjunctival flap (modified Hughes) covered by local flap or FTSG	Lower eyelid bridging full-thickness flap (Cutler-Beard) Lower eyelid switch flap (Mustarde)

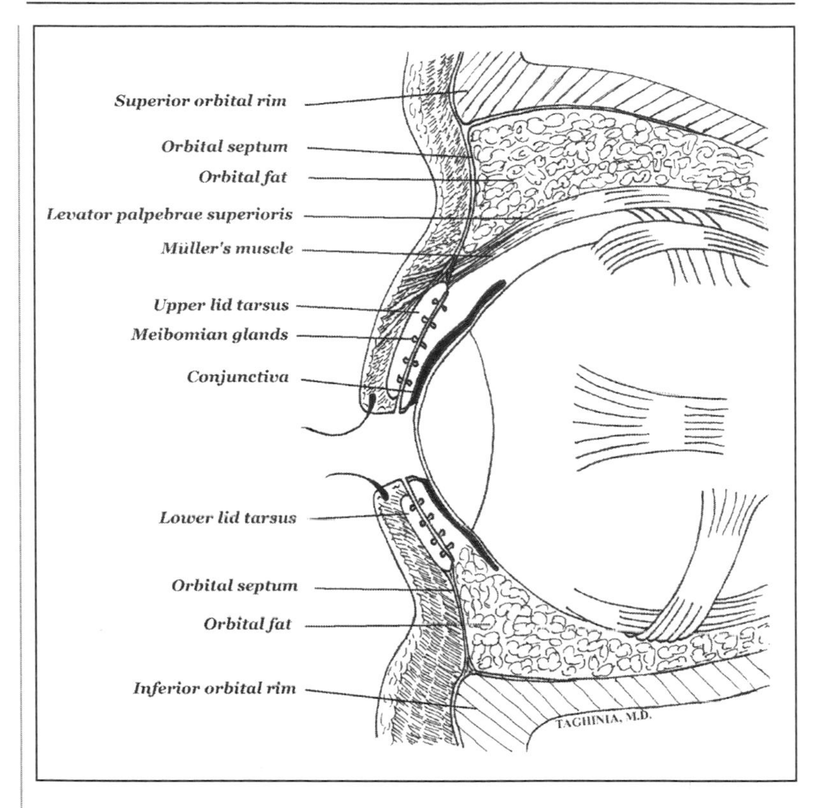

Figure 31.1. Cross-sectional anatomy of the upper and lower eyelids. Note that the levator palpebrae superioris and Müller's muscle are intimately associated. Müller's muscle attaches to the tarsal plate. The levator gives rise to distal fibers termed the levator aponeurosis. These fibers insert on the tarsal plate and the pretarsal orbicularis.

These fibers synapse in the superior cervical ganglion and intertwine as they ascend along the internal carotid artery into the cranium.

Eyelid Lesions

The skin and glands of the eyelids are susceptible to development of benign and malignant neoplasms. There is a higher preponderance of eyelid tumors in the lower eyelids (90% vs. 10% in upper eyelids). The most common malignant eyelid tumor is basal cell carcinoma. Squamous cell carcinoma of the eyelid is rare. Basal cell carcinoma has a high recurrence rate after excision and should be treated aggressively. Malignant melanoma in situ (lentigo maligna), though rare, also warrants wide excision; 5 mm margin is recommended. Eyelid reconstruction after surgical excision of tumors should be delayed until final pathological diagnosis and evaluation of margins is completed. The surgeon treats the open wound with dressings during this time.

Lower Eyelid Reconstruction

Partial-thickness defects of the lower eyelid may involve just skin or a combination of skin and orbicularis muscle. The reconstructive options for these defects include primary closure, local flap, full-thickness graft and split-thickness graft. Small defects can usually be closed primarily thereby avoiding excessive tension that may lead to ectropion (eversion of the eyelid). Larger defects may require one of a wide variety of local flaps that are available for lower eyelid reconstruction (Fig. 31.2). The bipedicled Tripier flap and medial/lateral skin-muscle flaps rely on ipsilateral upper eyelid tissue that provides excellent color match. The cheek flap also provides a good match in color and quality. The Fricke temporal brow flap, nasolabial flap and midline forehead flap are occasionally useful but provide mediocre color and texture matches.

If local flap coverage of a defect is not possible, the surgeon may use a full-thickness or split-thickness graft. Full-thickness grafts are superior to split-thickness grafts for eyelid reconstruction; they provide little or no contracture with better color and texture match. A full-thickness graft from the upper lid can cover small defects with excellent graft take and minimal donor site scar. For larger defects, skin from both upper eyelids can be used. For defects that involve skin and orbicularis muscle, a composite graft from the upper eyelids usually heals well. Full-thickness grafts may also be harvested from behind the ear and above the clavicle.

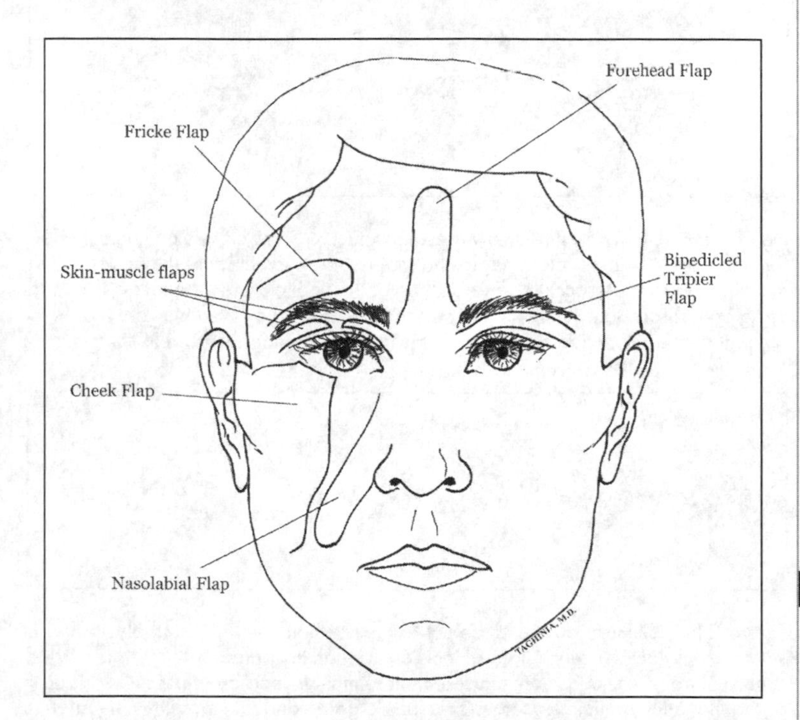

Figure 31.2. Local facial flaps for eyelid reconstruction. The medial and lateral skin-muscle flaps and the bipedicled Tripier flaps are mainly used for lower eyelid reconstruction.

Full-thickness defects of the lower eyelid involve the skin, orbicularis muscle, tarsus and conjunctiva. The reconstructive hierarchy for these defects is as follows:

1. Primary closure
2. Lateral canthotomy (division of the canthus) with cantholysis (division of the canthal ligament) and primary closure
3. Lateral extension with semicircular flap (Tenzel flap)
4. Lateral extension with Z-plasty (McGregor flap)
5. Upper lid tarsoconjunctival flap (modified Hughes flap) with skin graft or local flap
6. Cheek advancement flap (Mustarde flap) with or without a chondromucosal graft

Primary closure can usually be achieved for defects one-fourth to one-third of the lid margin. Parallel excision of the tarsal plates should be performed to avoid notching (Fig. 31.3). Absorbable sutures approximate the tarsus and nonabsorbable sutures approximate the skin. If a gap of a few millimeters prevents primary closure, a lateral canthotomy with cantholysis and closure is required. The lateral palpebral fissure is incised and the lower limb of the lateral canthal ligament is detached, thus allowing relaxation of the lateral portion of the lower lid. The lateral canthotomy incision is often carried superolaterally to create a semicircular flap for additional mobility (Fig. 31.4). An extra few millimeters can also be gained by adding a Z-plasty

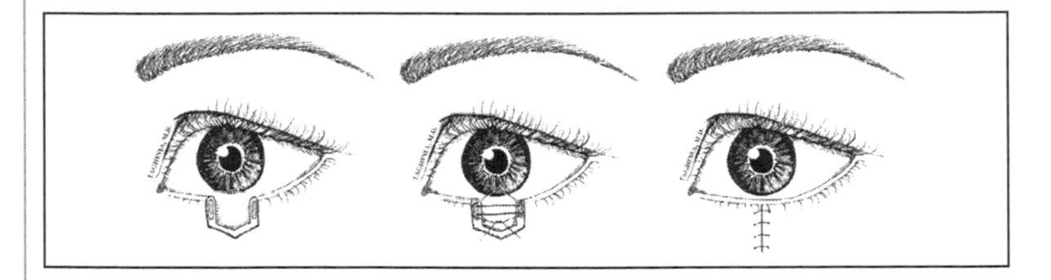

Figure 31.3. Primary closure of lower eyelid defect. Primary closure can be achieved for small defects. Tarsal plate edges that are not perpendicular to the upper eyelid margin tend to cause buckling or notching of the final closure. The tarsus is closed using fine absorbable sutures. The knots must be tied away from the conjunctiva to prevent corneal irritation. The skin is closed with fine nonabsorbable sutures.

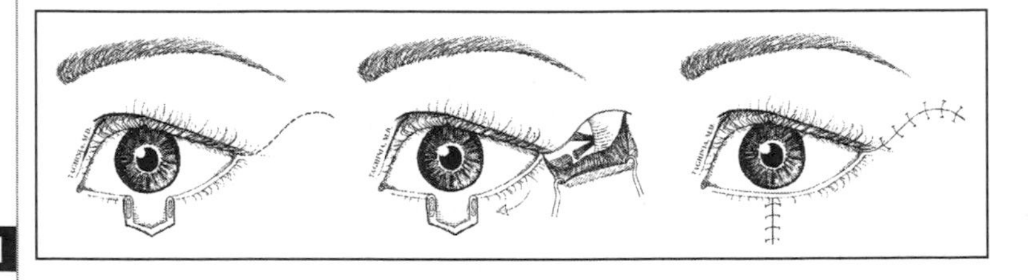

Figure 31.4. Closure of eyelid defect using canthotomy with cantholysis and a semicircular flap (Tenzel). A lateral incision through the palpebral fissure is carried superolaterally. Although Tenzel used a full semicircle incision, most surgeons now agree that a full semicircle is not necessary to gain adequate mobility. The inferior limb of the canthal ligament is divided (middle illustration). The tissues are mobilized medially and the wound is closed. The tarsus is approximated by the method outlined in Figure 31.3.

to the lateral incision (Fig. 31.5). All of these described techniques provide lining, structural support and cilia-bearing skin cover.

These maneuvers are usually not adequate for large defects (>50%). In such cases, lining, support and cover are required via other means. Two options exist for lining and support: the upper eyelid tarsoconjunctival flap (modified Hughes) and chondromucosal composite grafts. The tarsoconjunctival flap (modified Hughes) involves harvesting a flap of conjunctiva with attached tarsus from the upper eyelid (leaving 3-4 mm of inferior tarsal edge intact and separating the levator and Müller's muscles) and advancing this superiorly-based flap to the lower eyelid defect (Fig. 31.6). These flaps can cover up to 50% of lower eyelid posterior lamella defects. Coverage is obtained using a skin graft, local skin-muscle flap, or large cheek advancement flap (Mustarde). Larger lower lid defects prohibit borrowing from the upper lid for fear of upper lid distortion.

Cartilage, mucosal, or composite (chondromucosal) grafts are a better option for lining and support in larger defects. Options for these grafts include auricular cartilage grafts, mucosal grafts from the mouth (hard palate or buccal mucosa) and nasal septal chondromucosal grafts. Choice of graft is dictated by size of defect and the surgeon's familiarity with the technique. Ear cartilage grafts are commonly used

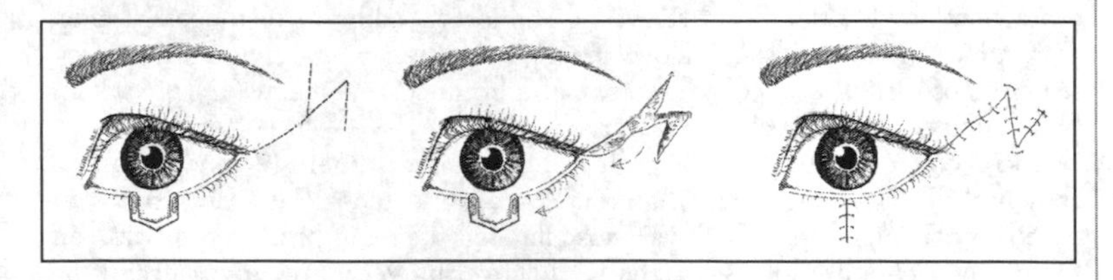

Figure 31.5. Lateral extension with Z-plasty (McGregor). In the original description, the lateral extension was aimed superolaterally (shown) in the curve of the lower eyelid to minimize the risk of cicatricial ectropion. However, this Z-plasty technique can also be used to gain additional mobility after making a semicircular flap as outlined in Figure 31.4. The limbs of the Z-plasty must be oriented vertically.

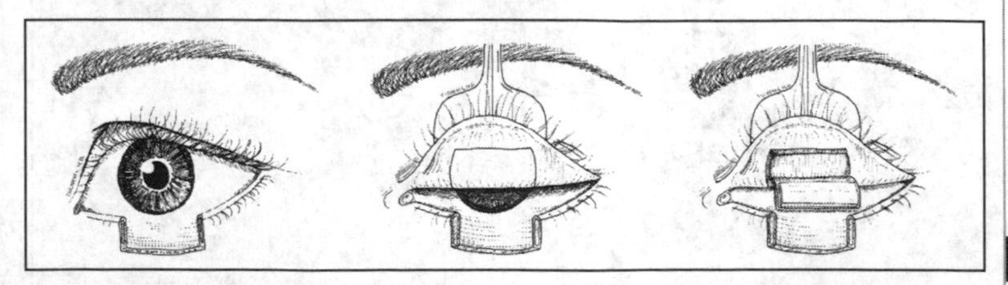

Figure 31.6. Tarsoconjunctival flap (modified Hughes). Although used mainly for lower eyelid reconstruction, this flap can also be applied for upper eyelid reconstruction. The distal end of the flap is designed 3-4 mm above the upper eyelid margin to leave enough tarsus for upper eyelid stability. The superiorly-based composite tarsoconjunctival flap is raised (leaving Müller's muscle behind) and sutured into the defect.

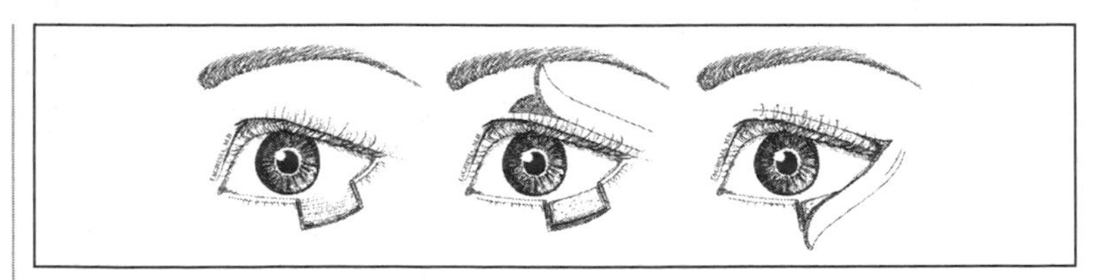

Figure 31.7. Cartilage-mucosa graft for lower lid reconstruction. Large lower lid defects usually require replacement of missing tarsoconjunctiva with free cartilage with or without attached mucosa. The graft can be covered with a laterally-based skin-muscle flap (shown) or by other means (see text for detail).

in larger defects. Thin cartilage from the scaphoid fossa is strong yet pliable, donor site morbidity is low, and the exposed surface of the cartilage becomes epithelialized by surrounding conjunctiva over several weeks. Hard palate mucosal grafts are advantageous because of intrinsic support and mucosal lining. In contrast, buccal mucosal grafts are too thin for support and their primary uses are for lining and lid margin reconstruction. Nasal septal chondromucosal grafts are advantageous because they provide significant structural support in addition to mucosal lining. Availability and ease of access allows these grafts to be commonly used for reconstruction of total or near-total lid loss. Chondromucosal grafts need flap coverage (with an adequate vascular bed) to survive.

Skin cover options for large lower lid defects include medial/lateral skin-muscle flaps (Fig. 31.7), large cheek advancement flaps (Fig. 31.8) or bipedicled Tripier flaps. These well-vascularized flaps can provide nourishment for primary chondromucosal grafts while covering very large lower eyelid defects (>75%).

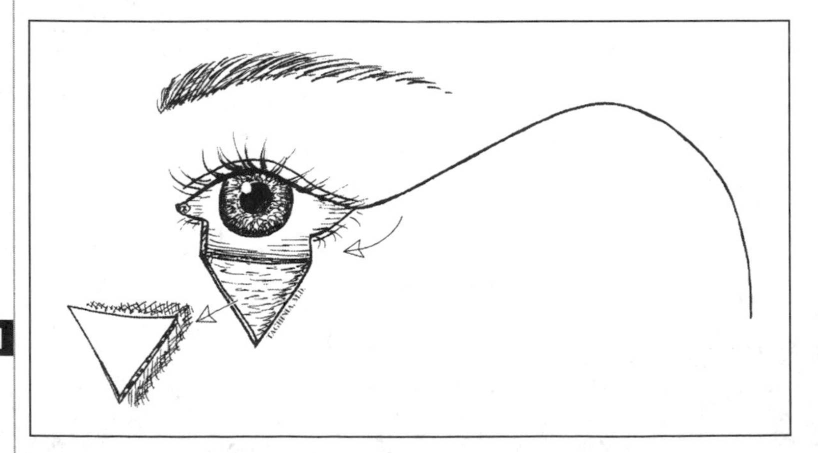

Figure 31.8. Advancement cheek flap (Mustarde). A large cheek flap can be raised to cover large defects of the lower eyelid. A triangular excision is necessary prior to flap inset. Support and lining are provided by a graft (see text for detail).

Upper Eyelid Reconstruction

Small partial defects of the upper eyelid can be closed primarily with a skin graft or a local flap. Local flap choices are somewhat limited but include a V-Y flap from lateral eyelid or temporal skin (best choice) and a midline forehead flap (poor choice). If local flaps are not available, skin grafts may provide adequate coverage. The best donor site is the contralateral upper eyelid; however it provides limited tissue. Other choices include grafts from behind the ear and the inner upper arm. Additional skin graft length should be used to account for graft contracture.

Full-thickness defects of the upper eyelid are addressed by:

1. Primary closure
2. Lateral canthotomy and cantholysis with primary closure
3. Lateral extension with semicircular flap (Tenzel)
4. Lateral extension with Z-plasty (McGregor)
5. Sliding upper eyelid tarsoconjunctival flap (modified Hughes) covered by skin graft or local flap
6. Lower eyelid full-thickness switch flap (Mustarde) or bridge flap (Cutler-Beard flap)

These methods are similar to methods for correction of lower eyelid defects. However, because constant blinking causes significant surface interaction between the upper eyelid and the cornea, inner irregularities in the upper eyelid are less forgiving. Sutures in the conjunctiva or thick cartilage grafts can scratch the cornea and lead to keratitis.

Full-thickness defects of the upper eyelid are addressed by primary closure, a semicircular flap or the Cutler-Beard flap. These methods are similar to the methods for correction of lower eyelid defects described above. However, because constant blinking causes significant surface interaction between the upper eyelid and the cornea, inner irregularities in the upper eyelid are less forgiving. Sutures in the conjunctiva or thick cartilage grafts can scratch the cornea and lead to keratitis.

Primary closure of full-thickness upper eyelid defects is feasible for defects up to 25% of the upper eyelid. As in the lower eyelid, the incision through the tarsus is made perpendicular to the lid margin to avoid buckling of the tarsus. Absorbable sutures approximate the tarsus and nonabsorbable sutures approximate the eyelid skin (Fig. 31.9). If primary closure cannot be achieved, a lateral canthotomy and

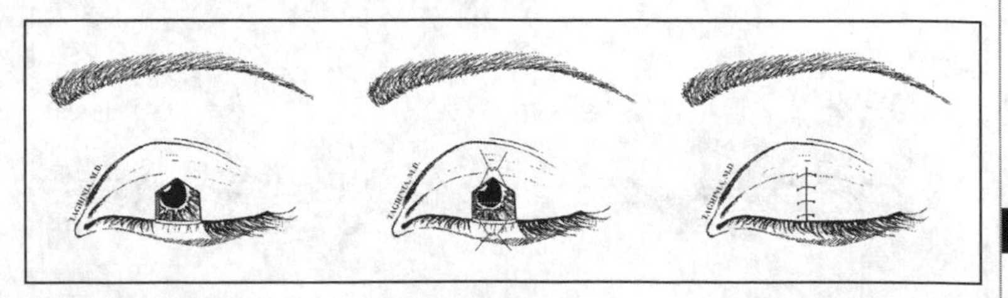

Figure 31.9. Primary closure of upper lid defect. Primary closure can be achieved for defects smaller than 25% of the lid. Closure method is similar to closure in small lower lid defects (see Fig. 31.3). However, because of its repetitive gliding action on the cornea, the upper eyelid is less forgiving of inner lining irregularities.

canthoysis can be performed. A lateral incision is made in the canthus, and a relaxing incision is made in the upper limb of the canthal tendon.

If a larger defect is present, additional mobility is usually obtained by performing a semicircular flap. Defects up to 50% can be closed in this manner. With the canthotomy and cantholysis already performed, the lateral incision of the canthotomy is extended inferolaterally in a semicircular fashion. The flap is then advanced into the defect (Fig. 31.10). Similar to lower lid reconstruction, a Z-plasty (McGregor) gives the semicircular flap additional mobility. The Cutler-Beard flap may be utilized for defects greater than 50% of the upper eyelid. This flap uses full-thickness tissue from the lower eyelid to reconstruct the upper eyelid. A full-thickness horizontal incision is made just inferior to the lower tarsus. Vertical incisions are made inferiorly from the lateral edges of the horizontal incision. The flap is then advanced superiorly under the intact lower tarsal bridge (Fig. 31.11). The flap is divided after 6-8 weeks. Table 31.1 provides a summary of the above-mentioned procedures and serves a guide to eyelid reconstruction. Although many other eyelid reconstructive techniques have been described, they are beyond the scope of this limited chapter.

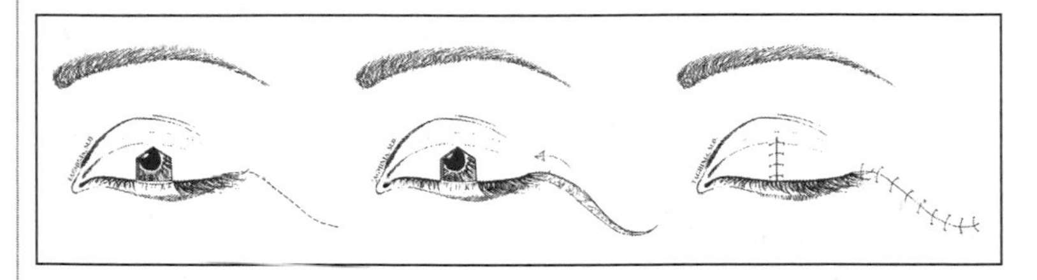

Figure 31.10. Semicircular flap (Tenzel) for upper eyelid defects. After canthotomy and cantholysis (not shown), the lateral incision is extended inferolaterally. Additional mobility can be obtained with a Z-plasty (McGregor) in the manner shown in Figure 31.5.

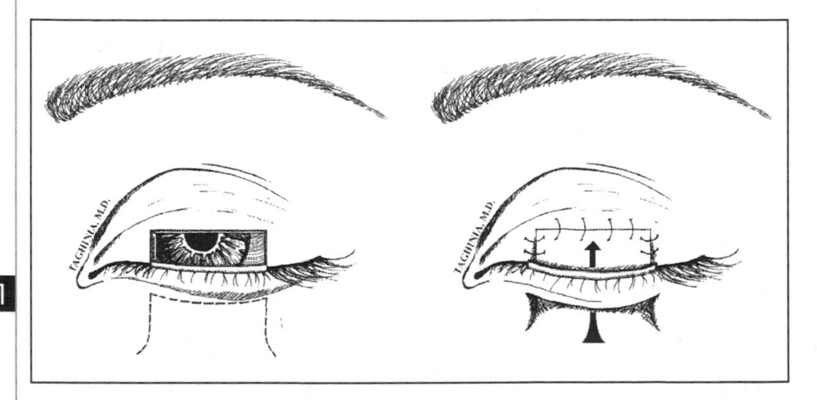

Figure 31.11. Lower eyelid bridging flap (Cutler-Beard). Large (>50%) defects of the upper eyelid can be closed with lower eyelid tissue as shown. The proposed flap is tunneled under a lower eyelid bridge to provide closure. The base of the flap is then divided at a second stage.

Complications

Some of the more common complications seen in eyelid reconstruction include:

- Asymmetry
- Lower lid laxity
- Lagophthalmos
- Ectropion
- Intropion
- Infection
- Conjunctivitis
- Trichiasis (turning inward of the lashes)

Pearls and Pitfalls

- Defects less than 25% of the eyelid can almost always be repaired by a layered primary closure.
- Small, medial, partial-thickness defects can be allowed to granulate instead of primary closure.
- Only use mucosa for reconstructing conjunctival defects, never skin. Remember that the graft will significantly contract in size so harvest a graft much larger than the defect.
- Convert a wedge shaped defects into a pentagonal shape before closure. This will help minimize tension that can lead to lid retraction.
- Always approximate the grey-line as a first step in closure of upper lid defects.
- A useful technique for lid support is to turn over a laterally-based strip of periosteum that can substitute for tarsus.

Suggested Reading

1. Achauer BM, Eriksson E, Guyuron B et al. Plastic Surgery: Indications, Operations, and Outcomes. St. Louis: Mosby Inc., 2000.
2. Aston SJ, Beasley RW, Thorne CHM. Grabb and Smith's Plastic Surgery. 5th ed. Philadelphia: Lippincott-Raven Publishers, 1997.
3. Evans GRD. Operative Plastic Surgery. New York: McGraw-Hill Companies, 2000.
4. Della Rocca RC, Bedrossiam EH, Arthurs BP. Ophthalmic Plastic Surgery: Decision Making and Techniques. New York: McGraw-Hill Companies, 2002.
5. Larrabee Jr WF, Sherris DA. Principles of Facial Reconstruction. Philadelphia: Lippincott-Raven Publishers, 1995.

Chapter 32

Nasal Reconstruction

Clark F. Schierle and Victor L. Lewis

Introduction

Reconstruction of the nose poses a particularly visible and unforgiving challenge for the reconstructive surgeon. Its central location in the face makes it a natural focal point, and contours, scars and textures must be precisely planned. The nose's prominent location also subjects it to more than its fair share of ultraviolet radiation, and by far the most common reason for nasal reconstruction is a post-surgical defect from the removal of a skin cancer, typically basal cell carcinoma. All layers of full-thickness defects must be reconstructed, and aesthetic subunit principles should be obeyed whenever possible. Reconstructive options range from skin grafts to complex free-tissue transfer.

Anatomy and Aesthetic Considerations

The nose is comprised of an inner mucosal lining, an osteocartilagenous skeleton, and an external layer of skin. Thus any reconstructive effort must ensure that all three of these elements are restored. The nasal skeleton consists of the paired nasal bones in the upper third, upper lateral cartilages in the middle third, and the lower lateral cartilages in the lower third. The nasal septum provides midline support and consists of the quadrangular cartilage, the perpendicular plate of the ethmoid and the vomer. The caudal edge of the nasal bones overrides and attaches to the upper lateral cartilages, suspending them above the nasal cavity. The internal nasal valve is the opening between the caudal end of the upper lateral cartilage and the nasal septum. The external nasal valve is the region caudal to this, consisting of the nasal alae laterally and the septum and columella medially. The lower lateral cartilages are divided into medal and lateral crura. The medial crura meet the caudal septum in the midline while the lateral crura attach to the pyriform aperture helping provide further support to the nasal vault. The chapter on nasal and NOE fractures has two useful figures of the septal and bony anatomy.

The surface anatomy of the nose is generally divided into a series of aesthetic subunits, first described by Millard and refined by Burget. These are the tip, dorsum, sidewalls, alae and soft triangles (Fig. 32.1). The borders of the subunits represent natural points of inflection which can serve to conceal scars quite satisfactorily. Partial replacement of a subunit results in scars lying within rather than between adjacent subunits and consequently a far more visually discordant light reflex. Furthermore, one must take into account the natural contour of the different subunits when choosing the tissue with which to replace it. Skin grafts will tend to contract in a flat manner and are ideally suited for replacing sidewall defects. Convex subunits lend themselves to replacement with full-thickness flaps, which naturally evolve into spherical shapes as they heal through centripetal contraction.

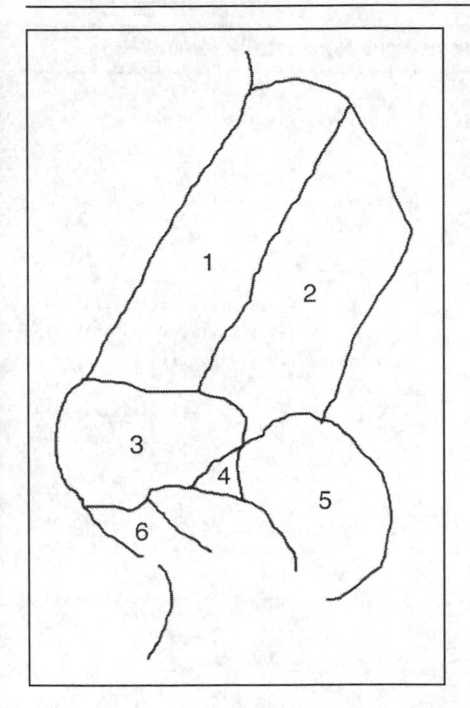

Figure 32.1. Surface anatomy of the nose: the aesthetic subunits. (1) Dorsum, (2) sidewall, (3) tip, (4) soft triangle, (5) alar-nostril sill, (6) columella.

It must also be noted that the thickness of the skin of the nose varies considerably. The skin of the upper dorsum and sidewalls of the nose (zone I) is smooth, thin, relatively nonscbaceous and moves fairly easily over the underlying skeleton. The skin of the supratip, tip and alae (zone II) is thick, dense and sebaceous. Finally the skin of the soft triangles, alar magins, infratip and columella (zone III) is smooth, thin and relatively nonsebaceous, but unlike the dorsum and sidewalls is densely adherent to the underlying cartilaginous skeleton and does not move easily.

Preoperative Considerations

As with all surgery, routine preoperative risk stratification should be undertaken, particularly when planning an extensive multi-staged reconstruction. A plan should be outlined for "replacing like with like." A functional nose must possess three basic elements: lining, support and cover. Options for replacing these tissues are outlined in Table 32.1. Several authors advocate the use of a preoperative or intraoperative template to plan the design of flaps. Classically the aluminum foil wrapper from a chromic suture is used to create a three dimensional template of the nasal subunits which need to be replaced. This template is then flattened out to reveal the actual size of the flap or skin graft which needs to be harvested. Skin defects comprising greater than 50% of a given aesthetic subunit should be enlarged to encompass the entire subunit to avoid noticeable scars and contour deformities within individual subunits. One must be cautious when analyzing the defect. Factors such as edema, scarring, previous attempts at repair, wound contracture, secondary healing, gravity and skin tension can all distort the true size and shape of the defect, and one must take any or all of these factors into account.

Table 32.1. Autogenous tissue options for nasal reconstruction

Lining
Full-thickness skin graft
Turnover flap
Nasolabial lining flap
Bipedicle alar margin ribbon flap
Contralateral mucoperichondrial flap
Septal pivot flap
Microvascular free flap
Support
Septal cartilage
Conchal cartilage
Costal cartilage (6th through 9th ribs)
Costocondral junction graft (usually 8th rib)
Cranial bone graft
Iliac crest bone graft
Costal bone graft
Surface Coverage
Local advancement flap
Preauricular skin graft
Bilobed flap
Nasolabial flap
Paramedian forehead flap
Scalp flap
Microvascular free flap

Operative Technique

Primary Closure

The relatively mobile nature of zone I skin can allow for primary closure of some small defects. As always, lines of relaxed skin tension should be utilized when able. The relatively thick, immobile nature of zone II and III skin makes for difficult mobilization. If primary approximation of the wound results in unacceptable tension or deformity, a skin graft or local flap should be used.

Skin Graft

Skin grafts can be of use in reconstruction of fairly superficial defects of the nose, particularly the nasal sidewalls which are planar subunits and well approximated by the flat contraction of a skin graft (as opposed to the convex contraction of a flap reconstruction). Appropriate donor sites in terms of color and texture match include preauricular and supraclavicular skin. Full-thickness skin should be used to minimize contraction and provide the best match for the depth of the defect. When relevant, perichondrium and periosteum at the recipient site should be preserved to facilitate skin graft take. Full-thickness defects including some nasal cartilage can also be addressed through an appropriately designed composite graft including auricular skin and cartilage.

Locoregional Flaps

The relatively mobile skin of the nasal dorsum and sidewalls can be used in a typical V-Y advancement fashion for small defects. The bilobe and rhomboid flaps can be used to address small defects of the nasal dorsum and sidewall, but in practice often generate distorting dog ears which must be carefully planned so as not to distort the normal contours of the nasal surface. The skin of the glabellar region can be mobilized in an advancement, V-Y, or transposition fashion to address defects of the upper third of the dorsum or sidewall.

The nasolabial flap has been used for reconstruction of defects of the nasal alae since the earliest descriptions of facial plastic surgery. The flap can be advanced or rotated into place based on an inferior or superior pedicle respectively, relying on random extensions of an axial blood supply derived from the angular branch of the facial artery. The flap provides reliable coverage, and the donor defect is easily concealed in the natural crease of the nasolabial fold. The superiorly based flap generally requires secondary revision of the cone of tissue generated by rotation of the flap into place. The inferiorly based flap results in a donor defect which can often be closed primarily and requires revision only to correct any excessive distortion of lip height.

The paramedian forehead flap is the workhorse for larger full-thickness defects of the lower two-thirds of the nose. Forehead skin is the ideal donor for the thick, sebaceous skin of zone II, and convex contracture of the flap results in an ideal contour match for the nasal tip and alae. The flap is based on random extensions of axial blood supply from both the supratrochlear and supraorbital arteries. The flap is designed over the contralateral supratrochlear artery to allow for greater ease of rotation. The base should include approximately 1.5 cm of width, with incisions designed to fall naturally into the procerus and corrugator skin creases. The distal portion of the flap is shaped based on a foil suture-package pattern designed to match the nasal subunits requiring replacement, taking care to accurately account for shortening of the flap with rotation. The distal flap is elevated in the subdermal plane to better approximate the depth of the defect it will be filling. The remainder of the flap is transitioned to a submuscular plane to optimize the vascular pedicle. The flap is divided after a period of three weeks allowing for inosculation of the distal flap.

Composite Flaps and Free Tissue Transfer

Significant loss of underlying structural elements and nasal mucosal lining require adequate replacement. Conchal, septal or rib cartilage may be harvested and shaped into structural support grafts to provide stability for overlying soft tissue reconstructions. Nasal lining may be provided by skin grafts, locoregional flaps or free microvascular tissue transfer. Contralateral mucoperichondrial flaps and facial artery musculomucosal flaps have been described for nasal lining. Recent reports have described the use of radial forearm skin as a thinned free flap for replacement of nasal lining in extreme defects.

Postoperative Care

The extent of postoperative care depends on the complexity of the repair. Head elevation, cold compresses and avoidance of nose blowing are recommended. Incisions should be washed daily to avoid crusting which makes suture removal very difficult. Intranasal saline spray should be used when needed. Skin grafts can be bolstered using cotton soaked in mineral oil and covered with Xerform® which is held in place with nylon sutures. Splints may be employed if significant osteocartilagenous reconstruction was undertaken. Use of devascularized tissues (cartilage, skin, or bone grafts) generally indicates some period of postoperative antibiotic coverage. Patients should be adequately counseled as to the multi-stage nature of more extensive reconstructions.

Pearls and Pitfalls

- Most surface contours of the nose are either flat (such as the dorsum) or convex (such as the alae). Reconstruction of flat surfaces is best done with a skin graft that contracts in a linear fashion, whereas convex surfaces should be reconstructed with a flap that contracts in a spherical manner.
- When designing a flap for nasal reconstruction, for example the forehead flap, it is critical to account for loss of flap length that results from the arc of rotation. A Raytek® sponge can be used to determine the designed length of the flap. By holding one end of the sponge over the base of the flap and rotating the other end into the defect, the amount of extra length needed to overcome the arc of rotation can be determined.
- There are exceptions to the rule of replacing "like with like" tissue. For example, the alar rims normally have a convex shape even though they do not contain cartilage. However, their convexity can best be restored using cartilaginous support.

Suggested Reading

1. Burget GC, Menick FJ. The subunit principle in nasal reconstruction. Plast Reconstr Surg 1985; 76(2):239.
2. Burget GC, Menick FJ. Nasal reconstruction: Seeking a fourth dimension. Plast Reconstr Surg 1986; 78(2):145.
3. Menick FJ. Artistry in aesthetic surgery. Aesthetic perception and the subunit principle. Clin Plast Surg 1987; 14(4):723.
4. Millard Jr DR. Aesthetic reconstructive rhinoplasty. Clin Plast Surg 1981; 8(2):169.
5. Singh DJ, Bartlett SP. Aesthetic considerations in nasal reconstruction and the role of modified nasal subunits. Plast Reconstr Surg 2003; 111(2):639.

Chapter 33

Lip Reconstruction

Amir H. Taghinia, Edgar S. Macias, Dzifa S. Kpodzo and Bohdan Pomahac

Introduction

The lips are not only a major aesthetic component of the face, but are also important for facial expression, speech and eating. Goals in lip reconstruction are to restore normal anatomy, oral competence and contour. These goals are easily attained following repair of small lip defects. However, restoring these characteristics of the lips in large defects remains a more arduous task. Although many different methods of lip reconstruction have been described in the literature, a few of the important and more commonly utilized methods are outlined in this chapter.

Anatomy

The subunits of the surgical upper and lower lips are shown in Figure 33.1. The surgical upper lip includes the entire area from one nasolabial fold to the other, and all structures down to the oral orifice. It extends intraorally to the upper gingivolabial sulcus. It is divided into the vermilion, one central and two lateral aesthetic subunits. The lower lip includes all structures superior to the labiomental fold including the vermilion and continuing intraorally to the inferior gingivolabial sulcus.

Extending from the nasal base are bilateral philtral columns flanking the centrally located philtrum (Fig. 33.2). The philtral columns extend downward to meet the vermilion-cutaneous junction (also known as the 'white roll') of the upper lip. Cupid's bow is the portion of the vermilion-cutaneous junction located at the base of the philtrum. The tubercle is the fleshy middle part of the upper lip from which the vermilion extends bilaterally to meet the commissures. The vermilion of the lower lip is bisected by the central sulcus which is prominent in some individuals. The lower lip is considered less anatomically complex than the upper lip because it lacks a definitive central structure.

The vermilion is made of a modified mucosa with submucous tissue and orbicularis oris muscle underneath. The large number of sensory fibers per unit of vermilion is reflected in its comprising a disproportionately large part of the cerebral cortex. It has a high degree of sensitivity to temperature, light touch and pain. The natural lines of the vermilion are vertical, thus scars on the vermilion should be placed vertically if possible.

The muscular anatomy of the lips is shown in Figure 33.3. The primary muscle responsible for oral competence is the orbicularis oris muscle. This muscle functions as a sphincter, puckering and compressing the lips. The fibers of the orbicularis oris muscle extend to both commissures and converge with other facial muscles just lateral to the commissures at the modiolus. The major elevators of the upper lip are the levator labii superioris, levator anguli oris and the zygomaticus major.

Practical Plastic Surgery, edited by Zol B. Kryger and Mark Sisco. ©2007 Landes Bioscience.

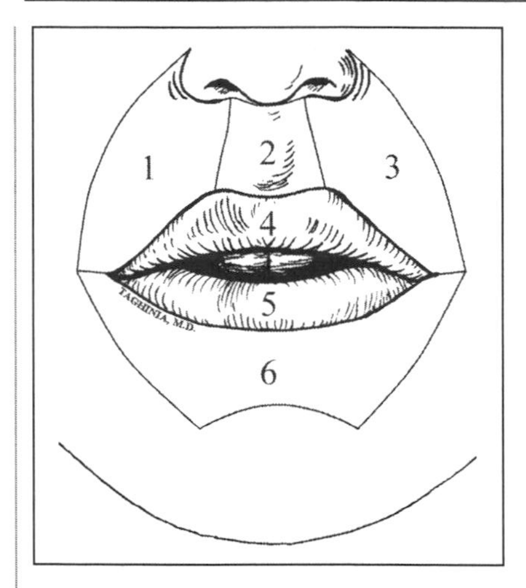

Figure 33.1. Subunits of the surgical upper and lower lips. The nasolabial folds on either side comprise the lateral borders of the upper and lower lips. The upper lip is made of the upper vermilion (4), two lateral subunits (1,3) and one central subunit (2) and the philtrum. These subunits are separated by the philtral columns and the white roll. The lower lip is made of the lower vermilion (6) and a large central unit that ends inferiorly at the labiomental fold.

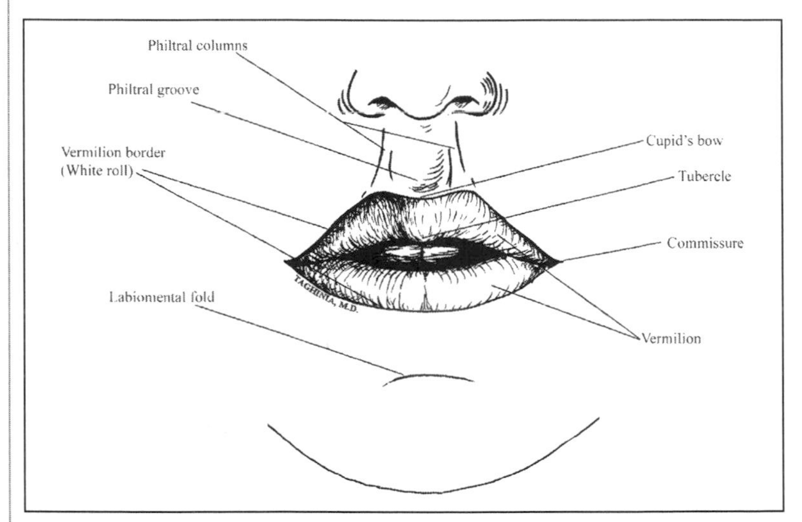

Figure 33.2. Topographic anatomy of the lips.

The mentalis muscle elevates and protrudes the middle portion of the lower lip. The major depressors of the lips are the depressor labii inferioris and depressor anguli oris. The risorius muscle pulls the commissures laterally.

The blood supply to the lips comes from the superior and inferior labial arteries, which in turn are branches of the facial arteries (Fig. 33.3). The paired superior and inferior labial arteries form a rich network of collateral blood vessels, thus providing a dual blood supply to each lip. These vessels lie between the orbicularis oris and the buccal mucosa near the transition from vermilion to buccal mucosa. There are no specific veins; instead there are several draining tributaries that eventually coalesce

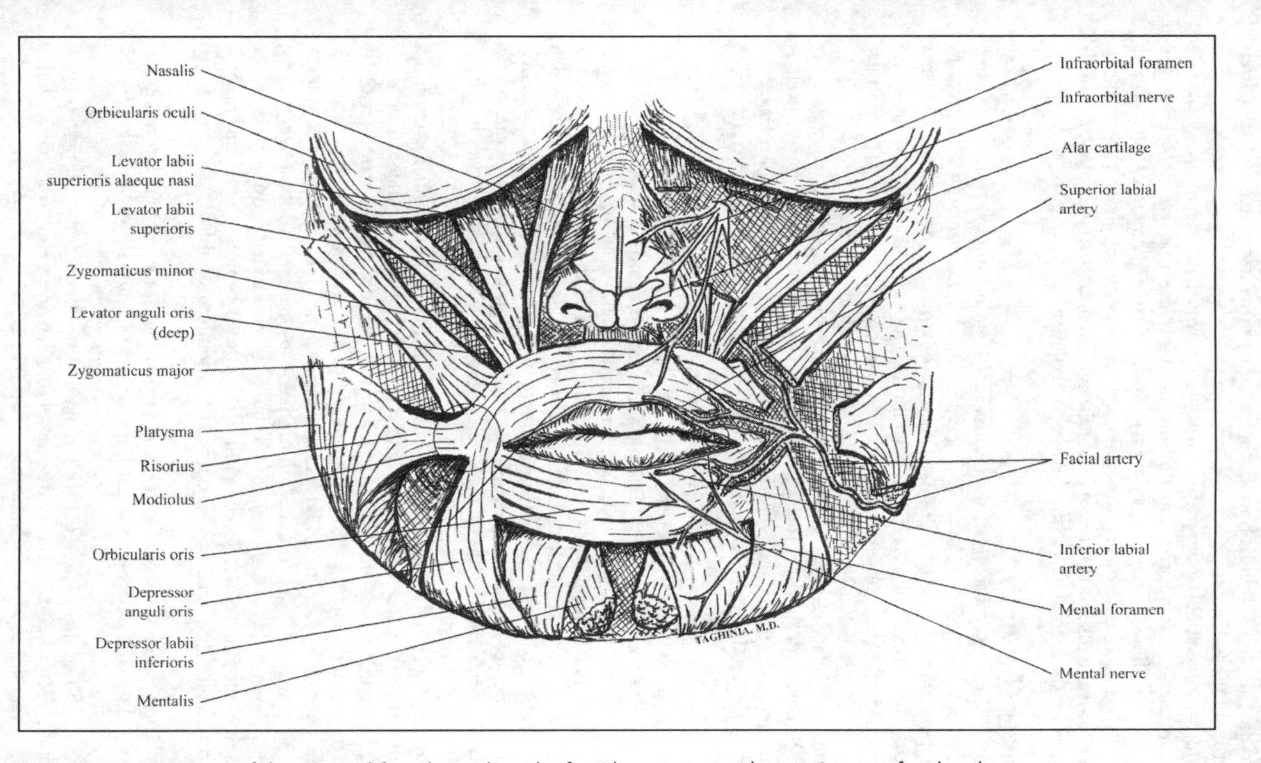

Figure 33.3. Anatomy of the perioral facial muscles. The facial nerve is not shown. See text for details.

into the facial veins. The lymphatic channels of the upper lip and lateral lower lip drain into the submandibular nodes; whereas, the central lower lip lymphatics drain into the submental nodes.

Motor innervation of the perioral muscles is from facial nerve branches. The buccal branches of this nerve supply motor input into the lip elevators; whereas, the marginal mandibular branches supply the lip depressors. The motor nerve enters each individual muscle on its posterior surface. Sensory supply to the upper lip comes from the infraorbital nerve (second trigeminal branch) and the lower lip is supplied by the mental nerve (third trigeminal branch).

Primary Closure

Lip lesions are typically due to trauma, infection, or tumors. Defects less than one-fourth to one-third of the total lip length can be closed primarily. This involves the apposition of the lateral margins of the wound on both sides and direct layered closure. The muscle is approximated with interrupted deep absorbable sutures. The white roll is closely approximated and then the labial mucosa and vermilion are closed. Finally, the skin is closed with fine nonabsorbable sutures.

Ideally, primary closure should cause minimal aesthetic and functional deformity; however, it can sometimes result in reduction of the oral aperture as well as asymmetry of the involved lip. Furthermore, primary closure in the upper lip can be problematic because opposing the edges of a large wound may create unfavorable distortion of the philtrum.

Vermilion Reconstruction

The vermilion spans the entire length of the oral aperture, becoming increasingly narrow and tapering laterally as it approaches the commissure on both sides. It forms the transition zone between skin and mucosa of the inner mouth. Defects involving the vermilion can range from superficial, such as leukoplakia in which there is limited compromise of the integument, to significant, in which tissue deficit extends to deeper muscle and mucosal tissue. Although small defects of the vermilion can be primarily closed or left alone to heal by secondary intention, larger defects require reconstruction.

Precise alignment of the vermilion-cutaneous margin on both sides ensures a curvilinear appearance of the border and avoids step-offs or lip notches after healing. The traditional labial mucosal advancement flap can replace vermilion resections that span the entire length of the lower lip. The mucosa on the buccal surface of the lower lip is undermined and advanced to the previous mucocutaneous junction. Maximal use of blunt undermining helps to preserve sensory innervation of this vermilion-to-be. Additional advancement can be achieved using a transverse incision in the gingivobuccal sulcus and in the process creating a bipedicled mucosa flap based laterally (Fig. 33.4). Extensive flap mobilization usually results in an insensate flap.

A notched appearance of the vermilion can result from scar contractures or vermilion volume deficiency (due to previous surgery or trauma). Scar contractures can be released with a Z-plasty. This procedure recruits vermilion tissue on either side of the scar to functionally lengthen the scar in the antero-posterior and supero-inferior direction. A notched appearance due to volume deficiency can be corrected with a local musculomucosal V-Y advancement flap (Fig. 33.5).

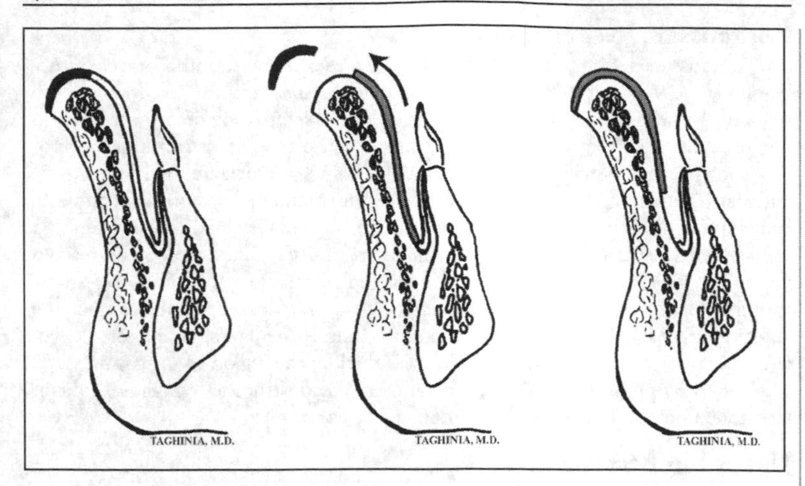

Figure 33.4. Vermilion reconstruction using labial mucosal advancement flap—cross-sectional view.

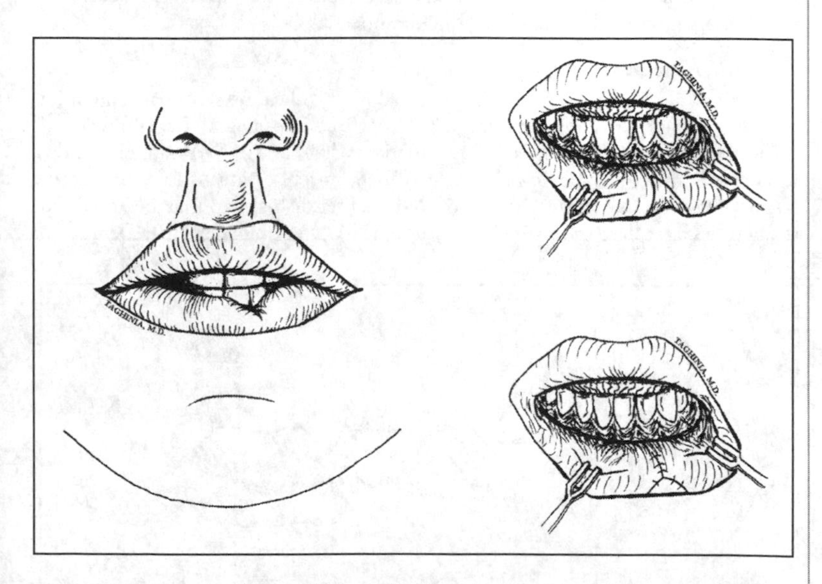

Figure 33.5. Repair of lower lip vermilion notch using V-Y advancement flap.

The next option of donor tissue is a flap from the ventral surface of the tongue but it is less than ideal because of color mismatch. Pribaz described the facial artery musculomucosal (FAMM) flap, which is a based on the facial artery and is used to reconstruct defects involving vermilion, lip, palate and a host of other oral structures. Labia minora grafts can also be used to reconstruct the vermilion.

Commissure Reconstruction

Commissure deformities often result from electrical burns, trauma, or reconstructive lip surgery. For post-burn commissure contractures, splinting techniques have reduced the need for surgical correction. Nevertheless, repairing deformities that do not respond to conservative measures remains complex. The intricate network of adjoining perioral muscle fibers at the modiolus (which is crucial for oral competence and facial animation) is nearly impossible to reconstruct. Furthermore, the contralateral commissure is the gold standard of comparison when evaluating the results of a unilateral reconstruction, thus leaving little room for discrepancy. Various approaches attempt to repair mucosal defects involving the commissure including the simple rhomboid flap, in which intraoral mucosa is advanced to reconstruct the commissure angles after an incision is made to widen the commissure laterally. The tongue flap also may be used when the mucosal defect is thick in the region of the commissure. Despite many proposed techniques, commissure reconstruction remains a difficult task and attempts at reconstruction often yield poor results.

Upper Lip Reconstruction

Upper lip cancers are usually basal cell carcinomas that spare the vermilion. The central aesthetic subunit of the upper lip, the philtrum, makes upper lip reconstruction more challenging than lower lip reconstruction. Upper lip defects can be divided into partial-thickness and full-thickness defects.

Partial-thickness Defects

Partial-thickness philtral defects can be allowed to heal by secondary intention or skin grafting. The triangular fossa skin-cartilage composite graft is well-described for reconstructing the philtrum in burn patients (Fig. 33.6). Partial-thickness defects of the lateral subunits can be repaired by a variety of means (Fig. 33.7). For larger lateral subunit defects, an inferiorly-based nasolabial flap may be employed (sometimes to replace the entire lateral subunit). Upper lip defects that are next to

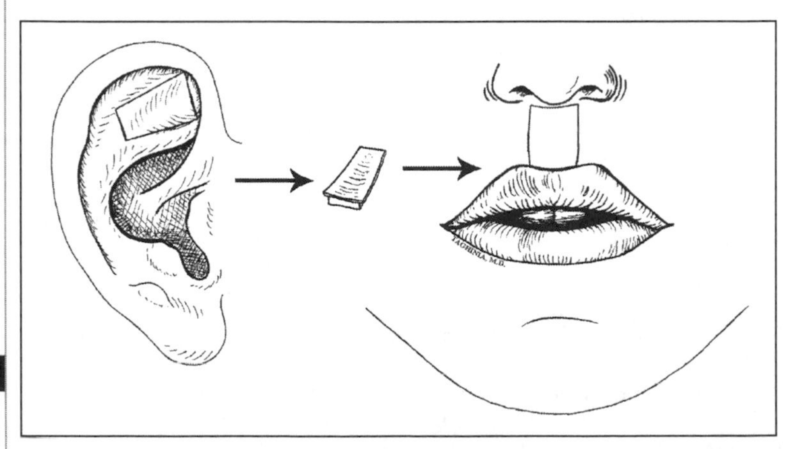

Figure 33.6. Conchal skin-cartilage composite graft to repair the philtrum in burn patients.

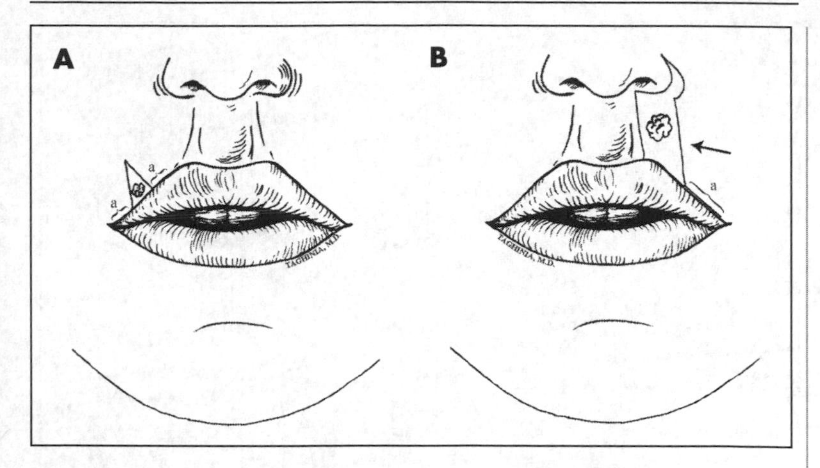

Figure 33.7. Repair of partial-thickness upper lip defects. In (A) the lesion is excised as a partial-thickness wedge. Lateral incisions along the white roll (*a*) allow the edges of the wound to be closed primarily. In (B) the lesion is excised as a partial-thickness section that incorporates a perialar crescent. A similar incision along the white roll (*a*) and undermining of the lateral flap allows the edges of the wound to come together.

the nasal ala may also be reconstructed with the nasolabial flap (Fig. 33.8A,B). This reconstructive method may not be ideal in men, however, because the nasolabial flap is not hair-bearing. Primary closure may be achieved for men by advancing adjacent lip and cheek tissue (Fig. 33.8C).

Full-Thickness Defects

For full-thickness defects, the choice of reconstructive option depends on the size of the defect. Defects of one-quarter to one-third of the upper lip can be closed primarily (Fig. 33.9). Larger defects of the upper lip require flaps from the lower lip or recruitment of adjacent cheek tissue. If these larger defects involve the central portion of the upper lip, perialar crescentic excisions may provide additional mobility if needed (Fig. 33.10).

Defects measuring one-third to two-thirds of the upper lip may be closed with the Abbe flap, the Karapandzic flap, or the Estlander flap (see below for description of each method; also see Figs. 33.12-33.15). The Abbe and Karapandzic flaps are used for central defects whereas, the Estlander flap is used for lateral defects that involve the commissure. The Abbe flap may also be used for lateral defects that do not involve the commissure.

Defects greater than two-thirds of the upper lip can be closed with the Bernard-Burow's technique if sufficient cheek tissue is available (Fig. 33.17). However, if sufficient cheek tissue is not available, most surgeons choose a free flap for reconstruction. The aforementioned reconstructive methods are described later in this chapter. Often, these methods can also be applied to closure of lower lip defects as well. Accordingly, for simplicity and ease of explanation, reference is often made to lower lip reconstruction.

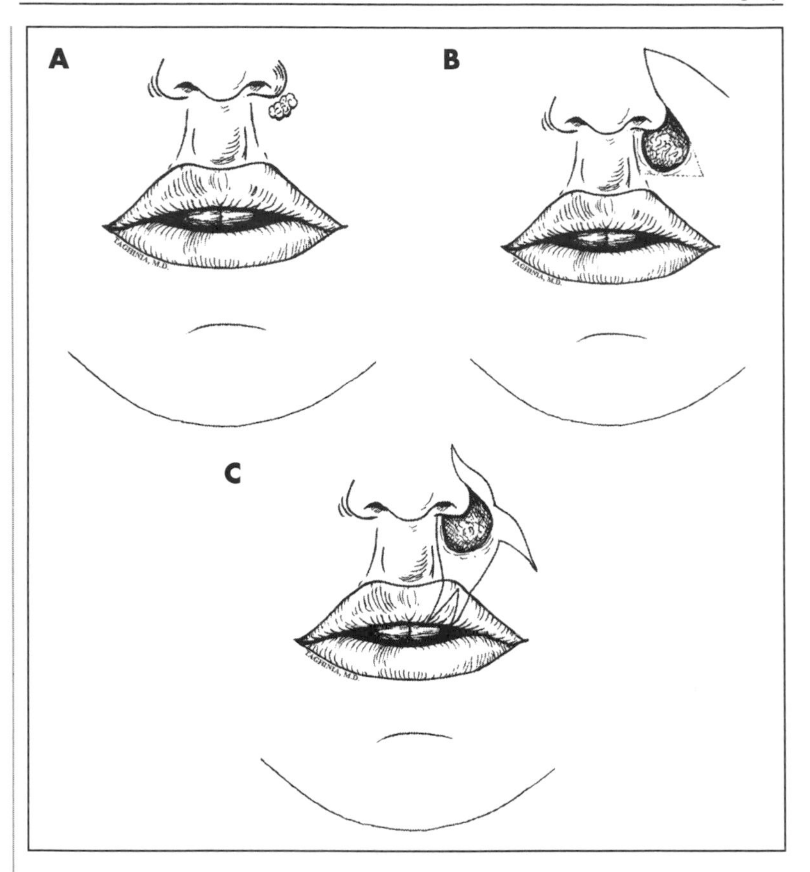

Figure 33.8. Repair of partial-thickness upper lip defects. The lesion (A) is excised leaving a circular partial-thickness defect. The defect can be closed with an inferiorly-based nasolabial flap (B) or using advanced tissue from the cheek and lip (C). The nasolabial flap is less ideal in men because the flap is not hair-bearing.

Lower Lip Reconstruction

In contrast to the upper lip, lower lip reconstruction tends to be simpler. This advantage is due to the greater laxity of the soft tissues and lack of a separate central aesthetic unit. Since oral competence is mainly mediated by the lower lip, function and sensation tends to be more important than aesthetics.

Partial-Thickness Defects

Partial-thickness defects of the lower lip are treated differently based on whether the defect involves skin and subcutaneous tissue or vermilion. Skin and subcutaneous defects of the lower lip subunit can be left to heal by secondary intention or skin grafted. More commonly, however, a local advancement flap, rotation flap or transposition flap is employed for reconstruction. Careful planning and

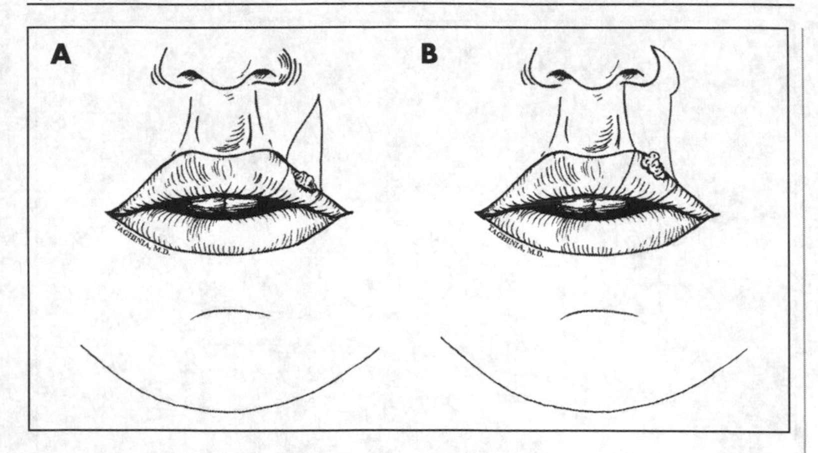

Figure 33.9. Full-thickness excisions of the upper lip. Defects up to one-third of the upper lip can be excised and closed primarily. Lateral defects often require wedge excision (A); whereas, defects that are closer to the philtrum can be excised with the help of perialar crescentic excisions for additional mobility (B).

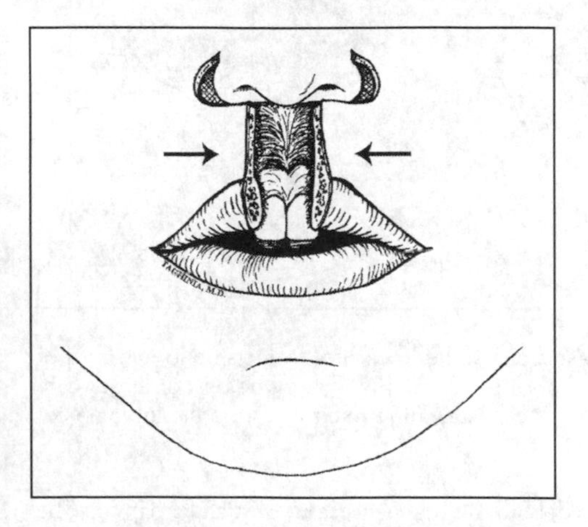

Figure 33.10. Perialar crescentic partial-thickness excisions for primary closure of full-thickness upper lip defects.

execution should allow the final scars to lie parallel to the natural skin tension lines. As previously mentioned, the white roll should be realigned as closely as possible.

Full-Thickness Defects

Many of the reconstructive methods used for upper lip reconstruction can also be used for lower lip reconstruction. As in the upper lip, reconstructive options for full-thickness defects depend on the size of the defect. Defects up to one-third of the lower lip can be closed primarily as described earlier (Fig. 33.11). Larger defects measuring one-third to two-thirds of the lower lip width may be closed with the Karapandzic, Abbe or Estlander flaps (see below).

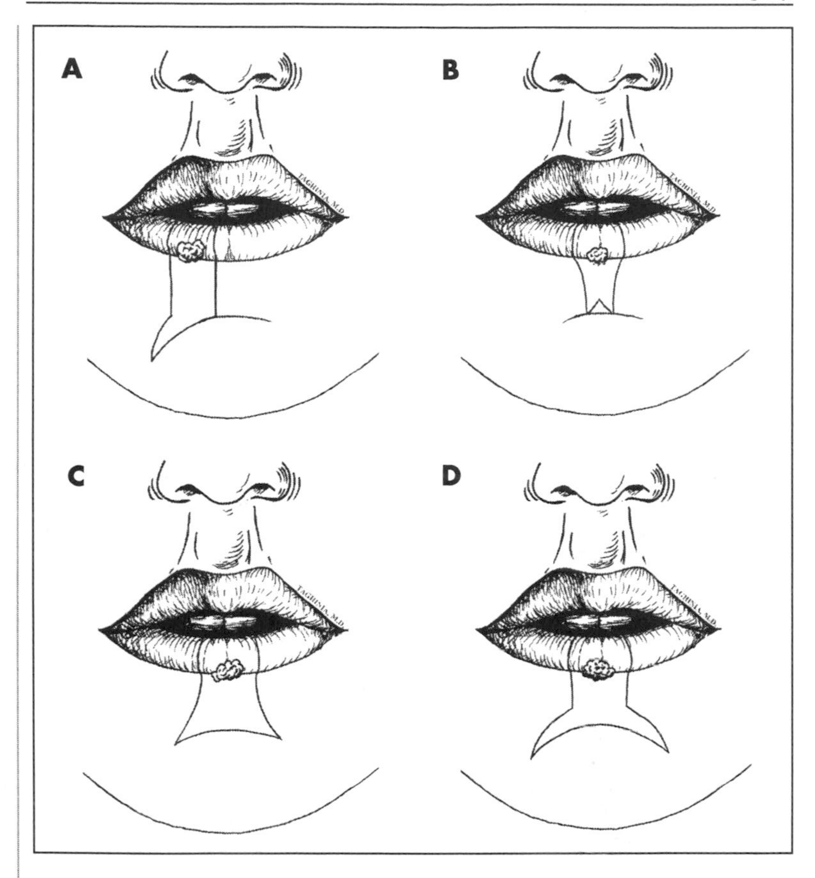

Figure 33.11. Full-thickness excisions of the lower lip. Defects up to one-third of the lower lip can be excised and closed primarily. Lateral defects or larger central defects may require partial-thickness wedge excisions from the labiomental fold (A-D).

If the commissure is involved, both the Karapandzic and Estlander flaps may be used; however, the Karapandzic is probably the better choice because it is better at maintaining oral competence. If the commissure is not involved, the Karapandzic or the Abbe flaps may be used. The Abbe flap is insensate; however it does provide a better cosmetic result.

In the case of larger lower lip defects (more than two-thirds of the lip), if there is sufficient adjacent cheek tissue, the surgeon may employ the Karapandzic (Fig. 33.13) or the Bernard-Burow's (Fig. 33.16) techniques. The Karapandzic flap may be used for defects up to three-fourths of the lower lip width whereas, the Bernard-Burow's can be used to reconstruct the entire lower lip. If enough cheek tissue is not available, distant or free flaps may be used for reconstruction.

Abbe Flap

Application: Upper and lower lip reconstruction

Defect size: One-third to two-thirds of the lip width

Donor site: Opposite lip

Blood supply: Medial or lateral labial artery

Comments: Ideal for reconstruction of the philtrum; often used with other methods for reconstruction of large defects; insensate.

This flap is often the first option in reconstruction of medium-sized upper and lower lip defects that do not involve the commissures. A full-thickness mucomusculocutaneous flap based on the medial or lateral labial artery is transposed from the opposite lip into the defect (Fig. 33.12). It may be used alone or in conjunction with other reconstructive measures such as perialar crescentic excisions. Typically done in two stages, the Abbe flap is set in place in the first stage and divided 14-21 days later in a second-stage procedure. One-fourth to one-third of the lower lip can be taken without significant loss of function. Studies have demonstrated evidence of muscle function in the transferred flap at its recipient site. Although this technique can be utilized for either lip, it is best for upper lip reconstruction because the lower lip has greater laxity and can contribute more tissue without disturbing a major central structure. Furthermore, the Abbe flap can be used to replace the entire philtral subunit.

The Abbe flap does not recruit new lip tissue; it simply transplants tissue from the lower (or upper) lip to its counterpart. Thus, the size of the oral aperture remains the same as if the lip defect is closed primarily. The goal is to recruit enough unaffected lip tissue to balance the discrepancy in lip lengths after a medium-sized excision.

A wedge-shaped pedicle flap is harvested from the opposite lip. At minimum the width of the flap should be one-half the size of the defect. The height of the flap should match the height of the defect, and the flap should be designed with sufficient tissue to permit a 180° arc of rotation into the defect. Because contralateral labial arteries form robust anastomotic connections in the midline, the flap can be based medially or laterally. Starting at the apex, an incision is made through skin, muscle and mucosa and is extended toward the vermilion border. As the vermilion border is approached, careful scissor dissection will avoid injury to the labial artery which can be found between the deep layers of orbicularis oris muscle and the mucosa approximately at the level of the vermilion border. Initial division of the nonpedicle side of the flap can locate the position of the labial artery and aid in its identification on the pedicle side. The pedicle should be at least 1 cm in width in order to maintain adequate venous drainage. The flap is rotated upon its pedicle, and a stay suture is placed after exact approximation of the vermilion border. The flap is secured with a three-layer closure approximating mucosa, muscle and skin, and the donor site is closed primarily or with the aid of crescentic excisions (labiomental or perialar depending on the donor site). The pedicle is usually divided 14-21 days later.

The most common complication is flap loss due to inadequate blood supply. Careful dissecting technique, an adequate soft tissue envelope around the artery, and ample flap width minimize flap ischemia. Careful attention should be paid to the accurate approximation of the vermilion border of both donor and recipient sites before and after pedicle division. Since the lower lip vermilion can be significantly thicker than that of the upper lip, resection of the vermilion can be undertaken in a secondary procedure for improved aesthetic result. Excessive pulling while

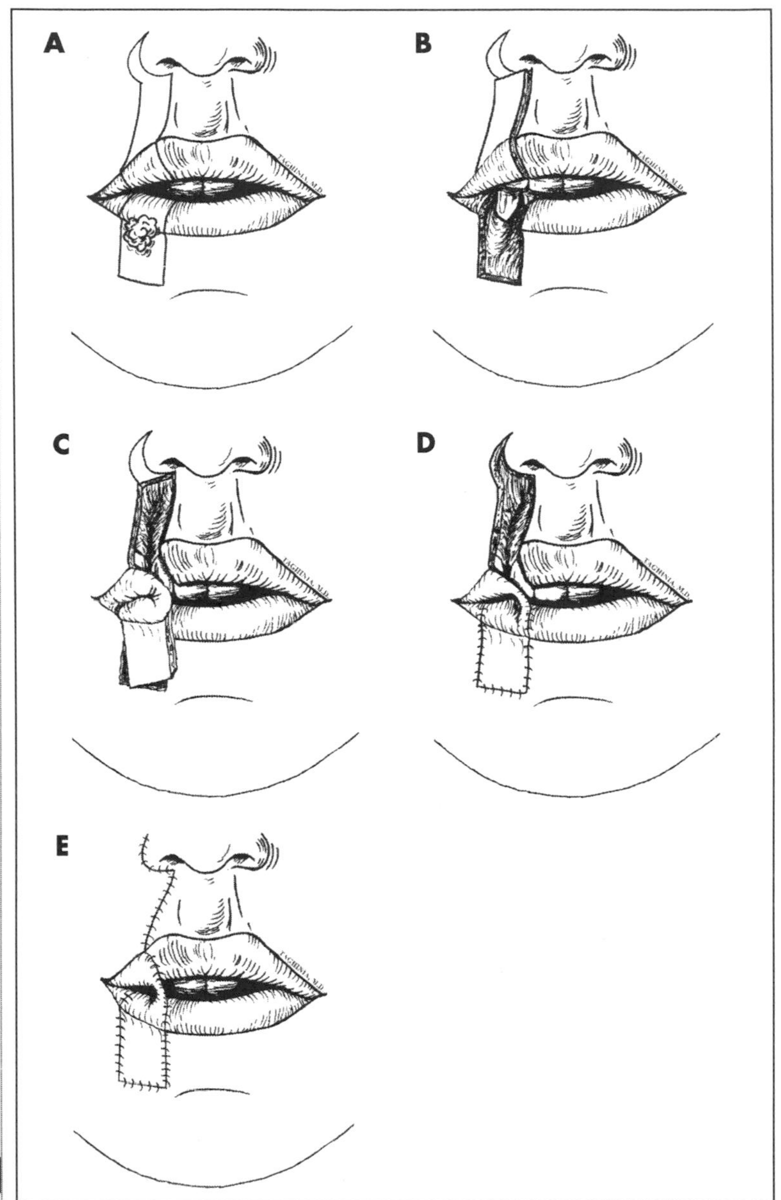

Figure 33.12. The Abbe flap. A lower lip lesion is excised and reconstructed with an Abbe flap from the upper lip. Perialar crescentic excision helps to close the donor defect. See text for full details.

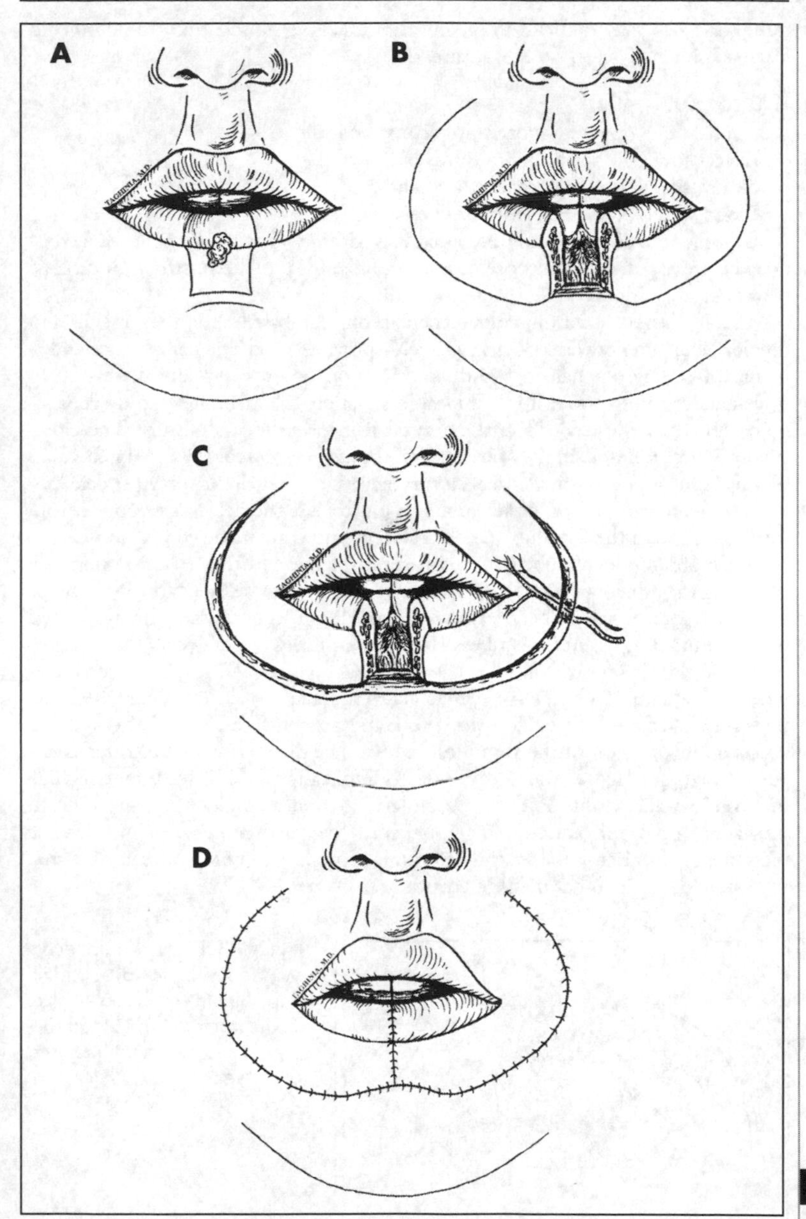

Figure 33.13. Lower lip reconstruction with the Karapandzic flap. A central lower lip defect is reconstructed using bilateral Karpandzic flaps. Avoiding deep dissection laterally helps to properly identify and avoid injury to the neuromotor and blood supply (C). See text for details.

raising the flap may result in the removal of excessive muscle from the donor lip leaving a notched defect on the closure.

Karapandzic Flap

Application: Upper and lower lip reconstruction
Defect size: One-third to two-thirds of lip width
Donor site: Cheek and lip advancement
Blood supply: Preserved labial arteries
Comments: A sensate and functional flap with poor aesthetic results; oral competence preserved at the expense of microstomia; ideal for reconstruction of large defects in the midline.

This is a sensate axial musculomucocutaneous flap based upon the superior and inferior labial arteries (Figs. 33.13, 33.14). It provides good oral competence and is useful for closing one-half to two-third defects of the upper lip and defects up to three-quarters of the lower lip. It is ideal in situations where no new lip tissue is required in central defects or lateral defects that involve the commissure. The blood supply is more robust than the Abbe flap, but the aesthetic outcome is inferior. Because new lip tissue is not recruited, microstomia may result after closure of larger defects.

A semicircular incision of adequate length to close the defect is extended from the defect toward the commissures. The skin incisions are made with a scalpel, and careful mobilization of subcutaneous tissues is achieved using electrocautery. By spreading the orbicularis oris muscle longitudinally along the line of the incision, or on a plane parallel to the fibers, separation from the adjacent musculature is attained while maintaining the nerves and vessels intact. Laterally, at the level of the commissures, the skin is incised only down to subcutaneous tissue. Careful dissection is needed to identify and preserve the labial arteries and buccal nerve branches. The flaps are rotated medially to close the defect, and a stay suture is placed after meticulous reapproximation of the vermilion border. The defect is closed in three-layers approximating mucosa, muscle and skin. Complications of this technique include microstomia and visible scarring. Secondary revision of the commissure is often indicated to prevent oral crippling in feeding, hygiene maintenance and denture placement. The circumoral scarring after this procedure is more noticeable because the scars do not lie in natural skin creases.

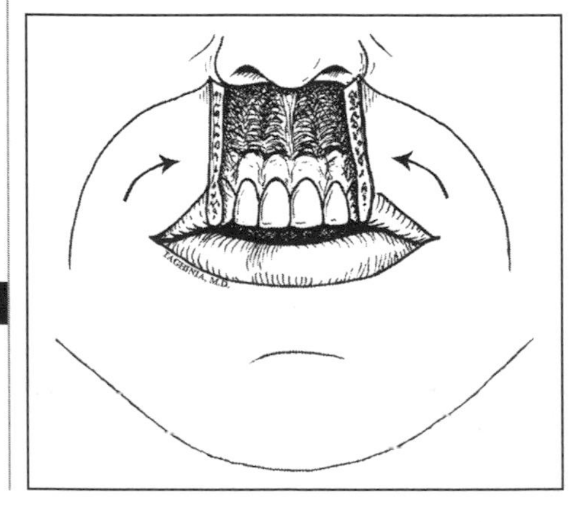

Figure 33.14. Upper lip reconstruction with the Karapandzic flap. Similar to the lower lip, the neuromotor and blood supply should be identified and preserved.

The Estlander Flap

Application: Upper and lower lip reconstruction

Defect size: One-third to two-thirds of lip width

Donor site: Opposite lip

Blood supply: Medial labial artery

Comments: Insensate but oral competence is preserved; one-step procedure that results in a rounded neo-commissure; frequently requires revision.

The Estlander flap is similar to the Abbe lip switch flap, but it is modified for use around the corner of the mouth (Fig. 33.15). It is a one-step procedure but sometimes requires future revision to improve the commissure. Continuity of the orbicularis oris ensures adequate oral competence; however, the modiolus functional region is distorted leading to altered oral animation. This alteration is compounded by a rounded neo-commissure which lacks definition. The flap is designed to be about half the width of the defect to cover. It is based on the opposite lateral lip. The vascular pedicle is within the pivoting point, supplied by the contralateral labial artery. It is rotated into the defect, and the donor site is closed primarily.

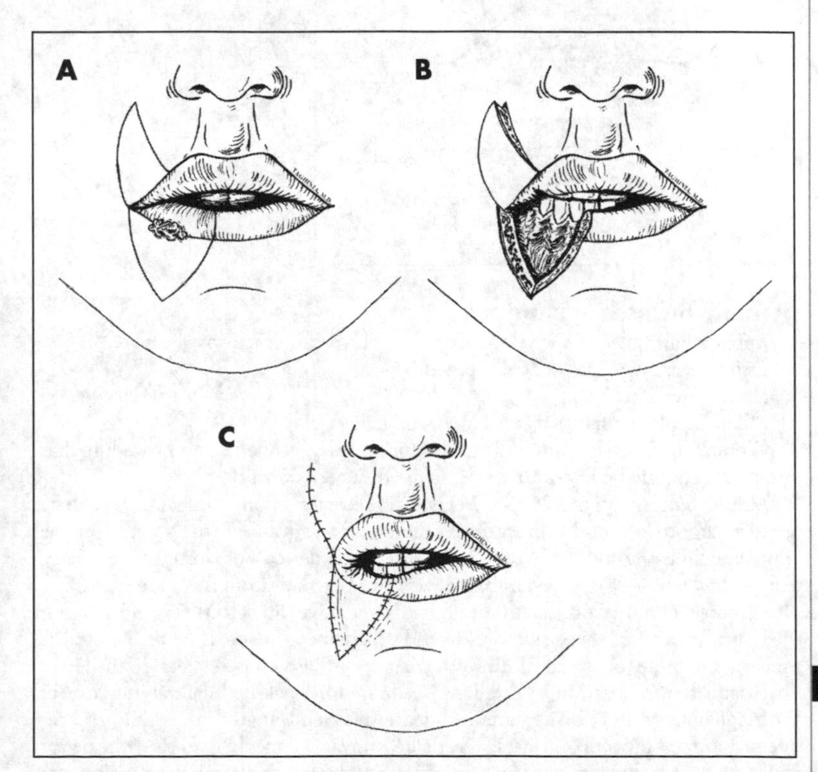

Figure 33.15. The Estlander flap. Similar to the Abbe flap, this flap is better suited for defects close to or involving the commissure.

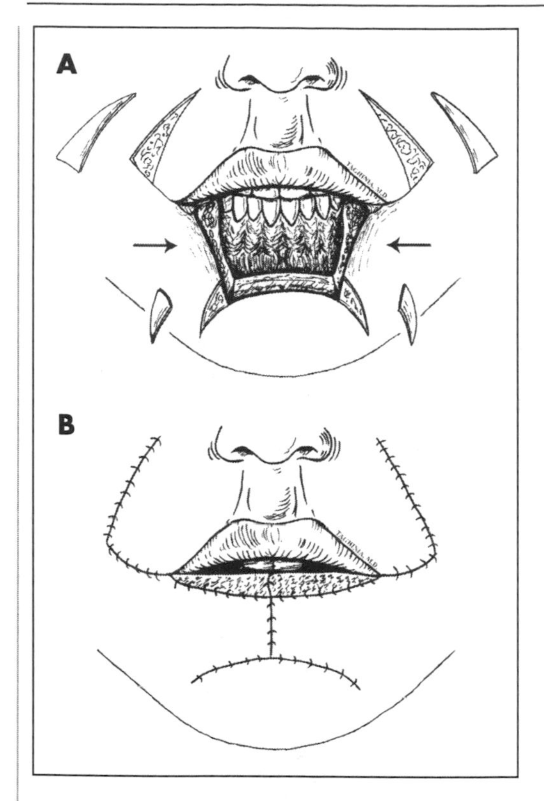

Figure 33.16. Bernard-Burow's technique for reconstruction of large lower lip defects. Partial-thickness Burow's excisions laterally in the cheek and labiomental fold help to close the defect (A). The neo-vermilion is constructed using buccal mucosa (B).

Bernard-Burow's Technique

Application: Lower lip (mainly) and upper lip reconstruction
Defect size: Two-thirds to full lip width
Donor site: Cheek
Blood supply: Labial and facial artery branches
Comments: Insensate but oral competence is preserved; one-step procedure that results in a rounded neo-commissure; frequently needs revision.

Although most commonly used in lower lip reconstruction, this technique can be useful in large defects of the upper lip as well. This is an advancement flap utilizing the remaining lip tissue and the cheeks for closure of the defect. For closure of very large defects this technique can be combined with an Abbe flap (from the opposite lip).

Closure technique is different for upper lip and lower lip defects (Fig. 33.17). For the upper lip, a perialar excision of skin and subcutaneous tissue is performed in the shape of a triangle (or crescent). Burow's triangles are also excised lateral to the lower lip. Adequate mobilization of the flaps is achieved by making bilateral incisions in the gingivobuccal sulci being careful to leave sufficient gingival mucosa for subsequent closure of the mucosal layer. The tissue is advanced medially to close the defect and is sutured in three layers. The skin and subcutaneous tissue perialar incisions are closed in a single layer. Perialar crescentic excisions are more aesthetically pleasing but may not provide enough mobility. The vermilion is reconstructed using cheek

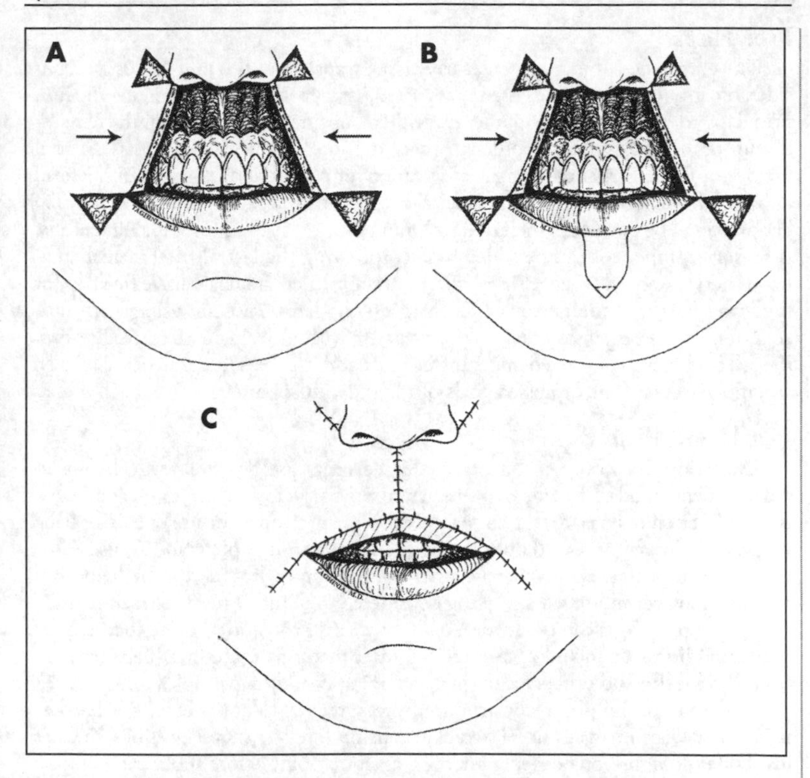

Figure 33.17. Bernard-Burow's technique for reconstruction of large upper lip defects. Partial-thickness triangular Burow's excisions in the perialar and commissure area help to advance the cheek tissue medially (A). The vermilion is reconstructed using buccal mucosa (C). Additional lip tissue can be recruited from a lower lip Abbe flap (B) if needed.

buccal mucosa. The resulting insensate, nonfunctional upper lip does not usually lead to oral incompetence. This is because gravity charges the lower lip with the responsibility of oral competence. The most common complication of this procedure is microstomia, which can sometimes be improved by combining this technique with an Abbe flap. This technique can also cause some excessive tension on the upper lip and cheek resulting in distortion of the nasolabial fold.

In the Bernard-Burow's technique for the lower lip, four Burow's triangles are excised lateral to the nasolabial folds and in the labiomental groove to allow relief space for advancement of bilateral lower cheek flaps medially to fill the defect. Excision of these triangles avoids a typical tight lower lip and excess upper lip, and can vary in size as long as closure is achieved without tension. A minor modification in the originally proposed procedure preserves innervation and function by avoiding deep dissection through perioral muscles. Although a bulkier upper lip and poor anterior projection at the vermilion is common, this procedure remains a suitable option for reconstructing very large defects.

Free Flaps

The emergence of microvascular free tissue transfers in the mid-1980s has considerably influenced methods used to repair massive facial defects involving the lips. Free flaps are often used in conjunction with an advancement flap from the remaining lip or adjacent cheek in order to meet the ideal reconstructive goals. Several methods of reconstruction using a wide variety of potential donor sites in the head and neck have been described. The radial forearm-palmaris longus tendon free flap has proven to be one of the preferred techniques for repairing substantial lip defects. A sensory component can be added by incorporating the lateral antebrachial cutaneous nerve. Recently, Lengele described a prefabricated gracilis muscle free flap for the lower lip that simultaneously reconstructs the labial muscular sling with mucosa, tendinous suspension and a skin cover. As with other reconstructive options, the selected method of free tissue transfer must address the soft tissue needs of each specific defect and the expressed goals of the individual patient.

Pearls and Pitfalls

Every lip reconstruction must be evaluated on a case by case basis. The lips of older patients tend to be more conducive to primary closure due to greater laxity. Reconstruction of facial defects in male patients may require the use of hair-bearing tissue. For women, when reconstructing the vermilion it is preferable to use tissue that will accept lipstick adequately, since application of lipstick can be helpful in camouflaging vermilion scars. Having considered the unique needs of each patient, the surgeon's options can be divided into three main categories: those that employ remaining lip tissue including primary closure; methods using local flaps (such as from the cheek); and others techniques involving distant flaps.

When performing lip reconstruction, the surgeon should adhere to a few key principles. Place incisions in relaxed skin tension lines whenever possible. In cases involving the vermilion border, mark the transition point before application of local anesthesia. Always align the markings of the white roll as the first step in closure. Finally, use deep absorbable sutures to oppose orbicularis oris fibers so that the closure scar does not widen or indent.

Suggested Reading

1. Behmand RA, Rees RS. Reconstructive lip surgery. In: Achauer BM, Eriksson E, Guyuron B, Coleman IIIrd JJ, Russell RC, Vander Kolk C, eds. Plastic Surgery: Indications, Operations, and Outcomes. St. Louis: CV Mosby, 2000:1193-1209.
2. Kroll SS. Repair of lip defects with the Abbe and Estlander flaps. In: Evans GRD, ed. Operative Plastic Surgery. New York: McGraw-Hill, 2000:289-297.
3. Kroll SS. Repair of lip defects with the Karapandzic flap. In: Evans GRD, ed. Operative Plastic Surgery. New York: McGraw-Hill, 2000:298-307.
4. Kroll SS. Staged sequential flap reconstruction for large lower lip defects. Plast Reconstr Surg 1991; 88(4):620-625.
5. Pribaz J, Stephens W, Crespo L et al. A new introral flap: Facial artery musculomucosal (FAMM) flap. Plast Reconstr Surg 1992; 90(3):421-425.
6. Spira M, Hardy SB. Vermilionectomy: review of cases with variations in technique. Plast Reconstr Surg 1964; 33:39-46.
7. Wechselberger G, Gurunluoglu, Bauer T et al. Functional lower lip reconstruction with bilateral cheek advancement flaps: Revisitation of Webster method with a minor modification in the technique. Aesthetic Plast Surg 2002; 26(6):423-428.
8. Zide BM. Deformities of the lips and cheeks. In: McCarthy JG, ed. Plastic Surgery. New York: WB Saunders Co., 1990.

Chapter 34

Mandible Reconstruction

Patrick Cole, Jeffrey A. Hammoudeh and Arnulf Baumann

Introduction

The mandible is of critical value to the functional and aesthetic integrity of the face. As the strongest bone of the face, the mandible significantly contributes to the lower third of the face, the structural continuity of the temporomandibular joint (TMJ), functions in deglutition and houses the lower dentition. The goal of mandibular reconstruction is to restore form and function following tumor resection, trauma, or secondary to congenital abnormalities. Reconstruction of the mandible is often both a soft tissue and bony problem. Though often technically demanding, precise mandibular reconstruction is necessary to optimize oral competence, unimpeded mastication, proper dental occlusion, intelligible speech and intraoral sensation.

Anatomy

Mandibular reconstructive procedures can be grouped according to the principal anatomic regions of the mandible: the condyle and ascending ramus, the horizontal ramus and the symphyseal region (Fig. 34.1). The condyle with the coronoid process and the ramus constitute the vertical portion of the mandible. Therefore this region is important to restore the vertical height of the face. Connecting the vertical and horizontal regions is the angle of the mandible. The horizontal region (the body of the mandible) then continues on in a curvilinear fashion to incorporate the symphyseal regions.

Each part of the mandible poses unique reconstructive challanges. The condyle is the basis of the TMJ joint. It should allow for rotation within the glenoid fossa to achieve adequate mouth opening and at the same time have the appropriate configuration to reestablish vertical facial height. Ankylosis in this region results in limited mouth opening and also pain in this region. The angle region should be restored for reestablishment of continuity and to help achieve an acceptable aesthetic result. The horizontal part of the mandible and the symphyseal region are important for dental rehabilitation (occlusion) and also for facial aesthetics.

Timing of Reconstructive Procedures

The timing of reconstruction depends on the etiology of the underlying bony defect, the size of the defect and various patient factors. For most cases, we advocate early primary reconstruction, especially as a single stage procedure, to minimize the deleterious effects that follow loss of hard and soft tissue.

In the trauma patient, primary reconstruction of the mandible and occlusion are achieved by precise reduction of the bone followed by stabilization with osteosynthesis plates and screws. Delayed reconstruction is of particular importance in gunshot wounds. It is advisable to wait until any sepsis or bacteremia has resolved, the soft tissue demonstrates that it is viable and tissue availability and quality is sufficient.

Practical Plastic Surgery, edited by Zol B. Kryger and Mark Sisco. ©2007 Landes Bioscience.

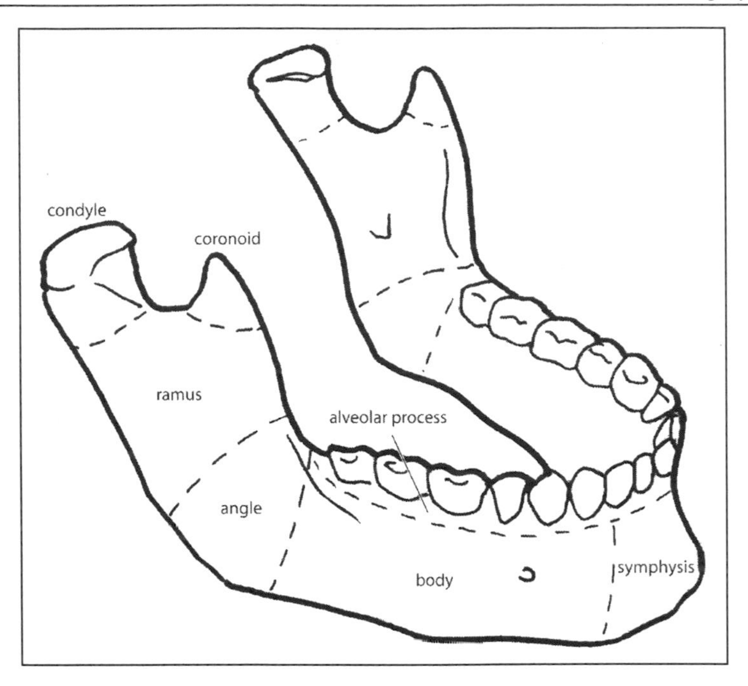

Figure 34.1. Anatomy of the mandible.

When mandibular reconstruction is needed after tumor resection, the timing of the reconstruction should consider the tumor characteristics, the amount of tissue to be resected and the possibility of achieving clear margins. If the resection margins are questionable, a two-stage procedure after several months would be more preferable.

Choices for reconstruction include composite free flaps, nonvascularized autologous bone grafts, or synthetic material. Immediate reconstruction with allograft materials or with autogenous bone may be associated with a 15-47% infection rate and loss of the transplant/implant. Therefore, proper preoperative dental evaluation should identify any potential decaying teeth that might serve as a source of infection. Such teeth should be addressed either by restoration or extraction.

Techniques for Bony Reconstruction

Nonvascularized Autogenous Bone Grafts

Bone grafts harvested from the iliac crest, rib, or calvaria are used for nonvascularized bone grafts. Autogenous bone grafts are more resistant to infection than allogenic grafts. They are usually of good quality and provide osteoconductive and osteoinductive properties. The rib is used for condylar reconstruction and can be shaped according to the requirements. Iliac bone grafts and calvarial bone grafts can be used to reconstruct alveolar defects or bone defects less than 5 cm. The graft is shaped and its cortex is perforated at multiple sites to enhance vascularity and eventual resorption. While rib grafts have proven ideal for condylar reconstruction, iliac crest grafts are better suited for the maintenance of dental implants.

The introduction of osteosynthetic plates markedly improved the outcomes of mandibular reconstruction. These plates are relatively easy to apply, provide a firm platform for healing, and are associated with low placement site morbidity. In addition, rigid internal fixation eliminates the need for external or intermaxillary fixation (IMF), maintains the appropriate dental relationships, reduces operative time and can provide effective condylar replacement.

Nonvascularized Alloplastic Bone Grafts

Allografts for mandible reconstruction are mostly comprised of freeze-dried bone. Such bone grafts are only suitable for small defects of the mandible where the continuity of the mandible is intact. The advantages of allografts include the relative ease of availability without a donor defect. The major disadvantage of allograft material is that it is prone to infection and only provides a matrix for osteoconduction. Alloplastic materials are available in paste, powder, or block form which may be easily contoured to fit the required shape. The lack of osteoinductive properties secondary to the absence of vascularity and cellular components limits the use of such material in radiated or poorly vascularized tissue. Significant failure rates (48%) are associated with implantation at radiated mandibular sites, compared to nonirradiated tissues (30%). Allogeneic bone cribs have been called the ideal vehicles for particulate cancellous bone marrow grafts. Bioresorbable and biocompatible, allogeneic cribs are easily adaptable and less prone to the poor graft regeneration seen with the use of alloplastic materials.

Vascularized Bone Grafts

With the development of the modern microsurgical technique, the reconstructive surgeon now has the ability to use a composite free flap for mandibular reconstruction. In many centers, this has become the standard of care for reconstructing large mandibular defects.

Fibula Free Flap

The fibula flap provides a large amount of cortical bone, allowing for multiple osteotomies to achieve the curvature of the mandible. This flap is very versatile and can be harvested as an osteomyocutaneous flap by incorporating a portion of the soleus muscle to provide additional soft tissue bulk. The vascular pedicle is the peroneal artery and vein (up to 8 cm length) and a skin island measuring approximately 10 × 30 cm. The free fibular graft provides up to 24 cm in length of bony material. There usually is sufficient bone height in the newly reconstructed mandible for future placement of a dental prosthesis. This flap provides osteocutaneous coverage that is reliable, durable and aesthetically acceptable to most patients. Disadvantages include poor donor site cosmesis, the limited amount of cutaneous tissue and a potential donor site neurapraxia. A detailed discussion of the design and harvest of this flap is provided in Appendix I.

Iliac Crest Free Flap

The iliac crest flap provides a curved, cortical section of bone that can be used to reconstruct the mandibular symphysis and the curved body region; 10-14 cm of corticocancellous bone can be harvested. The vascular pedicle consists of the deep circumflex iliac artery and vein providing up to 6 cm of length. The soft tissue island can measure as large as 16 cm in length, and it can be harvested with the internal oblique muscle. Drawbacks may include donor site morbidity, abdominal wall weakness or herniation, potential injury to the lateral femoral cutaneous nerve, and delayed postoperative ambulation.

Radial Forearm Free Flap

The radial forearm free flap is used mostly for soft tissue coverage, for example, in reconstruction of the anterior floor of the mouth after a cancer resection. This flap has limited use in mandibular reconstruction due to the limited length of radius that can be harvested, as well as the relatively short height of the harvested bone. It is based on the radial artery and venae comitantes in association with the cephalic vein. The pedicle can measure up to 20 cm in length, making it very versatile for use in sites distant from the recipient vessels. The skin paddle can be harvested as a sensate skin island receiving innervation from the lateral antebrachial cutaneous nerve (C5-C6). Disadvantages in using this flap include a visible donor site with potential skin graft loss over the flexor tendons and potential intra-oral hair growth. The nondominant forearm should be used, and adequate ulnar artery flow in the hand should be verified prior to the procedure. Preoperative laser hair removal should also be considered. A detailed discussion of the harvest of this flap is provided in Appendix I.

Temporomandibular Joint (TMJ) Reconstruction

The TMJ region poses unique reconstructive challenges. Historically TMJ reconstruction has been plagued with failed attempts using prosthetic joint devices. Recently, an implantable titanium glenoid fossa and/or condylar prostheses have been used at many institutions; however long term outcomes are still unknown. Whenever feasible, autogenous rib reconstruction in this region provides the best form and function. Costochondral grafts consisting of rib and associated cartilage allow for the tissue remodeling in response to stress and can provide good functional results despite the lack of an articular disc. Futhermore, in the pediatric patient the rib graft will continue to grow along with facial skeleton. Another reconstructive option is an extended free fibula flap with soft tissue arthroplasty. This flap uses a periosteal sleeve which improves subsequent TMJ reconstruction.

Distraction Osteogenesis

Mandibular defects of more than 5 cm are best reconstructed with an autologous vascularized flap. For smaller defects, several centers have studied the use of distraction osteogenesis (DO) as an alternative procedure. DO is probably best suited for small segmental defects. Recent reports describe success with DO in replacing defects up to 5 cm in the angle, body, and symphyseal regions. DO has an excellent track record in the treatment of craniofacial anomalies involving mandibular insufficiency.

The process of DO involves the gradual distraction of the bony segments flanking the defect, and the synthesis of new bone within the gap. The younger the patient, the less distraction time is required. Adults require a slightly longer period of distraction because their bone regenerative capabilities are slower than those of adolescents or infants. The advantages of DO are that it eliminates the need for bone grafts and the morbidity of donor sites. The entire procedure can usually be done intraorally without additional facial incisions. When adequate healing has been achieved, the device can be removed in a short, office-based procedure. In summary, DO is best suited for cases involving small defects, and proper patient selection is important.

Pearls and Pitfalls

Tissue Engineering

The future of mandibular reconstruction is headed the way of tissue engineering. In 2004, the first significant human mandibular replacement with engineered osseous tissue was described in a patient following subtotal mandibulectomy resulting in a 7 cm defect. After the computer-aided production of a titanium cage customized to the patient's defect, the structure was filled with bone mineral blocks and infiltrated with a mixture of recombinant human bone morphogenetic protein and autologous bone marrow. The transplant was implanted into the latissimus dorsi muscle. Following 7 weeks of in vivo maturation, a free bone-muscle flap based on the thoracodorsal vasculature was transplanted into the mandibular defect. Skeletal scintigraphy demonstrated bone remodeling and mineralization within the graft both before and after transplantation. Clinically, both aesthetic form as well as excellent functional capacity was regained with no unforeseen complication. This novel graft is expected to accommodate eventual dental implantation, and potentially the removal of the titanium scaffold. Currently, stem cell research is advancing in manner that eventually may eliminate all of the aforementioned therapies. We anticipate that the cutting edge research in stem cell biology will eventually lead to human trials and ultimately a reliable technique for providing soft and hard tissue replacement.

Suggested Reading

1. Coleman IIIrd JJ. Mandible reconstruction. Oper Tech Plast Reconstr Surg 1996; 3(4):213.
2. Herford AS, Ellis E. Use of a locking reconstruction bone plate/screw system for mandibular surgery. J Oral Maxillofac Surg 1998; 56:1261.
3. Hidalgo DA. Aesthetic improvements in free-flap mandible reconstruction. Plast Reconstr Surg 1989; 84:71.
4. Ilizarov GA. The principles of the Ilizarov method. Bull Hosp Jt Dis Orthop Inst 1988; 48:1.
5. Mathes S, Nahai F. Reconstructive surgery principles, anatomy, and techniques. New York: Churchill Livingstone, 1997:1353-1370.
6. Roumanas ED, Markowitz BL, Lorant JA. Reconstructed mandibular defects: Fibula free flaps and osseointegrated implants. Plast Reconstr Surg 1996; 99:346.
7. Swanson E, Boyd JB, Manktelow RT. The radial forearm flap: Reconstructive applications and donor-site defects in 35 consecutive patients. Plast Reconstr Surg 1990; 85:258.
8. Taylor GI. Reconstruction of the mandible with free iliac bone grafts. Ann Plast Surg 1986; 9:361.
9. Terheyden H, Behrens E. Growth and transplantation of a custom vascularised bone graft: In a man. Lancet 2004; 364:766.
10. Weinzweig N, Weinzweig J. Current concepts in mandibular reconstruction by microsurgical free flaps. Surg Technol Int 1997; VI:338.

Chapter 35

The Facial Nerve and Facial Reanimation

Zol B. Kryger

Relationship with the Superficial Musculoaponeurotic System (SMAS)

An understanding of the SMAS is critical in order to avoid injuring the facial nerve during facial surgery. The SMAS is a fascial layer that attaches the facial muscles to the overlying dermis. In the lower face, the SMAS is continuous with the platysma. Superiorly, the SMAS extends to the level of the zygomatic arch where its fibers are anchored. Above the arch, the temporoparietal fascia is the equivalent of the SMAS. It joins the deep temporal fascia which extends into the scalp region.

In the lower face and neck, the facial nerve runs deep to the platysma and the SMAS. Toward the midline, the nerve continues deep to the SMAS and runs superficial to the masseter muscle. In the midface and cheek region, the nerve runs within the SMAS, but as it passes over the zygomatic arch, it becomes more superficial. In the temporal region, the facial nerve travels within the temporoparietal fascia. Thus, it becomes apparent why facelifts are performed in the subcutaneous or sub-SMAS planes.

Course of the Extracranial Facial Nerve

The facial nerve enters the face upon exiting the stylomastoid foramen. At this point it is purely a motor nerve. The nerve then travels 15-20 mm sandwiched between the digastric and stylohyoid muscles before entering the parotid gland. Prior to its entrance into the parotid, it sends off a small branch to the posterior digastric and stylohyoid muscles. It also gives off the posterior auricular nerve which travels to the posterior auricular and occipitalis muscles.

Locating the Main Trunk

During total parotidectomy, it is essential to locate the facial nerve. This is done by mobilizing the parotid superiorly and the sternocleidomastoid laterally, which will reveal the posterior belly of the digastric. A key landmark in identifying the main trunk is the cartilaginous tragal pointer. It is located by following the posterior belly of the digastric towards its insertion at the mastoid, and releasing the parotid attachment to the cartilage of the external auditory canal. The main trunk lies about 1 cm deep and slightly inferior and medial to the tragal pointer. A branch of the occipital artery often can be found in close proximity, lateral to the nerve.

Intra-Parotid Anatomy

Within the substance of the parotid, the facial nerve travels in a fibrous plane between the deep and superficial lobes. At the pes anserinus, it divides into two major divisions. One division travels superiorly and the other inferiorly. These major divisions become five branches that exit the parotid: temporal (frontal), zygomatic, buccal, marginal mandibular and cervical. The buccal and zygmotic branches have a

Practical Plastic Surgery, edited by Zol B. Kryger and Mark Sisco. ©2007 Landes Bioscience.

number of interconnections; however the temporal and marginal mandibular branches are usually terminal endings that don't arborize with other branches. In addition, the temporal and marginal mandibular branches are at the greatest risk of injury during surgery. The incidence of injury during rhytidectomy is less than 1%.

Temporal Branches

The temporal branches run within the SMAS up to the level of the zygomatic arch. Cranial to this point the temporal (frontal) branch becomes more superficial and enters the temporal region within the temporopariatel fascia. The frontal branch runs roughly along an upward sloping line extending from 5 mm below the tragus to 15 mm above the lateral aspect of the eyebrow as described by Pitanguy et al. After crossing the zygomatic arch, the frontal branch travels in the superficial layer of the deep temporal fascia (temporopariatel fascia). It penetrates the undersurface of the frontalis muscle. To avoid injuring the nerve during elevation of the facial flap, one should dissect in the subcutaneous plane (superficial to the frontal branch) or in the sub-SMAS, "deep" plane (deep to the nerve).

Marginal Mandibular Branches

These branches originate from the mandibular division that runs along the inferior border of the body of the mandible in 80% of cases, or within 1-2 cm below the mandible in the remaining cases. These branches run deep to the platysma and course more superficially about 2 cm laterally to the corner of the mouth.

In children the facial nerve anatomy is not as predictable as in the adult, and the described landmarks may not be accurate. In fact, the facial nerve may be located in a more superficial plane then in adults.

Muscles of Facial Expression

There are 17 main paired muscle groups in the face (Fig. 35.1). The facial nerve innervates these muscles from their deep surface, with the exception of three muscles: the buccinator, levator anguli oris and mentalis. The important muscles of facial expression, their function and their innervating branches are shown in Table 35.1.

Reanimation of the Paralyzed Face

Causes of Facial Paralysis

The etiologies of facial paralysis are quite varied. Intracranial causes include congenital abnormalities, malignancies, degenerative diseases, trauma, vascular conditions and other rare causes. Intratemporal causes include malignancy, trauma, infections, Bell's palsy, osteopetrosis and iatrogenic causes. Extracranial causes include malignancies (parotid gland as well as tumors of adjacent structures), trauma and iatrogenic injury. Bell's palsy, or idiopathic facial nerve palsy, is the most common cause of facial nerve paralysis even though 85% of people with Bell's palsy will recover spontaneously.

Several important physical finding can help to distinguish intracranial (upper motor neuron) from extracranial causes of facial paralysis. Both supranuclear areas provide contributions to the frontal and upper orbicularis occuli muscles. Therefore, these muscles may be partially spared if the etiology is intracranial, creating an ipsilateral "lower face" paralysis. In addition, during periods of intense emotion, facial movements may appear on the affected side. It is important to remember that the extracranial facial nerve is a purely motor nerve; therefore extracranial paralysis

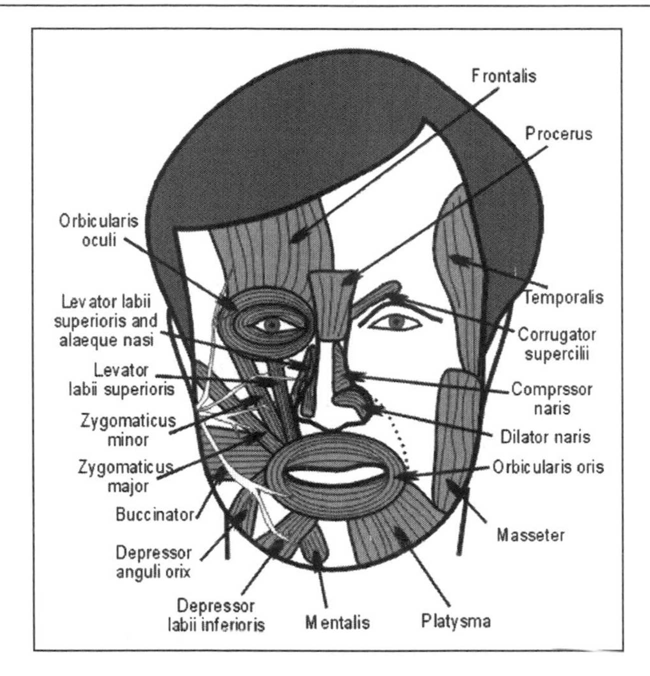

Figure 35.1. The major muscles of the face. Reprinted with permission from Microsurgeon.org.

should not involve decreased lacrimation (superficial petrosal nerve), changes in hearing (nerve to stapedius) or changes in taste (chorda tympani).

Preoperative Planning

It is essential to evaluate the patient carefully, in order to determine the cause and extent of paralysis and the status of the muscles involved. A history is obtained, focusing on the onset and duration of weakness. A complete physical exam of the head and neck including a cranial nerve exam is performed. The muscles of facial expression are evaluated for bulk, symmetry and function—both statically and dynamically. In addition, electrical testing is performed to determine the physiologic status of the facial nerve branches and the muscles of the face. Such tests, however, are not entirely accurate and tend to overestimate the extent of functional loss. High resolution helical CT is of value in localizing the precise site of pathology. A number of grading schemes for facial nerve paralysis have been described. The House-Brackmann Grading System is most widely used:

Grade	**Description**
I	Normal facial muscle function
II	Mild dysfunction
III	Moderate dysfunction
IV	Moderately severe dysfunction (symmetry at rest)
V	Severe dysfunction (asymmetry at rest)
VI	Total paralysis

Table 35.1. The primary muscles of facial expression, their innervation (branch of facial nerve) and function

Nerve Branch	Muscle	Action
Temporal	Anterior auricular	Pulls ear forward
	Superior auricular	Pulls ear up
	Corrugator	Pulls eyebrows inferomedially
	Procerus	Pulls eyebrows downward
	Occipitofrontalis	Moves scalp forward
Temporal-zygomatic	Orbicularis oculi	Closes lids, squinting
Zygomatic-buccal	Zygomaticus major	Elevates corner of mouth
Buccal	Zygomaticus minor	Elevates upper lip
	Buccinator	Pulls mouth laterally ("smile")
	Levator labii superioris	Elevates upper lip and nasolabial fold
	Orbicularis oris	Closes and compresses lips
	Nasalis	Flares nostrils
	Levator anguli oris	Elevates corners of mouth
Buccal-marginal	Depressor anguli oris	Pulls corners of mouth downward
Mandibular	Depressor labii inferioris	Pulls lower lip downward
Marginal mandibular	Mentalis	Pulls chin upward
Cervical	Platysma	Pulls corners of mouth downward
Posterior auricular	Posterior auricular	Pulls ear backwards
	Occipitofrontalis	Moves scalp backwards

Long-standing paralysis (greater than 2-3 years) will result in atrophy and fibrosis of the facial muscles and the inability to regain function purely by reinnervation. In these cases a muscle transposition or transplant procedure is required.

The goal of the patient is important to consider. Older patients may be content with achieving static facial symmetry at rest, whereas younger patients usually desire a dynamic repair that will allow them to smile.

Direct Nerve Repair

This is the most effective procedure for reanimating the paralyzed face. It is contingent on the adequate function of the target muscles. One should not attempt to restore function to a muscle that has been paralyzed for over 3 years solely by reinnervating it. In the past, many surgeons advocated waiting at least 3 weeks prior to nerve repair. It is now known that immediate repair of an injured facial nerve yields the best results. In direct nerve repair, an attempt is made to align the fascicles. Once proper orientation of the two stumps is achieved, the perineurium is sewn together followed by the epineurium using 9-0 silk. Smaller nerves in the distal branches can be repaired with a single full-thickness suture. If the stumps of the nerve have a neuroma or appear crushed, the nerve ends should be "freshened" until normal appearing nerve is evident. Direct repair should be undertaken only if a tension free repair is possible. Outcomes are directly correlated to the age of the individual, with younger patients faring far better than older ones.

Nerve Grafting

Autogenous sensory nerve grafts should be used when there is a gap in the facial nerve that cannot be primarily repaired. The length of the graft should be about 20% longer than the gap. The graft must also be placed in a tissue bed that is free of scar. Ipsilateral cervical plexus nerves are the first choice, followed by the contralateral cervical plexus. These usually can provide adequate length (about 10 cm) when several nerves are sewn together. If greater length is needed (e.g., for cross-face grafting described below), the sural nerve can provide up to 40 cm of length. A number of "sleeves" have been designed to cover the suture lines. These range from simple silastic tubes to collagen tubes lined by Schwann cells. Most of these have not demonstrated any significant benefit.

The classic teaching is that peripheral nerve axons regrow at a rate of about 1 mm per day. However, this does not take into account the time required for the reinnervated muscle to regain tone and function. For most patients, return of facial movement takes 1-2 years depending on the length of the graft. Movement usually begins at the oral commissure followed by motion around the eyes and the cheek. The muscles of the forehead and lower lip, however, usually do not regain much movement.

Nerve Transfer

This technique is employed when direct repair or grafting is not possible. This may be due to the absence of the main trunk of the facial nerve or in cases of intracranial nerve damage. It requires adequate mimetic muscle function and an intact peripheral nerve stump. It involves transferring one of the other cranial motor nerves, most commonly the hypoglossal nerve. Other nerves that have been used include the phrenic, accessory or glossopharyngeal nerve. Nerve transfer will be successful in most patients, even though it often results in mass movement. Following the cross-over procedure, the donor nerve target muscle loses some of its bulk and function. This is usually not a major issue since the tongue receives innervation from several nerves. One disadvantage of this technique is that eating and speaking can produce involuntary motion in the face.

Cross-Facial Nerve Grafting

This technique employs a nerve graft (typically the sural nerve) that acts as a conduit for motor axons from the normal, contralateral facial nerve (Fig. 35.2). Usually a single graft is used; however some surgeons will use multiple grafts from the intact contralateral divisions to the corresponding paralyzed ones. Cross-face grafting can be performed in a single stage or as a two-stage procedure. The advantage of the two-stage procedure is that it allows the surgeon to verify that the axons have successfully grown to the opposite side before connecting the nerve to the injured side. In the initial procedure, the desired branch is identified on the normal side and confirmed with a nerve stimulator. The sural nerve graft is sewn to it and then tunneled subcutaneously either above the upper lip or below the lower lip. A clip is attached to the end of the nerve graft. The second procedure is performed about 12 months later, after a Tinel's sign is present at the distal end of the nerve graft. Any neuroma evident should be resected prior to sewing the nerve graft into the injured nerve stump. The disadvantages of the cross-face graft include an additional donor site in the leg, violating the normal side of the face, two or more suture lines for the axons to cross, a long interval until return of function and reduced motor output from the donor side.

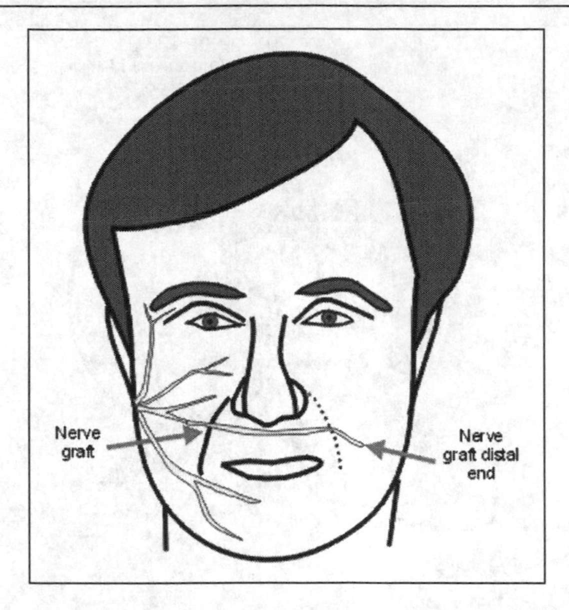

Figure 35.2. Cross-face nerve graft. Segments of the cervical plexus or the sural nerve are used as a conduit for motor axon regrowth from the normal side. Reprinted with permission from Microsurgeon.org.

Static Suspension Procedures

These procedures involve suspending structures in the face into static symmetry with the contralateral side. They provide no dynamic return of function. Suspension of the eyelids, nares, oral commissure and lower lip has been described. Traditionally, fascia lata slips were used, however recently suture suspension (e.g., contour threads), and slings made of Goretex® and Alloderm® have been described. Suspension procedures can be performed alone or in combination with muscle transfers.

Local Muscle Transposition

This technique is employed when there has been longstanding paralysis and the muscles of facial expression have atrophied and fibrosed. In addition, local muscle transposition should be considered if additional mimetic function at a specific site is required. The masseter and temporalis muscles are the two most commonly used for transfer. Transposition of the platysma and sternoclidomastoid muscles have also been described; however they have poor excursion compared to the muscles of mastication.

The temporalis may be transposed by transecting its origin on the skull along with a rim of epicranium which will serve as an anchor for sutures (Fig. 35.3). The muscle can be split longitudinally creating several slips. These may be transposed to the upper and lower eyelids, the ala, the mesolabial fold, and the upper and lower lips. Overcorrection should be performed by sewing the slips under tension. The depression left from removal of the temporalis can be repaired using a silastic block. The temporalis can also be detached from its insertion into the condyle and sewn into the oral commissure or nasolabial fold.

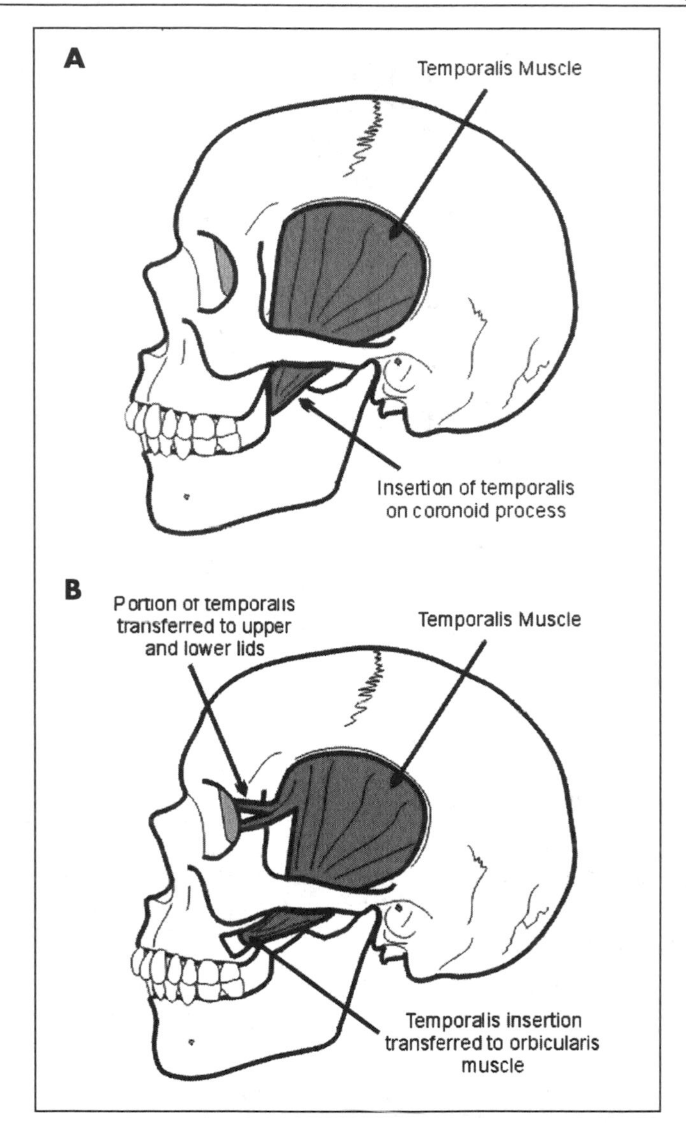

Figure 35.3. Transfer of the temporalis muscle. Reprinted with permission from Microsurgeon.org.

The masseter is transferred partially or entirely, and its medial end is split allowing it to sandwich the oral commissure (Fig. 35.4). The superior portion is sutured to the dermis of the mesolabial fold and to the underlying orbicularis oris, and the inferior portion is sutured to the lower lip.

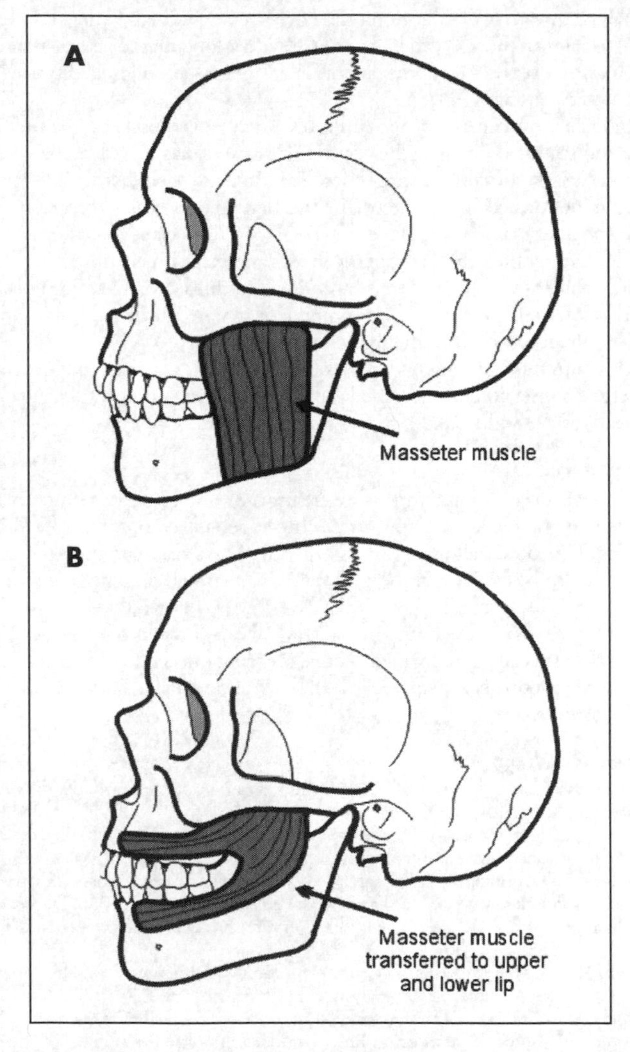

Figure 35.4. Transfer of the masseter muscle. Reprinted with permission from Microsurgeon.org.

Free Microneurovascular Transplant

The advantage of using a microneurovascular free muscle transplant is that it can be transferred with its nerve supply, which can be connected to the contralateral facial nerve using a cross-facial nerve graft. This will enhance voluntary control of the transferred muscle. Free flap transfer is suited primarily for patients who are paralyzed in the buccal distribution due to loss of the facial muscles after tumor resection. It is also appropriate for those with intracranial or congenital causes of facial paralysis. A variety of donor muscles have been described. Some of these, such as the biceps femoris muscle, can be harvested with a long neural pedicle that can be directly anastomosed to the contralateral facial nerve eliminating the need for a cross-facial nerve graft.

The gracilis is the muscle most commonly used. This muscle can be split longitudinally and trimmed down to the appropriate size. It has a predictable neurovascular pedicle. Harvest of this muscle is described in detail elsewhere in this text. The procedure is performed in two steps. In the first step, a cross-face nerve graft is performed as described above. Once the Tinel's sign indicates that the axons have grown the length of the sural nerve graft, the second stage is performed. The muscle is harvested and trimmed down to its anterior one-third. It is sewn into place in a manner that attempts to recreate the zygomaticus major. The origin is sewn into the zygoma and the insertion into the oral commissure.

Roughly one-third of patients will require a revision after free muscle transfer. These include reinsertion of a detached muscle, tightening of the muscle, debulking of the cheek, or the addition of a suspension procedure.

Pearls and Pitfalls

Nerve repair or grafting involves the creation of one or more suture lines. It is imperative that the repair be tension free. Intraoperatively, one must be sure that movements of the head, talking or eating will not place tension on the repair. Undue tension must also be avoided in the postoperative period. Some surgeons will even prohibit the patient from talking for several weeks. It is known that scar tissue at the site will interfere with axonal growth. Starting about 6 weeks after surgery, gentle massage of the area can be performed. Active muscle contraction aids in the speed of recovery, making postoperative physical therapy and facial exercises an important part of the process.

Suggested Reading

1. Baker DC, Conley J. Regional muscle transposition for rehabilitation of the paralyzed face. Clin Plast Surg 1979; 6:317.
2. Baker DC. Reconstruction of the paralyzed face. Grabb and Smith's Plastic Surgery. 5th ed. Philadelphia: Lippincott-Raven, 1997:545.
3. Baker DC. Reanimation of the paralyzed face: Nerve crossover, cross-face nerve grafting, and muscle transfers. Head and Neck Cancer. Philadelphia: B.C. Decker, 1985.
4. Braam MJ, Nicolai JP. Axonal regeneration rate through cross face grafts. Microsurg 1993; 14(9):589.
5. Harii K et al. Free gracilis muscle transplantation with microneurovascular anastomosis for the treatment of facial paralysis. Plast Reconstr Surg 1976; 57:133.
6. Manktelow RT. Free muscle transplantation for facial nerve paralysis. Microreconstruction of nerve injuries. Philadelphia: Saunders, 1987:607.
7. Pitanguy I, Ramos AS. The frontal branch of the facial nerve: The importance of its variations in face lifting. Plast Reconstr Surg 1966; 38(4):352.
8. Simpson RL. Anatomy of the facial nerve. The Paralyzed Face. St. Louis: Mosby, 1991.
9. Terzis JK, Noah ME. Analysis of 100 cases of free-muscle transplantation for facial paralysis. Plast Reconstr Surg 1997; 99:1905.

Chapter 36

Frontal Sinus Fractures

Joseph Raviv and Daniel Danahey

Introduction

Frontal sinus fractures can have serious consequences due to the proximity of the sinus to the intracranial cavity and the potential for nasofrontal duct obstruction with its long-term sequelae. Delayed or improper management of frontal sinus fractures can result in complications including meningitis, mucopyocele, pneumocephalus and brain abscess. Frontal sinus fractures comprise approximately 10% of facial fractures. Males are injured more frequently than females (8:1). The incidence of fractures of the frontal sinus is greatest in the third decade of life. Motor vehicle accidents are the most common cause. Other causes include physical altercations (including gunshot wounds), sports, industrial accidents and falls.

Relevant Anatomy

The paired frontal sinuses develop separately and are frequently asymmetric. The frontal sinus begins to develop early in childhood and is rarely visible on radiographs earlier than the second year of life. The sinus invades the frontal bone by about 5 years of age and slowly grows to reach adult volume of 6-7 ml by late adolescence.

The sinus is roughly pyramid-shaped with its apex inferiorly and its base superiorly. An intrasinus septum is usually present and the distal borders of the sinus often spread to form an irregular pattern, which makes mucosal removal difficult during frontal sinus obliteration. The anterior wall is the strongest of the sinus walls and is twice as thick as the posterior wall. Each sinus wall has an anterior and posterior table. The posterior wall separates the frontal sinus from the anterior cranial fossa. The floor of the frontal sinus is the thinnest of the three walls and is therefore the most convenient location for tapping an infected sinus (trephination). The floor of the sinus also functions as the supraorbital roof, and the drainage ostium is located in the posteromedial portion of the sinus floor. The frontal infundibulum is a more narrow area within the sinus that leads to the ostium.

The blood supply of the frontal sinus is via the internal carotid system through the supraorbital branch of the ophthalmic artery as well as through some branches of the anterior ethmoidal artery. Venous drainage is through two communicating routes. External drainage is through the angular and anterior facial veins; the deep drainage is via transosseous venous channels through the posterior wall of the sinus known as the foramina of Breschet. The nerve supply is mainly from the supraorbital branch of the ophthalmic division of the trigeminal nerve.

Biomechanics of Frontal Sinus Fracture

The amount of force necessary to fracture the frontal sinus is two to three times greater than that necessary to fracture other facial bones. The anterior wall is thicker than the posterior wall and can withstand between 800 to 2,200 pounds of force. For this reason, damage to the posterior wall must be suspected in all case of anterior wall frontal sinus fracture. Moreover, patients with frontal sinus fractures have frequently sustained serious concomitant injuries, which should be appropriately addressed prior to management of the frontal sinus fracture.

Diagnostic Assessment

Clinical Presentation

The patient suffering from a frontal sinus fracture has a characteristic history. A traumatic episode to the forehead has occurred involving considerable force, and the patient usually reports loss of consciousness. Initial management is directed at life-threatening conditions and stabilizing the patient's condition. Physical findings may include but are not limited to frontal swelling, pain, lacerations over the forehead skin and numbness over the forehead. A bony defect may be palpated over depressed anterior wall fractures. Epistaxis is often present and may be mixed with cerebrospinal fluid (CSF) from a dural tear caused by either a depressed posterior wall fracture or coincidental fracture of the anterior cranial fossa floor. A fracture of the superior orbital rim can be present, causing the globe to be displaced or trapped. A fracture of the nasoethmoidal complex can manifest as flattening of the pyramid and telescoping of the nose.

Radiographic Evaluation

The clinical picture in most instances does not allow clear differentiation between the fracture of the various frontal sinus walls or across the frontonasal duct. Radiographic evaluation, in particular computed tomography (CT), is clearly the most valuable diagnostic tool in frontal sinus fractures. CT clearly depicts fractures, the amount of depression, and the nature of the contents of the sinus cavity, adjacent brain and overlying soft tissue. Fine axial sections are useful for evaluating anterior and posterior table fractures of the frontal sinus and intracranial injuries. "True" coronal scans are useful for assessing the floor of the frontal sinus, the frontonasal duct and cribiform plate. Coronal images that are reconstructed from axial scans do not provide adequate resolution to assess these structures.

Classification

For ease of diagnosis and formulation of a treatment algorithm, frontal sinus fractures can be divided into two types: anterior table fractures and posterior table fractures. Each type of fracture can further be divided into displaced or nondisplaced fractures.

Management

The goals in management of frontal sinus fractures are: (1) prevention of intracranial infection; (2) prevention of frontal sinus disease (such as sinusitis and mucocele); and (3) a cosmetically acceptable outcome. Antibiotic prophylaxis is generally recommended. Antibiotics with high CSF penetration, such as ceftriaxone, together with metronidazole for anaerobic coverage are good choices when there is risk of intracranial sepsis. For Gram-positive coverage in cases of

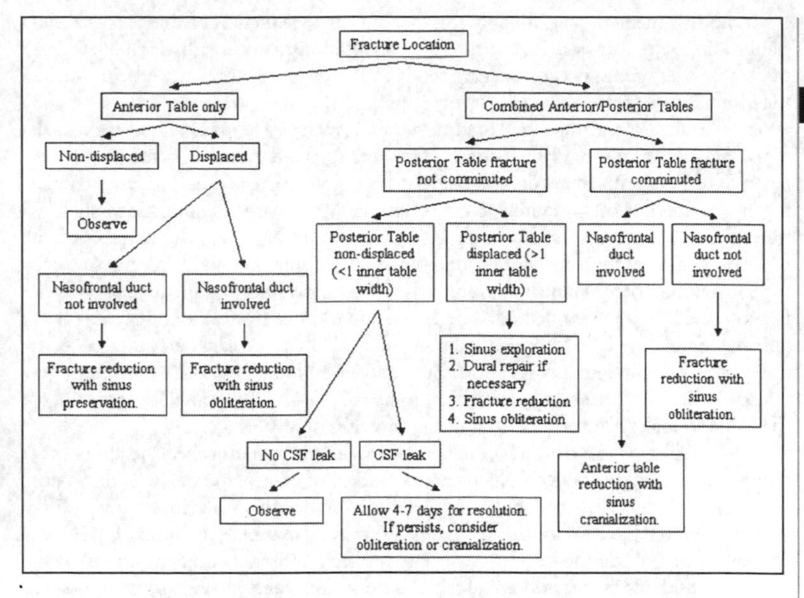

Figure 36.1. An algorithm for the management of frontal sinus fractures.

skin contamination, cefazolin can be added. For compound fractures, antibiotic therapy is maintained for 2 weeks. In cases of closed or isolated fractures, antibiotic use is controversial.

An algorithm outlining the management of frontal sinus fractures is shown in Figure 36.1. Although this simplifies frontal sinus fractures according to the involved walls, the surgeon is often faced with a patient with multiple fractures. A useful approach is to have a treatment scheme that addresses each specific site and then apply those principles to each particular case.

Anterior Table Fracture

Nondisplaced Anterior Table Fractures

Linear, nondisplaced fractures involving the anterior wall with no cosmetic deformity can be managed conservatively. No complications have been reported for nonoperative treatment of such fractures. Persistent opacification of the frontal sinus may indicate frontonasal duct obstruction or even CSF leakage and may mandate endoscopic evaluation.

Displaced Anterior Table Fractures

Displaced anterior wall fractures should always be surgically explored and repaired. The objectives of surgery for displaced anterior table fractures are: (1) aesthetically acceptable reconstruction of the anterior table; (2) removal of damaged sinus mucosa; and (3) direct inspection of the nasofrontal duct for injury. Displaced anterior table fractures are ideally reconstructed within 7 to 10 days to prevent a forehead deformity. Surgical access can be gained with a bicoronal approach, a supraorbital brow incision, or extension of an existing laceration.

In nonfragmented or minimally fragmented cases, reduction of the fractures and stabilization with 1.0 or 1.3 mm titanium adaptation plates is sufficient. Plating is preferable over wiring because wiring tends to flatten the normal arched contour of the frontal bone. In the management of severely comminuted fractures with bone loss, an effort is made to achieve maximal bone preservation. The painstaking process of replacing the comminuted fractures is necessary to avoid cosmetic deformity. Gaps larger than 4 or 5 mm are reconstructed with bone grafts. The use of synthetic material for reconstruction of the anterior table has been met with complications such as mucopyocele, infection, flap breakdown and extrusion of implant material. This is probably related to the direct communication of the frontal sinus with the nasal cavity.

The relation between obstruction of the frontonasal duct and formation of frontal mucoceles is well established. The best evaluation of frontonasal duct integrity and patency is made intraoperatively. Patency of the duct can be evaluated with fluorescin, benzylpenicillin solution, or methylene blue. Attempts to repair or reconstruct the duct have not been proven to be reliable and, therefore, obliteration of the sinus to make it nonfunctional is recommended.

Obliteration of the frontal sinus involves meticulous removal of the sinus mucosa while leaving the bony walls intact. Removal of all mucosa of the sinus is undertaken with a curet first, and then followed by drilling with an otologic drill. Obliteration has been accomplished using a variety of materials including fat, fascia, muscle, pericranium and cancellous bone. Obliteration eradicates the air-filled sinus cavity and nasofrontal duct making the sinus nonfunctional. This prevents subsequent infection of the sinus and mucocele formation.

Posterior Table Fractures

Nondisplaced Posterior Table Fractures

Posterior table fractures which are nondisplaced, or minimally displaced less than the width of the posterior table without CSF leakage, can be observed safely with prophylactic antibiotic treatment. If a CSF leak is present, then patients are put at bed rest with head elevation and given the opportunity for spontaneous resolution. A lumbar drain may be considered if the leak is profuse. If a leak persists after 5 to 7 days, a cranialization procedure should be considered to prevent intracranial complications including meningitis and pneumocephalus.

Cranialization of the frontal sinus involves excision of the posterior sinus wall. Cranialization is approached using a bicoronal frontal craniotomy, preserving an anterior pericranial flap for separating the nasal cavity from the intracranial space. Once any intracranial injury and dural lacerations are addressed, removal of all sinus mucosa is carried out, including the inner cortex of the anterior table. Removal of any residual posterior sinus wall and intersinus septum is accomplished as well. The nasofrontal duct orifices are obliterated with temporalis fascia, muscle, or bone in order to prevent retrograde spread of infection. Finally, the anterior pericranial flap is placed along the floor of the sinus. The anterior wall is then reconstructed using plates and bone grafts as necessary.

Displaced Posterior Table Fractures

Fractures displaced greater than one table width without nasofrontal duct involvement or CSF leak are explored, and fracture reduction and stabilization is performed. If the nasofrontal duct is injured but no dural tears or CSF leaks are present, sinus obliteration with occlusion of the nasofrontal duct is carried out. In the presence of persistant CSF leak or a significantly comminuted posterior wall fracture, cranialization of the sinus is required (as described above).

Long Term Follow-Up

Another component to managing fractures is the importance of long-term follow-up. It is necessary to emphasize to patients that a mucocele may form even after an obliteration procedure. One common approach is to have a CT scan done at 6 weeks after surgery, which will provide a new base line for the radiologist. Repeat scans are ordered at 6 and 12 month intervals. If the scans remain clear, then the patient can be reimaged at 2 years postoperatively.

Pearls and Pitfalls

The first controversy that often arises in frontal fractures with a CSF leak is whether to use antibiotics or not. The literature provides no clear answer, and our approach has been to treat these patients with antibiotics for at least one week. The controversy becomes more important if the leak persist beyond one week. Some feel that using antibiotics for a CSF leak over a long period of time actually promotes bacterial resistance that can be particularly problematic to treat. Fortunately, most CSF leaks will resolve with conservative management over 7 to 10 days, and neurosurgical intervention is reserved for cases that fail conservative management.

Determining the appropriate surgical approach is a common question in surgical repair of frontal sinus fractures. The three approaches mentioned in the above chapter are through an existing laceration, an open sky incision, or a bicoronal incision. For those patients that have a significant laceration, we usually do not hesitate to use this wound and even extend it if necessary. Of course, this is much more difficult to do if there is posterior wall fracture involvement, in which case a bicoronal approach is indicated. I have seen the open sky approach used in the past and have been very unhappy with the way those patients have healed. I have never personally utilized this approach because of the untoward scarring that results from this approach.

With respect to anterior frontal sinus wall fractures that are closed and have minimal displacement, there are some new techniques that are currently being utilized which involve an endoscopic approach—much like the approach for an endoscopic brow lift. Some surgeons have found that by providing the same exposure as with an endoscopic brow lift, they are able to inject hydroxyapatite into the small minimally displaced areas and to use a periosteal elevator through one of the portal incisions to sculpt the hydroxyapatite into the appropriate frontal configuration. For displaced anterior wall fractures, traditional open fixation repair was performed with titanium plates. Since the frontal sinus wall is not a support buttress, titanium fixation is not a necessity. We have been using resorbable plates for over two years with great success, with the advantage that after 18 months the plates are no longer palpable.

Posterior wall fractures are usually more complicated and should always involve a neurosurgical consultation. When the posterior wall is displaced we always ask the neurosurgeons to do a bifrontal craniotomy and to take down the posterior wall. Our role is to then drill out all the mucosa present along the anterior lateral walls and to plug the floor of the frontal duct. We use a bicoronal incision; however we back elevate all the way to the occipital area and then harvest a very long pericranial flap as a separate layer from the scalp tissue. This flap is based on the supraorbital and supratrochlear vessels. The cases should be reconstructed with titanium mesh, and not resorbable plates, because the pressure of the brain against the anterior wall will require more significant rigidity than the resorbable plates have to offer. The nasofrontal ducts are usually packed with pieces of temporalis muscle that are harvested from either side. There are a variety of options for frontal sinus obliteration. The first option is to use calvarial bone from the parietal area. Another option might be to use abdominal fat, which has been the traditional material of choice. More recently, we have been using hydroxyapatite cement because it saves time by avoiding a second operative sight.

Suggested Reading

1. Donald PJ. Frontal sinus fractures. In: Donald PJ, Gluckman JL, Rice DH, eds. The Sinuses. New York: Raven Press, 1995; Chapter 26.
2. Rohrich RJ, Hollier LH. Management of frontal sinus fractures. Changing concepts. Clin Plast Surg 1992; 19(1):219.
3. Stanley RB. Maxillofacial Trauma. In: Cummings CW, ed. Otolaryngology—Head and Neck Surgery. 3rd ed. St. Louis: Mosby Year Book, 1998; Chapter 24.
4. Wallis A, Donald PJ. Frontal sinus fractures: A review of 72 cases. Laryngoscope 1988; 98:593.
5. Xie C, Mehendale N, Barrett D et al. 30-Year retrospective review of frontal sinus fractures: The Charity Hospital experience. J Craniof Maxil Trauma 2000; 6(1):7.

Chapter 37

Orbital Fractures

John Nigriny

Anatomy

The orbit is composed of seven bones: the zygoma, sphenoid, frontal, ethmoid, maxilla, lacrimal and palatine bones (Fig. 37.1). Residing inside the bony orbit are two fissures, the superior and inferior orbital fissures. Through the superior orbital fissure (SOF) pass the occulomotor (CN III), trochlear (CN IV), ophthalmic division of the trigeminal (CN V1) and the abducens (CN VI) nerves. It is located in a superolateral position relative to the optic foramen and divides the greater and lesser wings of the sphenoid bone. The inferior orbital fissure (IOF) resides between the lesser wing of the sphenoid and the maxilla and is oriented in an inferolateral direction. Traversing the IOF are the second division of the trigeminal nerve (CN V2), branches off of the sphenopalatine ganglion and the inferior ophthalmic vein.

The optic foramen which transmits the optic nerve is located 42-45 mm posterior to the infraorbital rim. A tendinous ring, the annulus of Zinn, is located just anterior to the foramen and serves as the common origin of the four rectus muscles, superior oblique and the levator muscles. Whitnall's tubercle is the bony attachment of the lateral canthal tendon, and a check ligament for the lateral rectus, the suspensory ligament of Lockwood, and the lateral extension of the levator aponeurosis. The tubercle is located 1 cm inferior and 3 mm posterior to the frontozygomatic suture. Lockwood's ligament is actually a hammock-like system that suspends the globe. It has contributions from muscular septae, Tenon's capsule and the lower eyelid retractors. It arises from the fibrous attachments of the inferior rectus posteriorly and continues as the capsulopalpebral fascia anteriorly. It attaches to the lacrimal crest medially and Whitnall's tubercle laterally.

A thorough understanding of the four orbital walls (superior, inferior, medial and lateral) is important for describing fracture patterns and predicting associated injuries.

The superior wall, or orbital roof, is moderately resistant to trauma and infrequently fractured, comprising only 1-5% of orbital fractures.

The lateral wall is also moderately strong and made up of the zygomatic and frontal bones. This is the least common area for an isolated orbital fracture, although it is commonly fractured through the frontozygomatic suture line in ZMC fractures.

The medial wall is partially formed from the thin lamina papyracea of the ethmoid and lacrimal bones. As such it is more prone to fracture; however, these medial wall fractures are often observed nonoperatively. Problematic sequelae occur when there is a medial wall fracture in conjunction with injury to other nearby structures. The lacrimal sac is located in the lacrimal crest and may be lacerated or obstructed by bony fragments requiring immediate or late dacrocystorhinostomy. The medial

Practical Plastic Surgery, edited by Zol B. Kryger and Mark Sisco. ©2007 Landes Bioscience.

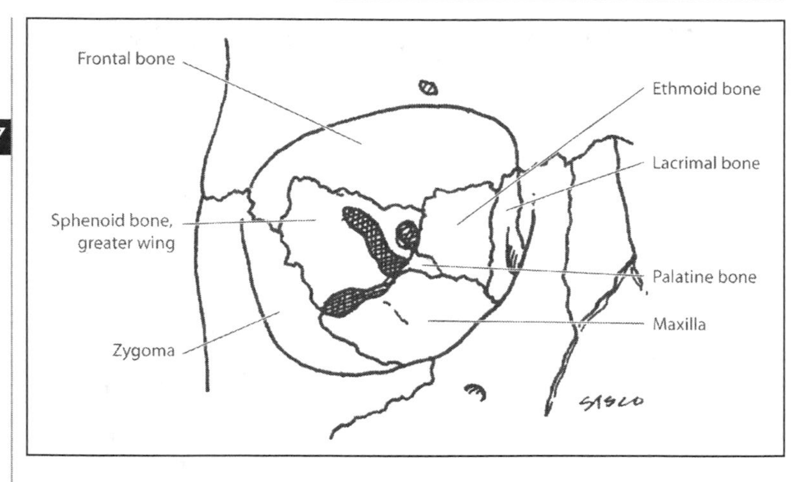

Figure 37.1. Bony anatomy of the orbit.

canthal tendon which inserts on the anterior and posterior lacrimal crests may be torn or avulsed creating traumatic telecanthus. The medial rectus muscle can become entrapped or attenuated resulting in late medial gaze abnormalities.

The inferior orbital rim, composed of the maxillary and zygomatic bones, is the structure most often involved in orbital fractures. A direct force applied to the inferior orbital rim is transmitted to the orbital floor. Since the orbital floor is a relatively weak structure, it is prone to fracture, resulting in a blow-out fracture of the floor in which the orbital contents (mostly fat) herniated downwards. As the medial floor is thinnest portion, floor fractures typically occur medial to the infraorbital groove.

Diagnosis

Clinical

The diagnosis of orbital fractures is based on clinical and radiographic findings. Although many signs and symptoms have been described, the common signs include periorbital ecchymosis and edema, the presence of a palpebral or subconjunctival hemorrhage or hematoma ("the spectacle hematoma"), limitation of extraocular function (due to entrapment of orbital contents, edema, or neurologic sequelae), hypoesthesia in the distribution of the infraorbital nerve and enophthalmos.

Radiographic

Radiologic assessment of suspected orbital fractures should be performed with CT scanning as plain films are of little diagnostic benefit. Both axial and coronal views (1-3 mm) with reconstruction are preferable. Common CT findings of orbital floor fractures include an air-fluid level in the ipsilateral maxillary sinus, a trap door deformity of the orbital floor with herniation of intraorbital contents, and various other orbital rim or wall fractures (Fig. 37.2). There are two prevailing theories for the mechanism of orbital floor fractures. First, the hydraulic theory attributes the fracture to a sudden increase in intraorbital pressure. The elevation in pressure on the thinnest bone segment in the orbit, the floor, causes a blow-out fracture to

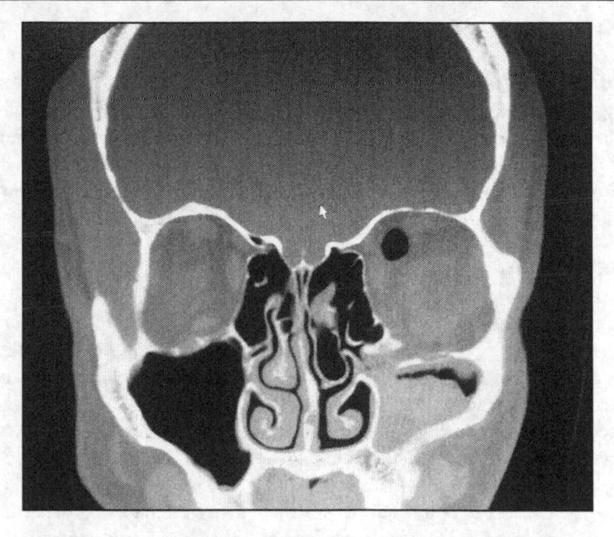

Figure 37.2. Coronal CT scan showing a fracture of the orbital floor with fluid in the maxillary sinus.

ensue. The buckling theory attributes orbital floor fractures to the force transmitted to the infraorbital rim causing acute deformation with or without fracture. The force is passed on to the thin orbital floor which cannot adequately resist the force and fractures.

Associated Injuries

A wide range of associated ocular and periocular injuries have been reported in the literature (2-93%). Interestingly, the reported rate of associated injury is much lower for nonophthalmologist (2-25%) compared to ophthamologists (9-93%). A recent study of 365 patients with orbital fracture had an associated ocular injury rate of 26%. The injuries which required immediate opthalmologic intervention were as follows: traumatic optic neuropathy (35%), elevated intraocular pressure >40 mm Hg (26%), hyphema (22%), traumatic iritis (9%) and ruptured globe (9%). As a general rule, all patients with orbital fractures require evaluation by an ophthalmologist.

Indications for Treatment

Indications for treatment of a subacute orbital fracture (within 2 weeks of injury) include entrapment, enophthalmos, exophthalmos and bony defects greater than 1 cm^2. The indications for immediate repair of acute orbital fractures are a nonresolving oculocardiac reflex, young patient (<18 yrs) with "white eye blow-out" associated with severe muscle entrapment, and a large fracture with globe prolapse into the maxillary sinus. Since orbital volume is roughly 20-30 ml, a 10% change in orbital volume (2-3 ml) may produce enophthalmos. As small volume changes can produce significant symptoms, a primary goal of treatment is anatomic reduction of the fracture to reestablish preinjury volume. If loss of osseous tissue interferes with anatomic realignment then autogenous or alloplastic grafts are necessary.

Graft Material for Reconstruction

Autogenous Grafts

Autogenous grafts described for orbital reconstruction include: bone (calvarium, ilium, rib, maxilla and mandible) or cartilage (nasal septal, costal and auricular). Calvarial grafts have been shown to be most beneficial in immediate reconstruction but tend to have problems with resorption when utilized in a delayed repair. Although described in the literature, cartilage grafts are infrequently used.

Alloplastic Grafts

Alloplastic implants can be divided into bioresorbable and nonbioresorbable types.

Allogeneic bone and cartilage, lyophilized dura, gelatin film, polydiaxone plates, polylactide plates, and polglactin mesh and plates are all examples of bioresorbable alloplasts used in orbital reconstruction. The advantage of all alloplastic implants is their inherent lack of donor site morbidity. The disadvantages of lyophilized dura and allogeneic grafts are the lack of cellularity and potential for disease transmission. The primary disadvantage of resorbable plate systems is the concern over long term stability especially in larger bony defects.

Nonbioresorbable implants include: silicone, methylmethacrylate, ceramics, polyurethane, polyethylene, porous polyethylene, titanium mesh and plates. All of these materials provide rigid support but have somewhat higher rates of infection than autologous tissues. Porous polyethylene has been shown to allow for fibrovascular ingrowth with pore sizes between 100-200 micrometers. Titanium mesh and porous polyethylene have been shown to mucosalize on the sinus-exposed surfaces of the implant.

Regardless of the method of reconstruction, patients with post-traumatic enophthalmos should be over corrected 2-3 mm to ensure adequate correction after resolution of edema. Over correction also helps to account for any volume loss secondary to fat necrosis. Opinions differ on the placement of orbital grafts. Some advocate rigid fixation, particularly of autogenous bone grafts, in an attempt to increase "take." Others merely place grafts to span the defect and allow the soft tissue to redrape over the graft.

Techniques

Techniques in orbital repair are mainly dictated by the severity of the fracture and the types of incisions used for gaining access to the fracture site. For example, multiple surgical approaches may be necessary for a four-wall reconstruction, while a single incision may be used for an isolated single wall fracture. Access incisions to the orbital floor and infraorbital rim include subciliary, subtarsal and transconjunctival. A lateral extension of the incision and a lateral canthotomy can be used to gain access to the lateral rim.

Subcilliary Approach

The subcilliary approach described by Converse is made a few millimeters below the lash line. The incision is carried down through the orbicularis muscle to the tarsal plate. The plane above the tarsal plate can then be followed to the orbital septum and subsequently the orbital rim periostium which is then divided anteriorly for access to the orbital floor. In a stair-stepped approach, a plane is dissected anterior to the orbicularis until the inferior tarsal margin. The dissection then proceeds through the muscle to the orbital septum and the orbital periostium. In the "skin only" subcilliary approach, the orbicularis muscle is divided at the level of the infraorbital rim, along with the orbital periostium. This variation of the subciliary approach is prone to skin flap necrosis, hematoma, ecchymosis and ectropion. Therefore, the skin/muscle flap techniques are preferred. The subcilliary incision extends from the punctum medially to the lateral canthus laterally. A lateral release of approximately 1.5 cm can be added to increase exposure of the lateral orbital wall, frontozygomatic suture and malar eminence. Advantages of the subcilliary approach include a well-camouflaged scar and ease of dissection. An important disadvantage is the higher incidence of scleral show than with a subtarsal approach.

Subtarsal Approach

The subtarsal approach, also popularized by Converse, is a version of a skin/muscle flap technique. The incision is made in the skin of the subtarsal fold, or if obscured by edema, 5-7 mm below the lash line directed inferolaterally, approximating the crease. The orbicularis muscle is divided a few millimeters below the incision, and then the dissection continues to the orbital septum and orbital rim periostium. The subtarsal approach offers the easiest dissection technically and an acceptable scar although it is the most conspicuous of the techniques described here. Also, there tends to be more lid edema with this approach.

Transconjunctival Approach

The transconjunctival approach for access to the bony orbit was described by Tessier. The technique involves scleral-corneal protection followed by a conjunctival incision just below the lower tarsal margin. Dissection then proceeds inferiorly between the orbicularis oculi and orbital septum to the infraorbital rim. The infraorbital periostium is incised to gain access to the orbital floor. A lateral canthal or paracanthal release can be performed to allow for wider exposure of the lateral orbital wall. The advantage of this approach over the others is the absence of a visible scar and enhanced medial exposure. Disadvantages include limited lateral exposure without lateral release and a risk of postoperative ectropion.

Regardless of which technique is utilized, an important point to decrease lower lid retraction is to not divide the orbital septum or risk shortening it. Rather, one should incise the orbital periostium inferior and anterior to the rim.

Endoscopic Approach

Another technique available for repair of isolated orbital floor fractures is the endoscopic approach. Although this technique requires a specialized skill set with a learning curve, it does not violate the eyelid soft tissues and thus is inherently less morbid. An upper gingivobuccal incision exposes the anterior maxillary wall. The technique proceeds as follows:

- A 1-2 cm antral bone flap is created to access the maxillary sinus and orbital floor. The sinus is evacuated and packed with oxymetazoline pledgets to aid in hemostasis.
- The size and fracture configuration are defined using a 30°, 4 mm endoscope. Complete removal of comminuted bony fragments is performed.
- Stable bony shelves are identified adjacent to the fracture. Either resorbable or nonresorbable bone grafts are cut slightly larger than the defect.
- The graft is introduced through the defect, rotated, and placed on the stable medial, lateral and anterior orbital shelves.
- Fixation is not required if there is adequate stability of the bony shelves. If not, direct screw fixation can be performed.

Complications

Many complications from orbital fracture repair have been reported. The most common include infection, visual disturbances, changes in visual acuity, diplopia, lid and canthal malposition (ectropion and entropion), lacrimal obstruction, resorption or malposition of implants, and infraorbital nerve dysesthesia. The most feared and dangerous acute postoperatve complication is retrobulbar hematoma with elevation of intraocular pressure. This is a surgical emergency which requires orbital decompression through immediate release of sutures and lateral canthotomy followed by control of hemorrhage. Steroids, diuretics and ice may also be of therapeutic benefit in this situation.

Pearls and Pitfalls

Orbital fractures with involvement of the medial orbital wall are significantly more likely to result in diplopia and exophthalmos than fractures without involvement of the medial wall.

Impacted fractures of the lateral orbital wall can be thought of as a "blow-in" fracture that may be accompanied by decreased visual acuity and ocular motility limitations. Early surgical treatment is warranted. Traumatic optic neuropathy will usually resolve with time.

Although extremely rare, retrobulbar hematoma following blunt orbital trauma is a serious complication since permanent loss of vision can ensue. It is due to postoperative or post injury orbital bleeding in the setting of an undisplaced orbital wall fracture. This results in increased intraocular pressure and ultimately ischemia of the optic nerve. The clinical presentation includes pain, exophthalmos with proptosis, internal ophthalmoplegia, and decreased or loss of the pupillary reflex. CT scan with thin-cuts is important in confirming the diagnosis. Any delay between the onset of symptoms and treatment can have a significant effect on functional recovery.

Suggested Reading

1. Cook Todd. Ocular and periocular injuries from orbital fractures. J Am Col Surg 2002; 195(6):831-834.
2. Manson Paul. The orbit after converse: Seeing what is not there. J Craniofac Surg 2004; 15(3):363-367.
3. Suga H, Sugawara Y, Uda H et al. The transconjunctival approach for orbital bony surgery: In which cases should it be used? J Craniofac Surg 2004; 15(3):454-457.
4. Rohrich, Rod, Janis J et al. Subciliary versus subtarsal approaches to orbitozygomatic fractures. Plast Reconstr Surg 2003; 111(5):1708-1714.
5. Manolides S, Weeks BH, Kirby M et al. Classification and surgical management of orbital fractures: Experience with 111 orbital reconstructions. J Craniofac Surg 2002; 13(6):726-737.
6. Jatla K, Enzenauer R. Orbital fractures: A review of current literature. Current Surgery 2004; 61(1):25-29.
7. Manson Paul. Facial fractures. In: Goldwyn R, Cohen M, eds. The Unfavorable Result in Plastic Surgery: Avoidance and Treatment. Philadelphia: Lippincott Williams and Wilkins, 2001:489-515.
8. Persons BL, Wong GB. Transantral endoscopic orbital floor repair using resobable plate. J Craniofac Surg 2002; 13(3):483-488.

Chapter 38

Fractures of the Zygoma and Maxilla

Zol B. Kryger

Zygomatic Fractures

Relevant Anatomy

The zygoma is comprised of two main components: the body, which creates the cheek prominence, and the arch, which articulates with the zygomatic process of the temporal bone. Three-dimensionally, the zygoma has a lateral (malar) surface and a medial (orbital) surface. From the standpoint of the orbit, the zygoma comprises the lateral orbital wall. The masseter muscle and fascia attach to the inferior border of the arch. During a fracture, this muscle produces a downward force on the zygoma. The temporal fascia attaches to the superior border of the arch.

Two points about zygomatic fractures are important to emphasize. First, the body of the zygoma bone itself does not usually fracture; rather one of the adjacent bones either fractures or separates at the suture line. Second, since the zygoma comprises the lateral orbital wall, displacement of the zygoma can significantly enlarge the orbital volume and produce enophthalmos.

Physical Exam

The following signs and symptoms are common and should be investigated:

- Infraorbital nerve distribution numbness
- Step deformity in the infraorbital region
- Depression of the malar eminencc
- Trismus due coranoid process impingement
- Enophthalmos
- Inferior globe displacement (hypoglobus)
- Visual disturbances

Imaging

Plain films are not usually needed. The Waters view is the most useful film for visualizing the zygomatic buttresses. Today, most centers will perform a CT scan for suspected facial fractures. Axial and coronal thin cuts (1.5 mm) will show all fractures. Three-dimensional reconstructions can be useful for characterizing and planning the treatment of complex fractures.

Treatment

As a general rule, undisplaced zygomatic fractures do not require surgical treatment. A soft diet for 6 weeks and protection of the malar region during sleep is sufficient. Displaced fractures usually require operative repair. In addition, fractures that are comminuted are more difficult to repair, and wider exposure is usually

Practical Plastic Surgery, edited by Zol B. Kryger and Mark Sisco. ©2007 Landes Bioscience.

required. Displaced fractures usually require several incisions for adequate exposure. A lower eyelid incision (subtarsal, subciliary or transconjunctival) exposes the infraorbital rim. A gingivobuccal sulcus incision exposes the zygomaticomaxillary buttress. The zygomaticofrontal suture can be exposed through the lower lid incision or through a lateral extension of an upper blepheroplasty incision. Coronal incisions should be used only when the arch must be visualized, such as in the case of a severely comminuted fractures. Once the fracture fragments have been adequately reduced, plate fixation is performed. The zygomaticofrontal suture is plated with a 1 mm plate. The zygomaticomaxillary buttress is plated with a heavier plate: either a 1.5 or 2 mm plate. Although two-point fixation is often adequate, some surgeons will also plate the infraorbital region with a 1 mm plate as well. This plate should be placed on superior surface of the rim so that it will not create a palpable step off.

Isolated zygomatic arch fractures require surgery in two cases: if there is a contour deformity or if trismus is present. Reduction is best achieved through a temporal hairline incision, termed the Gillies' temporal approach. The bone fragment is elevated using an elevator passed along the surface of the temporalis muscle, deep to the temporoparietal fascia. Severely comminuted and displaced fractures may require open reduction. Once proper reduction has been achieved, the fragment may require stabilization. This can be done with plates or K-wire fixation.

Complications

Diplopia

Temporary diplopia can last up to 6 months. It resolves in about half the cases. The incidence of permanent diplopia is 5%. It is usually only apparent on upward gaze. It is most often due to entrapment of the inferior rectus muscle or adjacent tissue; however it can also occur as a result of injury to a nerve.

Enophthalmos

Patients will often notice this complication themselves. The zygoma must be completely separated and repositioned with plate fixation, thus restoring normal orbital volume.

Descent of the Malar Soft Tissue

This complication can result in loss of the normal malar contour. It can be prevented by adequate periosteal suspension of the malar tissue to the orbital rim following fracture fixation.

Optic Neuropathy

It can range from mild changes in color perception to blindness and is due to traumatic injury or ischemia of the optic nerve. A short course of high dose steroids is given in acute cases; however many patients will not resolve their visual defects. All patients with post-traumatic optic neuropathy should undergo high resolution CT scanning, and an ophthalmologist should be consulted.

Bradycardia

It is often accompanied by nausea and syncope. These symptoms are part of the oculocardiac reflex which is mediated by the ophthalmic division of the trigeminal nerve's connections to the vagus nerve.

Maxillary Fractures

Classification

Most midface fractures occur along lines of bony weakness and fall into one of three patterns described by Le Fort (Fig. 38.1):

Le Fort I — A transverse fracture resulting in a floating palate.
Le Fort II — A pyramidal fracture that traverses the orbits and nasoethmoid region producing midface mobility.
Le Fort III — A fracture through the orbits and zygomas that results in craniofacial dysjunction.

The Le Fort II fracture is the most common, followed by Le Fort I and III patterns. It is unusual to diagnose a fracture that falls purely into one of these three classifications. Most midface fractures are more complex and have components of other fracture types. The nasal bones do not tolerate impact well and may be fractured along with the maxilla.

Clinical Examination

Facial fractures secondary to high speed motor vehicle accidents have a risk of associated cervical fractures or neurological injury. A cervical spine examination should be performed and cervical spine films obtained in order to rule out a fracture. A full cranial nerve and mental status exam should also be performed. Another commonly associated finding is ocular injury.

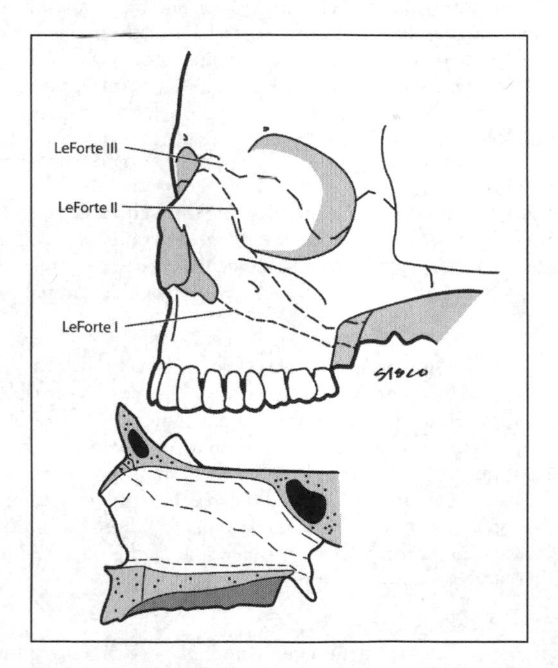

Figure 38.1. Le Forte fracture classification. The lower figure shows a sagittal view of the fracture patterns through the midface bones (see Fig. 39.2)

Maxillary fractures should be suspected whenever there is malocclusion following trauma to the face. An awake patient can usually tell whether or not his teeth alignment feels normal. In addition to the occlusion, mobility of the midface should be determined by grasping the premaxilla and eliciting movement while holding the head still with the other hand. Le Fort I fractures will demonstrate only movement of the lower maxilla, whereas both Le Fort II and III fractures will demonstrate additional movement at the nasal root. The Le Fort III fracture will also show lateral orbital rim motion.

The orbital rims and nasoethmoid regions should be palpated to help determine the level of the fracture. Le Fort I fractures can be subtle in their presentation and do not usually demonstrate periorbital findings. In contrast, Le Fort II fractures can present with periorbital and subconjunctival ecchymosis. The face will also appear lengthened, and there will usually be massive swelling of the middle third of the face.

Radiographic Examination

The standard of care in diagnosing facial fractures is a CT scan with thin cuts in both the axial and coronal planes. Plain films are not routinely needed for evaluating the midface. Three-dimensional reconstructions are useful for planning surgery but are not required in the diagnosis of these fractures.

Treatment

Several principles should be followed in treating midface fractures. Whenever possible, the repair should be performed in a single stage along with concomitant soft tissue injury management. All of the fracture fragments should be exposed and precise anatomic reduction with internal plate fixation should be attempted. If necessary, autogenous bone grafting should be performed during the initial repair. Restoring normal occlusion is essential and relies on maxillomandibular fixation (MMF). The mandible should be reduced and stabilized as an initial step in order to provide an occlusive platform for the maxilla. Since the midface lacks adequate sagittal buttresses, the mandible and frontal bone should be used as buttress the maxilla.

Exposure

Low maxillary fractures should be approached through an upper gingivobuccal sulcus incision. The infraorbital rims can be exposed via a transconjunctival, subtarsal or subciliary (least preferred) approach. Occasionally, an extended lateral canthotomy is required. The nasoethmoid region is approached through a coronal incision.

Plate Fixation

A number of studies support the use of resorbable plates and screws in the treatment of maxillary fractures; however many centers still use nonresorbable hardware. Rigid internal, three-point fixation is the current standard for treating maxillary fractures. Compression is not required and should never come at the expense of preserving occlusion or normal contours. Gaps less than 5 mm can be tolerated, although defects secondary to comminuted buttress fractures should be filled with bone grafts. Postoperative movement of the fragments will impair normal healing. At least two screws should be placed on either side of the fracture line. Buttress fixation requires at least a 2 mm thick plate.

Fractures in Adentulous Cases

Adentulous individuals are usually elderly, and the goals of restoration of normal chewing and facial appearance may be less of an issue. The indications for open reduction with internal fixation are less rigorous. If the fracture is not displaced, a soft diet for 2-3 weeks may suffice. A displaced fracture in a stable patient who is a good operative candidate should be treated with open reduction and internal fixation. For those patients who are unstable or in poor medical condition, minimal therapy is warranted. Le Fort I fractures should not be treated, and in Le Fort II and III fractures the focus should be on the nasoethmoid region and orbital rims. A Le Fort I osteotomy can be performed at a later time if necessary.

Fractures in Children

Maxillary fractures in children are unusual. The mandible and nasal bones are much more frequently fractured. Minimally displaced and greenstick fractures can usually be treated conservatively. Displaced fractures should be treated with a similar approach as in adults: early repair with open reduction and internal rigid fixation. Plates and screws should not be placed near the tooth buds. Many surgeons will also remove any hardware once the fracture has healed. Traditional MMF in young children without permanent teeth is risky. If it is required, it can be accomplished with acrylic splints and circummandibular wires in order to avoid damaging the tooth buds.

Fractures of the Palate

Maxillary fractures can occasionally involve the palate and alveolus. As is the case with most maxillary fractures, wide exposure and rigid internal fixation is usually the recommended treatment. The Hendrickson classification describes the six most common fracture types: anterior and posterolateral alveolar (Type I), sagittal (Type II), parasagittal (Type III), paraalveolar (Type IV), comminuted (Type V) and transverse (Type VI). The first four types are treated using internal rigid fixation followed by several weeks of MMF. Comminuted and transverse fractures (especially those with multiple small segments) are often not amenable to adequate reduction and internal fixation and should be treated with a palatal splint. Simple alveolar fractures can often be treated with closed reduction and immobilization for 4-6 weeks. Exposure of the fracture can be achieved through a transverse or longitudinal vestibular maxillary incision. Longitudinal incisions are less likely to devascularize the mucosa; however care must be taken to avoid the greater palatine artery.

Complex Fractures of the Face

High speed motor vehicle accidents and gunshot wounds often produce complex facial fractures. After the initial trauma assessment and treatment of any concomitant life-threatening injuries, the craniofacial injuries can be addressed. A high resolution facial CT scan is required for identifying the various fractures. Three dimensional reconstructions are occasionally useful in preoperative planning. Most authorities agree that early definitive treatment should be attempted. A tracheostomy may be necessary, especially if prolonged ventilatory support is expected.

Any ophthalmologic emergencies should be addressed first. The next step is to obtain wide exposure of the fractures through the various incisions described above. The order in which the various segments of the maxillofacial skeleton are addressed depends largely on surgeon preference. The goals in treating complex facial fractures are: (1) to restore the facial buttresses; (3) to restore normal occlusion; (4) to stabilize the various fracture segments; (5) to achieve the normal facial contours whenever possible.

The mandible is addressed first. Any fractures are reduced and rigidly fixed, and the mandible is stabilized relative to the cranial base. Zygomatic fractures are then reduced and plated since these may interfere with proper reduction of the maxilla. Attention is then turned to the maxillary fractures. The buttresses are restored, and the maxilla is stabilized to the mandible with MMF. Any intraoral lacerations are repaired. Any orbital rim fractures are then repaired. Reconstruction of the orbits may require bone grafting, especially in restoration of the supraorbital ridge. The nasoethmoid region is dealt with including reduction of any nasal fractures and reattachment of the medial canthal tendons. The frontal bone and frontal sinus are repaired when necessary. Finally, soft tissue lacerations and avulsions are repaired.

Complications of Facial Fractures

- Infection (sinusitis, meningitis, soft tissue infections)
- Brain damage
- Infraorbital nerve injury
- CSF rhinorrhea
- Ocular problems (e.g., blindness, diplopia, enophthalmos and telecanthus)
- Smell disturbances
- Nasal septal deviation
- Malocclusion
- Exposed or visible plates and screws
- Residual deformities in facial height and projection
- Malunion and nonunion

Pearls and Pitfalls

- The functional goals in midface fractures are restoration of orbital volume, movement and preinjury occlusion. The aesthetic goals include restoring the vertical height, the projection and the contours of the face. These goals should be kept in mind during any reconstruction.
- Maxillary fracture repair should begin with establishment of dental occlusion and placement of the patient in MMF, followed by restoration of the vertical buttresses using rigid fixation and bone grafting when indicated. Soft tissue repair comes last.
- Remember to resuspend the malar soft tissue after any subperiosteal dissection of the maxilla and zygoma. Failure to do so will result in a ptotic appearance of the cheek contour postoperatively. A cuff of periosteum should be preserved on the infraorbital rim to which the malar soft tissue can be reattached.
- Isolated zygomatic arch fractures are repaired when indicated with the Gillies' approach (as described above). A common pitfall is to pass the instrument used to "pop" the arch outward in the wrong plane. Usually, this is in a plane that is too superficial. The correct plane should follow the course of the temporalis muscle and allow the instrument to easily insert underneath the zygomatic arch without any resistance.

Suggested Reading

1. Hendrickson M, Clark N, Manson PN et al. Palatal fractures: Classification, patterns and treatment with rigid internal fixation. Plast Reconstr Surg 1998; 101:319.
2. Manson PN, Clark N, Robertson B et al. Subunit principles in midface fractures: The importance of sagittal buttresses, soft-tissue reductions and sequencing treatment of segmental fractures. Plast Reconstr Surg 1999; 103:1287.
3. Markowitz BL, Manson PN. Panfacial fractures: Organization of treatment. Clin Plast Surg 1989; 16(1):105.
4. Pollock RA, Gruss JS. Craniofacial and panfacial fractures. In: Foster CA, Sherman JE, eds. Surgery of Facial Bone Fractures. New York: Churchill Livingstone, 1987.
5. Rudderman RH, Mullen RL. Biomechanics of the facial skeleton. Clin Plast Surg 1992; 19(1):11.

Chapter 39

Nasal and NOE Fractures

Clark F. Schierle and Victor L. Lewis

Introduction

The central location and prominent nature of the human nose renders it particularly vulnerable to the forces of blunt facial trauma. The nasal bones are the third most commonly fractured bones in the body after the clavicle and the wrist. They are the most common injured facial bones accounting for roughly half of all facial fractures. Despite this they are often overlooked or missed entirely. The nasoorbitoethmoid (NOE) complex is comprised of the confluence of several critical facial structures. Failure to recognize and treat nasal bone and NOE complex fractures in the appropriate time frame can lead to severe functional and cosmetic problems which can be more difficult to treat at a later time point.

Anatomy

The bony pyramid of the nose is comprised of the paired nasal bones and frontal processes of the maxillary bones (Fig. 39.1). The nasal bones gradually thin as they meet the upper lateral cartilages caudally where they are most vulnerable to fracture. The upper lateral cartilage inserts on the undersurface of the nasal bone and contacts the quadrangular cartilage of the nasal septum anteriorly and the lower lateral cartilage caudally. The nasal septum is comprised of the quadrangular cartilage anteriorly, the perpendicular plate of the ethmoid bone posterosuperiorly, and the vomer bone posteroinferiorly (Fig. 39.2). The entire nasal structure rests on the groove of the maxillary crest terminating anteriorly at the nasal spine.

Fractures of the nasal bones also typically involve the bony or cartilaginous components of the nasal septum. Disruption of these structures can lead to inadequacy of the external nasal valve, formed by the angle of the upper lateral cartilages with the nasal septum, leading to chronic nasal airway obstruction. Cosmetic deformities include nasal asymmetry, inadequate dorsal support from the nasal septum resulting in a saddle nose deformity, or loss of tip support. This is due to deformation of the caudal septum or dissociation of the lower lateral cartilages from their supporting framework leading to a bulbous or drooping tip deformity.

The NOE complex is formed by the confluence of the nasal, orbital, ethmoid and cranial cavities and is comprised of bony contributions from the nasal, maxillary, frontal, ethmoid, lacrimal and sphenoid bones (Fig. 39.3). The vertical support buttress of the NOE complex is comprised of the maxillary process of the frontal bone, the nasal bones and the frontal process of the maxilla. Violation of this buttress leads to disruption of the more delicate structures of the medial orbital wall including the lacrimal bones and laminae payraceae of the ethmoid bones. The horizontal buttresses of the NOE complex are the supraorbital and infraorbital rims. Violation of these supports may compromise the integrity of the superior aspect of the ethmoid

Practical Plastic Surgery, edited by Zol B. Kryger and Mark Sisco. ©2007 Landes Bioscience.

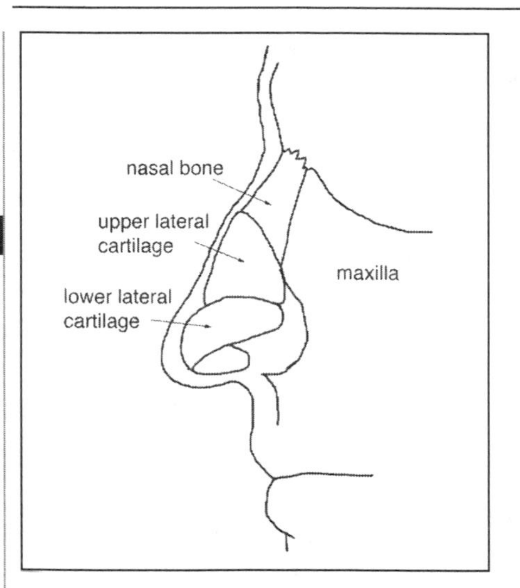

Figure 39.1. Lateral schematic view of the nasal bone and cartilages.

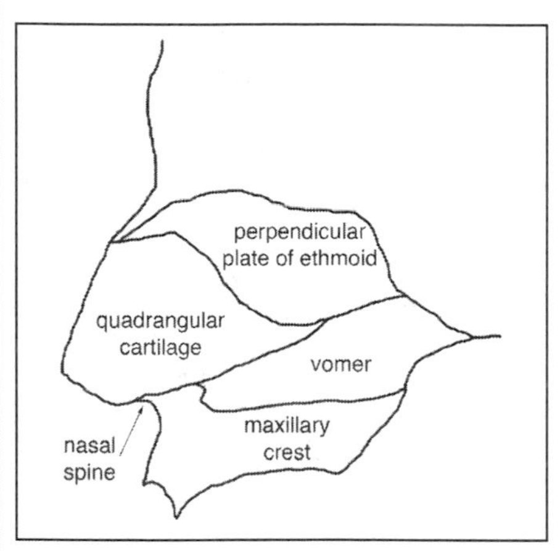

Figure 39.2. The components of the nasal septum.

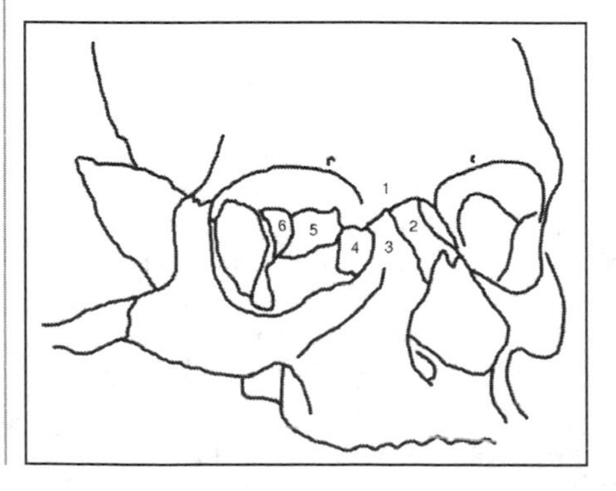

Figure 39.3. The nasoorbitoethmoid (NOE) complex: frontal bone (1); nasal bone (2); maxilla (3); lacrimal bone (4); ethmoid bone (5); sphenoid bone-lesser wing (6).

bone as it forms the cribiform plate at the base of the anterior cranial fossa resulting in a leak of cerebrospinal fluid (CSF) or transection of the arteries that span the anterior and posterior ethmoid foramina leading to a vision-threatening orbital hematoma.

The medial canthal tendon (MCT) is an important soft tissue structure with a critical role in the diagnosis and treatment of NOE fractures. The MCT arises from the anterior and posterior lacrimal crests, envelops the lacrimal sac within the lacrimal fossa, and diverges to form the orbital, preseptal and pretarsal heads of the orbicularis oculi muscle. The MCT provides a pumping action in the proper function of the lacrimal apparatus, as well as providing static support for the globe and dynamic support for the eyelid through the action of the orbicularis muscle. Fracture of the NOE complex leads to lateral displacement of the MCT, resulting in the characteristic telecanthus which should trigger recognition of an underlying NOE fracture. Failure to properly reposition and fix the MCT in an anatomically correct location can result in a cosmetic deformity, lagophthalmos, or dysconjugate gaze.

Nasoorbitoethmoid fractures are classified into three types of increasing severity and difficulty of repair:

- Type I fractures have no comminution and involve only the segment of the medial orbit containing the MCT.
- Type II fractures involve comminuted bone external to the insertion of the MCT, and typically the MCT is still attached to a relatively large segment of the fracture.
- Type III fractures are bilateral and involve extension of comminuted bone beyond the insertion of the MCT.

Preoperative Considerations

Preoperative imaging of any patient with a suspected fracture of the facial skeleton should include dedicated facial bone CT scans with thin (<2 mm) cuts to provide adequate detail. In addition to the standard axial views, true coronal images should be obtained to properly assess the fracture in three dimensions. The extent of displacement is usually best visualized on one of the two series of cuts. If true coronal images are not obtainable (as in the presence of a cervical spine injury) reconstructed images may be adequate if the axial source images are of sufficient resolution.

Operative Technique

Nasal Fractures

Reduction of nasal fractures should be performed either immediately or after at least three to five days have passed to allow for some resolution of the initial inflammation and swelling. Nondisplaced fractures do not require treatment. Comparison with preinjury photographs can help guide reduction. Nasal fractures may be reduced with open or closed techniques. Isolated nasal bone fractures or fractures which displace the nasoseptal complex by less than half the width of the nasal bridge are candidates for closed reduction. More extensive or complex fractures generally benefit from an open approach to maximize exposure. If local anesthesia with sedation is to be used, the infratrochlear, infraorbital, supratrochlear and anterior ethmoidal nerves are blocked by injection of the nasal root, dorsum, lateral walls and anterior septum; 4% cocaine-soaked pledgets are inserted to minimize hemorrhage.

For the closed technique a blunt instrument such as a Boise elevator, empty scalpel handle or Asch forceps are introduced into the nose and controlled force applied in the direction opposite the fracturing force. Once all fractures are adequately mobilized, the symmetry of the nasal pyramid is restored by direct manipulation. Depending on the extent of septal involvement and the mobility of the nasal segments, internal (e.g., silastic splints or Merocel packing) or external (e.g., Aquaplast or a Denver splint,) splints may be placed to maintain fixation for one week postoperatively.

Open reduction may be required if the nasoseptal deviation is too severe to be adequately addressed by the closed technique. A hemitransfixion incision at the caudal edge of the cartilaginous septum is used and mucoperichondrial flaps are elevated providing access to displaced fragments of the septum. Anchoring sutures may help in stabilization of the caudal septum to the periosteum of the nasal spine.

Nasoorbitoethmoid Fractures

NOE fractures necessitate an open approach. Choice of incision is determined by concerns for cosmesis, surgical exposure and location and orientation of any associated lacerations which may be exploited for access. Options include the traditional vertical Lynch incision in the medial canthal region, lower lid subciliary or transconjunctival incisions, bicoronal incisions and medial conjunctival incisions through or around the lacrimal caruncle. Whatever the approach, dissection must be sufficient to expose the fracture segment and adequately identify the MCT and severity of the NOE fracture. Adequate reduction of the MCT is the goal of any operation to treat NOE fractures. If the MCT remains attached to a large piece of bone, miniplates and monocortical screws or steel wire may be used to reduce the MCT through bony fixation. If the MCT has been completely avulsed, reattachment is accomplished through wire fixation of a sturdy portion of the MCT to a hole drilled in the posterior lacrimal crest. In type II fractures, where there is disruption or comminution of the anatomic point of fixation, transnasal fixation can restore the appropriate vector of pull of the MCT. In type III fractures, the MCT is secured using wire passing through the nasal septum to the contralateral MCT. As the wire is tightened the MCTs are drawn medially. Care must be taken to pass the wire through a point in the septum that is sufficiently superior and posterior to recreate an appropriate vector of pull.

Postoperative Care

Patients should be instructed to avoid nose-blowing and sneeze with an open mouth. Ice packs and head elevation help minimize swelling and discomfort. Splints and packing are removed at one week. Antibiotics are indicated if there are significant devitalized areas of cartilage or bone which must be treated functionally as grafts. Saline nasal spray can be used after packing is removed to aid in comfort and minimizing crusting. The role of antibiotics is not entirely clear; however many surgeons will prescribe 5-7 days of oral antibiotics. Steroids (e.g., methylprednisolone) can be given postoperatively to help minimize inflammation.

Pearls and Pitfalls

Although extremely rare, a septal hematoma must be ruled out in all nasal fractures. If untreated, the hematoma can lead to ischemic necrosis of the septal mucosa and consequent septal perforation. It is easy to diagnose and when present is quite striking. The mucosa is dark purple, dusky and bulging. Treatment consists of simply incising the mucosa and drainage of the blood.

The nose is extremely well-vascularized and tends to bleed a fair amount during nasal surgery. The use of lidocaine with epinephrine is essential, even when surgery is performed under general anesthesia. There is no point in using epinephrine if one does not wait at least 10 minutes for its vasoconstrictive effects to kick in. Additional efforts to minimize bleeding should include intraoperative control of blood pressure (ideally maintaining a systolic blood pressure below 120 mm Hg), direct application of pressure when bleeding is significant, and avoidance of retching and coughing during extubation. Finally, prior to extubation, an oral gastric tube should be used to suck out as much blood as possible from the stomach, esophagus and pharynx since ingested blood contributes to postoperative nausea.

Suggested Reading

1. Leipziger LS, Manson PN. Nasoethmoid orbital fractures. Current concepts and management principles. Clin Plast Surg 1992; 19(1):167.
2. Markowitz BL, Manson PN et al. Management of the medial canthal tendon in nasoethmoid orbital fractures: The importance of the central fragment in classification and treatment. Plast Reconstr Surg 1991; 87(5):843.
3. Mathog RH. Posttraumatic telecanthus. Facial Plast Surg 1988; 5(3):261.
4. Pollock RA. Nasal trauma: Pathomechanics and surgical management of acute injuries. Clin Plast Surg 1992; 19(1):133.
5. Rohrich RJ, Adams Jr WP. Nasal fracture management: Minimizing secondary nasal deformities. Plast Reconstr Surg 2000; 106(2):266.

Chapter 40

Mandible Fractures

Jeffrey A. Hammoudeh, Nirm Nathan and Seth Thaller

Introduction

Mandibular fractures are the second most frequently occurring facial bone fracture, second only to nasal fractures and are the primary maxillofacial injury requiring hospitalization. Interpersonal violence and motor vehicle accidents remain the most common etiology of mandibular fracture with anatomic distribution as follows: body 29%, condyle 26%, angle 25%, symphysis 17%, ramus 4% and coronoid 1% (Fig. 40.1).

Classification

Accurate classification of mandibular fractures is important for both medicolegal documentation and communication among clinicians. In addition, this often determines selection of the optimal method for surgical fixation. Basic classification necessitates a concise description including anatomic location, type of fracture (simple vs. complex), degree of displacement and whether it is open or closed (Fig. 40.2). Fractures that communicate intra- or extraorally are considered open fractures.

In the past fractures were also classified as either favorable or unfavorable (Fig. 40.3). This represents the relationship of the fracture segment to the vector that the associated musculature imposes on the bony segments. Favorable fractures are defined as those in which the muscle pull and the vector of displacement of the proximal fracture segment are in opposing directions, thereby preventing displacement. Unfavorable fractures occur when both muscle pull and the vector of displacement are in the same direction, resulting in bony displacement.

Diagnosis

Diagnosis of mandibular fractures can usually be made with an accurate and comprehensive history and physical exam. Initial questions should focus on the mechanism of injury and the location of impact. Predictable fracture patterns can help a clinician identify the correct diagnosis. For example, a chin laceration suggests an injury directly to the anterior mandible. This should also raise clinical suspicion of an associated contracoup injury to the contralateral condyle. Characteristic patterns of injury are based on a thorough knowledge of the underlying bony architecture and associated protective "fronts" of weakness within its horseshoe configuration. Common complaints include an abnormal bite, pain, swelling and altered sensation of the lower lip.

Physical findings generally include swelling, intra- and extraoral ecchymoses, associated soft tissue lacerations, trismus, fractures or displaced teeth and malocclusion. Bimanual palpation, which is a necessary component of all head and neck exams, may confirm fracture location as well as degree of mobility. Neurosensory function of the

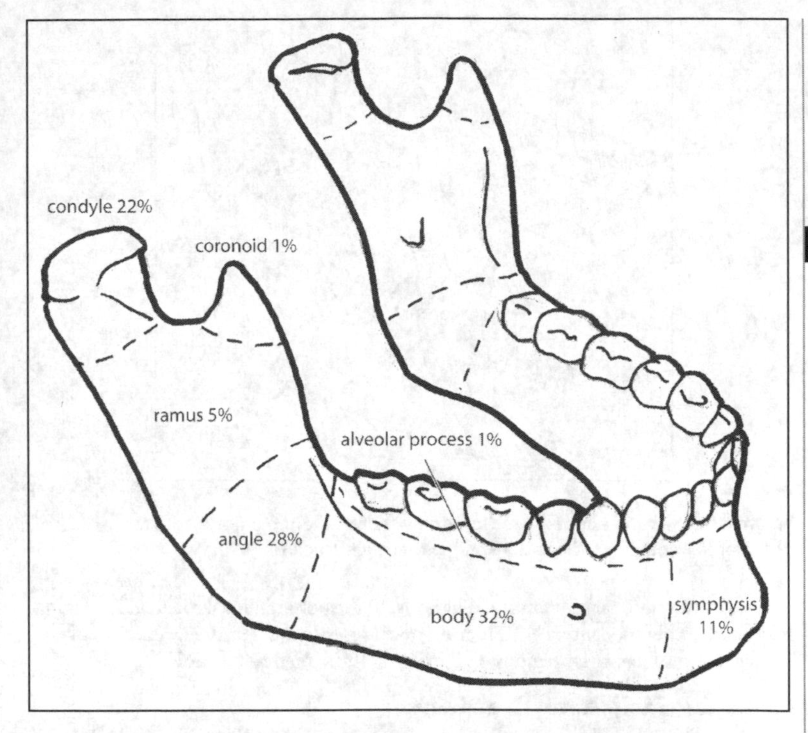

Figure 40.1. Mandible fracture pattern based on location and frequency.

inferior alveolar nerve using light touch, brush stroke and pinprick should be assessed. Fractures involving the angle or body may be associated with inferior alveolar nerve injury as it passes through the bony canal. It is imperative to document in the medical record completion of a neurosensory exam. The degree of eventual neurological return will vary depending on the location and displacement of the fracture. In approximately 85% of mandibular fractures, the neurosensory status was unchanged or improved after treatment. Fractures with greater than 5 mm displacement were associated with a six-fold increased risk for a decreased neurosensory score.

Imaging

Suspected mandibular fractures should be verified with a CT scan and a panoramic radiograph, or Panorex®. The Panorex® is especially useful in ruling out a fracture line that crosses the root of a tooth, since this can be missed even on a CT scan with thin cuts. The limitation of the panorex is that it necessitates the patient remain in the upright position for the duration of the radiography.

Since associated injuries are encountered in approximately 43% of mandibular fractures (11% of these injuries being cervical fractures), it is imperative for the clinician to rule out any associated cervical fractures via the appropriate radiological studies such as helical CT scans. Three-dimensional reformatted CT images can be helpful in visualizing the fracture and communicating management options to patients and families but

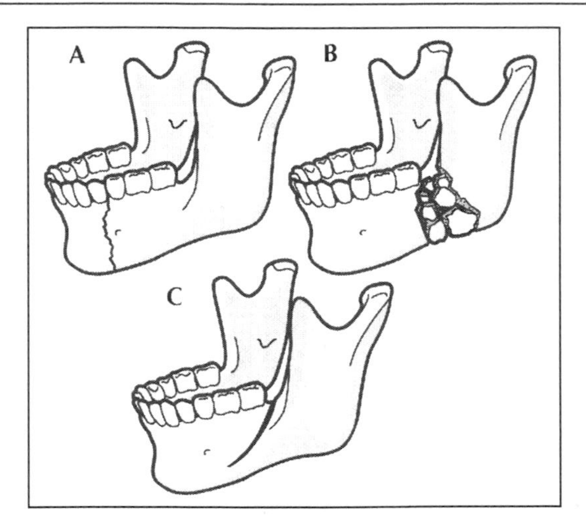

Figure 40.2. Fracture type based on complexity. A) Simple nondisplaced fracture. B) Complex comminuted fracture. C) Greenstick fracture.

are usually not necessary for clinical diagnosis. In the poly-trauma patient, CT imaging may be the only modality available for clinical assessment. For the evaluation of the condylar and subcondylar regions, CT imaging is the method of choice.

General Principles of Treatment

Immediate relief of pain is achieved by temporary immobilization of the fracture with Barton's head dressing, narcotic analgesics, topical ice therapy and a mechanically soft or pureed diet. Narrow spectrum antibiotics should be given immediately, especially in open fractures. Options include penicillin, clindamycin, or cefazolin. Definitive fracture management should be initiated as soon as practically possible, following stabilization of the patient's overall medical condition.

Two general treatment options are available: maxillomandibular fixation (MMF) alone or with open reduction internal fixation (ORIF). Closed treatment depends solely on the placement of MMF permitting secondary fracture healing by immobilization of the jaws. ORIF, on the other hand, will allow for primary bone healing. At the completion of ORIF, the MMF may be released, allowing for early mobilization. This is especially important for complex fractures involving the condylar region in which early motion will decrease the potential for development of temporomandibular joint (TMJ) dysfunction.

Condylar and Subcondylar Fractures

Condylar fractures are relatively common and constitute approximately 25% of all adult mandibular fractures (due to an inherent weakness of the condylar neck that helps prevent impaction of the condyles into the cranial vault). These fractures are often associated with a fracture at a second site, in particular parasymphyseal fractures. Definitive management options remain controversial.

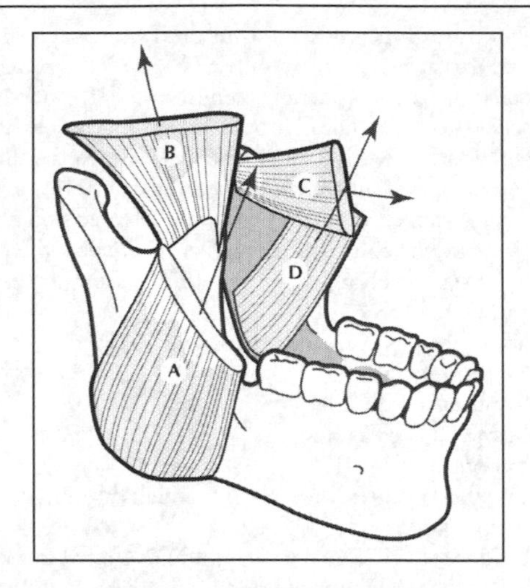

Figure 40.3. Displacement of fracture is based on the relationship of the muscle vector force with the angulation of the proximal/distal fracture segments. A) The masseter will try to displace the proximal fracture segment superolaterally. B) The temporalis muscle will try to displace the condylar segment superiorly. C) The medial pterygoid will try to displace the proximal fracture segment medially. D) The lateral pterygoid will try to displace the proximal sebment superiorly.

Classification

Classification of condylar fractures is based on location or displacement. A favorable condylar fracture is defined as minimally displaced or nondisplaced. Associated mild malocclusion can usually be treated through closed reduction technique. Our recommendation is 10 days of MMF followed by 10 days of elastics Ivy loops, a full MMF course, or four-point screw fixation. If the occlusion is stable, elastics may be removed and the patient should be continued on a mechanical soft diet. Reevaluation should be performed after 10 to 14 days and, if the occlusion is reproducible; a Panorex® should confirm a well-aligned, healing fracture. Arch bars may then be removed.

Unfavorable or displaced condylar fractures can be treated by either closed reduction or in certain specific instances with ORIF. Displacement of the proximal segment is caused by the insertion of the lateral pterygoid muscle resulting in one of the following: displacement anteromedially, laterally towards the zygoma, posteriorly towards the ear canal, or superiorly into the middle cranial fossa. Most cases, however, result in a medially displaced proximal fragment leading to a shortening of the ramus and an open bite deformity. Some institutions advocate initial closed reduction in this situation. This is based on data that evaluated the risks of ORIF vs. MMF which showed that adequate results can be obtained with closed reduction. However, as rigid fixation techniques have improved, and more critical data regarding ORIF vs. MMF has emerged, ORIF is now advocated by certain authors. In our opinion, the decision on when to treat with MMF vs. ORIF must incorporate other individual factors such as clinician experience, patient needs and associated facial features. One should further evaluate condylar fracture based on the specific anatomic location and the ability to actually position and fixate segments.

High condylar fractures which occur within the TMJ capsule itself may or may not involve the articular surface of the condylar head. If the proximal segment is small and the vascularity of the fragment is compromised, the fracture may eventually undergo avascular necrosis. These should be treated by closed reduction. In addition, open reduction is very difficult to complete due to the thin cortical bone at this site. On the other hand, subcondylar fractures that have a proximal segment that is adequate may be treated with rigid fixation. The decision tree on ORIF vs. MMF should incorporate the clinicians' operative experience, the patients' desires, anatomic feasibility of rigid fixation, and a risk stratification of the associated morbidity of either treatment.

Relative indications for ORIF include the following:

1. Patient compliance/preference
2. A history of recent seizure disorder
3. Poorly controlled psychiatric issues
4. Failed attempt at previous MMF
5. Bilateral subcondylar fractures with resultant malocclusion due to significant bony displacement

Absolute indications for ORIF of a unilateral subcondylar fractures include:

1. Displacement of the condyle into the middle cranial fossa or external auditory canal
2. Lateral extracapsular displacement of the condyle
3. The displaced condyle functionally blocks the opening and closing of the mandible
4. Severe anteromedial angulation of more than 45° with dislocation of the condylar head from the glenoid fossa
5. Open joint wound with presence of a foreign body

Severe facial trauma is the etiology of bilateral subcondylar fractures which are associated with malocclusion. An anterior open bite can be seen secondary to shortening of the ramus condyle unit (RCU) and associated clockwise rotation of the mandible. Comprehensive management is required, including a thorough work-up and algorithmic evaluation. Definitive management of bilateral subcondylar fractures remains controversial as well. In general, the approach depends on the surgeon's experience and institutional philosophy. Many centers advocate closed reduction of both condyles.

When open reduction is being considered, we recommend following these basic principles: If both condyles are accessible and capable of being plated, then bilateral ORIF will allow an early return of function. It is our recommendation to treat bilateral subcondylar fractures with ORIF of at least one condyle. ORIF of the more accessible RCU will reestablish the vertical dimension of occlusion in a manner that allows for adequate closed reduction of the contralateral RCU. It is the reestablishment if the proximal and distal segments of the RCU that will restore the vertical dimension of occlusion and thus allow for the possibility of closed manipulation and adequate realignment of the contralateral RCU. Again, once one RCU is treated by ORIF, the contralateral RCU should follow the same closed reduction protocol outlined above (10 days of MMF, followed by 10 days of guiding elastics). If a malocclusion develops, then the protocol should be extended by an additional 10 days of MMF followed by elastics.

In edentulous patients that have sustained either unilateral or bilateral subcondylar fractures, the recommended treatment is closed reduction. When a splint, existing dentures, or a gunning splint is available, then 10 days of closed reduction followed by a mechanical soft diet for 4-5 weeks is appropriate management. In cases where either the splint is unavailable or impossible to position because of alveolar ridge atrophy, we recommend a soft diet for approximately 6 weeks, followed by a general dental consult for refabrication of the dentures.

Inappropriate treatment of condylar fractures may lead to severe functional complications, including: avascular necrosis of the condylar head, persistent TMJ pain, arthritic changes to the TMJ (resulting in pain and decreased mobility of the jaw) and facial nerve injury. Injury to the branches of the facial nerve can occur during operative exploration and repair of subcondylar fractures with a reported incidence as high as 30%. Most nerve injuries result in only temporary paralysis. The search for a surgical approach that minimizes the risk to the facial nerve has led to the use of endoscopy-assisted ORIF of subcondylar fractures. This technique has a decreased risk of iatrogenic nerve palsy. Other advantages of the endoscopic approach over traditional ORIF include better access to the condyle through smaller incisions, and better outcomes with decreased morbidity and faster recovery. Specifically, there is less tissue dissection and manipulation, resulting in less tissue edema and postoperative pain and swelling. Disadvantages of the procedure include its higher cost and a learning curve that may initially result in a much longer surgical time. The subcondylar fracture can be approached intraorally or extraorally. The intraoral approach is of limited use, especially for the management of medially displaced and dislocated fractures.

Mandibular Body Fractures

Anatomically, the mandibular body is defined as the region bounded anteriorly by the parasymphysis and posteriorly by the angle. Body fractures represent approximately 30% of all adult mandibular fractures.

Favorable body fractures occur when the direction of the fracture and the pull of the masseter and medial pterygoid muscles resist bony displacement. In contrast, unfavorable body fractures are displaced at the fracture site due to the angulation of the masseter and medial pterygoid muscle pull.

In body fractures, both MMF (with closed reduction) and ORIF can be utilized. A common MMF technique is the use of prefabricated arch bars that are adapted and circumferentially wired to the teeth in each arch, maintained for 4-6 weeks.

Although MMF is more economical and lowers the overall risk of infection, it also possesses several disadvantages. Wiring of the teeth often leads to poor nutrition and oral hygiene. Patients cannot return to full preinjury function for a minimum of 4 weeks, thus requiring postoperative therapy to regain maximal incisal opening. Therefore, patient compliance is necessary. There is also a risk of aspiration in patients suffering from seizure disorders. Furthermore, patients with traumatic brain injuries have a tendency to generate forces that lead to the loosening of wire fixation.

ORIF is recommended for unfavorable and displaced body fractures or continued displacement of the body fracture fragments despite MMF. ORIF with rigid fixation of the body is a relatively simple surgical operation that can allow the patient to rapidly return to function. An incision is made at the depth of the vestibule. Intraoperatively, the surgeon should identify and protect the mental nerve. A rigid inferior border plate should be passively adapted and screwed into the bone. If the arch bars are to be removed at the end of the case, then a maxillomandibular tension band should be applied using monocortical screws. If the arch bars are to be kept on, they can act as the tension band. Although the arch bars can be left on, the patient should not be placed into MMF since the rigid plate fixation is adequate for fracture stabilization.

Angle Fractures

The mandibular angle is the triangular area defined by the anterior border of the masseter muscle and a line extending from the third molar region to the posterosuperior attachment of the masseter muscle. At the angle, the horizontal segment joins the vertical segment. It is the most common site of fracture in the edentulous mandible. Loss of alveolar bone due to lack of teeth weakens the mandible in the region of the angle, thus making it more susceptible to fracture. In addition, the third molars and long root of the premolars weaken the angle, causing fracture in dentate patients. Angle fractures constitute approximately 25% of all adult mandibular fractures.

In unfavorable angle fracture the proximal fragment is elevated and drawn forward by the action of the masseter and medial pterygoid muscles. Favorable fractures of the angle may have minimal or no displacement because the vector of fracture is opposite to the vector of muscle displacement or because there is bone stabilization of the proximal segment.

While angle fractures can sometimes be treated with closed reduction, more often they are unfavorable and displaced and accordingly require ORIF. In such cases, arch bars and MMF are not sufficient since they do not reach the displaced proximal segment. This can lead to fragment mobility and a risk of either malunion or nonunion.

Surgical intervention for unfavorable displaced angle fracture can employ several techniques. For minimally displaced fracture we advocate ORIF via an intraoral approach. In angle fractures that are significantly displaced, we advocate a rigid inferior boarder plate and a semirigid superior boarder plate. This is easily accomplished via an extraoral approach. Some clinicians will attempt this rigid fixation intraorally with a percutaneous trocar to facilitate screw placement. At the end of the case, the MMF and arch bars can be removed.

Symphyseal and Parasymphyseal Fractures

Symphyseal and parasymphyseal fractures usually result from direct trauma to the chin. With a simple isolated symphyseal fracture we recommend ORIF, since the operation is associated with minimal morbidity and allows for an early return to function. Some clinicians may elect a 4-6 period of MMF as opposed to ORIF.

Parasymphyseal fractures with coexisting condylar fractures require a different approach. Surgeons may elect to treat both fractures by ORIF, returning the patient to full function immediately. Some will treat the symphyseal fracture by ORIF and leave the arch bars on with the patient in MMF for 2 weeks to address the condylar fracture. This provides predictable, acceptable results and is the preferred treatment in the overwhelming majority of cases. In our opinion, the surgeon should not treat both fractures by closed reduction. This would require 4-6 weeks of immobilization for the symphyseal fracture, placing the patient at risk for ankylosis secondary to prolonged immobilization.

As a general rule, symphyseal and parasymphyseal fractures should be treated by compression plates and screws. Plates should be bent and molded to the angulation of the bone at the symphysis. Irregularities of the bone should be smoothened by burr to facilitate better placement of the plate and screws. Two screws should be placed on either side of the fracture line. Lag screws are often used to provide adequate compression at the fracture site. Screws should engage both anterior and posterior cortices. Fragmentation of the cortices should be carefully avoided.

Dentoalveolar Fractures

Fractures of the teeth and alveolus, while common, are often not treated in the emergency department. Frequently, the patient's dentist manages these injuries in an outpatient setting. Patients complain of loose teeth, bleeding from the gingiva, pain and malocclusion. Physical examination reveals an injury to the dentition resulting in chipped, fractured, displaced or avulsed teeth and possibly an associated alveolar bone fracture. Palpation reveals mobility of the alveolar segment and associated dentition. Imaging recommendations include a panoramic radiograph and dental spot or occlusal films. Most of these injuries can be treated by a composite splint or arch bar applied to the adjacent dentition to achieve immobility of the fractured dentoalveolar segment.

If a tooth is avulsed, attempt to replace it into the socket and contact a dentist or oral surgeon for definitive management. Antibiotic coverage is indicated and the patient should be urged to follow up with the dental or oral surgical service for definitive management. Also consider prescribing an antibacterial mouth rinse such as 0.12% chlorhexidine.

Fractures of the Edentulous Atrophic Mandible

Fixation of fractures of the edentulous atrophic mandible remains controversial because of two issues concerning the quantity and quality of the bone involved. First, due to resorption of the alveolar bone following the loss of teeth, the vertical height of the area is significantly reduced. Secondly, the edentulous atrophic mandible usually becomes osteopenic and weaker with thin cortical plates. Edentulous atrophic mandible fractures normally occur amongst the elderly, age-related diseases and other chronic, debilitating diseases (e.g., diabetes). These common comorbidities contribute to less than optimum results.

Many edentulous patients with fractured mandibles can be treated by wiring their own denture or by fabricating a gunning-type splint. Each side of the fracture is secured with circumferential mandibular wires. Adequate immobilization is difficult to achieve because the denture or splint may not sit well on the underlying soft tissue. Unstable edentulous mandibular fractures should be considered for ORIF. Advantages of the open method include better visualization of the fracture and early restoration of mastication.

Pediatric Dentoalveolar and Mandibular Injuries

Dentoalveolar injuries are more commonly seen in children than mandibular fractures and usually do not require hospitalization. Common causes of dentoalveolar and mandibular injuries in children include falls from bikes and less frequently from direct blunt trauma or motor vehicle accidents. Both types of injuries can lead to complications, especially malocclusion. When examining the patient, the clinician must rule out associated injuries (e.g., closed head injury or fractures of the face, skull, cervical spine and extremities).

The most common location of pediatric dentoalveolar trauma is the anterior maxilla. Children with protruding maxillary teeth are especially susceptible to dentoalveolar trauma, with associated lip and soft tissue injuries. After the anterior maxilla, the anterior mandible is the most common site of dentoalveolar injuries. Intrusion and avulsion of teeth is frequently seen in pediatric dentoalveolar injuries. A significant portion of intrusion injuries (25%) will involve the tooth bud which eventually will lead to enamel hypoplasia. A key prognostic factor in avulsion injuries is the status of the periodontal ligament cells at the time of reimplantation. Damaged cells can provoke an inflammatory response resulting in ankylosis or tooth loss. A tooth in isolated dentoalveolar injury should be stored in milk or saliva until attempted reimplantation.

Following completion of a thorough physical examination concentrating on associated injuries and stabilization of the child's condition, it is important to determine whether there are missing or chipped teeth involved in the injury. Foreign bodies may also be observed in the surrounding soft tissues. If teeth are missing upon examination, a plain film of the chest should be obtained to help locate them. Treatment of dentoalveolar injuries depends on the extent of injury and the association of primary or permanent dentition. In general, dentoalveolar fractures should be reduced and immobilized with either an acrylic bar, dental splints or a small arch bar for 3 to 4 weeks.

In contrast to mandibular injuries in the adult population, the most common site of pediatric mandibular fracture is the condyle. Diagnosis of mandible fractures in children is multifaceted. Associated cervical spine injuries should be ruled out with a thorough physical examination and radiographic studies (especially children under the age of two). Children may experience pain in the TMJ region or malocclusion. Pediatric mandibular fractures are most often treated with closed reduction because of the rapid bone growth rate and mixed dentition.

An anterior open bite and associated retruding jaw are usually indicative of a bilateral subcondylar fracture. A contralateral open bite and ipsilateral crossbite with prematurity is generally indicative of a unilateral subcondylar fracture. Panorex®, X-rays and CT scans are often useful in assessing the nature and extent of the injury.

Treatment should be instituted as soon as possible. Rapid rate of bony repair in the pediatric population makes mandibular fractures difficult to reduce even one week after the injury. There are additional considerations when treating pediatric subcondylar fractures: (1) it is easy to pull a wire through the mandible, because the cortex is much thinner in children; and (2) placement of wires and arch bars on the crowns of developing teeth is difficult and Ivy loops are preferred.

Fractures of the body and symphysis are rarely displaced and require only periodic observation. If the fracture is displaced, it should be treated with closed reduction. In children under two years of age, splints should first be attempted to immobilize the fracture. If the splints fail, MMF should then be utilized. Parasymphyseal fractures should be treated with a lingual splint.

Fractures of the condyle have the greatest risk of causing growth abnormalities in the pediatric mandible. It is critical to maintain the function of the condyle and ensure that normal ramus height is achieved. If these two tasks are accomplished, growth usually proceeds normally. Children with intracapsular subcondylar fractures are at particular risk for growth disturbances.

In summary, the majority of pediatric condylar fractures can be treated with noninvasive means, such as observation, simple exercise, or wires. Open reduction is only necessary if significant mandibular dysfunction is observed or if the child has permanent dentition and experiences persistent malocclusion after the undergoing noninvasive alternatives.

Pearls and Pitfalls

- It is important to preserve any teeth in the posterior fracture fragment; the tooth can provide stability, alignment, and can function as an occlusal stop with the corresponding tooth on the maxilla.
- Avulsed or loose teeth should be reimplanted and splinted in place as soon as possible (ideally within 90 minutes of injury). In such cases, a dentist should be consulted. Loose teeth with extensive periodontal disease should be removed.
- During the postoperative period of MMF, the average patient loses 15-20 pounds. Patients must be counseled on consuming a nutritious liquid diet during this period.
- A cuff of soft tissue should be preserved around the mental foramen to prevent traction injury to the mental nerve. Preserve at least 0.5 cm of tissue below the mental foramen since the nerve travels inferiorly.
- Proper reduction and stabilization of the fracture is more important than preservation of periosteal attachments of soft tissue. Whenever possible, a cuff of mentalis muscle attached to bone should be preserved and the muscle reapproximated at the time of soft tissue closure.
- The intraoral incision should be closed with a water tight seal by using a running locking stitch reinforced with interrupted sutures.
- Use monocortical screws adjacent to a tooth root or next to the inferior alveolar canal in order to avoid injury to these structures.
- Larger plates should be overbent so that they arc 2 to 3 mm off the fracture site. The overbent plate flattens against the outer border of the mandible after screw placement, anatomically reducing the lingual cortex in the process.
- Soft tissue interposed between the fracture fragments must be removed before proper alignment can be achieved. A lack of bone-to-bone contact will result in delayed union and potentially nonunion. In older fractures undergoing reduction, any extra callus in the fracture site must be removed to permit precise reduction in occlusion.

Suggested Reading

1. Baumann A, Troulis MJ, Kaban LB. Facial trauma II: Dentoalveolar injuries and mandibular fractures. In: Kaban LB, Troulis MJ, eds. Pediatric oral and maxillofacial surgery. Philadelphia: Saunders, 2004.
2. Halpern LR, Kaban LB, Dodson TB. Perioperative neurosensory changes associated with treatment of mandibular fractures. J Oral Maxillofac Surg 2004; 62:576.
3. Hammoudeh JA, Dodson TB, Kaban LB. Evaluation and acute management of maxillofacial trauma. In: Sheridan RL, ed. The Trauma Handbook of the MGH. Lippincott, Williams and Wilkens, 2003.
4. Haug RH, Assael LA. Outcomes of open versus closed treatment of mandibular subcondylar fractures. J Oral Maxillofac Surg 2001; 59:370-375.
5. Marciani RD, Hill O. Treatment of the fractured edentulous mandible. J Oral Surg 1979; 37:569.
6. Newman L. A clinical evaluation of the long-term outcome of patients treated for bilateral fracture of the mandibular condyles. Br J Oral Maxillofac Surg 1998; 36:176.
7. Ochs MW, Tucker MR. Management of facial fractures. In: Peterson LJ, ed. Contemporary oral and maxillofacial surgery. 4th ed. St. Louis: Mosby, 2003.
8. Spina AM, Marciani RD. Mandibular fractures. In: Fonseca RJ, ed. Oral and maxillofacial surgery. Philadelphia: Saunders, 2000.
9. Troulis MJ, Kaban LB. Endoscopic Open reduction and internal rigid fixation of subcondylar fractures. J Oral Maxillofac Surg 2004; 62:1269-1271.

Chapter 41

Breast Disease and Its Implications for Reconstruction

Kristina D. Kotseos and Neil A. Fine

Overview of Breast Cancer

Breast cancer is the second leading cause of cancer death and the most common cancer among women in the United States with over 200,000 new cases diagnosed each year. The most common risk factors include female gender, family history, increasing age, a prior breast cancer history, first live childbirth after age 30, early menarche, late menopause, previous radiation to the chest wall, prolonged use of hormone replacement therapy and genetic mutations such as BRCA1 and BRCA2. Women with a strong family history should be screened according to the National Comprehensive Cancer Network (NCCN) Genetics/Family Screening Guidelines. Risk reduction strategies may then be considered for women with a greater than 1.67% 5-year risk of breast cancer. High risk women who take tamoxifen for five years reduce their risk of breast cancer by 50%.

Abnormalities in proliferation of either ductal or lobular epithelium lead to breast cancer. These abnormalities exist over a proliferative continuum ranging from hyperplasia, atypical hyperplasia and carcinoma in situ to invasive carcinoma. Over 85% of breast cancers are ductal in origin and may include several variants such as tubular or colloid carcinomas.

Breast Cancer Staging

Breast cancer is staged according to the TNM staging system based on tumor size, axillary node status and identification of features of locally advanced breast cancer or distant metastases. To conceptualize treatment, it is best to categorize breast cancer according to stage. Stage 0 includes the noninvasive carcinomas such as ductal carcinoma in situ (DCIS) and lobular carcinoma in situ (LCIS). Stage I, stage II and some stage IIIA breast carcinomas include the primary operable locoregional invasive cancers. Stage I breast cancer is limited to tumors <2 cm in diameter and lack nodal metastases. Stage II breast cancer includes tumors >2 cm without nodal involvement or tumors <5 cm with positive nodes. Stage IIIB, stage IIIC and some stage IIIA breast cancers include the inoperable locoregional invasive tumors. Stage III breast cancer is a heterogeneous group of tumors greater than 5 cm with nodal involvement. Stage IV breast cancers includes metastatic or recurrent disease. Axillary nodal status is the most important determinant of outcome in early-stage breast cancer, followed by tumor size.

Breast Cancer Treatment

In women with LCIS, observation is recommended. If surgery is considered, a bilateral mastectomy would be necessary as the risk of an invasive breast cancer is equal in both breasts and approximates 21% over 15 years. Five-year treatment with tamoxifen in these patients reduces the risk of developing invasive breast cancer by 56%.

Practical Plastic Surgery, edited by Zol B. Kryger and Mark Sisco. ©2007 Landes Bioscience.

In women with DCIS contained to one quadrant, excision with breast-conserving therapy (BCT) is recommended. If DCIS is present in more than one quadrant, then total mastectomy is the treatment of choice. Local failure is reduced by half with the use of radiation after excision of DCIS and is equivalent to mastectomy; however this does not effect overall survival. Tamoxifen treatment should be considered in these women, especially in those with estrogen receptor positive DCIS.

Operable locoregional invasive cancers may be treated with mastectomy, sentinel lymph node biopsy (SLNBx) +/- axillary dissection, or breast-conserving therapy with lumpectomy, SLNBx +/- axillary dissection followed by breast irradiation. These two treatments have been shown to be equivalent primary treatment options in multiple randomized trials. If the sentinel lymph node is positive for metastasis, then a formal axillary lymph node dissection including level I and II lymph nodes is performed. For those under age 70, the current guidelines recommend adjuvant systemic therapy for tumors greater than 1 cm in diameter and for node-positive disease. If adjuvant chemotherapy is indicated, then radiation therapy should be given after chemotherapy is completed. In women undergoing mastectomy, tumors greater than 5 cm or positive surgical margins necessitate postmastectomy chest wall radiation due to the high risk of local recurrence. Furthermore, if greater than three axillary lymph nodes are involved, then regional lymph node radiation is also recommended. All tumors should be tested for hormone receptor content and level of HER2/neu expression to help guide adjuvant therapy. In some patients with large clinical stage tumors, preoperative chemotherapy may allow for breast-conserving therapy if the tumor responds favorably, otherwise a mastectomy with lymph node dissection is required.

Contraindications to breast-conserving therapy include previous irradiation to the chest wall, pregnancy, multicentric disease, positive margins after reexcision, active connective tissue diseases and tumors greater than 5 cm.

Inoperable locoregional invasive cancers are initially treated with chemotherapy. Local therapy after treatment includes mastectomy with lymph node dissection or lumpectomy with lymph node dissection. Regardless of surgical treatment option, post-surgical chemotherapy and radiation are also warranted.

Stage IV metastatic disease is treated with multiple modalities including chemotherapy, endocrine therapy and radiation. Women with recurrent local disease who have received breast-conserving therapy should undergo a total mastectomy whereas mastectomy treated women should undergo surgical resection of the recurrence plus radiation therapy. Following treatment of the local recurrence, chemotherapy or endocrine therapy should then be offered.

Overview of Breast Reconstruction

Breast reconstruction is now often considered as part of the breast cancer treatment when mastectomy is required or in women undergoing lumpectomy in cases where the specimen represents a large portion of the breast. The increasing number of women undergoing reconstruction after breast cancer surgery is related to several factors including increased patient awareness and screening, the number of mastectomies performed, greater awareness of reconstruction options, and change in insurer reimbursement. In 1998, Congress passed the Women's Health and Cancer Rights Act, guaranteeing insurance reimbursement for breast reconstruction or external prostheses, contralateral procedures for symmetry and treatment for any sequelae of mastectomy.

Patients should be educated at the plastic surgery consultation concerning the potential for multiple procedures required to ablate the breast cancer, fashion the breast mound and create a projecting nipple and areola along with the potential complications that may occur with each technique. This should especially be discussed in the context of immediate reconstruction where avoiding any delay to adjuvant treatment is an important consideration. On physical exam of the breast cancer patient it is important to note size, ptosis, asymmetry and scars, palpable axillary lymph nodes and whether previous radiation has been given, and if so, the quality of the breast and chest wall. The abdomen, back and other possible donor sites should be carefully inspected taking note of scars, overall fat content and strength.

A detailed description of the various breast reconstructive procedures can be found in the upcoming chapters of this section.

Immediate Breast Reconstruction

The previous fears that immediate breast reconstruction might delay recurrent cancer detection or have a negative impact on prognosis of breast cancer have been alleviated.

Furthermore, there is no evidence that immediate reconstruction increases the rate of local or systemic relapse. Overall, women benefit from the reduction in psychological trauma experienced after mastectomy by undergoing immediate breast reconstruction.

One of the major advantages of immediate breast reconstruction includes a reduction in the number of operations ultimately necessary to create a reconstructed breast mound following extirpation of the breast. In addition, the emotional benefit of having begun reconstruction at the time of breast removal may reduce the psychological impact of losing the breast. Disadvantages of immediate reconstruction include extended surgical time and potential complications of the mastectomy, such as skin loss or infection, which may adversely affect the reconstruction. Perhaps the greatest disadvantage of immediate reconstruction is the inability to predict (in many cases) who will need adjuvant radiation.

All patients undergoing immediate reconstruction should be marked preoperatively by the plastic surgeon. Markings include the inframammary folds, the midline, the sternal notch and the donor site in cases of autogenous reconstruction. In addition, many surgical oncologists will allow the plastic surgeon to design the incision through which the mastectomy will be performed. This usually involves an oval, periareolar incision for removal of the nipple-areola complex and a linear extension of the oval incision. This extension need not always be directed towards the axilla. For example, if no axillary dissection is indicated, the incision can be designed vertically towards the inframammary fold, simulating a tear-drop shaped mastopexy-like incision. Finally, in certain cases a nipple-sparing mastectomy may be planned. In such instances, the mastectomy can be performed through a horizontal, inframammary incision, an approach that is familiar to the plastic surgeon as it is commonly used for breast augmentation.

Delayed Breast Reconstruction

Delayed reconstruction may be performed several days to years after mastectomy; however waiting three to six months is generally recommended. Advantages to delayed reconstruction include increased time to allow for adequate skin flap healing making the tissues more mobile and pliable, as well as increased time to allow for patient recovery. Some surgeons feel it allows them to achieve better symmetry than with a single stage approach. Disadvantages include multiple procedures to obtain the same result as with immediate reconstruction and increased psychological trauma from losing the breast.

Intraoperative Considerations

The plastic surgeon should be present during the prepping and draping of the patient, since not all breast surgeons are familiar with the exposure required for the reconstruction. For instance, both breasts should always be prepped even in unilateral procedures.

Although cancer surgery should never be compromised for the sake of the reconstruction, there is a way to perform a mastectomy that is oncologically sound, yet ideal for the reconstructive procedure. The skin flaps should be handled gently during the mastectomy without excessive retraction. They should be kept as thick as possible, preserving all of the subcutaneous fat external to the breast capsule. In cases of DCIS, prophylactic mastectomy and small tumors distant from the chest wall, there is no need to remove the pectoralis fascia along with the breast specimen. Preservation of this fascia will make the expander/implant reconstruction much easier. Finally, if autogenous reconstruction is being planned in the setting of an axillary dissection, it is imperative that the thoracodorsal pedicle remain untouched, and that the branches arising from these vessels be ligated rather than cauterized to avoid thermal injury to the main artery and vein.

Chemotherapy and Reconstruction

As mentioned previously, adjuvant systemic therapy in the form of chemotherapy or hormonal therapy is routinely administered to women under the age of 70 if they are node positive and to the majority of women with tumors > 1 cm. Chemotherapy can be delivered to patients with implants, tissue expanders or flaps as soon as the wound has healed and there are no sign of underlying infection. There is no significant difference in postoperative complication rates between patients undergoing mastectomy alone followed by chemotherapy or mastectomy with immediate reconstruction followed by chemotherapy. Regardless of the reconstructive method, four weeks is usually the maximum time elapsed between reconstruction and the start of chemotherapy. Most studies have shown that a delay in the initiation of chemotherapy following immediate breast reconstruction happens only 1% of the time.

Radiation and Reconstruction

When discussing reconstructive options, special consideration should be given to women who will require adjuvant radiation and women who have received chest radiation in the past. For example, some women will require adjuvant radiotherapy following mastectomy for locoregional control and some women will have local recurrence following lumpectomy and radiation therapy, which would require treatment with a mastectomy. Both of these scenarios pose unique reconstructive challenges. Regardless of timing, radiation forever compromises the quality of the skin and underlying muscle, resulting in a higher incidence of complications, unsatisfactory expansion and a poorer aesthetic result.

Many surgeons will not use expanders/implants in patients with previous chest wall irradiation. However, studies of irradiated patients who undergo subsequent implant reconstruction have shown that, although the complication rates are much higher compared to autogenous reconstruction, the majority of women are satisfied with the final outcome. Therefore, all options should be presented to a patient with a prior history of radiation, with a detailed discussion of the risks and benefits of each type of reconstruction. Any woman undergoing radiation to the breast after implant reconstruction has a high risk of developing capsular contracture and tissue firmness ultimately requiring a salvage procedure using autogenous tissue. Women with obvious radiation-induced skin changes prior to reconstruction should undergo autogenous tissue reconstruction. Not surprisingly, myocutaneous flap reconstruction does not appear to be significantly affected by previous radiotherapy, mainly because nonirradiated distant skin is brought into the area of reconstruction as part of the flap. Furthermore, this skin and underlying muscle tolerate post reconstruction radiotherapy well. Loss of volume of the reconstructed breast will often occur.

Summary

With increased detection and evolving treatment options, breast cancer therapy has become complex and varied, encompassing several different therapeutic and reconstructive options. Post mastectomy and reconstruction complications such as difficulties in wound healing, compromised flap viability and infection can occur with any type of surgery, which may lead to prolonged healing time and delay in adjuvant therapies. If the patient is at high risk for complication, reconstruction may be delayed until all subsequent treatment modalities have been completed. Ultimately, the patient, in consultation with her reconstructive surgeon, must opt for a type of reconstruction which will allow optimization of the rehabilitation process following surgical treatment of breast cancer. Above all, one must always remember to not jeopardize an oncologically sound operation and adjuvant treatment for a better aesthetic result.

Pearls and Pitfalls

- Always send any excised tissue to pathology during secondary or revision procedures. For example, during implant replacement long after reconstruction, if capsulectomy is performed, the implant capsule should be sent to pathology to rule out any evidence of recurrence.
- A large number of patients who undergo implant/expander reconstruction will require a subsequent autogenous reconstruction due to complications related to the prosthesis (e.g., capsular contracture). Therefore, before performing any procedure that might compromise a future revision with autogenous tissue (such as an abdominoplasty), a full discussion of the ramifications should take place.
- Often after a more "extensive" mastectomy there is inadequate muscle or fascia to cover the tissue expander. One option is to use a strip of thin Alloderm® as a bridge of tissue from the pectoralis major edge to the chest wall.

  Although radiated skin often bleeds vigorously and appears hypervascular, this increased vascularity is disorganized and prone to fibrosis and impairs rather than improving blood flow to the tissue.

Suggested Reading

1. Cordeiro PG, Pusic AL et al. Irradiation after immediate tissue expander/implant breast reconstruction: Outcomes, complications, aesthetic results, and satisfaction among 156 patients. Plast Reconstr Surg 2004; 113(3):877.
2. Corral CJ, Mustoe TA. Controversy in breast reconstruction. Surg Clin North Am 1996; 76(2):309.
3. Downes KJ, Glatt BS et al. Skin-sparing mastectomy and immediate reconstruction is an acceptable treatment option for patients with high-risk breast carcinoma. Cancer 2005; 103(5):906.
4. Fisher B, Anderson S et al. Twenty-year follow-up of a randomized trial comparing total mastectomy, lumpectomy, and lumpectomy plus irradiation for the treatment of invasive breast cancer. N Engl J Med 2002; 347(16):1233.
5. Sandelin K, Wickman M et al. Oncological outcome after immediate breast reconstruction for invasive breast cancer: A long-term study. Breast 2004; 13(3):210.
6. Soong IS, Yau TK et al. Post-mastectomy radiotherapy after immediate autologous breast reconstruction in primary treatment of breast cancers. Clin Oncol (R Coll Radiol) 2004; 16(4):283.
7. Wilson CR, Brown IM et al. Immediate breast reconstruction does not lead to a delay in the delivery of adjuvant chemotherapy. Eur J Surg Oncol 2004; 30(6):624.

Chapter 42

TRAM Flap Breast Reconstruction

Amir H. Taghinia, Margaret L. McNairy and Bohdan Pomahac

Introduction

The transverse rectus abdominis myocutaneous (TRAM) flap is a popular technique for breast reconstruction using autogeneous tissue from the rectus abdominis muscle and overlying subcutaneous fat and skin (Fig. 42.1). Although it is most commonly used for breast reconstruction, the TRAM flap can also be used for lower extremity, groin, or craniofacial soft tissue reconstruction.

The TRAM flap can be harvested as either a free or a pedicled flap. In a pedicled TRAM, the entire length (or a large section) of the rectus abdominis muscle along with a transverse section of subcutaneous tissue and skin is tunneled to the location of the mastectomy defect where it is then molded into a breast. The pedicled TRAM flap maintains its native blood supply from the **superior epigastric vessels**.

In a free TRAM, only part of the rectus muscle is used and the flap (which includes its attached paddle of subcutaneous fat and skin) is totally detached from its surrounding tissues and transferred to its new location based on the deep **inferior epigastric vessels**, termed the pedicle. The pedicle is then anastomosed to recipient vessels in the axilla (thoracodorsal vessels) or chest wall (internal mammary or intercostal vessels).

Anatomy

The paired rectus abdominis muscles run vertically from the lower edge of the 5th to 7th costal cartilages down to the ipsilateral pubic symphysis and the pubic crest. The muscle is divided vertically by the linea alba in the midline and separated into three or four horizontal sections by tendinous intersections that divide the muscle into separate contracting elements (Fig. 42.2). The muscle is encased in fascial sheaths both anteriorly and posteriorly. The anterior rectus sheath is formed by two layers: the external oblique aponeurosis and the anterior leaflet of the internal oblique aponeurosis. The posterior rectus sheath is composed of the posterior leaflet of the internal oblique aponeurosis and the transversus abdominis aponeurosis. Deep to the posterior sheath lies a filmy thin layer of fascia termed the transversalis fascia. Deep to this fascia is the peritoneum.

Another important anatomical landmark is the arcuate line. This line refers to a level about half way between the umbilicus and the pubic symphysis where the posterior sheath ends. Inferior to the arcuate line, the rectus muscle lies directly on transversalis fascia. Below this level, the anterior rectus sheath is formed by three layers: the external oblique aponeurosis, the internal oblique aponeurosis (the two leaflets fuse to form one), and the transversus abdominis aponeurosis (it shifts anteriorly from its posterior position). Thorough knowledge of these fascial layers is crucial for abdominal wall closure after flap harvest (Fig. 42.3).

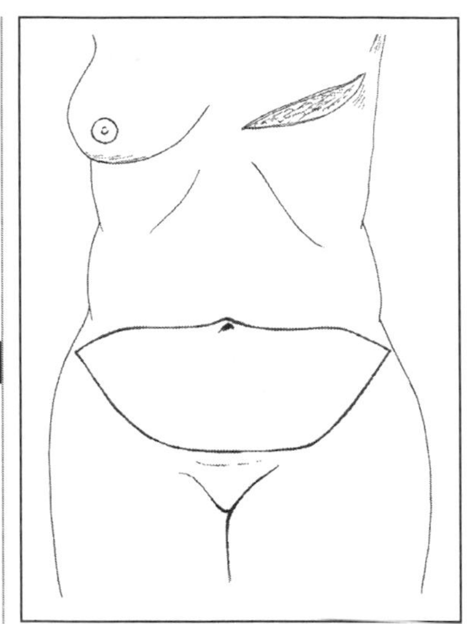

Figure 42.1. Skin markings in designing the TRAM flap. The marked area on the abdomen outlines the skin and subcutaneous tissue paddle that will be harvested with the rectus abdominis muscle.

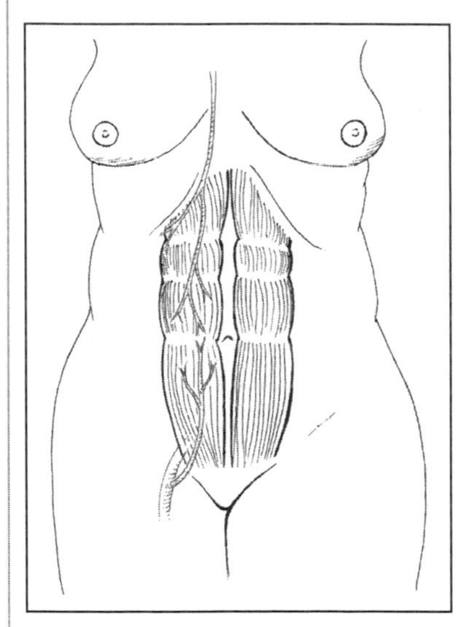

Figure 42.2. Anatomy of the rectus abdominis muscles. The rectus abdominis muscles originate at the pubic crest and pubic symphysis and insert into the ipsilateral lower costal cartilages. Each muscle is supplied by two dominant pedicles: the superior epigastric and the deep inferior epigastric vessels.

According to the Mathes and Nahai classification, the rectus abdominis muscle is a Type III muscle flap. It is supplied by two dominant pedicles, the superior and deep inferior epigastric vessels, and either one can supply the muscle in its entirety (Fig. 42.2). Thus, although the inferior epigastric vessels provide the more robust blood flow, the pedicled TRAM flap can be based on the superior epigastric vessels. The superior epigastric vessels extend from the internal mammary vessels, inserting deep to the muscle at the costal margin. The deep inferior epigastric vessels come off the external iliacs and enter the muscle at the lateral edge approximately 4 cm superior to the fibers of origin. In most cases, the anastomotic connections between the

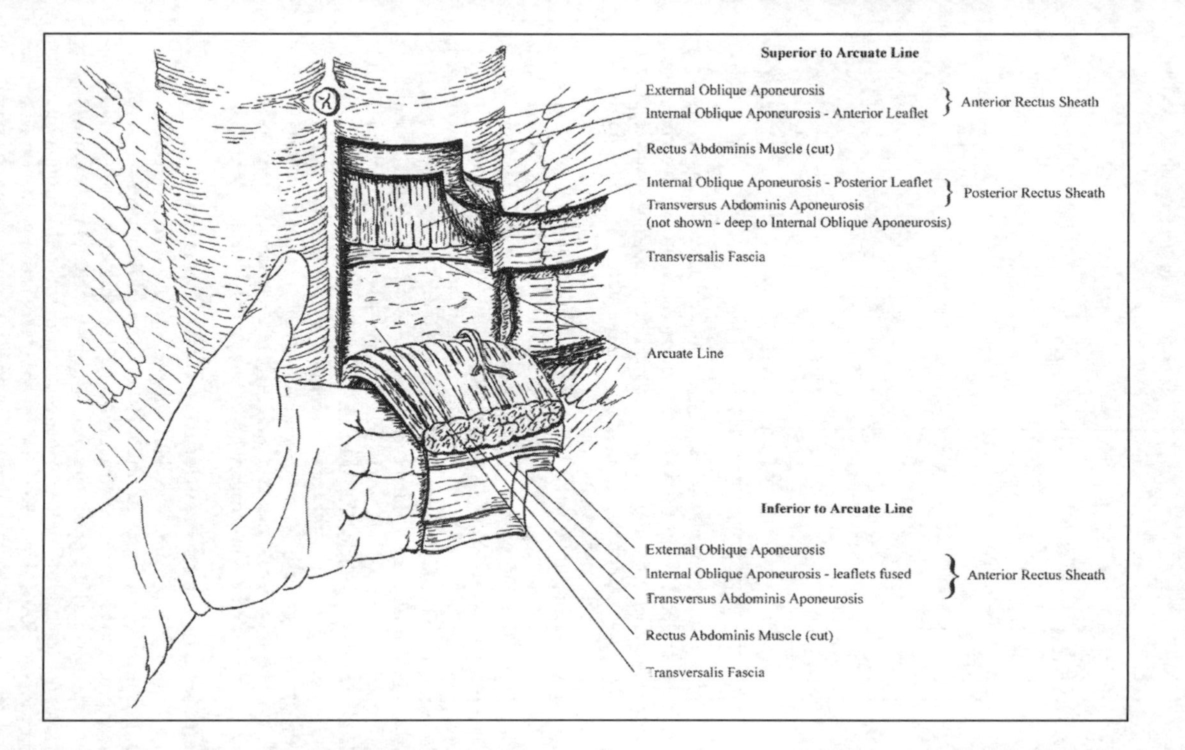

Figure 42.3. Layers of the abdominal wall. The abdominal wall layers adjacent to the rectus abdominis muscle have been exposed. Below the arcuate line, the posterior rectus sheath is not a discreet layer; rather it is fused with the components of the anterior rectus sheath.

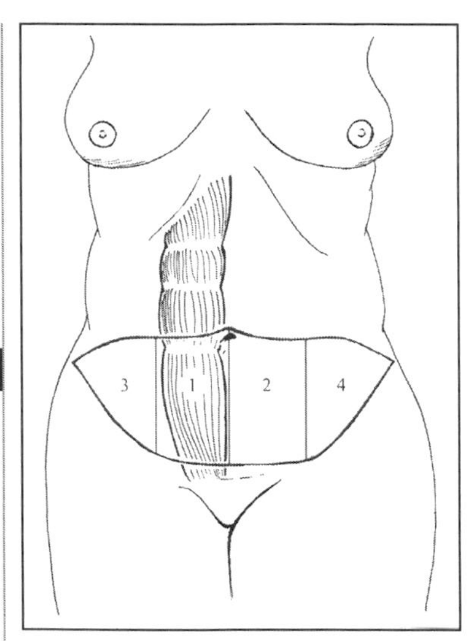

Figure 42.4. Skin zones of the TRAM flap. Despite the traditional ordering of the zones, it is now believed that zone 3 receives more robust blood flow than zone 2.

superior and deep inferior epigastric vessels are microscopic, via 'choke' vessels in the muscle that are located superior to the umbilicus. Along the length of the muscle, perforator vessels pierce the muscle and enter the subcutaneous tissue to provide blood supply to the superficial tissues including the skin. The largest perforating vessels are located in the periumbilical area.

The transverse skin paddle of the TRAM flap has been divided into several zones (Fig. 42.4). Zone 1 is immediately superficial to the rectus abdominis muscle. Zone 2 is immediately adjacent to zone 1 on the contralateral rectus. Zone 3 and zone 4 are lateral to zone 1 and zone 2, respectively. Ordering of the zones is somewhat misleading because the blood supply was not fully elucidated at the time of coinage. The zones are listed in order of most to least blood flow: 1, 3, 2, 4. Zone 3 receives axial blood flow from zone 1. However, zone 2 is primarily supplied by midline crossover at the subdermal and fascial levels. Zone 4 is the most ischemic and is usually discarded in a pedicled TRAM flap.

The motor innervation of the rectus abdominis is via the segmental motor nerves from the 7th through 12th intercostals nerves which enter the deep surface of the muscle at its lateral aspect. The corresponding lateral cutaneous intercostals nerves provide sensation to the skin territory of the anterior abdomen.

Patient Selection and Preoperative Considerations

The TRAM flap provides excellent autogenous tissue reconstruction of the breast. It is ideal for women with moderate abdominal wall excess tissue who do not want prosthetic breast reconstruction with implants. It is particularly advantageous for women whose oncologic treatment plan includes radiation as the rate of capsular contracture around implants is much higher in this group of patients. In contrast, the TRAM flap is less ideal for thin, very active women in whom the donor defect may hamper their lifestyle. In addition, smokers and the very obese patient are less than ideal candidates due to their relatively compromised blood flow to the abdominal skin.

Prior to the operation, a thorough history and physical examination must be performed. Patients should have enough abdominal wall tissue to allow the surgeon to make a breast and to close the donor defect. Abdominal scars from prior operations may modify operative plans. For example, a patient with an oblique subcostal scar should not have a pedicled TRAM harvested on that side; the superior blood supply of the flap was likely sacrificed in the previous surgery. Major complications can occur in smokers, diabetics and overweight patients. Smokers should be advised to quit. Patients with significant cardiovascular disease or other major comorbidities should not undergo this lengthy operation.

The Pedicled TRAM

The most commonly used approach is the pedicled TRAM. The pedicled TRAM can be harvested from the ipsi- or contralateral side of the breast defect (Fig. 42.5). The flap maintains the original rectus blood supply from the superior epigastric vessels. Once the flap is dissected to the desired pedicled length, it is transposed through an epigastric subcutaneous tunnel to the breast defect site. Although the superior epigastric vessels are adequate to keep the muscle well-perfused, they do not provide enough blood supply to perfuse the entire elliptical skin and subcutaneous paddle. Consequently, the least perfused part of the flap in zone 4 is usually sacrificed. If bilateral reconstruction is needed, both the rectus muscles can be used in conjunction with half-elliptical subcutaneous tissue and skin paddles.

The Free TRAM

The free TRAM flap constitutes approximately 5% of breast reconstructions. The free TRAM requires less rectus muscle and thus creates a smaller donor muscle defect (Fig. 42.6). The flap is nourished via the deep inferior epigastric vessels which provide a more robust blood supply than their superior counterparts (Fig. 42.7).

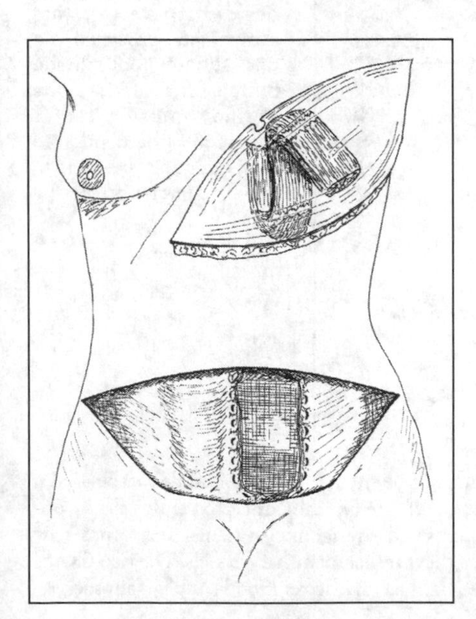

Figure 42.5. The ipsilateral TRAM flap. After raising an ipsilateral TRAM flap, the flap is inset into the breast defect. The flap can then be molded into the shape of a breast. The abdominal wall defect is usually repaired with mesh.

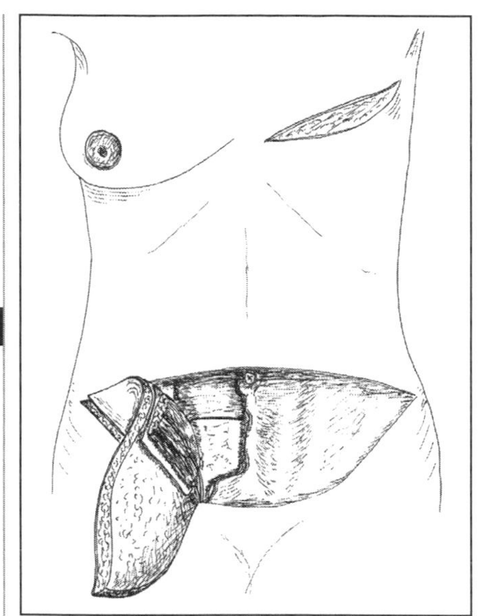

Figure 42.6. Harvesting the free TRAM Flap. The free TRAM flap has been fully raised and is ready for ligation of its inferiorly based pedicle, the deep inferior epigastric vessels.

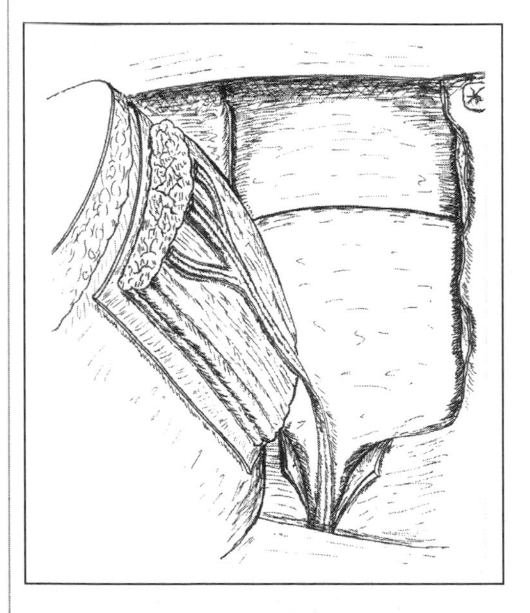

Figure 42.7. Details of the free TRAM flap pedicle. The deep inferior epigastric artery gives rise to a medial and lateral row of perforating vessels (not shown) that supply the skin and subcutaneous tissue paddle. The inferior edge of the anterior rectus sheath must be divided to gain adequate exposure to the proximal part of the pedicle.

Accordingly, the entire elliptical skin and subcutaneous tissue paddle can be used if needed. The deep inferior epigastric vessels are typically dissected close to their origins for maximal pedicle length. Venous and arterial microvascular anastomoses are performed using the thoracodorsal or the internal mammary vessels. The free TRAM may be considered in smokers whose flaps receive better blood supply from successful anastomoses rather than relying on collateral circulation between the superior and inferior blood supply.

TRAM Flap Delay

If TRAM vascularity is questionable, as in a smoker, the surgeon can take steps to improve chances of the skin flap survival. One technique to increase blood supply to the flap is **flap delay**. The pedicled TRAM flap can be delayed by ligating the superficial and deep inferior epigastric vessels a few weeks prior to breast reconstruction. This procedure enhances the superiorly based blood supply to the inferior aspect of the flap. Furthermore, the periumbilical perforating vessels on the side of the TRAM can be enhanced by ligating the opposing periumbilical perforators.

Supercharged TRAM

Supercharging refers to the use of microvascular anastomoses to increase the blood supply of a pedicled TRAM flap. Thus the flap will maintain a duel blood supply from both the superior and inferior vessels. For example, a pedicled TRAM flap (based on the superior epigastric vessels) can be supercharged by anastomosing its inferior epigastric vessels to the thoracodorsal vessels. Yamamoto et al (1996) reviewed 50 patients with TRAM flaps and found the incidence of flap complications significantly decreased with the use of supercharged pedicled TRAMs.

Operative Technique

Skin Paddle Design and Incision

Initially the skin paddle is marked, and the skin and subcutaneous tissue is raised up to the lateral and medial edges of the rectus. The largest perforators lie in the periumbilical region so most surgeons position some part of the skin island over the dense pack of these large vessels. The superior skin margin is marked from the upper umbilicus toward each anterior superior iliac spine, and the inferior skin margin is marked along the suprapubic crease to meet the lateral edges of the superior skin markings (Fig. 42.1). For a unilateral pedicled TRAM, the flap can be designed on the same or opposite side of the breast defect. The skin flap is elevated laterally until the lateral border of the rectus is reached on one side and until the midline is reached on the other side. Sacrificing the perforators of the nonharvested rectus allows the surgeon roughly determine the location of the perforators on the flap side. Once the skin and subcutaneous tissue paddle has been elevated, the remaining superior skin flap is raised off the anterior fascia in a manner similar to an abdominoplasty.

Muscle Harvest

Next, the rectus muscle (and its associated skin and subcutaneous tissue paddle) is raised from the abdominal wall. Most surgeons harvest the entire rectus muscle. Partial harvesting of the rectus abdominis muscle (e.g., leaving a lateral band of muscle intact) does not preserve muscle function. In addition, it may jeopardize the arterial supply and venous drainage of the skin and subcutaneous tissue paddle. The rectus sheath is divided lateral to the perforators, and the incision is extended inferiorly to the lower edge of the corresponding skin incision. Incising the rectus sheath at the lateral rectus abdominis muscle border allows for a cuff of anterior sheath to be left for subsequent closure. The cut is then extended medially to the midline. The deep inferior epigastric vessels are found and protected as the distal edge of the muscle is divided. The deep inferior epigastric vessels are then followed inferiorly for another few centimeters and ligated, keeping as much length as possible should future anastomosis be needed. The medial edge of the rectus sheath is then cut along the midline. Care should be taken to avoid muscle injury during dissection of the sheath off the tendinous inscriptions.

Continuing along the medial and lateral edges of the rectus muscle, the rectus sheath is incised. The muscle is then gently dissected off the posterior sheath and the lateral intercostals vessels and motor nerves are ligated or clipped. The dissection is carried superiorly to the subcostal margin where the superior epigastric artery and vein can usually be identified. A tunnel is created in the medial inframammary fold, and the flap is brought through and transposed into the breast pocket (Fig. 42.5). The least vascularized zone of tissue (zone 4) is usually discarded prior to final transposition.

For a bilateral TRAM flap, a similar procedure is performed on both sides. The skin and subcutaneous tissue paddle is split along the midline and each TRAM flap is used to reconstruct each ipsilateral breast defect.

Other Considerations in the Free TRAM

In a free TRAM flap, the entire subcutaneous tissue and skin paddle can be based on the deep inferior epigastric vessels (Figs. 42.6, 42.7). Similar steps as mentioned previously are followed; however the muscle and fascia are cut superiorly, usually at the level of the superior skin margin. Thus a section of muscle is removed with its attached anterior sheath and subcutaneous tissue/skin paddle. The deep inferior epigastric vessels are dissected towards their origin to give maximum length for microvascular anastomosis. The harvested flap is then transferred to the site of the breast defect and anastomosed to the prepared internal mammary vessels or the thoracodorsal vessels found at the level of the third intercostals space.

Donor Site Closure

The anterior rectus sheath defect is often closed with mesh (Fig. 42.5). The new location of the umbilicus is determined on the superior skin flap, and a small ellipse of skin and fat are excised at this point. The umbilicus is brought out through the resulting hole and is sutured to the dermis. When performing a unilateral pedicled TRAM, the umbilicus will be pulled laterally after the fascial defect is closed. Some surgeons will medialize the umbilicus with several tacking sutures that anchor it to the anterior fascia prior to bringing it through the skin flap.

The abdominal incision is then closed in layers over two Jackson-Pratt drains placed in the subcutaneous space. The first layer should consist of interrupted deep dermal, buried stitches. The skin layer should be a running intradermal suture. Some surgeons prefer absorbable nonreactive sutures, while others prefer a nonabsorbable one. Monocryl and Prolene are two commonly used sutures.

Recipient Site Closure

The TRAM flap is trimmed to the desired shape and secured into the mastectomy defect site with interrupted Vicryl sutures. Since most cancer resections are skin-sparing, much of the skin paddle can be deepithelialized, leaving only a small ellipse of skin to fill the postmastectomy skin defect. A suture can be placed on this paddle, identifying a site at which a Doppler tone can be obtained. Shaping the flap into an aesthetically pleasing breast mound that matches the contralateral side requires a great deal of experience. Medial and superior fullness are desirable in the final outcome. Furthermore, it is common to perform a symmetry procedure on the contralateral breast, such as reduction, augmentation or mastopexy.

Postoperative Considerations

Most patients usually require about five days of in-hospital stay and another 4-6 weeks for full recuperation. Postoperatively, the TRAM flap should be assessed for temperature, color and capillary refill. In addition, the free TRAM flap should be monitored by Doppler to assess perfusion across the anastomosis. The chapter on free flap monitoring addresses the details of postoperative free flap care. The bed should be adjusted to place the patient in a semi-Fowler position; the patient may ambulate with hips flexed. After most free TRAM flaps, the patient remains NPO for the first 24 hours in case the need arises to return to the operating room. Patients may be discharged once pain is under control and they can ambulate and tolerate oral intake. Typically, they are discharged with drains to be removed in the office once the drainage is minimal.

Several months after TRAM reconstruction, patients may undergo nipple-areola reconstruction using various techniques. These are discussed in detail in the "Nipple reconstruction" chapter of this book. The dennervated rectus abdominis muscle usually undergoes some degree of atrophy. Patients often have more difficulty with tasks (such as getting out of bed) that require an intact abdominal musculature. The abdominal skin on the breast is insensate and its autonomic protective mechanisms are nonfunctional. Accordingly, exposure to cold, heat, or solar radiation may cause injury. Patients should be counseled about these potential risks.

Complications

Overall, complication rates are reported between 16-28%. Most of these are minor complications including seroma, infection (usually at the donor site) and fat necrosis. More serious complications include large seromas, hematoma, dehiscence, abdominal hernias and flap necrosis. Long term complications include flap atrophy or necrosis. Since most of these patients have cancer and are minimally active, there is a small yet real risk of developing deep vein thrombosis (DVT). Unless contraindicated, all patients should be given prophylactic doses of heparin or low molecular weight heparin, as well as sequential compression devices (SCDs) in the postoperative period.

Total TRAM flap loss, independent of technique (pedicled or free TRAM), is reported between 1-6%. Khouri et al reported a 3% failure rate for a series of 250 free TRAM flaps. In a review of 48 pedicled and 20 free TRAMs, Schusterman et al reported a 10% incidence of abdominal herniation in both techniques. Others have reported a range of abdominal hernia rates between 0.3 and 13%. Obesity plays a significant role in complication rate. Kroll et al reported a 41.7% incidence of complications in morbidly obese patients verses a 15.6% rate in the nonobese. Serletti et al studied cost and outcomes after free and pedicled TRAMs. On average, operating room time and hospital reimbursement were similar in pedicled and free TRAM flap breast reconstruction. Length of stay was slightly longer for free TRAMs and cost approximately $1500 more. In conclusion, postoperative recovery, abdominal wall integrity and time to return to work are still debated without a clear difference in the literature between these two approaches.

Alternatives to the Pedicle or Traditional Free TRAM

Over the past few years, there has been growing enthusiasm for alternatives to the traditional free TRAM. These consist primarily of the muscle-sparing free TRAM, the deep inferior epigastric perforator (DIEP) flap, and the superficial inferior epigastric artery (SIEA) flap. The muscle-sparing TRAM is similar to the regular TRAM, except that a portion of either the medial or lateral rectus is preserved. Care must be taken to avoid dennervating this muscle. The DIEP flap is a perforator flap that involves dissection of one or more musculocutaneous perforators through the rectus to the origin of the deep inferior epigastric artery. Almost the entire rectus muscle is preserved, and only a small slip of anterior rectus fascia is taken with the flap. The SIEA flap is based entirely on the superficial inferior epigastric system, and therefore the rectus fascia is never violated.

There has been extensive debate over which technique of breast reconstruction offers the greatest advantages to the patient. There has been no definitive study to date that has demonstrated that any one technique is superior. The donor site morbidity (hernia, bulge, abdominal wall strength, etc) has not been shown to be significantly different between the various flaps. Long-term outcomes and patient satisfaction are not dependent on the method of reconstruction but rather on the surgeon's ability to achieve an excellent result with his preferred method.

Pearls and Pitfalls

- Adjuvant radiation is the enemy of the reconstructive surgeon. It can convert a beautifully reconstructed breast into a shrunken, fibrotic mound of tissue. In most cases, it is not possible to predict which patients will require postoperative chest wall radiation (this is discussed further in the chapter on breast cancer). However in a subset of patients, the likelihood for adjuvant radiation is extremely high. There is nothing wrong with delaying the reconstruction in these patients until after the radiation treatment has been completed.
- Patients who smoke deserve special mention. Their risk of developing complications is significantly higher than nonsmokers. Every effort should be made to maximize blood flow to the flap. This includes cessation of smoking as early as possible, consideration of performing a delay procedure, and using either a free or supercharged TRAM.
- In free TRAM reconstruction, never divide the superior epigastric pedicle until an adequate recipient vessel has been identified, since division of the superior system commits the surgeon to a free flap procedure.
- If a problem arises during the stage of microsurgical anastomosis, the risk of ischemic injury to the flap can be decreased by keeping it cold in an ice bucket on the back table.
- After tunneling the pedicled TRAM flap into the breast pocket, it is essential to examine the pedicle under direct vision, ensuring that there is no undue tension, kinking or compression.
- Postoperative abdominal bulge is a common long-term complication after TRAM flap harvest. The use of mesh in reconstructing the abdominal wall donor site should decrease the risk of bulge formation. One useful technique is to use a large prolene onlay mesh that is sutured to the anterior fascia under tension. This can be done by first securing one half of the mesh far lateral to the rectus sheath, then pulling the mesh tightly as it is sutured to the other side.

Suggested Reading

1. Arnez ZM, Bajec J, Bardsley AF et al. Experience with 50 free TRAM flap breast reconstructions. Plast Reconstr Surg 1991; 87(3):470.
2. Bajaj AK, Chevray PM, Chang DW. Comparison of donor-site complications and functional outcomes in free muscle-sparing TRAM flap and free DIEP flap breast reconstruction. Plast Reconstr Surg 2006;117(3):737.
3. Carlson G et al. Breast reconstruction. In: Achauer BM, Eriksson E, Guyuron B, Coleman IIIrd JJ, Russell RC, Vander Kolk CA, eds. Plastic Surgery: Indications, Operations, and Outcomes. St. Louis: Mosby, Inc., 2000:587.
4. Chevray PM. Breast reconstruction with superficial inferior epigastric artery flaps: a prospective comparison with TRAM and DIEP flaps. Plast Reconstr Surg 2004; 114(5):1077.
5. Grotting JC, Urist MM, Maddox WA et al. Conventional TRAM flap versus free microsurgical TRAM flap for innate breast reconstruction. Plast Reconstr Surg 1989; 83(5):828.
6. Hartrampf Jr CR, Bennett GK. Autogenous tissue reconstruction in the mastectomy patient: A critical review of 300 patients. Ann Surg 1987; 205(5):508.
7. Hartrampf CR, Scheflan M, Black PW. Breast reconstruction with a transverse abdominal island flap. Plast Reconstr Surg 1982; 69(2):216.
8. Khouri RK, Ahn CY, Salzhauer MA et al. Simultaneous bilateral breast reconstruction with the transverse rectus abdominus musculocutaneous free flap. Ann Surg 1997; 226(1):25.
9. Khouri RK, Cooley BC, Kunselman AR et al. A prospective study of microvascular free-flap surgery and outcome. Plast Reconstr Surg 1998; 102(3):711-721.
10. Kroll SS, Baldwin B. A comparison of outcomes using three different methods of breast reconstruction. Plast Reconstr Surg 1992; 90(3):455.
11. Kroll SS, Netscher DT. Complications of TRAM flap reconstruction in obese patients. Plast Reconstr Surg 1989; 84(6):886.
12. Nahabedian MY, Tsangaris T, Momen B. Breast reconstruction with the DIEP flap or the muscle-sparing (MS-2) free TRAM flap: is there a difference? Plast Reconstr Surg 2005; 115(2):436.
13. Schusterman MA, Kroll SS, Miller MJ et al. The free transverse rectus abdominis musculocutaneous flap for breast reconstruction: One center's experience with 211 consecutive cases. Ann Plast Surg 1994; 32(3):234.
14. Serletti JM, Moran SL. Free versus the pedicled TRAM flap: A cost-comparison and outcome analysis. Plast Reconstr Surg 1997; 100(6):1418.

Chapter 43

Latissimus Flap Breast Reconstruction

Roberto L. Flores and Jamie P. Levine

Introduction

The latissimus dorsi musculocutaneous flap was originally described almost a century ago by Iginio Tansini as a method to cover radical mastectomy defects which included a wide skin excision. Its use soon fell out of favor after the mastectomy described by Halsted became the standard of care, as much more skin was spared in this procedure. The flap was revisited in the 1970s as a means to reconstruct the breast mound after mastectomy. Shortly following this, the transverse rectus abdominis musculocutaneous (TRAM) flap was described. The advantage of complete autologous breast reconstruction provided by the TRAM flap versus the latissimus flap which required a prosthetic implant for additional volume, made the TRAM the flap of choice for breast reconstruction. In recent years, there has been a resurgence of the latissimus flap. Improvements in tissue expansion and implant design, as well as the ability to perform complete autologous reconstruction in selected patients has, once again, put the latissimus flap in the forefront of breast reconstruction.

Indications

Although the latissimus flap is a reliable flap with a robust blood supply based on the thoracodorsal vessels, for most surgeons it functions as an alternative flap when a TRAM cannot be used. Examples include patients who do not have adequate abdominal tissue for acceptable TRAM reconstruction, those who have undergone previous TRAM surgery and patients who prefer the back as a donor site for reconstruction. Patients in need of additional reconstruction after previous TRAM reconstruction with partial flap loss or necrosis can obtain excellent results with a latissimus flap reconstruction. Due to its reliable blood supply, the latissimus dorsi may be the preferred reconstruction in heavy smokers as there has been a documented increase in complications in TRAM flaps performed on smokers.

Overall, this flap provides a broad, thin and well-vascularized flap with a volume of soft tissue that usually falls short of the patient's requirement. For this reason, an implant is usually required as part of the reconstruction. For small-breasted patients, complete autogenous reconstruction may be possible. In addition, there is a role for use of the latissimus muscle flap in breast reconstruction after extensive lumpectomies that leave a very noticeable defect.

Anatomy

The latissimus dorsi, the broadest muscle of the back consists of two triangular shaped muscles with fascial origins from the spinous processes of the lower six thoracic, lumbar and sacral vertebrae, and from the iliac crest. Additionally, there are

Practical Plastic Surgery, edited by Zol B. Kryger and Mark Sisco.

muscular origins from the anterolateral aspect of the lower four ribs as well as the external oblique and tip of the scapula. The fibers converge superiolaterally and twist 180° before inserting into the intertubercular groove of the humerus. The muscle, which is largely expendable, functions to extend, adduct and medially rotate the arm.

The thoracodorsal artery, a terminal branch of the subscapular artery, provides the vascular supply to the muscle. One or two venae comitantes and the thoracodorsal nerve accompany the artery. The mean length of the vascular pedicle is 11 cm. The neurovascular bundle enters the deep surface of the muscle approximately 10 cm distal from the insertion. Once inside the muscle, the vessels divide into two branches which run a course parallel to the muscle fibers. Musculocutaneous perforators branch from these vessels, supplying the overlying skin. The thoracodorsal artery also supplies a branch to the serratus anterior muscle, which can be divided to increase the length of the vascular pedicle. Several branches of the lower intercostal and lumbar arteries contribute segmental minor pedicles to this flap.

Further discussion of the characteristics and anatomy of the latissimus flap, as well the technique for harvesting it, can be found in the "Reconstruction" section of this book.

Operative Technique

Breast reconstruction utilizing the latissimus dorsi flap can be performed in an immediate or delayed fashion. An implant can be placed immediately, or a tissue expander can be inserted with subsequent implant exchange once tissue expansion is complete. Alternatively, complete autogenous reconstruction may be performed in select patients. Immediate breast reconstruction with a prosthetic implant is described below.

Patient Marking

While the patient is in the upright position, draw the borders of the latissimus dorsi. Draw a line from the tip of the scapula to the top of the posterior axillary fold, identifying the superior margin. The lateral margin lies along a line connecting the anterior margin of the posterior border of the axilla to the iliac crest. The inferior and medial margins are the iliac crest and midline, respectively. The skin paddle is drawn obliquely along the line of the musculocutaneous perforators. Placing the skin paddle laterally, where the skin is more redundant, will facilitate closure. The skin paddle can also be oriented in a transverse fashion within the patient's bra line. The various incision options should be discussed with the patient.

Flap Harvest

After the breast surgeons have finished their dissection, the mastectomy site is prepared to receive the flap. The skin pocket is dissected up to 1-2 cm inferior to the inframammary crease. Medially, the skin pocket is mobilized to the sternal edge. The pectoralis muscle is dissected from the chest well in preparation for the implant. Care is taken not to dissect too far laterally as the tunnel through which the flap is passed may be entered, leading to lateral and posterior displacement of the prosthesis.

The patient is then placed in a prone or lateral decubitus position for harvesting of the flap. After the skin incision is made, the overlying skin and fat are mobilized away from the fascia overlying the latissimus dorsi. Wide skin flaps are dissected to the superior, inferior, medial and lateral extent of the muscle. The skin paddle is preserved, and temporary sutures are used to affix the skin paddle to the underlying muscle in order to minimize shearing injury during the procedure. Once this is done, the superior margin of the latissimus dorsi is identified. Dissection continues medially to mobilize the overlying trapezius muscle away from the latissimus dorsi. The dissection then continues superiolaterally, towards the axilla, taking care to separate the superiorly placed teres major from the latissimus muscle. Once the superior margin is mobilized, attention is drawn to the medial aspect of the muscle. Starting at the superiomedial aspect, the fascial origins are released from the paraspinous processes. As one proceeds inferiorly, the fascial attachments joining the lower border of the serratus to the latissimus can be easily identified and divided, preventing elevation of the serratus muscle with the latissimus dorsi. There can be a fair amount of interdigitation between these muscles but the cleavage plane is usually clearly identifiable. At this point, the intercostal perforating branches will be encountered. Careful control of theses vessels is crucial in preventing postoperative hematoma. Dissection is then continued laterally along the inferior margin of the muscle. The fibers of the external oblique and the intercostal muscles fuse with the latissimus dorsi at it inferior aspect and must be divided. Once the lateral edge is reached, dissection is continued superiorly toward the axilla.

Tunneling the Flap

After full mobilization of the latissimus muscle off the chest wall, the flap is placed into the axilla and the rest of the dissection is continued through the mastectomy incision. The tunnel through which the flap is passed should be large enough to accommodate the pedicle but not so small as to compress the vessels. One should be able to insert four fingers into the tunnel, without difficulty, depending on the size of the skin paddle and the flap being harvested. When the flap is placed on the chest, care is taken not to kink or twist the pedicle. There is no need to identify or to skeletonize the thoracodorsal vessels during this part of the procedure. If further length is needed, the serratus branch from the thoracodorsal artery may be cut. Additionally, the latissimus dorsi may be detached from the humerus, creating an island flap. If one chooses to detach the origin of the latissimus dorsi, the vascular pedicle must first be visualized entering the muscle, about 10 cm distal from the point of insertion. Once the pedicle is identified and protected, the proximal muscle can be divided. The back wound is closed over suction drains.

Creation of the New Breast

The patient is then placed in supine position, and the remainder of the surgery is performed through the mastectomy incision. Ideally, the latissimus is sutured to the pectoralis muscle superiorly, the sternum medially, and inferiorly to the chest wall, just below the inframammary crease or at the margin of the inframammary fold in order to create a new crease and provide complete muscular coverage over the implant. The implant is then inserted under the pectoralis and latissimus muscles. To prevent lateral migration of the prosthesis, the lateral edge of the latissimus dorsi can be sutured to the chest wall. The axillary tail can be reconstructed with a deepithelialized portion of the skin paddle or the insertion of the latissimus muscle, once detached. To maintain muscle volume in this area, the thoracodorsal nerve can be left intact innervating the muscle. If the resultant muscular contraction is unsightly or not desired surgically, this can be divided at the time of the operation.

The donor site is closed in layers over suction drains. Some surgeons advocate quilting sutures to tack down the overlying skin and subcutaneous tissue. It is thought that this maneuver may lower the incidence of postoperative seroma or hematoma formation. The breast is also closed in layers, usually over a suction drain placed in a manner that will drain fluid that collects in dependent positions. The drains are removed after several days, once the drainage has decreased below 30 ml per day.

Complications

As previously stated, the latissimus flap contains a vigorous blood supply. Flap necrosis is rare, even in smokers and diabetics and usually it is due to surgical injury, tension or torsion to the pedicle. The most common complications are seroma at the donor site, which usually occur after the drains have been removed. Needle aspirations address these collections until they resolve. Infection and hematoma are prevented by appropriate surgical technique and occur in equal frequency as other plastic surgery procedures. Transient brachial palsy can occur with improper patient positioning. This complication usually resolves after several weeks. Although shoulder weakness is not appreciated in most patients, physically active patients may complain when lifting themselves from a chair, when swimming competitively, or prolonged overhead activity. Finally, use of a prosthetic device in the chest lends to the development of capsular contracture. The development of newer shaped textured expanders and implants have decreased the incidence of the more severe capsular contracture.

Pearls and Pitfalls

1. It is critical during the preoperative phase to discuss the size and shape of the reconstructed breast with the patient.
2. Skin paddle orientation should also be discussed with the patient because this determines the location and orientation of the scar.
3. There is no need to preoperatively locate the pedicle by Doppler.
4. Dissection on the outer surface of the muscle insertion does not risk pedicle injury. Care should be taken when dissecting along the undersurface and towards the axilla.
5. Dissect up to the muscle insertion only as much as is needed to allow the flap to rotate into the mastectomy pocket. Further dissection will only risk injury to the pedicle.
6. Suture the muscle edges of the latissimus flap to the musculature of the chest wall to provide complete coverage of the implant. If the implant is not providing the desired shape/appearance for the reconstruction, then these sutures can be released or repositioned to obtain the desired reconstructive result.

Suggested Reading

1. Maxwell GP. Inginio Tansini and the origin of the latissimus dorsi musculocutaneous flap. Plast Reconstr Surg 1980; 65:686.
2. Hammond DC, Fisher J. Latissimus dorsi musculocutaneous flap breast reconstruction. In: Spear SL, ed. Surgery of the Breast: Principles and Art. Philadelphia: Lippincott-Raven, 1998:477-490.
3. Bartlett SP, May Jr JW, Yaremchuk MJ. The latissimus dorsi muscle: A fresh cadaver study of the primary neurovascular pedicle. Plast Reconstr Surg 1981; 67:631.
4. Bostwick J, Nahai F, Wallace JG et al. Sixty latissimus dorsi flaps. Plast Reconstr Surg 1979; 63:31.
5. Griffin JM. Latissimus dorsi musculocutaneous flap. In: Strauch B, Vasconez LO, Hall-Findlay EJ, eds. Grabb's Encyclopedia of Flaps. 2nd ed. Philadelphia: Lippincott-Raven, 1998:1295-1299.

Chapter 44

Tissue Expander Breast Reconstruction

Timothy W. King and Jamie P. Levine

Introduction

While many feel that autogenous breast reconstruction offers superior results to alloplastic reconstruction with tissue expansion/implants, many patients choose this method over autogenous options. Tissue expansion offers a faster, less complicated operation, decreased hospitalization, no donor site, and more rapid recovery than autogenous reconstruction.

Currently, most plastic surgeons prefer to perform prosthetic reconstruction as a two-stage technique. Stage one is the placement of the tissue expander and stage two is the removal of the tissue expander and exchange with a permanent implant. At the time of publication, only saline-filled implants are available to the general plastic surgeon. However, silicone gel implants are available to plastic surgeons involved in clinical trials for breast reconstruction.

Indications for Surgery

Indications for prosthetic breast reconstruction include patients undergoing a modified radical mastectomy or with significant congenital deformities who desire this technique for reconstruction for their breast deformity. It is also indicated for patients who do not qualify for autogenous reconstruction secondary to obesity, scars, lack of available tissue, or comorbidities. Commonly, if being performed for breast cancer reconstruction, the first stage is completed at the time of the modified radical mastectomy.

While not an absolute contraindication, this technique is relatively contraindicated in patients who will receive perioperative radiation therapy. In general, these patients should undergo an autogenous or delayed reconstruction.

Preoperative Patient Discussion

In addition to the standard discussion of risks, benefits and alternatives of the procedure, the patient should know that utilization of a tissue expander is a minimum of a two-stage procedure. It is important in the preoperative assessment to discuss with the patient their satisfaction with the contralateral (unaffected) breast. Depending on the size and ptosis of that breast, it may not be possible to appropriately match it with the reconstructed breast. Thus, during the second stage procedure additional surgery on the contralateral breast may be required to create more symmetric breasts. The patient should also understand that when she wakes up from the surgery, she will not have a breast mound, as the tissue expander is not fully expanded at the time of the surgery. She will have to undergo weekly injections of the tissue expander for weeks to months until the breast skin envelope is appropriately expanded. Finally, she should realize that there is a limited lifespan to the

Practical Plastic Surgery, edited by Zol B. Kryger and Mark Sisco. ©2007 Landes Bioscience.

implants and maintenance (i.e., possible replacement) is recommended approximately every 10 years.

Marking the Patient

The patient should be marked in the preoperative holding area with a thick felt tip marking pen. The patient should be sitting upright facing forward with her arms at her sides. The lateral border, medial border, midline and inframammary fold are marked. Once in the operating room, remarking, tattooing, suturing, stapling, or lightly scratching the original marks with an 18-gauge needle will help preserve them. The skin incision should be planned with the surgical oncologist in order to ensure proper oncologic resection while preserving as much skin as possible.

As the goal of this surgery is to provide a reconstructed breast that is symmetrical in size and shape to the contralateral breast, the tissue expander size is established by the base diameter of the contralateral breast. In bilateral procedures, choice of expander size should be based on an attempt to recreate the patient's own breast size. In addition, the patient's desires for larger or smaller breasts should be considered.

Intraoperative Considerations

Prior to induction, sequential compression device (SCD) boots should be placed on the patient's bilateral lower extremities. The patient should be placed under general anesthesia in the supine position with the arms abducted. The arms should be padded and secured to the armboards to allow the bed to be inclined fully during the case. The buttocks should be placed at the break of the bed so the patient can be placed in a sitting position during the surgery. If the surgery is a combined case with the oncologic surgeon, it is important to communicate this positioning with the breast surgeon prior to prepping the patient.

Once the mastectomy is completed, any nonviable or marginal skin and muscle should be excised. Capillary refill and bleeding dermal edges can be used to assess the viability of the skin flaps and, if in doubt, intraoperative, intravascular fluorescein can aid in the assessment. If major defects in the muscle exist, consideration should be given to the use of a latissimus muscle flap (discussed elsewhere in this book).

Placement of the Expander

The tissue expander should be placed either subpectorally or in a complete submuscular pocket. The complete submuscular pocket is used routinely by some or when the integrity of the soft-tissue envelope is in question. Elevating the lateral border of the pectoralis major muscle and proceeding medially to the level of the presurgical markings creates the subpectoral pocket. The development of this pocket may include elevating the pectoralis minor and possibly dividing the pectoralis major muscle fibers medially. The pectoral fascia should not be divided or the implant can migrate across the midline. The superior border of the pocket should only be dissected as much as needed to fit the tissue expander. Too aggressive dissection can cause undesired migration of the expander superiorly. The inferior border is created by entering the subcutaneous or subfascial plane inferior to the pectoralis muscle and by elevating the inferior muscular attachments to establish the inframammary fold (IMF) based upon the presurgical markings. Often, the tissue expander is placed at a slightly lowered IMF position to allow for expansion of this area. This may create ptosis and, if necessary, the IMF can be reestablished during the second stage procedure.

If a complete submuscular pocket is required, the pectoralis major is elevated just medial to its lateral border. The dissection continues laterally to the serratus anterior, which is also elevated to the level of the preoperative markings. Care must be taken to avoid creating a pneumothorax during dissection of these muscles. The inferior border of the pectoralis major is elevated in continuity with the rectus abdominus fascia and is carried down to the level of the inframammary fold. Elevating the pectoralis and the serratus alone will give approximately 80% coverage of the expander and can limit lateral migration of the expander if this is a concern.

One or two suction drains are placed in the newly created pocket and after ensuring its integrity, the tissue expander is inserted with the injection valve facing outward. The muscle is closed over the implant with an absorbable suture and a layered closure is used to close the skin. The tissue expander is filled with 100-300 ml of saline while ensuring that there is not too much tension on the closure. This amount of expansion will usually depend on the viability of the skin flaps created by the oncologic surgeon. The expander should be filled as much as possible in the operating room, as it will expedite the overall expansion process and give a fuller appearance immediately postoperatively.

Implant Exchange

During the second stage, the old tissue expander is removed and replaced with a permanent breast implant. Any adjustments needed to the ipsilateral IMF and the contralateral breast to match size and shape of the reconstructed breast can also be performed at this time.

An incision is made and the old scar is removed. Using Bovie electrocautery, the subcutaneous tissues and muscle are divided until the expander is encountered. The expander is then removed. The expander can be emptied to confirm the total volume and verify the size. Usually tissue expansion is performed to a size approximately 20% greater than the desired permanent implant size. This volume, along with anatomic measurements, including the chest wall/breast diameter, and desired breast projection will guide the surgeon in selecting the appropriate implant. If needed, a capsulotomy is performed and the inframammary fold is further defined with a capsulorraphy. A capsulectomy can be performed at the inferior section of the pocket. A drain is placed, and the permanent implant is inserted into the pocket and filled. Once the implant is in place, temporary sutures or staples are used to approximate the skin, and the patient is placed in an upright position. Careful attention is paid to the symmetry of the breasts. Final adjustments are made and the patient is placed back in the supine position. The skin incision is closed in layers consisting of deep dermal interrupted absorbable sutures followed by a running subcuticular absorbable suture.

Postoperative Considerations

Initial Tissue Expander Procedure

The patient remains in the hospital for one to two days. Initially, pain control can be managed with either intravenous patient controlled anesthesia (PCA) or oral narcotics. If a PCA is utilized, it can usually be converted to oral narcotics on postoperative day one. The suction drains are removed when their output drops below 30 ml/day. Many surgeons will leave a patient on oral antibiotics (e.g., Keflex) for a few days postoperatively or while the drains remain in place and an antispasmotic (e.g., flexeril) for postoperative muscle spasm.

The patient returns to the outpatient office in weekly intervals, initially for postoperative checks and then for filling the expander. The injections usually begin three to four weeks postoperatively and continue until the desired volume is instilled. Each injection is usually 50-100 ml. In most cases, the expander is overfilled by 10-20% to help create a little laxity in the skin envelope.

The expander is kept in place for 3-6 months at which time the patient undergoes exchange to a permanent implant. The expander does not interfere with chemotherapy, but radiation therapy should not be performed during the expansion phase. As most patients receive chemotherapy followed by radiation therapy, this is usually not an issue. If early radiotherapy is required, methods for rapid expansion have been described in the literature and should be utilized prior to the initiation of radiation therapy.

Implant Exchange Procedure

This stage is usually performed on an outpatient basis. The patient follows up for postoperative checks in the outpatient office. A drain, if needed, is removed when the output drops below 30 ml/day. A sports bra should be worn day and night for 2 weeks. Patients should avoid heavy lifting and strenuous activity for 1 month after their surgery. A new baseline mammogram should be performed approximately 6 months postoperatively in all patients over the age of 35.

Pearls and Pitfalls

- Tissue expander sizes are more limited than permanent implants. Usually, the best choice is to match the tissue expander diameter to the hemi-chest wall width. The expanders can tolerate overexpansion so the surgeon shouldn't feel completely limited by the volume fill number of the expander.
- The initial submuscular pocket formation is very important for the final result of the implant reconstruction. Although changes can be made at the second-stage procedure, if the expander is too high (superior) on the chest wall, the final position and reconstruction will never look natural. Correction at the second stage should be more related to pocket refinement, achieving symmetry, IMF reconstruction, capsulotomies, etc. and not to pocket repositioning. If true pocket repositioning is required to achieve an acceptable result, then either overexpansion (for minor malpositioning) or formal operating room repositioning of the expander may be required.
- There are many planes in which one can place the expander, and it is very surgeon-specific on which method is used. One reliable method is the elevation of the lower and mid-portion of the pectoralis major muscle and, at the same level, elevation of the serratus muscle slips to attach to the lateral border of the pectoralis muscle and contain the implant laterally. Anatomically, the implant will be subcutaneous (~20%) inferiorly, but this is contained at the IMF level and is where most of the expansion is needed.
- Mastectomy flap edges are trimmed, if needed, to grossly viable tissue and the trimmed skin is sent with the mastectomy specimen to pathology as additional skin margins. With appropriate muscular coverage, even if some skin edge breakdown occurs, exposure of the expander is unlikely and utilization of secondary healing or a minor revision, if needed, can be performed without any difficulties.
- At the time of implant exchange utilize as much of the original incision as needed to place the saline or silicone implant.

- If IMF elevation is needed, utilize a monofilament suture such as a 2-0 PDS. The suture should be placed from the subcutaneous or dermal tissue of the newly marked IMF to the chest wall, preferably on the rib periosteum. Place sutures as needed and they can remain 3-4 cm apart. Sutures can also be bought out transcutaneously and tied to a xeroform or telfa bolster, but these sutures can only remain in about one week.

Suggested Reading

1. Alderman AK, Wilkins EG, Kim HM et al. Complications in postmastectomy breast reconstruction: Two-year results of the Michigan Breast Reconstruction Outcome Study. Plast Reconstr Surg 2002; 109(7):2265.
2. Bostwick III J. Tissue expansion reconstruction. Plastic and Reconstructive Breast Surgery. 2nd ed. St. Louis: Quality Medical Pub., 2000:811.
3. Spear SL, Beckenstein MS. Breast reconstruction with implants and tissue expanders. In: Evans GRD, ed. Operative Plastic Surgery. New York: McGraw-Hill Professional (Appleton and Lange), 2000:635.
4. Spear SL, Majidian A. Immediate breast reconstruction in two stages using textured, integrated-valve tissue expanders and breast implants: A retrospective review of 171 consecutive breast reconstructions from 1989 to 1996. Plast Reconstr Surg 1998; 101(1):53.
5. Spear SL. Spittler CJ. Breast reconstruction with implants and expanders. Plast Reconstr Surg 2001; 107(1):177.

Chapter 45

Nipple Reconstruction and Tattooing

Kristina D. Kotseos and Neil A. Fine

Introduction

The final stage in breast reconstruction is creation of the nipple-areolar complex (NAC), which carries aesthetic and psychological importance to patients with congenital and acquired breast absence, whether from trauma, burns, or after mastectomy for cancer treatment. Nipple-areola reconstruction (NAR) began 40 years ago with the initial creation of both the nipple and areola from distant grafts. Subsequently, this approach transformed into a combination of local flaps for nipple reconstruction and distant grafts for the areola. Many of the earlier methods for nipple reconstruction are no longer used and have been relegated to historical significance including nipple banking due to spread of cancerous cells, nipple sharing due to insult on the contralateral nipple and free composite grafts of tissue from distant sites due to donor site morbidity. The current trend is the use of local dermal flaps alone with tattooing. These methods have proven to be cost-effective and carry low morbidity. These evolving techniques and modifications of NAR are based on simplicity and reliability; however, all are hampered to some extent by loss of long-term nipple projection. The goal of reconstruction is to create an aesthetically pleasing nipple areola complex with maintenance of nipple projection, symmetry and color. Although there have been numerous articles published regarding patient satisfaction, overall nipple projection and optimal color match of different methods, no one technique has proven to be superior.

Indications and Timing

A patient undergoing NAR can have nipple creation and tattooing performed simultaneously or as two separate procedures. Many patients and plastic surgeons do not consider a breast reconstruction complete until the NAR and tattooing are performed. It is recommended to delay NAR for approximately 3-6 months after breast mound creation in order to achieve stable breast volume, overlying skin and contour.

Contraindications

Nearly all patients, with the exception of those with active or progressive malignancy, should be offered NAR. As with all forms of postablative reconstruction, the first and foremost goal is adequate treatment of the specific malignancy, and this goal should never be compromised. Some consider radiation a relative contraindication due to its detrimental effects on wound healing. However, in patients with minimal breast skin changes, NAR may be performed with satisfactory results at approximately 6-8 wks following the completion of radiation treatment. Much of nipple projection is based on underlying subcutaneous fatty tissue; therefore patients who have undergone tissue expander reconstruction with thinning of this layer will need to be educated on realistic goals regarding nipple projection.

Practical Plastic Surgery, edited by Zol B. Kryger and Mark Sisco. ©2007 Landes Bioscience.

Anatomy

In unilateral breast reconstruction, the contralateral breast determines placement of the NAC and represents a comparison for normal anatomy. In bilateral breast reconstruction, the surgeon must rely more on measurements based on anatomical landmarks, namely the midpoint of the sternum, sternal notch, midclavicular line and inframammary fold. The distance from both the sternal notch and midclavicular line to the mid-nipple position should each measure approximately 19 to 21 cm. The distance from the midpoint of the sternum to the mid-nipple position should measure 9 to 11 cm. The inframammary fold lies 7 to 8 cm from the mid-nipple position, or 6 cm from the inferior areolar border. If discrepancies arise, it is best to remember that the NAC is the primary focus of the breast mound and aesthetically should be centered at the point of maximal projection. Other equally important anatomic considerations include the color, texture and size of the areola. The average nipple diameter is 8 mm, while the average areolar diameter is 42 to 45 mm. The areola is darker in color than the surrounding breast mound skin and is further affected by race and parity. Projecting a millimeter or so above the breast mound, the areola hosts several Montgomery glands contributing to its rough texture. Nipple projection is one of the most important considerations and is often overcompensated at the time of NAR in anticipation of loss of projection over time. When unstimulated, the nipple normally projects approximately 4 to 6 mm above the areola.

Preoperative Considerations

There are a vast number of operative techniques available for creation of the NAC. The method selected depends on the anatomic characteristics and projection of the contralateral nipple. Both the patient's desires and surgeon's preferred technique must be considered.

The patient should be marked in an upright position as the NAC may shift when supine. Furthermore, to avoid unwanted superior placement of the NAC, the patient's shoulders should be completely relaxed. Positioning should be determined based on previously described normal breast dimensions or the contralateral breast in unilateral breast reconstruction. Nipple-areolar positioning becomes difficult when there is a discrepancy between the normal anatomic measurements and the actual appearance of the NAC, typically as a result of asymmetric breast mounds. Preoperative antibiotics are typically not necessary, especially if the patient lacks an underlying implant. This procedure may be performed in an office under local anesthesia. Conversely, the procedure can be carried out in the operating room under conscious sedation or monitored anesthesia care if accomplished simultaneously with port removal or final touch-up procedures of previous autologous breast reconstructions.

Operative Technique

Approximately 100 different techniques and modifications of NAR have been described in the literature; however no single procedure can create the optimal NAC in one stage. Local flap techniques are the most popular since fears of donor site morbidity have largely been alleviated through modifications in previous flap designs allowing for direct donor site closure without the need for skin grafting. The most popular techniques are described in this chapter.

The Skate Flap

The Skate flap has proved to be reliable but is one of the techniques which may require skin grafting to cover the donor sites. A small circle (the base) measuring the exact size of the proposed new nipple is drawn at the appropriate position on the breast mound. The wings, elevated as partial-thickness skin flaps, depend on intradermal blood flow and a well-vascularized subcutaneous pedicle at the base or the proposed new nipple. They are elevated at the level of the deep dermis from the periphery inward until the width of the base is reached. The base is then used to line the flap vertically. This linear portion is elevated full-thickness including underlying deep fat. The wings are wrapped around this central portion to create the nipple. The areola is created by skin grafting the donor sites and the remaining circular de-epithelialized tissue, allowing the nipple to protrude through a circular hole in the center of the graft. Skin graft donor sites include the contralateral areola, labia, upper inner thigh or lateral mastectomy dog ear.

The Modified Star Flap

The modified star technique is similar to the skate flap except skin grafting is not required, and the lateral flaps or wings are raised full-thickness instead of just elevating the dermis. Primary donor site closure, increased blood supply and improved nipple projection are some of the significant advantages offered by the modified technique. The base may be positioned superiorly, laterally or inferiorly and measures approximately 1 to 1.5 cm in diameter, or the desired width of the nipple. Resulting nipple projection is directly related to the width and length of the laterally diverging flaps. The flaps should measure 1.5 times the desired height to account for loss in projection with time. The limbs are typically 1.5 to 2.0 cm in length. The longer the skin flaps the greater the projection. As previously mentioned, the lateral limbs and apex flap are elevated full-thickness, including the underlying fat; 5-0 plain or chromic gut sutures are used to secure the limbs after they are rotated and interdigitated. The donor sites are undermined slightly and closed primarily.

The S-Flap

The S-flap for nipple reconstruction uses an S-shape design to create two equal skin flaps with opposing bases. The center of the S represents the desired location of the new nipple. If a preexisting mastectomy scar exists, then the central limb of the S should be drawn along the scar to secure blood supply. Nipple projection is determined by the length of the flaps, which are raised full-thickness and sutured to each other. Circular de-epithelialization and subsequent placement of a skin graft for areolar reconstruction have been described in conjunction with the S-flap.

The Modified Double Opposing Tab Flap

The double opposing tab flap allows direct closure of the donor site without skin grafting. The tabs are raised as full-thickness skin flaps and the original back cuts first described have been eliminated to improve blood supply to the flap. Two equal semicircular flaps with opposing bases are drawn extending from a shared line measuring approximately three times the desired height of the final nipple. A round nipple projection of approximately 10-12 mm is accomplished after opposing the 18 mm wide flaps; 4-0 and 5-0 chromic sutures are used to primarily close both the donor and skin sites. This technique is similar to the S-flap and also allows incorporation of the mastectomy scar without compromising skin flap viability. This method may be used for nipple reconstructions requiring a larger broad-based nipple.

The C-V Flap

The C-V flap produces an excellent reconstruction, but it is dependent upon the underlying subcutaneous fat to provide the bulk of the reconstructed nipple. In most cases, the subcutaneous tissue is adequate. However, in cases of tissue expansion, the fat may be insufficient to produce a nipple of adequate bulk. This design uses two V flaps and a C flap to create the nipple. The widths of the V flaps determine projection, whereas the diameter of the C flap determines the diameter of the new nipple. This flap is quite similar in design to the modified star flap.

The Bell Flap

This bell-shaped flap design incorporates a unique purse string suture allowing for a tension free closure of the donor site. A circle, representing the final positioning of the new NAC, is drawn slightly larger than the contralateral areola, or approximately 45-50 mm in bilateral reconstructions. Next, the superiorly-based bell flap is elevated full-thickness toward the handle of the bell design and the entire circumference of the circle is incised in the subcutaneous plane. The defect created by flap elevation may be approximated after folding the flap on itself and securing the two sides of the bell around the handle. To produce final NAC projection, a subdermal nylon suture is placed around the circumference of the circle in a purse string fashion.

Tattooing

Tattooing is an optical effect used to color the areola and nipple approximately 6-8 weeks after nipple reconstruction is complete. It has several advantages, namely it is simple, reliable, easily correctable and has no effect on eventual nipple projection. Precise color matching to the contralateral NAC can be readily performed due to the many different pigments available. Intradermal pigmentation is generally performed with iron oxide and titanium dioxide pigments. The patient is encouraged to choose a pigment that is slightly darker than her normal areolar color, as the color often fades, requiring one or more touch up applications.

Postoperative Care

Following NAR, antibiotic ointment is applied to the nipple, and it is covered with sterile dressings for 48 hours. A small hole for the nipple can be cut in a few 4 x 4 gauze pads to avoid compression of the new nipple. In addition, steri-strips® can be used along the donor site regions at the base of the nipple. After 48 hours, patients may shower avoiding direct pressure to the nipple. They should be advised to wear loose-fitting, noncompressible bras for at least 2 weeks. Healing normally occurs in approximately 1 to 2 weeks. Following tattooing, impregnated-gauze dressings are applied and the patient is instructed to avoid direct water pressure to the area for 2 days.

Complications

Nipple necrosis is one of the feared and most serious complications of NAR, which can result in wound separation. Flap dehiscence with unfolding of the nipple may also occur. Smokers must be encouraged to quit smoking, at least temporarily, in order to avoid compromising the blood supply and healing of tissues. One of the more common undesirable outcomes of NAR is loss of nipple projection, occurring most significantly in the first 6 months. It is safe to assume that approximately 60%

of the projection achieved at surgery will be reliably maintained at one year follow up. Incorporating subcutaneous tissue allows for volume and projection, however fat is absorbed over time, thus causing flap flattening. Complications of tattooing can include infection, rash, slough and discoloration. Some patients may require retattooing for fading of pigment, poor take of pigment or revision of the shape of the pigmented area.

Pearls and Pitfalls

- Not all patients desire NAR. It should be presented as an option and not a necessity. The patient can always change her mind at a later time.
- When dressing the newly created nipple, a hole can be cut in the center of two gauze eye-patches. The patches surround the nipple and provide support and protection.
- In unilateral cases, the newly created nipple should match the contralateral side in terms of diameter only. Its height should be at least 1.5 times greater than the other side.
- In patients with extremely large, projecting nipples, always consider nipple reduction (e.g., wedge excision) of the normal side, and offer this to the patient. If she chooses to keep her normal nipple intact, the Skate flap is a good option for very projecting nipples.
- Beware of the donor site closed with excessive tension. A spread scar will not be easily concealed by tattooing, as the scarred dermis does not incorporate the pigment very well.

Suggested Readings

1. Cronin ED, Humphreys DH et al. Nipple reconstruction: The S flap. Plast Reconstr Surg 1988; 81(5):783.
2. Eng JS. Bell flap nipple reconstruction—a new wrinkle. Ann Plast Surg 1996; 36(5):485.
3. Eskenazi L. A one-stage nipple reconstruction with the "modified star" flap and immediate tattoo: A review of 100 cases. Plast Reconstr Surg 1993; 92(4):671.
4. Few JW, Marcus JR et al. Long-term predictable nipple projection following reconstruction. Plast Reconstr Surg 1999; 104(5):1321.
5. Hartrampf Jr CR, Culbertson JH. A dermal-fat flap for nipple reconstruction. Plast Reconstr Surg 1984; 73(6):982.
6. Kroll SS, Hamilton S. Nipple reconstruction with the double-opposing-tab flap. Plast Reconstr Surg 1989; 84(3):520.
7. Spear SL, Convit R et al. Intradermal tattoo as an adjunct to nipple-areola reconstruction. Plast Reconstr Surg 1989; 83(5):907.
8. Wellisch DK, Schain WS et al. The psychological contribution of nipple addition in breast reconstruction. Plast Reconstr Surg 1987; 80(5):699.

Chapter 46

Reduction Mammaplasty

Timothy W. King and Jamie P. Levine

Introduction

The goal of breast reduction is to reduce the overall volume of the breast while maintaining an aesthetically pleasing shape and the viability of the nipple areola complex. When considering the different techniques of breast reduction, it is important to understand that the two components of the surgery—the skin incisions and the pedicle—are independent of each other. Thus, you can have any type of skin incision with any type of pedicle (Table 46.1). This being said, there are commonly used approaches to breast reduction, and this chapter will focus on these combinations, namely the Wise pattern/inferior pedicle, the vertical scar/medial pedicle and the free nipple graft. The surgeon, however, should not be as limited in his thinking and should be open minded when selecting the best combination for each individual patient.

Preoperative Considerations

Indications for reduction mammaplasty are symptoms secondary to the large breast volume. These symptoms include chronic back, shoulder and neck pain, headaches, upper extremity neuropathy, postural changes, bra strap grooves in the shoulders and dermatological disorders in the inframammary folds.

In addition to the standard discussion of risks, benefits and alternatives of the procedure, the patient should know that there may be some loss of sensation, change in color, or even death/necrosis of the nipple-areola complex. Patients should also be warned that since some of the breast parenchyma is being removed, the ability to breastfeed might be altered with a breast reduction (although several studies show no significant differences in breast-feeding between women who have and haven't undergone breast reduction). The patient should be shown where and how long the scars will be on her breasts. She should realize that although everything will be done to avoid it, there will likely be some asymmetry between the breasts. A complete history and physical exam should be performed, particularly focusing on any history of breast disease. If age appropriate, the patient should have a screening mammogram within the past year.

Marking the Patient

Patient marking is a critical step in the planning of any breast reduction. The patient should be marked in the preoperative holding area with a thick felt tip marking pen. The patient should be sitting or standing upright facing forward, with her arms at her sides. The lateral border, medial border, midline and inframammary fold are marked. Once in the operating room, key midline and angle markings can be reinforced by remarking, tattooing, suturing, stapling, or lightly scratching the marks with an 18-gauge needle.

Practical Plastic Surgery, edited by Zol B. Kryger and Mark Sisco. ©2007 Landes Bioscience.

Table 46.1. The commonly used pedicles, their indications and the various skin incisions available for breast reduction

Type of Pedicle	Indications for Its Use	Types of Skin Incisions
Inferior pedicle	Good for almost any breast size and shape	Wise pattern/inverted T
Superior pedicle	Best for small to moderate breast reductions	Vertical
Medial pedicle	Reductions less than 1500 grams per side	Lateral
Partial breast amputation	Massive reductions (>40 cm from sternal notch to nipple)	Circumareolar
Liposuction	Minimal reduction, little to no ptosis	None

Note that any skin incision can be used with any type of pedicle.

Wise Pattern/Inferior Pedicle

A tape measure is required for marking, and a wire keyhole pattern of 14-15 cm long can be helpful to match a 42 mm cookie cutter. The sternal notch and midclavicular points are marked. The midclavicular point, for reasons of standardization and symmetry is normally set at 6 or 7 cm from the sternal notch. A meridian line from the midclavicular point through the nipple-areolar complex is made and then transposed to the inframammary fold and marked. The inframammary fold is marked and is translated to the anterior surface of the breast at the intersection of the meridian line. The new nipple position should be 0.5-1 cm above the anatomic inframammary fold in the meridian line. The wire keyhole is placed on the breast and adjusted for the individual patient. If the breasts are large and tense, the keyhole should be made narrower to reduce the amount of skin removed. On the other hand, if the breasts are lax and ptotic, the keyhole should be widened to allow more skin excision. Two lines are drawn from the end of the keyhole circle angling away from the meridian of the breast. The angle selected determines the amount of skin excised. The height of the vertical limbs should be 5-6 cm from the bottom of the keyhole circle. The keyhole pattern is then traced onto the skin. At the base of the vertical keyhole limbs a line creating a 90° angle is drawn and extended to the line marking the inframammary fold to complete the horizontal limb (Fig. 46.1). These horizontal lines define the future inframammary fold. These points can end in the breast folds. Some surgeons use a template for the entire markings and some complete the keyhole/areolar markings in the operating room.

Vertical Scar/Medial Pedicle

As with the Wise pattern, the meridian line is established and drawn. With this technique, the new nipple should be placed 2 cm lower than in the Wise technique. This will place the new nipple at or just below the level of the existing inframammary fold. A mosque-shaped areolar pattern is used and is marked on the skin (Fig. 46.1). Lines similar to the vertical limbs of the Wise pattern are drawn. These lines can be drawn in an individualized fashion by using the breast displacement method. The breast is moved medially and a straight line is drawn reapproximating the meridian. When the breast is returned to its normal resting position, a curved line remains as

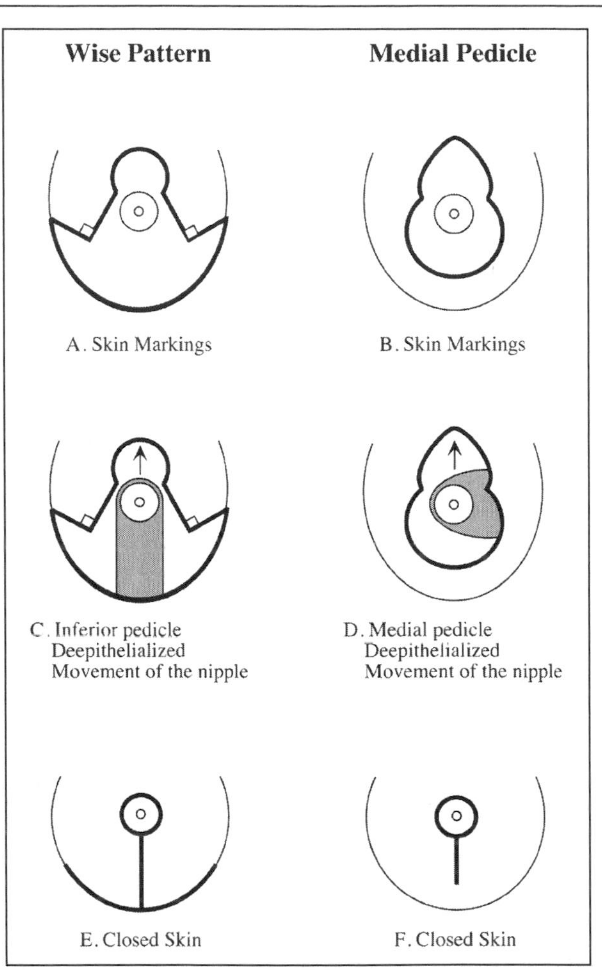

Figure 46.1. Comparison of the Wise pattern/inferior pedicle (A,C,E) with the vertical scar/medial pedicle (B,D,F). Illustrated from top to bottom are the initial skin markings (A,B), the de-epithelialized pedicle with the intact nipple-areola complex (C,D) and the final scars shown in bold (E,F).

the lateral extent of excision. This process can be repeated for the medial limb by lateral displacement of the breast. However, at the point where one makes a 90° angle, the line continues inferiorly and curves at the base, making a "U" shape. The distance from the inferior edge of the "U" to the inframammary fold should be 2-6 cm dependent upon the amount of ptosis of the breast. If the base of the "U" is too close to the inframammary fold, the vertical scar will extend down onto the chest/ abdominal wall when the incision is closed since this technique can raise the inframammary fold position.

Free Nipple Graft

As with the Wise pattern, the sternal notch, midclavicular points and the meridian line are determined and drawn. The markings for this procedure are similar to the Wise pattern. However, in this case there will not be a pedicle carrying the nipple/areola complex. The nipple/areola complex will be removed as a free, full-thickness graft and replanted onto the breast at its new location. The apex of the areola is marked and the vertical limbs are drawn in a similar fashion to the Wise pattern. The inframammary fold is marked and is translated to the anterior surface of the breast at the intersection of the meridian line. The height of the vertical limbs should be 5-6 cm. At the end of the vertical limbs, a line creating a 90° angle is drawn and extended to the line marking the inframammary fold, completing the horizontal limb (Fig. 46.1). These horizontal lines define the future inframammary fold. These markings should be placed while supporting the weight of the breast or the nipple-areola will end up too high on the final reduced breast.

Intraoperative Considerations

Wise Pattern/Inferior Pedicle

The inferior pedicle is marked vertically below the nipple-areolar complex and should be 8-9 cm wide at the base centered on a line from the nipple to the inframammary fold. This base can be widened with longer pedicles. The superior portion should extend in a curvilinear fashion 1 cm around the nipple-areolar complex. With the breast under moderate tension, a 38 or 42 mm cookie cutter is centered over the nipple and used to mark the areola for reduction. This should match with the keyhole pattern chosen. An incision is then made along this mark to the level of the dermis. Another incision is made along the marks for the inferior pedicle and the skin over the pedicle is deepithelialized (Fig. 46.1).

The breast tissue is resected in a systematic fashion. Tissue will be removed from the medial, lateral and superior aspects of the breast. Starting with the medial aspect of the pedicle, the breast tissue is dissected just superficial to the pectoralis fascia. Beveling the dissection away from the pedicle will aid in preserving the vascularity of the pedicle. A good base should remain attached to the chest wall. A skin incision is made along the lines previously marked in the inframammary fold and the medial aspect of the breast. The breast parenchyma is dissected down to the same level as performed previously. Once all the dissection has occurred, a small medial wedge of breast tissue is excised and weighed.

In a similar fashion the lateral breast tissue is excised. In general, more tissue is removed from the lateral aspect of the breast than the medial aspect since medial fullness is aesthetically more desirable. Attention is then turned to the superior aspect of the breast. The skin of the vertical limbs of the keyhole pattern is incised. Some surgeons will not incise the curvilinear pattern (or center) initially. Rather, they recommend extending the vertical limbs to form a triangle at the apex of the keyhole. This allows the surgeon to sit the patient up and determine the exact location of the nipple-areolar complex prior to committing to a specific location. A superior flap approximately 2 cm thick is created and dissected to the level of the pectoralis fascia. The remaining superior breast tissue is then removed from the breast. This creates the pocket allowing the pedicle to be placed in the more superior position. All tissue is weighed together from each breast and sent to pathology for evaluation.

Applying similar techniques, the contralateral breast undergoes the same procedure. Once both breasts have been reduced, temporary sutures or staples are used to approximate the skin and the patient is placed in an upright position. Careful attention is paid to the symmetry of the breast and position of the nipple. Final judgments and adjustments are made, and the patient is placed back in the supine position. If not performed previously, the curvilinear keyhole incision creating the opening for the nipple areolar complex is completed. A buried 2-0 vicryl suture is placed to bring together the junction of the "T". Some surgeons will place a 10 mm Jackson-Pratt (JP) drain in each breast prior to closure. The skin incisions are closed in layers using interrupted buried 2-0 vicryl deep, 3-0 monocryl in the dermal layer and a subcuticular running 3-0 or 4-0 monocryl or prolene. Steri-strips are applied to the incision lines, followed by sterile 4x8 fluffs or Telfa/Tegaderm and a sports bra.

Vertical Scar/Medial Pedicle

Although it is entitled a medial pedicle, the pedicle is often superomedially-based and full-thickness down to just above the pectoralis fascia. The flap is designed so that one-half to one-third of the flap is in the areolar opening and remainder is along the vertical skin incision line. The pedicle width should be equal to the pedicle length, which is usually about 8 cm. The pedicle is de-epithelialized including a rim around the areola. The pedicle is beveled superiorly to create a platform for the nipple-areolar complex and left attached to the chest wall.

The parenchymal breast tissue and skin to be removed are excised in one piece. The tissue laterally is beveled out extensively while there should be minimal beveling in the inferior-medial pole. Dissection continues with undermining down to the inframammary fold. Removing the inferiorly-based, subcutaneous tissue can reduce skin puckering on closure. If the inferiorly-based, subcutaneous tissue is not removed, puckering of the skin will occur. The tissue at the inframammary fold must be excised. Little to no breast tissue should be excised from the superior pole. All tissue is weighed together from each breast and sent to pathology for evaluation. The pedicle is rotated up into position, and a suture is placed at the base of the areola. The inferior border of the pedicle becomes the medial pillar, and this is sutured to the lateral pillar, starting at the base and progressing superficially. Skin closure and dressings are performed as described above.

Free Nipple Graft

An incision is made around the diameter of the areola and the nipple-areola is harvested as a full-thickness graft approximately 40 mm in diameter. The inframammary fold incision is made and the breast tissue is dissected off of the fascia to the level of the nipple. The horizontal limb incisions are made, beveling superiorly, and the inferior breast tissue is removed. An additional wedge of breast tissue is removed from the deep lateral segment of the breast to allow the breast to narrow during closure. The medial and lateral flaps are brought down to the inframammary fold and stapled in place. Excess skin can be removed. The patient is placed in the sitting position and the new nipple position is determined. The patient is lowered to a supine position and the skin around the site of the new nipple-areola is de-epithelialized. The free nipple graft is sutured in place with an absorbable suture and bolstered in place. All tissue is weighed together from each breast and sent to pathology for evaluation.

Applying similar techniques, the contralateral breast undergoes the same procedure. Once both breasts have been reduced, temporary sutures or staples are used to approximate the skin, and the patient is placed in an upright position. Careful attention is paid to the symmetry of the breast and position of the nipple. Final judgments and adjustments are made, and the patient is placed back in the supine position. Skin closure and dressings are performed as described above.

Postoperative Considerations

The patient may remain in the hospital overnight or be discharged the same day. Pain control is managed with oral narcotics. If JP drains are used, they are removed when the drainage drops below 30 ml/day. Many surgeons will leave a patient on oral antibiotics (Keflex) for a few days postoperatively. If drains are used, they may continue the oral antibiotics until the drains are removed. The sports bra should be worn day and night for 8 weeks. Patients should avoid heavy lifting and strenuous activity for 1 month after their surgery. A new baseline mammogram should be performed approximately 6 months postoperatively in all patients over the age of 35.

Pearls and Pitfalls

Markings of the breast meridian, medial and lateral extent of excision, etc., must be individualized to each breast since patients commonly have asymmetry. The preoperative discussion with the patient is essential. She should clearly understand all of the risks, benefits and alternatives, incisions, scars, etc. prior to arriving in the operating room. Breast evaluation with mammography and physical exam needs to be performed preoperatively, and all breast tissue should be sent to pathology for evaluation.

In dissection of either the inferior pedicle or medial pedicle, the pectoral fascia should be left intact. This will decrease postoperative pain. For the Wise pattern and even the vertical pattern, the final areolar placement can be made after the skin and parenchymal excision by sitting the patient up and examining symmetry. Nipple-areolar symmetry is extremely important in the overall aesthetic outcome.

In the inferior pedicle technique, the medial pocket should be left fuller since this gives a more aesthetic postoperative result. Maintaining a wide pedicle base and the attachments to the chest wall is essential for maintaining vascularity of the nipple-areola complex.

In the medial pedicle design, maintain a wide base of attachment of the entire pedicle to the chest wall. The excision of parenchyma should be greatest in the inferior and lateral pockets. Excision superiorly should only be enough to rotate the pedicle into its new position. Pillar suture placement helps to shape the breast and lift it superiorly. Do not be afraid to reopen incisions several times to remove more tissue for symmetry purposes and to obtain meticulous hemostasis.

Suggested Reading

1. Bostwick IIIrd J. Reduction mammaplasty. Plastic and reconstructive breast surgery. 2nd ed. St. Louis: Quality Medical Pub, 2000:371-498.
2. Chadbourne EB, Zhang S et al. Clinical outcomes in reduction mammaplasty: a systematic review and meta-analysis of published studies. Mayo Clinic Proceedings 2001; 76(5):503-510.
3. Cohen BE, Ciaravino ME. Reduction mammaplasty. In: Evans GRD, ed. Operative Plastic Surgery. New York: McGraw-Hill Professional (Appleton and Lange), 2000:613-630.
4. Cruz-Korchin N, Korchin L. Breast-feeding after vertical mammaplasty with medial pedicle. Plast Reconstr Surg 2004; 114(4):890-4.
5. Hall-Findlay EJ. A simplified vertical reduction mammaplasty: Shortening the learning curve. Plast Reconstr Surg. 1999; 104(3):748-759, (discussion 760-763).
6. Hall-Findlay EJ. Vertical breast reduction. Seminars Plast Surg 2004; 18(3):211-224.

Chapter 47

Sternal Wounds

Jonathan L. Le and William Y. Hoffman

Incidence

The majority of sternal wounds seen by the plastic surgeon are infections secondary to dehiscence of a median sternotomy following a postoperative infection. The incidence of median sternotomy dehiscence is reported in up to 5 percent of open heart procedures. The patients at highest risk are diabetics undergoing coronary revascularization using the internal mammary artery since this procedure results in impaired blood supply to the healing sternal incision.

Initial attempts to manage this complication conservatively with open drainage, debridement, and packing often resulted in bypass graft exposure, desiccation of wound margins, osteomyelitis and even death. Mortality due to sternotomy wound dehiscence is higher in patients with septicemia, perioperative myocardial infarction, or those requiring an intraaortic balloon pump. Mortality rates previously reached 50 percent; however, the development of wound reconstruction techniques using wide debridement and muscle or musculocutaneous flap transposition has reduced mortality rates to less than 5%.

Wound Classification

Pairolero's classification scheme identifies three types of sternal wound infections:

- **Type I (acute)** infected sternotomy wounds occur one to three days postoperatively and present with serosanguinous drainage. Wound cultures are negative and cellulitis, costochondritis and osteomyelitis are absent. These wounds require reexploration, minimal debridement and rewiring of the sternum. Reconstructive surgeons are typically not consulted.
- **Type II (subacute)** wounds present two to three weeks postoperatively and involve purulent mediastinitis with positive wound cultures, cellulitis, costochondritis and osteomyelitis.
- **Type III (chronic)** infected sternotomy wounds occur months to years after cardiac procedures and display draining sinus tracts as a result of chronic costochondritis and osteomyelitis.

Type II and III wounds require thorough debridement and may depend on temporary open-wound dressing changes for adequate infection control. Figure 47.1 provides an algorithm for the most appropriate management based on the presentation of the wound.

The majority of patients with postoperative mediastinitis have monomicrobial infections although virtually any organism can be responsible. Approximately 50 percent of infections are due to Gram positive bacteria, such as *Staphylococcus aureus*, with the remainder due to Gram negative bacilli, such as *Pseudomonas*

Practical Plastic Surgery, edited by Zol B. Kryger and Mark Sisco. ©2007 Landes Bioscience.

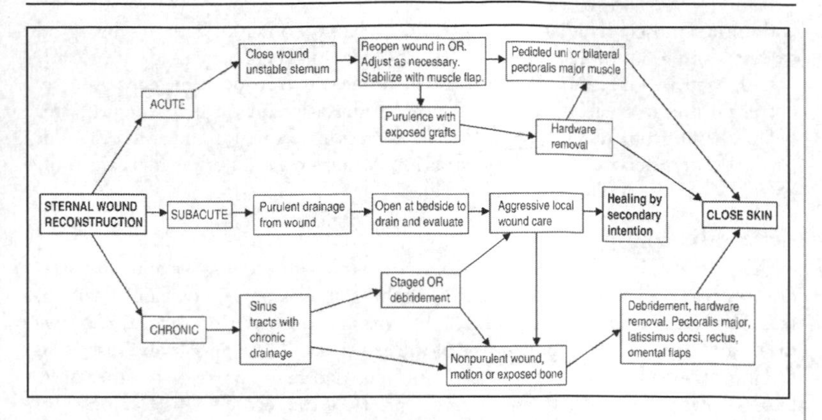

Figure 47.1. An algorithm for management of sternal wounds.

aeruginosa. Initial empiric antibiotic therapy should consist of broad coverage against these organisms. Rarely, postoperative mediastinitis is due to unusual organisms such as fungi, Legionella, or *Mycoplasma hominis.*

Preoperative Evaluation

Careful examination of the patient with a sternotomy wound dehiscence is extremely important. The primary contraindication to reconstructive surgery is a wound with active purulence. Such a wound requires thorough debridement prior to flap coverage to avoid possible flap failure. Patients who are hemodynamically unstable, have poor pulmonary or cardiac reserve, or are terminally ill should not be considered for surgery.

Appropriate laboratory and radiographic studies must be performed prior to operating on patients with infected sternotomy wounds. Easily accessible fluid collections should be aspirated and sent for culture. If the patient clinically deteriorates or further signs of wound breakdown are observed (such as increased erythema, drainage, dehiscence, or systemic signs of infection) then wound cultures should be obtained and sent. Wound cultures should include a quantitative microbiology count, a tissue specimen for analysis and sternum biopsies for culture.

Plain chest radiographs should be examined to evaluate the condition of the lung fields, to check for the presence of effusions, and to determine the number, position and condition of the sternotomy wires. Computed tomographic scans are usually not helpful unless the patient has an elevated fever and white blood cell count and the previously opened wound does not correspond with the illness. The assessment of possible osteomyelitis using a bone scan may have limited value for the acute wound due to the presence of inflammation and tracer uptake. Diagnosis of osteomyelitis is most accurate when bone biopsies are obtained rather than relying solely on clinical information.

Operative Approaches

Numerous surgical options are available for closure: the unilateral pectoralis major muscle turnover flap, the unipedicle pectoralis major muscle rotation advancement flap, bilateral myocutaneous pectoralis major muscle flaps, rectus abdominus muscle flap, latissimus dorsi muscle flap, omental flap, microsurgical free flap and vacuum-assisted closure. Although skin flaps were used historically, vascularized regional muscle and muscluocutaneous transposition flaps are now preferred due to their greater blood flow, obliteration of dead space and faster healing time from quicker resolution of infection. Omental and microvascular free tissue transfer are reserved for situations when no local muscle flaps are available or when other alternatives have failed.

Pectoralis Major Flap

The pectoralis major muscle is a common choice for coverage of a median sternotomy wound because of the proximity of the donor tissue, relative ease of harvest, versatility and its wide arc of rotation. The muscle originates on the clavicle, sternum, six upper ribs and aponeurosis of the external oblique muscle and inserts on the intertubercular sulcus of the humerus. This flap may be based on either of its two vascular pedicles, the thoracoacromial artery or the segmental perforators from the internal thoracic artery (see Appendix I).

Pectoralis Turnover Flap

By basing the flap medially on the segmental parasternal perforators from the internal thoracic artery, the muscle can be turned over on itself. The muscle is dissected from the chest wall lateral to medial to avoid injury to the perforators. The thoracoacromial pedicle is ligated, and the humeral end of the muscle is disinserted and placed in the sternal defect. The limitations of this approach are that one wastes part of the flap to allow the muscle to turnover and possibly creates a raised contour deformity of the anterior chest wall that would put tension on the skin closure.

Pectoralis Transposition/Advancement Flap

When the internal thoracic arteries are sacrificed bilaterally, the flap can be based on the thoracoacromial pedicle alone. The muscle is dissected off the sternum, disinserted from the humerus and advanced into the defect. This technique affords greater muscle mass to fill upper mediastinal dead space but without a wide arc of rotation the inferior aspect of the wound may remain inadequately covered. Wide undermining of skin flaps provides easier skin closure.

Bilateral myocutaneous medial advancement of both pectoralis major flaps may be considered for coverage of large defects. The flaps are elevated in the relatively avascular plane just deep to the pectoralis major muscles. This approach is best used in the absence of purulent mediastinitis as it does not fill the mediastinal dead space.

Rectus Abdominus Flap

The rectus is effective for covering inferior sternal defects and is used to supplement a pectoralis flap or when pectoralis muscle is unavailable. It may be used as a turnover flap based on the superior epigastric artery (see Appendix I). If only one internal thoracic artery is available, the flap may need to be harvested from the contralateral side. If both internal thoracic arteries have been sacrificed, then the flap can be based on the eighth anterior intercostal perforator to the rectus muscle; however, the distal third of the rectus muscle may become ischemic using this approach.

The rectus originates at the pubic crest and inserts at the xiphoid and fifth, sixth and seventh costal cartilages. The rectus muscle is harvested through a midline or paramedian skin incision. After the anterior rectus sheath is incised the muscle is divided at the appropriate level and elevated from the posterior rectus sheath towards the sternum. Once adequate mobilization has been achieved, the flap is transposed into the wound.

Latissimus Dorsi Flap

This flap is best used for smaller defects and should be reserved as a salvage flap in the event of pectoralis muscle flap failure. The muscle originates at the posterior iliac crest and spinous processes L7 to S1 and inserts at the crest of the lesser tubercle of the humerus. It is based on its major pedicle, the thoracodorsal artery, when it used for sternal defects (see Appendix I). When adequately mobilized, it can extend over the anterior chest to cover the mediastinum.

Omental Pedicle Flap

The broad, pliable, fatty nature of the omentum allows it to fill the deep spaces of large wounds. In addition the flap has a generous source of lymphatics and immune cells that support the clearance of infection. It can provide reliable coverage for deep sternal infections such as osteomyelitis, costochondritis, or mediastinitis. Its blood supply can be based on either the right or left gastroepiploic artery.

A midline laparotomy is the usual approach followed by lysis of adhesions from any previous abdominal surgery. After the pedicle from either the left or right gastroepiploic artery is identified, added mobilization may be achieved by dividing the short gastric vessels along the greater curvature of the stomach. This should be done cautiously as gastric outlet obstruction is associated with excessive cranial traction on the antrum of the stomach during mobilization and flap inset. Omental flaps heal without intraabdominal problems or chest wall instability 95% of the time. Complications include ventral hernia formation, wound infection and bowel injury. Laparoscopic harvest of the omentum is well described in the literature and should be considered for patients who are poor candidates for a laparotomy.

Microvascular Free Flaps

Free tissue transfer is usually reserved for conditions where local flaps are unavailable or have previously failed. Contralateral latissimus dorsi or tensor fascia lata are the most popular for this purpose. There are many potential recipient vessel sites for anastomosis of the donor pedicle including the axillary, thoracodorsal, subscapular, or internal mammary artery.

Postoperative Complications

Complications of infected sternal wound reconstruction include hematoma, seroma, dehiscence and sternal necrosis with osteomyelitis.

The formation of a hematoma is infrequent, but this complication generally requires reoperation. The most common cause of hematoma is premature reinstitution of anticoagulation therapy. Careful attention to hemostasis, meticulous pedicle dissection, and closure over suction drains will also help prevent this complication from occurring.

Seroma formation is infrequent but the risk may increase when two muscle flaps are used simultaneously. Closure of flaps over drains is an effective preventative measure. This minimizes dead space and provides for early flap adherence.

Dehiscence is more likely to occur in obese patients, older patients with chronic obstructive pulmonary disease, patients on prolonged ventilatory support, diabetics and patients with sepsis. In women with large, pendulous breasts the use of surgical bras and tapes will help prevent distraction on the medial chest and separation of the flaps.

Sternal necrosis and osteomyelitis may occur in patients with profound sepsis, Gram-positive infections, and when debridement is not adequate.The sternum should be debrided back to viable, bleeding bone. Surgery performed in the absence of frank purulence will also help lower the risk of infection recurrence.Reinfected wounds generally heal spontaneously when reopened, since foreign bodies and nonviable tissues have already been removed.The application of dedicated local wound care and/or vacuum-assisted closure therapy usually results in wound closure by secondary intention. Secondary flap reconstruction is only necessary in selected severe cases.

Pearls and Pitfalls

First and foremost, the timely recognition of an infected sternal wound is critical. Despite extensive data verifying the utility of flap closure of these wounds, it is not uncommon to see patients with mediastinitis treated with rewiring of the sternum, irrigation-drainage systems or some other alternative. Our experience has been favorable with simultaneous debridement and flap closure of the mediastinal wound, and this is done routinely at our institution unless the patient is medically unstable due to sepsis. Often the left sternum is more affected due to sacrifice of the left IMA for coronary revascularization. In these cases, we have found that debridement of the left side only and preservation of at least a portion of the right is still effective.

The manubrium is often unaffected by the infectious process, probably due to its more proximal blood supply. The xyphoid is often involved in severe cases. If the pectoralis muscles is detached from the clavicle as well as the humerus it can usually reach the xyphoid. Finally, new technology should be noted. There are recent reports that rigid fixation of the sternum can markedly reduce mediastinal infections and even be used in some infected cases. Laparoscopic harvest of omental flaps has also been reported for treatment of sternal wounds and might actually be less morbid than muscle flaps.

Suggested Reading

1. Jurkiewicz MJ et al. Infected median sternotomy wound: Successful treatment by muscle flaps. Ann Surg 1991; 738:1980.
2. Nahai F et al. Primary treatment of the infected sternotomy wound with muscle flaps: A review of 211 consecutive cases. Plast Reconstr Surg 1989; 84:434.
3. Obdeijin MC et al. Vacuum-assisted closure in the treatment of poststernotomy mediastinitis. Ann Thorac Surg 1999; 68:2358.
4. Pairolero PC, Arnold PG, Harris JB. Long-term results of pectoralis major muscle transposition for infected sternotomy wounds. Ann Surg 1991; 213:583.
5. Serry C et al. Sternal wound complications. Management and results. J Thorac Cardiovasc Surg 1980; 80:861.
6. Wening JV et al. Repair of infected defects of the chest wall by transposition of the greater omentum. Br J Clin Pract 1990; 44:311.

Chapter 48

Chest Wall Defects

Jason Pomerantz and William Hoffman

Introduction

As is true for reconstructive surgery in general, successful repair of chest wall defects involves a sequence of steps, each of which must be accomplished in order to proceed appropriately and safely to the next step. The precise nature of the defect will dictate the particular reconstructive requirements and options. With respect to function, chest wall defects may require repair of the skeletal support system in addition to soft tissue coverage in order to restore normal respiratory mechanics.

Preoperative Considerations

The evaluation of a patient requiring chest wall reconstruction begins with an assessment of the overall medical and functional status. Many patients with chest wall defects have significant comorbid conditions. Prior to definitive defect reconstruction, the patient's neurological, cardiovascular, and respiratory function should be stable and optimized. Definitive wound closure is often deferred until the patient's operative risk is acceptable, and interim management may include mechanical ventilation, local wound care, debridement and wound VAC placement for partial-thickness defects. With respect to the timing of repair, defects should be repaired as soon as possible after major organ system stabilization in order to decrease the incidence of wound and other related complications. In addition to impact on survival, chest wall defects must be evaluated in terms of their effect on quality of life, taking into account the patient's functional status, motivation and prognosis.

Chest Wall Defect Analysis

The specific etiology of a chest wall defect will influence the subsequent reconstructive plan. The underlying cause of the defect can be classified into one the following: trauma, tumor, infection, radiation, or congenital. A number of specific considerations arise depending on the etiology of the defect. For example, in any trauma case it is important to fully evaluate concomitant injuries. Reconstructive surgery is performed after life-threatening problems are recognized and addressed. With respect to operative planning, in traumatic wounds it is imperative to consider the "zone of injury" which may extend well beyond apparent borders of viable tissue. Generally, only tissue outside of the zone of injury should be used for wound coverage. Infected or traumatic wounds may require extensive debridement, potentially resulting in a significant alteration of the operative plan. This should be anticipated before beginning the operation.

Irradiated tissue poses additional issues. Previously irradiated tissue is often fibrotic and may also be poorly vascularized. Such tissue is less amenable to mobiliza-

Practical Plastic Surgery, edited by Zol B. Kryger and Mark Sisco. ©2007 Landes Bioscience.

tion as opposed to healthy tissue and is generally not a good source of tissue for wound coverage. If future irradiation is planned at the site of the defect, then reconstructive options utilizing bulkier, better vascularized tissue (flaps as opposed to skin grafts) are preferred.

Defect analysis requires an accurate anatomical description of the site. The location of the defect on the chest wall will dictate the reconstructive options. The chest wall can be divided into four regions:

- The **anterior chest** is located (medial to lateral) between the parasternal line and anterior axillary line and (superior to inferior) between the clavicle and superior costal margin. Breast defects may occur along with chest wall defects in this region, and breast reconstruction should be considered in conjunction with planning of chest wall reconstruction.
- The **superior chest** is located (medial to lateral) between the base of the neck to acromioclavicular joint and (anterior to posterior) between the deltopectoral groove and the spine of the scapula. In this region, absence of the clavicle does not impair chest wall stability, but exposure of the subclavian vessels and brachial plexus may occur.
- The **lateral chest** is located (medial to lateral) between the anterior and posterior axillary lines and (superior to inferior) between the apex of the axilla and inferior costal margin. Wounds in this region may involve exposure of the axillary vessels or the musculocutaneous, median or ulnar nerves.
- The **posterior chest** is located (medial to lateral) between the posterior midline and the posterior axillary line and (superior to inferior) between the spine of the scapula to the posterior costal margin at L1. Central posterior chest wall defects (spinal wounds) and central anterior chest wall defects (sternal wounds) are discussed in separate chapters in this text.

After location, chest wall defects are described in terms of depth: **partial-** or **full-thickness**. Partial-thickness defects may involve loss of skin and varying amounts of subcutaneous fat and muscle. The wound base may contain muscle or bone. Full-thickness defects involve loss of soft tissue as well as bone. In full-thickness chest wall defects, the size of the bony defect in centimeters, as well as the number of consecutive ribs missing, dictates the requirement for restoration of chest wall stability because of exposure of vital organs as well as the potential for paradoxical chest wall motion during respiration. It is important to note that abnormal chest wall motion may not be obvious during mechanical ventilation, and observation of spontaneous, unassisted respirations is required.

Reconstructive Options

The goal of chest wall defect closure is to perform adequate debridement followed by safe wound coverage that restores form and function. In addition, intrathoracic dead space should be obliterated. Reconstructive options include direct wound closure, healing by secondary intention, skin grafting, local flap options, microsurgical tissue transfer and tissue expansion. Accurate defect analysis permits development of a suitable plan for wound coverage or reconstruction. For partial-thickness wounds, including those devoid of muscle at the wound base, skin grafting is a viable option if there is intact periosteum on any exposed bone. However, skin grafting is not ideal because of the resulting contour deformity and because it generally is more prone to breakdown. In addition, if radiation is planned or possible in the future, a muscle flap is indicated.

If a defect involves the full-thickness of the chest wall, a muscle flap is always indicated, and the decision must be made whether skeletal stability needs to be restored. In addition to exposing the thoracic viscera, defects encompassing at least four consecutive ribs and greater than 5 cm in size are generally thought to have a greater likelihood of involving a flail segment and in these cases skeletal stabilization is indicated. Often, muscle or musculocutaneous flaps will result in a stable chest wall without significant paradoxical motion. Fasciocutaneous or omental flaps do not provide adequate structural support and require simultaneous rib reconstruction. More extensive full-thickness defects require skeletal stabilization with autogenous bone or synthetic materials in addition to soft tissue coverage. Split rib grafts are a good option in elective clean cases. However, given the wealth of available synthetic materials, the most commonly used of which are prolene mesh or composite methyl methacrylate/marlex (used for larger defects), autogenous bone grafts are being used less frequently. Synthetic materials are sutured to intact bony structures on either side of the defect, followed by coverage with a well-vascularized flap.

Muscle flaps commonly used in chest wall reconstruction all have adequate arcs of rotation, major vascular pedicles and adequate bulk. Particular flaps are chosen based on defect location and size. Table 48.1 lists the commonly used muscle flaps in chest wall reconstruction. It is important to note any prior procedures in the region, (e.g., thoracotomy) that may have sacrificed the pedicle of an otherwise suitable flap. In addition to flap transposition, tissue expansion and microsurgical free flaps are options as well. Tissue expansion offers the advantage of providing similar skin characteristics and avoids donor site morbidity, but is contraindicated in cases involving chronic infection or radiation injury. If tissue expansion is chosen, the defect may be temporarily covered with a skin graft during the 2-3 month period of expansion. Free flaps are rarely needed for chest wall reconstruction unless there is associated injury to the pedicles of other flaps, and more distant flap transpositions do not provide adequate arcs of rotation. However, with improvement in microsurgical technique and outcomes, free tissue transfers are being used more frequently, especially for the largest and most complex cases. Typical free flaps for chest wall reconstruction include the contralateral latissimus dorsi, tensor fascia lata or rectus abdominis.

A general treatment algorithm for managing chest wall defects is presented in Figure 48.1.

Postoperative Considerations

Suction drains placed during surgery are important for preventing postoperative fluid accumulation at the defect or donor site. Any violation of the thoracic cavity usually requires a chest tube in the immediate postoperative period until there is no evidence of an air leak. Most surgeons continue antibiotic therapy while drains are in place in order to decrease the incidence of infection, especially if synthetic materials such as mesh or methyl methacrylate are used.

Chest wall defects, especially those involving rib resection can be very painful, and aggressive postoperative pain control is important for optimal wound healing. Depending on the location, an epidural may be beneficial. Intercostal blocks may also be useful in addition to the standard narcotics PCA and oral pain medications.

Table 48.1. Commonly used local flaps for chest wall defect repair

Muscle Flap	Pedicle	Origin	Insertion	Defect Site
Latissimus dorsi	Thoracordorsal A. post. perforators	Vertebral spines post iliac crest	Humerus	Anterior, posterior lateral, superior intrathoracic
Pectoralis major	Thoracoacromial A. IMA perforators	Sternum, ribs, clavicle	Humerus	Anterior, lateral
Trapezius	Transverse cervical A.	Occiput, ligamentum nuchae, vertebrae	Acromion, scapula, clavicle	Superior, posterior
Serratus anterior	Lateral thoracic A. thoracodorsal branches	Ribs 7-10	Medial scapula	Anterior, intrathoracic
Rectus abdominis	Superior epigastric A	Pubis symphisis	Costal cartilage 5-7	Anterior

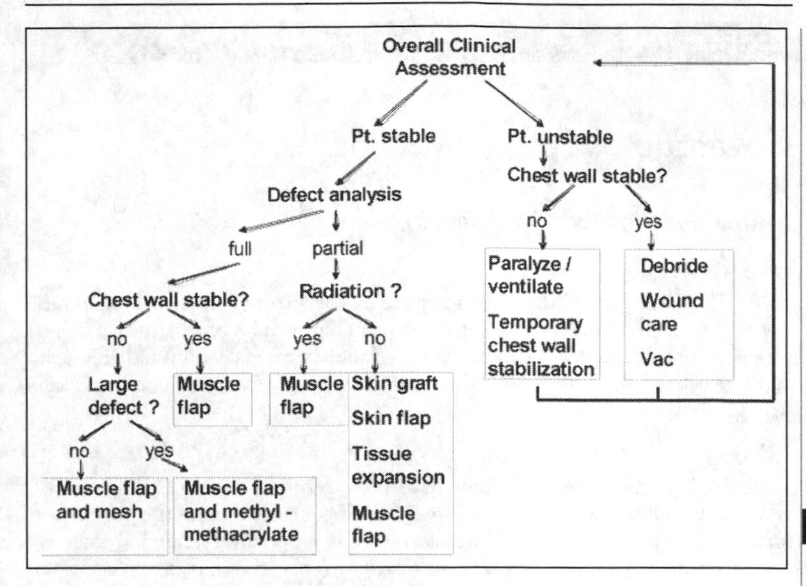

Figure 48.1. A management algorithm for the treatment of chest wall defects.

Pearls and Pitfalls

It is important to differentiate between traumatic chest wall wounds, usually the result of shotgun injuries, and surgical wounds due to resection of tumors or osteoradionecrosis. The trauma patient must be stabilized and serial debridements performed to establish the limits of viable tissue, while the surgical wound should be closed at the time of resection. It is useful to separate the reconstruction of the skeletal defect from the soft tissue problem; the former can almost always be addressed with polyethylene mesh and methyl methacrylate as a sandwich that is then sutured to adjacent remaining ribs. In smaller defects mesh alone can be used to stabilize the chest wall and support the overlying flap. Flap closure must be robust and airtight. The latissimus dorsi is really the workhorse of this region, either with a skin island or with an overlying skin graft. Intrathoracic defects, usually empyema cavities, pose a special challenge, requiring rib resection and control of bronchopleural fistulae for successful wound closure. Multiple muscle flaps may be needed in these cases.

Suggested Reading

1. Arnold PG, Pairolero PC. Chest-wall reconstruction: An account of 500 consecutive patients. Plast Reconstr Surg 1996; 98(5):804.
2. Chang RR, Mehrara BJ, Hu QY et al. Reconstruction of complex oncologic chest wall defects: A 10-year experience. Ann Plast Surg 2004; 52(5):471.
3. Cordeiro PG, Santamaria E, Hidalgo D. The role of microsurgery in reconstruction of oncologic chest wall defects. Plast Reconstr Surg 2001; 108(7):1924.
4. Losken A, Thourani VH, Carlson GW et al. A reconstructive algorithm for plastic surgery following extensive chest wall resection. Br J Plast Surg 2004; 57(4):295.
5. Mansour KA, Thourani VH, Losken A et al. Chest wall resections and reconstruction: A 25-year experience. Ann Thorac Surg 2002; 73(6):1720.
6. Mathes SJ. Chest wall reconstruction. Clin Plast Surg 1995; 22(1):187.

Chapter 49

Coverage of Spinal Wounds

Jason Pomerantz and William Hoffman

Overview

As with other parts of the body, adequate management of wounds in the posterior midline of the trunk is based on fundamental principles of reconstructive surgery. These include general patient assessment (multisystem evaluation) and thorough, accurate defect analysis that guide the subsequent plan for timely, durable and safe wound closure.

Etiology

Wounds in the posterior midline have a few common etiologies. Most spinal wounds in adults occur after spine surgery and frequently involve the presence of foreign material (hardware). Midline back wounds, especially postsurgical wounds, are often infected, adding increased complexity to their management. Aside from infection, other factors such as a history of irradiation, chronic illness, malnutrition, and use of systemic steroids all complicate the management of spinal wounds. Pressure is also a contributing factor either solely or in combination with surgical manipulation. Any postsurgical spinal wound complication, such as dehiscence or purulent drainage, should be explored and debrided as soon as it is noted. Other causes of spinal wounds include traumatic soft tissue injuries and congenital anomalies such as spina bifida comprise.

Spina bifida is the most common birth defect of the central nervous system and involves incomplete fusion of the vertebrae dorsally. Types of spina bifida include meningocele (meninges only in the defect), myelomeningocele (meninges and spinal cord), syringomyelocele (meninges and spinal cord with increased fluid and pressure in central canal) and myelocele (absence of epithelial covering of spinal cord). Myelomeningocele is the most common, and initial treatment involves coverage of the spinal cord and meninges with well vascularized soft tissue in order to preserve spinal cord function and prevent future neurological sequelae. In contrast to post-surgical spinal wounds, closure of myelemeningocele defects is often straightforward and accomplished with skin flaps. In more difficult cases, other reconstructive options including muscle flaps are required in a cooperative effort between neurosurgeons, pediatric surgeons and plastic surgeons. In addition, prenatal surgical treatment of spina bifida is under investigation and may ultimately provide the best long term outcome by virtue of early correction of the defect.

Defect Analysis

After the etiology is determined, a description of the defect is made. The location (superior, middle or inferior) is the most important for determining which muscle flaps are most suitable for coverage. The spine can be divided into thirds based on the original scheme described by Casas and Lewis. The superior region extends from C3 to T7, middle from T7 to L1 and inferior from L1 to S5. The size of the wound should be noted as well as the presence of infection, necrotic tissue, exposed hardware, bone or dura.

Management

A general treatment algorithm for treating spinal wounds is shown in Figure 49.1. As with defects in other regions of the body, definitive closure of spinal wounds should take place as soon as the patient is medically stable. The patient's nutritional status should be optimized prior to reconstruction. Adequate control of infection may require multiple debridements prior to wound closure. Treatment with systemic and local (beads) antibiotics is often implemented. Debridement in conjunction with the orthopedic or neurosurgery teams may include removal of spinous processes and various amounts of vertebral tissue in order to permit eradication of infection. Wounds should be cultured intraoperatively and antibiotic therapy tailored to treat specific organisms. During debridement, effort should be made to leave functioning hardware in place, as the ultimate goal is to achieve stable wound and hardware coverage, often attainable with the transposition of well-vascularized muscle flaps. The wound VAC offers a suitable option as a bridge to wound closure, either in systemically unstable patients or in situations that require multiple debridements. The VAC is discussed in detail in the "Wound VAC" chapter.

Closure of postsurgical spinal wounds usually requires muscle as well as skin. Table 49.1 lists some of the commonly used muscle flaps in spinal wound coverage. Healing by secondary intention, skin grafting or skin advancement are generally

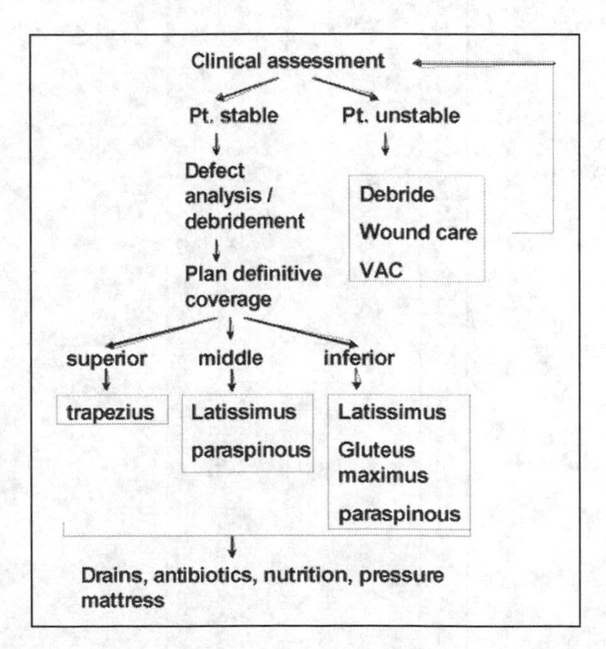

Figure 49.1. A treatment algorithm for spinal wounds.

Table 49.1. Commonly used local flaps for coverage of spinal wounds

Muscle flap	Pedicle	Origin	Insertion	Wound Site
Latissimus Dorsi	thoracordorsal A. post. perforators	vertebral spines post. iliac crest	humerus	upper, middle lower
Paraspinous	lumbar perforators	spinous process, Iliac crest	posterior-medial ribs	middle, lower
Trapezius	transverse cervical A.	occiput, ligamentum nuchae, vertebrae	acromion, scapula, clavicle	superior
Gluteus maximus	superior or inferior gluteal A.	sacrum, ileum	greater trochanter	lower
Omentum	left or right gastroepiploic A.	N/A	N/A	lower

considered inadequate for providing a durable repair, except in the most superficial wounds that do not contain hardware. Local muscle flaps that have adequate bulk and are well vascularized are available for all regions of the spine. Free tissue transfers are rarely required. In addition, there are few convenient recipient vessels in this region, and free tissue transfers often require pedicle extension with vein grafts.

Middle and Inferior Spinal Wounds

Several muscle flaps are used most frequently for coverage of spinal wounds. Flap choice is usually based on location of the wound. The **paraspinous muscle flap** can be used in the middle and inferior regions. This flap avoids a significant donor site defect and is based on posterior perforating vessels. The paraspinous flap is mobilized by releasing the fascia laterally, on either side of the wound. The muscle and fascia are then turned over or advanced into the midline to cover the wound and exposed hardware. Healthy skin flaps are advanced and closed on top of the muscle. Potential disadvantages of the paraspinous flap are that the muscle is often small and may have been partially debrided because of proximity to the wound. Furthermore, there is sometimes a need to debride overlying skin, making the skin closure difficult. In these cases, a skin graft may be required. Inferior spinal wounds, including those overlying the sacrum, can be covered with the **gluteus muscle flap** based on the superior and/or inferior gluteal vessels. Alternatively, the superior gluteal artery perforator (SGAP) flap can be used in order to minimize donor site morbidity in ambulatory patients.

Superior Spinal Wounds

The paraspinous muscles are particularly small and difficult to mobilize in the superior part of the back. In the superior region, the **trapezius muscle flap** is a good option. A skin paddle is designed over the distal portion of the trapezius, and the muscle is elevated off the deeper tissues and then divided lateral to its pedicle. The flap is then rotated into the wound. The donor site can usually be primarily closed. Division of the superior portion of the trapezius should be avoided to prevent postoperative shoulder drop. The **latissimus dorsi flap** is a commonly used muscle flap for spinal wound closure, appropriate for all regions of the spine. The latissimus can be used with a skin paddle or covered with a skin graft. This muscle can be rotated or advanced, based on the thoracodorsal pedicle. Alternatively, the latissimus flap may be based on posterior perforators after release from its attachment to the humerus and division of the thoracodorsal pedicle. This large muscle can be successfully transposed to cover most spinal wounds.

Finally, the **omental pedicle flap** has been used for coverage of spinal wounds, but is not a first choice given the availability of the muscle flaps described above, as well as the necessary creation of a lumbar hernia in order to pass the omentum from the abdominal cavity to the back.

Postoperative Considerations

Suction drains are usually placed at the time of surgery to prevent seroma formation at both the donor and recipient sites. These are left in place for variable periods of time depending on drain output as well as the surgeon's preference. As a general rule, drains should not be removed until their output is almost zero.

Excess pressure on the flap is an important issue postoperatively, as many spinal wound patients are bedridden. Low pressure mattresses should be used to avoid flap ischemia. On the other hand, some surgeons prefer some mild pressure in line with the thinking that it decreases seroma formation. Usually, patients are kept prone or in the lateral decubitus position for an extended period of time (up to 3 weeks).

For infected spinal wounds, especially those with hardware in place, a prolonged course of IV antibiotics is generally used. Although not well studied for spinal wounds in particular, 6 weeks of organism-specific IV antibiotics after debridement and muscle flap coverage is a common practice, in accord with the antibiotic treatment of osteomyelitis in other regions of the body.

Pearls and Pitfalls

In cases involving bone, often with exposed hardware, muscle coverage is required to eliminate dead space and to combat infection. The most difficult area of the spine for obtaining muscle coverage is the lumbosacral region; the gluteus muscle is below and the latissimus above but neither reaches this area easily, and the paraspinous muscles end at this level as well. The gluteus remains the best choice for this area.

Hardware is always a difficult issue. In today's complex spine cases, the hardware may be necessary to stabilize the spine after scoliosis repair or multiple level fusions. Any loose hardware must be removed and/or replaced. If there is severe infection (gross purulence) and loose hardware present, a body cast may be required while the wound heals. In some cases we have actually kept patients on antibiotics until some degree of fusion and spine stability has been achieved and the hardware can be removed.

Functional deficits must be considered in these cases when muscle flaps are used. In particular, the latissimus muscle may be important in using crutches or a wheelchair, and this fact argues against its use in myelomeningocele patients as well as spinal patients with lower extremity weakness. The gluteus muscle is used commonly in paraplegic patients for coverage of the lower lumbar spine and sacrum but may cause a problem with ascending stairs in an ambulating patient.

Suggested Reading

1. Casas LA, Lewis Jr VL. A reliable approach to the closure of large acquired midline defects of the back. Plast Reconstr Surg 1989; 84(4):632.
2. Dumanian GA, Ondra SL, Liu J et al. Muscle flap salvage of spine wounds with soft tissue defects or infection. Spine 2003; 28(11):1203.
3. Foster RD, Anthony JP, Mathes SJ et al. Flap selection as a determinant of success in pressure sore coverage. Arch Surg 1997; 132(8):868.
4. Hochberg J, Ardenghy M, Yuen J et al. Muscle and musculocutaneous flap coverage of exposed spinal fusion devices. Plast Reconstr Surg 1998; 102(2):385.
5. Manstein ME, Manstein CH, Manstein G. Paraspinous muscle flaps. Ann Plast Surg 1998; 40(5) :458.
6. Mathes SJ, Stevenson TR. Reconstruction of posterior neck and skull with vertical trapezius musculocutaneous flap. Am J Surg 1988; 156(4) :248.
7. Ramasastry SS, Schlechter B, Cohen M. Reconstruction of posterior trunk defects. Clin Plast Surg 1995; 22(1):167.
8. Stahl RS, Burstein FD, Lieponis JV et al. Extensive wounds of the spine: A comprehensive approach to debridement and reconstruction. Plast Reconstr Surg 1990; 85(5):747.

Chapter 50

Abdominal Wall Defects

Mark Sisco and Gregory A. Dumanian

Background

Acquired abdominal wall defects can occur as the result of postoperative wound complications, trauma, or surgical resection. Several options exist to manage each of these scenarios. Congenital defects comprise a distinct subset of abdominal wall defects; their treatment is the purview of pediatric surgeons and is not covered here. The objectives of abdominal wall reconstruction, in order of priority, are protection of the abdominal contents, the restoration of visceral support and the production of a natural body contour. Knowledge of abdominal wall anatomy as it relates to the defect is paramount in planning appropriate management. Reconstruction of the myofascial layer restores visceral support and structural stability. Reconstruction of the cutaneous layer provides wound closure and provides an aesthetic outcome.

Preoperative Considerations

Relative vs. Absolute Defects

The plastic surgeon is typically consulted when closure of an abdominal defect cannot be accomplished primarily. Relative defects arise when an increase in the size of the viscera or contraction of the soft tissues of the abdominal wall precludes wound closure. Absolute defects result from loss of soft tissues, as in trauma, necrotizing infection, or tumor resection. Absolute defects are more difficult to treat and are more likely to require flap mobilization.

Layers of the Abdominal Wall

Defects involving the abdominal wall may affect the skin and subcutaneous tissue or the myofascia. A defect of the skin and subcutaneous tissue leads to an open wound. A myofascial defect leads to loss of structural support for the viscera. Finding a suitable surgical solution depends on an analysis of which component is missing and the most straightforward means of replacement or repair.

Location the Defect

Several classifications of abdominal wall defects have been proposed. Mathes et al have devised a classification of abdominal wall defects based on location.

Midline defects with a myofascial deficiency, such as incisional hernias, are usually amenable to meshes, grafts, or the components separation technique, which uses advancement of adjacent myofascial tissue.

Lateral defects, often due to extirpation or trauma, are best treated with local and distant flaps based on the size of the defect.

Practical Plastic Surgery, edited by Zol B. Kryger and Mark Sisco. ©2007 Landes Bioscience.

Timing of Wound Closure

Emergent Wound Closure

In critically-ill patients such as those who have undergone damage-control laparotomies for trauma, reconstruction of the abdominal wall may become an issue when edematous intraabdominal contents prohibit wound closure. In these cases, the priority is to restore the protective function of the abdominal wall such that the viscera are protected from dessication and evisceration. Temporary closure methods are suitable because of the likelihood that such patients may need to be reexplored and because prompt resolution of the edema is the rule. Such efforts are usually performed by the trauma surgeon. These methods include sewing nonabsorbable materials, such as IV bags, into the defect and using various improvised and commercial vacuum devices.

Urgent Wound Closure

In unstable patients or in those with acute wound failure complicated by gross contamination or bacterial infection, urgent wound closure is indicated. The primary goal in such situations is to prevent evisceration. Large defects may be simply and quickly closed with absorbable mesh which is then covered with saline-soaked dressings. Definitive reconstruction in such cases is delayed and performed as a second-stage operation. In stable patients, definitive reconstruction with autologous tissue may be considered at that time.

Definitive Wound Closure

When possible, definitive reconstruction should be performed in a single stage. Definitive closure may also be performed following procedures such as those described above, when the patient is stable and the wound is more amenable to reconstructive intervention. When possible, most authors suggest a delay of at least six months following the initial surgery to reduce the likelihood of iatrogenic bowel injury. The key to effective reconstruction of the abdominal wound is tension-free closure at the healing edge. Primary closure, mesh and graft reconstruction, component separation, locoregional flap reconstruction and distant flap reconstruction represent four approaches to the increasingly complex abdominal wound.

Simple Skin Defects

Simple skin defects, when small, may be closed with local skin elevation and primary closure. Larger skin defects can be treated with a negative pressure wound therapy (NPWT) device, skin grafts, local skin flaps, or fasciocutaneous flaps. Local skin flaps include the groin flap, superficial inferior epigastric artery flap, iliolumbar flap, or thoracoepigastric flap. Alternatively, tissue may be expanded while a skin graft or NPWT device is in place.

Musculofascial Defects

Musculofascial defects with intact skin coverage are synonymous with ventral hernias. Primary repair may be attempted for small (<3 cm) defects; however this may be associated with a high recurrence rate. Medium and large defects should be reconstructed with grafts that have sufficient tensile strength to restore the structural integrity of the abdominal wall. They are treated with prosthetic materials or autologous tissue such as fascial grafts, local flaps, or advancement flaps.

Open Wounds

Open wounds with exposure of the viscera can be closed when possible by elevating and closing the skin, which allows for subsequent repair as a myofascial defect. When primary closure of the skin is impossible, patients can be reconstructed immediately using local flaps in combination with skin grafts, distant flaps, combinations of prosthetic materials and flaps, or free tissue transfer. Unstable patients with large open wounds can be managed by placing an absorbable mesh and dressing changes.

Mesh and Graft Reconstruction

Mesh is often used for the repair of ventral hernias in which a pure fascial deficit exists. A variety of synthetic materials have been used to reconstruct the fascia successfully. Other materials such as acellular dermal matrix (AlloDerm®) and sodium hyaluronate and carboxymethylcellulose coated mesh (Sepramesh®) are also in early clinical use.

Permanent mesh is most commonly used to close the abdomen. Since there are risks of infection, extrusion and fistula, permanent meshes should be used in clean wounds in which there is enough soft tissue to provide wound coverage. Permanent meshes include polypropylene, polytetrafluoroethylene (PTFE) and polyester. Polypropylene (Prolene®, Marlex®). It has excellent strength at the suture line but may be associated with adhesions. PTFE (Gore-Tex®) demonstrates less in-growth of surrounding tissue than polypropylene, which confers a theoretical increased risk of hernia recurrence since there is less strength at the suture line. Polyester mesh (Mersilene®) is strong and pliable. It causes a significant inflammatory response in surrounding tissues.

As described above, polyglactin (Vicryl®, Dexon®) absorbable meshes can be used when urgent wound closure is required and in the settings of wound contamination/sepsis or bowel fistula. This mesh provides a base for granulation tissue formation and is retained until infection and critical illness have resolved. The intestines adhere to each other and to the fascial edges, becoming "frozen." The mesh is then removed and skin grafts are placed. The use of absorbable mesh virtually guarantees the formation of a large ventral hernia and requires second-stage reconstruction, usually six months later.

Nonvascularized fascial grafts have excellent incorporation. As autologous tissue, they are better suited to contaminated wounds than permanent mesh. However, fascial grafts have less tensile strength than PTFE or polypropylene and require creation of a donor defect. They are associated with a significant rate of dehiscence and recurrence.

Flap Reconstruction

Locoregional flaps can replace both the fascial and cutaneous components of the abdominal wall in large open wounds. The rectus abdominis muscle flap is the most commonly used flap in abdominal wall reconstruction. It can be used with or without a skin paddle to reconstruct any area of the abdominal wall, especially in the middle and lower thirds. It can be based on the superior epigastric or the deep inferior epigastric arteries. The tensor fascia lata (TFL) myocutaneous flap, based on the lateral circumflex artery, is useful to reconstruct defects inferior to the umbilicus. The rectus femoris can be used to cover defects involving the lower quadrants and the umbilical and the epigastric areas. Table 50.1 summarizes the local and distant flaps commonly used for abdominal wall reconstruction.

Table 50.1. Local and distant flaps commonly used for abdominal wall reconstruction

Local Flaps	Arc of Rotation	Notes
Rectus abdominus pedicle	Entire abdomen	May be based on the superior or inferior pedicles Can use skin paddle
Component separation (separation of parts)	Midline defects	Reconstruction of choice for large midline myofascial defects
External oblique	Middle and upper thirds	Limited arc
Internal oblique	Lower third and groin	Technically difficult
Distant Flaps		
Tensor fascia lata	Middle and lower thirds	Donor site may require skin graft
Latissimus dorsi	Upper third	Excellent for upper lateral defects Requires harvesting of the pregluteal fascia
Rectus femoris	Middle and lower thirds	Good in chronically infected wounds May be associated with weak knee extension
Vastus lateralis	Lower third	Limited by lack of fascial component
Gracilis	Lower third	Small

Component Separation

The components separation procedure, also known as the "separation of parts," involves the longitudinal release of the medial edge of the external oblique aponeurosis and occasionally the release of the posterior rectus fascia. This procedure is especially useful for closing large midline musculofascial defects up to 30 cm transversely. It can also be performed as a salvage procedure when defects recur or prosthetic materials become exposed or infected. It allows medial advancement of the anterior rectus fascia and the internal oblique and provides dynamic support of the abdominal wall.

After entering the abdomen, dissection along the semilunar lines if achieved via 6 cm transverse incisions located at the inferior aspect of the rib cage. These lateral incisions allow for preservation of the abdominal skin blood flow. The muscle and fascia of the external oblique are then incised 1 cm lateral to the semilunar line. This incision is extended superiorly to the costal margin and inferiorly to the inguinal ligament. This release allows for approximation of the rectus sheaths at the midline. Drains are left at each semilunar line and at the midline.

Postoperative Considerations

Complications related to abdominal wall reconstruction are manifold and depend partly on the indication. Urgent abdominal wall reconstruction using simple cutaneous flaps or absorbable mesh are associated with a high rate of ventral hernia formation. Repair of ventral hernias with permanent mesh may be complicated by mesh extrusion, infection, enterocutaneous fistula and hernia recurrence.

Pearls and Pitfalls

1. Think of abdominal wall reconstruction as the solution to two related problems: How to deal with the structural support and how to deal with the skin.
2. Preservation of skin blood flow decreases local wound complications.
3. Often, complex abdominal wounds with infected mesh and fistulae can be excised "en bloc", allowing for the reconstruction to be performed with noninflamed tissues. In such cases, the best repair discards the most tissue.
4. Leaving the skin open to be managed with NPWT after repair of the abdominal wall is an effective management strategy in the obese patient with contaminated wounds.

Suggested Reading

1. Mathes SJ, Steinwald PM, Foster RD et al. Complex abdominal wall reconstruction: a comparison of flap and mesh closure. Ann Surg 2000; 232(4):586-96.
2. Ramirez OM, Ruas E, Dellon AL."Components separation" method for closure of abdominal-wall defects: an anatomic and clinical study. Plast Reconstr Surg 1990; 86(3):519-26.
3. Rohrich RJ, Lowe JB, Hackney FL et al. An algorithm for abdominal wall reconstruction. Plast Reconstr Surg 2000; 105(1):202-16.
4. Stone HH, Fabian TC, Turkleson ML, Jurkiewicz MJ. Management of acute full-thickness losses of the abdominal wall. Ann Surg 1981; 193(5):612-8.
5. Szczerba SR, Dumanian GA. Definitive surgical treatment of infected or exposed ventral hernia mesh. Ann Surg 2003; 237(3):437-41.

Chapter 51

Pelvic, Genital and Perineal Reconstruction

Mark Sisco and Gregory A. Dumanian

Preoperative Considerations

The ideal reconstruction of any major pelvic defect includes reestablishment of the pelvic floor, obliteration of pelvic dead space, closure of the perineum and recreation of the genitalia while maintaining ambulatory function. As with any reconstructive effort, the etiology, size and anatomic features of the defect determine the surgical options available. Separate procedures may be necessary to satisfy each of these imperatives. Patient comorbidities and expectations also play a major role in deciding the extent of reconstruction to pursue.

In defects associated with malignancy, the use of pre- and postoperative radiation should be considered. Preoperative radiation often presents the surgeon with tissues that are scarred and relatively ischemic, making the use of locoregional and distant flaps more likely to be necessary. Reconstruction of the pelvic floor restores structural support to the viscera to prevent herniation and injury. In patients who are likely to receive adjuvant radiation therapy, exclusion of the intestines from the pelvis is especially important to prevent radiation enteritis.

Soft Tissue Defects of the Genitalia

Reconstruction of soft tissue defects of the perineum, including the vagina, penis and scrotum, is guided by the extent of tissue loss and by the functional and cosmetic goals of the patient.

Defects of the Vulvo-Perineal Surface and Vagina

Defects resulting from superficial skin cancer excisions, such as Bowen's disease, can often be treated with a skin graft alone, provided that a satisfactory, nonirradiated donor bed exists. Split-thickness skin grafts (STSG) are especially useful when margin status is questionable or reexcision is likely. In the setting of prior radiation therapy, locoregional skin flaps, such as rhomboid flaps, may be used for superficial defects. Small defects may be covered with flaps from the lateral and posterior perineum, where skin is most lax. Larger defects may require larger local flaps, such as the gracilis myocutaneous flap or the pedicled groin flap. When reconstructing the vagina, it is important to provide a solution that is durable and prevents herniation of intestines into the pelvis. The most commonly used flap for this purpose is the rectus abdominus myocutaneous flap. Options for flap reconstruction of the vagina are presented in Table 51.1.

Defects of the Scrotum and Penile Skin

The scrotum has a robust blood supply and is highly elastic. As such, large scrotal flaps can be used to reconstruct most defects. Complete scrotal loss can be treated with STSGs. Subcutaneous medial thigh pouches should be used as cover for the

Table 51.1. Options for vaginal reconstruction

Flap	Notes
Oblique rectus abdominis myocutaneous (ORAM)	Good for total and partial reconstruction Fills dead space Insertion does not usually need to be divided Elliptical skin paddle is tubularized to create pouch and delivered transpelvically to the perineum
Gracilis myocutaneous	Useful for partial defects in small women; bilateral flaps necessary for complete reconstruction Tunneled proximally into the defect Skin islands sewn together edge-to-edge
Posterior thigh fasciocutaneous	Useful when large amounts of skin are necessary and rectus is unavailable May be tunneled or the skin bridge can be transected Remember to divide posterior femoral cutaneous nerve
Omentum/STSG	Useful in obese women or small pelvis where other flaps are too bulky Flap secures at the introitus and at the pelvic inlet Remaining omentum is sutured to the pelvic inlet STSG is held in place by a vaginal stent

testicles. In the interim, they may be replaced back into the neoscrotum to maintain spermatogenesis. Skin loss on the penis can be managed using a nonexpanded meshed STSG, being sure to tumesce the penis prior to grafting in order to prevent constriction during subsequent erection. The graft is arranged such that the mesh slits are transversely oriented. The suture lines between grafts should be made obliquely or in a zigzag pattern to prevent unidirectional scar contraction.

Defects of the Penis

The three aims of penile reconstruction are to create the ability to urinate normally, restore sexual function and achieve acceptable cosmesis. In order to meet these demands, the reconstructed phallus must have sufficient rigidity and sensation. The **radial forearm** free flap is a versatile flap that can be used for this purpose. The antebrachial cutaneous nerve can be anastomosed to the pudendal nerve to achieve sensation. The "cricket bat" phalloplasty employs a narrow "handle" of skin to recreate the urethra inside of the attached "blade" of skin that recreates the phallus itself. The use of a prosthesis is required to attain rigidity; this is best inserted as a second procedure, after protective sensation has developed. The **free fibula** osteocutaneous flap, based on the perineal artery, has the potential advantage of rigidity without a prosthesis. The lateral sural cutaneous nerve is used to achieve sensation. Urethral reconstruction is performed using skin grafts.

Reconstruction of the Pelvic Floor and Perineum

Pelvic defects may involve the pelvic floor, the perineum, or both. The choice of reconstruction depends on the relative involvement of each of these structures. Three solutions for flap reconstruction of these defects are presented in Table 51.2.

When a pelvic floor defect exists with minimal perineal soft tissue loss, the omentum or mesh can be used. The omentum brings with it excellent blood flow and immunologic properties. It can fill small cavities in the pelvis but may be too elastic to

Table 51.2. Options for pelvic floor and perineal defects

Flap	Pedicle	Advantage	Disadvantage	Application
Gracilis myocutaneous	Medial femoral circumflex a.	Cutaneous segment can resurface vaginal wall or central perineum	Vascular spasm may cause necrosis, especially skin paddle	Low intrapelvic defects Small perineal defects Suprapubic area Groin Vagina
ORAM	Deep inferior epigastric a.	Large local flap available for pelvic and perineal reconstruction	Requires laparotomy for lower pelvic inset Partial harvest of rectus muscle	Iliac crest Perineum Groin Internal pelvic defects, including floor
Posterior thigh	Inferior gluteal a.	Less bulky than gracilis Can incorporate posterior cutaneous nerve of the thigh for sensation	No muscle component	Perineal skin

provide durable suspension of the viscera. The rectus abdominis muscle flap or de-epithelialized posterior thigh flaps can also be used for pelvic defects in this setting.

If a large perineal defect coexists with loss of the pelvic floor, bulky flaps such as the oblique rectus abdominis myocutaneous (ORAM) flap can be used to restore the pelvic floor while filling pelvic dead space and providing a large skin island. The ORAM, based on the deep inferior epigastric artery, employs a skin flap based on periumbilical perforators that extends superolaterally to the anterior axillary line toward the tip of the ipsilateral scapula. The skin paddle can be up to 7 cm wide.

Finally, free tissue transfer using large flaps such as the anterolateral thigh flap may be necessary for extremely large defects or patients who have few local flap options because of radiation or prior surgeries. Vein grafts to the vascular pedicle are often required.

Pearls and Pitfalls

- When adjuvant radiation therapy is planned, it is important to exclude the small bowel from the pelvis in order to prevent radiation enteritis. The ORAM flap is an excellent solution as it is a large soft tissue flap that can be raised with little muscle loss.
- When an abdomino-perineal resection is planned, the right ORAM can be harvested through a paramedian incision which is then used to access the abdomen. The ostomy can then be brought through the left rectus muscle. This approach is especially useful in patients who have undergone preoperative radiation and/or chemotherapy.
- The gracilis muscle flap is especially useful for obliterating draining perineal sinuses. Its small size precludes it from filling large amounts of dead space.
- We have noted preliminary success using pedicled inferior gluteal artery perforator (IGAP) flaps to fill large pelvic defects from below.

Suggested Reading

1. Achauer BM, Braly P, Berman ML et al. Immediate vaginal reconstruction following resection for malignancy using the gluteal thigh flap. Gynecol Oncol 1984; 19(1):79-89.
2. Black PC, Friedrich JB, Engrav LH et al. Meshed unexpanded split-thickness skin grafting for reconstruction of penile skin loss. J Urol 2004; 172(3):976-9.
3. Friedman J, Dinh T, Potochny J. Reconstruction of the perineum. Semin Surg Oncol 2000; 19(3):282-93.
4. Giampapa V, Keller A, Shaw WW et al. Pelvic floor reconstruction using the rectus abdominis muscle flap. Ann Plast Surg 1984; 13(1):56-9.
5. Kusiak JF, Rosenblum NG. Neovaginal reconstruction after exenteration using an omental flap and split-thickness skin graft. Plast Reconstr Surg 1996; 97(4):775-81.
6. Lee MJ, Dumanian GA. The oblique rectus abdominis musculocutaneous flap: Revisited clinical applications. Plast Reconstr Surg 2004; 114(2):367-73.
7. McCraw JB, Massey FM, Shanklin KD et al. Vaginal reconstruction with gracilis myocutaneous flaps. Plast Reconstr Surg 1976; 58(2):176-83.
8. Semple JL, Boyd JB, Farrow GA et al. The "cricket bat" flap: A one-stage free forearm flap phalloplasty. Plast Reconstr Surg 1991; 88(3):514-9.
9. Sengezer M, Ozturk S, Deveci M et al. Long-term follow-up of total penile reconstruction with sensate osteocutaneous free fibula flap in 18 biological male patients. Plast Reconstr Surg 2004; 114(2):439-50.
10. Tobin GR, Pursell SH, Day Jr TG. Refinements in vaginal reconstruction using rectus abdominis flaps. Clin Plast Surg 1990; 17(4):705-12.
11. Vincent MP, Horton CE, Devine Jr CJ. An evaluation of skin grafts for reconstruction of the penis and scrotum. Clin Plast Surg 1988; 15(3):411-24.

Chapter 52

Lower Extremity Reconstruction

Mark Sisco and Michael A. Howard

Introduction

The etiologies of wounds in the lower extremity include trauma, oncologic resection, diabetes, radiation, peripheral vascular disease and chronic osteomyelitis. The relative paucity of soft tissue that is available for reconstruction requires a creative approach to filling various deficits. The distal leg in particular can be problematic due to its lack of tissue elasticity, the need for a durable reconstruction, and the frequent presence of neuropathy, ischemia, osteomyelitis and edema. The advent of free-tissue transfer and the increased familiarity with local flaps have revolutionized the treatment of difficult lower extremity wounds.

The evaluation of defects of the lower extremity must include a determination of what tissue and structures are missing and an evaluation of the potential functional outcome. Treatment must then be geared toward achieving this outcome. As such, initial approaches to management may include a recommendation for amputation when overwhelming injury has occurred or jumping to the top of the reconstructive ladder with the immediate use of free-tissue transfer when other options are insufficient.

Preoperative Considerations

Compared to the upper extremity, the leg has relatively simple functions: it must bear weight and contribute to ambulation. Function can often be restored by achieving wound closure, recreating structural stability and preserving basic motor and sensory function.

The arterial blood supply to the distal leg is provided by the popliteal artery, which is a continuation of the superficial femoral artery. Below the knee, the popliteal artery trifurcates into the anterior tibial, posterior tibial and peroneal arteries. The anterior tibial artery (which becomes the dorsalis pedis) and the posterior tibial artery cross the ankle. In general, the anterior tibial artery supplies the dorsum and distal foot; the posterior tibial artery feeds the heel and posterior plantar foot; and the peroneal artery branches supply the lateral foot and ankle and the calcaneus.

In acute trauma, attempts for limb salvage should be guided by the likelihood for survival and functional recovery. Long-term extremity function is poor in patients with sciatic or posterior tibial nerve injury. Scoring systems such as the Mangled Extremity Severity Score (MESS) may be helpful (Table 52.1). Amputation should be considered in any patient with a MESS score greater than 7.

Compartment Sydrome

Lower extremity compartment syndrome occurs when increased soft tissue pressure within an enclosed fascial space causes ischemic necrosis of nerves and muscles.

Practical Plastic Surgery, edited by Zol B. Kryger and Mark Sisco. ©2007 Landes Bioscience.

Table 52.1. The Mangled Extremity Score (MESS)

Skeletal and soft tissue injury	Low energy	1
	Medium energy (open fractures)	2
	High energy (military gunshot wound)	3
	Very high energy (gross contamination)	4
Limb ischemia (double score for > 6 h ischemia)	Near-normal	1
	Pulseless, decreased capillary refill	2
	Cool, insensate, paralyzed	3
Shock	Systolic BP always > 90 mm Hg	0
	Transient hypotension	1
	Persistent hypotension	2
Age (y)	< 30	0
	30-50	1
	> 50	2

Note: Amputation should be considered for any patient with a score above 7.

This increased pressure is usually due to direct local trauma or ischemia and reperfusion seen with proximal vascular occlusion. Compartment syndrome occurs in up to 10% of open tibial fractures.

There are four fascial compartments in the leg: anterior, lateral, superficial posterior and deep posterior (Table 52.2). A high index of suspicion is the single most important factor in recognition of compartment syndrome, especially in the obtunded patient.

In awake patients, the classic "5 Ps," pain, paresthesia, paresis, pressure and pulselessness suggest the diagnosis. Pain out of proportion to the injury is probably the most useful symptom; paresthesias and pulselessness are usually late findings. Measurement of compartment pressures is more accurate than the history and physical exam. A compartment pressure greater than 30-35 mm Hg or a pressure within 30 mm Hg of the diastolic blood pressure are indications for fasciotomy. Clinical suspicion may also warrant fasciotomy even in the setting of a normal compartment pressure.

Fasciotomy is performed using two incisions, each of which releases two compartments. The anterior and lateral compartments are released using a longitudinal incision over the anterior intramuscular septum. A second incision posteromedially releases the superficial and deep posterior compartments.

Table 52.2. The compartments of the leg

Comparment	Muscles	Nerve	Vessels
Anterior	Extensor digitorum longus Extensor hallicus longus Anteroior tibialis	Deep peroneal	Anterior tibial a.
Lateral	Peroneus longus Peroneus brevis	Superficial peroneal	
Superficial posterior	Gastrocnemius Soleus	Tibial	
Deep posterior	Flexor hallicus longus Tibialis posterior Flexor digitorum longus	Tibial	Peroneal a. & v.

Nerve Injuries

Nerve injuries in the lower extremity are often more challenging than in the upper extremity because of the force involved in lower extremity trauma and the fact that most injuries are relatively proximal. When a complete motor and sensory deficit exists, there is likely to be complete transection or severe traction injury. Partial neurologic deficits related to penetrating trauma suggest partial transection. Progressive neurologic dysfunction, meanwhile, may indicate the presence of compartment syndrome, ischemia, or an expanding hematoma. Exploration should be considered in any of these cases. Partial deficits related to blunt trauma may improve spontaneously and may warrant a delay in exploration. Immediate repair should be performed when a sharp transection of the nerve is found without major contamination or nearby soft tissue injury. Traction and crush injuries may benefit from the use of vascularized nerve grafts.

Reconstructive Options

The operative approach to the traumatized extremity should proceed in the following order:

- Examination (often under anesthesia)
- Debridement
- Fracture fixation
- Vascular repair
- Tendon repair
- Nerve repair
- Wound closure (temporary vs. definitive)

A wide array of flaps has been described for reconstruction of the lower extremity. These include local flaps, pedicle flaps (Table 52.3) and free tissue transfer (Table 52.4).

Hip and Proximal Femur

The thigh can typically be reconstructed using local or pedicle flaps. Common workhorses for thigh reconstruction include the vastus lateralis and gracilis. Fasciocutaneous flaps in the region include tensor fascia lata, medial thigh, rectus femoris and lateral posterior thigh. Finally, the inferior-based rectus abdominis with our without a skin paddle can reach the thigh.

Knee

The medial or lateral gastrocnemius flap is useful for coverage of ipsilateral and inferior defects around the knee. Other pedicle flap options include the distally-based vastus lateralis or vastus medialis. A wide variety of fasciocutaneous flaps, including the saphenous, sural, V-Y retroposition, posterior thigh and lateral genicular flap may be used for small defects. In some cases, free-tissue transfer may be warranted.

Proximal Tibia

The medial or lateral gastrocnemius flaps are most commonly used to reconstruct soft tissues about the proximal tibia. Alternatively, the saphenous or lateral/medial fasciocutaneous flaps may be used alone or in combination with the gastrocnemius muscle flap.

Middle Tibia

Free flaps are commonly necessary to cover defects of the middle tibia. The soleus muscle flap can be used to cover a variety of wounds in the middle tibia. Smaller

Table 52.3. Pedicle flaps for leg reconstruction

Location	Flap	Pedicle	Comments
Thigh	Vastus lateralis	Lateral femoral circumflex a.	Good for trochanteric and femoral head defects
	Rectus abdominis muscle	Deep inferior epigastric a.	Can be used with skin paddle Can reach knee if extended approach used
	Gracilis muscle	Medial femoral circumflex a.	Good for smaller defects
	Tensor fascia lata	Transverse branch of lateral femoral circumflex a.	May cause lateral knee instability May need delay procedure to ensure distal flap survival
Knee	Gastrocnemius	Medial and lateral sural a.	Take care to preserve the peroneal n. when mobilizing lateral head Medial head usually larger
	Vastus lateralis (distally-based)	Superficial branch of lateral superior genicular a.	Using mid-muscle belly only Good for popliteal fossa
	Vastus medialis	Minor pedicle from SFA	Upper knee
Proximal tibia	Gastrocnemius	Medial and lateral sural a.	Medial or lateral head usually sufficient
Middle tibia	Soleus	Posterior tibial a. perforators Peroneal a. perforators	Good for small wounds without significant dead space
Distal tibia	Dorsalis pedis fasciocutaneous	Dorsalis pedis perforators Posterior tibial a. perforators	Donor site morbidity limits utility
	Soleus	Peroneal a. perforators	

Table 52.4. Free flaps used for leg reconstruction

Flap	Pedicle	Comments
Gracilis	Medial femoral circumflex a.	Good for small areas; round muscle may be and teased out flattened
Rectus abdominis muscle	Deep inferior epigastric a.	Can be used with skin paddle Can reach knee if extended approach used
Latissimus dorsi	Thoracodorsal a.	Large
Fibula	Peroneal a.	Good source of vascularized bone
Tensor fascia lata	Transverse branch of lateral femoral circumflex a.	Good for large defects Lateral femoral cutaneous may allow sensory reinnervation

defects may be covered using a turnover flap of the anterior tibialis muscle. Again, fasciocutaneous flaps may also be useful.

Distal Tibia

Free flaps are commonly used in the distal fibia. The dorsalis pedis fasciocutaneous flap can cover relatively large defects about the distal tibia. The lateral supramalleolar flap and the extensor brevis flap may be used to cover somewhat smaller defects. Distally-based fasciocutaneous flaps may also be used.

Foot

Extensive wounds of the foot may require free-tissue transfer for closure (Table 52.5). The sural artery flap may be of use for wounds about the ankle and heel. Plantar defects in ambulatory patients must be reconstructed with tissue that is durable enough to bear weight and shear forces. Underlying bony abnormalities that might create pressure points should be corrected.

Replantation may be considered when there is a single simple, transection without associated blunt injury and the warm ischemia time is less than 6 hours. It is most suited to healthy, young individuals without significant comorbidities. Lower extremity replantation is relatively rare due to several factors including: the likelihood of significant concomitant injury; the difficulty in achieving useful neurologic function; and the widespread acceptance and utility of prosthetics.

Conclusions

Attempts at lower extremity reconstruction should be geared toward providing the most functional outcome in light of the extent of injury and comorbid factors. In some cases, amputation may provide the quickest road to meaningful recovery.

Pearls and Pitfalls

Essential principles for lower extremity wound management include: prompt evaluation, debridement of nonviable tissue, patience (awaiting infection resolution and signs of granulation) and prompt coverage when ready. Utilization of negative pressure therapy often temporizes the situation, reducing the need for urgent coverage and often down-staging the wound allowing for simpler closure method.

Flaps are required for bone and tendon coverage of the lower extremity. Usually, a local flap (muscle or fasciocutaneous) can be designed to cover small defects (<10

Table 52.5. Free flaps used for coverage of the foot

Type	Flap	Pedicle	Comments
Cutaneous	Groin	Superficial iliac circumflex a.	Large flap can be raised
	Scapular or parascapular	Circumflex a. of the scapula	Thick flap
Fasciocutaneous	Radial	Radial a.	Widely used Can be re-innervated or combined with bone
	Lateral arm	Branches from brachialis profunda a.	Small, thin flap
Muscle	Latissimus dorsi	Thoracodorsal a.	Large, bulky flap Can be myocutaneous
	Gracilis	Medial femoral circumflex a.	May be myocutaneous Easy dissection
	Anterior serratus	Branch of thoracodorsal a.	Difficult dissection
Osteocutaneous	Iliac crest	Deep and superficial iliac circumflex a.	Double pedicle Useful for calcaneal reconstruction
	Fibula	Peroneal a.	Useful for metatarsal loss Can include soleus

cm^2) of the foot and ankle; the additional soft tissue and flap donor site defects can be skin grafted. Larger wounds require more distant tissue transfer.

Suggested Reading

1. Arnez ZM. Immediate reconstruction of the lower extremity—an update. Clin Plast Surg 1991; 18(3):449.
2. Arnold PG, Irons GB. Lower-extremity muscle flaps. Orthop Clin North Am 1984; 15(3):441.
3. Attinger C. Soft-tissue coverage for lower-extremity trauma. Orthop Clin North Am 1995; 26(2):295.
4. Byrd HS, Cierny IIIrd G, Tebbetts JB. The management of open tibial fractures with associated soft-tissue loss: External pin fixation with early flap coverage. Plast Reconstr Surg 1981; 68(1):73.
5. Heller L, Levin LS. Lower extremity microsurgical reconstruction. Plast Reconstr Surg 2001; 108(4):1029.
6. Johansen K, Daines M, Howley T et al. Objective criteria accurately predict amputation following lower extremity trauma. J Trauma 1990; 30:568.
7. Park S, Han SH, Lee TJ. Algorithm for recipient vessel selection in free tissue transfer to the lower extremity. Plast Reconstr Surg 1999; 103(7):1937-48.
8. Zenn MR, Levin LS. Microvascular reconstruction of the lower extremity. Semin Surg Oncol 2000; 19(3):272.

Chapter 53

Basic Dental Concepts

Mark Sisco and Jeffrey A. Hammoudeh

Introduction

Dental anatomy, physiology and occlusion are important aspects of plastic and reconstructive surgery. They allow clinicians to communicate consistently and are the underpinning of surgical intervention for many pediatric craniofacial syndromes.

Nomenclature

The 20 deciduous (primary) teeth, starting at the midline, are named central incisor, lateral incisor, canine, first molar and second molar. The 32 permanent teeth, starting at the midline, are named central incisor, lateral incisor, canine (cuspid), first premolar (bicuspid), second premolar (bicuspid), and first, second and third molars.

The Universal system is most commonly used to describe the permanent teeth. The maxillary teeth are numbered starting at the upper right third molar (no. 1) and ending at the left upper third molar (no. 16). The mandibular teeth are numbered in the opposite direction, from the left lower third molar (no. 17) to the right lower third molar (no. 32).

1 2 3 4 5 6 7 8 | 9 10 11 12 13 14 15 16
32 31 30 29 28 27 26 25 | 24 23 22 21 20 19 18 17

The deciduous teeth are denoted by letters in the same order, from A to T. A through J represent the maxillary teeth, starting at the upper right second molar. K through T represent the mandibular teeth, starting at the lower left second molar.

A B C D E | F G H I J
T S R Q P | O N M L K

Anatomy

The normal tooth is comprised of the crown and the root. The majority of the tooth consists of dentin. The crown, which is exposed to the oral cavity, is covered by a layer of enamel. The root, which interfaces with the alveolar socket, is covered by cementum. The pulp chamber and canal are found in the middle of the crown and root, respectively. The pulp tissue supplies the tooth with its sensory and blood supply.

The maxillary teeth are supplied by the posterior, middle and anterior superior alveolar nerves, which are branches of the maxillary division of the trigeminal nerve (CN V). The latter two superior alveolar nerves are supplied by CN V via the infraorbital nerve. The mandibular teeth are supplied by the inferior alveolar nerve from the mandibular division of the trigeminal nerve.

Table 53.1. Age of eruption of the deciduous teeth

Age (Months)	Deciduous Teeth
6-7	Central incisors
7-9	Lateral incisors
12-14	First primary molars
16-18	Canines
20-24	Second primary molars

Pattern of Eruption

The deciduous teeth begin to erupt at six months of age. Maxillary and mandibular teeth tend to erupt at the same times (Table 53.1). The primary dentition eventually exfoliates and is replaced by the permanent detention in a predictable sequence (Table 53.2).

Pathologic Descriptions

Occlusion describes the relationship, or fit, between the upper and lower teeth when in contact. The occlusal plane is the curvilinear plane along which the teeth erupt and meet each other.

Overbite refers to the amount of vertical overlap between the maxillary and mandibular incisal edges when in occlusion. **Anterior open bite** describes a situation where these incisal edges do not overlap, resulting in a negative overbite.

Overjet refers to the forward projection of the upper incisors beyond the lower incisors when the teeth are in occlusion.

Crossbite exists when there is an abnormal buccolingual relationship among the upper and lower molars. In normal or neutral occlusion, the buccal cusps of the maxillary teeth overlap those of the mandibular teeth. Buccal crossbite occurs when the entire maxillary tooth is buccal to the mandibular tooth. Lingual crossbite occurs when the buccal cusps of the mandibular teeth overlap the maxillary teeth.

Table 53.2. Age of eruption of the permanent teeth

Age (Years)	Permanent Teeth
6	Mandbular first molar followed by maxillary first molar
6	Mandibular central incisor
7	Mandibular lateral incisors
7	Maxillary central incisors
8	Maxillary lateral incisors
10	First maxillary and mandibular premolars
10	Mandibular canines
11	Second maxillary and mandibular premolars
11	Maxillary canines
12	Second molars
17	Third molars

Common Tooth Injuries

Subluxation refers to intrusion and extrusion. Most intruded deciduous teeth require no treatment as they will erupt spontaneously. However, if the intruded tooth fails to resorb or erupt in an appropriate timeframe, it must be extracted to allow for eruption of the permanent tooth bud. Intruded permanent teeth are electively repositioned by the orthodontist.

Extruded primary teeth are usually not replanted; the patient should instead be referred to an orthodontist for fabrication of a space maintainer. An attempt may be made to replant an extruded permanent tooth within 1-2 hours of injury. Prior to any attempt, the socket should be irrigated. After one hour, any clot or granulation in the socket should be aggressively debrided. After two hours, replantation will generally fail.

Tooth fractures that involve the crown are typically treated by the endodontist. Fractures involving the root usually require removal of the tooth.

Pearls and Pitfalls

The importance of accurate identification of teeth cannot be overstated. Learn the standard nomenclature. If you have any doubt about which tooth number to use, refer to the proper name, such as "right mandibular second molar." You do not want to write the wrong number in a progress note or extract the wrong tooth in the line of a fracture.

It is important to treat extruded primary teeth with a space maintainer, which is fabricated by an orthodontist or pediatric dentist. Loss of primary teeth may lead to shifting of the remaining teeth. The resultant loss of arch space can ultimately cause crowding and malocclusion of the permanent dentition.

Suggested Reading

1. Bischoff RJ, Simon WJ. The occluding system of the teeth. A review of the occluding system of the mouth—from nomenclature to malocclusions. Dent Assist 1975; 44(1):17.
2. Goodman P. A universal system for identifying permanent and primary teeth. J Dent Child 1967; 34(5):312.
3. Lunt RC, Law DB. A review of the chronology of eruption of deciduous teeth. J Am Dent Assoc 1974; 89(4):872.
4. Natiella JR, Armitage JE, Greene GW. The replantation and transplantation of teeth. A review: Oral Surg Oral Med Oral Pathol 1970; 29(3):397.
5. Wilson KS, Hohmann A. Dental anatomy and occlusion. Otolaryngol Clin North Am 1976; 9(2):425.

Chapter 54

Cephalometrics

Matthew Jacobsen and Jeffrey A. Hammoudeh

Introduction

The analysis of a patient with a skeletal deformity can be complex. However, there are many diagnostic tools available that help the surgeon understand the etiology of the deformity, including the history and physical, radiographic exams, videocephalometric analysis and cast dental models. The integration of this information provides the appropriate diagnosis and guides treatment. The goal of this chapter is to provide the clinician with the basic working principles for detecting and diagnosing maxillofacial deformities.

Clinical Exam—Dental Classification

Terminology:

- **Open bite**—the tips of the incisors are not directly opposed
- **Overbite**—the vertical distance between the tips of the incisors
- **Overjet**—the horizontal distance between the tips of the incisors
- **Neutral occlusion (Class I)**—the maxillary first molar mesiobuccal cusp fits in the buccal groove of the mandibular first molar
- **Distal occlusion (Class II)**—the maxillary first molar mesiobuccal cusp is anterior to the buccal groove of the mandibular first molar
- **Mesial occlusion (Class III)**—the maxillary first molar mesiobuccal cusp is posterior to the buccal groove of the mandibular first molar

The frontal smiling exam documents symmetry of the smile as well as the amount of maxillary and mandibular dentition and gingival exposure. The normal amount of exposed gingiva during smiling is 1-2 mm. Particular attention should be paid to the amount of maxillary tooth show from central incisor to canines. In addition, during the intraoral exam the clinician should look for an **anterior** or **posterior open bite**. It is important to determine the maxillary and mandibular midpoints in order to understand if the maxilla and mandible are coincident with the facial midline or if there is a deviation of one or both. Finally the intraoral exam is used to assess for the presence of **overjet**, **overbite** and **malocclusion**.

Facial Relationships

Aside from the standard history and physical, the clinical exam incorporates subjective numerical data. With the patient seated at eye level across from you, evaluate the face into its proportions of facial thirds. Facial proportions are be considered within the context of the following normal relationships:

- The intercanthal width is roughly equal to the alar base width.
- The lower eyelid rests at or above the most inferior position of the iris. Measure any scleral show as it may be a sign of exophthalmos or infraorbital hypoplasia.

Practical Plastic Surgery, edited by Zol B. Kryger and Mark Sisco. ©2007 Landes Bioscience.

- The width of the nasal dorsum is half the intercanthal width.
- Facial midline, nasal tip, maxillary and mandibular midlines, and chin point are in line.
- Upper face height (glabella to subnasale) should be equal to lower face height (subnasale to menton).

Inspect for **symmetry** and **size** of the forehead, orbits, eyes, ears and nose. Useful measurements include:

- **Interpupillary width**: normal is 65 mm
- **Intercanthal width**: normal is 32 mm for whites and 35 mm for African Americans
- **Upper lip length**: normal is 22 mm for males and 20 mm for females. Measured from subnasale to upper lip stomion.
- **Upper tooth to lip relationship**: 2.5 mm of incisal edge lips the lips relaxed
- **Lower lip length**: normal is 42 mm for males, 38 mm for females. Measured from lower lip stomion to menton.

At the end of the physical exam, take high quality digital photographs. These photos will be crucial to videocephalometric predictions, correlation with model surgery, and will be referred to intraoperatively. The following photos are recommended:

- Frontal photograph relaxed
- Frontal photograph smiling
- Lateral profile—right and left
- 45° oblique
- Intraoral—central, right, left

The fabrication of dental casts aids in diagnosis as well as treatment. For example, in orthognathic surgery it is essential to fabricate surgical splints. The casts must be properly mounted using the facebow, bite record and articulator.

The orthognathic workup and cephalometric analysis are important since they determine many of the surgical movements. A key point to an accurate workup is the careful positioning of the mandible in centric relation rather than in centric occlusion during the bite registration. **Centric relation** denotes position of mandible where the condyles are in the most superior, posterior position in the mandibular fossa. **Centric occlusion** is the position with maximal intercuspation of the teeth. During surgery, the mandible is positioned in centric relation as this depicts the skeletal defect and allows for a reproducible anatomic position.

Radiographic Analysis

After the clinical exam is completed, radiographic analysis helps to further define the nature of the patient's maxillofacial deformity. The **Panorex**, **lateral cephalogram** and **AP cephalogram** are used. The Panorex is inspected for pathology of the sinus, joints, mandible, maxilla and dentition. Examine the Panorex closely for condylar morphology and position. The **Ramus Condyle Unit** is measured to determine if there are any asymmetries in condylar growth. This may be indicative of conditions such as idiopathic condylar resorption or condylar hyperplasia. The lateral and AP cephalograms are the two principle radiographs of othognathic surgery and cephalometric analysis. Classical cephalometric analysis required tracing the cephalometric film by hand. This allowed for identification of the hard and soft tissue landmarks used in the cephalometric analysis. Digital radiographs and computer cephalometric analysis are replacing traditional radiographs and hand tracings. Using a computerized tracing program, the surgeon selects several landmarks on the digitized lateral cephalogram. The computer then produces the cephalometric measurements and the digital tracing.

The relationship between the cranial base, nasomaxillary complex, mandible and maxillomandibular dentition is determined by the resulting angular and linear measurements. Notably, the surgeon may measure the distance from **sella to posterior nasal spine (PNS)**. This value determines the position of the posterior maxilla in relation to the cranial base. The normal value is between 45-50 mm.

Cephalometric Analysis

The cephalometric analysis is a valuable tool that will assist the clinician with the diagnosis of a facial deformity. The diagnosis and treatment plan are determined during the clinical and cephalometric analysis. There are a variety of cephalometric analyses in clinical practice; the Singer and Harvard methods are most commonly used. Normal values for cephalometric measurements in adults are listed in Table 54.1.

Important cephalometric measurements include the following:

1. **Porion**: midpoint of upper contour of the external auditory canal
2. **Sella**: midpoint of the sella turcica
3. **Orbitale**: most inferior point along the bony orbit
4. **Pterygomaxillary fissure**: most superior posterior point of the pterygomaxillary fissure
5. **Nasion**: most anterior point of the frontonasal suture
6. **Basion**: Most inferior anterior point of the foramen magnum
7. **Anterior nasal spine (ANS)**: most anterior point of the anterior nasal spine
8. **Posterior nasal spine (PNS)**: most posterior point of the anterior nasal spine
9. **A point**: most posterior point along the bony premaxilla between ANS and maxillary incisor

Table 54.1. Cephalometric measurements for adults (in mm)

Parameter	Male	Female
Maxillary length	114 4)	105 (3)
Mandibular length	127 (5)	119 (4)
Total facial height	137 (8)	123 (5)
Upper face height	80 (6)	55 (2)
Lower face height	80 (6)	69 (5)
Ethmoid point—PNS	55 (4)	50 (3)
Sella—PNS	56 (4)	51 (3)
Posterior face height	88 (6)	79 (4)
Palatal plane—Menton	76 (6)	67 (4)
Palatal plane—upper molar	28 (3)	25 (2)
Palatal plane—upper incisor	33 (3)	30 (3)
PNS—ANS	62 (4)	57 (4)
Mandibular plane—lower incisor	49 (3)	42 (3)
Manibular plane—lower molar	38 (3)	33 (3)

10. **B point**: most posterior point along the bony contour of the mandible between mandibular incisor and pogonion
11. **Pogonion**: most anterior point along the contour of the bony chin
12. **Menton**: most inferior point along the mandibular symphysis
13. **Gnathion**: point along bony chin between menton and pogonion
14. **Gonion**: angle of the mandible at the intersection of the tangents drawn from the posterior ramus border and the lower ramus
15. **Articulare**: point of intersection between the cranial base and posterior ramus
16. **Condylion**: most superior posterior point of the bony condyle

Important cephalometric planes and angles include the following:

1. **Sella-Nasion-A point**: SNA
2. **Sella-Nasion-B point**: SNB
3. **A point-Nasion-B point**: ANB
4. **Frankfort horizontal (FH) plane**: Pogonion—Orbitale
5. **Palatal plane**: ANS-PNS
6. **Occlusal plane**: plane from mesial cusp of maxillary molar through point bisecting overbite
7. **Mandibular plane**: tangent along lower border of mandible
8. **Gonial angle**: Articulare-Gonion-Menton

SNA and SNB provide an assessment of the maxillary and mandibular relationship to the cranial base. However, in order to utilize SN as the normal inclination of the anterior cranial base, the surgeon must first normalize the actual SN position to the normal SN-FH of 6 degrees. In most patients, the sella turcica is normally positioned and thus the sella to Frankfort angle is normal. However, in patients with craniofacial deformities, congenital syndromes, or sequences, it is imperative to identify the relationship of sella-to-FH. Altering the normal position of sella will, by default, alter the SNA and SNB. In order to assess the true SNA/SNB one must determine the SN-FH correction.

1. First measure the SN-Frankfurt angle (normal is 6°)
2. If the SN-F angle is abnormally **acute** due to a shallow SN plane, the measured SNA and SNB will be too **obtuse**. Therefore, we must subtract the difference ([$SN\text{-}FH_{actual}$-$SN\text{-}FH_{normal}$]) to generate the "corrected" SNA and SNB.
3. If the SN-F angle is abnormally **obtuse** due to a steep SN plane, the measured SNA and SNB will be too **acute**. Therefore, one must add the difference ([$SN\text{-}FH_{actual}$-$SN\text{-}FH_{normal}$]) to generate the "corrected" SNA and SNB.

PA Cephalogram

The PA cephalogram illustrates transverse and vertical skeletal relationships and evaluates facial symmetry. The first step in analyzing the PA film is to analyze the transverse dimension using a J point analysis. A horizontal measurement is made from the lateral aspects of the maxilla at the level of the pyriform rims. Similarly, a horizontal measurement is made from antegonial notch form right to left. The difference between these two values is used to identify an excessively narrow maxilla or enlarged mandible. The normal J point value is between 20 and 23 mm. Next, the PA cephalogram is drawn on tracing paper. Horizontal lines are drawn at the infraorbital rim, pyriform rim, occlusal plane and gonial angle. These marks are compared to a vertical midline mark in order to determine symmetry and cant of the orbits, zygomas, maxilla and mandible.

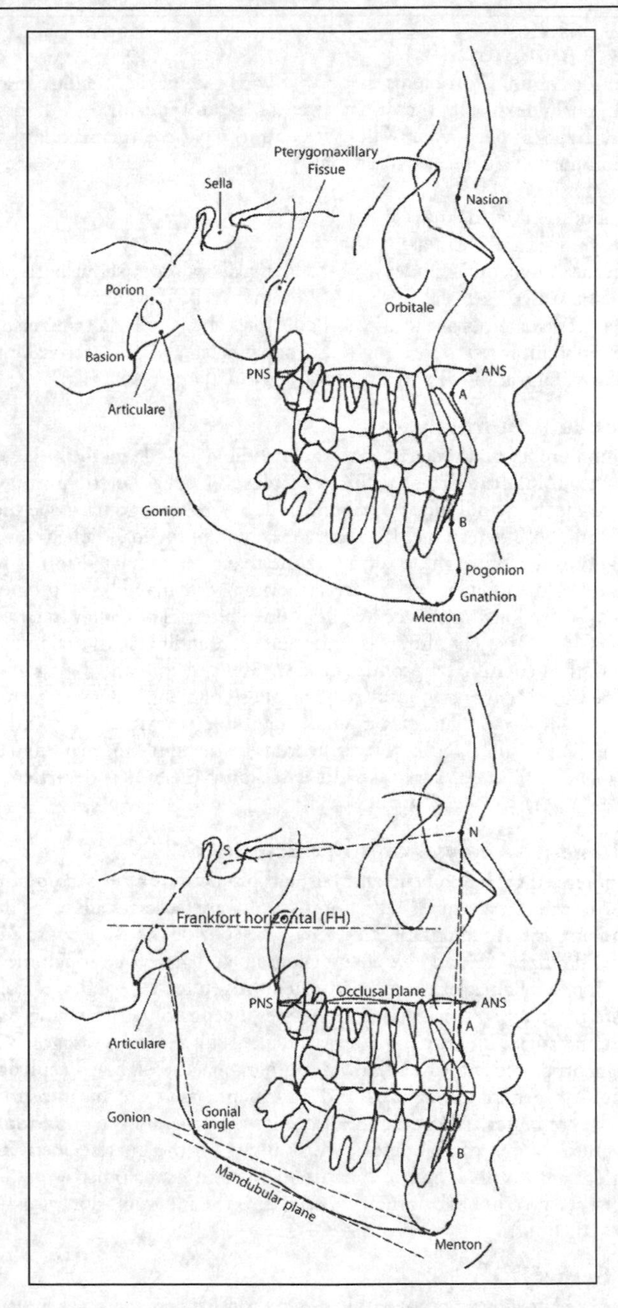

Figure 54.1. Common cephalometric points and their relationships.

Maxillary Abnormalities

The most common problems associated with the maxilla include **maxillary sagittal hypo/hyperplasia, maxillary vertical hypo/hyperplasia** and **maxillary transverse hypo/hyperplasia**. When evaluating a patient for maxillary surgery the surgeon must evaluate the following:

1. Vertical position of the maxilla
2. Sagittal position of the maxilla
3. Transverse width of the maxilla

Useful measurements include: PNS-Sella, maxillary tooth show, occlusal plane angle, J point analysis, nasolabial angle, SNA and ANB. The LeFort I osteotomy is used to correct deformities of the maxilla in the vertical, sagittal and transverse planes. It is commonly employed to treat sagittal hypoplasia (small SNA), vertical maxillary excess (excess gingival show) and anterior open bite (large PNS-Sella).

Mandibular Abnormalities

Common problems associated with the mandible include **mandibular sagittal hypoplasia** and **mandibular sagittal hyperplasia**. Cephalometric measurements used to evaluate the mandibular position include SNB, ANB, gonial angle and mandibular length. SNB determines the relative position of the mandible to the cranial base, and ANB illustrates the position of the mandible as related to the maxilla. Large values of mandibular length are associated with mandibular prognathism. The patient with a long lower face height or open bite will often have a large gonial angle. The Class III patient often has a diagnosis of mandibular sagittal hyperplasia. Most often, the patient has a prominent lower jaw and chin and a long lower face height. The Class II patient typically has a diagnosis of mandibular sagittal hypoplasia. Most often, the patient has a small mandible, retrusive chin and obtuse cervicomandibular angle. Surgical treatment of the mandible for mandibular prognathism includes **bilateral sagittal split osteotomy (BSSO)** and **vertical ramus osteotomy (VRO)**.

Cephalometric Analysis and Predictions

Computerized cephalometric analysis and predictions have become the standard in orthognathic surgery. Many popular software packages allow for incorporation and analysis of radiographic images. They enable the clinician to pinpoint angles on a digitized AP and lateral cephalogram and generate cephalometric calculations. These calculations can be adjusted for gender, race and age. Following cephalometric analysis, the software generates a cephalometric tracing based on the points and angles identified by the physician. The clinical photograph is overlaid and incorporated into the digitized radiograph. The digitized cephalometric data is used to generate diagnosis and treatment based on the measurements. Overall, the computerized cephalometric software program is a good tool in the rapid diagnosis of dentofacial deformity. It allows for the interface between digitized radiographs, cephalometric measurements and cephalometric predictions. Ultimately, the goal in cephalometric imaging is the incorporation of 3-D imaging and predictions.

Model Surgery

As discussed previously, othognathic surgery consists of mandibular surgery, maxillary surgery and double jaw procedures. Surgeons use model surgery to aid in planning of the operation. Model surgery facilitates precise surgical movements. In general, the patient presents for diagnostic evaluation when the skeletal malformation is first identified. At that time, diagnostic records, radiographs, images and dental models are fabricated. This process is repeated immediately before the operation. Appropriate images, measurements, radiographs and models are generated prior to the operation. Specifically, dental casts are fabricated from dental impression and stone. A face-bow record of the patients Frankfort horizontal plane to occlusal plane is generated. The dental casts are articulated using a bite record in centric relation and these are mounted on an anatomic semiadjustable articulator. Reproducible vertical reference lines are placed at the midline, canine tip, mesiobuccal cusp of the first molar and posterior retromolar region. Horizontal reference lines are placed at 10 and 20 mm from the articulator mounting plate. The cast is separated from the mounting ring at 15 mm, centered between the two horizontal reference lines.

Next, model surgery is performed on the mandible, maxilla, or both by positioning the maxillary and mandibular arches into an ideal occlusal relationship. The horizontal and vertical reference lines keep the models in the appropriate vertical, anteroposterior and transverse relationship. A final surgical splint is fabricated using methyl methylacrylate resin. In the case of double jaw surgery, the maxilla is first positioned in the predicted horizontal and vertical position. A splint is fabricated using methylmethacrylate that preserves this "intermediate" maxillomandibular relationship. Next, the mandibular cast is moved into an ideal occlusal relationship with the maxilla, creating the final splint.

Pearls and Pitfalls

An inexperienced surgeon who takes a bite registration in centric occlusion rather than centric relation will end up with inaccurately mounted modes, inaccurate splints, and eventually a nonreproducible, unstable result with a malocclusion intraoperatively.

The SN-FH correction must be taken into consideration in order to have an accurate cephalometric prediction and treatment plan. Most patients have a normal SN-FH and thus do not require any correction factors; however children with craniofacial anomalies often require close analysis of the corrected FH, otherwise the cephalometric analysis will not be accurate.

Suggested Reading

1. Ash M. Wheelers Dental Anatomy, Physiology and Occlusion. Philadelphia: Saunders, 2000.
2. Betts N, Turvey T. Orthognathic surgery. In: Fonseca R, ed. Oral and Maxillofacial Surgery. Philadelphia: Saunders, 2000.
3. Ferraro JW. Cephalometry and cephalometric analysis. In: Ferraro JW, ed. Fundamentals of Maxillofacial Surgery. New York: Springer, 1997.
4. Kaban LB, Troulis MJ. Pediatric Oral and Maxillofacial Surgery. Philadelphia: Saunders, 2004.
5. Peterson LJ, Ellis E, Hupp JR et al. Contemporary oral and maxillofacial surgery. St. Louis: Mosby, 1998.
6. Profitt WR, Thomas PM, Camilla-Tulloch JF. Contemporary orthodontics. St Louis, Mosby: 1986.
7. Proffit WR, White RP, Sarver DM. Contemporary treatment of dentofacial deformity. St. Louis. Mosby: 2002.

Chapter 55

Craniofacial Syndromes and Craniosynostosis

Zol B. Kryger and Pravin K. Patel

Craniosynostosis

Introduction

Craniosynostosis is the premature fusion of the sutures of the skull. Although the cause of this condition is not known, TGF-beta has been strongly implicated as playing a major role. Craniosynostosis can occur as an isolated event or in the context of a craniofacial syndrome. The sporadic nonsyndromic cases are more common (incidence of 1 in 2000 live births) than the syndromic cases, many of which are related to defects in the FGF receptor (FGFR). Virchow's law can help predict the developing skull shape. It states that growth is restricted perpendicular to the fused suture, and compensatory growth occurs parallel to the affected sutures.

Affected Suture

One or more sutures can be affected in craniosynostosis. Sagittal synostosis is the most common form, accounting for over half of all cases. Table 55.1 summarizes the various involved sutures and the characteristic appearance of the skull:

Deformational (Nonsynostotic) Plagiocephaly

This condition is more common than posterior plagiocephaly due to craniosynostosis with an incidence of 1 in 300 live births. The incidence of this condition increased significantly after recommendations by pediatricians that infants sleep supine in order to decrease the risk of SIDS. Deformational plagiocephaly occurs as a result of supine positioning during the first few weeks of life. Features that help distinguish it from synostotic plagiocephaly (lambdoid synostosis) can be seen in Table 55.2.

Table 55.1. Various involved sutures and characteristic appearances of the skull

Involved Suture	Skull Appearance
Unicoronal	Anterior plagiocephaly
Bicoronal	Turribrachycephaly (bitemporal widening)
Sagittal	Scaphocephaly (biparietal narrowing)
Metopic	Trigonocephaly (triangular forehead)
Lambdoid	Posterior plagiocephaly (flat posterior skull)
Coronal, lambdoid, metopic	Kleeblattschadel (clover leaf skull)

Table 55.2. Comparison of features between synostotic and deformational plagiocephaly

Feature	Synostotic Plagiocephaly	Deformational Plagiocephaly
Sutures	Fused	Patent
Eyebrow	Elevated on affected side	Eyebrow lower on affected side
Ear	Rotated anterosuperior	Rotated posteroinferior
Nose	Deviated to opposite side	Deviated to affected side
Chin	Deviated to opposite side	Deviated to affected side
Cheek	Forward on affected side	Flattened on affected side
Torticollis	Contralateral side	Ipsilateral side

Associated Symptoms

Elevated intracranial pressure (ICP)—occurs in 13% of single suture synostosis and 42 % of multiple suture synostoses.

Strabismus—seen most often in unilateral, coronal synostosis. It is due to paresis of the superior oblique muscle.

Torticollis—seen in about 15% of cases of anterior plagiocephaly, usually on the unaffected side.

Cognitive deficits—seen most commonly with metopic synostosis.

Treatment

The goal of treatment is foremost to allow adequate space for the brain to grow. Of lesser importance is the creation of an aesthetically normal skull and forehead.

Molding (orthotic cranioplasty)

Molding of the skull, or orthotic cranioplasty, is done with a helmet worn up to 23 hours a day for 2-4 months. It should begin around the age of 6 months and always before the age of 14 months. It is the recommended treatment for deformational plagiocephaly and can also be used as an adjunct to surgery.

Surgery

Since the most rapid phase of skull growth occurs within the first year of life (Fig 55.1), early treatment is required. Surgical intervention is most commonly performed between the ages of 3 to 12 months of age. There is some evidence that earlier intervention results in fewer learning disabilities and emotional problems in later years. However, these findings are disputed by those who believe that operating on infants with craniosynostosis will not have an effect on their future cognitive abilities. Regardless of the timing, surgical treatment must be tailored to the individual's deformed structures.

Unilateral coronal synostosis is managed with unilateral or bilateral fronto-orbital advancement. Bilateral coronal synostosis is managed with bilateral fronto-orbital advancement. If there is evidence of brachycephaly, total calvarial reconstruction may be required. Sagittal synostosis is treated with strip craniectomies and partial-wedge osteotomies. If biparietal expansion is required, anterior and posterior parietal wedges may be required. Lambdoid synostosis is managed with excision of the lambdoid suture. Metopic synsotosis is treated with removal of the supraorbital bar, corticotomy and correction of the midline angle with bone grafts

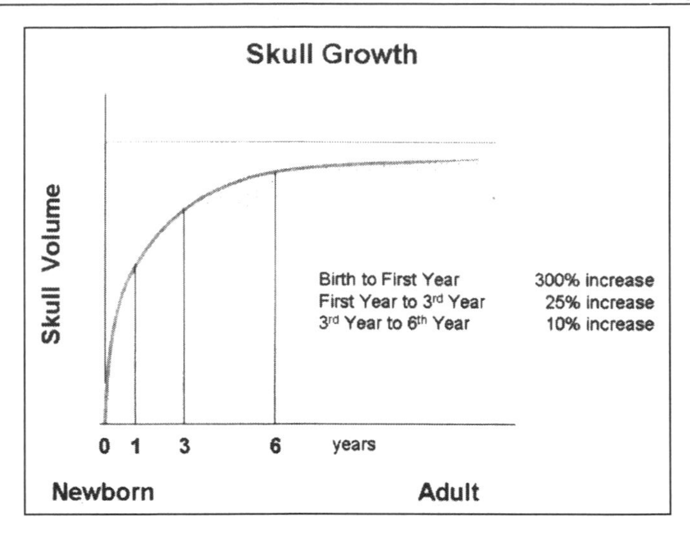

Figure 55.1. The growth curve of the skull.

or miniplates. Kleeblattschadel is treated after life-threatening conditions are addressed (e.g., hydrocephalus, airway obstruction) with anterior calveriectomy and fronto-orbital advancement. Excess constricting bands of bone are removed.

Complications

As might be expected from the magnitude of the procedures, the complications can be devastating. Early complications specific to these procedures include: sagittal sinus tears with venous infarction, subdural hematoma, cerebral edema, excess vasopressin production (SIADH), nerve injuries, injury to the orbits, infections, elevated ICP and dural leaks. Late complications include: incomplete advancement, palpable hardware, alopecia, asymmetry, orbital deformities, pseudomeningocele and increased ICP.

Craniofacial Syndromes

Many craniofacial syndromes have been described in the literature. However, with a few exceptions, these are rare syndromes seen primarily in specialized centers. A discussion of the more commonly encountered craniofacial syndromes follows with an emphasis on the similarities and differences between them.

Apert Syndrome (Acrocephalosyndactyly)

- Autosomal dominant
- Bicoronal synostosis
- Hypertelorism or exorbitism
- Strabismus
- Maxillary hypoplasia
- Acne and coarse skin

- Large ear lobes
- Syndactyly of all fingers (mitten hands) and toes
- Variable mental retardation

Crouzon Syndrome (Craniofacial dysostosis)

- Autosomal dominant
- Bicoronal (occasionally sagittal) synostosis
- Hypertelorism or exorbitism
- Maxillary hypoplasia
- Micrognathia
- No abnormalities of the hands or feets
- Normal intelligence

Pfeiffer Syndrome (Acrocephalosyndactyly type V)

- Autosomal dominant
- Craniosynostosis of multiple sutures
- Maxillary hypoplasia
- Broad thumbs and great toes
- Brachydactyly with partial syndactyly
- Normal intelligence

Saethre Chotzen Syndrome (Craniocephalosyndactyly)

- Autosomal dominant
- Bicoronal synostosis
- Maxillary hypoplasia
- Strabismus
- Vertebral anomalies
- Low hairline
- Palatal abnormalities (often a cleft)
- Brachydactyly with syndactyly
- Normal intelligence

Carpenter Syndrome

- Autosomal recessive
- Craniosynostosis
- Deafness
- Brachydactyly with syndactyly
- Polydactyly (preaxial)
- Mental retardation

Treacher-Collins Syndrome (Mandibulofacial dysostosis)

- Autosomal dominant
- Prominent eyelid abnormalities (e.g., pseudocoloboma and absence of lashes)
- Downward sloping palpebral fissures
- Poorly developed orbital rims
- Midface (zygomaticomaxillary) hypoplasia
- Macrostomia and resulting airway distress
- Cleft palate

Velocardiofacial Syndrome

- Autosomal dominant
- Cleft palate or submucus cleft
- Cardiac abnormalities
- Ectopic carotid arteries
- Malar flattening
- Prominent broad nose
- Epicanthal folds
- Vertical maxillary excess
- Retrognathia
- Developmentally delayed

Pierre Robin Sequence

- Retrognathia (occasionally described as retromicrogenia)
- Glossoptosis
- Airway obstruction
- Variable cleft palate

Although not a true craniofacial syndrome, Pierre Robin sequence is an important constellation of findings. This condition can be life-threatening due to airway compromise. Initial treatment consists of prone positioning, while surgical management is reserved for more severe cases.

Pearls and Pitfalls

- Treatment of craniosynostosis and craniofacial syndromes requires a multidisciplinary team approach. Early involvement of the other relevant specialties (neurosurgery, otolaryngology, oral surgery, speech therapy, a prosthodontist and a geneticist) is essential for achieving a good outcome.
- Prevention of deformational plagiocephaly is quite simple. It consists of turning the infant's head from side to side several times throughout the day, and placing the infant prone while he is awake and under observation by the caretaker.
- In any infant suspected of having a craniofacial abnormality, evaluation of the airway is the most important initial step in the work-up.
- Distraction osteogenesis of the facial skeleton is an alternative to bone grafting that offers excellent long-term outcomes. It should be considered whenever greater than 10-15 mm of advancement of the facial bones is required.
- Crouzon's syndrome can be distinguished from Apert and Pfeiffer syndromes by the absence of hand abnormalities.
- Almost all of the craniofacial syndromes are transmitted via an autosomal dominant pattern with variable expression. An important exception is Carpenter syndrome.
- The FGFR is known to be mutated in the majority of the craniofacial syndromes involving craniosynostosis. Future therapeutic approaches may target these mutations.

Suggested Reading

1. Anderson FM, Geiger L. Craniosynostosis: A Survey of 204 cases. J Neurosurg 1965; 22:229.
2. Breugem CC, van R, Zeeman BJ. Retrospective study of nonsyndromic craniosynostosis treated over a 10-year perior. J Craniofac Surg 1999; 10:140.
3. Tessier P. The definitive plastic surgical treatment of the severe facial deformities of craniofacial dysostosis. Crouzon's and Apert's diseases. Plast Reconstr Surg 1971; 48:419.
4. Huang MH, Gruss JS, Clarren SK et al. The differential diagnosis of posterior plagiocephaly: True lambdoid synostosis versus positional molding. Plast Reconstr Surg 1996; 98:765.
5. Kapp-Simon KA. Mental development and learning disorders in children with single suture craniosynostosis. Cleft Palate Craniofac J 1998; 35:197.
6. Losken HW, Pollack IF. Craniosynostosis. In: Bentz ML, ed. Pediatric Plastic Surgery. 1st ed. Stamford: Appleton and Lange, 1998.
7. Marchac D, Renier D, Broumand S. Timing of treatment for craniosynostosis and faciocraniosynostosis: A 20-year experience. Br J Plast Surg 1994; 47:211.
8. Mulliken JB, Vander Woude Dl, Hansen M et al. Analysis of posterior plagiocephaly: Deformational versus synostotic. Plast Reconstr Surg 1999; 103:371.
9. Noetzel MJ et al. Hydrocephalus and mental retardation in craniosynostosis. J Pediatr 1985; 107:885.
10. Persing JA, Jane JA, Edgerton MT. Surgical treatment of craniosynostosis. In: Persing JA, Edgerton MT, Jane JA, eds. Scientific Foundation of Surgical Treatment of Craniosynostosis. Baltimore: Williams and Wilkins, 1989.

Chapter 56

Craniofacial Microsomia

Zol B. Kryger

Introduction

Craniofacial microsomia, also termed hemifacial microsomia, is defined as a congenital hypoplasia of the facial skeleton and soft tissues. There is no known genetic cause. It is thought to be due to an in utero vascular insult in the developing first or second branchial arches. It is unilateral in 90% of cases. After cleft lip and palate, this condition is one of the more common craniofacial congenital abnormalities. The incidence in the U.S. is estimated to be 1 in 5,000 births.

Clinical Findings

Any of the skeletal, nervous, and soft tissue structures derived from the first and second branchial arches can be affected. Consequently, there is a wide spectrum of presenting signs and symptoms.

Microtia and inner ear abnormalities are usually present on examination. One may notice abnormalities of the auricle with preauricular skin tags on the affected side. Examination of the lower face reveals a cant of the occlusion plane sloping downward and away from the hypoplastic side. The chin is deviated towards the affected side, and the distance from the oral commissure to the ear may be shortened.

A full examination of oral cavity, including the teeth and dental occlusion, should be performed. The muscles of mastication are often dysfunctional; however, their involvement is not always proportional to the degree of mandibular hypoplasia. This dysfunction may be manifested in difficulty opening the mouth. The lateral pterygoid is also usually involved, causing an inability to deviate the chin to the contralateral side. Soft tissue and palatal clefts may also be present.

A complete cranial nerve examination is essential since patients may also demonstrate neurologic abnormalities. The marginal mandibular branch of the facial nerve is most commonly involved. Cerebral abnormalities are rare.

Radiographic evaluation has traditionally included cephalograms and a panorex. Computed tomography with 3-D reconstructions will give the most precise information to evaluate the skeletal abnormalities and assist in the preoperative planning.

Skeletal Abnormalities

The mandible is most commonly involved in craniofacial microsomia. The degree of mandibular hypoplasia has been classified by Pruzansky:

- **Type I**. Mild hypoplasia of the condyle and ramus; the body is unaffected. The TMJ is functional. There are minimally noticeable morphological changes.
- **Type IIA**. The condyle and ramus are severely hypoplastic. The coranoid process can be absent. The condyle maintains a normal position relative to the glenoid fossa.

Practical Plastic Surgery, edited by Zol B. Kryger and Mark Sisco. ©2007 Landes Bioscience.

- **Type IIB**. Similar to Type IIA, yet the condyle and glenoid fossa are not in the normal position and plane.
- **Type III**. The condyle and ramus are absent and there is no TMJ.

Other bones of the craniofacial skeleton can be hypoplastic as well. The maxilla is reduced in the vertical plane, and contributes to the cant of the occlusion plane. The zygoma and zygomatic arch are often decreased in length and the arch may be entirely absent. The temporal and frontal bones are affected less; however the orbits are often reduced in all dimensions leading to microphthalmos.

Classification of Unilateral Craniofacial Microsomia

There are a number of classification schemes. The two that are common and easy to use are the OMENS classification (Table 56.1), and the system proposed by Munro and Lauritzen in 1985 (Table 56.2).

Treatment by Age

There is no uniform treatment plan that is appropriate for everyone. Children under the age of two with craniofacial microsomia do not usually undergo major reconstructive surgery. Preauricular skin tags or other cartilaginous remnants may be excised. Commissuroplasty can correct those patients with macrostomia at this young age.

After the age of two, distraction osteogenesis can correct mandibular ramus hypoplasia as seen in Pruzansky Type I and IIA. This will both lengthen the mandible and improve the function of the muscles of mastication. Bilateral cases can be treated with bilateral distraction, allowing closure of tracheostomies in many children.

At the age of four, children with Pruzansky Type III mandibular hypoplasia undergo costochondral rib-graft reconstruction of the mandible combined with a LeFort I osteotomy and sagittal split of the contralateral mandibular ramus. The zygomatic arch and glenoid fossa are also reconstructed using a rib graft and a cap of costochondral cartilage, respectively.

After the age of six, and microtia repair may be performed and orthodontic treatment initiated. Augmentation of the facial soft tissue with muscle flap transfers is also occasionally necessary at this age. Once skeletal maturity has been reached in the early teenage years, residual deficiencies should be addressed.

Teenagers who have not undergone any prior treatment will require alternative types of reconstruction. Interpositional bone grafts can be used to correct mild mandibular length deficiencies. More severe microsomia can be treated with combined LeFort I osteotomy, bilateral sagittal split of the mandible, and genioplasty. Micrognathia can be treated with bilateral mandibular advancement.

56

Table 56.1. The "OMENS" classification for craniofacial microsomia

	Involved Structure	Description
O	Orbit	Size and position of the orbit
M	Mandible	Degree of mandibular hypoplasia
E	Ear	Extent of microtia
N	Facial nerve	Which branches are involved
S	Soft tissue	Degree of muscular and subcutaneous deficiency

Table 56.2. The Munro and Lauritzen classification of craniofacial microsomia. There is a progressive addition of abnormalities, ranging from Type IA, the least severe, to Type V, the most severe.

Type	Description
IA	Mild craniofacial skeletal hypoplasia, normal occlusion plane
IB	Mild craniofacial skeletal hypoplasia, canted occlusion plane
II	Absent condyle, part of the ramus
III	Absent condyle, part of the ramus, glenoid fossa, zygomatic arch
IV	Absent condyle, part of the ramus, glenoid fossa, zygomatic arch, hypoplastic zygoma, lateral orbital wall displaced
V	Absent condyle, part of the ramus, glenoid fossa, zygomatic arch, inferior orbit displacement with loss of orbital volume

Pearls and Pitfalls

- Roughly 25% of patients with craniofacial microsomia have obstructive sleep apnea (OSA). Infants with more severe mandibular and orbital deformities appear at a greater risk for OSA. Macrostomia can also result in airway obstruction. Therefore, all children with craniofacial microsomia should undergo evaluation for airway difficulties and OSA prior to undergoing any other types of reconstructive surgeries.
- The importance of treating children with craniofacial microsomia is underscored by the fact that facial asymmetry is progressive in this condition. Untreated children will develop worsening degrees of bony and soft tissue asymmetry. Furthermore, the more severe the deformity, the worse the asymmetry will become if left untreated.
- Treatment of the soft tissue defects in this condition range from injection of fillers to free tissue transfer. The amount of soft tissue that is missing will dictate the treatment. The soft tissue defect should be addressed after the major skeletal reconstruction has taken place. For mild defects, fillers (e.g., Sculptra®) and fat injections work well. Medium-sized defects can be addressed with Alloderm® or silicone implants. Larger defects may require a dermal fat graft or even a free tissue transfer of fat (e.g., DIEP flap).

Suggested Reading

1. Kearns GJ et al. Progression of facial asymmetry in hemifacial microsomia. Plast Reconstr Surg 2000; 105:492.
2. McCarthy JG, Schreiber JS, Karp NS et al. Lengthening of the human mandible by gradual distraction. Plast Reconstr Surg 1992; 89:1.
3. McCarthy JG. Craniofacial microsomia. Grabb and Smith's Plastic Surgery. 5th ed. Philadelphia: Lippincott-Raven, 1997:305.
4. Mulliken JB, Kaban LB. Analysis and treatment of hemifacial microsomia in childhood. Clin Plast Surg 1987; 14:91.
5. Munro IR, Lauritzen CG. Classification and treatment of hemifacial microsomia. In: Caronni EP, ed. Craniofacial Surgery. Boston: Little, Brown and Co., 1985:391.
6. Vento AR, LaBrie RA, Mulliken JB. The O.M.E.N.S. classification of hemifacial microsomia. Cleft Palate-Craniofac J 1991; 28:68.

Chapter 57

Microtia Repair

Zol B. Kryger

Introduction

Microtia ranges from a very mild form in which the ears are grossly normal but prominent, to complete absence of the ear, termed anotia. Severe abnormalities occur in roughly 1 in 8,000 births. There is a predisposition of microtia to occur in males, with a male: female ratio of about 2:1. The right side is also twice as likely to be affected as the left, and bilateral cases occur in 10% of cases. Microtia is most commonly described according to the classification described by Tanzer (Table 57.1).

Preoperative Considerations

There are a number of congenital conditions associated with microtia:

- Narrowing or atresia of the external auditory canal (very common)
- Middle ear abnormalities (very common)
- Combined ear canal and middle ear abnormality
- Cleft lip or palate
- Facial nerve abnormality
- Hemifacial microsomia (in about half of microtia cases)
- Cardiac or urogenital defects

Middle ear abnormalities are more likely to occur the more severe the degree of microtia. The tragus is the structure that is most highly correlated with the presence of an adequate middle ear cleft.

Hearing

All patients with microtia should undergo hearing evaluation, either with an audiogram or auditory brainstem response testing. Hearing loss can be either conductive or sensorineural. Conductive deafness is more common in microtia. Children with unilateral microtia and normal hearing in the contralateral ear will develop normal speech. In bilateral microtia cases, a hearing aid is required for normal speech development. This will often be a bone conduction hearing aid that is fitted shortly after birth. Most commonly, a percutaneous, bone-anchored hearing aid is used. At

Table 57.1. Classification of microtia

Grade	Description
I	Anotia
II	Complete hypoplasia (± atresia of the external auditory canal)
III	Middle third auricular hypoplasia
IV	Superior third auricular hypoplasia
V	Prominent ears

Practical Plastic Surgery, edited by Zol B. Kryger and Mark Sisco. ©2007 Landes Bioscience.

about the age of four, a CT scan of the middle ear is useful to determine whether the middle ear is amenable to reconstruction by the otologist.

Timing of the Repair

Most experienced surgeons who treat microtia wait until the child is 5-7 years old before considering reconstruction. Some have reported auricular framework reconstruction as early as 2 to 3 years of age. There are a number of reasons to wait until the child is around 6 years old. First, this is the age that children will usually begin to tease other children with different appearances. The child is more likely to cooperate with the postoperative regimen. Second, by the age of 6, the normal ear has almost fully developed, reaching 85% of its full size. Third, the rib cartilage has sufficiently developed by this age, and it can provide an adequate framework for auricular construction.

In general, when microtia is unilateral, middle ear atresia repair is not indicated since unilateral hearing is sufficient. In bilateral cases, repair of the atresia on at least one side is usually required. If middle ear surgery is indicated, it should be postponed until after completions of the first stage of the auricular reconstruction-insertion of the costal cartilage framework. This is due to the fact that atresia surgery preceding auricular reconstruction can result in scarring and impaired vascularity in the mastoid region.

The size of the auricular framework is controversial. According to Brent, it should be matched to create an ear that is as close to the size of the normal one as possible. The rationale for this is that the cartilaginous framework continues to grow at roughly the same pace as the cartilage of the normal ear. Some surgeons, however, will create a framework that is a few millimeters larger than the other ear. They feel that the other side will "catch up" with the reconstructed ear.

Intraoperative Considerations

Microtia repair is a multi-stage procedure. Although the timing and sequence of the reconstruction is variable, a number of steps are generally followed. Brent and Tanzer each advocate following these steps:

Step I—cartilage framework construction and placement
Step II—transposition of the lobule
If indicated—atresia repair
Step III—construction of the tragus and conchal cavity
Step IV—creation of the auriculocephalic sulcus

A number of the steps can be combined, such as performing tragal reconstruction in combination with the initial cartilage framework placement, or repair of the lobule during step I. The approach that is chosen must be tailored to the individual's wishes and unique anatomy. If atresia repair is undertaken, an oval shape of skin is excised when creating the conchal cavity. This will serve as the opening of the reconstructed auditory canal. The canal is lined with a skin graft.

In contrast to the multi-step microtia repair, Nagata has described a two-stage approach that encompasses the steps described above. The initial procedure consists of fabrication of the costal cartilage framework, rotation of the lobule, conchal accentuation and fashioning of the tragus. The second stage focuses primarily on ear elevation and creation of the auriculocephalic sulcus.

The Auricular Framework

The framework can be either **autologous**, using costal cartilage, or **alloplastic**, composed of a synthetic material.

Alloplastic Framework

A number of materials have been used to create auricular frameworks. Currently, the most commonly used substances are silastic or porous polyethylene mold. Allopastic frameworks have a higher risk of erosion and exposure compared to autogenous ones. Factors contributing to this high rate of extrusion are scar tissue, excessively thin skin, tension over the implant, trauma and infection. Nevertheless, with adequate soft tissue coverage, such as the temporoparietal fascial flap-described below, alloplastic frameworks can be used successfully. Most authors feel that they are a second choice to costal cartilage.

Autogenous Framework

Costal cartilage can be taken from the 6th, 7th, 8th and 9th ribs. A large piece is cut out from the 6th and 7th ribs and used for the body of the framework. A smaller adjacent wedge is removed and will be banked for future use during ear elevation. Two thin, smaller pieces are taken from the 8th and 9th ribs and used to create the helix and a strut for the tragus (Fig. 57.1). The cartilage can be removed along with the perichondrium, or the dissection can be subperichondrial, as advocated by Tanzer, leaving the perichondrium behind. Some authors prefer to use the ipsilateral ribs, whereas others use the contralateral side (as shown in Fig. 57.1). After removal of the cartilage and closure of the donor site, the pieces of

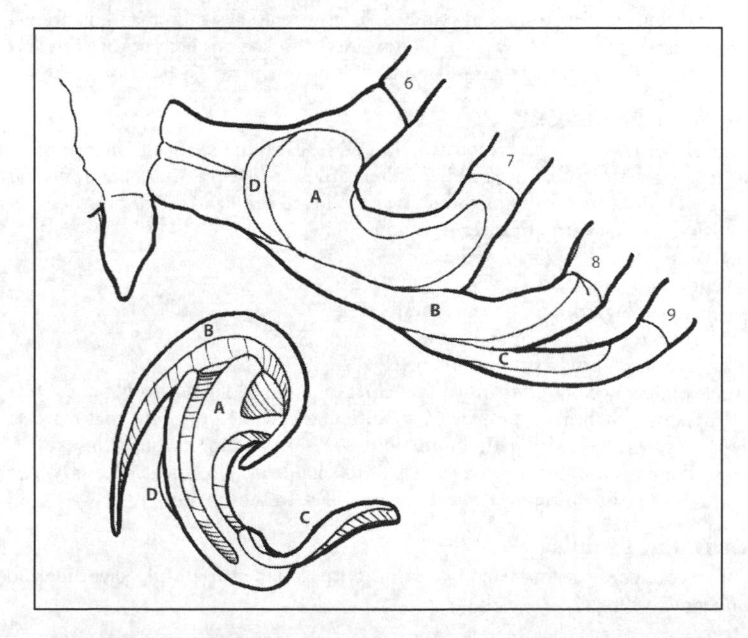

Figure 57.1. The auricular framework. Harvesting the cartilage framework from the contralateral costal cartilage (above). Segment 4 is banked for use in the final procedure of ear elevation by creating greater projection. Creating the framework from the segments labeled 1 (main body), 2 (helical rim) and 3 (tragus). The pieces of cartilage are sutured together using 4-0 and 5-0 clear nylon sutures (below).

cartilage are carefully carved into the desired shapes with a scalpel and chisels. In adults, the rib cartilages are often fused and the entire framework must be sculpted as a single unit.

The framework is inserted into a pocket in the desired auricular region. The pocket must be much larger than the framework in order to have a tension free closure. Suction tubing attached to vacuum test-tubes is used to adhere the skin envelope to the framework. Problems with the hairline or inadequate tissue for coverage are described below.

Postoperative Considerations

Most children will be hospitalized for one to two days. The suction test tubes are changed daily for 5 days postoperatively and then removed. This system provides adequate compression; therefore the external dressing does not need to be compressive. Sports are restricted for 4-6 weeks, for protection of the ear as well as the chest wound. Patients are not instructed to avoid sleeping on the operative site, because most children turn in their sleep and will not be able to comply with this instruction.

Complications

As mentioned previously, alloplastic frameworks have a higher extrusion rate than for costal cartilage frameworks. Extrusion requiring removal occurs in 5-30% of silastic frameworks, compared to 1-2% for costal cartilage. Other complications include infection, hematoma and skin loss. These are usually minor, and the framework can almost always be salvaged. Donor site complications include unacceptable chest scars, mild to severe retrusion and flattening of the rib cage contour.

Auricular Prosthesis

An alternative to surgical reconstruction of the ear is to use an auricular prosthesis. In select patients, it is an excellent alternative. Its use precludes any inner ear surgery. The following patients should be considered for an auricular prosthesis:

- Major auricle loss after cancer resection
- Absence of the lower half of the ear
- Poor quality of local tissue
- Patients at high risk for general anesthesia
- Poorly compliant patients
- Salvage after unsuccessful reconstruction

Osseointegrated titanium implants are first implanted in the mastoid bone. Once the implants have healed completely, an auricular silicone prosthesis that matches the other side is created. The titanium abutments protruding through the skin attach to the prosthesis by one of a variety of mechanisms. No glue is necessary. The prosthesis can easily be removed and the area cleaned thoroughly.

Pearls and Pitfalls

Two commonly encountered problems in microtia repair are a low hairline and inadequate coverage.

Low Hairline

A low hairline is a frequently encountered problem. A number of techniques for dealing with this problem have been described. If hair covers only the upper helix, it can be removed by electrolysis. If it covers the entire upper third of the ear, the hair-bearing skin can be excised, and the defect covered with a graft from the contralateral postauricular region. Perhaps the most precise and least morbid technique is to use a laser for hair removal as a first step. This can create the ideal hairline prior to insertion of the framework.

Inadequate Coverage

The Temporoparietal Fascial Flap

The temporoparietal fascial flap is very effective for obtaining additional soft tissue coverage over the auricular framework in cases in which the skin is overly thin, poorly vascularized, or the pocket is too small. This fascial flap is based on the superficial temporal artery. It is usually elevated from the ipsilateral side. It is inferiorly based, and raised off the underlying deep temporal fascia. It is turned over to cover the framework and coapted to it by suction drainage. A thick split-thickness or full-thickness skin graft is used to cover it. If this fascial flap dehisces and the framework becomes exposed, a salvage procedure has recently been described using the deep temporal fascia.

Tissue Expansion

Tissue expansion has also been described as an attempt for obtaining addition soft tissue in cases of severe skin shortage, or a low hairline. The long-term results with expansion have been disappointing. The skin in this region is relatively inelastic. In response to expansion, a thick capsule forms which can obscure the details of the auricular framework. In addition, the expanded skin contracts and can compromise the framework.

Suggested Reading

1. Aguilar IIIrd EF. Auricular reconstruction of congenital microtia (grade III). Laryngoscope 1996; 106(82):1.
2. Brent B. Auricular repair with autogenous rib cartilage grafts: Two decades of experience with 600 cases. Plast Reconstr Surg 1992; 90:355.
3. Brent B. Technical advances in ear reconstruction with autogenous rib cartilage grafts: Personal experience with 1200 cases. Plast Reconstr Surg 1999; 104(2) :319.
4. Cronin TD, Ascough BM. Silastic ear reconstruction. Clin Plast Surg 1978; 5:367.
5. Edgerton MT. Ear reconstruction in children with congenital atresia and stenosis. Plast Reconstr Surg 1969; 43:373.
6. Hackney FL. Plastic surgery of the ear. Selected readings in plastic surgery 2001; 9(16):9.
7. Nagata S. Modification of the stages in total reconstruction of the auricle: Part I-IV. Grafting the three-dimensional costal cartilage framework for lobule type microtia. Plast Reconstr Surg 1994; 93:221.
8. Tanzer RC. Congenital deformities of the auricle. In: Coverse JM, ed. Reconstructive Plastic Surgery. 2nd ed. Philadelphia: WB Saunders, 1977.
9. Tanzer RC. Total reconstruction of the auricle. The evolution of a plan of treatment. Plast Reconstr Surg 1971; 47:523.

Chapter 58

Cleft Lip

Alex Margulis

Introduction

A cleft lip is more accurately described as a cleft lip, nose and alveolar deformity because all of these anatomic structures are commonly affected. As the deformity worsens, the effect on the lip, nose and alveolar structures become more apparent.

There is no agreement on the ideal timing and the technique of repair. Advocates of different methods may demonstrate results that are comparable, underscoring the fact that more than one treatment plan is acceptable. Total familiarity with the details and limitations of a technique is as important as the type of repair chosen.

Brief History

Modern repairs have in common the use of a lateral lip flap to fill a medial defect, a concept that can be credited to Mirault. The LeMesurier repair involves the lateral quadrilateral flap, whereas the Tennison repair employs a lateral triangular flap.

In 1955, Millard described the concept of advancing a lateral flap into the upper lip combined with downward rotation of the lower segment. The benefits are two-fold: the incision lines follow the natural anatomic position of the philtral column, and placement of scars across the philtrum in the lower part of the lip is avoided. This technique has become popular because of its aesthetic advantages.

More recently, emphasis has shifted away from skin flap design and has been placed on accurate and functional reconstruction of the orbicularis oris muscle and on primary nasal reconstruction. The concept of differential reconstruction of the orbicularis oris muscle was emphasized by Miller. McCoomb, Anderl, Salyer and others have championed primary nasal tip-plasty. Primary nasal repair can achieve long-lasting improvements that can be achieved without detrimental effects on the growth and development of nasal tip cartilages.

Classification

Accurate analysis and reporting of the cleft lip and palate deformity should be done in a standardized manner. Many different classification systems were introduced over the years. In 1971, Kernahan introduced a simple classification scheme that may be reported on a diagrammatic Y-shaped symbol with the incisive foramen represented at the focal point (Fig. 58.1). This was subsequently modified by Millard, and other versions were later proposed to allow easier analysis, reporting and surgical planning.

Embryology

The branchial arches are responsible for the formation of the face, neck, nasal cavities, mouth, larynx and pharynx. The first branchial arch contributes to the maxillary and mandibular prominences. The paired maxillary and mandibular prominences form the lateral and caudal borders of the stomodeum (primitive mouth)

Practical Plastic Surgery, edited by Zol B. Kryger and Mark Sisco. ©2007 Landes Bioscience.

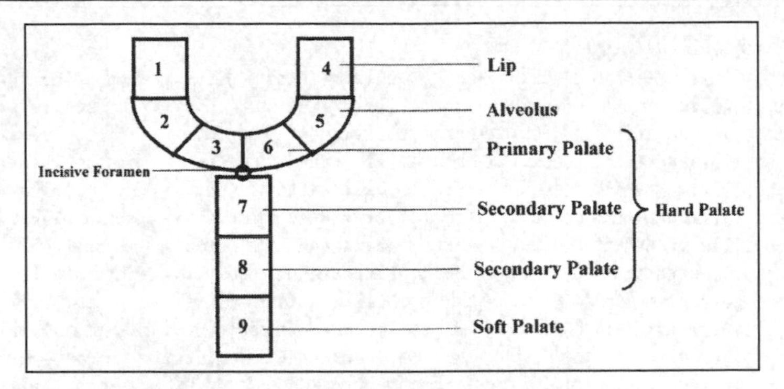

Figure 58.1. The Kernahan Y classification system for cleft lip and palate.

respectively. The frontonasal prominence, a central process formed by the proliferation of the mesenchyme ventral to the forebrain, forms the cranial boundary of the stomodeum by merging with the first arch derivatives. These five facial prominences are responsible for the development of adult facial features.

The mesenchyme of all five facial prominences that border the stomodeum is continuous; therefore mesenchymal migration may occur freely between the facial prominences. The fusion of the medial nasal, lateral nasal and the maxillary prominences produces continuity between the nose, upper lip and palate. Facial development occurs between the fourth and eight weeks. By the age of 10 weeks the face has a clearly human appearance.

Unilateral cleft lip results from failure of fusion of the medial nasal prominence and maxillary prominence on one side. A bilateral cleft lip results from failure of fusion of the merged medial nasal prominences with the maxillary prominences on both sides.

Etiology

Clefting is multifactorial, with both genetic and environmental causes cited. The observation of clustered cases of facial clefts in a particular family indicates a genetic basis. Approximately 33% to 36% of cases have a positive family history for clefting. Clefting of the lip and/or palate is associated with more than 150 syndromes. The overall incidence of associated anomalies (e.g., cardiac) is approximately 30% (more common with isolated cleft palate). Environmental causes such as viral infection (e.g., rubella) and teratogens (e.g., steroids, anticonvulsants, alcohol and smoking) during the first trimester have been linked to facial clefts. The risk also increases with advanced parental age, especially when older than 30 years, with the father's age appearing to be a more significant factor than the mother's age. Nevertheless, most presentations are of isolated patients within the family without an obvious etiology.

Incidence and Epidemiology

Ethnic variations exist in the incidence of clefting. The incidence is approximately 2.1 per 1000 live births in Asians, 1:1000 in Caucasians and 0.41:1000 in African-Americans. Isolated clefts of the lip occur in 21% of the affected population, while 46% of cases involve clefts of the lip and palate and 33% are isolated clefts of the palate. Clefts of the lip are more commonly left sided and unilateral (6: 3: 1 left: right: bilateral) and show a male predominance.

Clinical Findings

Unilateral clefts are placed in one of three categories for the purpose of treatment planning: microform cleft lip, incomplete cleft lip or complete cleft lip. The associated nasal deformity is categorized as mild, moderate or severe.

The unilateral complete cleft lip involves a full-thickness defect of the lip and alveolus (primary palate) and often is accompanied by the palatal cleft (secondary palate). The premaxilla typically is rotated outwardly and projects anterior in relation to a relatively retropositioned lateral maxillary alveolar element. The nasal structures of the ala base, nasal sill, vomer and septum are distorted significantly. The lower lateral cartilage on the cleft side is positioned inferiorly, with an obtuse angle as it flattens across the cleft. The alar base is rotated laterally, inferiorly and posteriorly (acronym LIP). The developing nasal septum pulls the premaxilla away from the cleft, and the septum and the nasal spine are deflected toward the noncleft side. The cleft continues through the maxillary alveolus and palatal shelf, extending to the palatal bone and soft palate.

The bilateral cleft lip may be either complete or incomplete. The complete cleft lip involves the entire upper lip, with the cleft traversing the alar base and potentially involving the primary and secondary palates. The anatomic components of the bilateral cleft lip include widened alar bases with flared internal nasal valves; a shortened columella; excessively obtuse nasolabial angles; a hypoplastic prolabium; a vertically short upper lip; protruding premaxillary segment; absence of the orbicularis oris muscle in the prolabial segment; absence of the philtral dimple, columns and tubercle; absence of Cupid's bow; aberrant insertion of the lateral lip orbicularis oris muscle into the alar bases; and potential involvement of the primary and secondary palates.

Timing of Repair and Treatment Planning

The goals of reconstruction include restoring the normal morphologic facial form and function as they are related intimately for proper development of dentition, mastication, hearing, speech and breathing. A multidisciplinary team approach is ideal because it provides a setting in which parents can recognize that there is a plan that will be carried out over a long term in a coordinated and specialized fashion by experts who are interested, educated and experienced in the care of children with clefts.

Presurgical orthodontic treatment is initiated in the first or second week following birth, with the maximum response occurring during the first six weeks. The lip repair with the orbicularis oris muscle reconstruction and primary nasal repair are deferred until the patient is 2 to 3 months of age. Some centers will perform alveolar closure when the segments are ideally aligned and <2 mm apart (gingivoperiosteoplasty). In our center primary alveolar bone graft is performed as a separate procedure when the child is 6 to 8 months of age and the segments are aligned. Closure of the palatal cleft is accomplished when the patient is approximately 10 to 12 months of age.

When alveolar closure is not completed in the first year of life, a definitive two-layer closure of the alveolus with cancellous bone grafting is performed between 7 and 9 years of age. The timing of this closure is mitigated by presurgical orthodontic treatment to align the segments and the guideline of obtaining surgical closure and bone grafting before eruption of the permanent canine teeth. Further correction of the nasal deformity, if required, is deferred until late adolescence.

Relevant Anatomy

An understanding of normal lip and nasal anatomy is essential for achieving a satisfactory repair. The elements of the normal lip are composed of the central philtrum, demarcated laterally by the philtral columns and inferiorly by Cupid's bow and tubercle. Just above the junction of the skin-vermilion border lies a mucocutaneous ridge known as "the white roll." Within the red portion of the lip is the wet-dry junction demarcating the moist (inner) mucosa from the dry (outer) vermilion, the increased keratinized portion of the lip that is exposed to air.

The primary muscle of the lip is the orbicularis oris, and it has two well-defined components: the deep (internal) and the superficial (external). The deep fibers run horizontally or circumferentially from commissure (modiolus) to commissure and function as the primary sphincter of the mouth. The superficial fibers run obliquely, interdigitating with the other muscles of facial expression to terminate in the dermis. They provide subtle shades of expression and precise movements of the lip for speech. The superficial fibers of the orbicularis decussate in the midline and insert into the skin lateral to the opposite philtral groove forming the philtral columns. The resulting philtral dimple centrally is depressed as there are no muscle fibers that directly insert into the dermis in the midline. The tubercle of the lip is shaped by the pars marginalis, the portion of the orbicularis along the vermilion forming the tubercle of the lip with eversion of the muscle.

In the upper lip, the levator labii superioris contributes to the form of the lip. Its fibers, arising from the medial aspect of the infraorbital rim, sweep down to insert near the skin-vermilion junction. The medial-most fibers of the levator labii superioris sweep down to insert near the corner of the ipsilateral philtral column helping to define the lower philtral column and the peak of the Cupid's bow.

The nasal muscles are equally important. The levator superioris alaeque arises along the frontal process of the maxilla and courses inferiorly to insert on the mucosal surface of the lip and ala. The transverse nasalis arises along the nasal dorsum and sweeps around the ala to insert along the nasal sill from lateral to medial into the incisal crest and anterior nasal spine. These fibers join with the oblique fibers of the orbicularis and the depressor septi (nasalis), which arises from the alveolus between the central and lateral incisors to insert into the skin of the columella to the nasal tip and the footplates of the medial crura.

A unilateral or bilateral cleft disrupts the normal termination of the muscle fibers that cross the embryologic fault line of the maxillary and nasal processes. This results in asymmetric (or symmetric but abnormal) muscular forces between the nasolabial and oral groups of muscles. With an unrestrained premaxilla, the deformity accentuates with differential growth of the various elements. The alar cartilages on the cleft side are splayed apart and rotate caudally, subluxed from the normal position. Consequently, the nasal tip broadens, the columella is foreshortened and the alar bases rotate outwardly and cephalad.

Laboratory and Imaging Studies

Routine lab studies are not necessary in otherwise healthy infants with a cleft. Some centers obtain a blood count as a routine study before performing surgery on a child with cleft. At our institution, we do not find this necessary unless some other associated medical conditions coexist. The child's weight, oral intake and growth and/or development are of primary concern and must be followed closely. Routine imaging studies are also not needed in otherwise healthy infants who undergo cleft lip repair.

Diagnostic Procedures

Early collaboration with an audiologist and an otolaryngologist, including examination and early audiologic assessment; can prevent long-term hearing deficits in patients with cleft lip and palate. Patients with isolated cleft lip are not believed to have a higher incidence of middle ear disease. If, however, the palate is also clefted then there is a significant risk of inner ear infections due to eustachian tube dysfunction (see chapter on cleft palate repair).

Orthodontic Interventions

The goals of presurgical nasal and alveolar molding are the active molding and repositioning of the nasal cartilages and alveolar processes, and the lengthening of the deficient columella. This method takes advantage of the plasticity of the cartilage in the newborn infant during the first 6 weeks after birth. This high degree of plasticity in neonatal cartilage is due to elevated levels of hyaluronic acid, a component of the proteoglycan intracellular matrix. A description of the protocol for treatment of the patient with bilateral cleft deformity was introduced in 1993 by Grayson, Cutting and Wood. This combined technique has been demonstrated to have a positive influence on the outcome of the primary nasal, labial and alveolar repair.

Repair of Unilateral Cleft Lip

General endotracheal anesthesia with an oral Rae tube is used for all stages of cleft lip repair. A cursory description of a modified Millard operative technique is as follows:

Presurgical Marking (Fig. 58.2)

The key points that are identified and marked are as follows:

- Midline and bases of the columella (1, 6)
- Alar base
- Peak and midpoint of Cupid's bow on the noncleft side (2, 3)
- Proposed point of Cupid's bow on the cleft side (4)

Two key elements are involved in the markings: the placement of the final position of the new Cupid's bow peak and the vertical length of the philtral column to be created on the cleft side. Referring to the diagram, Point 3 is determined as the mirror image of Point 2 based on the distance from the midpoint to the peak of the Cupid's bow on the noncleft side. The peak on the cleft side, Point 4, is not determined as

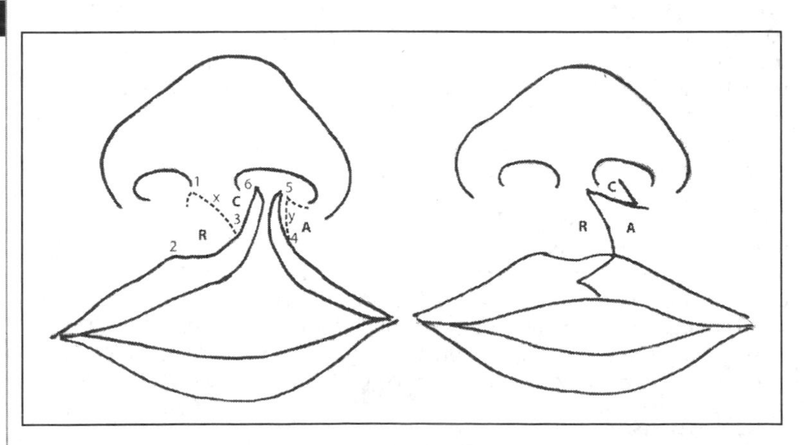

Figure 58.2. Presurgical marking.

easily but typically is placed level with Point 2, where the dry vermilion is widest and the white roll above is well developed. The white roll and the dry vermilion taper off medial to this point. It is unreliable to determine the peak on the cleft side, using the distance between the peaks of the Cupid's bow from the commissure on the noncleft side because of unequal tension of the underlying orbicularis muscle.

Once the anatomic points are marked, draw incision lines that define the five flaps involved in the lip reconstruction. These are the inferior rotation flap (R) of the medial lip element, the medial advancement flap (A) of the lateral lip element, the columellar base flap (C) of the medial lip element, and the two pared mucosal flaps of the medial and lateral lip elements. An additional flap that refines the repair is the vermilion triangular flap to allow for a smoother transition at the vermilion cutaneous junction and at the vermilion contour.

The essential marking is the line that determines the border between the R and C flaps. This line becomes the new philtral column on the cleft side. For the vertical lengths of the philtrum on the cleft side and noncleft side to be symmetric, the length of the rotation advancement flap (y) should equal the vertical length of the philtral column (x) on the noncleft side (distance between alar base and Cupid's bow peak). For the two lengths, x and y, to be equal, the path of y must be curved as illustrated. In marking the curve, take care to avoid a high arching curve that comes too high at the columellar base to create a generous philtrum, as this significantly diminishes the size of the C flap.

Description of the Repair

After markings, 0.5% lidocaine with epinephrine (1:200,000) is injected into the lip and the nose. In the region of the vermilion-cutaneous junction, incise the muscle for approximately 2-3 mm on either side of the cleft paralleling the vermilion border to allow development of vermilion-cutaneous muscular flaps for final alignment.

Develop the R and C flaps by incising the line (x) between the flaps to allow inferior rotation of the R flap so that it lies horizontally tension free with Point 3, level with Point 2. For this to occur, release must be at all levels (skin, subcutaneous tissue, muscle, fibrous attachments to the anterior nasal spine, labial mucosa). Correspondingly free the C flap with the medial crus of the alar cartilage and allow it to be repositioned, creating a large gap to be filled by the A flap.

Develop the A flap from the lateral lip element for advancement into the gap between the R and C flaps. In developing the A flap, keep the incision along the alar base at a minimum; it rarely is required to extend much beyond the medial-most aspect of the alar base. A lateral labial mucosal vestibular release is also required to mobilize the A flap medially and to avoid a tight-appearing postoperative upper lip deformity.

As part of the mobilization of the ala, make an incision along the nasal skin-mucosal vestibular junction (infracartilaginous) where the previously developed L flap may be interposed if needed. Widely undermine the nasal tip between the cartilage and the overlying skin approaching laterally from the alar base and medially from the columellar base. While the A flap can be inserted as a mucocutaneous flap incorporating the orbicularis, the author repairs the muscle separately to allow for differential reorientation of its vectors. Dissect the muscle from the overlying skin and the underlying mucosa to accomplish this and divide it into bundles that can be repositioned and interposed appropriately.

Once all the flaps are developed and the medial and lateral lip elements are well mobilized, begin reconstruction. Typically, this begins with creating the labial vestibular lining from superior to inferior and then proceeding to the junction of the wet-dry vermilion with completion of the remainder of the vermilion after the cutaneous portion of the lip is completed. At this point, the labial mucosa can be advanced as needed, with additional lengthening and a back cut to allow for adequate eversion of the lip and to avoid a tight-appearing lip postoperatively. Approximation of the muscle bundles must be complete. Appropriately reorient the nasolabial group of muscles toward the nasal spine. Follow this by approximating the orbicularis, interdigitated with its opposing element along the full length of the vertical lip.

Inset the C flap to create a symmetric columellar length and flare at its base. Millard originally described the C flap to cross the nasal sill to insert into the lateral lip element as a lateral rotation-advancement flap. Millard later refined the C flap as a medial superior rotation flap to insert into the medial lip element, augmenting the columellar height and creating a more natural flare at the base of the medial footplate. The latter method occasionally results in a nexus of scars at the base of the columellar with unfavorable healing if the flaps are not well planned. However, the author continues to use the C flap in either position as needed. Set the ala base in place. As the C and A flaps and the ala are inset, take care to leave an appropriate width to the nasal sill to avoid a constricted-appearing nostril, which is nearly impossible to correct as a secondary deformity.

Approximate the vermilion-cutaneous junction and inset the vermilion mucocutaneous triangular flap. Use dermal sutures to approximate the skin edges. Final approximation is with nylon sutures, ideally removed at 5 days. If the cutaneous edges are well approximated with dermal sutures alone, one may occasionally use a cyanoacrylate-type adhesive. Reposition the cleft alar cartilage with suspension/transfixion sutures and a stent. Further shape the ala with through-and-through absorbable sutures as needed.

Repair of Bilateral Cleft Lip

Originating on either side of the columellar base, vertical lines are marked ending in a triangular base such that Cupid's bow is 6 to 8 mm wide. Lateral forked flaps are also outlined prior to making the skin incisions. All philtral-based flaps are elevated from the surrounding vermilion. The prolabial mucosal vermilion complex is thinned before being sutured together, creating the midline posterior labial sulcus.

The lateral lip segments are incised vertically down from the medial alar base, analogous to the originally made prolabial incisions. Medially-based buccal mucosal flaps are rotated from the alar base horizontally. The alar cartilages are freed via an intercartilaginous incision, originating from the piriform aperture, and secured together at the domes and to the upper lateral cartilages. The buccal mucosal flaps are then sutured into the inferior intercartilaginous incision to increase length for the nasal floor reconstruction. The mucosal orbicularis flaps are sutured together to create the anterior labial sulcus, with the most superior suture secured to the nasal spine to prevent inferior displacement. Finally, the inferior white roll-vermilion-mucosal flaps are apposed to create Cupid's bow and tubercle complex.

Postoperative Considerations

For the child who is breastfed, the author encourages uninterrupted breastfeeding after surgery. Some centers will allow bottle-fed children to resume feedings immediately following surgery with the same crosscut nipple used before surgery, while others have the child use a soft catheter-tip syringe for 10 days and then resume normal nipple bottle feeding.

The author uses velcro elbow immobilizers on the patient for 10 days to minimize the risk of the child inadvertently injuring the lip repair. The parents are instructed to remove the immobilizers from alternate arms several times a day in a supervised setting. For the child with sutures, lip care consists of gently cleansing suture lines using cotton swabs with diluted hydrogen peroxide and liberal application of topical antibiotic ointment several times a day. This is continued for 10 days. If cyanoacrylate adhesive is used, no additional care is required in the immediate postoperative period until the adhesive film comes off. The parents are told to expect scar contracture, erythema and firmness for the first 4-6 weeks postoperatively, and that this gradually begins to improve 3 months after the procedure. Typically, parents are also instructed to massage the upper lip during this phase and to avoid placing the child in direct sunlight until the scar matures.

Pearls and Pitfalls

There is no agreement on the ideal timing and the technique of the repair. It is important for the surgeon to view the various repairs as principles of repair—a guideline to be followed, not a rigid design to which the surgeon must strictly adhere. Cleft lip repair is one of the few procedures that has a lot of room for modifications and innovations on the part of the surgeon.

Occasionally, an additional 1- to 2-mm back cut just medial to the noncleft philtral column is required along with a mucosal back cut to allow for adequate inferior rotation of the rotation (R) flap.

The current trend in cleft surgery is toward a more aggressive mobilization and repositioning of the lower lateral cartilages of the nose as an integral part of the cleft lip repair.

It is important to recall that the maxillary alveolar arches typically are at different heights in the coronal plane, and the ala must be released completely and mobilized superomedially to achieve symmetry, although ultimately its maxillary support is inadequate until arch alignment and bone grafting can be accomplished.

After approximation of the vermilion-cutaneous junction and inset of the vermilion mucocutaneous triangular flap, the lip may appear to be vertically short. One solution is to inset a small, 2- to 3-mm triangular flap into the medial lip just above the vermilion.

Suggested Reading

1. Afifi GY, Hardesty RA. Bilateral cleft lip. Plastic Surgery, Indications, Operations and Outcomes. St. Louis: Mosby, 2000:769-797.
2. Byrd HS. Unilateral cleft lip. Grabb and Smith Plastic Surgery. 5th ed. Philadelphia: Lippincott-Raven Publishers, 1997:245-253.
3. Cutting CB. Primary bilateral cleft lip and nose repair. Grabb and Smith Plastic Surgery. 5th ed. Philadelphia: Lippincott-Raven Publishers, 1997:255-262.
4. Grayson BH, Santiago PE. Presurgical orthopedics for cleft lip and palate. Grabb and Smith Plastic Surgery. 5th ed. Philadelphia: Lippincott-Raven Publishers, 1997:237-244.
5. Kernahan DA. On cleft lip and palate classification. Plast Reconstr Surg 1973; 51:578.
6. LaRossa D. Unilateral cleft lip repair. Plastic Surgery, Indications, Operations and Outcomes. St. Louis: Mosby, 2000:755-767.
7. LaRossa D, Donath G. Primary nasoplasty in unilateral and bilateral cleft lip nasal deformity. Clin Plast Surg 1993; 29(4):781.
8. Mulliken JB. Primary repair of the bilateral cleft lip and nasal deformity. In: Georgiade GS, ed. Plastic, Maxillofacial and Reconstructive Surgery. 3rd ed. Philadelphia: Williams and Wilkins, 1997.
9. Salyer KF. Primary correction of the unilateral cleft lip nose: A 15-years experience. Plast Reconstr Surg 1986; 77:558.

Chapter 59

Cleft Palate

Alex Margulis

Introduction

A cleft palate has tremendous aesthetic and functional implications for patients in their social interactions, particularly on their ability to communicate effectively and on their facial appearance. The treatment plan focuses on two areas: speech development and facial growth. Speech development is paramount in the appropriate management of cleft palate. Many surgical techniques and modifications have been advocated to improve functional outcome and aesthetic results. The most controversial issues in the management of cleft palate are the timing of surgical intervention, speech development after various surgical procedures and the effects of surgery on facial growth.

Classification

Numerous classifications have been suggested over the years. The most common classification scheme is that of Kernahan (Fig. 59.1). This "striped Y" classification has been almost universally adopted for its simplicity and usefulness. A modification of the Kernahan classification was introduced several years ago by Smith et al and it uses an alphanumeric system to describe all cleft varieties.

Embryology

The embryogenesis of the palate has two separate phases: the formation of the primary palate followed by the formation of the secondary palate. Palatal development begins at approximately day 35 of gestation with the emergence of facial processes. In formation of the primary palate, the fusion of the medial nasal process (MNP) with the maxillary process (MxP) is followed by the lateral nasal process (LNP) fusing with the MNP. Failure of fusion or breakdown of fusion of the processes results in a cleft of the primary palate. The genesis of the secondary palate begins at the completion of primary palate formation. The secondary palate arises from the bilateral shelves that develop from the medial aspect of the MxP. The two shelves meet in the midline, and the fusion process begins as the shelves move superiorly. Interference in the fusion leads to clefting of the secondary palate.

Incidence

The incidence of cleft lip/palate (CL/P) by race is 2.1/1000 in Asians, 1/1000 in whites and 0.41/1000 in blacks. Isolated cleft palate shows a relatively constant ratio of 0.45-0.5/1000 births. The foremost type of clefting is a bifid uvula, occurring in 2% of the population. The second most frequent type is a left unilateral complete cleft of the palate and prepalatal structures. Midline clefts of the soft palate and parts of the hard palate are also common. Complete clefts of the secondary palate are twice as common in females as in males while the reverse is true of velar clefts. About

Practical Plastic Surgery, edited by Zol B. Kryger and Mark Sisco. ©2007 Landes Bioscience.

7-13% of patients with isolated cleft lip and 11–14% of patients with CL/P have other anomalies at birth.

Inheritance Patterns

In 25% of patients, there is a family history of facial clefting, which does not follow either a normal recessive or dominant pattern. The occurrence of clefting deformities do not correspond to any Mendelian pattern of inheritance, and it would appear that clefting is inherited heterogeneously. This observation is supported by evidence from studies of twins that indicate the relative roles of genetic and nongenetic influences of cleft development. For isolated cleft palate and combined CL/P, if the proband has no other affected first- or second-degree relatives, the empiric risk of a sibling being born with a similar malformation is 3-5%. However, if a proband with a combined CL/P has other affected first-degree relatives, the risk for siblings or subsequent offspring is 10-20%.

Etiology

The causes of cleft palate appear to be multifactorial. Some instances of clefting may be due to an overall reduction in the volume of the facial mesenchyme, which leads to clefting by virtue of failure of mesodermal penetration. In some patients, clefting appears to be associated with increased facial width, either alone or in association with encephalocele, idiopathic hypertelorism, or the presence of a teratoma. The characteristic U-shaped cleft of the Pierre Robin anomaly is thought to be dependent upon a persistent high position of the tongue, perhaps associated with a failure or delay of neck extension. This prevents descent of the tongue, which in turn prevents elevation and a medial growth of the palatal shelves.

The production of clefts of the secondary palate in experimental animals has frequently been accomplished with several teratogenic drugs. Agents commonly used are steroids, anticonvulsants, diazepam and aminopterin. Phenytoin and diazepam may also be causative factors in clefting in humans. Infections during the first trimester of pregnancy, such as rubella or toxoplasmosis, have been associated with clefting.

Clinical Findings

The pathologic sequelae of cleft palate can include airway issues, feeding and nutritional difficulties, abnormal speech development, recurrent ear infections, hearing loss and facial growth distortion.

Airway Problems

The infant with Pierre Robin sequence or other conditions in which the cleft palate is observed in association with a micrognathia or retrognathic mandible may be particularly prone to upper airway obstruction. Prone positioning is the initial step in management.

Feeding Difficulty

The communication between the oral and nasal chamber impairs the normal sucking and swallowing mechanism of the cleft infants. Food particles can reflux into the nasal chamber. Although a child with cleft palate may make sucking movements with the mouth, the cleft prevents the child from developing adequate suction. However, in general swallowing mechanisms are normal. Therefore, if the milk or formula can be delivered to the back of the child's throat, the infant feeds effectively. Breastfeeding is usually not successful unless milk production is abundant.

Speech Abnormalities

Speech abnormalities are intrinsic to the anatomic derangement of cleft palate. The facial growth distortion appears to be, to a great extent, secondary to surgical interventions. An intact velopharyngeal mechanism is essential in production of nonnasal sounds and is a modulator of the airflow in the production of other phonemes that require nasal coupling. The complex and delicate anatomic manipulation of the velopharyngeal mechanism, if not successfully learned during early speech development, can permanently impair normal speech acquisition.

Middle Ear Disease

The disturbance in anatomy associated with cleft palate affects the function of the eustachian tube orifices. Parents and physicians should be aware of the increased possibility of middle ear infection so that the child receives treatment promptly if symptoms arise. The abnormal insertion of the tensor veli palati prevents satisfactory emptying of the middle ear. Recurrent ear infections have been implicated in the hearing loss of patients with cleft palate. The hearing loss may worsen the speech pathology in these patients. Evidence that repair of the cleft palate decreases the incidence of middle ear effusions is inconsistent. However, these problems are overshadowed by the magnitude of the speech and facial growth problems.

Facial Growth Abnormalities

Multiple studies have demonstrated that the cleft palate maxilla has some intrinsic deficiency of growth potential. This intrinsic growth potential deficiency varies from isolated cleft of the palate to complete CL/P. This growth potential is further impaired by surgical repair. Any surgical intervention performed prior to completion of full facial growth can have deleterious effects on maxillary growth. Disagreement exists as to the appropriate timing of surgery to minimize the harmful effects on facial growth and on what type of surgical intervention is most responsible for growth impairment. The formation of scar and scar contracture in the areas of denuded palatal bones are most frequently blamed for restriction of maxillary expansion. The growth disturbance is exhibited most prominently in the prognathic appearance during the second decade of life despite the normal appearance in early childhood. The discrepant occlusion relationship between the maxilla and the mandible is usually not amenable to nonsurgical correction.

Associated Deformities

The surgeon must always keep in mind that in as many as 29% of patients, the child with cleft palate may have other anomalies. These may be more commonly associated with isolated cleft palate than with CL/P. High among the associated anomalies are those affecting the circulatory and skeletal systems.

Surgical Goals and the Benefits of Repair

The broad goal of cleft palate treatment is to separate the oral and nasal cavities. Although this is not absolutely necessary for feeding, it is advantageous to normalize feeding and decrease regurgitation and nasal irritation. More important than repairing the oral and nasal mucosa is the repositioning of the soft palate musculature to anatomically recreate the palate and to establish normal speech. Another goal of palate repair is to minimize restriction of growth of the maxilla in both sagittal and transverse dimensions.

Palate repair with repositioning of the palatal musculature may be advantageous to eustachian tube function and ultimately to hearing. Because the levator and the tensor veli palatini have their origins along the eustachian tube, repositioning improves function of these muscles, improves ventilation of the middle ear and decreases serous otitis, which further decreases the incidence of hearing abnormality. Palate repair alone does not usually completely correct this dysfunction and additional therapy frequently includes placement of ear tubes as necessary.

Relevant Anatomy

The bony portion of the palate is a symmetric structure divided into the primary and secondary palate based on its embryonic origin (Fig. 59.1). The premaxilla, alveolus and lip, which are anterior to the incisive foramen, are parts of the primary palate. Structures posterior to it, which include the paired maxilla, palatine bones and pterygoid plates, are part of the secondary palate. The severity of the clefting of the bony palate varies from simple notching of the hard palate to complete clefting of the alveolus. The palatine bone is located posterior to the maxilla and pterygoid lamina. It is composed of horizontal and pyramidal processes. The horizontal process contributes to the posterior aspect of the hard palate and becomes the floor of the choana. The pyramidal process extends vertically to contribute to the floor of the orbit.

Even though the bony defect is important in the surgical treatment of cleft palate, the pathology in the muscles and soft tissues has the greatest impact on the functional result. Six muscles have attachment to the palate: levator veli palatini, superior pharyngeal constrictor, musculus uvulae, palatopharyngeus, palatoglossus and tensor veli palatini. The three muscles that appear to have the greatest contribution to the velopharyngeal function are the musculus uvulae, levator veli palatini and superior pharyngeal constrictor.

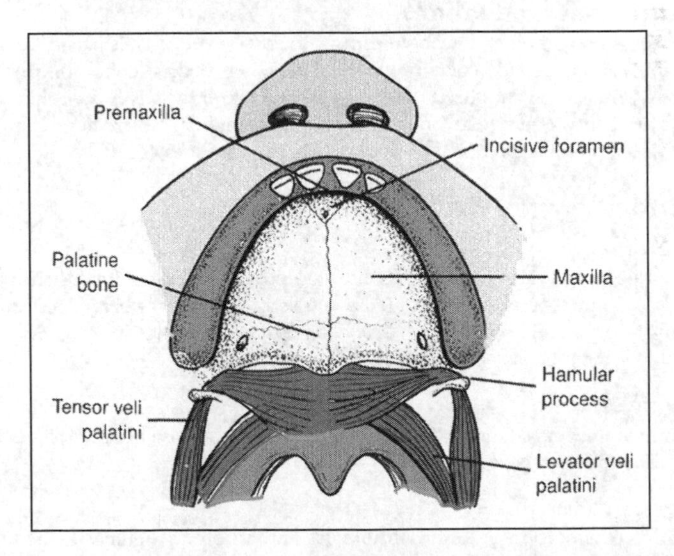

Figure 59.1. Anatomy of the palate. (Reprinted from emedicine.com with permission.)

The musculus uvulae muscle acts by increasing the bulk of the velum during muscular contraction. The levator veli palatini pulls the velum superiorly and posteriorly to appose the velum against the posterior pharyngeal wall. The medial movement of the pharyngeal wall, attributed to the superior pharyngeal constrictor, aids in the opposition of the velum against the posterior pharyngeal wall to form the competent sphincter. The palatopharyngeus displaces the palate downwards and medially. The palatoglossus is mainly a palatal depressor that plays a role in the production of phonemes with nasal coupling by allowing controlled airflow into the nasal chamber. The tensor veli palatini does not contribute to the movement of the velum. The tensor veli palatini's tendons hook around the hamulus of the pterygoid plates and the aponeurosis of the muscle inserts along the posterior border of the hard palate. The muscle originates partially on the cartilaginous border of the auditory tubes. The function of the tensor veli palatini, similar to the tensor tympani with which it shares its innervation, is to improve the ventilation and drainage of the auditory tubes.

In a cleft palate, the aponeurosis of the tensor veli palatini, instead of attaching along the posterior border of the hard palate, is attached along the bony cleft edges. All the muscles that attach to the palate insert onto the aponeurosis of this muscle. Thus, the overall length of the palate is shortened. The abnormality in the tensor veli palatini increases the incidence of middle ear effusion and middle ear infection. The muscle sling of the levator veli palatini is also interrupted by the cleft palate. The levator does not form the complete sling. The medial portion of each side attaches to the medial edge of the hard palate. Thus, in patients with cleft palate, the effectiveness of the velar pull against the posterior pharyngeal wall is impaired. Of the six muscles, the prevailing theory attributes most of the contribution to the velopharyngeal competence to the levator veli palatini.

Relative Contraindications

There are no absolute contraindications for the repair of cleft palate. Relative contraindications include concurrent illness or other medical condition that can interfere with general anesthesia, possible compromise of the airway in a child with a preexisting airway problem (such as severe micrognathia), severe developmental delay, or a short life expectancy because of other severe illnesses.

Preoperative Considerations

Lab Studies

- Routine lab studies are noncontributory in otherwise healthy infants with cleft palate. Some centers obtain a blood count as a routine study before performing surgery on a child with cleft palate. The author does not find this necessary unless some other associated medical condition coexists.

Imaging Studies

- Routine imaging studies are noncontributory in otherwise healthy infants who undergo primary cleft palate repair.

Diagnostic Procedures

- Early collaboration with an audiologist and an otolaryngologist, including examination and early audiologic assessment, can prevent long-term hearing deficits.

Orthodontic Intervention

The available data suggest that to optimize speech development some degree of facial growth distortion may need to be accepted. One role of orthodontic intervention is to minimize the severity of the growth disturbance. Interventions vary according to the type of cleft. Many types of orthodontic appliances have been used in the treatment of cleft palate patients. In CL/P, orthodontic appliances can be used to realign the premaxilla into a normal position prior to lip closure. Orthodontic interventions in patients with cleft palate are frequently aimed at maxillary arch expansion, correction of malocclusion and correction of a developing class III skeletal growth pattern. The maxillary dental arch contracture may become significant, requiring the surgical repair of the hard palate.

Orthodontic interventions may be started early or delayed for several years. When orthodontic manipulation is initiated early, difficulties may occur. Maintaining orthodontic appliances in the infant population may present a challenge unless these appliances are fixed in position. The benefit of these orthodontic interventions has also been questioned, especially in patients with isolated cleft palate. The most beneficial period for orthodontic interventions in isolated cleft palate may be during the mixed dentition period.

At the age of 6-8 years, the permanent incisors begin erupting. At this age, the presence of grossly misaligned teeth and severe malocclusion can lead to social isolation. The incisor relation can be corrected and maintained with relatively simple interventions. Patients who undergo palatal arch expansion during this period can benefit from the rapid growth phase. The orthodontic intervention can also proceed with more cooperation from the patient in this age group. Orthodontic management of arch deformities after the permanent dentition has erupted is more limited. The established malocclusion and asymmetry between the maxillary arch and mandibular arch usually require orthognathic surgery.

Surgical Therapy

Timing of Palatal Closure

The timing of cleft palate closure remains controversial. The goals of palatal repair include normal speech, normal palatal and facial growth and normal dental occlusion. Early palate repair is associated with better speech results, but early repair also tends to produce severe dentofacial deformities. Several studies have consistently shown that children whose palates were repaired at an earlier age appeared to have better speech and needed fewer secondary pharyngoplasties compared to those whose surgery had been delayed beyond the first 12 months of life.

Noordhoff and associates found that children undergoing delayed palatoplasty for cleft palate had significantly poorer articulation skills before the hard palate closure compared to children of the same age who did not have clefts. These benefits of early cleft palate repair, from the standpoint of speech and hearing, must be weighed against the increased technical difficulty of performing the procedure at a younger age and possible adverse effects on maxillary growth. Numerous studies failed to demonstrate an observable difference in underdevelopment of the palatal arch among children undergoing operations at various ages. The surgical intervention appears to interfere with midfacial growth without regard to the age of the patient at the time of repair.

Bifid uvula occurs in 2% of the population. Although this can occur in association with a submucous cleft palate, most infants with bifid uvula do not have this problem. The recommended management of a bifid uvula is close observation to ensure that speech develops normally.

Sequence of the Procedures

Multiple protocols for the management of CL/P have been suggested over the years by various authors. Today, the mainstream of cleft repair calls for closure of the lip at an early age (from age 6 weeks to 6 months) followed by closure of the palate approximately 6 months later. This protocol has little impact on facial development. When managing a residual alveolar defect and an associated oronasal fistula, the primary goal of surgery is to allow subsequent development of a normal alveolus. Optimal eruption of teeth at the cleft site and development of normal periodontal structures of the teeth adjacent to the cleft occur when bone grafting and final fistula closure are performed prior to eruption of the permanent canine teeth at the cleft site.

Choice of Repair

The list of surgical techniques used in palatal cleft closure is extensive. The repairs differ depending upon whether the cleft is an isolated cleft palate or part of a unilateral or bilateral cleft lip and palate. The three main categories include: (1) simple palatal closure; (2) palatal closure with palatal lengthening; and (3) either of the first two techniques with direct palatal muscle reapproximation.

The Von Langenbeck Procedure

The simple palatal closure was introduced by von Langenbeck and is the oldest cleft palate operation in wide use today (Fig. 59.2). The bipedicle mucoperiosteal flaps are created by incising along the oral side of the cleft edges and along the posterior alveolar ridge from the maxillary tuberosities to the anterior level of the cleft. The flaps are then mobilized medially with preservation of the greater palatine arteries and closed in layers. The hamulus may need to be fractured to ease the closure. The von Langenbeck repair continues to be popular because of the simplicity of the operation. This technique can successfully close moderate-sized defects. Modern critics of the von Langenbeck technique cite the unnecessary anterior fistulas it promotes, the insufficiently long palate it produces and the poor speech result associated with it.

Trier and Dreyer combined primary von Langenbeck palatoplasty with levator sling reconstruction (intravelar veloplasty). The author has observed better speech and superior velopharyngeal function following intravelar veloplasty with muscle reconstruction and recommended careful reconstruction of the levator sling at the time of palate repair.

Palatal Lengthening or V-Y Pushback

Veau's protocol for closure of cleft palate stressed the need for (1) closure of the nasal layer separately, (2) fracture of the hamular process, (3) staged palatal repair following primary lip and vomer flap closure, and (4) creation of palatal flaps based on a vascular pedicle. Kilner and Wardill devised a technique of palatal repair in 1937 that was more radical than Veau's and that ultimately became the V-Y pushback. It includes lateral relaxing incisions, bilateral flaps based on greater palatine vessels, closure of the nasal mucosa in a separate layer, fracture of the hamulus, separate muscle closure and V-Y palatal lengthening.

The 4-flap technique is similar to the Wardill-Kilner 2-flap technique, except the oblique incisions are more posterior to create four unipedicle flaps. The flaps are again mobilized medially and closed. These pushback techniques achieve greater

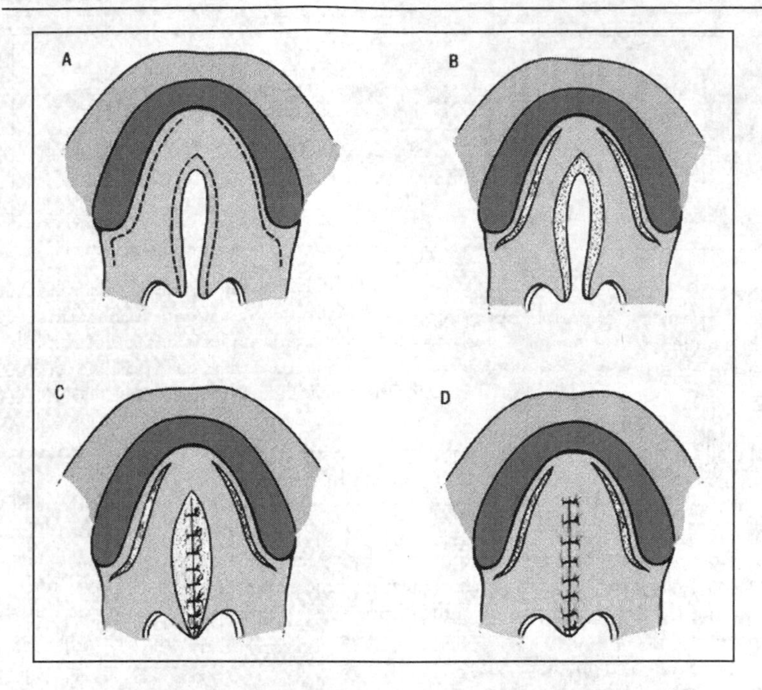

Figure 59.2. The von Langenbeck repair. Two bipedicle mucoperiosteal flaps are created by incising along the oral side of the cleft edges and along the posterior alveolar ridge from the maxillary tuberosities to the anterior level of the cleft. The flaps are then mobilized medially with preservation of the greater palatine arteries and closed in layers. The hamulus may need to be fractured to ease the closure. (Reprinted from emedicine.com with permission.)

immediate palatal length but at the cost of creating a larger area of denuded palatal bone anterolaterally. The gain in the length of the palate has not been demonstrated to be permanent nor has it translated to improved velopharyngeal function. This approach has been associated with a higher incidence of fistula formation.

Intravelar Veloplasty

Several studies have emphasized the necessity of realignment of the muscle in the soft palate. The approach was designed to lengthen the palate as well as to restore the muscular sling of the levator veli palatini. Improved velopharyngeal function was sporadically reported. Marsh et al conducted a prospective study of the effectiveness of primary intravelar veloplasty and found no significant improvement in velopharyngeal function.

Double-Opposing Z-Plasties

In 1986, Furlow described a single-stage palatal closure technique consisting of double opposing Z-plasties from the oral and nasal surfaces (Fig. 59.3). Use of the double Z-plasty minimized the need for lateral relaxing incisions to accomplish closure. The palate was also lengthened as a consequence of the new position of the

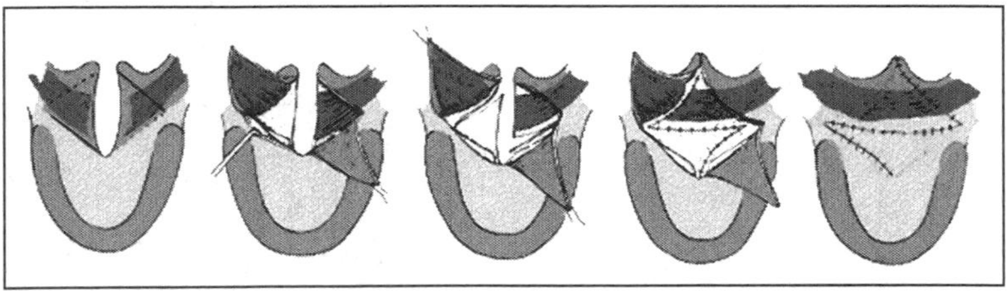

Figure 59.3. Double-opposing Z-plasties. Furlow's single-stage palatal closure technique consisting of double opposing Z-plasties from the oral and nasal surfaces. The double Z-plasty minimizes the need for lateral relaxing incisions to accomplish closure. The palate is lengthened as a consequence of the new position of the velar and pharyngeal tissues. (Reprinted from emedicine.com with permission.)

velar and pharyngeal tissues. Preliminary data revealed that speech development was excellent, with 86% exhibiting normal speech in Furlow's study.

Others have confirmed the improvement in speech development. The closure of the hard palate in Furlow's technique avoids the use of lateral relaxing incisions. The mucoperiosteal flaps are mobilized from the bony hard palate and the palatal defect closed by tenting the flaps across and creating a moderate empty space between the flaps and the bony hard palatal vault. Furlow's technique appears to be quite successful in clefts of limited size. In moderate-size clefts, lateral relaxing incisions may still be required to obtain closure.

Two-Flap Palatoplasty

Bardach (1984) and Salyer independently modified the 2-flap palatoplasty to combine elements of other operations with some innovative details (Fig. 59.4). The main goals are complete closure of the entire cleft without tension at an early age (<2 mo) with minimal exposure of raw bony surfaces and the creation of a functioning soft palate. The authors believe that a muscle sling within the soft palate, not velar lengthening, is essential to adequate speech. Morris and colleagues note that 80% of patients treated with this method developed velopharyngeal function within normal limits, although 51% required speech therapy before normal speech production was achieved.

Velar Closure—Delayed Hard Palate Closure

Schweckendiek (1978) closed the soft palate early (at age 6-8 mo) but left the hard palate open, albeit occluded with a prosthetic plate, until the age of 12-15 years. In unilateral clefts the soft palate is closed first, followed by lip surgery 3 weeks later. In bilateral clefts one side of the lip is closed first in conjunction with primary veloplasty, with repair of the other side of the lip and the alveolar cleft 3 weeks later. Schweckendiek reported normal jaw development subsequent to this protocol. Many European surgeons now use Perko's (1991) approach of two-stage palatal closure. Repair of the soft palate occurs at age 18 months and of the hard palate at 5-8 years. Perko found that the remaining cleft in the hard palate does not disturb speech development to a significant degree. Several long-term assessments of patients who undergo the Schweckendiek approach or the Perko (Zurich) approach disclosed an unusually high incidence of short palate and poor mobility of the soft palate, with a correspondingly high degree of velopharyngeal insufficiency (VPI). Conversely, facial growth was judged to be quite acceptable in most patients.

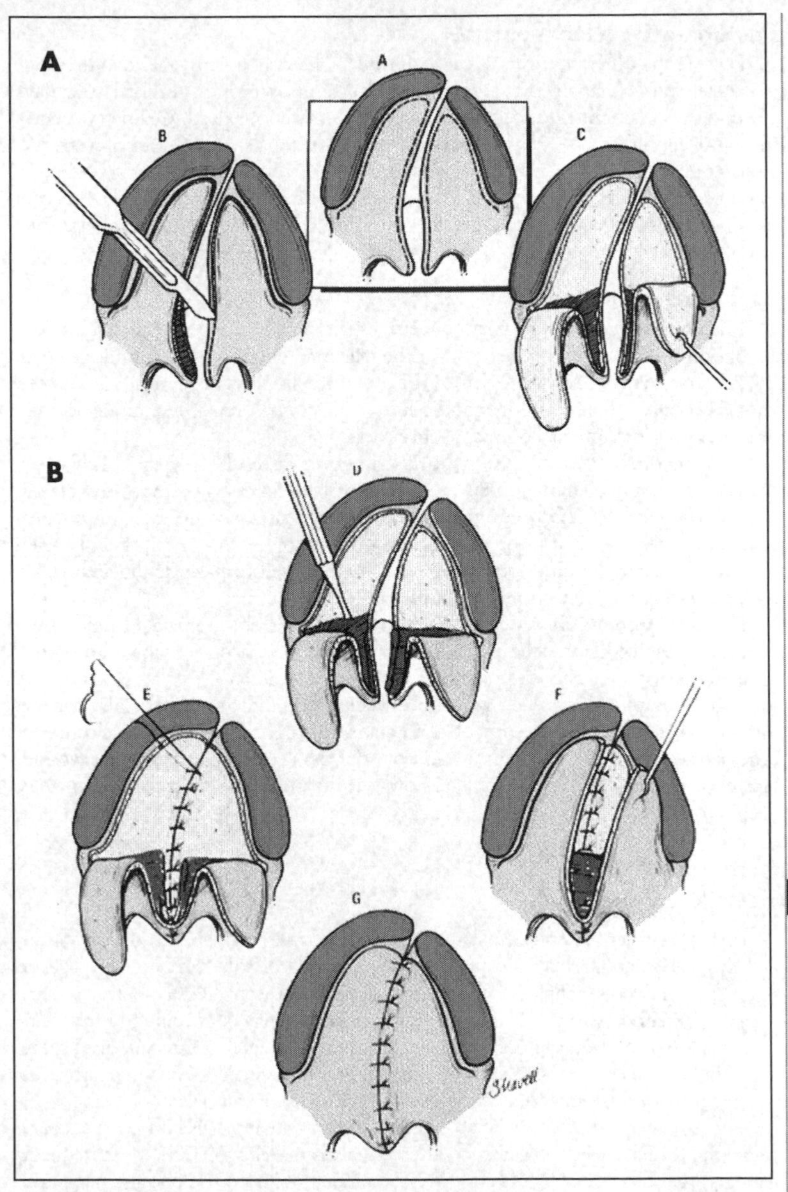

Figure 59.4. Two-flap palatoplasty. A) After lateral relaxing incisions are performed, bilateral flaps are elevated based on greater palatine vessels. B) Closure of the nasal mucosa is performed. The hamulus may be fractured, the muscle is repaired, and the oral mucosa is closed as a separate layer. (Reprinted from emedicine.com with permission.)

Postoperative Management

Despite the difference in surgical technique, a general postoperative routine exists. After surgical repair, the child is given nothing by mouth until the next day. Hydration is maintained during this time with intravenous fluid. Oximetry is continuously monitored for 24-48 hours. Arm splints are also applied to prevent the child from disrupting the wound by placing his fingers in his mouth. Oral feeding is initiated by syringes or drinking from cups. Nipple feeding is avoided. Patients can usually be discharged the day after the operation with extension arm splints. The liquid diet is continued for 7-10 days with solid food to follow.

Early Complications

The complications of greatest concern in the immediate postoperative period are bleeding and respiratory distress, yet the true incidence of these complications is difficult to determine from a review of the literature. Reports of surgical experiences with CL/P typically mix children and adults, type of cleft, repair technique, timing of the surgery, or sequence of operations.

Some reports suggest that the Wardill-Kilner repair results in greater morbidity than other methods. This technique typically involves increased postoperative bleeding following division of the anterior branch of the greater palatine artery. Epinephrine is routinely injected prior to the incision to allow better visibility and easier control of bleeding. Hemostatic agents can also be used to pack denuded areas of the palate to minimize the amount of bleeding.

Respiratory compromise secondary to obstruction from the palate lengthening or sedation can be life threatening. Airway obstruction was considerably more common after a von Langenbeck procedure with pharyngeal flap.

Other complications, such as wound dehiscence and oronasal fistula, can be difficult to manage. Dehiscence of the palatal closure, as with wound closure in other parts of the body, is usually a result of poor tissue quality and excessive wound tension. The incidence of dehiscence is low, but the incidence of oronasal fistula has been reported at 5-29%.

Long-Term Complications

Palatal Fistula

Fistula treatment after cleft palate repair is a difficult problem. Fistulas are classified as prealveolar, alveolar and postalveolar. A fistula of sufficient size can lead to significant problems, ranging from food passing into the nasal chamber to speech difficulties secondary to nasal air emission. Factors that may contribute to fistula formation are type of cleft, type of repair, wound tension, single-layer repair, dead space below the mucoperiosteal flap and maxillary arch expansion. The usual management strategy is to avoid closure of a fistula until arch expansion can be completed.

The management of a fistula secondary to cleft palate repair is limited in success, with a high incidence of recurrence after initial fistula closure. The most frequent technique used in palatal fistula closure is local flap mobilization. Bone and periosteal grafts have been reported to improve the results. Tongue flaps and microvascular tissue transfers are used for difficult palatal fistulas and large palatal defects, respectively.

Velopharyngeal Incompetence

Morris, in his review of the literature, reported an incidence of velopharyngeal competence of 75%, as defined by the absence of consistent evidence of VPI. No differentiation was made on the type of cleft or the technique of repair. Peterson-Falzone (1991) reported 83.4% competence based on the same criteria. However, when using the criterion of no nasal emission or hypernasality, the incidence of velopharyngeal competence decreases to 60%. The analysis of velopharyngeal competence after various techniques is difficult to interpret in the different studies. The anatomy of the cleft has a great degree of variability that is usually not controlled.

Growth and Morphology

The severity and laterality of the clefts, as well as the choice of cephalometric measurements used in the assessment, account for much of the variability in the reported effects of clefting on facial growth. Grayson et al studied the net effect of palatal clefts on the facial skeleton as viewed by lateral cephalograms and determined by mean tensor analysis. The authors note reduced facial bone growth in all directions but principally in the horizontal dimension. The effect was most pronounced at the level of the palate and slightly less so in height of the mid face. Vertical facial growth was most restricted in subjects who had clefts of both the primary and secondary palate compared with those who had clefts of the secondary palate alone.

Pearls and Pitfalls

The management of a patient with cleft palate is complex. No current universal agreement exists on the appropriate treatment strategy. Several main points should be emphasized. Normal speech should be the most important consideration in the therapeutic plan. Growth disturbance should be minimized but not at the expense of speech impairment because facial distortion can be satisfactorily managed with future surgery, whereas speech impairment is often irreversible. Repair of cleft palate with the goal of establishing a competent velopharyngeal sphincter should be completed at age 6-12 months. At the present time, there are widely divergent claims of superior results from various techniques. There remains a need for well-controlled, prospective studies to determine the optimal technique of cleft palate repair. Until then, cleft patients should be managed in a center with an experienced, multidisciplinary team

Suggested Reading

1. Bardach J, Morris HL, Olin WH. Late results of primary veloplasty: The marburg project. Plast Reconstr Surg 1984; 73(2):207-18.
2. Dufresne CR. Oronasal and nasolabial fistulas. In: Bardach J, Morris HL, eds. Multidisciplinary Management of Cleft Lip and Palate. 1st ed. WB Saunders Co, 1991:425-436.
3. Furlow Jr LT. Cleft palate repair by double opposing Z-plasty. Plast Reconstr Surg 1986; 78(6):724-38.
4. Hodges PL, Pownell PH. Cleft palate surgery and velopharyngeal function. Plast Surg 1994; 7(23):1-36.
5. Kaufman FL. Managing the cleft lip and palate patient. Pediatr Clin North Am 1991; 38(5):1127-47.
6. Lindsay WK. Surgical repair of cleft palate. Clin Plast Surg 1975; 2(2):309-18.
7. Nguyen PN, Sullivan PK. Issues and controversies in the management of cleft palate. Clin Plast Surg 1993; 20(4):671-82.
8. Perko M. Two-stage palatoplasty. In: Bardach J, Morris HL, eds. Multidisciplinary Management of Cleft Lip and Palate. 1st ed. WB Saunders Co, 1991:311-320.
9. Rohrich RJ, Byrd HS. Optimal timing of cleft palate closure. Speech, facial growth, and hearing considerations. Clin Plast Surg 1990; 17(1):27-36.
10. Schweckendiek W, Doz P. Primary veloplasty: Long-term results without maxillary deformity. A twenty-five year report. Cleft Palate J 1978; 15(3):268-74.

Chapter 60

Rhytidectomy

Stephen M. Warren and James W. May, Jr.

Introduction

Aesthetic facial surgery is intended to rejuvenate the cervicofacial contour. Recognizing the elements of an aging face and neck are a prerequisite to planning any procedure. Common stigmata of facial aging include: ptotic malar pads, heavy nasolabial folds, nasojugal creases, marionette lines, jowls, geniomandibular grooving, cheek and neck skin laxity, platysmal banding, lateral orbital wrinkling, submental lipodystrophy and salivary gland ptosis. While all of these structures may be affected by the pull of gravity, repetitive contraction of the underlying muscle, and cellular/subcellular aging, each can be improved by facelifting techniques. Other problems such as forehead and glabellar lines, eyelid bulges and excess skin, fine facial wrinkles, lip atrophy, cheek fat atrophy, senile nasal dysmorphia cannot be corrected with a facelift. Since it is not possible to design a universal technique for all patients, facelifting must be preceded by a sound knowledge of the anatomy and a thorough understanding of the elements to be corrected. Careful planning and good technique are necessary to precisely remove redundant skin, resuspend or resect fat and repairing lax musculature and fascia.

Anatomy

There are five important anatomic levels in the face and neck: skin, subcutaneous fat, the superficial musculoaponeurotic system (SMAS)/muscle layer, fascia and the facial nerve. While these layers are consistent throughout the face and neck, in some area such as over the zygomatic arch, the layers are highly compressed. In addition to these tissue planes, the surgeon must be familiar with the folds, retaining ligaments, glands , blood supply and fat pads of the face.

Skin

As we age, the skin changes in its appearance and characteristics. Skin aging is accelerated by sunlight; this process is known as dermatohelisosis, solar elastosis, or photoageing. Photoageing is accelerated by long and short wavelength ultraviolet radiation (UVA and UVB) injury to the epidermis and dermis. Studies suggest that UV light can activate enzymes that degrade collagen and elastin in skin. Repetitive solar damage can cause fine lines and wrinkles, telangiectasias, solar comedones, dryness and actinic lentigines (diffuse or mottled brown patches). Signs of skin aging are accelerated by smoking. Facelifting cannot directly improve the quality of photoaged skin, but it can improve the appearance.

Subcutaneous Tissues

The subcutaneous plane provides a relatively safe plane for dissection. This layer contains innumerous fine ligaments passing from the subjacent SMAS/muscle layer to the overlying dermis of the skin. These ligaments transmit mimetic movements into facial expressions but also contribute to facial lines and wrinkles.

SMAS/Muscle Layer

Below the skin and subcutaneous tissues, is the SMAS/muscle layer. The SMAS/muscle layer is a continuum from neck to scalp. It is composed of fibrous, muscular, or fatty tissues. In the neck, the platysma represents the most inferior portion of the SMAS/muscle layer. In the face, the SMAS is a tough fibrofatty layer over the parotid. Medial to the parotid, the muscles of facial expression (e.g., zygomaticus major/minor and orbicularis oculi) are contiguous with the SMAS layer. Above the zygoma, the SMAS is contiguous with the frontalis muscle and the superficial temporal fascia (or temporoparietal fascia). The temporoparietal fascia blends into the galea as it reaches the scalp. Collectively, this layer may be thought of as the platysma-SMAS-temporoparietal-galea layer.

Fascial Layer

Between the SMAS/muscle layer and the facial nerve is a fascial layer. In the neck, the layer is termed the cervical fascia. Over the parotid, it exists as a filmy, areolar layer called the parotideomasseteric fascia. This thin, nearly transparent layer lies immediately superficial to the facial nerve. The fascia continues cephalad passing over the zygoma. In the upper third of the face, the layer becomes the innominate fascia that blends into the subgaleal fascia over the scalp. The innominate fascia lies between the SMAS/muscle extension (i.e., temporoparietal fascia) and the superficial layer of the deep temporal fascia. The anatomy here is critical because the frontal branch of the facial nerve (see below) pierces the innominate fascia at the level of the zygomatic arch and travels along the undersurface of the temporoparietal fascia. Collectively, this fascia may be thought of as the cervical-parotideomasseteric-innominate-subgaleal layer.

Facial Nerve

The main facial nerve trunk emerges from the stylomastoid foramen to provide motor innervation to 20 paired muscles of facial expression as well as the posterior belly of the digastric, stylohyoid and stapedius muscles. In addition, the facial nerve provides sensory innervation to the anterior two-thirds of the tongue, external auditory meatus (nerve of Jacobsen), soft palate and pharynx. The motor portion of the facial nerve divides into five major branches (Fig. 60.1). The branches of the facial nerve travel just deep to the cervical-parotideomasseteric fascia to innervate all muscles of facial expression from their deep surface with three exceptions: (1) mentalis, (2) buccinator, and (3) levator anguli oris. These muscles lie deep to the facial nerve branches and are, therefore, innervated on their superficial surfaces.

The frontal branch of the facial nerve leaves the parotid gland immediately beneath to the zygomatic arch. As it crosses the superficial surface of the zygomatic arch, the frontal branch pierces the innominate fascia to travel along the undersurface of the temporoparietal fascia (superficial temporal fascia). At this point, the frontal branch is extremely susceptible to inadvertent injury. The path of the frontal branch can be approximated by connecting a line 2 cm lateral to the edge of the eyebrow to the lower edge of the earlobe, termed Pitanguy's line. The frontal branch innervates the muscles of the upper part of the face including the upper orbicularis oculi, frontalis and corrugator muscles. Transection of the frontal branch leads to brow ptosis.

The zygomatic branch provides motor fibers to the lower orbicularis oculi, procerus, some lip elevator and some nasal muscles. The buccal branch has tremendous overlap with the zygomatic branch and sends fibers to similar muscles, as well

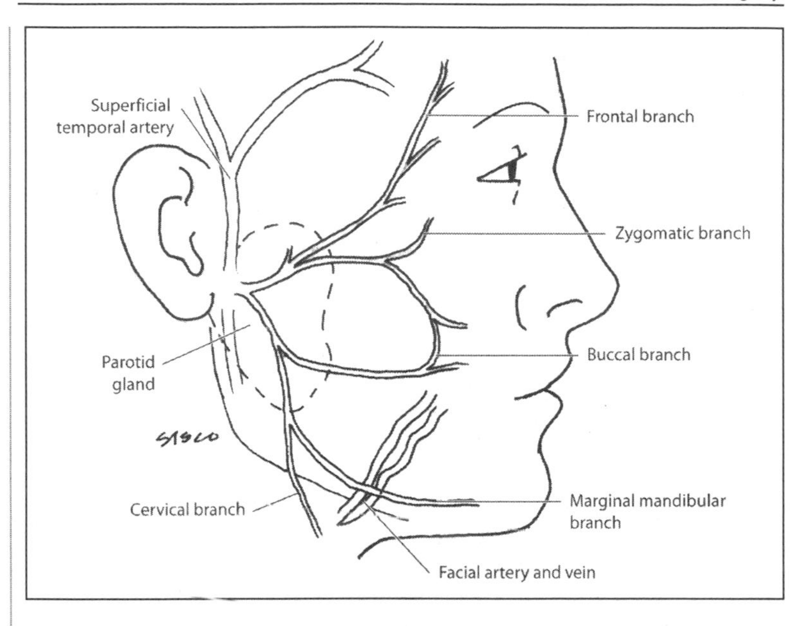

Figure 60.1. Facial nerve anatomy.

as the buccinator, orbicularis oris, depressor anguli oris and risorius muscles. As the zygomatic and buccal branches exit the medial portion of the parotid, they travel along the superficial surface of the buccal fat pad, just below the SMAS. This position makes them susceptible to injury during facelift procedures, particularly at the lateral edge of the zygomaticus major. As discussed below, when dissecting in the sub-SMAS plane, the surgeon must change to a subcutaneous plane at the lateral border of the zygomaticus major muscle in order to avoid interrupting the zygomatic and buccal branches. Transection of the zygomatic and buccal branch leads to unpredictable defects because muscular innervation in the mid face is variable. The buccal branch is the most commonly injured branch of the facial nerve.

The marginal mandibular nerve emerges from the inferior border of the parotid gland and crosses the inferior border of the mandible deep to the platysma to reach the face. Studies suggest that posterior to the facial artery, the marginal mandibular branch may dip as far as 2 cm below the border of the mandible. Anterior to the facial artery, the marginal mandibular nerve nearly always lies above the mandibular border. The marginal mandibular nerve has little cross-innervation as it enters the orbicularis oris and lip depressors. Transection of this nerve results in paralysis of the muscles that depress the corner of the mouth; therefore, the paralyzed side of the mouth will appear higher than the innervated side!

The cervical branch travels on the undersurface of the platysma. The platysma acts synchronously with other muscles of the lower lip to draw the oral commissure and lower lip downward. Transection of the cervical branch is uncommon, but it does not result in significant functional or cosmetic deficits.

Additional discussion of this anatomy can be found in Chapter 35.

Tear Trough, Nasojugal, Malar, Nasolabial and Labiomental Folds

The tear trough is a depression near the medial palpebral fissure formed by the separation of the orbicularis oculi and levator labii superioris. The nasojugal fold extends inferiorly and laterally from the tear trough onto the cheek. The malar fold runs inferiorly and medially from the lateral palpebral fissure towards the inferior extent of the nasojugal fold.

The cutaneous insertion of the zygomaticus major/minor and levator labii superioris muscles determines the nasolabial fold. In a sense, the nasolabial fold may be considered a fasciocutaneous ligament necessary for lip elevating muscles to initiate a smile. Laxity of this fasciocutaneous ligament causes the malar fat pad to travel inferomedially over the crease to deepen the nasolabial fold. The depressor anguli oris superiorly and the mandibular ligaments inferiorly determine the labiomandibular crease, which similarly is converted into a fold as a result of the laxity of the masseteric ligaments that occurs with age.

Parotid and Submandibular Salivary Glands

Invested by the deep fascia, 80% of the parotid gland lies between the mastoid process and the posterior border of the mandible. About 20% of the gland extends convexly forward over the masseter muscle occasionally as far as the zygomaticus major. The parotid duct (Stensen's duct) and branches of the facial nerve emerge from the anterior border of the parotid, beneath the parotideomasseteric fascia. The parotid duct (4-6 cm in length) travels parallel to the zygomatic arch, 1.5 cm (approximately 1 finger breadth) below its inferior border, passing over the masseter muscle and then turns medially 90° to pierce the buccinator muscle at the level of the second maxillary molar where it enters the oral cavity. Using surface landmarks, Stensen's duct lies midway between the zygomatic arch and corner of the mouth along a line between the upper lip philtrum and the tragus. The buccal branch of the facial nerve parallels the parotid duct.

The submandibular glands, often referred to as the submaxillary glands because of the tendency of British anatomists to refer to the mandible as the submaxilla, lie in the submandibular triangles formed by the anterior and posterior bellies of the digastric muscles and the inferior border of the mandible. The marginal mandibular branch of facial nerve courses superficial to the submandibular gland and deep to the platysma. The submandibular ducts (Wharton's ducts) exit the medial surface of each gland and run between the mylohyoid (lateral) and hyoglossus muscles along the genioglossus muscle to empty into the oral cavity lateral to the lingual frenulum. The lingual nerve wraps around Wharton's duct, starting lateral and ending medial to the duct, while the hypoglossal nerve parallels the submandibular duct, just inferior to it. The identification of the hypoglossal and lingual nerves as well as Wharton's duct is important prior to resecting portions of the submandibular glands.

Retaining Ligaments

The retaining ligaments of the face support soft tissues in their youthful anatomic positions. Furnas described four retaining ligaments that support the soft tissues of the face. The platysma-auricular and the platysma-cutaneous ligaments are aponeurotic condensations attaching platysma to dermis. Of greater significance are the osteocutaneous zygomatic and mandibular retaining ligaments. The zygomatic ligaments (McGregor's patch) anchor the skin of the cheek to the inferior border of the zygoma just posterior to the origin of the zygomaticus minor muscle. With age, these ligaments become lax, leading to inferomedial migration of the malar fat pad and formation of the nasolabial fold. The mandibular retaining ligaments arise from the parasymphysial mandibular body and insert into the skin inferior to the insertion of the depressor anguli oris. The mandibular ligaments define the anterior extent of the jowls. The zygomatic and mandibular ligaments are obstacles to surgical maneuvers intended to lift the skin flap and, therefore, both are usually divided.

Malar and Buccal Fat Pad

In a youthful midface, the superior border of the triangular shaped malar fat pad lies along the orbital rim and extends laterally to the zygoma. The lateral border can be identified by drawing a line from the lateral canthus to the lateral commissure. The malar fat pad is located beneath the skin and subcutaneous fat, but it is superficial to the SMAS. It is fibrous and fatty, and it is readily distinguishable from the overlying subcutaneous fat. With advancing age, the malar fat pad slides downward and medially, over the SMAS. Ptosis of the malar fat pad also empties the midface, producing a crescent-shaped hollow at the lower lid-cheek junction. The malar fat pad descent also contributes to the nasojugal and nasolabial folds. To a lesser extent, this displacement also results in the formation of labiomandibular folds (marionette lines) and jowls.

The buccal fat pad lies over the masseter and buccinator muscles, deep to the plane of the parotid duct and facial nerve branches. Medially, it may reach into the pterygopalatine space. It can be approached from a sub-SMAS dissection plane by separating the buccal branches of the facial nerve. Alternatively, it can be approached through the mouth by penetrating the mucosa and buccinator muscle. There are few indications to remove this fat pad because it tends to hollow the cheek giving an aged appearance.

Preoperative Considerations

All patients should receive a complete medical examination by the appropriate specialist, including complete blood counts, metabolic chemistries, EKG and, if indicated, a chest roentgenogram. Patients with diabetes mellitus, hepatic, cardiovascular, renal, or thyroid disorders must have preoperative medical clearance. Patients should be instructed to stop taking alcohol or tobacco products 3 weeks prior to surgery. Aspirin, nonsteroidal anti-inflammatory agents, anticoagulants, vitamin E, multivitamins, Alka Seltzer® and homeopathic remedies should also be discontinued 3 weeks prior to surgery.

One of the dreaded complications of the facelift is hematoma. Postoperative nausea and vomiting (PONV) and hypertension are believed to be contributing factors. All patients, unless contraindicated, should receive preoperative antiemetic therapy. For example, preoperative odansetron (Zofran®), 4 mg IV, has been shown to significantly decrease the incidence of PONV. Other less expensive antiemetics are also available. In addition, patients with hypertension should take their medications the morning of surgery. Any patient with even mild hypertension the morning of surgery, should be considered for antihypertensive therapy. Oral clonidine (0.1-0.2 mg) is a commonly used medication for this purpose. The night before surgery, a benzodiazepine can be given (e.g., lorazepam 2 mg) to prevent preoperative anxiety-induced hypertension.

The patient should refrain from using cosmetics, perfumes, aftershave, and moisturizers on the morning of surgery. Hair coloring should not be performed within 10 days of surgery. Make-up should be removed the night before surgery, and the patient should be instructed to wash their face and shampoo their hair with an antimicrobial soap. It is standard to administer a single intravenous dose of preoperative antibiotics. Antibiotics are generally not required postoperatively.

Surgical Techniques

Skin/Subcutaneous Facelift

Incisions vary and depend on the technique, patient anatomy and hairline and surgeon preference. The temporal incision is generally marked in a curvilinear fashion, just within the temporal hairline and superior to the ear. This avoids any loss or elevation of the temporal hairline. The preauricular incision lies in the natural crease at the junction of the auricle and the face, following the curve of the helical root. The incision can then be continued in either the pretragal crease or behind the tragus. The inferior aspect of the incision is located at the junction of the earlobe and cheek. Curving posteriorly and superiorly around the lobe, the incision is placed in the postauricular crease. The incision then curves tangentially into the occipital hairline at the level of the inferior crus of the antihelix. This incision placement helps prevent a step-off deformity of the posterior hairline.

Flap elevation proceeds in a subcutaneous plane with care taken to avoid hair follicles. Transilluminating the flap can help to maintain the proper dissection plane. Pre- and postauricular flaps are extended into the neck over the sternocleidomastoid muscle. The great auricular nerve emerges from the anterior border of the sternocleidomastoid muscle 6.5 cm inferior to the external auditory meatus. A separate submental incision may be used to elevate the anterior portion of the cervical flap in a preplatysmal plane. The dissection is limited superiorly by the inferior border of the mandible and inferiorly by the hyoid bone. The preplatysmal plane serves to protect the marginal mandibular nerve as it courses below the mandible. Once elevated the skin is redraped, tailored and inset under limited tension.

The subcutaneous facelift technique is simplest to perform, with the least risk of injury to the facial nerve branches. The skin-only facelift produces good results for thin women with good skin tone and underlying bone structure. It is difficult to obtain a natural look in patients with heavier faces because high skin tension produces a pulled-appearance, wider scars and alopecia. The skin-only facelift has limited application.

SMAS/Muscle Facelift

The SMAS facelift begins with the incisions and skin flap elevation as described above. Classically, the SMAS is elevated in the preauricular area, from 1 finger breadth below the zygoma to the lower border of the mandible. The parotideomasseteric fascia is left intact just below the dissection plane, protecting the facial nerve branches. Dissection continues anteriorly to the nasolabial fold, remembering to change the level of dissection at the lateral border of the zygomaticus major muscle. The dissection plane remains superficial to the zygomaticus major muscle and extends inferiorly to the oral commissure. Sharp division of zygomatic and mandibular retaining ligaments allows full mobilization of the skin and soft tissue, facilitating redraping.

In the neck, subplatysmal dissection can be performed to expose the triangular shaped subplatysmal fat pad. After resecting this fat pad under direct vision, the medial edges of the platysma can be trimmed and the diastasis closed. The muscle sling should be securely plicated in order to correct the platysmal banding.

After completing the dissection, the SMAS is lifted in a vector parallel to the zygomaticus major, trimmed and inset. The skin is then redraped in a vector perpendicular to the nasolabial fold. It is critical that the skin is inset tension-free. The advantage of a SMAS/muscle facelift over a skin only lift is the ability to independently control the vectors of the deep tissues and skin. Moreover, since the lift is based on the SMAS, the skin can be trimmed and inset without tension. There are many variations of this technique which include the limited SMAS, extended SMAS and lateral SMASectomy.

Deep Plane Facelift

Deep plane facelift refers to sub-SMAS dissection without significant undermining in the subcutaneous plane. The subcutaneous dissection is carried approximately 2-3 cm in front of the tragus, from zygoma to the jaw line. The sub-SMAS plane is dissected beyond the nasolabial fold, exposing the orbicularis and zygomatic muscles (the SMAS is transected at the level of the zygomaticus major muscle and the dissection continued in a subcutaneous plane). This maneuver frees the SMAS from the attached mimetic muscles, allowing the pull on the skin to be transmitted to the fold. The cheek fat is dissected free from the underlying mimetic muscles and is elevated with the skin/SMAS flap. The technique is said to diminish the appearance of the nasolabial fold. The risk of nerve injury may be greater with the more extensive dissection. There is improved vascularity compared to the subcutaneous plane facelift. However, the major drawback to the deep plane operation is the development of persistent infraorbital and midface ecchymosis and edema that greatly prolong the convalescence.

Composite Facelift

The composite facelift is a modified deep plane facelift designed to additionally address the orbicularis oculi muscle. With the addition of a lower blepharoplasty incision, the orbicularis oculi is elevated off the malar prominence. This frees the muscle of its attachments to the malar eminence, allowing mobilization and repositioning. As originally described, this dissection plane is then connected to the deep-plane dissection by an incision made between the inferior lateral border of the orbicularis oculi and the zygomaticus minor muscle. A distinct division between these muscles is not always present because they lie in the same plane. This maneuver prevents inadvertent elevation of the zygomaticus minor muscle into the composite flap. The inferior aspect of the orbicularis oculi muscle is trimmed, and the muscle is repositioned in a superomedial vector.

Subperiosteal Lift

The subperiosteal lift is another type of deep plane facelift. Through a coronal approach, subperiosteal undermining is carried out around the orbital rims, over the zygomatic arch and body, over the maxilla and down to the piriform aperture. After undermining, the tissue is advanced superiorly and sutured to the temporal fascia. In older patients with skin laxity, the procedure is combined with a preauricular incision. Patients frequently have marked facial edema for several weeks after surgery and a mask effect for several months. Risk of injury to the frontal branch of the facial nerve was high in the initial series but has been minimized with a deep approach to the zygomatic arch. Many surgeons prefer this technique for patients 45 and under who desire facial implants. There is more swelling with the subperiosteal lift than with more SMAS lifts.

Postoperative Care

In the recovery room following surgery, the patient should be evaluated for pain, nausea, or vomiting. If present, pain medication and antiemetics should be administered. The blood pressure must be frequently monitored and precisely controlled with antihypertensives. Patients should rest, but need not stay in bed. While in bed, the patient's head should be elevated. Drains are usually removed the morning after surgery. For at least 2 weeks after surgery, the patient should refrain from physical exertion, bending or heavy lifting, sexual activity, driving and flying. The patient should continue to abstain from alcohol and tobacco products, aspirin, nonsteroidal anti-inflammatory agents, anticoagulants, vitamin E, multivitamins, Alka Seltzer® and homeopathic remedies for 3 weeks. A shower and hair washing are permitted on the day after surgery, but no hair brushing or make-up applications are permitted for 10 days. The patient should avoid sun exposure until the scars are mature. Preauricular/temporal sutures are generally removed in 5-7 days and postauricular sutures are removed in 2 weeks.

Complications

Hematoma

Men generally have twice the incidence of hematoma after facelift surgery as women (8% versus 4%). This may to be due to the hair follicles in a male's beard. Secondary facelifts have a lower incidence of bleeding.

Skin Slough

Skin slough occurs most often in the postauricular region, and it is more common in patients who smoke. Patients should refrain from smoking at least 3 weeks before and 2 weeks after the operation. Another risk factor for skin slough is acne scarring. The subdermal acne scar is hypothesized to compromise blood flow to skin flap. Good judgment is necessary to determine the amount of skin undermining that can be safely performed in higher risk patients. Skin slough is usually treated by allowing the wound to heal by secondary intention.

Nerve Injury

The most commonly injured nerve during a facelift is the greater auricular nerve. Patients undergoing a subcutaneous rhytidetomy have a facial nerve injury risk of 0.5-2% (mean of 1%). Patients who undergo a SMAS-based lift have a facial nerve injury risk of 2-9% (mean of 4%).

Alopecia

Hair loss is uncommon during a face lift (1.2%). Suture line alopecia tends to occur in areas of inappropriately high tension. Elevation of the sideburn or notching of the postauricular hairline is more common, particularly during a secondary lift. Both of these complications can be avoided by careful planning.

Scarring

Scarring is present in every facelift. Well-designed incisions closed without tension produce the best scars. Patients with a family or personal history of hypertrophic scarring or keloid formation or risk factors for excessive scarring after a facelift should be counseled preoperatively.

Infection

Infections occur very uncommonly (0.18%) during facelifting due to the robust blood supply of the face.

Pearls and Pitfalls

A thorough understanding of the nerves encountered during rhytidectomy in the face and neck is essential for avoiding the most dreaded complications of this procedure.

The frontal division of the facial nerve lies within the temporoparietal/SMAS fascia. Dissection in the vicinity must be either extratemporoparietal/extra-SMAS or subtemporoparietal/sub-SMAS. In the subperiosteal approach, dissection should proceed deep to the deep layer of the temporal fascia. The zygomatic branch of the facial nerve lies deep to the zygomaticus major muscle. Sub-SMAS dissection at this point causes trauma to the nerve, as does blind incision of the zygomatic ligaments. The marginal mandibular branch of the facial nerve usually is not visualized during facelifting. In sub-SMAS dissection in the lower face, it is safer to stay above the mandible posterior to the facial vessels. Use appropriate caution with electrocautery hemostasis around vessels in the SMAS since electricity may be transmitted to nerves causing injury.

In the neck the cervical branch of the facial nerve lies deep to the platysma muscle and is in no danger in a supraplatysmal dissection. The great auricular nerve lies deep to the superficial layer of the deep investing fascia on the sternocleidomastoid muscle as it traverses from posteroinferior to anterosuperior to emerge in the vicinity of the infra-aural region, where the skin is firmly attached to the sternocleidomastoid muscle. Caution with the infraorbital nerve must be exercised during dissection in the subperiosteal plane in the region.

Suggested Reading

1. Baker DC, Conley J. Avoiding facial nerve injuries in rhytidectomy: Anatomical variations and pitfalls. Plast Reconstr Surg 1979; 64:781.
2. Barton Jr FE. Rhytidectomy and the nasolabial fold. Plast Reconstr Surg 1992; 90:601.
3. Furnas DW. The retaining ligaments of the cheek. Plast Reconstr Surg 1989; 83:11.
4. Hamra ST. The deep-plane rhytidectomy. Plast Reconstr Surg 1990; 86:53.
5. Mitz V, Peyronie M. The superficial musculo-aponeurotic system (SMAS) in the parotid and cheek area. Plast Reconstr Surg 1976; 58:80.
6. Pitanguy I, Silveira Ramos A. The frontal branch of the facial nerve: The importance of its variations in face-lifting. Plast Reconstr Surg 1966; 38:352.
7. Stuzin JM, Baker TJ, Gordon HL. The relationship of the superficial and deep facial fascias: Relevance to rhytidectomy and aging. Plast Reconstr Surg 1992; 89:441.

Chapter 61

Browlift

Clark F. Schierle and John Y.S. Kim

Introduction

The term browlift is generally used to describe a family of procedures aimed at the rejuvenation of the upper third of the face. A number of different incisions, planes of elevation, vectors of pull and methods of anchoring may be employed, depending on the underlying anatomic pathophysiology at work in the individual patient. Techniques continue to evolve and must be tailored to each patient taking into account the patient's sex, age, facial features and expectations.

Anatomy and Aesthetics

Traditionally, the ideal forehead is thought of as occupying one-third of the height of the face when viewed from the front. The aesthetically ideal brow is generally thought of as a graceful arc occupying the space just superior to the orbital rim, ending at a point along a line drawn from the lateral nasal ala and the lateral canthus of the eye. The zenith of the arc should lie above a point between the lateral limbus and lateral canthus in females, while in males, it may lie more directly above the pupil. The soft tissues of the brow and forehead are basically comprised of five layers, often remembered with the aid of the mnemonic SCALP: Skin, subcutaneous tissue, Aponeurosis, Loose areolar tissue and Periosteum. The skin of the forehead is quite thick with many fibrous connections to the underlying facial muscles. There is also a relative paucity of fat compared with other regions of the face. The strong connection of the skin to the dynamic muscles of facial expression, coupled with the lack of subcutaneous fat to act as a filler contribute to the vulnerability of this facial region to the stigmata of aging.

The arterial supply to the forehead derives from the supraorbital and supratrochlear arteries medially (tributaries of the internal carotid system) and the superficial temporal artery laterally (a terminal branch of the external carotid). This dual arterial system forms a rich and robust blood supply with many anastomotic connections. Venous drainage, as with most of the skin of the face is supplied primarily by an extensive subdermal venous plexus rather than discrete named vessels. The region is innervated by all three divisions of the trigeminal nerve with the supratrochlear and supraorbital nerve branches of the first division supplying the brow medially, the zygomaticotemporal branch of the second division supplying the medial temple, and the auriculotemporal nerve supplying the lateral aspect of the temple.

The brow is home to several muscles of facial expression whose function can lead to the development of deep rhytides over time. The frontalis muscle serves to elevate the brow, while the actions of the orbicularis oculi, corrugators and procerus all depress the brow. The frontalis muscle is the anterior half of the epicranius muscle and is not attached to bone. Its action over time contributes to the formation of

Practical Plastic Surgery, edited by Zol B. Kryger and Mark Sisco. ©2007 Landes Bioscience.

deep horizontal rhytides in the forehead. The orbicularis oculi close the eyes, but their action over time contributes to crow's feet, brow ptosis and hooding, particularly laterally where their action is less well opposed by the more attenuated frontalis. The corrugator supercilii muscles lie deep to the frontalis and orbicularis muscles. They originate from the medial orbital rim and insert into the dermis overlying the supraorbital foramen, producing the vertically oriented glabellar frown line. The procerus muscle originates from the inferior portion of the nasal bones and inserts into the dermis above the glabella and creates a horizontal rhytid between the eyes.

In addition to the neurovascular supply and musculature of the brow and forehead, a knowledge of the fascial planes is critical to understanding the anatomy of this region. The superficial temporal fascia, also known as the temporoparietal fascia, lies immediately deep to the dermis and is contiguous with the galea aponeurotica above and the superficial musculoaponeurotic system (SMAS) below. The superficial temporal artery, vein, and temporal branch of the facial nerve all lay within the temporoparietal fascia. The temporal branch of the facial nerve can be found along Pitanguy's line which runs from 0.5 cm inferior to the tragus to 1.5 cm above the lateral aspect of the eyebrow. Deep to the temporoparietal fascia lies the fascia of the temporalis muscle, known as the deep temporal fascia, which is contiguous with the periosteum of the skull at the conjoint tendon. The deep temporal fascia splits into superficial and deep layers above the zygomatic arch to envelop the superficial temporal fat pad.

Preoperative and Anesthetic Considerations

Preoperative workup consists of standard screening for risks of anesthesia with laboratory and cardiac workup tailored to the age and comorbidities of the patient and the type of anesthesia planned. Aspirin, other blood thinning medications and herbal remedies are discontinued. Frontal, lateral and oblique photographs should be obtained in a standardized fashion. Most browlift techniques can easily be performed under intravenous conscious sedation in conjunction with effective local anesthesia although some still favor general anesthesia. Local anesthesia should be infiltrated along all incision lines as well as performing blocks of the supraorbital and supratrochlear nerves.

Operative Technique

Coronal Browlift

Although minimally invasive techniques are rapidly gaining popularity, the full coronal browlift is still employed by many and does offer some advantages. The technique offers full exposure of the frontalis, corrugator and procerus muscles with an incision that is concealed within the hair-bearing scalp. Care must be taken to select patients whose hairlines will tolerate the inevitable elevation associated with this technique. The incision is performed such that the resultant scar will lie approximately 3 cm posterier to the hairline following the excision of excess scalp. The incision is scyved in alignment with hair follicles, and electrocautery is used judiciously to minimize the region of alopecia associated with the scar.

Dissection takes place within the relatively bloodless subgaleal plane. In the temporal region, dissection is carried out between the superficial temporal fascia and the superficial layer of the deep temporal fascia, protecting the temporal branch of the facial nerve. Medially, the trunks of the supratrochlear and supraobital nerves are identified and preserved. The corrugator supercilii and procerus are easily visualized and may be divided to address horizontal or vertical rhytides in the medial orbitital region, the so called "frown lines." Finally, 1-2 cm of excess scalp is excised from the incisional edge of the flap and a layered closure is performed.

Pretrichial and Trichophytic Browlift

These techniques seek to minimize or eliminate disturbance to the hairline. The pretrichial incision is carried out at or just anterior to the hairline while the trichophytic incision lies a few millimeters posterior to the hairline. Careful beveling of the incision and minimal use of electrocautery are critical to avoiding unsightly incisional alopecia. The plane of dissection and remainder of operative technique are essentially those of the coronal browlift; however great care must be taken with closure as the scar is far more likely to be visible, especially if the patient's hairline recedes with time.

Temporal Lift

A temporal approach can be of great benefit in patients with isolated lateral brow ptosis or hooding. The incision is performed in the temporal region, running anterosuperior to posteroinferior across the temporalis muscle, similar to the Gillies approach to zygomatic arch repair. Dissection is carried out in a similar fashion to the coronal technique, in the plane between the temporoparietal fascia and the superficial layer of the deep temporal fascia, protecting the frontal division of the temporal branch of the facial nerve which courses through the superficial temporal fascia. Dissection continues inferomedially to the supraorbital rim and arcus marginalis. The superficial temporal fascia is then anchored superolaterally to the deep temporal fascia achieving the desired degree of elevation. Excess skin is excised, and skin is closed in a layered fashion.

Midforehead Browlift

This technique places the incision directly in the middle of the forehead, concealed within an existing forehead crease. It is useful for the patient seeking correction of severe brow ptosis or asymmetry for whom the appearance of horizontal forehead rhytides is of secondary concern. Either a single incision extending the full length of the forehead or two fusiform incisions above each brow are performed centered on a prominent forehead crease. Asymmetric positioning of these incisions can assist in their camouflage as natural wrinkles. In contrast to the coronal, trichophytic and pretrichial techniques, the plane of dissection is subcutaneous, superficial to the frontalis muscle and is carried down until the orbicularis oculi are visualized. The galea may be incised 2-3 cm superior to the orbital rim and excess galea excised or redraped and anchored to underlying periosteum. Excess skin is excised and closed in a layered fashion.

Direct Browlift

Direct browlift refers to a skin and subcutaneous tissue-only technique which directly addresses positioning of the brow. A fusiform incision is made superior to each brow. Dissection is carried out in the subcutaneous plane with preservation of underlying muscular and neurovascular structures. Long term fixation is achieved through placement of sutures anchoring the superior aspect of the incision to the periosteum. Meticulous skin closure is essential as the scar is located in a very visible location.

Endoscopic Browlift

Minimally invasive surgical techniques have revolutionized all aspects of surgical practice, and aesthetic plastic surgery is no exception. Endoscopic browlift techniques allow results comparable to traditional incision techniques without the risks of incisional anesthesia and alopecia associated with longer scars. Although techniques are rapidly evolving, the typical approach utilizes one midline coronal incision located along the traditional coronal browlift course and two temporal incisions placed along the traditional temporal browlift incision line (along a line drawn from the lateral ala throught the lateral limbus).

The dissection is similar to the open coronal and temporal browlift techniques. The central dissection may proceed in the traditional subgaleal plane or alternatively in the subperiosteal plane, which some find to provide a more bloodless dissection and a better optical cavity. The supraorbital notch serves as the landmark to identify and preserve the neurovascular bundles. The corrugators and procerus musculature may be disrupted as in the open technique. Methods include sharp or blunt transection, thermal injury through electrocautery, or laser ablation. A superolateral orbicularis myotomy may be performed to maximally release the brow for elevation.

Laterally the dissection through the temporal ports proceeds along the traditional plane between the superficial temporal fascia and the superficial leaf of the deep temporal fascia until one reaches the supraorbital rim and arcus marginalis. Proceeding superomedially, the two dissections are joined at the conjoint tendon where the lateral dissection is transitioned to the deeper subperiosteal plane of the central dissection. Care is taken in this area to identify the so called "sentinel vein" which identifies the course of the temporal branch of the facial nerve that lies directly above the vein in the superficial temporal fascia.

Fixation of the soft tissues along vectors providing ideal superolateral elevation of brow structures is a matter of rapidly evolving debate. Techniques and materials include soft tissue suture fixation, permanent or absorbable cortical screws with suture or staple fixation, cortical bone tunnels and various tissue glues. Regardless of fixation method, proper mobilization of the soft tissues during dissection is critical, and the ideal fixation method is likely of secondary importance so long as it provides a reliable result and is easily accomplished through the limited incisions of the endoscopic technique.

Postoperative Care

Most browlift procedures are well tolerated and can be performed as an outpatient with a responsible family member on visiting nurse to provide reliable observation in the first 24-48 hr postoperatively. Ice packs may be used judiciously to limit postoperative swelling. Patients typically do well with a few days of mild oral narcotic analgesia, transitioning to over the counter pain medications as tolerated. Sutures and staples should be removed one week after surgery. Patients may advance their activity as tolerated with most returning to normal daily activities within 3-5 days. Aerobic and other strenuous activities should be avoided for two weeks postoperatively.

Complications

Swelling and bruising should be expected, although the incidence of both can be minimized through judicious use of electrocautery and careful adherence to relatively avascular tissue planes. Hematomas may occur in the setting of inadequate hemostasis or in the failure to recognize a potential bleeder due to intraoperative vasospasm or vasoconstriction from the use of local anesthetics containing epinephrine. Patients should be warned of the risk of incisional alopecia and anesthesia. Both can be minimized by limiting thermal injury associated with electrocautery and taking care to minimize tension along the suture line during closure. Standard perioperative antibiotics may be administered to limit the incidence of wound infection although the rich vascular supply of facial skin makes this a rare complication even in the absence of such prophylactic measures. In addition to incisional anesthesia, risk of injury to major sensory or motor nerves must be related to the patient. Careful dissection along safe tissues planes and avoidance of excessive tissue traction minimize these risks. In the absence of complete transection, most nerve injury is transient, and patients may be reassured that partial or complete return of function is typical. Finally, as in any elective aesthetic procedure, the patient should be counseled on the inherent unpredictability of the final outcome and the very real possibility of under- or overcorrection, asymmetry, hypertrophic scarring and other aesthetic considerations which are a function of the body's inherent response to the surgery out of the control of the surgeon.

Pearls and Pitfalls

- The highest brow peak in women is between the lateral limbus and lateral canthus, whereas in men it is more directly above the pupil, less of a peak and roughly at the level of the orbital rim.
- Vertical glabellar wrinkles are due to the corrugators, whereas horizontal wrinkles are primarily due to the procerus.
- Excessive resection of the procerus, corrugator and frontalis muscles can result in visible depressions in the center of the forehead and glabellar regions.
- For patients who have a very high hairline preoperatively, the pretrichal approach will preserve the hairline without elevating it further.
- For patients with any lateral ptosis, insufficient release of the orbital retaining ligaments (dermal to periosteal adhesions) will result in under correction of the ptosis and a likely dissatisfied patient.
- If an upper blepharoplasty is planned along with the browlift, much of the dissection can be performed through the upper bleph approach. After dissection above the supraorbital rim and development of the subperiosteal plane, the periosteum can be released from the bone and the muscles readily divided under direct vision.

Suggested Reading

1. Chajchir A. Endoscopic subperiosteal forehead lift. Aesthetic Plast Surg 1994; 18(3):269.
2. Core GB, Vasconez LO, Graham IIIrd HD. Endoscopic browlift. Clin Plast Surg 1995; 22(4):619.
3. Freund RM, Nolan IIIrd WB. Correlation between browlift outcomes and aesthetic ideals for eyebrow height and shape in females. Plast Reconstr Surg 1996; 97:1343.
4. Ramirez OM. Endoscopic subperiosteal browlift and facelift. Clin Plast Surg 1995; 22:639.

Chapter 62

Otoplasty

Clark F. Schierle and Victor L. Lewis

Introduction

The pinna of the human auricle serves primarily an aesthetic function. This is in contrast to other members of the animal kingdom where active control of the pinna allows for focusing of sound waves for optimal sound localization, providing a distinct survival advantage in certain species. Congenitally prominent ears can be the source of great emotional and psychosocial distress for patients in all stages of life, especially young school age children.

Anatomy and Aesthetics

The auricle is a complex fibroelastic cartilage structure which normally measures approximately 6 cm in the vertical axis. The auricle is comprised of several cartilaginous landmarks which serve to make up the normal appearing ear (Fig. 62.1). These can be divided into the various fossae of the auricle and the cartilaginous ridges which separate them. The outermost cartilaginous ridge of the ear is the helical rim. This graceful arch forms the outermost boundary of the auricle and terminates anteriorly in the helical crus. Proceeding centripetally, the next ridge encountered is the antihelix, a fold of cartilage which bifurcates anterosuperiorly into the superior and inferior crura. The fossa between the helix and antihelix is the scaphoid fossa, or scapha. The antihelix defines the border of the conchal bowl, which is divided into the concha cymba superiorly and concha cavum inferiorly by the helical crus. Within the concha cymba the superior and inferior crura of the antihelix define the triangular fossa. Two small cartilaginous protuberances inferiorly form the tragus anteriorly and the antitragus posteriorly at the end of the antihelical rim (the cauda helicis). These are separated by the incisura intertragica. The antitragus and cauda helicis are separated by a small notch known as the fissura antitragohelicina.

The blood supply to the ear derives from the posterior auricular and superficial temporal arteries which supply the posterior and anterior surfaces of the auricle respectively, and are both terminal branches of the external carotid artery. The sensory nerve supply arises anteriorly from the auriculotemporal nerve (branch of CN V3) and posteriorly from the great auricular nerve arising from the cervical plexus, with small contributions to the ear canal from Arnold's nerve (auricular branch of CN X), Jacobsen's nerve (tympanic branch of CN IX) and minor fibers of CN VII. Reflecting its more prominent role in focusing sound waves in other species, several rudimentary muscles are variable present in the human auricle. Among these are extrinsic muscles including the superior and posterior auricular muscles which insert on the posterior auricle and triangular fossa respectively, as well as several intrinsic muscles including the major and minor helical, tragal, antitragal, transverse and oblique auricular muscles located on the anterior and

Practical Plastic Surgery, edited by Zol B. Kryger and Mark Sisco. ©2007 Landes Bioscience.

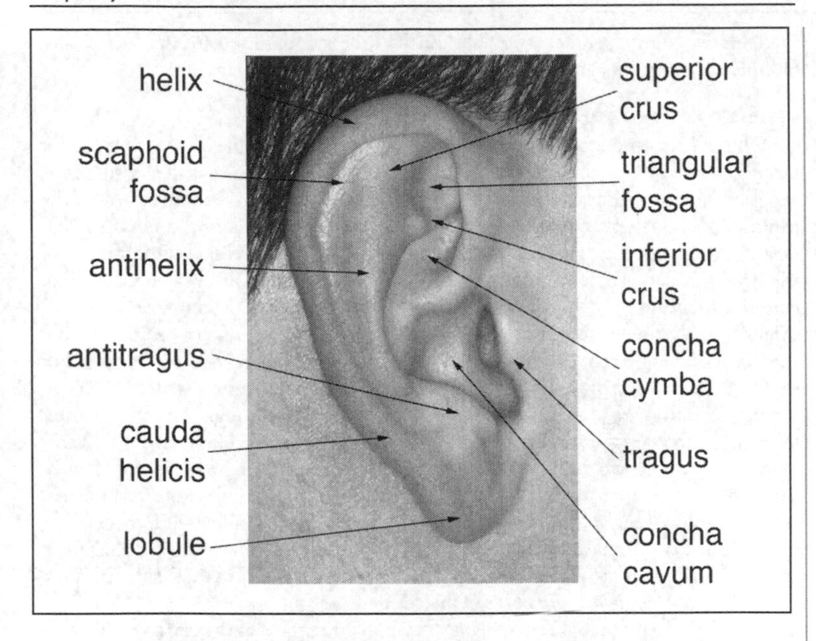

Figure 62.1. Surface anatomy of the human ear.

posterior surfaces of the auricular cartilage. When present, these muscles or structures are supplied by rudimentary branches of the facial nerve (CN VII).

The auricle typically measures between five and six centimeters in vertical dimenstion with a long axis which deviates approximately 20° from the vertical when viewed from the side. This parallels the angle of the profile of the nose. The auricle should be about half as wide as it is long with a helix to mastoid distance ranging from 1 cm superiorly to 2 cm inferiorly.

A commonly encountered ear deformity is the congenitally prominent ear, which is characterized by an increase distance from the helical rim to the mastoid scalp. It is often associated with a deep conchal bowl and may lack and antihelical fold altogether. The scaphoconchal angle is increased with flattening of the superior antihelical crus. The cup ear deformity, also referred to as constricted ear or lop ear involves cosmetically unacceptable overhang of the helix, associated with widening of the concha, flattening of the antihelix, and compression of the scapha an fossa triangularis. Stahl ear, also referred to as "Spock," "Vulcan," or Satyr ear, is characterized by the presence of a third antihelical crus, resulting in flattening of the antihelix and angular distortion of the scapha and superior helical rim.

Preoperative Considerations

By the age of five or six years, the pinna has attained 85% of its adult height and otoplasty may be considered. In adults, surgery on the ear can generally be performed under intravenous conscious sedation supplemented with local nerve blockade or local anesthesia alone. Blood thinning medications should be discontinued including careful screening for prescription, over-the-counter and herbal medications.

Preoperative photodocumentation should include standardized frontal, lateral and oblique views with magnified views of any particular areas of interest.

Operative Technique

Due to the complex three dimensional nature of the ear, the ideal technique for otoplasty must be tailored to the particular patient's deformity and the surgeon's experience with various techniques. The traditional excision for correction of the prominent ear removes a fusiform ellipse of skin just lateral to the postauricular crease, although some delay committing to both limbs of the ellipse until the end of the procedure after determining precisely how much skin will need to be excised. Anterior incisions may also be employed to provide access to the anterior perichondrial surface. Skin excision alone is generally insufficient to hold an ear in position. Cartilage shaping is achieved through the placement of mattress sutures to reanchor the cartilage, scoring of the cartilage to facilitate reshaping and/or excision of excess cartilage. Sutures are placed in a mattress fashion and go through the full thickness of cartilage and perichondrium and should be fairly wide based (at least 1.5 to 2 cm) to avoid cutting through or creating an antihelix that appears too sharp. Conchoscaphoid mattress sutures (Mustarde sutures) serve to restore the normal conchoscaphoid angle of 90° and enhance the antihelical ridge while decreasing projection. Conchomastoid mattress sutures serve to "pin" back the concha. Both these sets of sutures are placed in a radial fashion and tied down gradually until the desired effect is achieved with evenly distributed tension.

The effect is augmented by scoring the anterior surface of the cartilage. Scoring of a cartilaginous or perichondrial surface will generally result in expansion of that surface as an assist in bending of the cartilage away from the side being scored. Patients with excessively prominent conchal bowls may benefit from resection of an ellipse of conchal cartilage. Care and experience allow the proper amount of cartilage to be removed to recreate a graceful fossa without creating an unnatural stepoff in the conchal contour. A mattress suture from the triangular fossa to the temporalis fascia can help correct any residual prominence of the superior pole of the auricle.

Prominent lobules can be addressed by extending the incision inferiorly and closing this portion of the flap in a V-Y manner resulting in reduction of the frontal appearance of the lobule. Alternatively, if the prominence of the lobule is due to an excessively long cauda helicis, simple exision of the excess cartilage and redraping of the skin will improve the deformity.

Finally the skin is redraped, excess skin is excised from the posterior surface of the ear and the wound is closed. Meticulous hemostasis is essential as hematoma formation can have disastrous consequences and some authors favor the placement of a small penrose type rubber-band drain to facilitate drainage. A compressive nonadherent dressing is placed and wrapped with a large turban style dressing.

Postoperative Care

In teenagers or adults, otoplasty can easily be performed as an outpatient procedure with a responsible relative or friend available for overnight observation. After the dressing is removed at the first postoperative visit, patients should be instructed to continue to wear lightly supportive tennis or skiing headbands or ear wraps to take tension off of the reconstruction for an additional 6-8 weeks. Hypertrophic scarring or keloid formation can be treated with injection of steroids with or without excision as dictated by the severity of the scar.

Complications

Along with the ubiquitous risks of bleeding and wound infection, the most common sequelae involve an unsatisfactory cosmetic result. Over-correction or, more commonly, under-correction or reversal of the correction with time are the most common complaints. Avoid asymmetry by measuring the helix to mastoid distance. Over-correction of the conchal prominence without addressing superior pole projection and an excessive lobule results in the "telephone ear" deformity. Conversely, over-correction of superior pole and lobule projection with insufficient correction of conchal excess results in the so called "reverse telephone ear." Failure to anchor conchomastoid sutures in a posterior vector can lead to compromise of the external auditory meatus by occlusion with redundant conchal cartilage. Inadequate hemostasis and failure to recognize a postoperative hematoma can lead to catastrophic loss of the cartilaginous architecture of the auricle, requiring far more complex reconstruction, potentially requiring rib cartilage. Infection should be promptly treated with systemic antibiotic therapy and surgical drainage.

Pearls and Pitfalls

- The biggest disaster is ischemia of the ear. This can lead to cartilage necrosis and a resulting misshapen ear. It can be caused by folding of the ear from too tight a dressing or poor placement of the ear in the dressing due to numbness from anesthesia. Severe postoperative pain is an indication for removal of the dressing and evaluation of the ear.
- Tie the Mustarde mattress sutures first and then look and measure the two ears to achieve as much symmetry as possible. Ears are typically not identical preoperatively.
- Over-correction or an ear that lies flat against the head will not typically improve with time.
- Correction of the antihelix without treating a deep conchal bowl results in an odd deformity requiring revision.
- There are very few occasions when both auricles are seen simultaneously in the course of normal daily activities. Therefore, perfect symmetry from side-to-side is less critical than the correct anatomical harmony of each auricle viewed alone. As an example, this can be very useful when repairing traumatic defects of the ear which require removal of tissue from the helix. The precise size matching of the two auricles is of secondary importance to achieving a balanced, harmonious, natural appearing ear.

Suggested Reading

1. Kelley P, Hollier L, Stal S. Otoplasty: Evaluation, technique, and review. J Craniofac Surg 2003; 14(5):643.
2. Stenstrom SJ, Heftner J. The Stenstrom otoplasty. Clin Plast Surg 1978; 5(3):465.
3. Dingman RO, Peled I. Corrective cosmetic otoplasty: A simple and accurate technique. Ann Plast Surg 1979; 3(3):250.
4. Elliott Jr RA. Otoplasty: A combined approach. Clin Plast Surg 1990; 17(2):373.

Chapter 63

Blepharoplasty

Robert T. Lancaster, Stephen M. Warren and Elof Eriksson

Introduction

Aesthetic eyelid surgery is intended to brighten and refresh the eyes. By removing baggy skin, an ellipse of muscle and protruding fat, the surgeon has a chance to erase the signs of periorbital aging and endow a youthful appearance. Functional eyelid surgery is largely geared towards correcting congenital or acquired ptosis, ectropion (eversion of the eyelid) and epiblepharon (inversion of eyelashes against globe). While aesthetic and functional eyelid surgery may be nosologically divided, both have a role in every operation.

Eyelid surgery must be preceded by a sound knowledge of the anatomy and a thorough understanding of the deformity to be corrected. Moreover, patient selection may be as important as the technical aspects of resecting/resuspending the periorbital tissues or controlling the lateral canthus. Goals for eyelid surgery include the creation of crisp upper lids, correction of fatty protrusions without hollowing the eyes, reestablishment of lower lid tone, control of the lateral canthus, preservation of aperture length and height, avoidance of scleral show and lagophthalmos, and correction of deep groves, all while maintaining the illusion of symmetry.

Anatomy

The palpebral fissure (aperture of the eye) measures 12-14 mm vertically and 28-30 mm horizontally. The eye is almond-shaped with the lateral canthus slightly more superior than the medial canthus: typical superior elevations at the lateral canthus are 2 mm for men and 4 mm for women. The distance from the lateral canthus to the orbital rim is about 5 mm. The upper lid fold in Caucasians is approximately 8-11 mm. The lower lid crease is about 5-6 mm. The high point of the brow is superior to the lateral limbus. The upper lid rests 2 mm below the superior limbus of the iris and the lower eyelid rests at the inferior limbus.

Upper Eyelids

The skin of the eyelid is less than 1 mm thick. With aging and loss of elasticity, wrinkling and sagging of the eyelid occurs. Beneath the skin and subcutaneous tissues lies the orbicularis oculi muscle, which is divided into an outer orbital portion and an inner palpebral portion (Fig. 63.1). The palpebral portion is further subdivided into preseptal and pretarsal parts. Collectively, the orbicularis oculi closes the eye, shortens and milks the canniliculi, and expands the lacrimal sac. Beneath the orbital and preseptal portions of the orbicularis oculi is the preseptal fat, known as the retroorbicularis oculi fat (ROOF). This fat pad lies over the orbital rim extending outward toward the tail of eyebrow. Resecting the ROOF decreases the heaviness of the lateral brow and upper lid, but the ROOF isn't the primary culprit in

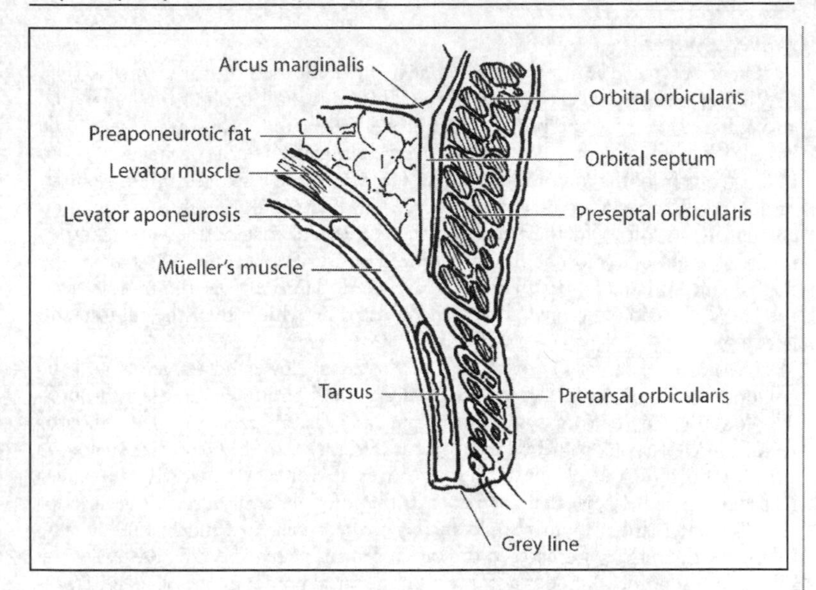

Figure 63.1. Upper eyelid anatomy.

baggy upper lids. The orbital septum lies deep to the orbicularis. It hangs from the superior orbital rim and joins the levator aponeurosis at the superior border of the tarsal plate. Weakening of the septum with aging, hereditary predisposition, or trauma may cause protrusion of the orbital fat. The upper lid contains two fat pads: medial and central (the lateral space is occupied by the lacrimal gland). The medial and central compartments are separated by the superior oblique muscle. The medial fat pad is lighter in color (similar to butterscotch), firmer in consistency, and usually requires more local anesthesia during resection than the central compartment. The medial compartment also contains the terminal branch of the ophthalmic artery; we find that attending surgeons will take an extra second to ensure that this vessel is satisfactorily coagulated during fat resection.

The upper lid tarsus is a fibrous plate that is approximately 10 mm wide in the central upper lid, narrowing medially and laterally. The tarsal plates extend from the lateral commissure to the punctum, and it contains numerous meibomian glands that empty into the ciliary border. There are two muscles responsible for opening the upper eyelids: the levator muscle (primary lid elevator) and Müller's muscle. The levator muscle originates at the apex of the orbit, just superior to the superior rectus. At the orbital aperture, it is supported by Whitnall's ligament, which functions to translate the horizontal force of this muscle into the posterior and vertical motion necessary for lid elevation. The levator muscle becomes aponeurotic as it passes Whitnall's ligament. The anterior interdigitation of the aponeurosis with the orbicularis muscle fibers leads to the formation of the supratarsal fold. Müller's muscle originates from the posterior aspect of the levator aponeurosis and travels inferiorly, closely adherent to the conjunctiva, to insert on the superior border of the tarsus. Müller's muscle is sympathetically innervated. The conjunctiva is the inner most layer of the upper lid.

Lower Eyelids

The lower eyelid is often described as being composed of three lamellae (Fig. 63.2). The anterior lamella consists of the skin, orbicularis oculi (orbital, preseptal and pretarsal) muscle, and preseptal suborbicularis oculi fat (SOOF). The middle lamella is the orbital septum. The septum is a continuation of the orbital periostium that extends from the inferior orbital rim (arcus marginalis) to the inferior border of the tarsus. It provides the anterior border of the medial, middle and lateral fat compartments. Although these compartments are more imaginative than anatomic, the inferior oblique (most commonly injured during blepharoplasty) separates the medial and middle compartments; the arcuate expanse divides the middle from the lateral fat pad. The medial fat pad tends to be whiter than the middle and lateral pads.

The posterior lamella is composed of the tarsus, lower lid retractors and the conjunctiva. The lower lid tarsal plate is only about 4.5 mm wide at the mid-pupil. The lower eyelid retractor system originates as a fascial extension of the inferior rectus muscle (capsulopalpebral head). This fascial system splits to encapsulate the inferior oblique muscle and then reunites to form a dense fibrous sheet (capsulopalpebral fascia) that inserts onto the inferior tarsal border. The inferior tarsal muscle is the smooth muscle analog of the upper lid Müller's muscle. The inferior tarsal muscle originates in the inferior fornical area and extends toward the inferior tarsal border but does not insert on the tarsal border as does its counterpart in the upper eyelid. The inferior tarsal muscle provides sympathetic innervation to the lower eyelid, and interruption of its innervation results in a slightly elevated position of the lower eyelid margin, as observed in Horner syndrome. Otherwise, the inferior tarsal muscle has little pathologic significance.

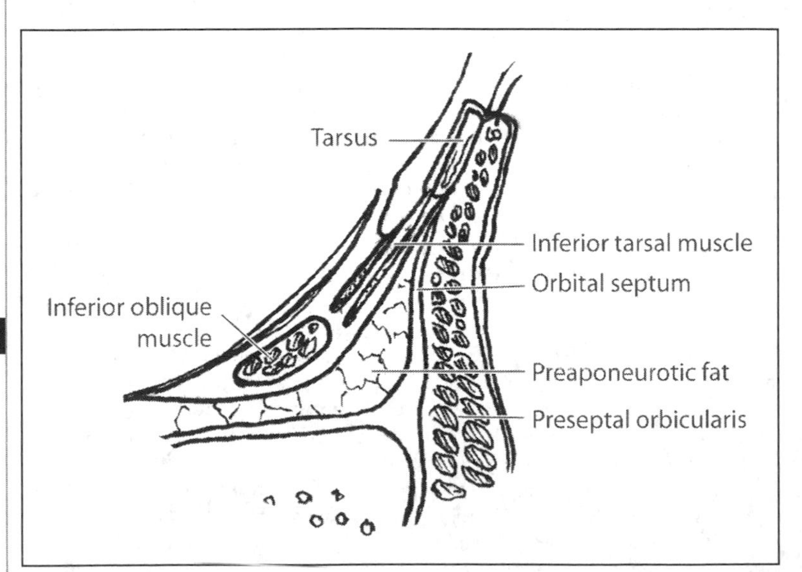

Figure 63.2. Lower eyelid anatomy.

Asian Eyelids

There are a number of anatomic differences between Asian and Caucasian eyelids. In the Asian upper eyelids, the skin and subcutaneous fat tend to be thicker. The ROOF is heavier. The orbital septum is weaker and thinner and it tends to drape below the upper border of the tarsal plate. The levator aponeurotic dermal extensions are weak or nonexistent. The paucity of levator-dermal extensions is responsible of the low-lying or absent supratarsal fold. Epicanthal folds cover the medial angle and lacrimal caruncle. Trichiasis is caused by the overhanging upper eyelid skin pushing down on the eyelashes. Collectively, these features give the Asian eyelid more fullness, a lower lid crease and a narrower palpebral fissure.

Canthal Tendons

Medially, the superficial heads of the upper and lower pretarsal muscles join to form the medial canthal tendon (Fig. 63.3). This tendon is firmly attached to the anterior lacrimal crest. The superficial heads of the preseptal muscles attach to the medial canthal tendon. The deep heads of the preseptal and pretarsal muscles attach to the posterior lacrimal crest, just behind the lacrimal sac. Laterally, the upper and lower pretarsal muscles join to form the lateral canthal tendon, which inserts just posterior to the orbital tubercle. The upper and lower preseptal muscles join laterally to form the lateral palpebral raphe, which is attached to the skin.

Preoperative Considerations

During the complete medical history, one should inquire about a history of diabetes, hypertension, coagulopathy, hypothyroidism, hyperthyroidism, renal disease, cardiopulmonary disease or glaucoma; each of these diseases can have a role in eye symptomatology and effect postoperative recovery. Patients with a history of collagen vascular diseases, such as scleroderma, systemic lupus erythematous, periarteritis nodosa, Wegener's granulomatosis, Stevens-Johnson syndrome, rosacea, rheumatoid arthritis, or secondary Sjögren syndrome have a higher risk of postoperative dry eye syndrome. Finally, elicit an ophthalmologic history. This includes previous eyelid surgery, eyelid trauma, eyelid infection, eyelid allergy, eyelid swelling, the use of glasses or contact lenses, and changes in visual fields and/or visual acuity.

The physical examination first includes looking at the full face. Note facial appearance, asymmetry and wrinkles. Examine the periorbital area for crow's feet, fine wrinkles (at rest and with smiling), and the appearance of the globes, infraorbital

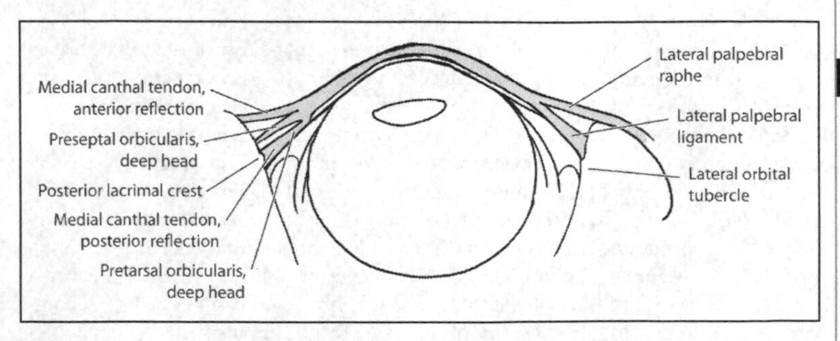

Figure 63.3. Canthal tendons.

rims, cheeks and malar bags. Assess for **negative vector** (in the lateral view: anterior most projection of the globe, the lower eyelid margin and the malar eminence). A negative vector is one which angles posteriorly and indicates an absence of support for the lower lid). Patients with a negative vector are at higher risk for postoperative dryness. Note any pseudoherniated orbital fat and hypertrophied orbicularis muscle. Pressure on the upper eyelid and globe causes pseudoherniated orbital fat in the lower eyelid to be more evident. Having the patient look upward can help to delineate the lower eyelid fat pockets. Examine the eyelids for discoloration, hypertrophied skin and skin lesions. Excessive skin produces a crepe-like quality in the lower eyelid skin. Also, note the position of the lacrimal gland, in particular whether or not it has fallen from the lacrimal fossa. Assess for Bell's phenomenon. The Schirmer's I test may be performed by placing a 5 x 35 mm strip of #41 filter paper between the lower lateral lid and globe for 5 minutes. This test measures both basic and reflexive tear production. Less than 10 mm of moisture on the paper suggests that the patient may be at risk for postoperative dry eyes. The Schirmer's II test measures only basic tear secretion by blocking reflex secretion with a topical anesthetic. The snapback test is performed by grasping the lower eyelid skin and pulling the lid away from the globe. The rapidity with which the lid snaps back against the globe gives an estimate of the probability of postoperative ectropion.

Surgical Technique

Upper Blepheroplasty

Begin the procedure by marking the lower border of the skin resection with the patient in the upright position. This curvilinear line is usually located 10 mm above the lid margin in the natural eyelid crease. The upper border of the skin resection is usually estimated by a pinch test; there should be 15-18 mm of skin between the upper border of the skin resection and the eyebrow. Infiltrate local anesthetic evenly throughout the upper eyelid. Excise the skin and achieve hemostasis with pinpoint electrocautery. Resect a sliver of orbicularis oculi muscle to reveal the preaponeurotic fat. The preaponeurotic fat is resected as desired, and the orbital septum is opened in defined points at the medial and central fat pads or along its length. Excess orbital fat is estimated by applying gentle pressure to the globe. This excess fat is resected with care taken to ensure hemostasis. The skin then is reapproximated, and the procedure is complete.

Lower Lid Blepharoplasty, Subciliary

At the start of the subciliary blepharoplasty approach, a Frost suture can be placed in the gray line lateral or medial to the limbus to facilitate retraction and protect the globe. Alternatively, a lubricated corneal shield may be inserted. An incision is made just lateral to the lower lid margin. A scissors is used to develop a subcutaneous plane across the subciliary margin and complete the skin incision. Care is taken to protect the lashes. A skin-only or skin-muscle flap can be created to gain access to the orbital fat. If a skin-only flap has been elevated, the orbicularis is opened by incisions over the medial, central and lateral compartments. If a skin-muscle flap is chosen, all three fat compartments are in plain view. The inferior oblique muscle is identified between the medial and middle fat pads, and it is protected easily. The lateral compartment is slightly higher than the middle and should be identified carefully because it is the most common compartment to be overlooked. After resecting the orbital fat, a conservative skin excision is performed and the skin is closed.

Lower Lid Blepharoplasty, Transconjunctival

Two approaches to transconjunctival blepharoplasty are used: retroseptal and preseptal. The retroseptal approach is taken by incising from the caruncle to the lateral canthal area at a level half way between the inferior margin of the tarsal plate and the fornix. A traction stitch can be placed through the upper conjunctiva. The orbital fat pads can be assessed and excess fat resected to the level of the orbital rim. Once the fat is resected, some surgeons will close the transconjuctival incision with absorbable sutures, but most will realign the incision without suturing.

In the preseptal approach, the incision is made through the conjunctiva below the tarsus, and a dissection plane is developed between the orbicularis muscle and the orbital septum. The orbicularis is opened by incisions over the medial, central, and lateral compartments, and the fat is resected. Closure is the same as above.

Lateral Canthopexy

There are many techniques to suspend the lateral border of the lower tarsal plate. Classically, the lateral canthus is sutured to Whitnall's tubercle, which is located about 1 cm inside the lateral orbital rim. The purpose of any type of lateral tarsal suspension is to simply tighten the lower lid, improve its coaptation against the globe, and reduce the incidence of postoperative ectropion.

Kunt-Simonowsky Procedure

A small composite wedge resection of the lateral lower lid is an alternative to lateral suspension procedures. This procedure is designed to take up slack in the lower lid and improve lid position. However, some surgeons argue that wedge resection shortens the aperture of the eye and causes a rounding of the lateral palpebral fissure.

Postoperative Care

The patient should stay in an observational area after surgery for at least 1-2 hours. Ice water-soaked gauze compresses should be applied continuously for 24-48 hours to decrease swelling, as well as to reduce postoperative pain. Additionally, the patient's head should be elevated greater than 45° to decrease edema and the collection of sanguinous fluid. Ocular lubrication with artificial tears should be prescribed, particularly if the patient has a preexisting history of dry eyes or if postoperative lagophthalmus is present. The patient should avoid any strenuous activity. Stitches are removed in clinic in 3-7 days.

Complications

Retrobulbar Hematoma

The most concerning postoperative complication is a retrobulbar hematoma (bleeding posterior to the orbital septum). The incidence is about 0.04%. This complication is more likely to occur in hypertensive patients with intraoperative hypertension. Thus, adequate pain control during the procedure becomes equally as important as precise cauterization of vessels. Significant retrobulbar hemorrhage will lead to visible protrusion of the globe initially and then to a rapid increase in intraocular pressure to greater than 30 mm Hg (normal = 10-20 mm Hg). As the intraocular pressure approaches diastolic levels, the risk of total vision loss increases dramatically. Clinically, the signs and symptoms include acute pain, proptosis, chemosis and opthalmoplegia. The globe and lid may become hard to palpation. Acute orbital hemorrhage is a medical and surgical emergency. An emergency ophthalmology consult should be obtained, but treatment should not be delayed. Treatment includes immediate suture removal, wound exploration and lateral canthotomy. Mannitol, acetazolamide and steroids may be administered to reduce intraocular pressure. Anterior chamber paracentesis and bony orbital decompression are rarely used.

Diplopia

Diplopia in the early postoperative period is not uncommon. It may be caused by edema or anesthesia infiltration of the extraocular muscles. Long-lasting diplopia is extremely rare and may be caused by damage to the inferior oblique muscle, which is especially vulnerable during the transconjunctival approach. Management is supportive.

Ptosis

Post-blepharoplasty ptosis occasionally occurs secondary to edema and ecchymosis. In these cases the ptosis is temporary and usually resolves after 2-3 weeks. However, the most frequent cause of postoperative ptosis is the failure to recognize it preoperatively. Early postoperative asymmetry is best managed by time and gentle massage of the higher crease. Ptosis lasting longer than 3 months requires reexploration.

Scleral Show

The incidence of scleral show is reported to be as high as 15% in some series. Many surgeons suggest the principle cause of lower eyelid malposition is unrecognized laxity in the tarsoligamentous sling. In addition to scleral show, ligamentous laxity tends to cause rounding of the lateral palpebral fissure. Treatment is conservative with massage therapy for 2-3 months. The round eye appearance may improve slowly with time and gentle upward massage. Lateral canthopexy can be done if symptoms persist. It is important to remember that minimal skin should be excise in a lower blepharoplasty.

Ectropion

Although ectropion is one of the most commonly discussed complications of lower blepharoplasty, its incidence is estimated to occur in less than 1% of patients. The best treatment is prevention by attention to lower lid laxity and conservative skin and orbicularis excision. Ectropion is usually treated by massage therapy and taping for at least 3 months. If the ectropion persists, tightening procedures or skin grafts can be performed.

Lagopthalmos

Lagopthalmos is the inability to completely close the eyes. It occurs immediately after surgery due to swelling and local anesthesia impairment of orbicularis oculi function. Lagopthalmos requires treatment with eye lubricants to protect the cornea and reduce irritative symptoms. Persistent lagopthalmos is probably related to the amount of skin excised from the upper lid and not the amount of muscle excised. In most cases, lagopthalmos resolves as the wound matures.

Dry Eyes

Mild dry eye syndrome secondary to lagopthalmos is a common transient problem. However, it can produce corneal ulcerations that may threaten vision. All patients should be provided with ocular lubricants.

Pearls and Pitfalls

It is important to identify ptosis preoperatively. Possible causes of ptosis include trauma, chronic progressive external ophthalmoplegia, Horner's syndrome, myasthenia gravis, levator dehiscence and upper lid tumors. Pseudoptosis is excess skin that causes hooding and depression of the upper lid. Pseudoptosis can be differentiated from true ptosis by elevating the excess skin with a cotton-tipped applicator. Ptosis, particularly asymmetric ptosis, should be highlighted for the patient preoperatively. Operative correction is dictated by the cause of ptosis.

Preoperative testing for dry eyes (e.g., Schirmer's tests) may have a poor positive predictive value. The best predictors of postoperative dry eyes are abnormal ocular history or abnormal orbital anatomy. Abnormal ocular anatomy includes: scleral show, lagopthalmos, lower lid hypotonia, proptosis, exopthalmos and maxillary hypoplasia. Patients with one or more of these anatomic findings should be provided with additional preoperative warnings and postoperative ocular protection.

Although ectropion is uncommon, the surgeon must have a high preoperative index of suspicion. Ectropion may be prevented by a variety of lateral canthal tightening procedure. The senior author prefers a Kunt-Simonowsky lid shortening procedure for lower risk patients and a lateral canthopexy for higher risk patients. Correcting an ectropion is more difficult than preventing an ectropion.

Suggested Reading

1. Jelks GW, Jelks EB. Preoperative evaluation of the blepharoplasty patient: Bypassing the pitfalls. Clin Plast Surg 1993; 20:213.
2. Rees TD, LaTrenta GS. The role of the Schirmer's test and orbital morphology in predicting dry-eye syndrome after blepharoplasty. Plast Reconstr Surg 1988; 82:619.
3. Zarem HA, Resnick JI. Operative technique for transconjunctival lower blepharoplasty. Clin Plast Surg 1992; 19:351.
4. Zide BM. Anatomy of the eyelids. Clin Plast Surg 1981; 8:623.

Chapter 64

Rhinoplasty

Ziv M. Peled, Stephen M. Warren and Michael J. Yaremchuk

Introduction

Nasal surgery requires careful analysis of the patient's problems, detailed planning and meticulous operative technique. Specific nasal problems must be detailed in order to understand the patient's likes and dislikes about the nose. External and internal examinations are necessary to understand the complex three-dimensional anatomy and define the aesthetic and functional problems. Rhinoplasty is a procedure in which subtle changes in anatomy often make profound differences in appearance, making this one of the most challenging yet elegant procedures in the plastic surgery. Although rhinoplasty is a complex procedure, according to the American Society of Plastic Surgeons (ASPS), over 350,000 rhinoplasties were performed in the United States in 2003, making it one of the most common cosmetic surgical procedure.

Anatomy

To begin to understand the complexities of rhinoplasty, one must first become familiar with nasal anatomy. This understanding is often hampered by the plethora of terms used to describe the anatomy of the nose. This chapter will attempt to underscore the key components of nasal anatomy, how these components interact functionally, and how they are affected during nasal surgery. In the process, we will try to maintain a consistent nomenclature to avoid confusion.

Beginning cephalically, the nose consists of a pair of nasal bones that extend caudally from the frontal bone (Fig. 64.1). These bones function as a structural support extending from the skull. Posteriorly, the nasal bones are supported by the anterior edge of the frontal process of the maxilla on either side. Caudally, the nasal bones fuse with the dorsal septum forming a support structure that extends along the length of the nasal dorsum. Moving towards the tip, the caudal aspect of the nasal bones overlaps the cephalic portion of the upper lateral cartilages for a distance of 2-4 mm. The upper lateral cartilages comprise the distal nasal sidewalls and extend towards the tip of the nose. At their caudal free edge, they are overlapped by the cephalic portions of the lower lateral cartilages.

The lower lateral cartilages are also known as the alar cartilages. They provide the structural support to the soft tissues of the lower third of the nose, also known as the tip or lobule, which is also the portion of the nose which has the greatest projection from the facial skeleton. The lower lateral cartilages are subdivided into three parts (Fig. 64.2). The portions of the lower lateral cartilages closest to the midline on either side are known as the medial crura. They comprise the structural support for the columella. At their most posterior aspect, they flare slightly laterally to form the medial footplates which project into the nasal vestibule. More anteriorly, are the

Practical Plastic Surgery, edited by Zol B. Kryger and Mark Sisco. ©2007 Landes Bioscience.

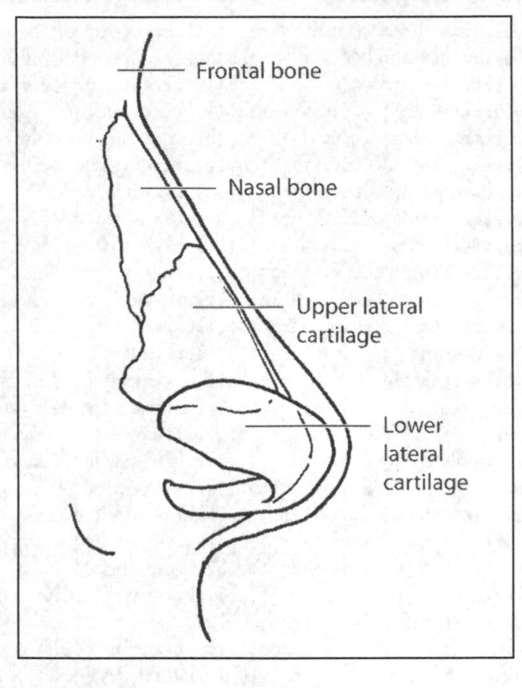

Figure 64.1. Lateral view of the nose illustrating the relationship between the frontal bones, nasal bones, and the upper and lower lateral cartilages. Note also the superior location of the lateral extent of the lateral crus on the patient's right side demonstrating that the lateral aspect of the alar lobule is devoid of cartilage. (Reproduced with permission from "Aesthetic Rhinoplasty" by J.H. Sheen, 1998, Quality Medical Publishing.)

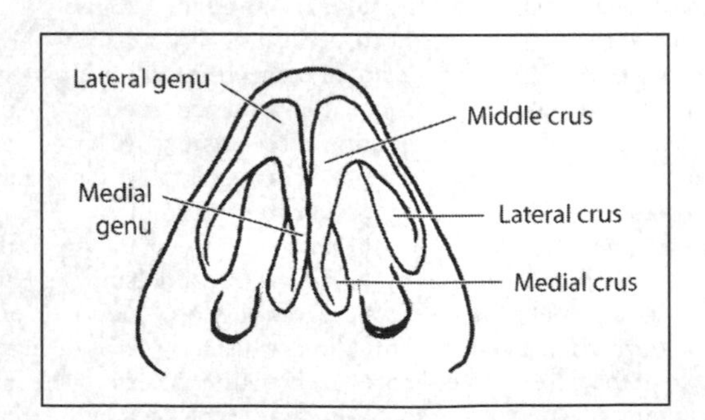

Figure 64.2. Components of the lower lateral cartilages. Note the slight projection of the posterior-most portion of the medial crura into the nasal vestibule. This portion of the medial crura is also known as the medial footplate. (Reproduced with permission from "Aesthetic Rhinoplasty" by J.H. Sheen, 1998, Quality Medical Publishing.)

middle crura which continue from the most anterior portion of the medial crura towards the tip of the nose and begin the turn laterally to form part of the genu (i.e., curve) of the lower lateral cartilages. The lateral crura comprise the remainder of the genu by continuing laterally, posteriorly and slightly cephalically. It is important to note that as one traces the lateral crura from the tip posterolaterally, the crura project more cephalically (Fig. 64.1). Hence, the lateral crura provide structural support to the nasal rim predominantly at their medial portions, leaving the very lateral portions of the nasal alae devoid of cartilage. This portion of the nose is comprised of fibrofatty tissue covered by overlying skin.

The nasal septum is composed of three structures: the perpendicular plate of the ethmoid posteriorly and cephalically, the vomer posteriorly and caudally and the quadrangular (i.e., cartilaginous) septum anteriorly. The septum functions as a support structure for the mid-portion of the nose and it also comprises the medial component of the internal nasal valve (completed posteriorly by the nasal floor and laterally by the upper lateral cartilages). The average internal valve angle is 12°.

The arterial anatomy of the nose is important to consider for several reasons. While the blood supply to the nose is abundant making tissue necrosis a rare complication, the potential for clinically significant bleeding exists. Bleeding can be significant in that it can compromise tissue due to compression (e.g., septal hematoma) or compromise visualization during the rhinoplasty. Again, beginning cephalically, the blood supply to the dorsal nose is derived from the dorsal nasal artery and the external nasal branches of the anterior ethmoidal artery. The lateral nasal artery which arises from the angular artery supplies blood flow to the nasal sidewalls and the caudal nasal dorsum and tip. The columellar branches of the superior labial artery anastomose with the distal branches of the lateral nasal artery to supply the nasal tip from below. The blood supply to the septum comes from the anterior and posterior ethmoidal arteries, the sphenopalatine artery and the posterior septal artery. The convergence of the anterior and posterior ethmoidal plexuses in the anterosuperior septum is known as Little's area and is the most common site of injury causing epistaxis. Because of its location, the anterior ethmoidal artery is the most often injured in nasal trauma.

The cutaneous nerve supply to the nose is important because adequate local analgesia can allow the plastic surgeon to perform a rhinoplasty without general anesthesia. Since the local anesthetics contain epinephrine, appropriate infiltration can also reduce blood loss. Furthermore, the local anesthetics can be used to hydrodissect delicate nasal tissues facilitating subsequent sharp dissection. For example, local anesthesia infiltrated submucosallly in the septal area, not only provides excellent hemostasis, but also separates the septal mucosa from the underlying cartilaginous septum. Beginning cephalically, the nasal branches of the supraorbital nerve are infiltrated to anesthetize the radix and proximal nasal dorsum. Sensation along the nasal sidewalls, alae and columella is blocked using local anesthesia on the nasal branches of the infraorbital nerve. The middle and distal thirds of the nasal dorsum as well as the tip of the nose are anesthetized by blocking the external nasal branches of the anterior ethmoidal nerve. The anterior septum is blocked by anesthetizing the medial and lateral branches of the anterior ethmoidal nerve. If a septoplasty will be performed in addition to the rhinoplasty, the nasopalatine and posterior nasal nerves, which supply sensation to the posterior septum medially and laterally, respectively, should be blocked. Once the nose is adequately blocked, the rhinoplasty procedure can begin.

Surgical Technique

Whether to perform an open or a closed rhinoplasty remains a controversial topic. To be sure each technique has its advantages and disadvantages and these, along with the requests of the patient, should be used to guide the surgical approach. The open rhinoplasty involves any of a variety of mid-columellar incisions to expose the nasal anatomy much like one would expose the engine of a car by opening the hood. Clearly, the advantages of this approach are the excellent visualization which facilitates the operative procedure and teaching. Disadvantages include the external scars, the longer operative time and the increased postoperative swelling secondary to more aggressive manipulation of the nasal tissues. During a closed rhinoplasty, no external incisions are made and access to the nasal framework is obtained via any number of internal incisions (e.g., inter-, intra- or infracartilage, or rim). Advantages with this technique are the lack of external scarring and the relative expeditiousness of the procedure. Its primary disadvantage is the limited visualization which therefore limits the manipulations than can be performed. This approach is best suited for patients requiring minor tip work, straight-forward resection of a prominent dorsal hump, or those with a wide alar base.

Dorsal Hump

One of the most common complaints is a prominent dorsal hump. In consideration of the anatomy just reviewed, one can understand that this prominence can be caused either by projecting nasal bones, upper lateral cartilages, cephalic dorsal cartilaginous septum or some combination thereof. Resection of this dorsal hump can be performed either with the closed or open rhinoplasty technique. Once the overlying soft tissues have been dissected in the subcutaneous plane from the underlying bony and cartilaginous anatomy, the reason(s) for the projecting dorsum can be determined. While many techniques are used to achieve a more balanced nasal dorsum, the general principles are to sharply resect the cartilaginous components of the dorsal hump and to use either an osteotome or rasp to resect the bony dorsum. The rasp is generally used for more subtle bony adjustments. Two points about dorsal hump resection deserve emphasis. First, conservative resection is best. One can always resect more and as noted earlier, small changes often make pronounced differences in appearance. Secondly, it is important to avoid resection of the mucoperichondrium as this tissue layer provides support to the upper lateral cartilages and can lead to an inverted-V deformity.

Tip Projection

Under-projection of the nasal tip is another common problem. Tip projection can be enhanced in a number of ways. A transdomal suture bringing the genua closer together will add a mild degree of projection. Placement of an onlay tip graft can enhance tip projection as well as widen the nasolabial angle and increase lobular volume. Other ways of achieving the same effect are placement of a columellar strut graft and suturing the medial footplates together. In addition, lowering a dorsal hump will often give the impression of a cephalically oriented tip. Each of these techniques has its unique degree of tip enhancement and differing effects on surrounding nasal structures (e.g., columella, lobule, upper lip).

The over-projecting tip (Pinocchio nose) is due to excess length of the medial and lateral crura. Resection of the medial or lateral crura or scoring of the dome-medial crura junction with or without resection of the lateral crura can be combined with a transfixion incision to reduce tip support to allow immediate posterior settling of the tip.

Boxy Tip

A boxy nasal tip is due to either an increased angle of divergence of the medial crura or a wide arc in the dome segment of the lateral crura.. This problem is commonly treated by resection of the cephalic portion of the lateral alar cartilages and interdomal suturing.

Widened Alar Base

A wide alar base is a masculinizing feature. Since most patients seeking rhinoplasty are female, a wide alar base is a commonly encountered problem. The simplest way to narrow a wide alar base is through resection of a portion of the fibrofatty tissue making up the lateral-most portion of the alar rim (e.g., Weir excision). This technique has the disadvantage of placing visible (albeit small and well-hidden) scars on the external nasal skin. Transdomal sutures often used to augment tip projection can also pinch the lateral crura together thus slightly narrowing the alar width.

Wide Nasal Bridge

A wide nasal bridge is also a frequent complaint. This problem can be dealt with directly and indirectly. Clearly, nasal osteotomies will allow medialization of the upper and middle nasal vaults. These osteotomies can be performed via either open or closed approaches, using a continuous or perforated technique. This maneuver is often performed when a prominent dorsal hump has been resected, leaving the dorsal edge of the nose with a widened, open roof, appearance. Preoperatively, the patient's internal nasal valve angle must be evaluated because in-fracturing will narrow the nasal passage. If the internal nasal valve <12°, the patient will have obstructed airflow. Dorsal augmentation also gives the appearance of a narrower nose, again highlighting the interplay of the various nasal components.

Complications

Retracted Ala

A retracted ala occurs in patients who have had overresection of the cephalic portions of the lateral crura in an attempt to improve tip definition. As healing progresses, wound contraction rotates the lateral crura cephalically and with it the alar rim. To avoid this pitfall, it is recommended that at least 6-9 mm of lateral crus remain after resection and that one make every attempt to leave as much vestibular mucosa as possible during the procedure. To fix this problem, a free cartilage graft can be used to augment the remaining lateral crura in minor cases. If a severe retraction is encountered, composite grafts from the contralateral ear can be used to provide, lining, coverage and support.

Parrot Beak Deformity

A parrot beak deformity refers to excessive supratip fullness following rhinoplasty. This problem can be caused by either underresection of the supratip dorsal hump or overresection of the nasal dorsum. If the cause is the former, further resection to achieve a proper tip-to-supratip proportion is mandated. If the cause is the latter, a dorsal graft is appropriate. When the parrot beak deformity occurs in the early postoperative period it is attributed to edema and wound contraction. In this situation, a trial of conservative management with taping and steroid injection is appropriate. Careful technique with frequent assessments of the appearance of the nose during the procedure is the best way to avoid this problem.

Saddle-Nose Deformity

The saddle-nose deformity appears as a disproportionately flattened dorsum, akin to a boxer's nose. The most common cause is over-resection of the cartilaginous septum leaving less than 15 mm of support. Treatment in mild to moderate cases involves placing additional graft material over the depressed areas to restore the lateral and frontal profile. In severe cases, cantilevered bone grafts suspended from the frontal bone may be required.

Open Roof Deformity

An open roof deformity occurs after resection of a dorsal hump without adequate in-fracture of the nasal bones. The dorsal bony edges become visible within a flattened area of the dorsum. Treatment is intuitive and consists of adequately in-fracturing the nasal bones.

Inverted-V Deformity

If the mucoperichondrium of the upper lateral cartilages is inadvertently resected, support to the upper lateral cartilages is lost. This problem causes the upper lateral cartilages to collapse inferomedially. On frontal view, the caudal edges of the nasal bones become visible. The best treatment is avoidance, but if it occurs, dorsal cartilage grafting to restore dorsal nasal balance is the preferred treatment.

Pearls and Pitfalls

Each case has it own challenges and requires a careful estimation of the deformity preoperatively, a clear understanding of the techniques available, a proposed plan of action and sequence, and a meticulous, uncompromising surgical technique. Every operation has a risk for complication, and only the surgeon who does not operate has no complications. Under-correction and over-correction of a preexisting deformity may lead to either persistence of the deformity or the introduction of a new one. A new deformity may introduce functional deficits. Some deformities are illusory and correction can only follow accurate diagnosis. For example, when a patient requests dorsal reduction, first examine the nose in thirds. If the nasal radix is too low, augment it, don't reduce the dorsum. The radix augmentation will give the illusion of a smaller nose. Furthermore, the nose must also be examined in relation to the face. For instance, a nose may appear large because the chin is small; a chin implant may be the best choice in some cases to achieve the illusion of facial harmony.

Suggested Reading

1. Becker DG. Complications in rhinoplasty. In: Papel I, ed. Facial Plastic and Reconstructive Surgery. New York: Thieme, 2002:87-96.
2. Daniel RK. Rhinoplasty. In: Aston SJ, Beasley RW, Thorne CHM, eds. Plastic Surgery. Philadelphia: Lippincott-Raven, 1997:651-669.
3. Dingman RO, Natvig P. Surgical anatomy in aesthetic and corrective rhinoplasty. Clin Plast Surg 1977; 4:111.
4. Guyuron B. Dynamics in rhinoplasty. Plast Reconstr Surg 2000; 105:2257.
5. Rohrich RJ, Muzaffar AR. Primary rhinoplasty. Plastic Surgery: Indications, Operations and Outcomes V. 2000:2631.
6. Sheen JH, Sheen AP. Aesthetic Rhinoplasty. 2nd ed. St. Louis: Mosby, 1998:1-1440.
7. Sheen JH. Rhinoplasty: Personal evolution and milestones. Plast Reconstr Surg 2000; 105:1820.
8. Zide BM, Swift R. How to block and tackle the face. Plast Reconstr Surg 1998; 101:840.

Chapter 65

Genioplasty, Chin and Malar Augmentation

Jeffrey A. Hammoudeh, Christopher Low and Arnulf Baumann

Introduction

The chin provides harmony and character to the face. A strong chin or prominent jaw line is considered to be aesthetically pleasing, especially in males. When chin surgery is indicated, whether by anterior horizontal mandibular osteotomy (AHMO) or by alloplastic implant augmentation, it can create an aesthetically pleasing facial contour and establish proportionate facial height. In addition, the AHMO can improve obstructive sleep apnea by elevating the hyoid bone.

Most genioplasty procedures are done to improve the mandibular profile in order to obtain a more natural profile. Genioplasty can shorten or lengthen the lower third of the face. Facial asymmetry may be corrected by rotation of the chin-point to coincide with the midline. The advantages of osseous genioplasty are versatility, reliability and consistency in correcting problems in the sagittal and vertical planes to achieve greater chin projection.

In order to be able to make an appropriate recommendation, the correct preoperative workup should be performed, including soft and hard tissue analyses. Ideally, cephalometrics and video cephalometric predictions would also be performed.

Anatomy and Analyses

It is important for the surgeon to be familiar with the classic soft tissue analysis and diagram of facial proportions. The size, shape and position of soft and hard tissue can enhance facial harmony and symmetry. The relationship between soft tissue and bone is important for planning the chin correction. For chin advancement, the bone to soft tissue proportion is 1:0.8, meaning that 1 mm of bony change is associated with 0.8 mm of soft tissue change.

The face can be divided into upper, middle and lower thirds. The upper third of the face spans from the hairline to the glabella (G); the middle third from glabella to subnasale (Sn); and the lower third from subnasale to menton (Me). The lower third of the face can be further divided into an upper half (Sn to vermilion of the lower lip) and a lower half (Me to vermilion of the lower lip). The face is "balanced" when the three thirds are of similar height. Cephalometric analysis ensures that skeletal and occlusal disparities are identified and can be corrected before or at the same time as a genioplasty.

Many patients that complain of a small chin truly do not have microgenia. They often have a true deficit of the mandible in the sagittal plane, which can be a class 2 malocclusion (retrognathia) or normo-occlusion (retrogenia). Retrognathia is ideally corrected with a bilateral sagittal split osteotomy (BSSO); however if the discrepancy is small, advancement genioplasty may sufficiently camouflage the facial profile into an orthognathic appearance. Retrogenia (chin point deficiency

Practical Plastic Surgery, edited by Zol B. Kryger and Mark Sisco. ©2007 Landes Bioscience.

in the setting of a class I occlusion) and mild retrognathia (≤3 mm) are ideal cases for a genioplasty. It is important to understand the relationship of the dentition to the chin point. The boney chin point should be about 2 mm posterior to the labial surface of the mandibular incisors. This will help maintain a natural labiomental fold. The position of the labiomental angle is paramount and profoundly influences the aesthetic outcome. Cephalometric analysis helps the surgeon to plan the operative procedure. The treatment plan is based on incorporation of these data into clinical assessment that will facilitate a postsurgical profile that is esthetically pleasing.

Perceived Chin Abnormalities Due to Anomalies of the Maxilla

When facial analysis identifies disharmony within a patient's profile, the surgeon must determine whether there is an underlying occlusal and skeletal deformity or merely a poorly or over-projected mentum. True maxillomandibular discrepancies should be addressed with orthognathic surgery. In the case where occlusion is stable and a small mandibular deficiency exists (retrogenia), an isolated mandibular sagittal deficiency may be a candidate for an AHMO.

To highlight the importance of the correct diagnosis, one can take the common occurrence of a patient complaining of a "small chin." A recessed chin may be retrogenia or microgenia. An over projected chin may be macrogenia or prognathia. Micrognathia and macrognathia are rare. Prognathia and retrognathia more commonly contribute to chin point abnormalities. In the setting of a patient complaining of a small chin, the lateral profile should be evaluated. Concavity or convexity in conjunction with the proportions of the middle and lower third of the face should be considered in the planning. The maxilla should be evaluated. If the maxilla is set appropriately in the sagittal plane and there is mild retrognathia (≤3 mm) or retrogenia, then a genioplasty is appropriate. However, if a maxillary developmental dysplasia is present, a formal orthognathic work-up should be done.

In contrast, patients complaining of a "prominent chin" often have pseudomacrogenia. These individuals may have maxillary sagittal hypoplasia, which manifests with a retruded upper lip, a midfacial concavity or deficiency, and a chin that may appear prominent in the sagittal plane. Since the true etiology is maxillary hypoplasia, the corrective procedure would be a Le Fort I osteotomy to advance the maxilla anteriorly to coincide with the chin point. A pitfall would be for a novice surgeon to perform a genioplasty to set the chin point back to coincide with the maxilla.

In maxillary vertical deficiency, the patient presents with pseudomacrogenia due to the counterclockwise rotation of the mandible. In this case, the chin is accentuated and appears larger than normal. Patients with this condition have a short lower third facial height and present with poor maxillary tooth show at rest and when smiling. When the mandible is placed in the normal centric relation, the chin point increases in the sagittal plane. Maxillary vertical height correction will allow for a more natural position of the chin and only then can a decision be made on the need for genioplasty.

Maxillary vertical excess may manifest as pseudomicrogenia due to the excessive downward growth of the maxilla causing a clockwise rotation of the mandible. In such cases, the rotation of the mandible results in the appearance of a small chin due to poor projection of the chin in the sagittal plane. The patient will likely have excess gingival show, a long lower third facial height and mentalis muscle strain from the forces needed to close the interlabial gap. The treatment for this type of facial anomaly may be to reposition the maxilla superiorly, particularly in the posterior area.

Maxillary sagittal hyperplasia is extremely rare. Patients may complain of a small chin as well. Once again, this is most likely a case of pseudomicrogenia, where the chin appears relatively small due to the prominence of the maxilla in the sagittal plane. These patients will have a convex facial profile associated with maxillary protrusion and an acute nasolabial angle. This form of microgenia can be corrected with repositioning of the maxilla, after which a decision can be made on the need for an adjunct genioplasty.

Evaluation of the Mandible

After a thorough investigation to rule out any maxillary discrepancies, the next step is to evaluate the mandible. For a patient with mandibular hypoplasia with either gross malocclusion or severe hypoplasia (greater than 4 mm), a formal orthognathic work-up is necessary. The risks of advancing or augmenting a chin greater than 5-6 mm include an unnatural appearance, a deep labiomental angle, and the risk of advancing the chin point past the lower central incisor. Severe mandibular sagittal hypoplasia is corrected with a BSSO, and genioplasty should be viewed as an adjunct procedure. Prognathia in the setting of class 3 malocclusion should be corrected by a setback with a BSSO. Isolated true macrogenia in the presence of a normal class 1 occlusion can easily be treated with a genioplasty, setting the chin point back and even reducing chin height if needed.

Indications for Isolated Genioplasty

After careful scrutiny of the skeletal, dental and soft tissue structures, there exist certain cases that are amenable to isolated genioplasty. An isolated genioplasty can be considered if functional occlusion is present and the lower third profile has mild hypoplasia or hyperplasia in the sagittal or vertical plane. A sagittal hypoplasia (3-4 mm) in the setting of functional normo-occlusion with acceptable facial proportions is an ideal candidate for AHMO with advancement. A variation of the standard sliding genioplasty is the "jumping" genioplasty. The "jumping" genioplasty is ideal for sagittal advancement when vertical reduction is needed.

Patients who are considered for isolated genioplasty should have a good overall profile and occlusion. The surgical goals for these patients include creating an aesthetically pleasing facial contour and establishing proportionate facial height. Ideal candidates for a genioplasty are: (1) retrogenia, i.e., recessed chin point with class I occlusion; (2) mild retrognathia (≤4 mm) with a functional occlusion; and (3) macrogenia.

For example, in a patient that may have a long lower third of the face, a reduction genioplasty is performed to reduce the vertical dimension of the chin. Vertical reduction is done by performing a second horizontal osteotomy that is parallel to the first osteotomy, and a segment of bone is removed. Another indication for an isolated genioplasty may be a mild asymmetry, when the chin does not coincide with facial midline. An oblique triangular wedge of bone can be removed from one side and transplanted to the other side to correct chin asymmetry. In all cases, the chin has to be rigidly fixed by miniplates, wire or screws. A variety of genioplasty is the "jumping" genioplasty. This type of procedure allows the surgeon to both increase the chin projection and shorten the vertical dimension of the chin simultaneously. After the osteotomy is completed, the basilar segment is elevated on top of the upper symphysis.

Surgical Technique

Genioplasty can be performed under local anesthesia with IV sedation or under general anesthesia. General anesthesia is more commonly used with this procedure. Lidocaine with epinephrine is infiltrated along the depth of the buccal vestibule. An incision is made in the buccal vestibule, initially perpendicular to the mucosa then perpendicularly to the muscle and bone. The dissection is continued in the subperiosteal plane to identify the mental foramen on both sides. After identification of the mental foramen, the mental nerves should be protected from both direct and traction injury. The osteotomy is done under the apices of the teeth and the mental nerve. Upon completion of the osteotomy, the chin is then rigidly fixated. A step-off (sharp edge) at the posterior part of the genioplasty should be avoided. The contour of the mandible should be smooth. There are a variety of techniques used for fixation of the chin including wires, resorbable or titanium bone plates. Closure should be done in multiple layers. The mentalis muscle must be reapproximated. The muscle layer can also be reattached to the chin using Mitek anchors. We prefer using two Mitek anchors to secure the mentalis muscle to bone. Alternatively, simple reapproximation with two horizontal mattress sutures is acceptable. This prevents ptosis of the mentalis muscle. Nonfixation may result in a "witches chin." A compressive chin dressing is worn for 5 days postoperatively. The oral mucosa is closed with a running 3-0 chromic suture. The advantage of this procedure is its versatility, reliability and reproducible correction of chin point discrepencies. The disadvantages, when compared to alloplastic augmentation, include increased operative time, bleeding and incidence of mental nerve hypoesthesia.

Chin Augmentation Using an Alloplastic Implant

An alternative to genioplasty in correcting chin hypoplasia is the use of a chin implant. The use of implantable biomaterials and devices plays a potential role in most forms of reconstructive surgery. The common locations for alloplastic augmentation are the chin and malar regions. There are numerous synthetic implantable materials that can be classified as carbon based polymers, noncarbon-based polymers, metals and ceramics. This wide selection allows the surgeon to choose a material tailored to the individual needs of the patient. Silicone (silastic), a noncarbon-based polymer, was one of the first materials used in facial implants. Silicone is resistant to degradation and when fixed against a bony surface its long term stability is very high. Carbon-based polymers include polytetrafluoroethylene (PTFE), polyethylene (PE) and aliphatic polyesters. PE is currently being used in its high density form (HDPE), which allows for contouring by the surgeon. The solid implantable form of HDPE (Medpor®) allows for increased fibrous ingrowth leading to long-term stabilization of the implant.

Alloplastic chin augmentation is a well-accepted technique used in the correction of mild retrognathia or true retrogenia. The advantages of alloplastic augmentation include the material being readily available without donor morbidity, shorter operating time and less blood loss. The disadvantages include bone resorption, infection, extrusion and displacement.

When an implant is used, either an intraoral or extraoral approach can be used. With the extraoral approach, the incision placement is in a natural crease location.

The extraoral approach has a lower rate of infection although it is more difficult to achieve precise placement of the implant. The intraoral technique introduces oral flora into the pocket of dissection, increasing the risk of infection and extrusion.

Malar Augmentation

The malar prominences contribute to aesthetic balance and beauty. Deficiencies or asymmetry in the malar region are usually secondary to trauma, congenital anomalies, such as Treacher Collins syndrome, cancer, or aging. Malar augmentation can be used in either aesthetic or reconstructive practices to achieve symmetry and balance of the face. For example, patients that have undergone repair of a cleft lip and palate may require a LeFort I osteotomy to correct midface deficiency and malocclusion. Correction of the occlusion can be done with the LeFort I osteotomy. The concave profile can be enhanced by malar augmentation (or doing a high osteotomy) to help reestablish overall facial harmony and proportions. In the aging patient, the face may have a tired appearance due to resorption of the maxilla and descent of the malar fat over the zygoma. Malar augmentation may help to create a more youthful appearance.

Preoperative assessment of the deficient or asymmetric malar region is important prior to any augmentation. The facial skeleton can be analyzed using three-dimensional CT generated models from which custom-made implants can be fabricated. Any asymmetries found should be pointed out to the patient and documented. Malar augmentation can be done with autologous bone or with alloplastic implants.

Technique

A number of approaches have been described, including the lower eyelid, coronal, temple and preauricular incisions. However, the most popular approach is by an intraoral incision. This approach leaves no visible scar and allows the procedure to be performed under local anesthesia if desired. The intraoral approach is done through an upper buccal sulcus incision followed by subperiosteal dissection of a pocket in the malar region. It is important to identify the infraorbital nerve and protect it. The implant is contoured and placed after creation of an adequate pocket. It should remain "uncontaminated "during its placement in order to decrease the risk of infection. In the no touch technique, the implant does not come in contact with gloved hands, skin, or oral mucosa in hopes to decrease any bacterial contamination load. Finally, the implant should be secured with either screws, nonabsorbable sutures, or held in place by a tight pocket of periosteum.

Pearls and Pitfalls

Over the long-term, genioplasty is much more durable than alloplastic augmentation of the chin. At some point, most implants will require replacement due to malposition, extrusion, infection or overlying soft tissue changes. Osseous genioplasty, however, can last a lifetime. It is important to discuss this with the patient.

When designing the proposed osteotomy for genioplasty, one has to pay particular attention to the apices of the teeth and the mental nerve. The intraoral approach provides a simple access without a visible scar. The mental foramen lies on the same vertical plane defined by the second bicuspid tooth, infraorbital foramen and pupil. The mental nerve should be dissected out, retracted superiorly and protected during the osteotomy. The three branches of the mental nerve exit the mental foramen and supply general sensation to the chin point. The osteotomy line should be about 3 mm below the mental canal to avoid the route of the inferior alveolar nerve.

Genioplasty by AHMO has also found a functional role in patients with obstructive sleep apnea. Advancement of the genioglossal muscle leads to indirect elevation of the hyoid, thus serving as an adjunct to bimaxillary surgery. A long distance (>15 mm) from the hyoid to the mandibular plane angle can contribute to a decrease in the posterior airway space. By advancing the genial tubercle and muscles, this will indirectly pull the hyoid closer to the mandibular plane and away from the posterior airway.

Suggested Reading

1. Chang EW, Lam SM, Karen M et al. Sliding genioplasty for correction of chin abnormalities. Arch Facial Plast Surg 2001; 3(1):8.
2. Constantinides MS, Galli SK, Miller PJ et al. Malar, submalar and midfacial implants. Facial Plast Surg 2000; 16(1):35.
3. Millard DR. Chin implants. Plast Reconstr Surg 1954; 13(1):70.
4. Spear SL, Kassan M. Genioplasty. Clin Plast Surg 1989; 16(4):695.
5. Spear SL, Mausner ME, Kawamoto Jr HK. Sliding genioplasty as a local anesthetic outpatient procedure: A prospective two-center trial. Plast Reconstr Surg 1987; 80(1):55.
6. Wolfe SA. Chin advancement as an aid in correction of deformities of the mental and submental regions. Plast Reconstr Surg 1981; 67:5.
7. Yaremchuk MJ. Facial skeletal reconstruction using porous polyethylene implants. Plast Reconstr Surg 2003; 111(6):1818.

Chapter 66

Augmentation Mammaplasty

Richard J. Brown and John Y.S. Kim

Introduction

Breast augmentation, also known as augmentation mammaplasty, is performed to balance a difference in breast size (developmental or involutional), to improve body contour, or for reconstruction after breast cancer resection. Clinical trials have shown that breast implants are a safe technique for breast augmentation. Attention to detail during patient evaluation and preoperative planning can significantly impact outcomes in breast augmentation. Women requesting the procedure are most commonly in their twenties or thirties. It is imperative to assess a patients' motivation for seeking augmentation mammaplasty in an effort to avoid dissatisfaction after the procedure. The best candidates for surgery are women who are looking for improvement, not perfection, in physical appearance, and who understand that augmentation only enhances breast size and will not change their social situation.

Anatomy

The female breast is a modified integumentary and glandular structure. The dimensions vary depending on the patients' body habitus and age. It covers the anterior chest wall from the second rib superiorly to the fourth or fifth rib inferiorly. The upper half overlies the pectoralis major muscle, the lower half overlies the serratus anterior, and part of the lateral breast overlies the axillary fascia. Blood supply to the breast is supplied via the internal mammary artery from the medial aspect, the lateral thoracic artery from the lateral aspect and the third through seventh intercostal perforating arteries. Deep venous drainage accompanies the arterial supply and superficial drainage arises from the subdermal plexus. Lymphatic drainage is primarily from the retromammary plexus located within the pectoral fascia. Sensory innervation is derived primarily from the intercostal nerves. Nipple sensation is derived from the third through fifth anterior cutaneous nerves and the fourth and fifth lateral cutaneous nerves. The anterior branch of the fourth lateral intercostal nerve provides the main sensation to the nipple-areolar complex.

Patient Selection

As with any elective surgery, certain severe systemic illnesses may preclude a patient from being considered an acceptable candidate for augmentation. Severe ptosis is a relative contraindication and may concomitantly require mastopexy. The one absolute contraindication to the subglandular placement of implant is in patients with a history of breast irradiation. Radiation impairs blood supply making the submuscular approach much safer. A strong family history of breast cancer is another relative contraindication to the subglandular approach because a small amount of breast tissue may be obscured during a mammogram when the implant is in the subglandular position. Relatively contraindicated is the patient with severe psychosocial issues, grossly unrealistic expectations, or body dysmorphic disorder.

Practical Plastic Surgery, edited by Zol B. Kryger and Mark Sisco. ©2007 Landes Bioscience.

Preoperative Planning

History and Physical

The importance of the initial consultation cannot be overemphasized. Determining breast volume, selecting the type and shape of the implant, deciding on placement (subglandular or submuscular) and choosing the surgical approach should all be discussed. Evaluation of anatomic features is an essential element when planning augmentation. Noting asymmetry is important because the patient may be unaware of the problem. In such cases, the surgeon should explain that efforts to correct asymmetry will be attempted, but that perfectly symmetric results are unrealistic. Attempting to meet the patient's expectations regarding size without compromising a natural appearance is very important. In young healthy patients, routine blood tests are not required. It is, however, important to rule out a personal or family history of clotting disorders. A urine pregnancy test should be performed in most cases.

Choice of Incision

The trend in plastic surgery is to minimize scarring by remote placement of access incisions. The position of the incision is important and should be individualized since no single incision works best for all patients. Most surgeons have their own preference as to the surgical approach used in the procedure. In addition, most patients are knowledgeable about the various incisions and have their own preferences as well. The choice of incision should be discussed with the patient in a nonbiased manner highlighting pros and cons of each approach (Fig. 66.1). Currently, there are four commonly used types of incisions: **periareolar**, **inframammary crease**, **axillary** and **umbilical**. Some patients are candidates for an endoscopic transaxillary or inframammary approach; however results depend in part on the surgeon's familiarity and experience with endoscopic techniques.

Type of Implant

Silicone gel-filled implants were the most common implants used prior to their removal from the market by the Food and Drug Administration (FDA) and are available only through FDA approved clinical trials. Currently, saline-filled implants are the type of implant most widely used. They can be classified as smooth or textured, round or contoured and as having a high or low profile.

Cancer Detection

It is important to educate patients about the potential long-term effects on breast cancer surveillance. A woman's risk of developing breast cancer is not affected by breast implants, and to date there is no evidence to support a relationship between silicone or saline implants and breast cancer in humans. Slight modifications in mammographic technique (Eklund displacement technique) may be required for adequate visualization of all breast tissue. Capsular contracture has the greatest impact on mammography. Severe contracture can reduce the accuracy of the mammographic findings. Current recommendations suggest women aged 30 or older should have baseline and postaugmentation mammograms and should follow the same schedule of routine mammography as all other women. They should also perform monthly self-exams and become familiar with the new feel of their breast tissue. Self-exams are much more likely to detect a breast cancer in augmented patients compared to nonaugmented women.

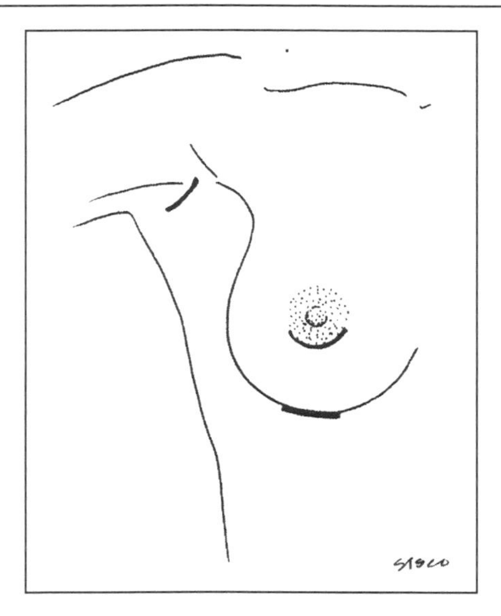

Figure 66.1. Commonly used incisions for breast augmentation.

Sensation

Breast and nipple sensation are usually compromised only temporarily (weeks to months) due to nerve stretching from aggressive lateral dissection. Accidental nerve division is uncommon. With the subglandular approach, the reported incidence of nerve injury is 10%, and even lower with the submuscular approach. The degree of sensation loss is directly proportional to the size and diameter of the implant due to the large pocket that must be dissected. Impaired breast sensation is more likely after secondary procedures, especially extensive capsulotomy which may subject sensory nerves to injury.

Pregnancy and Lactation

Many women seek breast augmentation prior to their childbearing years. Breast augmentation is extremely unlikely to impair lactation or breast-feeding, especially when implants are placed in the submuscular position. Many surgeons will advise patients to undergo breast augmentation a minimum of 6 months prior to pregnancy or at least 6 months after the termination of lactation. Breast appearance after childbirth varies in patients who have had augmentation prior to pregnancy. Some will return to their prepregnancy appearance while others do not.

Intraoperative Considerations

Inframammary Incision

This is the most common approach for placement of a breast implant. The incision is made just above the imframammary crease, and should be no more than 3-4 cm in length. The surgeon must estimate where the incision will sit in relation to the crease created by the larger enhanced breast. An incision placed too low may be obviously visible. The inframmary incision is a favorite among surgeons because it gives complete visualization of the subpectoral plane and offers a great deal of control over placement of the implant. If the patient needs revision surgery in the future, this incision can be reused without the need to create a new scar. A potential problem with this incision occurs when patients decide to undergo an implant exchange to a different size. This can cause migration of the inframammary fold, thus exposing the original scar.

Periareolar Incision

This approach involves an incision within the pigmented areolar tissue and often results in the least conspicuous scar. A medially placed incision avoids the fourth intercostals nerve, which supplies sensation to the nipple-areola complex. Dissection may be carried through the breast parenchyma or towards the imframammary fold subcutaneously around the lower pole of the breast. Since most breast ducts contain bacteria there is an increased risk of infection when dissecting through breast tissue. If a patient is undergoing simultaneous mastopexy, this is a favorable approach since the two procedures may share the same incision. This approach works well with every type of implant and for placement both above and below the pectoralis muscle.

Transaxillary Incision

The benefit of this approach is the inconspicuous scar. The incision is in the axilla, where it is only seen when raising the arm. This scar heals well with only slight discolorization, although there is a greater risk of forming a hypertrophic scar compared to incisions at the breast. The implant may be placed on, above or below the muscle with this technique. However, there are many disadvantages to the transaxillary incision. The main disadvantage with this approach is the poor exposure obtained during placement of the implant; however this can be improved using the endoscopic technique. It is difficult to create symmetric pockets so patience and skill are required. There is also an increased incidence of paresthesia to the nipple-areolar complex with this approach. Another downside to this incision is in the event of a complication or the need for future corrective surgery, removal of the implant would require conversion to one of the previously discussed incisions. Other potential complications that have been reported are damage to the intercostobrachial nerve and subclavian venous thrombosis.

Periumbilical Incison

Placement of the implant using this approach is restricted to a prepectoral plane, and this approach provides the worst control for dissection of the pockets. Dissection of the superior pole and symmetry of placement are difficult even in the most experienced hands. Complications of hematoma or infection require conversion to one of the previously discussed incisions. The periumbilical approach is the least utilized technique.

Location of the Implant Pocket

Implants are commonly placed in the **submuscular position**, the **subglandular position**, or in a position that combines the two, termed the **biplanar** approach. In the submuscular technique, the implant is placed in a plane below the pectoralis major muscle, whereas in the subglandular approach the implant is deep to the glandular breast tissue but superficial to the muscle. Advantages to submuscular placement include a decreased rate of capsular contracture, and reduced sensory changes in the nipple. In addition, with this technique there is a decreased incidence of hematoma formation since the plane below the muscle is relatively avascular.

Disadvantages to this approach include limitations on the size of the pocket that can be dissected, increased postoperative pain, potential lateral displacement of the implant and the inability to obtain significant medial fullness, or cleavage. Releasing the inferior portion of the pectoralis muscle from its medial sternal attachments can help improve cleavage appearance. Advantages to the subglandular approach include ease of dissection, the ability to use larger implants and a more predictable size and contour. Furthermore, breast ptosis may be better addressed with subglandular placement. Disadvantages include an increased risk of capsular contracture, higher risk of nerve injury, abnormal contour appearances such as rippling and a greater risk of rupture during a future breast biopsy.

Postoperative Considerations

For the first few hours while in the recovery room, the patient's incisions and breasts should be examined for evidence of hematoma. Oral narcotics are given for postoperative pain, and the patient should refrain from vigorous activity for the first 48 hours. A soft elastic bra may be worn after the operation for comfort, support and molding of the breast, but is not required. At the first postoperative visit, the patient is encouraged to begin massaging her breasts daily. Movement of the implant against the walls of its cavity helps maintain an expanded capsule and result in a softer breast. If used, patients are instructed to remove Steri-strips about two weeks postoperatively. At that time, unrestricted activity may commence.

Complications

Hematoma

One to two weeks prior to surgery patients should discontinue medications that may impair platelet activity. The frequency of hematoma formation is less than 2%. The best prevention is operative technique utilizing blunt dissection and meticulous hemostasis. A hematoma may form slowly with no symptoms or rapidly with symptoms such as unilateral pain, swelling and fever. Immediate postoperative hematomas should be taken back to the operating room for evacuation. Late-onset hematomas that are symptomatic should be drained due to the risk of infection and capsular contracture.

Sensory Changes

The incidence of diminished sensation in the nipple is 15%. Most patients experience temporary dysesthesia of the nipple that often resolves within several months. A small percentage of patients may experience long-term sensory loss in one or both nipples, a complication that is most common with the transaxillary approach. Avoidance of extensive lateral pocket dissection and use of the the submuscular position will decrease nerve injury.

Infection or Seroma

The incidence of infection is about 2% and usually manifests 7-10 days postoperatively, but may occur at any time. Either the wound or the periprosthetic space may be involved. Symptoms include fever, swelling, discomfort, pain, drainage and cellulitis of the breast. The two most common organisms are *Staphylococcus epidermidis* or *Staphylococcus aureus*. Uncomplicated wound infections should be treated with a course of antibiotics for 1-2 weeks. Wounds draining pus should be opened, and the implant will often need to be removed. If seroma fluid is clinically present, ultrasound-guided aspiration with culture and sensitivity should be performed. Posi-

Table 66.1. The Baker classification of capsular firmness in augmented breasts

Grade I	No palpable capsule	The augmented breast feels as soft as an unoperated one
Grade II	Minimal firmness	The breast is less soft and the implant can be palpated, but is not visible
Grade III	Moderate firmness	The breast is harder, the implant can be palpated easily, and it can be seen
Grade IV	Severe contracture	The breast is hard, tender, painful and cold. Distortion is often marked

tive culture results should prompt appropriate antibiotic therapy and consideration of implant removal. If cultures are negative, seromas may be followed clinically. If they do not resolve, then patients should be presented with management alternatives, including repeat aspiration or implant removal. It is important to remember that infection and seroma are both risk factors for subsequent capsular contracture.

Capsular Contracture

Capsular contracture is the most common cause of dissatisfaction after breast augmentation. It occurs due to the formation of a fibrous scar that may become thick and constrict a soft implant (Table 66.1). Etiology is unknown, but the two main theories are hypertrophic scar formation or infection-induced contracture. Saline-filled implants have a lower incidence of contracture compared to silicone-filled implants. Textured surface saline implants have a lower rate of contracture when placed in the subglandular position provided the patient has adequate tissue for coverage, while smooth surface saline implants have a lower rate of contracture in the subpectoral position. Steps to help minimize the risk of contracture include meticulous hemostasis and sterile technique. The implant should be soaked in antibiotic irrigation solution and handled as little as possible. In addition, creating an adequately sized pocket is important. The implant should fit into the pocket without being tight or firm, allowing mobility during postoperative massage exercises.

Treatment options include closed or open **capsulotomy** and **capsulectomy**. The goal of closed capsulotomy is to rupture the scar capsule by manually squeezing the breast until there is an audible pop without fracturing the implant. Complications such as hematoma, rupture of the implant and migration make this an unfavorable procedure. Open capsulotomy is best when capsular contracture is less severe, and involves stripping or scoring of the capsule. Capsulectomy, or complete capsule removal, should be reserved for a firm, thick, calcified capsule. Implants may be replaced and repositioned if necessary with open capsulotomy or capsulectomy. The patient may also choose to remove the implants without replacement.

Implant Displacement

Asymmetry and displacement are the second most common causes of dissatisfaction after breast augmentation. Implants can be placed too high, too low or excessively lateral. Displacement may also rarely occur after a capsulotomy procedure. Displacement can occasionally be treated conservatively with prolonged taping and closed manipulation. More commonly, however, persistent dissatisfaction with displacement usually culminates in surgical revision.

Rupture or Deflation

Implants begin to lose their integrity approximately 10 years after insertion. Magnetic resonance imaging is the most sensitive method for visualizing breast implants and determining their integrity. When saline implants rupture there is usually a noticeable change in size, shape, feel and appearance of the breast. Saline is absorbed, and the implant should be replaced to avoid the possibility of capsular contracture. Silicone implants may leak or rupture, and usually the gel remains within the breast capsule. Many are silent ruptures discovered at the time of routine mammograms or during implant replacement. If a significant portion of the gel moves outside of the implant capsule, it can then migrate into the breast or surrounding tissue. The body reacts by depositing collagen around the silicone, leaving a firm mass. Ruptured silicone implants need to be removed along with the extruded gel, which may involve simple implant replacement, or in the case of severe leaks, subcutaneous mastectomy with reconstruction.

Pearls and Pitfalls

Upward displacement of the implant can be caused by incomplete release of the pectoralis muscle. If the proper position of the implant is not judged correctly, upward displacement of the implant in a submuscular plane can also result in ptotic, "snoopy" deformity with the projection of the implant mismatched with the bulk of the breast tissue around the nipple-areolar complex. Advocates of the biplanar approach (in which dissection and release occurs in both a subglandular and submuscular plane) believe that a more natural curve displacement is possible with their approach.

It is essential to be aware of the dissection planes to avoid asymmetric inframammary folds. Similarly, care must be taken not to go too far medially with the dissection as symmastia can result.

In particularly thin patients, a pneumothorax may result from overly aggressive dissection through the intercostals (small leaks can be corrected with temporary placement of a small suction catheter).

Relative underfilling—especially in thin patients or with a subglandular approach—may result in rippling and is also believed to impair the overall structural integrity of the capsule.

While the subglandular approach will allow a modest lift of the nipple areolar complex, concomitant mastopexy may be necessary; however, if the skin integrity is weak it may be prudent to stage the procedures to avoid lowering of the nipple-areolar complex.

Consistent postsurgical massage is believed to ameliorate the onset of capsular contracture and yield softer breasts.

Finally, it is essential that a final intraoperative view is taken of the patient in an upright position after **all** surgical adjustments have been made.

Suggested Reading

1. Bostwick IIIrd J. Augmentation mammaplasty. Plastic and Reconstructive Breast Surgery. Vol. 1. 2nd ed. St. Louis: Quality medical publishing Inc., 2000:239-369.
2. Salomon JA, Barton JR FE. Augmentation mammaplasty. Selected Readings in Plastic Surgery. 2004; 28(8):1-34.
3. Spear SL, Elmaraghy M, Hess C. Textured-surface saline-filled breast implants for augmentation mammaplasty. Plast Reconstr Surg 2000; 105:1542-1552.
4. Tebbetts JB. A surgical perspective from two decades of breast augmentation. Clin Plast Surg 2001; 28(3):425-434.

Chapter 67

Gynecomastia Reduction

Richard J. Brown and John Y.S. Kim

Introduction

Gynecomastia is defined as a benign enlargement of the male breast due to proliferation of the glandular tissue. The term is derived from the Greek words **gyne** and **mastos** meaning female and breasts, respectively. Gynecomastia is the result of an imbalance between estrogens and testosterone in the male body whereby the stimulatory effect of estrogen on breast tissue exceeds the inhibitory effects of testosterone. The cause of this imbalance has many etiologies that will be discussed in this chapter. Often gynecomastia occurs at birth, but most cases are discovered during puberty, with the peak incidence between 14-15 years of age. When occurring during puberty the condition is usually self-limited and will regress within 2 years.

Gynecomastia can involve one breast; however 75% of the time it is bilateral. It may be secondary to hormonal imbalances, medications, illicit drug use, genetic conditions and exogenous hormone use. Frequently gynecomastia is misdiagnosed as pseudogynecomastia, which is an increase in male breast size that develops from fat deposition, not glandular proliferation. Medical management of gynecomastia is important in ruling out serious underlying pathology. Treatment of persistent gynecomastia itself is predicated on surgical removal.

Etiology

Gynecomastia can be classified as either physiological or pathological.

Physiologic Gynecomastia

It is usually seen in newborns, adolescents at puberty or aging men. In neonates, circulating maternal estrogens at birth stimulate neonatal breast tissue to hypertrophy. This condition usually resolves spontaneously within a few weeks. The average age of onset of adolescent gynecomastia is 14 years and it commonly disappears by 20 years of age. It often produces asymmetrical enlargement with accompanying breast tenderness. Declining testosterone levels in aging men can lead to mild gynecomastia.

Pathologic Gynecomastia

Pathological causes include estrogen excess, androgen deficiency, or drugs that interfere with the normal estrogen-testosterone balance.

Causes of deficient production or action of testosterone include: congenital anorchia, Klinefelter's syndrome, androgen resistance, defects in testosterone synthesis and secondary testicular failure (neurological conditions, renal failure, orchitis and trauma).

Causes of increased estrogen production include: estrogen secretion from neoplasms (testicular, lung, pituitary) or increased substrate for the actions of peripheral aromatase (cirrhosis, thyroid excess, adrenal disease and starvation).

Practical Plastic Surgery, edited by Zol B. Kryger and Mark Sisco. ©2007 Landes Bioscience.

Drugs that interfere with estrogen-testosterone balance include: estrogens, estrogen-like compounds (marijuana, heroin), gonadotropins, inhibitors of testosterone (spironolactone, cimetidine and alkylating agents) and several drugs with an unknown mechanism of action (isoniazid, methyldopa, D-penicillamine, captopril, diazepam and tricyclic antidepressants).

Evaluation and Diagnosis

History

The history should include the age of onset, laterality of disease, tenderness and symmetry of the deformity. A thorough assessment of any hepatic, testicular, pulmonary, adrenocortical, or thyroid dysfunction is important in ruling out an endocrine etiology. An abnormal exam should guide the surgeon to order focused tests and consult specialists. Gynecomastia is common in older men. Enlargement of breast tissue is usually central and symmetric arising from the subareolar position. Unilateral eccentric gynecomastia may be secondary to neurofibromas, hematoma, lipomas, lymphangiomas, or dermoid cysts. A careful review of systems focusing on medications, alcohol and drug use is important in revealing any conditions associated with gynecomastia. The most difficult condition to differentiate from gynecomastia is pseudogynecomastia. Patients with this condition are often obese, have bilateral enlargement and do not complain of breast pain or tenderness.

Physical Exam

Patients should be examined in the supine position. The examiner grasps the breast between the thumb and forefinger (pinch test) and gently moves the two digits toward the nipple. If gynecomastia is present, a firm, rubbery, mobile, disk-like mound of tissue arising from beneath the nipple-areolar region will be felt. When pseudogynecomastia is present it may be difficult to palpate this firm disk of tissue. Breast exam should also include evaluation of the axillary contents to rule out lymphatic involvement. Simon et al graded gynecomastia into four groups (Table 67.1).

Young patients with no previous medical history and new onset bilateral gynecomastia should have a testicular exam looking for atrophy, enlargement or abnormal masses. If indicated, an ultrasound of the testicles should be performed. If physical exam demonstrates characteristics of feminization, it is prudent to check the appropriate hormone levels (e.g., estradiol, leutenizing hormone, testosterone and DHEA). A marfanoid body habitus should prompt a karyotype to rule out Klinefelter's syndrome.

Findings such as axillary lymphadenopathy or a unilateral hard mass fixed to underlying tissues should prompt further evaluation. Skin dimpling, nipple retraction, nipple discharge and axillary lymphadenopathy are all associated with breast carcinoma. Breast cancer must be ruled out even though it accounts for less than 1% of cancers in men. If cancer is suspected, imaging of the breasts (mammography or MRI), and a coreneedle biopsy or fine-needle aspiration (FNA) should be performed

Table 67.1. Grading of gynecomastia

Grade I	Small enlargement, no skin excess
Grade IIA	Moderate enlargement, no skin excess
Grade IIB	Moderate enlargement with extra skin
Grade III	Marked enlargement with extra skin

to rule out carcinoma. Gynecomastia in conjunction with Klinefelter's syndrome carries a sixteen-fold increased risk of male breast cancer.

Finally, patients with gynecomastia and an otherwise normal history and physical exam may be observed if the condition has been present for less than 12 months. However, when the disease has been present for over a year, surgery should be considered since breast tissue may become irreversibly fibrotic as time progresses beyond this stage.

Treatment

Before considering treatment, it is important to keep in mind that gynecomastia may regress spontaneously. Although surgery is indicated as a diagnostic procedure, patients often request surgery as treatment for the physical discomfort or emotional distress that is common in men with this condition. Most patients who visit a plastic surgeon request treatment for psychological reasons.

Discontinuing offending medications or correcting any underlying imbalance between estrogens and androgens should result in spontaneous regression of new-onset gynecomastia. Medical treatment with androgens, anti-estrogens and aromatase inhibitors has been used with minimal efficacy. Surgery remains the accepted standard for management of gynecomastia, especially in patients with long-standing gynecomastia and fibrotic breast tissue. Surgical options can range from simple excision to a more complex, inferior pedicle breast reduction. The two most widely used surgical techniques are the subcutaneous mastectomy and liposuction-assisted mastectomy.

Subcutaneous Mastectomy

Several approaches may be used when performing an open subcutaneous mastectomy. The choice of incision should be guided by the degree of gynecomastia present. Patients with small or moderate gynecomastia may have an intra-areolar incision along the inferior hemisphere of the nipple (Webster incision). This incision can be extended medially or laterally for better exposure. An alternative incision that helps maximize exposure is the triple-V incision. It is made along the superior border, parallel to the nipple. Moderate or massive gynecomastia may require skin resection along with nipple relocation or nipple grafting. The most common incision for moderate gynecomastia is the Letterman incision. This approach allows for the nipple-areola complex to be rotated superiorly and medially after skin resection. With massive gynecomastia, en bloc resection of skin and breast tissue with free nipple grafting can be performed through an elliptical incision. In cases of severe gynecomastia, the dissection may be carried to the level of the pectoralis major fascia and may require the use of postoperative suction drains.

Liposuction-Assisted Mastectomy

Experienced surgeons may perform endoscopic-assisted mastectomy with liposuction through an axillary incision in lieu of open mastectomy. In the past, liposuction-assisted mastectomy was utilized after open excision to assist with breast contouring. In an effort to avoid large visible scars on the chest wall, liposuction-assisted mastectomy (suction lipectomy) is becoming the preferred surgical technique for most cases of gynecomastia. It is the most commonly used technique for correcting pseudogynecomastia. With this technique there is less compromise of the blood supply as well as a decreased risk of nipple distortion. Postoperative complications such as hemorrhage, infection, hematoma, seroma and nipple necrosis have been minimized with suction lipectomy.

This technique allows removal of glandular and fibrotic tissue from the breast. However, pure glandular gynecomastia may still require an open technique. More recently, ultrasound-assisted liposuction (UAL) has been introduced in conjunction with standard liposuction as a safe and effective method of treatment for gynecomastia, especially in cases where dense fibrous tissue is involved. The ultrasound probe is introduced through an axillary or inframammary incision and advanced through the dense parenchymal tissue. Energy from the ultrasound waves cavitates and emulsifies breast parenchyma that may be removed via suction lipectomy. UAL stimulates the dermis, allowing for postoperative skin retraction to occur. In the future, it may become standard treatment to utilize the UAL technique first, followed by an excisional procedure 6-9 months later (once maximal skin contraction has occurred) if excess skin or breast tissue persists.

Complications

Hematoma and seroma are the most common complications and can be avoided by judicious hemostasis and the coordinate use of pressure dressing and suction drains postoperatively. Appropriate care must be taken with liposuction to ensure viability of the overlying skin. Skin or nipple/areola necrosis can occur if the vascularity is compromised. Pigment changes in the areola have also been reported, especially in free-nipple grafts. Asymmetry and discontent with scars are frequent patient complaints.

Pearls and Pitfalls

The treatment of gynecomastia is predicated on the exclusion of potentially dangerous (or easily reversible) causes. As is the case with surgery of the female breast, the surgical approach should consider the magnitude of skin and volume excess as well as the quality of that excess. For instance, an elderly man's skin will not retract with liposuction as well as a young man's. Hence, an informed discussion of issues related to skin redundancy is important. In cases in which there is a fair degree of dense, fibrotic tissue, ultrasound-assisted liposuction may be the treatment of choice. For milder forms of gynecomastia, liposuction with mastectomy through a periareolar incision may be helpful. While gynecomastia may regress spontaneously, it is important to note that the longer the gynecomastia is present, the more fibrotic the breast tissue can become. Once fibrosis sets in, surgical removal of the tissue remains the optimal treatment option.

Suggested Reading

1. Bostwick IIIrd J. Gynecomastia. Plastic and Reconstructive Breast Surgery. Vol.1, 2nd ed. St. Louis: Quality medical publishing Inc., 2000:239-369.
2. Eaves IIIrd FF et al. Endoscopic techniques in aesthetic breast surgery. Clin Plast Surg 1995; 22:683.
3. Rohrich RJ, Ha RY, Kenkel JM et al. Classificaiton and management of gynecomastia: Defining the role of ultrasound-assisted liposuction. Plast Reconstr Surg 2003; 111(2):909.
4. Simon BB, Hoffman S, Kahn S. Classification and surgical correction of gynecomastia. Plast Reconstr Surg 1973; 51:48.
5. Spear SL, Little IIIrd JW. Gynecomastia. Grabb and Smith's Plastic Surgery. 5th ed. Lippincott-Raven Publishers, 1997.
6. Wilson JD. Gynecomastia. Harrison's Principles of Internal Medicine. 11th ed. New York: McGraw-Hill Book Co., 1987.

Chapter 68

Mastopexy

Richard J. Brown and John Y.S. Kim

Introduction

Mastopexy, or breast lift, is a surgical procedure that can help restore a more youthful and natural shape to sagging (ptotic) breasts. Gravity, pregnancy, nursing, weight gain and aging can all lead to ptosis and a loss of firmness. Breast implants in conjunction with mastopexy can increase breast firmness and their size. The goals of surgery are to create improved projection and a more youthful, uplifted appearance while minimizing visible scarring. In addition to reshaping the breast, mastopexy can also reduce the size of the nipple areola complex (NAC). Mastopexy can be performed in any size breast; however very large breasts may be more suited to a formal breast reduction procedure. Pregnancy and nursing will usually stretch breasts that have been previously lifted; therefore the best outcomes are seen in patients who are past their childbearing years. Ideal candidates for mastopexy are healthy, emotionally stable women who are realistic about what the surgery can accomplish. It is important to emphasize that the tradeoff for lifted, youthful breasts are the scars that remain after surgery. Patients with relatively small breasts and minimal ptosis may be candidates for modified procedures requiring less extensive incisions.

Anatomy

The relevant anatomy is discussed in the breast augmentation chapter.

Indications

In most instances mastopexy is performed primarily to improve an unaesthetic appearance of the breasts. However, certain cases, such as postmastectomy reconstruction or ptosis after implant removal, may require a mastopexy to restore symmetry.

Contraindications

There are no absolute contraindications to breast mastopexy. Planned future pregnancy is a relative contraindication because lactation and subsequent involution can change the shape of breast tissue. Capsular contracture after breast augmentation is another relative contraindication to mastopexy. In many of these patients, the breasts appear ptotic when in fact they truly are not. Therefore, removal and inspection of the implants while in the operating room is paramount prior to committing to mastopexy. Finally, women with a high risk of breast cancer should be evaluated carefully since surgery may alter the architecture of breast tissue making detection and treatment of cancer difficult.

Preoperative Considerations

Judicious care should be taken during patient assessment and selection to clarify expectations and ensure that desired results are obtainable. A complete physical examination should be performed which includes inspection as well as palpation of the breast parenchyma to rule out suspicious masses. All patients 40 years or older should have a baseline mammogram prior to surgery, a follow-up mammogram 6 months after surgery, and then follow the American Cancer Society recommendations for annual screening mammograms. Determining the degree of breast ptosis is central to planning mastopexy as it will guide which technique is best suited to achieve the optimal aesthetic appearance (Table 68.1).

Determining the correct level of the nipple areolar complex (NAC) is critical when planning a mastopexy. The nipple should be placed at or slightly above the inframammary fold taking care to avoid placing the nipple too high on the breast mound, which can be difficult to fix. Breast volume is important to consider when planning a mastopexy, and any parenchyma that falls below the inframammary crease should be reduced or elevated.

Next, the position, length and definition of the inframammary crease should be evaluated. When augmentation is used in conjunction with mastopexy, the implant pocket is used to define and retain the new inframammary crease. Breast mobility is directed by the firmness of glandular attachment to the underlying deep fascia and should be assessed prior to surgery. Skin and tissue quality should be assessed since women with ptosis have an excess of breast skin compared to the amount of underlying parenchymal tissue. The appearance of striations indicates a weakness in underlying dermis, and this skin usually has poor elasticity that will not support or shape the breast. Recurrence of breast ptosis in these patients is predictable; therefore planning skin removal and incision placement is an important preoperative consideration. These are planned on a continuum from periareolar to circumareolar to vertical scars and finally to horizontal scars. Should ptosis recur, additional breast tissue can be excised through old incisions.

Women with small breasts and upper pole flatness may benefit from simultaneous augmentation. The addition of an implant can enhance the size and contour while increasing the longevity of the uplifting effects of mastopexy. Simultaneous breast augmentation and mastopexy should be considered carefully since the two have somewhat conflicting goals. The goal of breast augmentation is to enlarge the breast, which involves stretching the skin and NAC, while mastopexy is designed to reduce the skin that envelopes the parenchymal tissue. Patients should be aware of

Table 68.1. Grades of Ptosis according to the Regnault classification

Minor Ptosis (Grade I)	Nipple at the level of inframammary fold, above lower above lower contour of gland
Moderate Ptosis (Grade II)	Nipple below level of inframammary fold, above lower contour of gland
Major Ptosis (Grade III)	Nipple below level of inframammary fold, at lower contour of gland
Pseudoptosis	Inferior pole ptosis with nipple at or above the inframammary fold
Glandular Ptosis	Nipple is above the fold but the breast hangs below the fold

the increased risk of poor scarring, implant-nipple misalignment and implant extrusion. The best scenario occurs when the implant fills out the excess skin envelope while leaving enough excess skin to reshape the breast. Depending on the complexity of the problem and quality of the skin support, it may be better to perform two separate, staged procedures.

Preoperative markings vary with surgical plan and are essential for obtaining optimum results. In most patients the nipple should be at or slightly above the inframammary fold. Once the proper nipple location has been determined, an indelible marker may be used to mark the remainder of the skin incision.

Intraoperative Considerations

Since scars are the greatest drawback to aesthetic breast surgery, it is best to choose techniques that minimize the length of incisions and place them in hidden areas. Intraareolar and periareolar incisions are tolerated best because they are less likely to become hypertrophic provided there is no tension on the incision. A median inferior vertical incision is also tolerated well compared to a horizontal inframammary incision. As a rule, incisions should be kept off of the superior hemisphere of the breast because women often wear clothing that exposes this area. There are several techniques available to correct breast ptosis, and no single technique is considered ideal. The degree of ptosis varies from patient to patient and treatment should be individualized. The primary focus is on altering breast volume and contour by removing excess skin and repositioning the NAC.

The common surgical options for ptosis correction are:

1. Augmentation with or without mastopexy
2. Periareolar scar technique
3. Circumareolar scar with periareolar purse-string closure (Benelli mastopexy)
4. Wise-Pattern mastopexy
5. Vertical mastopexy. Vertical mastopexy can be combined with the horizontal inverted T technique or the short horizontal scar technique.

Augmentation for Ptosis

Patients that are well suited for augmentation alone are those that have pseudoptosis or grade I ptosis. In these patients, minimal elevation of the NAC is required. Their breasts usually have flattened upper poles and are hypoplastic and involuted. It is important to be aware that if the nipple is below the inframammary fold, an implant may actually enhance the deformity giving a more ptotic appearance. Patients who seek a more elevated NAC may require circumareolar incisions with augmentation. When augmentation is used to correct breast ptosis, the implant is placed in either the submuscular position in the upper portion of the breast or in the subglandular position in the lower portion of the breast. When placing the implant in the submuscular position, it is important to maintain a loose submusculofascial pocket to avoid the appearance of a double silhouette (double bubble). This occurs when breast parenchyma descends over the implant while the implant remains fixed at the upper pole.

Periareolar Technique

This approach is best utilized in patients with a minor degree of ptosis who require minimal elevation of the NAC. The periareolar technique involves a crescenteric excision and lift of the NAC. It affords the shortest possible scar and is well hidden within the NAC. Patients that require a greater degree of elevation of the NAC should have a different technique performed since the risk of areola deformity is proportional to the amount of skin removed. Skin quality is important to consider for healing purposes as well as for assessing the risks of recurrent ptosis.

Circumareolar Scar (Benelli) Technique

Circumareolar mastopexy alone tightens the breast envelope without raising the NAC and may cause central breast flattening. Patients who have large areola or tubular breasts may benefit from this technique. Two incisions are required: an inner incision around the areola and a second parallel outer incision demarcating the area for skin excision. The final diameter of the new NAC should be 40-45 mm. A pursestring, nonabsorbable suture is placed around the outer dermal circumference in order to reduce tension on the suture line and limit the risk of scar widening. This procedure is called a "Benelli or "round block" mastopexy. The round block technique allows control of the diameter of the areola and maintains it in a fixed circular scar thus avoiding protrusion. Limiting the size of the outer diameter to three times that of the inner diameter helps minimize tension as well. In addition to a pursestring suture, a Benelli mastopexy may also include pexying the retroglandular surface of the breast parenchyma to underlying rib periostium in a crisscross fashion.

Vertical Scar Technique

Vertical mastopexy is needed to correct more severe breast ptosis, such as grade II or III, where the nipple is below the level of the inframammary crease. If the distance the nipple needs to be elevated is significant, and there is excess skin requiring excision, then a vertical limb is required. Removing skin in a vertical direction allows the medial and lateral breast skin to be moved toward the center and prevent flattening of the breast apex. The scar is usually minimally visible with time and is located inconspicuously on the lower portion of the breast out of view when low-cut clothing is worn.

The procedure begins by determining the new position of the NAC. With the patient standing or sitting upright, the apex and the width of the new NAC position are marked. An ellipse is drawn starting from the top of the new position of the NAC around the existing NAC and downward to the inframammary crease. Incisions are made along the lines of the ellipse as well as around the NAC, and skin is deepithelialized within the ellipse. If implants are being used, they are placed into a subpectoral pocket. Patients with upper pole flattening may benefit from a lower pole deepithelialized parenchymal flap turned beneath the NAC into a new position in the upper pole of the breast. Prior to making any further incisions through breast parenchyma, a technique called tailor tacking may be employed to help the surgeon predict the final outcome. This technique is useful for determining the position of the NAC. The deepithelialized areas are invaginated and the skin edges approximated with staples. The outer edges are marked, the staples are removed and the excess breast tissue is then excised.

Wise-Pattern Mastopexy

Patients with significant ptosis, very full lower breast poles, or those who require a long transposition of the NAC may opt for the Wise-pattern technique. A keyhole incision is made around the NAC with a vertical limb and a horizontal extension resulting in an inverted T type scar after removal of excess tissue. Different pedicles can be used with this approach depending on surgeon preference (if concomitant implants are to be placed, a superior or medial pedicle may be suitable; if significant lift of the NAC is required, an inferior pedicle may be preferable). Once the skin and pedicles have been incised and dissected, judicious undermining and removal of excess tissue can be performed. The vertical and horizontal limbs are approximated and the NAC is sutured in place with a tension-free subcuticular closure.

Postoperative Considerations

Postoperatively an elastic bandage or surgical bra is worn over gauze dressings. Several days later, a soft support bra can be worn continuously for 3-4 weeks. Lifting objects above the head should be avoided during the immediate postoperative period. Patients should be instructed about potential complications such as numbness and hematoma formation. Breastfeeding should be normal after some types of mastopexy; however other techniques increase risk of loss of lactation. Many of the same complications seen after breast augmentation apply to mastopexy as well. These include hematoma or seroma formation, infection, nipple sensory loss and implant contracture. There are several other complications worth mention. These are necrosis of the nipple-areolar complex, recurrent ptosis, nipple and breast asymmetry, upper pole flattening and unacceptable scarring.

Nipple-Areolar Necrosis

Adequate nipple-areolar microcirculation is imperative to the survival of the NAC. Patients who are heavy smokers or have predisposing vascular diseases such as diabetes or collagen vascular disease are at risk for nipple-areolar necrosis. All patients who smoke should quit smoking prior to and after surgery in order to help decrease nipple necrosis. Placing a subglandular implant or a periareolar pursestring suture may also increase the risk of nipple necrosis because the central parenchyma may be damaged to a point where the blood supply to the NAC is compromised.

Recurrent Ptosis

Many surgeons leave large amounts of lower pole breast tissue beneath their skin closure. Subsequently, many women return with lower pole ptosis, termed "bottoming out." Ptosis may recur when there is asynchrony between breast parenchyma and NAC descent. Correction during a secondary procedure requires removal of the lower pole parenchyma with simultaneous skin revision.

Nipple and Breast Asymmetry

The goal of breast surgery is to obtain perfect symmetry; however, most women have some degree of asymmetry preoperatively. Postoperative asymmetry of the NAC or patient dissatisfaction may be corrected during a follow-up procedure. Discussing this issue prior to surgery may help alleviate anxiety when there is postoperative asymmetry. Periareolar techniques may afford a modest correction of asymmetry; however in cases of significant asymmetry, a complete revision mastopexy may be necessary.

Upper Pole Flattening

Mastopexy alone may not be sufficient to correct breast ptosis. To avoid upper pole flattening the surgeon may place implants, rotate lower breast parenchymal flaps beneath the upper pole, or perform a reverse periareolar pursestring mastopexy.

Poor Scarring

Patients with the highest risk of poor scar outcome are cigarette smokers and any patients with comorbidities that predispose them to microvascular disease. Closing wounds in a tension free manner offers the best chance for optimal scar outcome. If the surgeon anticipates a tight skin closure below the NAC, then extrapigmented skin may be left inferiorly and excised at a later time. Horizontal incisions commonly heal with hypertrophy due to the amount of tension on these wounds. Efforts to minimize incision length combined with proper wound taping may help limit hypertrophic scarring. If hypertrophic scarring does occur, silicone gel therapy or steroid injections should be attempted prior to scar revision.

Pearls and Pitfalls

Choosing a periareolar technique for significant ptosis will result in poor scarring, deformity of the nipple and recurrence. The vertical mastopexy approach may allow a sufficient lift; however, the use of a variable length of horizontal incision in conjunction with this may allow for the necessary removal of excess skin. A modification of the vertical mammaplasty is the use of a curvilinear J-type incision. This may obviate the need for a short horizontal scar for more moderate forms of ptosis. An adjunctive internal suspension of the breast parenchyma may be necessary with these approaches depending on the quality of skin and its ability to be an active element in shaping and holding form. With the Benelli approach, care must be taken with preoperative counseling about the scar and the relatively long process of smoothening of contours. While concomitant implant placement may improve not only ptosis but also the overall aesthetic result, it is critical to judiciously evaluate the timing of implant placement in light of patient expectation and anatomy. In some instances it may be more prudent to stage the augmentation and mastopexy.

Suggested Reading

1. Bostwick IIIrd J. Augmentation mammaplasty. Plastic and Reconstructive Breast Surgery. Vol. 1. 2nd ed. St. Louis: Quality Medical Publishing Inc., 2000:239-369.
2. Benelli L. A new periareolar mammoplasty: The "round block" technique. Aesth Plast Surg 1990; 14(2):93.
3. Regnault P. Breast ptosis: Definition and treatment. Clin Plast Surg 1976; 3(2):193.
4. Rohrich RJ, Thornton JF, Jakubietz RG et al. The limited scar mastopexy: Current concepts and approaches to correct breast ptosis. Plast Roconstr Surg 2004; 114(6):1622.
5. Spear SL, Little III rd JW. Reduction mammoplasty and mastopexy. Grabb and Smith's Plastic Surgery. 5th ed. Lippincott-Raven Publishers, 1997.
6. Weinzweig J. Augmentation mammaplasty. Plastic Surgery Secrets. Philadelphia: Hanley, 1999:238.

Chapter 69

Abdominoplasty

Amir H. Taghinia and Bohdan Pomahac

Classification and Choice of Procedure

Procedures used to change the shape of the anterior abdomen include liposuction (suction-assisted lipectomy), mini-abdominoplasty (infraumbilical elliptical skin and fat excision), abdominoplasty and panniculectomy. Patient selection and preference determines which operation to perform. A thorough history and physical examination usually reveals the best operation for a given patient. The physical examination should determine the amount of excess skin, the amount of excess subcutaneous fat and the laxity of abdominal fascia. Classification schemes have been developed to assist the surgeon in decision-making:

- Type 1 Fat deposit with normal fascia and skin → liposuction only
- Type 2 Mild excess skin, normal fascia with or without excess fat → infraumbilical elliptical skin excision with liposuction for excess fat
- Type 3 Mild excess skin, infraumbilical fascia laxity with mild to moderate fat excess → infraumbilical elliptical skin resection, infraumbilical fascia tightening and liposuction or direct lipectomy
- Type 4 Mild excess skin, total fascia laxity with or without excess fat → infraumbilical elliptical skin excision, supra- and infraumbilical fascia tightening and liposuction or direct lipectomy for excess fat
- Type 5 Large excess skin, total fascia laxity with or without excess fat → **complete abdominoplasty with supra- and infraumbilical fascia tightening**

In general, abdominoplasty addresses excess skin and lax abdominal fascia whereas liposuction addresses only excess fat. The ideal abdominoplasty patient has excess skin, mild excess fat and mild to moderate fascia laxity (Type 5). In contrast, the ideal abdominal liposuction patient has excess fat with normal, taut fascia and taut skin (Type 1). Liposuction alone is not recommended for patients with lax or excess skin. In these patients, liposuction without concomitant skin excision results in an unsightly 'deflated balloon' look.

The mini-abdominoplasty is reserved for patients with mild excess skin with or without excess fat (Types 2, 3 and 4). An infraumbilical ellipitical skin and fat excision is performed. If the infraumbilical fascia is lax (Type 3), it may be tightened directly.

Panniculectomy is removal of excess abdominal wall tissue (skin and fat) without associated undermining or umbilical transposition. It is usually performed in previously massively obese patients who have lost weight. These patients have a large segment of overhanging abdominal skin and fat. The "pannus" can lead to problems with skin breakdown or panniculitis. These patients have poor vascularity of the abdominal wall tissue; therefore even limited flap undermining can lead to complications.

Patients with excess abdominal wall fat usually benefit from weight loss prior to an operation. Successful preoperative weight loss heralds a motivated patient and promises a pleasing outcome. Certain patients will present with a somewhat bulging abdomen with minimal to moderate excess abdominal wall fat, normal fascia and no excess skin. Typically these patients will have significant excess intraabdominal fat. Weight loss is encouraged in these patients as well.

Preoperative Considerations

In the preoperative setting, the surgeon should pay close attention to the unique needs, characteristics and concerns of the patient. A thorough history and physical examination is crucial—not only to determine the surgical strategy but also to uncover potential sources of future complications. Previous abdominal surgery should raise the possibility of incisional hernias. Scars from prior abdominal surgery may also indicate the need for a more conservative approach. For example, an oblique right upper quadrant incision (e.g., from a previous open cholecystectomy) indicates potential interruption of the superolateral blood supply to the abdominoplasty flap (Fig. 69.1). The astute surgeon may consider flap delay, minimal undermining,

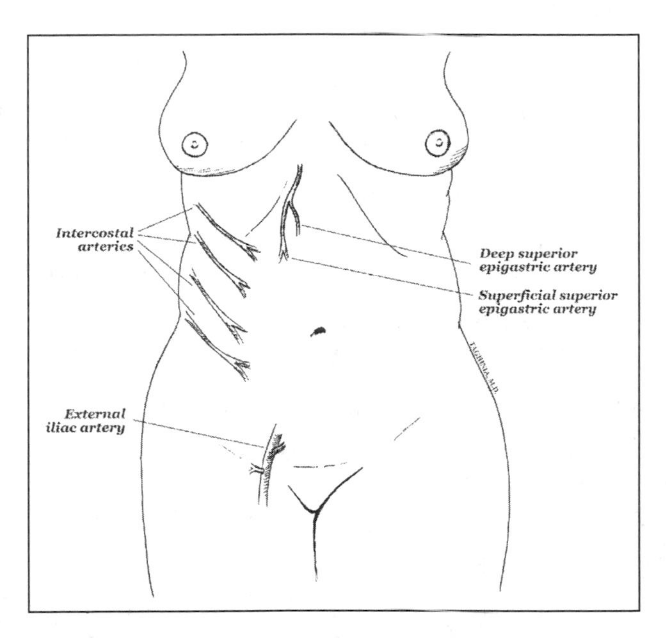

Figure 69.1. Blood supply to the anterior abdominal wall. The superficial epigastric artery perfuses the abdominal skin directly. The deep epigastric arteries (inferior and superior) perfuse the abdominal skin and subcutaneous tissue via perforators through the rectus abdominus muscle. Once fully raised, the abdominoplasty flap receives its main blood supply from the deep superior epigastric and intercostal arteries. A subcostal right upper quadrant incision (e.g., from an open cholecystectomy) can interrupt the superolateral blood supply to the abdominal wall. Liposuction of the lateral abdominoplasty flap can jeopardize blood supply from the lateral intercostal vessels.

restrained resection, or a combination of these approaches. Smoking should be stopped at least 1 month prior to surgery. Aggressive resection in smokers invites complications.

Hernia repair during abdominoplasty needs to be considered carefully. Occasionally, hernia repair requires extensive lysis of adhesions. Possible bowel injury during these operations may result in significant wound infections. In addition, these patients inevitably endure a longer hospital stay (for return of bowel function), thus providing a set up for other complications such as deep vein thrombosis and hospital-acquired infections.

Suction-assisted lipectomy during abdominoplasty can lead to feared complications of flap necrosis and infection. It is important to note that raising the abdominoplasty flap disrupts the blood supply from the deep superior and inferior epigastric perforators (Fig. 69.1). Thus, the predominant blood supply to the lower flap comes from the laterally-based segmental perforators of the intercostals vessels. Lateral liposuction of the abdominoplasty flap can injure these vessels, thus compromising blood supply of the flap, especially in the lower midline watershed area. Liposuction should not be performed in the lateral abdominoplasty flap. When done in the midline or inferior far-lateral flanks, it should be done very conservatively and in experienced hands.

Operations and Techniques

With the patient standing, the midline is determined. A pinch test is performed to determine the amount of excess skin. The lower border of the dissection is marked at a point 5-7 cm above the vulvar commissure. Lines are drawn laterally curving up to the anterior superior iliac spines. The superior borders of excision are then drawn laterally from a point just above the umbilicus. The superior lines serve only as useful guides for preliminary planning-rarely do they determine the final extent of excision (Fig. 69.2).

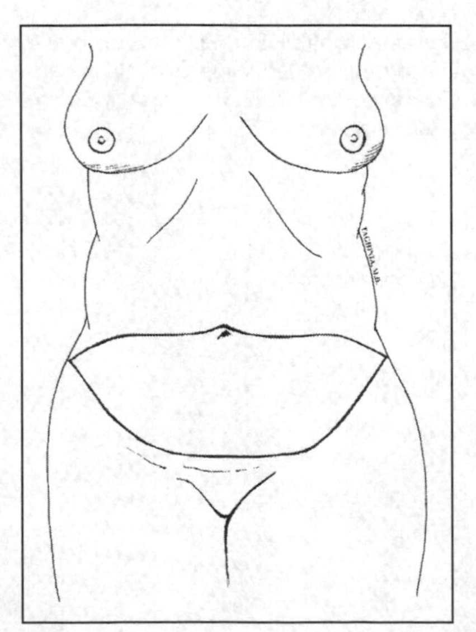

Figure 69.2. Markings for abdominoplasty. The lower incision is made and the flap is elevated. The superior markings serve as a guideline and may be modified once the flap is raised and redraped.

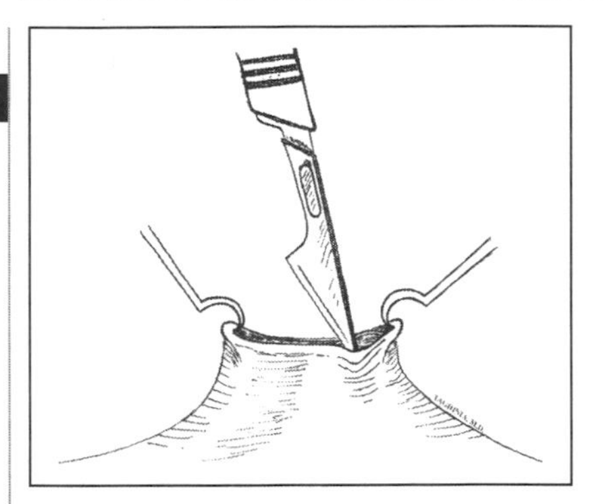

Figure 69.3. Umbilical incision. Two skin hooks elevate the umbilicus and provide tension for making the incision. An 11-blade incises the umbilicus circumferentially. The incision is then carried down to the fascia using heavy curved scissors to cut the fibrous bands that tether the umbilicus (not shown).

The patient is placed in the supine position with arms abducted. An intravenous antibiotic is administered, and pneumatic compression boots are fitted. After induction of adequate general anesthesia, a Foley catheter is inserted and the abdomen is prepped and draped from the xyphoid to the pubis. Abdominoplasty can also be performed under sedation and local anesthesia.

The umbilicus is sharply circumscribed and dissected down to the anterior abdominal wall fascia (Fig. 69.3). The inferior incisions are made and carried to the fascia. Injury to the lateral femoral cutaneous nerve is avoided by leaving a small amount of redundant fat attached to the anterior superior iliac spine. The dissection is directed cephalad towards the xiphoid along the plane anterior to the abdominal fascia. Perforating vessels must be identified and coagulated before being divided or they can retract and may cause delayed hematomas.

If the abdominal fascia is lax, it may be tightened with running or figure-of-eight interrupted nonabsorbable sutures (Fig. 69.4). This tightening is usually performed

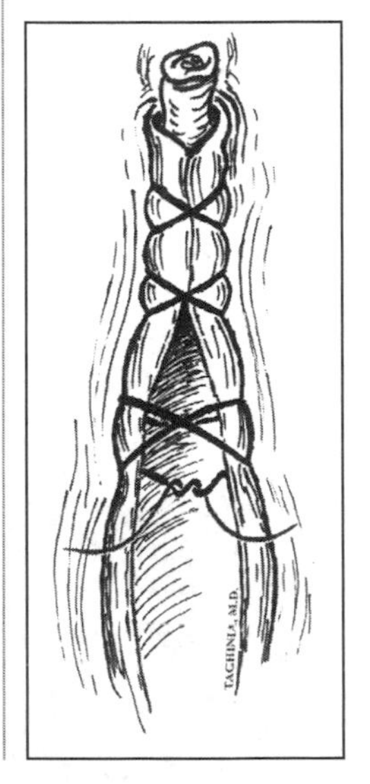

Figure 69.4. Fascial plication. Buried, nonabsorbable figure-of-eight interrupted sutures approximate the fascia in the midline below the umbilicus. A running, heavy gauge suture can also be used.

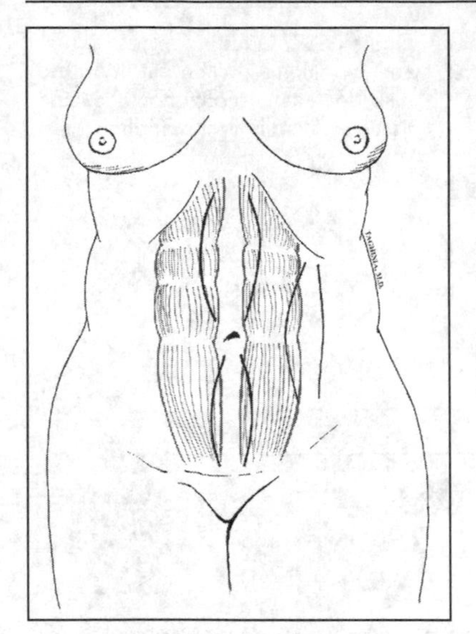

Figure 69.5. Tightening patterns for complete abdominoplasty. Vertical midline curves are drawn and approximated above and below the umbilicus. Alternatively, tightening can be performed at the linea semiluminaris (the intersection of rectus fascia and external oblique fascia).

at the midline above and below the umbilicus depending on the degree of laxity (see classification scheme). Arguing that the lax midline fascia may not support tightening in the long-term, some authors prefer to plicate the fascia laterally at the linea semiluminaris (the adjoining border of the external oblique and rectus fascia as shown in Fig. 69.5).

The patient's hips are flexed and the excess skin of the flap is excised (Figs. 69.6, 69.7). It is crucial to avoid excessive skin removal that could cause undue tension on the closure. The new site for the umbilicus is determined, and an opening is made to accommodate the umbilicus. In general, a smaller umbilicus looks more natural and less conspicuous. Most surgeons err towards a smaller, more inferiorly positioned umbilicus. Some prefer a vertical incision for the umbilical opening whereas others prefer a V-shaped incision (Fig. 69.8).

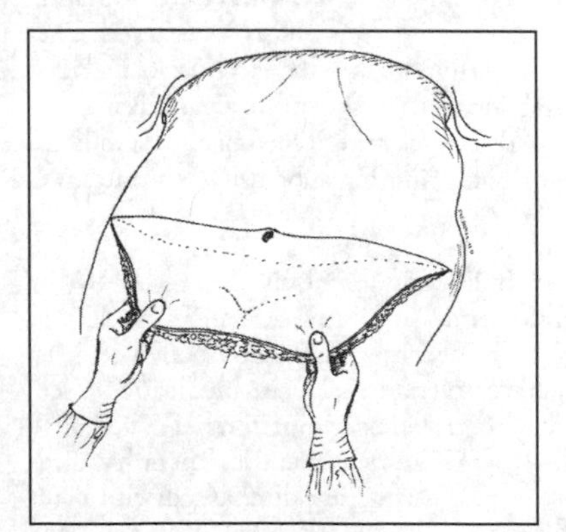

Figure 69.6. Gauging excess tissue. Once the flap is fully raised, the patient's hips are flexed and the flap is pulled inferiorly to assess the amount of excess tissue to be excised.

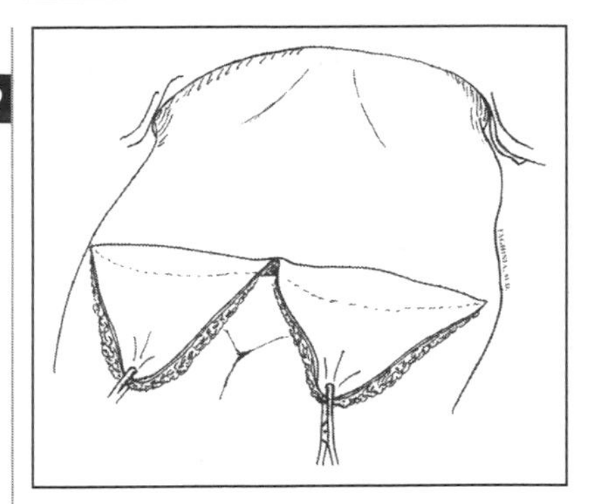

Figure 69.7. Gauging excess tissue. The lower flap can be cut in half to provide better visualization of excess tissue-especially in the midline.

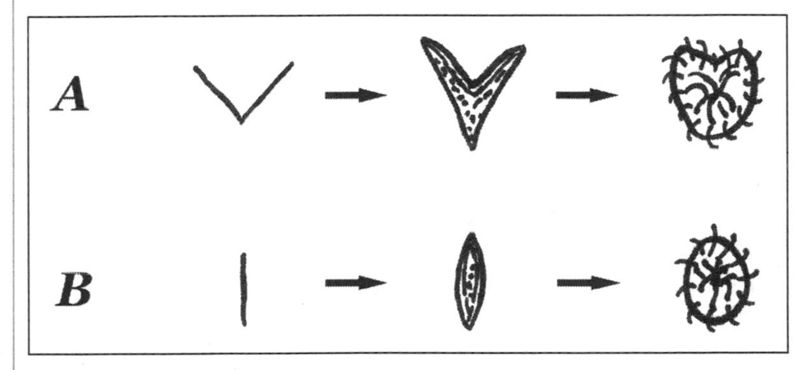

Figure 69.8. Techniques for umbilicoplasty. Some surgeons prefer a V-shaped incision (A) whereas others prefer a short vertical incision (B). The umbilical stalk may be sutured to the periumbilical anterior fascia (not shown).

If there is significant excess fat in the flap, it can be directly excised from below Scarpa's layer. To accentuate the midline, fat may be directly excised from the sub-Scarpa's layer in the midline. At this point, careful hemostasis is achieved. The flap is brought down, and the umbilicus is housed in its new location. Jackson-Pratt closed suction drains are placed through stab incisions in the pubic area. It is important to reapproximate Scarpa's fascia prior to dermal closure. The flap and umbilicus are then closed with interrupted deep dermal and running subcuticular sutures.

Postoperative Care

Patients are admitted to the hospital overnight and are discharged to home the following day. The hospital bed is kept in the semi-Fowler position and ambulation is encouraged with hips flexed. Intermittent intravenous analgesia usually provides adequate pain relief until patients are able to tolerate oral pain medications. Ice packs applied to the suprapubic region can also provide good pain relief. The Foley catheter is removed at midnight prior to discharge. Voiding difficulties can be handled early in the day thus allowing timely discharge. Patients are discharged with pain medication, antibiotics and instructions for Jackson-Pratt drain care. While at home, patients monitor and record drain output. The drains are usually removed after a week during the first office visit. At that point, activity is gradually increased until full return to activities, including sports and heavy lifting, at six to eight weeks.

Complications

Fortunately, worrisome complications of abdominoplasty rarely occur. These include deep vein thrombosis, pulmonary embolism and major skin necrosis. The more common complications of abdominoplasty include seroma, infections, hematoma and minor skin edge necrosis. Infections should be treated with antibiotics and debridement of nonviable tissue. Small hematomas self-absorb and may be observed. Larger hematomas should be treated more aggressively by open or percutaneous drainage. Observation is usually warranted for most seromas. Large seromas may be drained percutaneously, but often times the fluid reaccumulates. Minor skin edge necrosis is treated with conservative debridement and, if necessary, late scar revision to improve cosmetic appearance.

Pearls and Pitfalls

- Design of the skin incision should take into consideration patient preference. Some women wear very low-riding pants, and an incision that approaches the iliac crest may be visible.
- The larger caliber, periumbilical perforating vessels should be cauterized and transected well above the level of the fascia so that if one bleeds it will not fully retract into the rectus muscle below. If this occurs, a figure-of-eight suture should be used to achieve hemostasis.
- Any palpable bulge on the fascia will be a visible one on the surface when the patient stands and should be tightened. Occasionally this may involve horizontal tightening of the fascia.
- Plication of the fascia should span from xyphoid to pubis, otherwise a bulge will occur above or below the tight suture line.
- Many surgeons use tacking sutures to help adhere the abdominoplasty flap to the fascia in order to reduce dead space. This has the theoretical benefit of decreasing seroma formation.
- Never make the superior skin incision until there is no doubt that the skin will close without undue tension. It is better to have a small vertical scar (from the umbilicus incision) than central skin necrosis or hypertrophic scarring from excess tension.

Suggested Reading

1. Bozola AR, Psillakis JM. Abdominoplasty: A new concept and classification for treatment. Plast Reconstr Surg 1988; 82(6):983-93.
2. Grazer FM, Goldwyn RM. Abdominoplsty assessed by survey with emphasis on complications. Plast Reconstr Surg 59(4) :513-7.
3. Greminger RF. The mini-abdominoplasty. Plast Reconstr Surg 1987; 79(3) :356-65.
4. Hester Jr TR, Baird W, Bostwick IIIrd J et al. Abdominoplasty combined with other major surgical procedures: Safe or sorry? Plast Reconstr Surg 1989; 83(6):997-1004.
5. Lockwood TE. High-lateral-tension abdominoplasty with superficial fascial system suspension. Plast Reconstr Surg 1995; 96(3):603-15.
6. Lockwood TE. Maximizing aesthetics in lateral-tension abdominoplasty and body lifts. Clin Plast Surg 2004; 31(4):523-37.
7. Pitanguy I. Abdominal lipectomy. Clin Plast Surg 1975; 2(3):401-10.

Chapter 70

Liposuction

Zol B. Kryger

Indications

Liposuction, or suction-assisted lipectomy, is the most common plastic surgical procedure performed in the U.S. It is used for removal of fat from small to moderately localized deposits; however many surgeons now perform large-volume liposuction for more generalized fat removal. Ultrasound-assisted liposuction has a role in treating gynecomastia and other deposits of tough, fibrous fat. These include lipodystrophy due to protease inhibitors (submental and "buffalo hump") and benign lipomatosis. There are reports of treating lipomas, lymphedema and post-rhytidectomy hematomas or fat necrosis with liposuction.

Relevant Anatomy

Liposuction is performed on subcutaneous adipose tissue, which can be divided into superficial, intermediate and deep layers. The superficial layer is composed of denser fat with organized fibrous septa. The deep layer is more loosely arranged with areolar fat and minimal fibrous tissue. The intermediate layer is a transition between the two. Most liposuction is performed on the deep and intermediate layers. Removing fat from the superficial layer is associated with a risk of skin irregularities and contour deformities. Some surgeons will create several closely spaced tunnels using small diameter cannulas in the superficial fat layer without suctioning. This can help the skin retract and scar down to the underlying tissue in order to avoid irregularities in the skin. Superficial lipectomy should not be done on the buttocks.

A study by Rohrich et al described the term "zones of adherence" where liposuction should be avoided:

- Lateral gluteal depression (just superior to the "saddle bags")
- Gluteal crease
- Posterior thigh above the popliteal fossa
- Inferolateral iliotibial tract region
- Medial mid-thigh

In these areas, the skin is tightly adhered to the underlying fascia, and contour deformities are more likely to occur.

Preoperative Considerations

Liposuction is a significant surgical procedure; there are several potentially serious complications. Patients should be ASA class I and within 30% of their ideal body weight. Many patients have expectations that cannot be met by liposuction alone. A detailed discussion of the risks, including the likelihood of contour irregularities requiring a secondary procedure, is essential.

After consideration of the anatomical factors listed above, patients should be marked while they are standing upright. A common technique used for marking is to use to use a topographical map-like sytem in which the area to be suctioned is marked with concentric rings. An increasing number of rings indicates a greater thickness of fat to be suctioned. The areas to be avoided are marked as well. Prior to surgery, an intravenous antibiotic in given.

Tumescent Technique

Almost all liposuction is performed after injection of tumescent solution. Tumescent solution is typically composed of 0.9% saline containing 0.05–0.1% lidocaine and 1:1,000,000 epinephrine. Several liters of this solution are introduced into the area to be suctioned 10-20 minutes prior to starting. The local anesthetic effect of the lidocaine reduces the need for heavy sedation or general anesthesia. The epinephrine assists in hemostasis and slows the absorption of the lidocaine, allowing a dose of up to 35 mg/kg to be used. The typical volume of tumescent solution used is three times the volume of fat to be aspirated.

Liposuction Techniques

Syringe Aspiration. Small volume liposuction can be performed using a syringe with the plunger withdrawn. This technique is excellent for harvesting small amounts of fat for soft-tissue augmentation. It is a controlled technique and can be used to remove fat from the superficial layer.

Suction-Assisted Lipectomy (SAL). Often referred to as "traditional" liposuction, this technique relies on the energy generated by the longitudinal motion of the surgeon's hand. The speed of the movement, the size of the cannula, and the characteristics of the suctioned fat combine to determine the rate of liposuction. When the surgeon's hand is still, there is no effect on the tissue because no kinetic energy is generated.

Power-Assisted Liposuction (PAL). An external motor is used to generate reciprocating motion of the suction cannula, thus sparing the surgeon from the repetitive motion of manual liposuction. This significantly reduces surgeon fatigue at the expense of a small amount of control. Studies demonstrate equivalent cosmetic results with PAL compared to traditional liposuction.

Ultrasound-Assisted Liposuction (UAL). This technique relies on the delivery of ultrasonic waves that liquefy the fat, making suctioning more efficient. Since the cannula generates heat, one must constantly keep it in motion when on. Care must be taken not to burn the skin, resulting in "end hits." Since UAL removes fat quickly, most surgeons achieve the final contour with traditional or power-assisted liposuction. The second generation UAL device, the VASER system, emits pulsed waves instead of continuous waves of energy. This reduces the heat of the probe by about 50%, decreasing the risk of thermal injury.

Complications

Minor complications from liposuction occur in 10% of cases and include:

- Infection
- Contour irregularities
- Hypesthesia (which usually resolves within 6 months)
- Skin sloughing
- UAL skin burns
- Skin discoloration
- Seroma or hematoma

Major complications occur in 0.1–0.3% of patients and include:

- DVT or PE
- Cannula penetration of the abdominal or thoracic cavity
- Significant hemorrhage
- Hypovolemia due to inadequate resuscitation
- Fat emboli

Complications from the tumescent solution are rare but account for most of the deaths from liposuction. Most authors recommend overnight monitoring for patients undergoing large volume liposuction (greater than 5 liters of total aspirate). Complications include the following:

- Electrolyte abnormalities can occur in large volume liposuction due to massive fluid shifts and electrolyte losses. Intraoperative hypokalemia is common but usually corrects itself rapidly. Symptoms from electrolyte abnormalities are uncommon but should always be evaluated.
- Volume overload leading to pulmonary edema can occur in lengthy cases.
- Lidocaine toxicity is rare; some authors believe that a dose of up to 55 mg/kg is safe. Symptoms occur 8-14 hours after surgery and are typically neurologic: circumoral numbness, light-headedness and drowsiness followed by tremors and seizures. Cardiac and respiratory signs are late findings.
- Epinephrine toxicity has been reported. This stresses the importance of performing larger volume liposuction on patients with good cardiac function and reserve. Patients should always have basic cardiac monitoring with EKG leads during liposuction to detect any arrhythmias.

In light of these potential complications, the surgeon must be able to administer anticonvulsant medications and advanced cardiac life support.

Pearls and Pitfalls

In order to maximize the effect of the tumescent solution, at least 10 to 20 minutes should elapse prior to suctioning. This allows for the epinephrine-induced vasoconstriction and even distribution of the tumescence. In order to minimize intraoperative time, an area should be tumesced, and then one should move on to other areas or adjunct procedures rather than stopping while the tumescent takes effect. This can be difficult since it may involve turning the patient from side-to-side, or from the supine to the prone position multiple times. Nevertheless, the benefits of a nonbloody aspirate and shorter procedure time are worth this inconvenience.

Suggested Reading

1. De Souza Pinto EB, Erazo Indaburo P, da Costa Muniz A et al. Superficial liposuction: Body contouring. Clin Plast Surg 1996; 23(4):529.
2. Gasperoni C, Salgarello M. MALL liposuction: The natural evolution of subdermal superficial liposuction. Aesthetic Plast Surg 1994; 18:253.
3. Hughes IIIrd CE. Reduction of lipoplasty risks and mortality: An ASAPS survey. Aesthetic Surg J 2001; 21:120.
4. Klein JA. Tumescent technique for regional anesthesia permits lidocaine doses of 35 mg/kg for liposuction. J Dermatol Surg Oncol 1990; 16:248.
5. Klein JA. Tumescent technique for local anesthesia improves safety in large-volume liposuction. Plast Reconstr Surg 1993; 92:1085.
6. Lockwood TE. Superficial fascial system (SFS) of the trunk and extremities: A new concept. Plast Reconstr Surg 1991; 86:1009.
7. Rohrich RJ, Smith PD, Marcantonio DR et al. The zones of adherence: Role in minimizing and preventing contour deformities in liposuction. Plast Reconstr Surg 2001; 107:1562.
8. Zocchi ML. Ultrasonic assisted lipoplasty. Technical refinements and clinical evaluations. Clin Plast Surg 1996; 23(4):575.

Chapter 71

Laser Resurfacing

Keren Horn and Jerome Garden

Introduction

The acronym LASER (light amplification by stimulated emission of radiation) describes the physical process by which light is produced. All laser systems require four basic elements: a laser medium that can be excited to emit laser light by stimulated emission, an energy source to excite the medium, mirrors at the ends of the laser, forming the 'cavity' and a delivery system. Laser light is monochromatic; it is one discrete wavelength and color. Monochromicity is determined by the chosen lasing medium, such as **gas** (e.g., carbon dioxide {CO_2} or halide), **solid** (e.g., alexandrite or ruby) or **liquid**. Coherence describes energy waves that are in phase, both spatially and temporally. Collimation refers to the ability of laser light waves to travel in a parallel fashion and propagate over long distances without divergence. Laser beams are able to maintain a very high powered intensity over long distances traveled. All of these properties help to explain the clinical effectiveness of laser light in the skin.

Mechanism of Action

Light striking the epidermis may be reflected, transmitted to deeper tissues, scattered or absorbed. A clinical effect occurs when light is absorbed. The energy absorbed is measured in joules per square centimeter and is known as the energy density or fluence. A photon surrenders its energy to an atom or molecule (known as a **chromophore**) once absorption occurs. The photon then ceases to exist, and the chromophore becomes excited. Water, hemoglobin and melanin are the principle endogenous chromophores in the skin. Each chromophore is absorbed at a particular wavelength of light (Table 71.1). It is this selective absorption spectrum that allows for specific targets by laser. Three basic effects are possible once laser energy is absorbed in the skin: photothermal, photochemical, or photomechanical.

In the 1980s, Anderson and Parish introduced the theory of selective photothermolysis, revolutionizing cutaneous laser surgery. Selective photothermolysis occurs when a laser light wavelength reaches a specific target and is preferentially absorbed, thereby minimizing unwanted thermal injury. For this to occur, there must be selective light absorption and a short enough laser delivery time so that heat conduction is minimal. The duration of laser delivery must be shorter than the thermal relaxation time (TRT) of the target. TRT is the time needed for a given heated tissue structure to lose half its heat. Table 71.2 lists the TRT for common targets. The energy density, or fluence, supplied by the laser must also be satisfactory to attain the desired response of the target within the allotted time for selective photothermolysis to occur.

Types of Lasers

Several types of lasers are used to treat the skin including continuous-wave (CW), quasi-CW, and pulsed device lasers. CW mode lasers (e.g., CO_2 or older argon

Practical Plastic Surgery, edited by Zol B. Kryger and Mark Sisco.

Table 71.1. Various chromophores and the range of light that they absorb

Chromophore	Absorption Range
DNA, RNA, proteins	Ultraviolet
Hemoglobin	Blue-green and yellow
Hematoporphyrin derivative	Red
Melanin	UV > visible >> near infrared
Black tattoo ink	Visible and infrared
Water	Infrared

technology) supply a constant beam of laser light. These long exposure durations can result in nonselective tissue injury. These lasers also tend to have a limited peak power. Quasi-CW mode lasers (e.g., potassium-titanyl-phosphate [KTP], copper vapor, copper bromide, krypton and argon-pumped tunable dye [APTD]) break the CW beam into short segments. The pulsed laser devices emit very short pulses at very high peak powers with relatively long intervening time periods (0.1-1 second). These lasers can be either long-pulsed (e.g., pulsed dye laser) or very short-pulsed (e.g., quality-switched [QS] ruby, alexandrite, or neodymium: yttrium-aluminum-garnet [Nd:YAG]). Q represents a quality factor of energy storage in the lasing medium. This can be changed abruptly to produce a short, intense burst of light. The repetition rate for pulsed lasers is expressed in hertz.

Skin Cooling

Laser skin surgery requires precise control over placement, timing and temperature regulation (i.e., skin cooling) to prevent the amount of heat-induced skin injury. Skin cooling minimizes epidermal damage. This is especially important in the treatment of darkly pigmented skin which is at high risk for epidermal injury with visible or near infrared light at wavelengths of less than 1200 nm. There are three basic types of cooling methods: precooling, parallel cooling and postcooling, depending upon when in the process of laser exposure they are utilized. All cooling methods place a cold medium in contact with the skin surface, and the depth of cooling depends on contact time. The epidermis is cooled in tens of milliseconds, the papillary dermis in hundreds of milliseconds and the bulk of the dermis in seconds. Advantages of skin cooling are threefold: epidermal protection, ability to deliver higher fluences and anesthesia. Skin cooling methods range from ice packs and cooled aqueous gel to cold sapphire contact hand pieces and dynamic cooling devices.

Table 71.2. Thermal relaxation time (TRT) for common targets

Target	TRT
200 μm hair follicle	20 msec
100 μm port wine stain blood vessel	15 msec
50 μm blood vessel	1 msec
50 μm epidermis	1 msec
7 μm erythrocyte	20 μsec
1 μm melanosome	1 μsec
0.1 μm tattoo particle	10 nsec

Table 71.3. Ablative and nonablative resurfacing lasers

Laser Type	Wavelength
Ablative	
Carbon dioxide (CO_2) (CW)	10,600 nm
CO_2 (pulsed)	10,600 nm
Erbium:YAG (pulsed) CO_2/erbium	2490 nm
Nonablative	
Pulsed dye laser (PDL)	585-595 nm
Nd:YAG	1064 nm, 1320 nm
Diode	980 nm, 1450 nm
Erbium:glass	1540 nm
Intense pulsed light (IPL)	515-1200 nm

Laser Resurfacing

Laser resurfacing (also termed laser peel, laser rejuvenation and laser surgery) removes wrinkles, discolorations, age spots and sun damaged skin. It can be performed with either ablative or nonablative laser systems (Table 71.3). Success of ablative lasers (e.g., CW CO_2 laser) in treating dermatoheliosis, facial rhytides and atrophic scars has been well documented over the past decade. However, the unpredictable degree of thermal necrosis and scarring that can be seen has led to the development of nonablative, high-energy, pulsed systems for facial rejuvenation. Although these newer devices do not achieve as dramatic results, they do eliminate the high risk profile and extended healing time seen with the ablative systems.

Ablative Lasers

The CO_2 laser emits light at 10,600 nm, in the far infrared portion of the spectrum. The chromophore for this wavelength is intracellular and extracellular water. It produces the most effective results in facial resurfacing. Epidermal ablation occurs after one pass at standard treatment parameters, but collagen shrinkage and remodeling require an additional one to two passes. These latter two factors are likely responsible for the long-term clinical improvements seen. Thermal desiccation with concomitant collagen shrinkage is believed to be the mechanism by which resurfacing works. When dermal temperatures rise beyond 55-62°C, disruption of the interpeptide bonds within collagen's triple helix shrinks the moiety to one-third of its normal length. This procedure usually produces a 50% improvement. This laser also provides excellent hemostasis. Newer ablative systems exist which minimize nonspecific thermal damage. These include high-energy, short duration CO_2 laser and the erbium:YAG (Er:YAG) laser. Another less aggressive approach is to perform only a single pass with the CO_2 laser. This method leads to only partial desiccation of the treated area, leaving an intact layer of skin to serve as a biologic wound dressing.

Short-pulse Er:YAG laser systems at a wavelength of 2940 nm were developed in the hopes of minimizing thermal injury and tissue necrosis. The Er:YAG laser penetrates less deeply into the dermis than the CO_2 laser. This is due to its target chromophore, water. The Er:YAG laser is absorbed by water sixteen times more efficiently than the CO_2 laser. Water makes up over 90% of the epidermis, and it is here where most of the energy of the Er:YAG laser is absorbed. As a result, there is less thermal damage. This explains its more modest clinical improvement as well as its less intense side effect profile. This laser provides very poor hemostasis. Combined, sequential CO_2-Er:YAG laser resurfacing combines both systems in an attempt to have the greatest clinical response with the minimum amount of tissue necrosis.

Indications

Indications for resurfacing with CO_2 and Er:YAG lasers are photoaging (including dyschromias, facial rhytides, nonfacial rhytides (Er:YAG only) and facial solar lentigines), epidermal nevi, rhinophyma, atrophic acne scars, varicella or hypertrophic scars, adenoma sebaceum, sebaceous hyperplasia, seborrheic keratoses, syringomas, trichoepitheliomas and xanthelasma. Premalignant and malignant skin conditions that may be treated with these laser systems include actinic chelitis, actinic keratoses, Bowen's disease, erythroplasia of Queyrat and superficial basal cell carcinoma.

Relative and Absolute Contraindications

Caution must be taken in patients who have a tendency towards keloid formation and those with a reduced number of adnexal structures. This absence or decrease of appendageal structures may prolong healing time and/or prevent complete reepithelialization. Diseases which koebnerize, such as psoriasis and lichen planus, are considered relative contraindications and patients should be well-informed prior to resurfacing. Atypical scarring after dermabrasion or chemical peeling has been associated with isotretinoin therapy. This has been reported even in cases where isotretinoin therapy was given over one year prior to resurfacing. Skin necrosis and scarring have been associated with resurfacing done in close temporal sequence to face lifting or blepharoplasty. Resurfacing of the hands, neck and chest with a CO_2 laser must be done with great caution due to the high potential for scarring.

Complications

Side effects and complications of laser skin resurfacing are common. Oozing, crusting, and edema occur in the days and weeks that follow the procedure. Infectious complications in this acute period include bacterial, viral and/or yeast infections. Erythema, pruritus and skin tightness can be seen weeks to months later. Eczema and allergic contact dermatitis, along with acneiform eruptions, milia and perioral dermatitis can all occur in this time period as well. Hyperpigmentation is most commonly seen in the first six months after the procedure, whereas hypopigmentation (permanent, delayed) occurs between the sixth and twelfth month after laser resurfacing. Atrophic-, hypertrophic- and keloidal-scarring tend to occur after the first month.

Preoperative Considerations

The use of topical tretinoin, vitamins C and E and full-spectrum sunscreens have been used for preoperative regimens, although only the sunscreen appear to be of any proven benefit. All patients undergoing laser resurfacing receive prophylactic oral antivirals, antifungals and antibiotics with Gram-positive coverage to avoid secondary bacterial infection.

The CO_2 laser is generally painful. Although local anesthesia may be sufficient in patients undergoing treatment of one or two cosmetic units, a combination of topical, local and systemic anesthesia is generally used for full face CO_2 resurfacing. The Er:YAG laser is generally considered less painful than the CO_2 laser. Nerve blocks with lidocaine (1-2%) with epinephrine (1:100,000) are appropriate for the central face. Sensory blockade of the lateral surfaces of the face tend to be more challenging (Table 71.4).

Table 71.4. A regimen for sensory blockade of the lateral aspects of the face

Local infiltration is performed with a mixture of the following:
1. 2% lidocaine and 1:100,000 epinephrine
2. 0.5% bupivicaine
3. 1:10 dilution of 8.4% sodium bicarbonate (neutralizes the pH of the mixture and decreases pain during the injection)
4. Hyaluronidase 75U (improves tissue diffusion)

Intraoperative Technique

CO_2 Laser

Both patient and physician must have eye protection. Regardless of the type of laser being used, the treatment protocol is standardized to use single-pulse vaporization with minimal overlap of the pulses. If more than one-pass is to be performed, the desiccated tissue debris is wiped away with moist gauze after the entire surface has been covered. This is performed to enhance the laser-tissue interaction of the second pass. Repeat treatment of any area should be done after a minimum six month waiting period to allow for adequate new collagen formation, remodeling of collagen and normalization of pigment.

Er:YAG Laser

The skin is first cleansed with an antiseptic and then rinsed with saline. It is imperative that the operative field be kept dry due to the Er:YAG laser's high coefficient of absorption of water. In contrast to the CO_2 laser, wiping between laser passes is not always necessary. Early pinpoint bleeding is an indication of penetration into the papillary dermis. Heavier bleeding indicates ablation to the level of the midpapillary to reticular dermis.

Postoperative Considerations

All patients undergoing an ablative laser procedure will have at least one week of significant morbidity until complete reepithelialization occurs. Head elevation and ice application can help reduce erythema and edema. Cool compresses can be applied to the treated area for the first week to remove the serous exudates and any residual necrotic debris. Either a thick healing ointment to the open skin surface or bio-occlusive dressings should be used during reepithelialization.

The burning sensation experienced during the immediate postresurfacing period can be treated with acetaminophen with or without codeine phosphate or hydrocodone. Antihistamines or mild topical steroids can be used for pruritus. A several day course of oral corticosteroids may be necessary to help reduce edema. Strict sun avoidance should be emphasized to reduce postinflammatory hyperpigmentation and all patients should be counseled to comply with this measure.

Nonablative Lasers

Nonablative lasers have largely replaced ablative laser systems because of their low risk profile and decreased recovery time. Pulsed dye laser (PDL), Nd:YAG laser, diode, erbium:glass laser and intense pulsed light (IPL) are all examples of nonablative systems. These systems emit light between 500-1540 nm. Absorption by superficial water-containing tissue like the epidermis is relatively weak at these wavelengths. Cooling systems are used in conjunction with all nonablative lasers to ensure epidermal preservation. The epidermis is therefore not visibly disrupted. These systems work via their thermal effects on the dermis. The mechanism of action of nonablative lasers is thought to be through induction of collagen remodeling by creation of a dermal wound. Although not completely understood, fibroblast activation, collagen remodeling and subsequent increased pro-collagen III expression has been noted after use of these systems.

Nonablative rejuvenation is useful in the treatment of vascular markings including erythema and telangiectasias, pilosebaceous changes including pore size and skin smoothing, and pigmentary skin alterations including dyschromia, lentigines, mottled pigmentation and photoaging. Significant improvement of dermal and subcutaneous senescence including rhytides and lipodystrophy is more challenging with these devices. However, nonablative treatments can improve skin texture or surface irregularities. The more superficial wavelength systems are considered more effective in treating vascular, pigmentary and pilosebaceous irregularities, whereas the longer wavelengths may induce more dermal collagen and ground substance changes.

Practical Considerations

Multiple treatment sessions at 3 to 4 week intervals are necessary. Typically, 5 to 6 treatments are performed. Minimal erythema and edema are usually noted immediately after treatment and tend to resolve several hours soon thereafter. Patients are usually able to return to work the following day. Minimal anesthesia is required, and topical or local techniques are the mainstay of pain control.

Mild and limited pinpoint bleeding, transient erythema and edema, blistering, pinpoint scarring and post-inflammatory, transient hyperpigmentation are the most commonly reported adverse reactions. Purpura is characteristically seen with the PDL. The longer wavelength of the erbium:glass laser makes it the laser least absorbed by tissue melanin, providing an advantage when treating darker-skinned individuals.

Although their clinical efficacy does not yet meet that of the ablative laser systems, nonablative laser resurfacing has been shown to improve mild to moderate atrophic scars and rhytides with virtually no external wound. There is yet to be one in this class that stands out above the rest. As technological advances continue to refine these systems even more, nonablative lasers remain a popular choice for patients seeking noninvasive rejuvenation. Further trials and clinical experience are needed to help determine the overall efficacy of these systems in laser resurfacing.

Nonablative Radiofrequency Resurfacing

A new device has recently come to the facial rejuvenation market which appears to produce immediate collagen contraction with a single treatment. The nonablative device uses radiofrequency to produce an electric current which generates heat through resistance in the dermis. The epidermis is protected and preserved through a cooled electrode before and during the radiofrequency pulse via a cryogen-spray device. Intense, sustained, uniform heat is thought to cause shrinkage of collagen followed by a period of neocollagenesis and subsequent tissue remodeling.

Effects noted in early clinical trials report tissue tightening of the cheeks with improvement of the nasolabial fold, cheek contour, marionette lines and possibly the jaw line. Clinical changes appear approximately 4 to 6 months after the procedure. Benefits of this new system are believed to be gradual changes, no downtime and minimal risk profile. This procedure represents the possibility of a 'minimally invasive face lift, or a noninvasive, nonsurgical approach to tissue tightening in facial skin. Further technological advances, refinement of technique and studies to better determine clinical effect will help us to evaluate and determine the future success of this device.

Pearls and Pitfalls

Ablative resurfacing is deceptively easy to perform, and one must refrain from being too invasive around thin skin areas such as the periorbital region and the jawline as this has been associated with a higher incidence of scarring after resurfacing. In general, the deeper the ablation, the better the results will be. The caveat is that the risk of scarring will be greater.

Once the top layer of skin is denuded, the patient needs very close monitoring and care. Even on broad-spectrum infectious coverage, any pustulation or blistering during the healing period must be aggressively evaluated and treated. Furthermore, areas of persistent erythema may indicate early fibrosis and may need to be immediately treated with intralesional therapy and possibly nonablative lasers.

Hypopigmentation, the most common side effect of laser resurfacing, may not appear until a year after the procedures and this risk must be reviewed in detail with the patient prior to the procedure. It may be that the lighter skin-toned patient may actually have a greater tendancy to develop hypopigmentation.

Nonablative remodeling is much safer and better tolerated by the patient. With nonablative lasers, the epidermis remains intact and many of the concerns regarding resurfacing are minimized or eliminated. However, the patient must still be informed of the rare potential of dyspigmentation or skin texture changes. Furthermore, there is great variability of improvement from patient to patient. Some respond favorably while many may experience only modest response, especially for the treatment of rhytides and skin tissue laxity.

Radiofrequency energy treatment of tissue laxity may be enhanced by the addition of laser energy. Radiofrequency application is usually painful depending on the amount of deposited energy. This may dissuade patients from undergoing this procedure and must be addressed in the preoperative discussion.

Suggested Reading

1. Alster TS, Lupton JR. Are all infrared lasers equally effective in skin rejuvenation. Semin Cutan Med Surg 2002; 21(4):274.
2. Fitzpatrick R, Geronemus R, Goldberg D et al. Multi-center study of noninvasive radiofrequency for periorbital tissue tightening. Lasers Surg Med 2003; 33(4):232.
3. Fitzpatrick RE, Goldman MP. Carbon dioxide resurfacing of the face. In: Fitzpatrick RE, Goldman, MP, eds. Cosmetic Laser Surgery. St. Louis: Mosby, 2000.
4. Fitzpatrick RE, Goldman MP. Use of the Erbium:YAG laser in skin resurfacing. In: Fitzpatrick RE, Goldman MP, eds. Cosmetic Laser Surgery. St. Louis: Mosby, 2000.
5. Grema H, Greve B, Raulin C. Facial rhytides—subsurfacing or resurfacing? A review. Lasers Surg Med 2003; 32(5):405
6. Manuskiatti W, Fitzpatrick RE, Goldman MP. Long-term effectiveness and side effects of carbon dioxide laser resurfacing for photoaged facial skin. J Am Acad Dermatol 1999; 40(3):401.
7. Papadavid E, Katsambas A. Lasers for facial rejuvenation: A review. Int J Dermatol 2003; 42(6):480.
8. Tanzi EL, Alster TS. Side effects and complications of variable-pulsed erbium: yttrium-aluminum-garnet laser skin resurfacing: Extended experience with 50 patients. Plast Reconstr Surg 2003; 111(4):1524.

Chapter 72

Chemical Rejuvenation of the Face

Keren Horn and David Wrone

Introduction

Chemical peeling is a popular option for facial rejuvenation. A wide spectrum of chemical peels exist to produce varying effects on the skin, ranging from salon procedures which exfoliate only to the depth of the stratum corneum to peels which are comparable in efficacy to carbon dioxide laser ablation. Just like dermabrasion and laser resurfacing, chemical peeling is a form of controlled wounding resulting in skin rejuvenation. Different solutions are used to target a specific depth of injury with the consequent removal of damaged skin resulting in improved skin characteristics. The classification spectrum ranges from superficial to medium to deep (Table 72.1). Agents are generally classified based on the level or depth of skin injury. Although less efficacious than deep peels, superficial and medium depth peels are desirable due to their long-standing safety record, effectiveness, relative low-cost and quick recovery time. Chemical peels are generally performed on only the face. If applied to other sun exposed areas susceptible to photodamage such as the neck, chest and hands, great caution must be taken due to the decreased number of oil glands and lack of uniform penetration. Superficial peels such as salicylic acid and 30% glycolic acid are usually safe to use in these areas.

Pre-Procedure Considerations

Relative Indications

It is important to identify both at-risk and ideal patients for rejuvenation. The level of skin irregularities should be matched to the depth of injury from the peel. The major indications for chemical peel resurfacing are solar lentigo and other signs of photodamage including rhytides, scarring, preneoplastic lesions of epidermal origin (e.g., actinic keratoses), acne vulgaris, dyschromia (e.g., melasma) and demarcation lines secondary to other resurfacing procedures. As a guideline, early photodamage consisting of primarily textural (including benign epidermal growths) and pigmentary changes without wrinkles responds nicely to repeated superficial peels or one to two medium-depth peels. Lesions originating deeper in the skin such as actinic keratoses or melasma deserve treatment with one or more medium-depth peels. Permanent or sempermanent reduction in wrinkles necessitates deeper dermal injury that only a deep peel or similar wounding procedure can produce. Although temporarily useful (such as for an upcoming special event), superficial and medium-depth peels do not permanently improve or remove wrinkles.

Contraindications and Special Cases

Underlying skin problems and risk factors must be identified to avoid or limit complications.

Practical Plastic Surgery, edited by Zol B. Kryger and Mark Sisco. ©2007 Landes Bioscience.

Table 72.1. Classification of chemical peel rejuvenation methods

Superficial Wounding

- Very light: to depth of s. corneum or s. granulosum depth
  - Low potency formulations of glycolic acid or other α-hydroxy acids
  - 10%-20% TCA (weight-to-volume formulation)
  - Resorcin
  - Salicylic acid
  - Tretinoin
- Light: to stratum basale or upper papillary dermal depth
  - 25%-35% TCA, unoccluded, single or multiple frost
  - Solid CO_2 slush
  - Jessner's solution (Combes' formula)
    - 14 g Resorcinol
    - 14 g Salicylic acid
    - 14 g 85% lactic acid
    - 100 ml 95% ethanol
  - 70% glycolic acid

Medium-Depth Wounding

- Through the papillary dermis to the upper reticular dermis
  - Combination peels, single or multiple frost
    - Solid CO_2 + 35% TCA (most potent combination)
    - Jessner's + 35% TCA (most popular combination)
    - 70% glycolic acid + 35% TCA
  - 35-50% TCA, unoccluded, single frost (not recommended)
  - 88% full-strength phenol, unoccluded (rarely used)
  - Pyruvic acid

Deep-Depth Wounding

- To the mid-reticular dermis
  - Baker-Gordon formula, occluded or unoccluded
    - 3 ml 88% liquid phenol, USP
    - 2 ml tap water
    - 8 drops Septisol® liquid soap
    - 3 drops Croton oil
  - > 50% TCA concentration

- Oral isotretinoin therapy in the preceding six months precludes use of medium-depth and deep peels because studies have shown that the medication impairs wound healing.
- Burn or radiation treated patients with a complete absence of intact pilosebaceous units on the fact should not undergo medium-depth or deep peels.
- Past or active herpes simplex infection should be treated appropriately prior to either medium-depth or deep peels.
- When contemplating deep peeling, caution should be taken in patients with a propensity towards hypertrophic or keloid scar formation.
- Deeper peels in patients with any condition which may impair the immune system such as human immunodeficiency virus should be done with great care.
- Absolute contraindications to chemical skin rejuvenation for all types of peels include but are not limited to: poor physician-patient relationship; unrealistic expectations; lack of psychological stability; poor general health and nutritional status; and active infection or open wounds.

Table 72.2. Fitzpatrick classification system of skin types

Type	Skin Characteristics
I, extremely fair or freckled skin	Always burns, never tans
II, fair skin	Usually burns, sometimes tans
III, fair to olive skin	Sometimes burns, always tans
IV, olive to brown skin	Rarely burns, always tans
V, dark brown skin	Very rarely burns, always tans
VI, black skin	Never burns, always tans

- The main contraindication for superficial peels is the presence of active skin disease such as rosacea, seborrheic dermatitis, atopic dermatitis, psoriasis and vitiligo. Rosacea skin is particularly sensitive to peeling. Consideration for a very superficial peel such as salicylic acid should be given for patients with this skin condition due to the peel's anti-inflammatory properties.
- As a general rule, Fitzpatrick skin types I-III (Table 72.2) will tolerate all depths of resurfacing with little risk of pigmentary changes. In contrast, Fitzpatrick skin types IV-VI have a higher risk of hyper- and/or hypopigmentation, and patients should be aware of this possible outcome. Deep peels should not be performed on patients with these skin types. Superficial peels, especially salicylic acid peels, are still safe in even the most highly pigmented skin.

General Considerations

All patients should be adequately prepared for the procedure. This includes counseling and explanation of all the risks and benefits. It is controversial whether or not topical tretinoin should be used nightly preprocedure as it enhances peel penetration. Use of tretinoin is restricted in the postoperative period until complete reepithelialization and inflammation has subsided. Topical hydroquinone is another potential adjuvant therapy to be considered in the pre- and postprocedural periods, especially in darker skinned individuals. In the postoperative period, strict sun precautions and sun avoidance must be insisted upon to minimize pigmentary changes.

The need for intraoperative sedation and analgesia should be considered mainly for deep peels (Table 72.3). The patient's own pain threshold, anxiety level, comorbid medical conditions, and willingness to take on the financial burden of heavier sedation are all factors to be considered.

Superficial peels require only gentle face washing. For medium and deep peels, residual oils, debris and excess stratum corneum must be removed with vigorous cleansing prior to applying a chemical peel solution. Cleansing and degreasing assures uniform penetration of the peeling solution and even results without skip areas. If any residual oil is felt after this, the process should be repeated prior to peeling.

Specific markers help to determine the appropriate endpoint for a given chemical peel. This varies depending on the chemical agent used. For example, **frosting**, or the whitening of skin after the chemical is applied, is an important indicator of how evenly trichloroacetic acid (TCA) been applied. The extent of photodamage, the applicator utilized, and the adequacy and uniformity of defatting all impact the frost achieved. It is not a valuable marker of depth. Table 72.4 reviews the levels of frosting. The following sections will specifically discuss superficial, medium and deep peels and highlight one example in each class.

Table 72.3. Examples of anesthesia for facial resurfacing

Superficial Resurfacing Procedures
- No sedation or analgesia is required

Medium-Depth Peels
Limited degree of perioperative sedation and analgesia, such as:
- Topical amide anesthetics
- Regional blocks of the supraorbital, infraorbital and mental nerves
- Oral diazepam
- Intramuscular meperidine hydrochloride
- Intramuscular hydroxyzine hydrochloride

Deep Resurfacing Procedures
- Nerve blockade, local injections of lidocaine and mild sedatives
- Intravenous or general endotracheal tube anesthesia
- Tumescent anesthesia technique

Table 72.4. Levels of frosting of the skin during chemical peeling

Level I	Erythema and streaky white frosting
Level II	Erythema with a solid white enamel frosting
Level III	Solid white enamel frosting without erythema

Superficial Chemical Peeling

Multiple peels will usually be necessary to achieve optimal results. Superficial peels have the advantage of very little stinging or burning during the procedure with little recovery time afterwards. Multiple peels in some circumstances can give results similar to medium-depth or deep peels when the goal of the procedure is destruction of superficial photodamage. There are very light and light superficial peels. Very light peels slough the stratum corneum and stratum granulosum in a process known as **exfoliation**. Light peels extend deeper, through the entire depth of the epidermis. This stimulates regeneration of fresh, new epithelium. Examples of very light superficial peels include low potency formulations of glycolic acid and other alpha-hydroxy acids, salicylic acid, 10%-20% trichloroacetic acid (TCA) and resorcin. Light superficial peels include 25%-35% TCA, Jessner's solution and 70% glycolic acid.

30%-70% Glycolic Acid

Weekly or biweekly 40%-70% unbuffered glycolic acid is one method of light peeling. Patient's skin develops tolerance to glycolic acid peels. Therefore, a typical regime of six peels would start with a low strength glycolic peel of perhaps 30% applied for three minutes. Erythema and patient level of pain are pertinent factors in determining the appropriate endpoint for glycolic acid peels. The peels are spaced one month apart. At the next visit, the patient would be questioned about length of recovery, pain and possible benefits. Only the rare patient would have significant pain or erythema after such a light peel. Assuming she tolerated the first peel well, the next peel could either be 30% left on the skin longer, such as 5 or 6 minutes or the patient could be graduated up to a 40 or 50% peel. At each visit, the process of evaluating the patient and selecting the correct peel and peeling time would be repeated. Unlike TCA peels or salicylic acid peels which self-neutralize, glycolic acid must be rinsed off with water or neutralized with 5% sodium bicarbonate after 2-4 minutes. However, neutralization with bicarbonate may produce heat and some practitioners avoid this.

Medium-Depth Chemical Peeling

Controlled wounding in medium-depth chemical peels extends through the epidermis and papillary dermis to the upper reticular dermis. The most commonly used medium-depth chemical peels include combination peels such as solid CO_2 + 35% TCA, Jessner's + 35% TCA and 70% glycolic acid + 35% TCA. Acutely, there is epidermal necrosis and papillary dermal edema. Visible healing occurs over the ensuing 1 to 2 weeks. Patients frequently do not feel comfortable returning to work for several days because they may have light scabs or open areas on their face. Over the next several months there is increased collagen production and organization, which correlates clinically with wrinkle reduction. 50% TCA as a single agent is not used as widely given the frequency of scarring and pigmentary changes. The efficacy of 35% TCA in combination with several agents (see Table 72.1) reaches that of 50% TCA alone without the associated complications. The depth of the peel determines how efficacious it will be, which is determined in part by effective degreasing preprocedure as well as how much of each solution is used. Care must be taken to avoid ocular damage when treating the eyelids.

35% TCA

The combination of 35% TCA with glycolic acid, Jessner's solution, or solid CO_2 generates a white frost almost immediately. A uniform level II or III frosting is the desired endpoint. There should be at least a three to four minute break period before retreating any area to guarantee that maximal frosting has occurred. Retreating over frosted areas should be approached with great caution because penetration can be variable in the previously damaged skin. An acute sensation of burning occurs with peel application; this generally subsides once frosting is present.

Deep Chemical Peeling

Deep chemical peeling is a very safe procedure when performed by an experienced physician. Many newly trained physicians have replaced deep peels with skin laser resurfacing as the latter may provide more control over depth of dermal injury. With deep chemical peeling, injury extends to the midreticular dermis with consequent production of new collagen. Two options exist for deep chemical peeling: TCA solutions and phenol-based compounds. As is true for medium-depth peels, >50% TCA peels have associated complications and are used less frequently.

The Baker-Gordon Peel

This phenol-based peel must be freshly prepared and mixed well prior to use. It can be applied under occlusion to increase penetration and extend injury deeper into the midreticular dermis. The chemical agent is sequentially applied with 15 minute intervals between each of six aesthetic units: forehead, perioral region, right cheek, left cheek, nose and periorbital area after complete cleansing and degreasing. Care must be taken to avoid scarring and ectropion formation when treating periorbital skin. If there is ocular involvement, mineral oil should be used to flush the eyes, as water increases the penetration of phenol. Pain relief should be addressed appropriately. Intravenous hydration with lactated Ringer's solution prior to and during the procedure helps promote phenol excretion and prevent toxicity.

Surveillance during the perioperative period must include continuous electrocardiography, pulse oximetry and blood pressure monitoring.

Patient's undergoing deep chemical peels must be completely aware of the significant risk of complications and morbidities including scarring, textural changes and pigmentary disturbances. A key point to mention to patients is that new collagen formation takes months to occur. Therefore, they can expect significant skin tightening even 6 months after their procedure. In addition, phenol is cardiotoxic and is eliminated by the liver and kidneys. For this reason, preoperative evaluation including complete blood count, liver function tests, basic metabolic panel and a baseline electrocardiogram. Patients with known arrhythmias should not undergo phenol peels. Serious consideration should be given for any patients with known hepatic or renal insufficiency prior to the procedure. Appropriate patient selection cannot be over-emphasized.

Pearls and Pitfalls

It is prudent to start with a low strength concentration for peels to ensure the patient is not unusually sensitive to any ingredient. When a practitioner is just starting to use peels, salicylic acid is a good choice, since its penetration is superficial and it self neutralizes. Another key advantage of salicylic acid is that it is lipophilic and therefore penetrates into the hair follicle. For this reason, it may also be the best peel for acne and may help to at least temporarily shrink pores.

Regarding skin preparation, gentle face washing prior to superficial chemical peels is important. Before medium-depth peels, facial skin can be prepped with application of Jessner's solution. Deep peels should be preceded by a medium depth peel such as acetone, Jessner's, or 30% glycolic acid.

During application of any chemical peel, care should be taken to never pass over the patients' eyes and to very cautiously apply solution to the nasal root or bridge in order to avoid dripping and the solution running into the eyes. Don't use the spray applicator for rinsing because the partially neutralized solution can run into the eyes. Be careful with heavily soaked saline gauze around the eyes, as a trickle of saline can collect peeling agent. The first round of neutralization is done with moist gauze, and only then is more heavily soaked cool gauze used.

Determining the correct time to neutralize a glycolic peel is tricky, because the depth of injury relates both to the time on the skin as well as the concentration. The two key factors to follow are skin erythema and patient's pain perception. Tell the patients to alert you if any area develops a 5 out of 10 or higher pain threshold. The other endpoint is erythema. Ideally, you should see a light to medium pink color, almost like a wind burn. Any area that gets deeper red is immediately neutralized. In summary, neutralization times are different for each facial area, and areas that require early neutralization should be noted in the chart.

Many physicians use a fan for reducing the pain of the procedure. There is one exception: don't use the fan during glycolic peels until you make the decision to neutralize because a genuine assessment of the patient's pain threshold must be noted. A blowing fan in close proximity to the patient's face should be utilized once an estimated pain level of 5 out of 10 is reached.

Suggested Reading

1. Brody HJ. Chemical Peeling and Resurfacing. 2nd ed. St. Louis: Mosby, 1997.
2. Drake LA et al. Guidelines of care for chemical peeling. Guidelines/Outcomes Committee: American Academy of Dermatology. J Am Acad Dermatol 1995; 33:497.
3. Glogau RG. Chemical peeling and aging skin. J Geriatr Dermatol 1994; 2(1):30.
4. Mendelsohn JE. Update on chemical peels. Otolaryngol Clin North Am 2002; 35(1):55.

Chapter 73

Fat Injection and Injectable Fillers

Darrin M. Hubert and Louis P. Bucky

Overview

Contour deformities of soft tissues confront the plastic surgeon in both aesthetic and reconstructive endeavors. Atrophy of fat and subcutaneous tissue manifests itself in the characteristics of the aging face. In addition, chronic acne may result in destruction of subcutaneous tissues due to repeated inflammation. Other soft tissue defects include traumatic injuries, secondary cleft deformities, Romberg's disease (hemifacial atrophy), iatrogenic defects such as those caused by liposuction and mid-face hypoplasia. The ideal soft tissue filler should be safe, predictable, easy to use, readily available and have a reasonable duration. At present, there is no consensus as to a perfect single injectable material for soft tissue volume augmentation. Fillers can be divided into those that can produce volume augmentation and those that can correct fine lines.

Fat Grafting

Autologous fat grafting has a long history. Recently it has regained renewed popularity due to advances in technique that have led to more reliable results. Fat represents the ideal facial soft tissue filler for both aesthetic and reconstructive purposes. The advantages are several. The use of autologous fat obviates the concern for an immunogenic response. The donor tissue is often in abundant supply, even when fairly large volumes are required. This safe procedure is technically easy to perform and generally well tolerated. Patients easily understand the concept of volume loss and are receptive to replacement procedures. Lastly, the majority of facial volume loss over time is due to fat atrophy. Thus replacing similar tissues is preferable.

Variable graft survival with long-term unpredictable success represents the major shortcoming of fat grafting. Typically a percentage of the grafted fat will undergo resorption, which sometimes necessitates additional procedures. Some authors advocate over-correction at the initial procedure in anticipation of losing some volume over time. The literature reports widely disparate results with a two-year average of about 50% (range 30%-70%) persistence of the injected fat. These reports, however, are plagued by differences in the methods of fat harvesting, processing and injection.

Technique

Adipose tissue is extremely fragile. Therefore fat must be harvested with minimal trauma. The entire procedure should create only a minimum of physical and chemical disruption, and the fat specimen should have only limited exposure to the air. The specimen is at no point washed with saline or lactated Ringer's solution. Strict adherence to sterile technique must be observed. The majority of the time, small

Practical Plastic Surgery, edited by Zol B. Kryger and Mark Sisco.

volumes of fat may be harvested and injected under local anesthesia. For larger volumes and anxious patients, however, sedation may be required.

Typical harvest sites include the abdominal wall, flanks and hips. A small stab incision is made at the donor site. The fat is obtained with a 10-ml Luer-lock syringe attached to a blunt cannula. Low-pressure suction is applied manually by withdrawing the plunger while passing the two-holed cannula (Byron™) to minimize tissue trauma. Coleman and others popularized centrifugation of fat to facilitate the separation of the specimen into three layers. The authors, however, have modified the process and do not employ centrifugation. The syringe is allowed to stand in a vertical position for a period of minutes. Then the specimen is gently injected onto a Telfa® pad and rolled with the end of a blunt instrument, e.g., an empty scalpel handle. This atraumatic process is continued until the specimen reaches its purified, natural yellow color and achieves a meringue consistency. It is then immediately returned to the 10 ml syringe and subsequently transferred to 1 ml tuberculin syringes for injection with a blunt microinjection cannula (Byron™).

The recipient site is prepared in a sterile fashion, and a percutaneous puncture is made with iris scissors. Small aliquots (< 0.1 ml) of fat are injected in multiple subcutaneous planes. The puncture site may be closed with a simple 6-0 fast-absorbing plain gut suture. Because fat survival depends on the diffusion of nutrients, over-correction of no greater than 10% is preferred. Significant over-correction or traumatic harvest techniques lead to greater resorption, cyst formation and fibrosis, and should therefore be avoided.

Motion tends to diminish fat survival. Hence, static recipient areas usually result in greater graft survival than do dynamic ones. The malar, periorbital and parasymphyseal regions of the face are more reliable in terms of maintenance of volume. Results with the upper nasolabial fold appear to be better than in the lower marionette region, once again due to an absence of inherent motion. Fat injection in the glabellar region can be enhanced with preoperative botox injection in order to diminish motion and improve graft survival. Lip injection is complicated for two reasons. First, persistent motion in this area leads to unpredictable results, and second, fat grafting to the lips is associated with prolonged edema. However, use of a small 20-gauge cannula seems to decrease postoperative edema in this area. The same small cannula is effective in the nasojugal groove region in order to avoid over-correction and palpability. In addition, the thin skin of the lower eyelid region requires fat grafting to be performed deep to the orbicularis oculi muscle.

Other variables that impact fat survival in the face are the age of the patient, presence of subcutaneous scarring, and degree of the deformity. Secondary procedures typically should occur at six-month intervals in order to diminish inflammation and fibrosis. Other untoward sequelae are extremely rare but can include hematoma, infection, necrosis of overlying tissues and erythema. Ultimately fat grafts are living adipose tissue, and they tend to diminish concurrent with the normal aging changes to the face. Fat grafting, when performed utilizing appropriate care in harvest, processing and injection, can be an extremely useful technique for soft tissue augmentation. It is useful in isolation as well as in combination with other surgical procedures.

Other Injectable Fillers

Efforts have been made to develop synthetic fillers for volume augmentation. The advantage of these other fillers lies in the fact that they are essentially available "off the shelf." Radiesse™ (BioForm Medical, San Mateo, CA) represents a suspension of synthetic bioceramic microspheres (25-45 µm diameter) comprised of calcium hydroxylapatite in a gel carrier. Indications approved by the FDA include vocal cord augmentation and filling of maxillofacial defects. Although more common elsewhere, its off-label use for facial soft tissue augmentation is becoming increasingly more common within the United States. It is supplied in single-use, 1.0 ml and 0.3 ml syringes and can easily be injected through a 26-gauge needle. Metabolism is the same as that for endogenous bone fragments. The durability of Radiesse™, therefore, is significant and ranges from several months to years. It works best in regions with significant soft tissue cover, e.g., the malar area, nasolabial folds, marionette lines and the parasymphyseal region. Likewise, it should be avoided under thin tissues such as the eyelids and oral mucosa. The authors have found Radiesse™ to be particularly useful in patients with HIV-associated soft tissue atrophy or fat grafting failures. The most common complication is palpability.

Sculptra™ (Dermik Laboratories, Berwyn, PA) is an injectable suspension of microparticles composed of poly-L-lactic acid, a member of the alpha-hydroxy acid family. It is presently FDA-approved for the treatment of facial lipoatrophy in patients with HIV. Although this synthetic implant is biodegradable, volume effects may still be evident up to two years after injection. For this reason, over-correction is not recommended. However, multiple treatment sessions (3-6) at approximately two-week intervals may be required to achieve optimal effect. No skin test is required. It is supplied as a freeze-dried preparation that is reconstituted in sterile water and 2% lidocaine. Injection involves 0.1 to 0.2 ml aliquots in the deep dermis or subcutaneous planes of the cheek or nasolabial region with a 26-gauge needle. Side effects include bruising, edema and the delayed appearance of subcutaneous papules that are palpable in the region of injection.

Hyaluronic Acid

Other injectable fillers have been developed which are utilized primarily in the treatment of fine lines and wrinkles. Synthetic fillers such as hyaluronic acid are rapidly replacing collagen, long considered the "gold standard" of soft tissue injectable fillers. This is due to three main reasons. First, no skin test is required as is the case with collagen. Second, the longevity of hyaluronic acid is slightly greater than that of collagen. Third, it has an ease of injection similar to that of collagen. Chemically identical across all species, hyaluronic acid consists of a long-chain polysaccharide of repeating disaccharide units of glucuronic acid and N-acetyl glucosamine. Because of this structure, hyaluronic acid avidly binds water, which imparts its significant viscoelastic characteristics. It is ultimately degraded due to hyaluronidase activity although molecular cross-linking slows this degradation process.

Restylane® (Q-Med, Uppsala, Sweden) is a nonanimal stabilized (cross-linked) hyaluronic acid that can be used for volume augmentation. Derived from engineered bacteria, it has an attractive safety profile. Injection typically produces no palpability. However, its lack of longevity (3-6 months) and cost make it a better choice for sensitive areas like the lower eyelids and lips. Restylane® is supplied in prepackaged 0.5 ml and 1.0 ml sterile syringes at a concentration of 20 mg/ml. A 30-gauge needle is provided for injection into the middle dermis, and over-correction is not recommended.

A second hyaluronic acid gel, **Hylaform®** (Inamed, Santa Barbara, CA), is FDA-approved. Derived from rooster combs, it should be avoided in individuals with an allergy to any avian protein. It is indicated for facial volume augmentation in wrinkles and folds, where it is injected in a fashion similar to Restylane into the mid- to deep dermis. The longevity profile, compared to Restylane, has not been determined. However, its cross-linked nature is associated with less swelling, which may be of particular importance in the lips.

Collagen

The most abundant protein in the human body, collagen comprises roughly 30% of its dry weight. It is a triple-helix composed of polymers of primarily three amino acids: glycine, proline and hydroxyproline. The most favorable results with collagen are obtained in correction of fine to medium lines and soft, mobile scars. The main drawbacks of collagen include the potential for a skin reaction. Moreover, the duration of collagen is limited to 3-4 months, and it is removed by an inflammatory process which may result in erythema. Of critical importance is the technique of injection: collagen is injected in small volumes within the dermis. Most of the implanted volume is eventually degraded, and often several injection sessions, or "touch-ups," may be required to achieve long-term volume persistence. This can be prohibitively expensive and time-consuming. Many authors recommend over-correction of lesions whenever possible.

Commercially available collagen has historically been derived from both human (typically of cadaveric origin) and bovine sources. Injectable bovine collagen, FDA-approved since 1981, has been associated with hypersensitivity reactions in 1.3% to 5% of patients. Thus a skin test is recommended before initiation of therapy. Bovine collagen has long been available in two forms, marketed under the names **Zyderm®** (Inamed, Santa Barbara, CA) and **Zyplast®** (Inamed). Both are suspensions of dermal collagen in saline solution with lidocaine. These products undergo selective hydrolysis of the terminal peptides to decrease antigenicity. Despite this, however, hypersensitivity skin testing is positive about 3% of the time. The use of Zyderm® and Zyplast® is rapidly waning.

CosmoDerm® (Inamed) and **CosmoPlast®** (Inamed) are utilized in much the same fashion as Zyderm® and Zyplast®. They are comprised of collagen which is isolated and purified from human dermal fibroblasts grown in culture. As such, no skin test is required. They are dispersed in physiologic saline solution with 0.3% lidocaine. CosmoDerm-1 (35 mg/ml) is injected with a 30-gauge needle into the superficial dermis for shallow scars and fine lines and wrinkles, especially in the perioral region. Over-correction of approximately 1.5 to 2 times is recommended. CosmoDerm-2 is similar to CosmoDerm-1 but, at nearly twice the concentration, it requires less over-correction. CosmoPlast (35 mg/ml) is cross-linked with glutaraldehyde, which increases its longevity. Particularly effective for deeper lines such as the nasolabial folds, it is injected into the mid-to-deep dermis of furrows and scars. These products are supplied in disposable single-use only syringes that may be stored in a refrigerator.

Isolagen® (Isolagen, Inc., Houston, TX) is derived from autologous fibroblasts cultured from a 3-mm retroauricular skin biopsy. A period of approximately six weeks is required to produce roughly 1.0 to 1.5 ml for injection. No skin test is required. Living fibroblasts, the cells that produce collagen, are injected with a 30-gauge needle into the superficial dermis. Although Isolagen has not achieved widespread acceptance, investigational studies continue to evaluate this product.

Pearls and Pitfalls

The key to effective facial filling is to be able to differentiate between fillers that are used for volume augmentation from fillers that are used to obliterate fine lines. Fat grafting is the gold standard for volume augmentation. Careful harvest, purification and injection of fat are required around the face for volume augmentation. It is most useful in areas where there is minimal motion. Over-correction is routinely not successful, and if secondary corrections are required, waiting six months is typically appropriate. In addition, the use of fat around the eye or superficially to fill fine lines is often fraught with persistent lumps and bumps that are displeasing. Radiesse™ has become an effective filler in patients who fail fat; however, it should not be used in the more sensitive areas of the lips and lower lids.

The addition of hyaluronic acid-based fillers has been a significant improvement due to their increased longevity compared to collagen and their excellent safety profile; however, they are truly not cost effective at this point as a structural volume filler in areas such as the nasolabial folds or malar region. Lastly, the key to effective volume augmentation of the face is to differentiate between soft tissue atrophy or deflation versus gravitational changes and subsequent line development. One should not attempt to overcompensate for gravitational changes with volume augmentation.

Suggested Reading

1. Coleman SR. Facial recontouring with lipostructure. Clinics Plast Surg 1997; 24(2):347.
2. Kanchwala SK, Bucky LP. Facial fat grafting: The search for predictable results. Facial Plast Surg 2003; 19(1):137.
3. Billings Jr E, May Jr JW. Historical review and present status of free fat graft autotransplantation in plastic and reconstructive surgery. Plast Reconstr Surg 1989; 83(2):368.
4. Bergeret-Galley C. Comparison of resorbable soft tissue fillers. Aesthetic Surg J 2004; 24(1):33.
5. Narins RS, Brandt FS, Leyden J et al. A randomized, double-blind, multicenter comparison of the efficacy and tolerability of restylane versus zyplast for the correction of nasolabial folds. Dermatol Surg 2003; 29(6):588.

Chapter 74

Cosmetic Uses of Botulinum Toxin

Leonard Lu and Julius Few

Introduction

The use of botulinum toxin type A (Botox®) for facial aesthetic procedures has gained exponential popularity. In fact, more than 2.2 million Botox procedures were performed in 2003. The widespread use of Botox is attributed to its effectiveness in reducing facial rhytids and also to its short and long term safety record. It is important to note that Botox Cosmetic is currently approved by the FDA for the treatment of glabellar lines; however, its use continues to find new applications in dermatology and plastic surgery (for example, for hyperhidrosis).

Mechanism of Action

Botox is a neurotoxin derived from the bacteria, *Clostridium botulinum.* It blocks neuromuscular transmission through a three-step process. First, the toxin binds to presynaptic cholinergic motor nerve terminals. Next, the toxin is internalized into the nerve terminal by endocytosis, where it eventually enters the cytoplasm. Finally, Botox inhibits acetylcholine release by cleaving a cytoplasmic protein. The end result is that the muscle contraction is inhibited. However, Botox's action is not permanent because collateral axonal sprouting establishes new neuromuscular junctions, restoring muscle function.

Storage and Reconstitution

Botox is supplied in a vial with 100 units (U) of dried neurotoxin complex. It can be stored at 2 to 8°C for up to 24 months. The manufacturer recommends reconstitution of the toxin with 2.5 ml of 0.9% saline to yield a final concentration of 4.0 U/0.1 ml. Although the package insert states Botox should be used within 4 hours of reconstitution, studies have shown that at 4°C, the drug can be stored up to 6 weeks without a decrease in efficacy. Although there are no special handling precautions when using Botox, it is important to note that alcohol neutralizes the toxin. Therefore, if the practitioner uses alcohol to prepare the skin, it should be allowed to dry before injection.

Injection Technique

Injections should be performed with a tuberculin or insulin syringe (30 or 32-gauge needle). The top of the vial can be removed to avoid dulling the needle when drawing Botox into the syringe. Topical anesthesia with ice or other agents may be beneficial to decrease pain associated with injections but is not necessary. In general, men require higher doses of Botox because of a greater muscle mass. Thicker skin, such as that found in Asians, may also require higher doses. With the exception of the perioral and periocular areas, injections should be made into the muscle

Practical Plastic Surgery, edited by Zol B. Kryger and Mark Sisco. ©2007 Landes Bioscience.

belly **perpendicular** to the skin. It is useful to use facial animation to mark the areas that require treatment. Injections should be made with the patient reclined to 45°. In addition, the practitioner should prospectively look for superficial vessels in the skin and avoid injecting them in order to decrease the risk for ecchymosis. If a bruise starts to develop, the practitioner should hold 5-10 minutes of pressure to avoid a hematoma, which could lead to migration of the toxin. Although the clinician may gently massage the site of injection at the time of treatment, patients should be told not to massage the area as it may cause diffusion of the drug and result in weakness of unintended muscles. Patients should contract the treated areas as it may increase local uptake of the toxin.

Patient Counseling

Similar to other surgical procedures, treatment should be individualized. Clinicians should set realistic expectations and discuss if off-label use is planned. The effect of Botox may not manifest for 24-72 hours after injection, with optimal results seen at 2 weeks. Results typically last 2-3 months, but clinicians have reported results lasting 6 months or more. In general, assessment and possible retreatment can be done 14 days postinjection. To assess efficacy, photography is essential for guiding preinjection and postinjection treatment plans. The main side effects that a patient may encounter include: pain at the injection site; bruising (avoiding NSAIDs for 10-14 days prior to injection is helpful); and unexpected weakness of muscle groups (for example, eyelid ptosis).

Treatment Areas

FDA Approved Uses

Glabellar Complex/Vertical Frown Lines

The main muscles treated in the glabella include the corrugators and procerus. Fibers from the orbicularis oculi may also contribute to the glabellar complex. These muscles are brow depressors and are responsible for the vertical frown lines. The most common number of injection sites is five (two per corrugator and one to the procerus), and a total Botox starting dose of 20-30 U for women and 30-40 U for men can be used. This dose does not have to be divided equally among injection sites.

Off-Label Uses

Frontalis/Horizontal Forehead Lines

The frontalis elevates the brow and is responsible for horizontal forehead wrinkles. The number of injections varies from four to ten, depending on the severity of the rhytids. Injections are done 2 cm above the brow and should be lateral enough to avoid a "quizzical" eyebrow appearance. A total starting dose of 10-20 U for women and 20-30 U for men is used in equally divided doses. Complete paralysis of the frontalis should be avoided; rather, the goal should be to weaken the muscle. In addition, the forehead should not be treated without the glabella, as this will lead to a potential increase in rhytids due to loss of activity of the antagonist muscles.

Orbicularis Oculi/Crow's Feet

The orbicularis oculi functions for voluntary and involuntary closing of the eyelids. The usual number of injections is two to five per side. The injections should be superficial (to avoid bruising), lateral, and 1 cm from the orbit. The total starting dose is typically 12 U, divided equally per injection site. Similar to the treatment of the frontalis muscle, the goal should be to weaken the muscle rather than cause complete immobility.

Orbicularis Oris/Perioral Rhytids

The orbicularis oris surrounds the mouth and acts as a sphincter to close the lips. Wrinkles of the upper lip are often treated by multiple modalities such as with fillers or resurfacing, but Botox can aid in improving the appearance of the perioral area. In general, four sites are treated—one injection per lip quadrant. If more sites are treated, the injections should be symmetrical. Avoid the midline of the upper lip and the corner of the lips. The total starting dose is 4-10 U, divided equally among each injection. Injections should be started with low doses and repeated as necessary to avoid oral incompetence or an increase in dental show.

Platysma/Platysmal Bands

The platysma acts to depress the lower jaw and pull the lower lips and corners of the mouth down and sideways. Bands are treated with three to five injection sites per band at 1 cm intervals. Injections are done by grasping the band and injecting directly into the belly of the muscle. A total starting dose of 10-30 U for women and 10-40 U for men is used in equally divided aliquots. Treatment can be expected to last 3 to 4 months.

Pearls and Pitfalls

1. Glabellar complex: Assess preinjection brow position and symmetry because treatment can affect eyebrow shape and position. Injections should be directed outside the orbital rim.
2. Frontalis: Stay 2 cm above the brow and start with a low dose to avoid complete forehead immobilization.
3. Orbicularis oculi: Avoid the delicate veins around the eye to prevent bruising, and evaluate lid laxity prior to treatment because excessive laxity increases the risk of an ectropion.
4. Orbicularis oris: Avoid injections too far from the lip margin and avoid using Botox in patients who rely on their lips for their professions (for example, singers).
5. Platysma: Avoid injecting the strap muscles.

Suggested Reading

1. Carruthers J, Fagien S, Matarasso SL et al. Consensus recommendations on the use of botulinum toxin type A in facial aesthetics. Plast Reconstr Surg 2004; 114:1S.
2. Klein AW. Dilution and storage of botulinum toxin. Dermatol Surg 1998; 24:1179.
3. Allergan, Inc. Botox Cosmetic purified neurotoxin complex (package insert).
4. Carruthers JD, Lower NJ, Menter MA et al. Double-blind, placebo-controlled study of the safety and efficacy of botulinum toxin type A for patients with glabellar lines. Plast Reconstr Surg 2003; 112:1089.
5. Fagien S. Botox for the treatment of dynamic and hyperkinetic facial lines and furrows: Adjunctive use in facial aesthetic surgery. Plast Reconstr Surg 1999; 103:701.

Chapter 75

Dermabrasion

Zol B. Kryger

Introduction and Terminology

Dermabrasion has become a term that collectively encompasses microdermabrasion and dermabrasion. Dermaplaning and dermasanding are two additional techniques less commonly used. All these terms differ in both the technique that is used and thickness of the skin that is removed. This chapter deals primarily with dermabrasion.

Microdermabrasion

Microdermabrasion involves the use of a hand-held Dremel-like device with a burr embedded with tiny crystals. It removes microns of skin at a time. It is most suitable for removing the very most superficial layer of the dermis. It is often used for very fine wrinkles and scars. As opposed to traditional dermabrasion, it is quite painless, and can be performed without local anesthesia. It can be repeated multiple times in short intervals, and the recovery time is much shorter than for dermabrasion. It does, however usually require repeated treatments to achieve the desired outcome, and the results are quite variable.

Dermabrasion

Dermabrasion uses a similar hand-held motorized device that has either a wire brush or a diamond particle coated fraise or burr. It requires local anesthetic and involves the removal of the epidermis and most of the outer layers of the dermis. Steady, even pressure and an experienced operator are essential since it is easy to dermabrade too deep. This can result in new scarring.

Dermaplaning

Dermaplaning requires a hand-held dermatome, similar to the type used for harvesting split-thickness skin grafts. The dermatome is used to shave off layers of dermis to a depth that will remove the blemish without causing permanent scarring of the deeper dermis.

Dermasanding

Dermasanding uses sterile sandpaper and is a completely manual technique. It is obviously cheaper than dermabrasion but requires more time and effort to produce results. It also requires less specialized training. It has been shown to be useful for the treatment of some surgical scars.

Indications

The following are the common indications for dermabrasion:

- Unacceptable appearance of a scar
- Scarring from long-standing acne
- Fine wrinkles
- Tattoo removal
- Removal of seborrheic keratoses
- Removal of a number of other benign skin tumors
- Treatment of the rhinophyma of acne rosacea

Relative Contraindications

The following are conditions under which dermabrasion should be postponed:

- History of Accutane® use in the past 12 months due to the risk of hypertrophic scarring in those who have been taking isotretinoins
- Predisposition to keloid scarring
- Active herpes simplex infection at the site of dermabrasion
- Severe immunosupression

Preoperative Considerations

A number of preoperative steps can be taken that will minimize the risk of complications. Many of these deal with decreasing the risk of infection in certain higher risk patients:

- Smoking cessation for at least 2 weeks
- Avoidance of anticlotting medications for at least 10 days
- Face scrubbing the night before and morning of surgery
- Prophylaxis with valacyclovir in those with a history of oral herpes (Valtrex® for 10-14 days)
- Prophylaxis with antistaphlococcus antibiotics in patients with impetigo who have a positive nasal swab culture
- Daily Retin-A® application for two weeks is recommended by some dermatologists

Anesthesia

Some patients will undergo additional procedures requiring general anesthesia. Most, however, will undergo dermabrasion while receiving local anesthesia and intravenous sedation. A common routine is to provide an amnestic the night before surgery for anxious patients. A benzodiazepine such as diazepam can be given the morning of surgery, as well as an antiemetic such as Zofran®. Commonly used intraoperative sedation regimens are fentanyl and midazolam, or fentanyl and propofol. These medications are discussed in greater detail in the chapter on conscious sedation. Local anesthesia is administered, and additional anesthesia can be provided by regional nerve blocks.

Intraoperative Considerations

Some surgeons will pretreat the areas to be abraded with a cryoanesthetic right up to the time of dermabrasion. Allowing the skin to rewarm can result in vasodillation and consequently increased bleeding. In addition, a refrigerant, such as Frigiderm®, can be admininstered for 10 seconds following the cryoanesthesia. This creates a firm surface for dermabrading.

The area to be dermabraded should be under tension, ideally with three-point fixation. This obviously requires the hands of an assistant. The correct hand position is to grasp the dermabrader firmly, with the thumb extended along the shaft. The movement of the hand engine should be perpendicular to the direction of the rotating burr or wire brush. It is preferable to begin dermabrading centrally and move towards the periphery so that the blood will run away from the field in the supine patient.

The most important consideration is the depth of dermabrasion. It is easy to go too deep. In the superficial papillary dermis, small red dots will appear. White parallel lines of collagen can be seen in the reticular dermis. The appearance of sebaceous glands indicates a very deep plane.

Upon completion, gauze soaked in 1% lidocaine with epinephrine should be placed over the abraded area for 5-10 minutes. This will help both postoperative pain and hemostasis. A semipermeable dressing should then be used; however, if the area is small, some surgeons will use only ointment.

Postoperative Considerations

It is important to keep the abraded region moist and clean for the 7-10 days required for complete reepithelialization. Vitamin A&D ointment, Vaseline® or aquaphor® are some of the more commonly used ointments. Any crusty buildup should be removed, and the area washed with soap and water daily.

Many have advocated a short course of postoperative steroids to decrease the swelling, especially when dermabrasion has been performed near the eyes. Kenolog® or a Medrol Dosepak® are often used.

Redness persists for 1-2 weeks, and the skin then becomes more pinkish. This lasts for 2-3 months. Avoidance of strong sun and wind is advised during this period. Judicious use of sunscreen with the highest SPF value is recommended. Swelling will also persist during this 2-3 month postoperative period.

The patients should be advised not to resume Retin-A® use until a month after surgery. As mentioned above, valacyclovir treatment, if it was given preoperatively, should continue for 7-10 days postoperatively. For patients with a history of acne, some surgeons will give tetracycline for two weeks.

Complications

The most common complication from dermabrasion is hypo- or hyperpigmentation. The darker the skin, the greater the risk of developing permanent hypopigmentation. Hyperpigmentation is almost always reversible with sun avoidance and hydroquinone treatment.

Postoperative infections are unlikely but should be treated immediately when suspected. Valtrex® is used for a suspected herpes outbreak to prevent permanent scarring. Contact dermatitis is not unusual and can be caused by antibacterial ointments. Treatment consists of topical steroids.

Enlarged skin pores and milia formation can occur in the dermabraded area for several weeks. These are usually self-limited. Acne flare ups can also occur in acne-prone individuals. This too is temporary and will usually not cause new scarring.

Worsening of the dermabraded scar (e.g., hypertrophic scarring) or even new scarring is the most dreaded complications. Scarring is preceded by persistent erythema. If suspected early on during the hyperemia phase, topical steroids can be used. Intralesional Kenalog is useful once it becomes apparent that a scar is forming. Silicone sheeting has shown some promise in treating hypertrophic scars and can be tried in this setting.

Pearls and Pitfalls

- When applying traction on the skin to create a planar surface for dermabrading, avoid using gauze since it easily becomes entangled in the dermabrasion instrument.
- If the patient is under sedation and not general anesthesia, either the surgeon or the assistant should brace the patient's face in case of sudden movement.
- It is essential to continuously watch the dermabraded skin for the transition from flaking epidermis to punctate bleeding indicating entry into the upper dermis. From this point, it is easy to go too deep. Never keep the dermabrader in any one spot, but keep it moving continuously.

Suggested Reading

1. Coimbra M, Rohrich RJ, Chao J et al. A prospective controlled assessment of microdermabrasion for damaged skin and fine rhytides. Plast Reconstr Surg 2004; 113(5):1438.
2. Fulton Jr JE, Rahimi AD, Mansoor S et al. The treatment of hypopigmentation after skin resurfacing. Dermatol Surg 2004; 30(1):95.
3. Harmon CB. Dermabrasion. Dermatol Clin 2001; 19(3):439.
4. Hirsch RJ, Dayan SH, Shah AR. Superficial skin resurfacing. Facial Plast Surg Clin North Am 2004; 12(3):311.
5. Koch RJ, Hanasono MM. Microdermabrasion. Facial Plast Surg Clin North Am 2001; 9(3):377.
6. Poulos E, Taylor C, Solish N. Effectiveness of dermasanding (manual dermabrasion) on the appearance of surgical scars: A prospective, randomized, blinded study. J Am Academy Dermatol 2003; 48(6):897.
7. Shpall R, Beddingfield IIIrd FC, Watson D et al. Microdermabrasion: A review. Facial Plast Surg 2004; 20(1):47.
8. Szachowicz EH. Microepidermabrasion: An adjunct to medical skin care. Otolaryng Clin North Am 2002; (35)1:135.

Chapter 76

Hair Restoration

Anandev Gurjala

Introduction

Hair restoration is based on the concept of "donor dominance," in which hair from a hair-bearing area is transferred through a variety of techniques to an area of alopecia or thinning hair. The average scalp contains 110,000-150,000 hairs. Each hair's growth is influenced by age, weather, health, and genetic factors. Hair typically grows at a rate of 0.35 mm per day or roughly 5 inches per year. Approximately 100 hairs are lost per day as part of the normal hair growth cycle, in a resting period termed the telogen phase. The same number of hairs enter a growth phase each day, termed the anagen phase.

Permanent hair loss is thought to be due to, in large part, the testosterone metabolite, dihydrotestosterone (DHT). DHT acts to "turn off" genetically sensitive follicles. These follicles are located in predictable patterns, usually the frontal or frontoparietal scalp, and this hair loss is termed **male pattern baldness**. The follicles in the occipital and parietal regions of the scalp serve as the donor areas for restoration surgery. A balding person has on average between 5000 and 6000 follicles available for hair donation, having lost up to 30,000 follicles.

Female hair loss was classified by Ludwig as Grade I-III from least to most severe. The **Norwood classification** is often used to describe the severity of male hair loss:

Grade I	indistinguishable hair loss
Grade II	slight temporal hair line regression
Grade III	more prominent temporal regression
Grade III *vertex*	stage III combined with slight vertex hair loss
Grade IV	frontotemporal regression and prominent vertex hair loss
Grade V	marked frontotemporal regression with vertex hair loss
Grade VI	almost complete frontal-vertex hair loss
Grade VII	complete frontal-vertex hair loss

Preoperative Considerations

Three principal strategies have evolved for surgical hair restoration: follicular grafting, scalp reduction, and flap rotation. A thorough evaluation of the patient's hair pattern as well as his expectations and desires is essential, as these will determine the treatment method(s). For all three approaches, especially grafting, finer hair and less color contrast between the scalp and the hair allow for a more undetectable result. Prior to surgery patients may be asked to shampoo their hair with an antimicrobial shampoo (e.g., 4% chlorhexidine gluconate, Betadine shampoo, 3% chloroxylenol, or pHisoHex). Table 76.1 lists the indications, advantages and disadvantages of these three procedures.

Practical Plastic Surgery, edited by Zol B. Kryger and Mark Sisco. ©2007 Landes Bioscience.

Table 76.1. The three main surgical approaches for treating hair loss

Technique	Advantages	Disadvantages	Patient Characteristics
Follicle grafting	Most commonly used for fronto-temporal restoration Ideal for providing coverage but not hair density	Several sessions may be required to reach adequate density May produce a "pluggy" appearance in significant crown alopecia Can take 6-12 months to see incipient growth	Frontotemporal alopecia Satisfied with gaining hair coverage but not necessarily hair density Fears more extensive surgery Has adequate occipital fringe as a donor site
Alopecia reduction	Achieves immediate hair density after several sessions	Added morbidity compared to grafting May be difficult to conceal the scars	Crown or vertex alopecia (<12 cm) with stable hairline (age >40) Good scalp mobility Values hair density not just hair coverage
Flap rotation	Achieves immediate hair density in three sessions	Added morbidity compared to grafting	Frontotemporal alopecia Stable hairline not as important Values hair density not just hair coverage Nonsmoker

76

Intraoperative Considerations

Grafting

Anesthesia for hair grafting is achieved using IV sedation (e.g., midazolam) combined with local ring block of the recipient and donor areas (1% lidocaine with 1:100,000 epinephrine). Creation of the new anterior hairline is key; generally the central forelock should be no less than 10 cm from the orbital rim (or 8.5-10 cm above the root of the nose). The frontotemporal recessions should be no less than 12 cm lateral to the orbital rim. Following injection of either tumescent solution or saline, two rows of donor hair are harvested. A Tori-style scalpel with three No. 10 blades spaced 2 mm apart is directed parallel to the follicles at an upward angle of 20-30°. Donor areas can typically consist of two sites, one inferior to the occipital area and the other superior (and contralateral) to the occipital, parietal or temporal areas. Hair of different textures and colors is obtained in this way. During subsequent sessions, more strips can be taken from the analogous contralateral areas. Donor sites may be closed with running, locking 2-0 polypropylene sutures. Micrografts (1-2 hairs) and minigrafts (3-7 hairs) are then dissected from the donor strips. To create the most natural effect, micrografts are placed most anteriorly along the neo-hairline and micrografts behind them providing more density and volume posteriorly. Slits for the micrografts are created using a no. 15c Bard-Parker blade or a 16 gauge needle; holes for the minigrafts are created by a 1-2 mm punch or a 1.75 mm rotating hand drill. Atraumatic handling of the grafts and constant moisture are essential to maintaining graft viability.

Recreation of frontal forelocks on average will utilize 4000-4500 hairs, achieved in one session of 600-700 grafts or a single 1250 graft session. Care should be taken to transplant bald as well as thinning areas to prevent chasing an enlarging area of alopecia; 90-95% graft take can be expected.

Scalp Reduction

Following sedation, local anesthetic is administered in a ring block fashion with a 10 minute delay before incision to ensure adequate hemostasis. A “Mercedes” or inverted-Y incision is made at the vertex down to the galea and dissection is carried out laterally and posteriorly in the subgaleal plane. About 2 cm of midline advancement is possible for the parietal flaps and 1-2 cm of anterior advancement is available for the occipital flap. Following resection of appropriate bald tissue, a silastic band with titanium hooks running along each of its long sides is placed under tension beneath the wound edges of the skin flaps. Use of this extender device prevents “stretch back” of the skin during healing by absorbing tension along the suture line.

The above procedure achieves about 2-3 cm of scalp reduction, with successful treatment of 12 cm of vertex alopecia requiring about 4 sessions each separated by 1 month. Adequate treatment of greater than 13 cm of vertex alopecia is not usually possible. For the final reduction procedure, closure is performed using multiple Z-plasties to disguise the scar. At that point, hair direction can also be reoriented to create the crown-vertex “whorl.”

Scalp Rotation

The twice-delayed Juri flap has been found by some authors to be the ideal method for scalp rotation coverage of frontotemporal alopecia. The new frontal hairline is designed using the aforementioned guidelines. The Juri flap is designed as a single flap intended to arc the entire length of the new hairline and is based anteriorly off the posterior branch of the superficial temporal artery (identified by Doppler) and posteriorly off the occipital and postauricular arteries. The base of the flap is made 4 cm superior to the helical crus and extends anteriorly with a width of roughly 4 cm at an angle of 30-40° from the horizontal. The first stage of the procedure is a delay, in which the distal third of the flap is raised and then replaced with staple closure. In the second stage performed one week later, the proximal portion of the flap is delayed. After three weeks, the entire flap is elevated, the bald scalp is excised, and the flap is rotated into position to create the new hairline. The flap may be sutured with 6-0 polypropylene sutures anteriorly and 4-0 sutures posteriorly. The donor site is closed following considerable subgaleal undermining.

After completion of these stages, minor corrections including triangle reduction of the kink produced anteriorly at the base of the flap (although this often settles down on its own and does not require correction), and grafting to refine the anterior hairline at the incision.

Postoperative Considerations

Following surgery, no dressing is necessary and patients can begin shampooing their hair on the first postoperative day. Complications are not common and are usually due to poor surgical design. Misuse of donor site hair resulting in an inadequate supply for complete hair restoration is the most frequent complication and is best avoided by careful planning. Infection, bleeding, and graft failure occurs in fewer than 1% of cases. Epidermal inclusion cysts (1-2 mm pustules) may form in 2-5% of hair grafting cases but are easily treated with warm compresses and mechanical unroofing of the cysts. Hairs adjacent to transplants may also enter a telogen phase in 5-10% of cases, although this phase should last no more than 3 months for the majority of these hairs. In scalp reduction, the most common complication is a "slot deformity" scar centrally caused by stretch of the scar and telogen of the hair flanking the scar. This tendency can be minimized by use of the extender device described above. Scalp rotation methods are subject to distal tip flap necrosis caused by the conventional errors of excessive tension or kinking of the flap base. Other risk factors for distal necrosis are overly aggressive cautery and excessive galeal dissection. Early debridement of compromised areas of the flap may be postponed since hair that has fallen out may be in the telogen phase and regrow within 3-6 months. Micrografting may also be used to selectively treat areas that have healed by secondary intention.

Pearls and Pitfalls

Tips on micro- and minigrafting:

- Achieving a natural appearance when transplanting coarse black hair can be quite challenging. Even more so if the recipient site is a shiny, oily surface. A random distribution of implants should be used, and the temple recessions should be maintained.
- When harvesting a donor ellipse of skin, bevel the blade at an angle to avoid damaging the hair follicles.
- The epidermis should be preserved on harvested grafts.
- Prior to implantation, infiltration of the recipient scalp should be performed with a generous amount of tumescent solution. It should have a ballooned, marbled appearance prior to grafting.
- After a slit is made with a microsurgical blade, inserting the graft prior to removal of the blade can help with it "sticking" into place.

Suggested Reading

1. Epstein JS. Follicular-unit hair grafting: State-of-the-art surgical technique. Arch Facial Plast Surg 2003; 5:43.
2. Juri J. Use of parieto-occipital flaps in the surgical treatment of baldness. Plast Reconstr Surg 1975; 55:456.
3. Lam SM, Hempstead BR, Williams EF. A philosophy and strategy for surgical hair restoration: A 10-year experience. Dermatol Surg 2002; 28:11.
4. Martinick JH. The latest developments in surgical hair restoration. Facial Plast Surg Clin N Am 2004; 12:249.
5. Unger RH, Unger WP. What's new in hair transplants? Skin Therapy Letter 2003; 8(1):5.
6. Uebel CO. The use of micrograft and mingraft megasessions in hair transplantation. In: Nahai F, ed. The art of aesthetic surgery. St. Louis: Quality Medical Publishing, Inc., 2005.
7. Vogel JE. Advances in hair restoration surgery. Plast Reconst Surg 1997; 100(7):1875.

Chapter 77

Anatomy of the Hand

Zol B. Kryger

Introduction

The hand is a magnificently designed, complex organ with a large number of nerve endings, muscles, ligaments, bones and joints. These various structures function in a highly orchestrated manner, producing an organ capable of intricate motion and sensibility. This chapter will cover the anatomy of the fingers, hand and wrist, grouping structures based on their volar, dorsal or intrinsic anatomical location. Relevant clinical correlates will be given throughout. Much of the anatomy presented in this chapter is discussed elsewhere in the Hand Section of this text.

Intrinsic Anatomy

Bones and Ligaments of the Wrist

The wrist is composed of the distal radius and ulnar as well as the two rows of carpal bones (Fig. 77.1). The distal radius and ulna are joined by an interosseous membrane which is flexible enough to allow rotation of the radius around the ulna in pronation and supination. Pronator and supinator muscles course between the two bones, and the long flexor and extensor muscles of the wrist and fingers originate from these bones in the forearm. The radius articulates with the lunate and scaphoid at the lunate fossa and the scaphoid fossa. Ligaments join the scaphoid and lunate to the radius, and these articulations can also be the source of arthritic pain. The distal ulna articulates with the triquetrum through the triangular fibrocartilage complex (TFCC). The radius and ulna meet at the sigmoid notch and are joined by the TFCC forming the distal-radial-ulnar-joint (DRUJ), a frequent site of injury and arthritis. The ulna styloid lies most distally and laterally and is frequently involved in fractures of the distal radius. It is easily palpated on the dorsal surface of the wrist.

The proximal carpal row (from radial to ulnar) consists of the scaphoid, lunate, triquetral and pisiform. It has some motion about the radius, with the lunate serving as the fulcrum. The distal row consists of the trapezium, which articulates with the first metacarpal, trapezoid, capitate and hamate. The distal carpal row has very little motion and moves as a unit. The capitate is the fulcrum for this group of bones.

Bones and Ligaments of the Hand and Fingers

The fingers are comprised of three bones, the proximal, middle and distal phalanges. Each phalanx is further subdivided into a shaft, neck and head. The bones articulate at the proximal interphalangeal (PIP) and the distal interphalangeal (DIP) joints. The proximal phalanx articulates with the metacarpal bone at the metacarpophalangeal (MP) joint. There are five metacarpal bones that articulate proximally with the carpal bones at the carpometacarpal (CMC) joints. The first CMC joint (thumb) is very mobile. The second and third CMC joints have almost no mobility. The fourth and

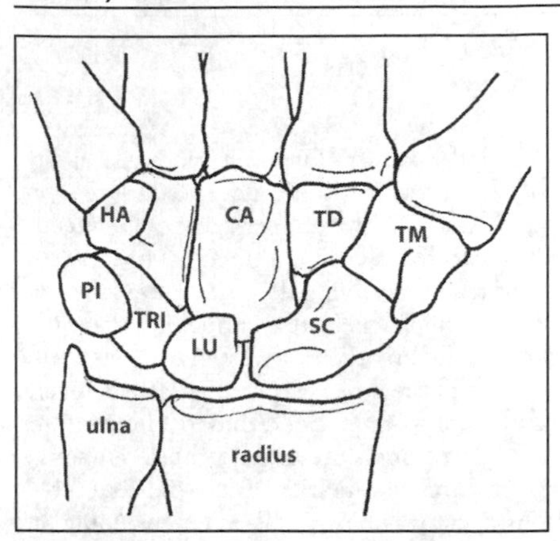

Figure 77.1. The carpal bones. HA, hamate. CA, capitate. TD, trapezoid. TM, trapezium. PI, pisiform. TRI, triquetrum. LU, lunate. SC, scaphoid.

fifth CMC joints have about 50° of mobility to aid in the power grip. The finger metacarpals are interconnected through the deep transverse metacarpal ligament. Lateral stability of the fingers is largely dependent on the radial and ulnar collateral ligaments located at the base of each digit. Rupture of a collateral ligament will result in instability and excess laxity on the side of injury. On the volar surface of these joints is the volar plate to which the collateral ligaments attach. During MP joint dislocations, the volar plate can slip into the joint space and interfere with anatomical reduction.

The thumb has only two phalanges, and thus only a single interphalangeal (IP) joint. The superior motion of the thumb over the digits is largely due to the extremely mobile CMC joint of the thumb. Unfortunately, this is also a common site for osteoarthritis to develop.

Intrinsic Hand Musculature

The intrinsic tendons are the dorsal interosseus muscles, the volar interossei, and the lumbricals. The four dorsal interossei insert onto the proximal phalanges, and abduct the fingers and weakly flex the proximal phalanx. The three volar interossei do not attach to bone but rather insert onto the lateral bands that unite with the lateral slips of the extensor tendons. They adduct the fingers and flex the PIP joint. The lumbricals arise from the tendon of flexor digitorum profundus (FDP) on the palmer side and insert onto the radial lateral band of each finger. They primarily extend the IP joints, as well as functioning as flexors of the MP joints. The intrinsic muscles are innervated primarily by the ulnar nerve, except for the first and second lumbricals, which are innervated by the median nerve. The intricate extensor, flexor and intrinsic systems are interconnected, aligned and stabilized with a system of ligaments found in each finger. The triangular, transverse retinacular and oblique retinacular ligaments perform a variety of complex actions. The pathology of these ligaments is addressed in the chapter on Dupuytren's disease.

Compartments of the Hand

There are ten compartments in the hand: four dorsal interosseous, three volar interosseous, one thenar, one hypothenar, and one adductor pollicis compartment. In the fingers, the ligaments (Cleland's, Grayson's, and transverse retinacular) can also segregate the digit into compartments, resulting in a compartment syndrome of the neurovascular bundle.

Volar (Palmar) Anatomy

Wrist

The most superficial tendon of the wrist is the palmaris longus. It is located roughly in the midline and is absent in 10-15% of patients. It is commonly used for tendon grafting procedures. Just radial to this tendon lies the palmar sensory cutaneous branches (PSCB) of the median nerve. It branches off 5-7 cm proximal to the wrist crease and terminates in the subcutaneous tissues overlying the thenar musculature.

Deep to these structures lies the carpal tunnel. The carpal tunnel is a space defined by the concave arch of the carpus enclosed by the transverse carpal ligament (TCL). The scaphoid, trapezium, and sheath of the flexor carpi radialis (FCR) make up its radial margin. The ulnar boundary consists of the triquetrum, the hook of the hamate, and pisiform. Ten structures course through the carpal tunnel. These include the median nerve, four flexor digitorum superficialis (FDS) and four FDP tendons, all ensheathed by the ulnar bursa, and the flexor pollicis longus, found at the radial side of the canal and surrounded by the radial bursa. The flexor carpi radialis tendon (FCR) travels through its own separate osteofibrous tunnel. The TCL itself attaches medially to the pisiform and laterally to the tuberosity of the scaphoid bone. The TCL can be severely thickened in carpal tunnel syndrome.

The median nerve is the most superficial structure within the carpal tunnel and is covered by a layer of cellulo-adipose tissue. The nerve lies directly under the TCL in the radiopalmar portion of the canal. The superficialis and profundus tendons of the index finger lie immediately dorsal to the median nerve. The nerve divides distally into five sensory branches and the recurrent motor branch. In 80% of cases, the motor branch arises from the radiopalmar region of the median nerve. In the remainder of patients the origin is in the central location, and a small percentage take-off from the ulnar aspect. Ten percent of patients have multiple motor branches. A sensory communicating branch between the median and ulnar nerves in the palm is present in 80% of patients, with roughly half lying within millimeters of the TCL. Injuries to this nerve have been reported during carpal tunnel release, leading to paresthesias or dysesthesias in the ulnar digits.

Guyon's canal is located on the ulnar aspect of the wrist and contains the ulnar artery and nerve. Its floor is the TCL and lies just radial to the pisiform.

Palm and Fingers

The skin of the palm is thick and tough due to the stratum corneum. The dermis is actually of similar thickness to the skin of the dorsum of the hand. The palmar skin is firmly anchored to the palmar fascia through a network of vertical fibers. The palmar fascia is also anchored to the underlying metacarpals. Thus edema fluid cannot easily collect in the palm; however an abscess (collar-button abscess) can develop in the tight space of the palm. The palmar fascia extends into the fingers, and a number of the ligaments of the fingers attach to this fascia. In Dupuytren's disease, this fascia can become thickened and diseased.

The flexor tendons course through the palm en route to the fingers. Superficial to the tendons lies the superficial palmar arch. This arch is the terminating anastomosis between the ulnar artery and the palmar branch of the radial artery. The digital arteries arise from the superficial palmar arch. Deep to the flexor tendons is the deep palmar arch. It too is a terminal anastomosis between the radial and ulnar circulation. The volar metacarpal arteries originate from the deep arch. The rich anastomosis between the radial and ulnar arteries accounts for expendability of the radial artery (e.g., for CABG or free flap) in close to 100% of patients without any long-term sequelae in the hand and fingers.

The tendon of FDS inserts on the volar aspect of the middle phalanx, and the FDP tendon inserts on the volar aspect of the distal phalanx. FDS flexes the PIP joint and FDP flexes primarily the DIP joint. In the digits, the flexor tendons travel in synovial-lined tunnels called flexor tendon sheaths. The sheaths are anchored to the bones by a series of five annular pulleys, numbered A1-A5 from proximal to distal. The odd numbered pulleys are located over the joints; the even pulleys lie over the bones. There are three thin cruciate pulleys, numbered C1-C3, that maintain tendon motion and collapse during flexion. The palmer aponeurosis lies proximal to the A1 pulley and is often referred to as the A0 pulley. It acts in unison with the first two annular pulleys. Proximal to the entrance into the digital sheath (A1 pulley), the FDS tendon lies palmer to the FDP tendon. At this point, the FDS tendon divides and becomes deep to the FDP tendon. The two portions reunite at Camper's chiasma and go on to attach to the middle phalanx. The FDP tendon, after passing through the FDS bifurcation, attaches to the distal phalanx.

Flexor pollicis longus (FPL) is the primary flexor of the thumb. It is the most radial structure in the carpal tunnel. It travels in its own fibrous sheath in the palm and inserts into the base of the distal phalanx of the thumb. The thumb, unlike the fingers, has two annular pulleys, A1 and A2, located over the MP and IP joints, respectively. Lying between them is an oblique pulley that is the most important of these three pulleys.

Fingertip

The fingertip is the end organ for touch, enabling the hand to relay the shape, temperature, and texture of an object. The skin covering the pulp of the finger is very durable and has a thick epidermis with deep papillary ridges. The glabrous skin of the fingertip is well-suited for pinch and grasp functions. Its volar surface consists of a fatty pulp covered by highly innervated skin. The skin of the fingertip is firmly anchored to the underlying terminal phalanx by multiple fibrous septa that traverse the fatty pulp. Hence an infection that develops within the pulp can result in a closed space infection, or felon.

Dorsal Anatomy

Wrist

The extrinsic muscles that extend the hand and fingers enter the wrist through six synovial-lined, dorsal compartments covered by the extensor retinaculum. At the wrist, the tendons are surrounded by a sheath, but this sheath is not present in the hand and fingers. The six compartments are numbered 1-6 from radial to ulnar:

1. Abductor pollicis longus (APL), extensor pollicis brevis (EPB)
2. Extensor carpi radialis brevis (ECRB), extensor carpi radialis longus (ECRL)
3. Extensor pollicis longus (EPL)
4. Extensor digitorum communis (EDC), extensor indices proprius (EIP)
5. Extensor digiti minimi (EDM)
6. Extensor carpi ulnaris (ECU)

The first compartment is commonly involved in stenosing tenosynovitis, termed de Quervain's disease. The second compartment contains the radial extensors of the wrist, and this compartment is located beneath the anatomic snuffbox. The hollow of the snuffbox (so named because it was a common site for the placement of snuff) is easily created by extending and abducting the thumb. The radial artery passes through the snuffbox, and the scaphoid bone is deep to it. Therefore, tenderness in the snuffbox can be seen in scaphoid fractures. The second and third compartments are separated by Lister's tubercle.

Hand and Fingers

The four fingers are extended by EDC; however the communis tendon to the little finger is present only 50% of the time. The index and little fingers also have independent extensor muscles—EIP and EDM, respectively. These tendons usually lie ulnar and deep to the communis tendons to these two fingers.

The EDC tendons are joined proximally to the MP joints by the juncturae tendinum. They are almost always present between the EDC of the middle, ring and little fingers. Thus, lacerations proximal to the juncturae may not impair digit extension due to the connection to the adjacent digits. The tendons inserts proximally into the MP joint volar plate through attachments known as the sagittal bands. Distal to the MP joint, the extensor tendons divide into one central and two lateral slips. The central slip inserts into the middle phalanx and extends the PIP joint. The lateral slips reunite distally and attach to the distal phalanx, extending the DIP joint.

The thumb is extended by three tendons: the first metacarpal by APL, the proximal phalanx by EPB, and the distal phalanx by EPL. However, the MP and IP joints of the thumb can both be extended by EPL due to the attachments of the dorsal apparatus. It is worth mentioning that the IP joint of the thumb is extended by the combined actions of all three major nerves: the radial nerve (EPL), the median nerve (thenar muscles) and the ulnar nerve (adductor pollicis).

Deep to the extensor tendons and proximal to the metacarpals lies the dorsal carpal arch. This is the dorsal anastomosis between the radial and ulnar circulation. The dorsal metacarpal arteries originate from this arch.

Fingertip

The fingernail protects the fingertip and has a major role in tactile sensation and fine motor skills. The nail complex, or perionychium, includes the nail plate, the nail bed, and the surrounding skin on the dorsum of the fingertip (paronychium). The fingernail is a plate of flattened cells layered together and adherent to one another. The nail bed lies immediately deep to the fingernail. The nail bed is composed of the germinal matrix, the sterile matrix, and the roof of the nail fold. The germinal matrix, which produces over 90% of nail volume, extends from the proximal nail fold to the distal end of the lunula. The lunula represents the transition zone of the proximal germinal matrix and distal sterile matrix of the nail bed. The sterile matrix (ventral nail) contributes additional substance largely responsible for nail adherence. The roof of the nail fold (dorsal nail), which includes the germinal matrix, is responsible for the smooth, shiny surface of the nail plate. The hyponychium is the area immediately below the fingernail at its cut edge which serves as a barrier to subungual infection, and also marks the terminal extension of bone support for the nail bed. The eponychium is the skin covering the dorsal roof of the nail fold. The paronychium is the skin at the nail margin, folded over its medial and lateral edges.

Innervation of the Hand and Fingers

The median, ulnar and radial nerves are the primary nerves of the hand and fingers. The first two have both motor and sensory fibers, whereas the radial nerve provides only sensory fibers to the hand. Its motor branches terminate in the arm and forearm.

Median Nerve

The median nerve travels in the forearm between the muscle bellies of FDP and FDS and provides motor input to most of the flexors of the forearm. Just proximal to the wrist, it gives off the palmar cutaneous branch which supplies sensation to the thenar region. At the wrist it enters the carpal tunnel, where it is the most superficial of the structures that traverse this tunnel as described above. It then gives off the motor branch which innervates the radial side of the thenar muscles: opponens pollicis, abductor pollicis brevis, the superficial part of flexor pollicis brevis, as well as the two radial lumbricals. Finally, it divides into sensory branches whose territory includes the palmar surface of the thumb, index, middle, and radial side of the ring fingers, and the radial side of the palm (palmar sensory branch). On the dorsal surface, it sends sensory branches to the distal third of the above-mentioned fingers.

The most consistent sign of median nerve injury is loss of skin sensibility on the palmer surface of the first three digits and loss of thenar opponens function.

Ulnar Nerve

The ulnar nerve travels in the forearm ulnar to the FDP muscle belly. It gives off its palmar sensory branch proximal to the wrist. At the wrist it travels in Guyon's canal, after which it begins to branch into many motor and sensory branches. It innervates all the intrinsic hand muscles except those mentioned above that are innervated by the median nerve. The last muscle innervated by the ulnar nerve is the first dorsal interosseus nerve. Its sensory territory includes both the palmar and dorsal sides of the little finger and the ulnar side of the ring finger.

A common sensory sign of ulnar nerve injury is the loss of sensibility to the small and ulnar side of the ring fingers. Motor signs include FCU paralysis, interosseous and thumb adduction loss with a weak "key pinch," and FDP paralysis of the small and ring fingers. A long-standing injury can present with the classic "claw hand" deformity.

Radial Nerve

The radial nerve provides all the motor innervation of the extensor muscles of the forearm. Only the superficial branch of the radial nerve reaches the hand. This purely sensory nerve travels over the radial side of the wrist. It then branches into the dorsal digital nerves that supply the skin on the dorsum of the thumb, index, middle, and radial side the ring fingers (with the exception of the distal third of each which is supplied by the median nerve).

Signs of injury include loss of sensibility on the dorsum of the hand and in the first web space. Patients will be unable to extend their fingers.

Pearls and Pitfalls

A resource that is highly recommended for understanding hand anatomy and anatomic relationships is *The Interactive Hand* CD-ROM (McGrouther DA, Colditz JC, Harris JM, Eds.) published by Primal Pictures, Inc. 2002. The upper extremity can be rotated and displayed from various angles, and it can be viewed layer by layer from bone to skin.

Suggested Reading

1. Bogumill GP. Functional anatomy of the flexor tendon system of the hand. Hand Surg 2002; 7(1):33-46.
2. Furnas DW. Anatomy of the digital flexors: Key to the flexor compartment of the wrist. Plast Reconstr Surg 1965; 36(3):315-9.
3. In: Green DP, Hotchkiss RN, Pederson WC, Wolfe SW, eds. Green's Operative Hand Surgery. 5th ed. New York: Churchill Livingstone, 2005.
4. Lister's The Hand. In: Smith P, ed. Diagnosis and Indications. 4th ed. New York: Churchill Livingstone, 2002.
5. Rockwell WB, Butler PN, Byrne BA. Extensor tendon: Anatomy, injury, and reconstruction. Plast Reconstr Surg 2000; 106(7):1592-603.

Chapter 78

Radiographic Findings

Zol B. Kryger and Avanti Ambekar

Introduction

Radiologic diagnosis of bony hand injury and deformity is often successfully achieved with conventional radiographs, despite advances in cross-sectional imaging. At least two orthogonal views should be performed routinely. Special views, such as the scaphoid view, can be helpful in selected cases. The images shown in this chapter were obtained and reprinted with permission from the website http://www.gentili.net.

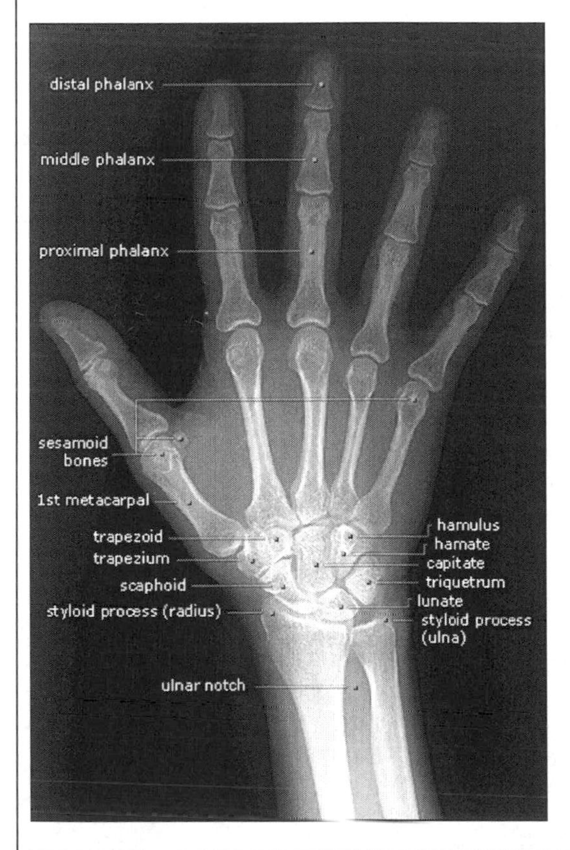

Figure 78.1. P-A view of a normal hand radiograph.

A Systematic Approach

An organized evaluation is required for all image interpretation. The patient's name, date of exam, correct body part, and laterality (right or left side) should be verified prior to analyzing the image. The exam should be assessed for quality and completeness. Attention can then be turned to film interpretation. The bones, joint spaces, and soft tissues should be sequentially inspected. Each feature should be evaluated in a systematic fashion (e.g., from proximal to distal). All abnormalities should be confirmed on a second view.

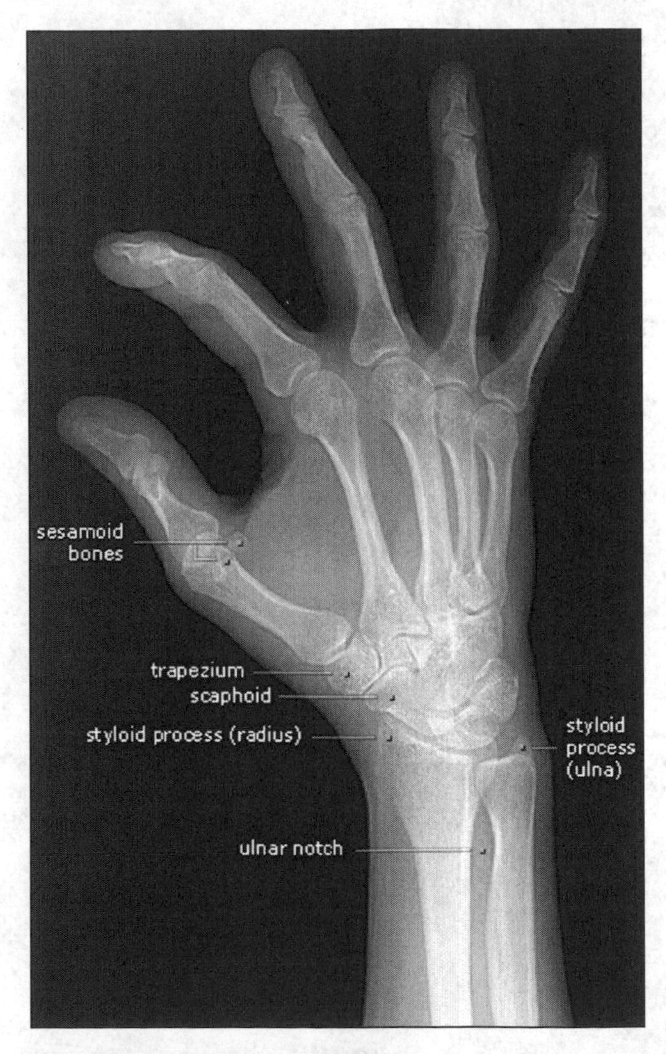

Figure 78.2. Oblique view of a normal hand radiograph.

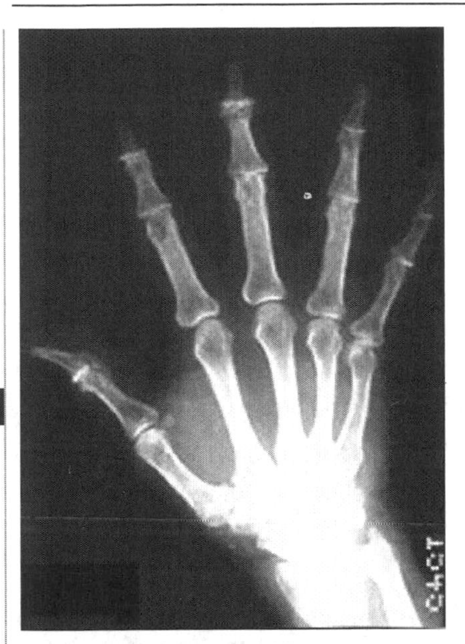

Figure 78.3. Osteoarthritis of the hand, with joint space narrowing and adjacent bony sclerosis. The DIP joints and 1st CMC joint (thumb) are characteristically involved.

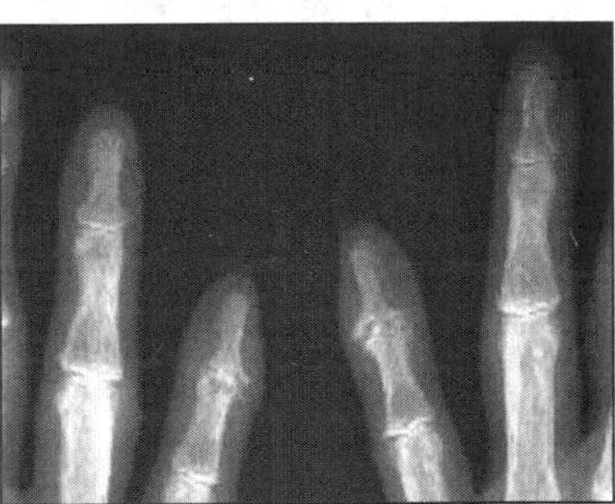

Figure 78.4. Osteoarthritis of the fingers. Note greater involvement of the DIP joints compared to the IP joints.

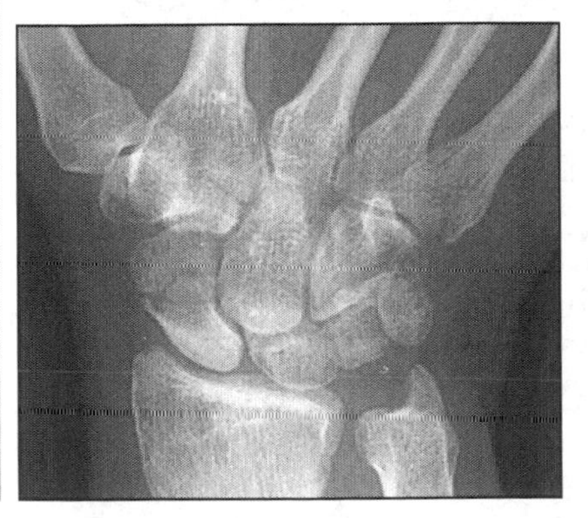

Figure 78.5. Fracture of the scaphoid bone. The scaphoid is the most commonly fractured carpal bone, accounting for 80% of all carpal fractures. It is often the result of a fall on an outstretched hand. Scaphoid fractures are at high risk for nonunion and avascular necrosis.

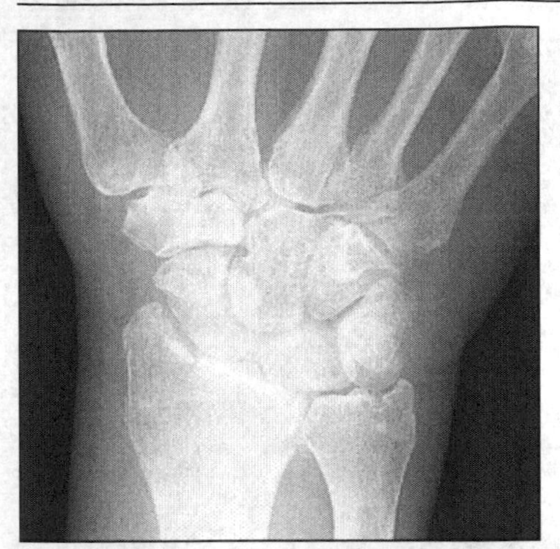

Figure 78.6. Fracture of the radial styloid, also termed a chauffeur's fracture. A scapholunate ligament tear often accompanies this fracture.

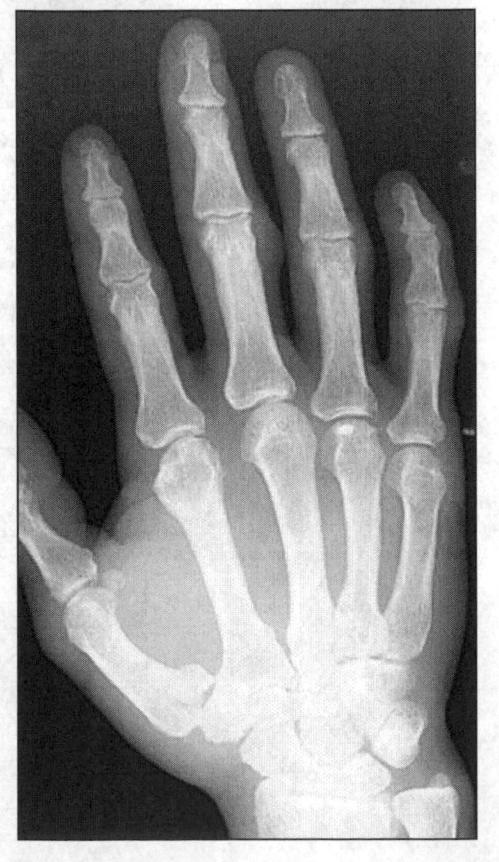

Figure 78.7. Fracture of the first metacarpal base, If it is intra-articular, it is referred to as a Bennett fracture. If it also demonstrates comminution, it is called a Rolando fracture.

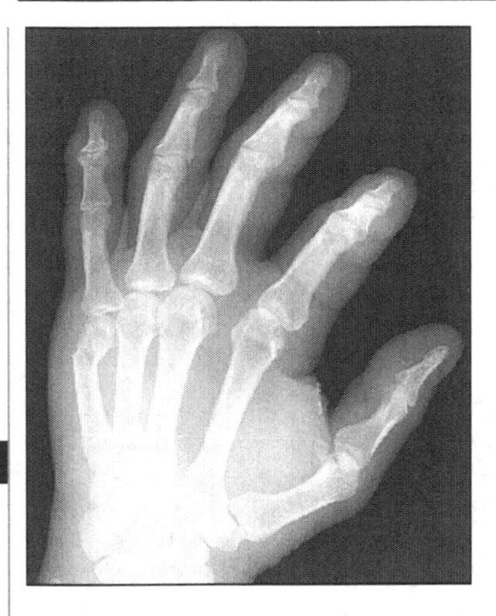

Figure 78.8. Fifth metacarpal fracture, commonly known as a boxer's fracture.

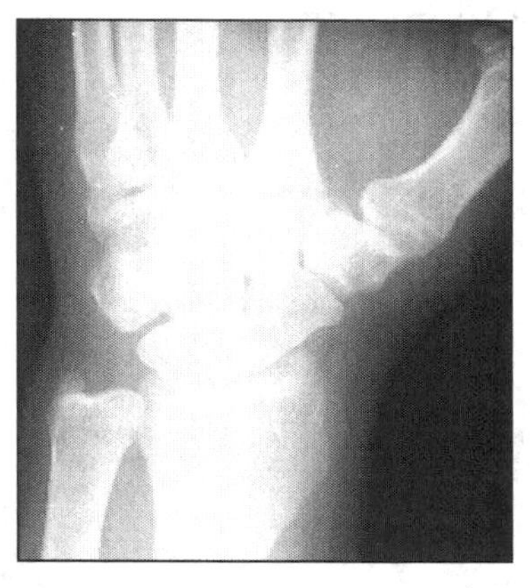

Figure 78.9. Fracture of the hamate bone. Hook fractures commonly occur in golfers, baseball players or construction workers with a complaint of a dull ache when gripping.

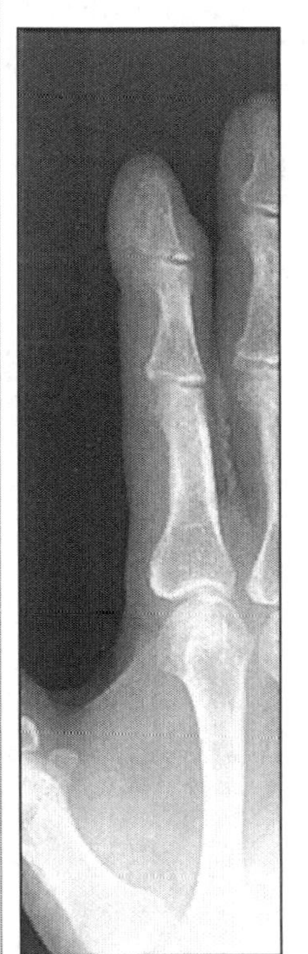

Figure 78.10. Fracture of the distal phalanx, or a mallet fracture. Distal phalanx fractures can occur due to avulsion of a bony fragment attached to the flexor or extensor tendon.

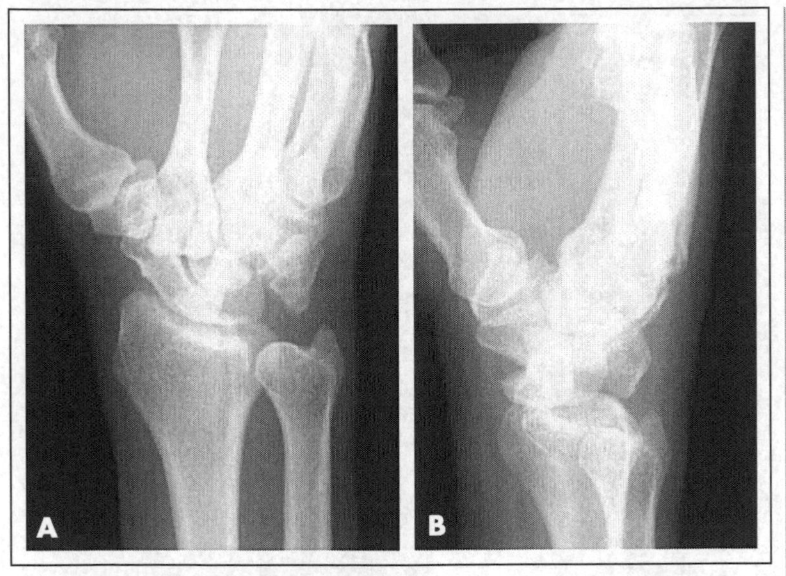

Figure 78.11. Dislocation of the lunate bone in the P-A (A) and lateral views (B). Note the normal position of the capitate bone with respect to the distal radius. In milder cases, the lunate can sublux dorsally after scapholunate ligament disruption (DISI) and will rarely sublux volarly after lunotriquetral ligament disruption (VISI).

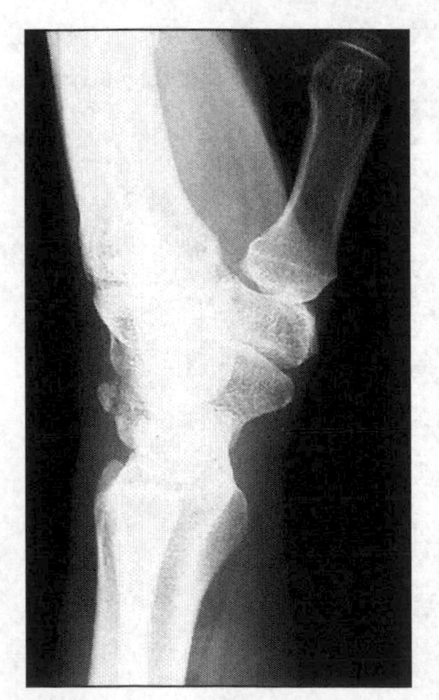

Figure 78.12. Triquetrum fracture. Usually associated with other wrist fractures after a fall on an outstretched arm. This fracture is usually a dorsal cortex chip fracture best seen on a lateral view.

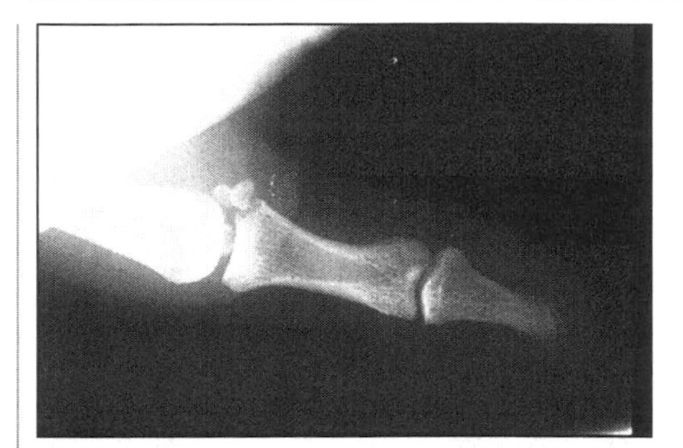

Figure 78.13. Avulsion fracture of the ulnar collateral ligament insertion at the base of the thumb, commonly referred to as a gamekeeper's or skier's thumb. If the fragment is widely displaced, a Stener lesion can result, requiring surgical intervention.

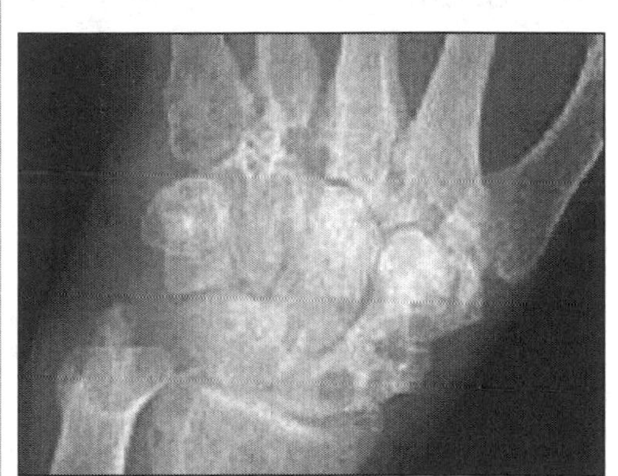

Figure 78.14. Rheumatoid arthritis of the wrist, demonstrating bony ankylosis of the carpus, ulnar deviation at the MP joints, and ulnar styloid erosions.

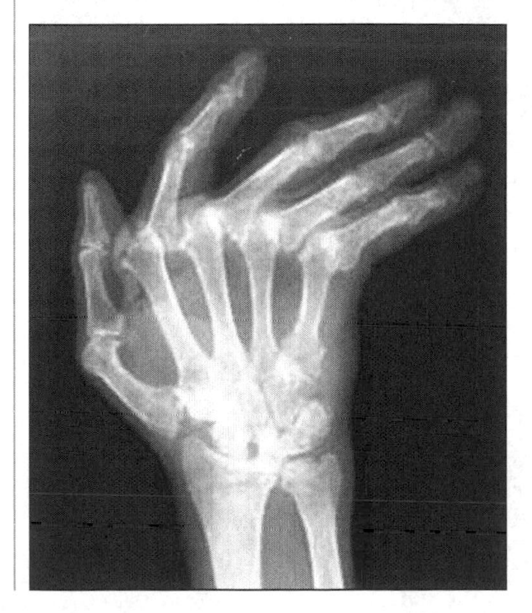

Figure 78.15. Late rheumatoid arthritis. There is marked MP joint destruction and narrowing. Weakening of the radial sagittal bands causes ulnar subluxation of the extensor mechanism and subsequent ulnar deviation of the MP joints.

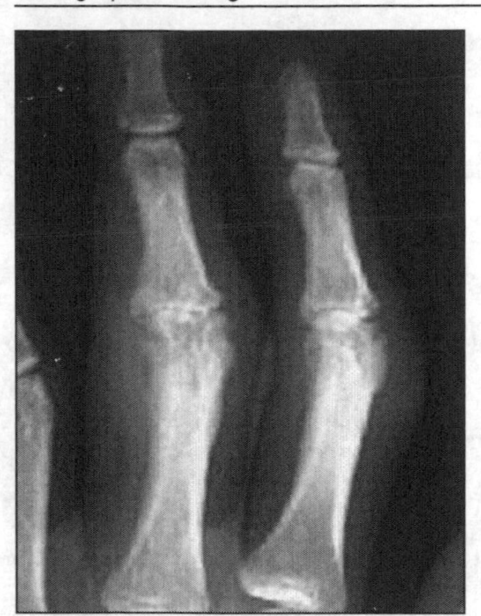

Figure 78.16. Rheumatoid arthritis in the fingers. In contrast to osteoarthritis, the IP joints are narrowed with characteristic erosions.

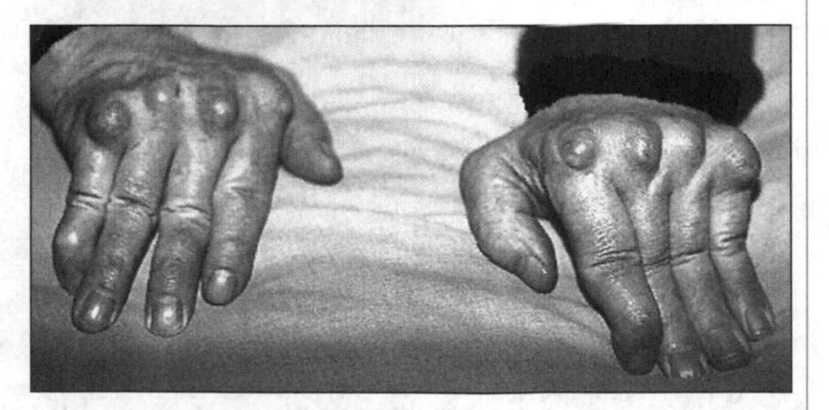

Figure 78.17. An illustration of several of the common clinical features of rheumatoid arthritis: symmetrical joint involvement, rheumatoid nodules, ulnar deviation of the MP joints, and swan neck deformity of the digits (IP hyperextension and DIP flexion).

Chapter 79

Examination of the Hand and Wrist

Zol B. Kryger

Observation

The hand should be observed in the resting position. This is termed the **position of function**. The wrist will be slightly dorsiflexed, the MP joints will be in 45-75° of flexion, the PIP joints in about 10° of flexion, and the DIP joints will be in 0-10° of flexion. Any alterations in the normal resting position should be noted. A single finger fixed in extension can be due to a flexor tendon laceration or rupture. A single finger in fixed flexion may be due to an extensor tendon injury. If the abnormal flexion is chronic, flexor tendon contracture must also be considered. This can be seen in Dupuytren's disease.

The patient should be instructed to slowly make a fist by flexing the fingers towards the palm. The fingertips should point towards the scaphoid. A finger fracture with rotational deformity can result in one finger overlapping another, or an alteration in the axis of convergence.

Sensory Exam

The hand receives sensory innervation from the median, ulnar and radial nerves. The most reliable method of testing normal sensory discrimination is the **two-point discrimination test**. The minimal distance at which the patient can distinguish two distinct points of pressure is recorded. Two-point discrimination should be 2-3 mm on the pulp of the fingers. This value can vary based on the individual's occupation. A measurement of greater than 5 mm is abnormal in most people.

To test median nerve sensation, skin sensibility on the palmer surface of the first three digits should be evaluated. Ulnar nerve injury will result in abnormal sensation of the little finger and ulnar side of the ring finger. Loss of radial nerve sensation will result in loss of sensibility on the dorsum of the hand and in the first web space. In addition, any areas of numbness, tingling or other sensory abnormalities reported by the patient should be noted as this can provide clues as to the nerve that may be involved.

Evaluation of Motor Nerve Function

The most consistent sign of median nerve injury is loss of thenar opponens function. Patients will make a fist without the thumb and index finger folded into the palm. If the flexor pollicus longus (FPL), flexor digitorum superficialis (FDS) and flexor digitorum profundus (FDP) of the index and middle finger are flexing normally, then the nerve injury has occurred distal to the take off of the anterior interosseous nerve.

Practical Plastic Surgery, edited by Zol B. Kryger and Mark Sisco.

Signs of ulnar nerve injury include flexor carpi ulnaris (FCU) paralysis, interosseous and thumb adduction loss with a weak "key pinch," and FDP paralysis of the small and ring fingers. A long-standing injury can present with the classic "claw hand" deformity. In a low lesion of the ulnar nerve just above the wrist, there will be more clawing of the 4th and 5th fingers compared to a high lesion at the elbow. This is known as ulnar paradox and is due to the fact that in a high lesion the flexors will also be paralyzed.

Patients with a radial nerve injury will be unable to extend their wrist or fingers. This can occur following humeral shaft fractures. Injury to the radial nerve in the axilla will result in paralysis of the extensor carpi radialis longus (ECRL) and brevis (ECRB), triceps and brachioradialis muscles. These muscles will atrophy in long-standing nerve injury. Loss of ECRL and ECRB function leads to loss of wrist extension.

Muscle power is assessed on a scale of 0-5, with a score of 5 indicating full power and a score of 0 being paralysis. 1 is given for just a flicker of muscle contraction, 2 for the ability to move with gravity eliminated, and 3 is the muscle strength necessary to lift a joint against the force of gravity. A score of 4 is for muscle power between 3 and 5.

Flexor Tendon Evaluation

Separate evaluation of both FDP and FDS function is important. Division of the FDS without injury to the FDP will not be noticeable in the resting posture. The FDS is evaluated by immobilizing the surrounding fingers in extension and having the patient flex the finger at the PIP joint. FDS to the index finger is evaluated by having the patient perform a firm pulp-to-pulp pinch with the thumb. An injured FDS will cause pseudo mallet deformity of the distal phalanx (flexed DIP, extended PIP), whereas an intact FDS will result in a pseudo boutonniere deformity of the distal phalanx (extended DIP, flexed PIP). FDP is evaluated by immobilizing the PIP and IP joints and evaluating flexion of the isolated DIP joint.

The palmaris longus is present in 85-90% of patients. It courses over the hamate and is identified by having the patient forcibly oppose the thumb and little finger. This tendon is expendable and can be readily used for tendon grafting procedures.

Extensor Tendon Evaluation

Extension of each individual digit at each joint should be examined. It is important to remember that the juncturae tendinum located at the level of the MP joints connects the extensor tendons of extensor digitorum communis (EDC). Therefore, lacerations proximal to the juncturae may not impair digit extension due to the connection to the adjacent digits. Limitation of extension of the DIP joint of the finger is most often due to rupture of the insertion of the extensor tendon into the distal phalanx. This type of injury will produce a mallet finger deformity. If the extensor tendon is disrupted proximal to the PIP joint, a boutonniere deformity may be produced due to hyperextension of the DIP joint by the taught lateral slips.

Evaluation of extensor pollicus longus (EPL) function is performed by asking the patient to place his palm flat on the table and to raise the thumb off of the table. Extension of the thumb alone is not an adequate test of EPL function since extensor pollicus brevis (EPB) and the intrinsic muscles of the thumb will contribute to its extension.

Intrinsic Hand Muscle Evaluation

The dorsal interossei are tested by asking the patient to abduct the fingers away from the middle finger, whereas the volar interossei are evaluated by adduction of the fingers towards to the middle finger. The lumbricals are tested by asking the patient to flex the fingers at the MP joints while the IP joints are held in extension.

Thumb Laxity

Instability of the MP joint of the thumb should be assessed. Laxity should be determined by abducting the thumb in both extension and flexion. Thumb laxity is normally greater in extension and should be compared between the two hands. Rupture of the ulnar collateral ligament (UCL) that connects the proximal phalanx of the thumb to the first metacarpal will result in excessive MP laxity. This injury is commonly seen in skiers, skateboarders, and anyone who falls on an extended thumb that is forcefully abducted. The acute injury is termed skier's thumb, and the chronic injury is referred to as gamekeeper's thumb.

Vascular Exam

The color of the digits should be observed, and any pallor, hyperemia or cyanosis noted. Capillary refill greater than 2 seconds is not normal. Arterial inflow to the hand is determined by palpating the radial and ulnar artery pulses. The **Allen test** is performed using the thumb and fingers to compress the radial and ulnar arteries at the wrist. The patient exsanguinates the hand by making a fist several times and then opens the hand so that the fingers are in a relaxed and gently extended position. The examiner then releases pressure from over the ulnar artery. Capillary refill time in the hand is noted. A normal Allen test is refill in less than 5 seconds, and greater than 5 seconds indicates an abnormal Allen test.

Wrist Stability and Motion

Pronation and supination of the wrist are examined with the elbow flexed to 90° and held firmly to the sides in order to stop rotation of the shoulder. From this position, there is usually about 90° of pronation and 90° of supination. Flexion, extension, radial and ulnar deviation should all be compared between the two wrists simultaneously, and any discrepancies should be noted.

Volar and dorsal stability are determined by the examiner exerting axial traction while holding the patient's forearm with one hand and the metacarpal heads with the other. Volar and dorsal displacement of the wrist in this position is minimal. If wrist laxity is present during volar pressure, midcarpal subluxation may be due to a flexion deformity of the first carpal row (the VISI deformity). If laxity is noted during dorsal pressure, a dorsal deformity (the DISI deformity) may be present.

Bony Landmarks

Patients will often present with focal pain in the hand or wrist. An understanding of the bony landmarks can help identify the structure that may be injured.

Volar Surface

The pisiform is located on the ulnar side of the palm just proximal to the palmar crease. Tenderness at the bony prominence should be assessed. The hook of hamate is 1-2 fingerbreadths towards the midline from the pisiform. The examiner places his thumb IP joint over the pisiform and directs his thumb towards the patient's index finger. The thumb pulp will rest over the hamate. The hook becomes more prominent with wrist flexion.

Dorsal Surface

The anatomic snuff box is located on the radial side of the wrist. Its radial border is EPB and abductor pollicus longus (APL; first dorsal compartment). The ulnar border is EPL (third dorsal compartment). This structure is important because the most commonly fractured carpal bone, the scaphoid, lies below the snuff box. The distal pole of the scaphoid can be palpated in the snuff box with the wrist in ulnar deviation. The scapholunate (SL) joint is the most common site of carpal dislocation. The examiner traces the third metacarpal proximally towards the wrist. The examiner's finger falls into a depression overlying the SL joint. The lunate is the second most commonly fractured carpal bone. It lies just ulnar to the SL joint. It is palpated with the wrist in flexion during which it is the most prominent area on the dorsum of the wrist. Lister's tubercle is an easily palpable prominence at the distal radius that lines up with the third metacarpal. It is easier to palpate it during mild wrist flexion. Recall that it separates the second and third compartments.

Important Provocative tests

Tinel's sign is a useful test for neuropathy. If percussion at the site of nerve entrapment produces tingling along the course of the nerve or in the digits supplied by the nerve, the test is positive. At the wrist, the volar carpal ligament can be percussed to test for median nerve neuropathy (carpal tunnel syndrome). The test has a low sensitivity (50-70%), but a high specificity (94%).

Phalen's test is used to provoke median nerve compression. It is performed by having the patient hold both wrists in flexion for a minute by opposing the dorsum of the two hands. If this reproduces the symptoms, the test is positive. It has a sensitivity of 70-80% and a specificity of 80%. The more rapidly the symptoms are produced, the higher the specificity of the test.

Finkelstein's test is used to test for tenosynovitis of the first dorsal extensor compartment (de Quervain's disease). The wrist is braced in ulnar deviation while the thumb is passively adducted and flexed. A positive test produces pain at the base of thumb. There may also be point tenderness over the radial styloid.

The grind test is used to assess for arthritis of the carpometacarpal (CMC) joint of the thumb, a common condition that produces pain similar to de Quervain's tenosynovitis. The examiner exerts axial pressure on the thumb and grinds it against the first metacarpal against the trapezium. A positive test produces pain. In addition, plain radiographs will show arthritic changes—a finding that will be absent in de Quervain's disease.

The scaphoid shift, or Watson test, is used to test for SL instability or scaphoid fracture, although it has a low sensitivity and specificity. The examiner's fingers are placed dorsally on the distal radius, while the thumb is placed firmly on the scaphoid tubercle (on the volar surface). The other hand holds the metacarpals, and the wrist is deviated ulnarly which places the scaphoid in extension. As the wrist is moved in radial deviation, the scaphoid is blocked from flexing by the examiner's thumb. If the SL ligament is injured, the scaphoid will move dorsally under the posterior margin of the radius inducing pain when it touches the examiners fingers. When pressure on the scaphoid is removed, the scaphoid goes back into position with a thunk or clunk. The test should be performed on the uninjured wrist for comparison.

The screwdriver test is used to examine the triangular fibrocartilage complex (TFCC). The examiner shakes the patient's hand and performs alternating supination and pronation of the wrist (i.e., screwing and unscrewing motion). This motion will usually cause ulnar-sided pain at the wrist if a TFCC injury is present. Decreased range of motion may also be apparent.

Pearls and Pitfalls

- Any patient with an acute injury should be asked the following questions:
  - When did the injury occur?
  - What was the mechanism of injury?
  - What was the position of the hand and/or fingers during the injury?
  - Was the environment clean or dirty?
  - Have any medications been administered by the ER staff such as a local anesthetic?
  - Is there a history of prior injury?
  - What is the tetanus status?
- Palpation for specific tender spots is often the most painful part of the physical assessment and should therefore be reserved for the end of the exam.
- A patient complaining of chronic hand or wrist pain should be asked to demonstrate the movements that most accurately replicate the pain.
- In order to avoid missing an important finding on the initial physical exam, it is useful to follow an examination routine such as:
  - Observation
  - Vascular: Radial and ulnar artery, capillary refill, skin color
  - Sensory: Median, ulnar and radial nerves
  - Motor: Median, ulnar and radial nerves
  - Flexor tendons
  - Extensor tendons
  - Intrinsic hand muscles
  - Wrist stability and range of motion
  - Palpation of bony landmarks
  - Provocative tests

Suggested Reading

1. Corley Jr FG. Examination and assessment of injuries and problems affecting the elbow, wrist, and hand. Emerg Med Clin North Am 1984; 2(2):295-312.
2. Daniels IInd JM, Zook EG, Lynch JM. Hand and wrist injuries: Part I. Nonemergent evaluation. Am Fam Physician 2004; 69(8):1941-8.
3. Daniels IInd JM, Zook EG, Lynch JM. Hand and wrist injuries: Part II. Emergent evaluation. Am Fam Physician 2004; 69(8):1949-56.
4. Kuschner SH, Ebramzadeh E, Johnson D et al. Tinel's sign and Phalen's test in carpal tunnel syndrome. Orthopedics 1992; 15:1297-1302.
5. Skvarilova B, Plevkova A. Ranges of joint motion of the adult hand. Acta Chir Plast 1996; 38(2):67-71.
6. Waylett-Rendall J. Sensibility evaluation and rehabilitation. Orthop Clin North Am 1988; 19(1):43-56.

Chapter 80

Soft Tissue Infections

Zol B. Kryger and Hongshik Han

Introduction

Infections of the hand can range from minor, superficial cases to infections of the deeper spaces of the hand that can potentially become limb-threatening and must be treated aggressively. Trauma is the major cause of hand infections, followed by human bites and animal bites. *Staphylococcus aureus* (*S. aureus*) accounts for about two-thirds of all hand infections.

Clinical Presentation

Diagnosis

A thorough history is important. The exact time and nature of the injury should be ascertained since this will guide treatment. For example, *S. aureus* is often implicated in home and industrial infections, whereas Gram-negative bacteria should also be considered if the infection occurred on a farm setting. Determination of hand position during and after the injury is also important. Accompanying symptoms help determine how extensive the infection has become. The age of the patient is important as well. Hand infections in children may involve different bacteria such as oral flora and Gram-negative rods (e.g., *H. Influenza* or pseudomonas). Table 80.1 lists the common hand infections and antibiotics used to treat them.

Differential Diagnosis

There are a number of conditions that can simulate a hand infection: acute calcific tendinitis (usually affecting flexor carpi ulnaris), gout or pseudogout, pyogenic granuloma, pyoderma gangrenosum, metastatic cancer, and the necrosis from the brown recluse spider bite, to name a few. Radiographs are required in all suspected hand infections to rule out gas in the soft tissue and occult fracture or osteomyelitis.

The early hand infection may masquerade as ischemia. Findings on exam might include areas of patchy discoloration, edema, slow capillary refill and a mottled, cyanotic appearance. Gross purulence and inflammation may not be observed. Late infections developing 7-10 days after the initial injury indicates an insufficient local host response against bacterial contamination. The wound usually becomes fluctulant with purulent discharge. Late infections are seen more commonly in diabetics or immunocompromised individuals (e.g., transplant recipients, chemotherapy recipients, HIV, and other immune disorders).

Treatment

The principles of treatment are similar in all hand infections. They are summarized in Table 80.2.

Practical Plastic Surgery, edited by Zol B. Kryger and Mark Sisco. ©2007 Landes Bioscience.

Table 80.1. Hand infections and the empiric antibiotics of choice

Infection	Common Bacterial Pathogen	Antibiotic
Human bite	*S. aureus*, streptococci, anaerobes *Eikenella corredens*	Cefazolin and penicillin or amoxicillin/clavulonic acid
Dog/cat bite	*S. aureus*, *S. viridans*, Bacteroides *Pasturella multocida*	Cefazolin and penicillin or amoxicillin/clavulonic acid
Tenosynovitis	*S. aureus*, streptococci, anaerobes	Cefazolin and penicillin or amoxicillin/clavulonic acid
Deep space infection	*S. aureus*, streptococci, anaerobes Gram-negative rods	Amoxicillin/clavulonic acid or cefazolin and penicillin
Cellulitis	*S. aureus*, *S. pyogenes*	Nafcillin, oxacillin, or cefazolin
Felon or paronychia	*S. aureus*, anaerobes	Cefazolin or dicloxacillin

Note: emerging bacterial resistance to cephalosporins may require the use of a second- or third-generation agent.

The most common infections and their management are discussed below:

Cellulitis

The dorsum of the fingers are most commonly affected. Streptoccoci and *S. aureus* are the likely offending organisms. Conservative therapy with oral antibiotics and elevation is often sufficient. However, if there are signs and symptoms of systemic infection (fevers, chills, increased WBC count, etc), ascending lymphangitis, lymphadenitis, or failure of oral antibiotics, the patient should be admitted for IV antibiotics and hand splinting in the intrinsic plus position.

Bites

Animal and human bites account for about a third of all hand infections. They not only can cause tissue damage but also can lead to serious infections if not treated properly. Human bites often cause infection due to the virulent microaerophilic Streptococci, *S. aureus*, and *Eikenella corredens* found in the human mouth.

Human Bites

Human bite wounds can be devastating. The human mouth contains mixed flora that can synergistically infect and destroy soft tissues, tendons, joints, and bones. For instance, a clenched fist that strikes a person's mouth can result in a

Table 80.2. Principles of hand infection management

- Incision and drainage of pus
- Debridement of necrotic tissue
- Antibiotics-first empiric (usually a two-drug regimen) then targeted based on cultures
- Tetanus prophylaxis when indicated
- RICE (rest, immobilization, cold, and elevation)
- Early hand therapy

tooth penetrating the extensor tendon, metacarpal head or joint space. In such cases it is important to examine the hand with the fist in the clenched position. Early infections may present with nothing more than minimal swelling and erythema. These can often be treated as an outpatient, with exploration, debridement, irrigation and oral antibiotics. Advanced infections, however, can be severe with purulence and accompanying lymphangitis or lymphadinitis. Untreated, irreversable damage to vital structures can occur. At the first sign of severe infection, the patient must be hospitalized and the infection treated aggressively with debridement, drainage, irrigation, antibiotics, immobilization, and elevation. Human bite wounds to the hand should not be closed, but rather left open to heal secondarily. The wound should be evaluated on a daily basis to ensure that it is healing.

Dog and Cat Bites

Dog bites account for 90% of all animal bites. Dog bite wounds are often lacerations or avulsions of soft tissue, whereas cat bites are usually deep puncture wounds. Not all bites become infected. Animal bites are less likely to progress to a clinical infection than human bites. An infected animal bite presents as rapidly progressing cellulitis and swelling with drainage from the bite wounds. The organisms frequently seen in dog bites are *S. aureus*, *S. viridans*, *Pasturella multocida* and *Bacteroides* species. Cat bites have similar organisms with a large number of *Pasturella multocida* often present. Treatment is similar to human bites. Antibiotic coverage against Pasturella (penicillin) should be included.

Paronychia

A paronychia is a soft tissue infection adjacent to the nail. Bacteria enter the surrounding tissue through the nail fold, usually after the patient pulls or bites a hang nail, or secondary to a foreign body at the site (such as a splinter). Consequently, the infection will often contain oral flora in addition to *S. aureus*. If discovered early, it can be treated with warm daily soaks, elevation and oral antibiotics. After the first 24 hours of infection, pus may develop under the nail fold. In such cases, treatment consists of elevating the nail fold off the nail to allow it to drain. Paronychia that fail this more conservative approach may require removal of the nail plate for adequate drainage to occur.

Chronic paronychia occur in the immunocompromised; *Candida albicans* is the most common organims involved. It usually involves nail bed destruction with the resulting nail plate fragments acting as a foreign body. This condition requires X-rays to rule out distal phalanx involvement. Treatment consists of excision of the proximal nail fold with healing by secondary intention or removal of the nail and topical antifungal therapy. Tissue specimens should be sent for culture and pathology to rule out an exotic infection or occult carcinoma/melanoma.

Eponychia

Eponychia is a soft tissue infection at the base of nail. It is usually caused by an infected hangnail or extension of an untreated paronychia. Conservative treatment is similar to that of a paronychia. More advanced infections may require paired incisions at the base of the nail. These are perpindicular incisions made at each lower corner of the nail.

Felon

A felon is an infection of the distal finger tip pulp, usually due to *S. aureus*. It can develop after a puncture wound or as the result of a proximally-extending paronychia. Patients often complain of dependent pain due to the pressure build up of fluid in the confined space of the pulp. Because of the fibrous septa in the volar pad, a felon is a closed space infection that must be incised and drained. If the infection has not pointed, the incision should be made over the site of greatest tenderness. Antibiotics and elevation of the hand are also important. When not treated adequately, a felon can spread into the phalanx, adjacent joint space, or flexor tendon sheath. X-rays should be performed to evaluate for the presence of a foreign body.

Deep Space Infections

The deep spaces in the hand include the thenar space, midpalmar space, web spaces, and subtendinous (Parona's) space. Deep space infections usually follow a deep puncture wound, and *S. aureus* is the most commonly cultured organism. They require incision and drainage, antibiotics, splinting and elevation. Care must be taken not to injure the nearby neurovascular structures. These infections are often contained in a single space, but if neglected and allowed to spread they can form a "horse shoe abscess," involving several deep hand spaces.

Infectious Tenosynovitis

Infections of the tendon sheath are most commonly due to Gram-positive, aerobic bacteria such as *S. aureus*. Human bites predispose to anaerobic organisms as well. Animal bites carry the additional risk of *Pasteurella multocida* infection. Gonococcal flexor tenosynovitis is due to hematogenous spread of disseminated gonorrhea.

Diagnosis

Dr. Kanavel described the cardinal signs of acute flexor tenosynovitis (Kanavel's signs):

1. Pain with passive finger extension
2. Tenderness over the entire length of the flexor sheath
3. Finger held in the semiflexed position
4. Swelling of the finger

Treatment

Acute flexor tenosynovitis is a serious condition that requires immediate treatment. In cases of nonsuppurative tenosynovitis, 24 hours of broad spectrum antibiotics, splinting and elevation can be tried. If the patient does not improve, tendon sheath irrigation is indicated. Suppurative tenosynovitis is a surgical emergency, and tendon sheath irrigation is the mainstay of treatment. A proximal incision is made at the level of the distal palmar crease proximal to the A1 pulley. A distal incision is made over the DIP joint distal to the A5 pulley. The sheath is exposed, opened, and vigorously irrigated with antibiotic solution. Postoperatively, a catheter can be left in place distally and proximally to allow irrigation of the sheath. Intravenous antibiotics are also used, although the irrigation of the sheath is primary treatment. Patients are usually hospitalized for 2-3 days.

Outcome and Complications

Two to three days following surgery, active motion exercises are initiated. Complete functional recovery usually occurs within 1-2 weeks. Complications of flexor tenonsynovitis include tendon necrosis, rupture, or adhesions, joint complications, osteomyelitis, deep space abscess and even amputation.

Pearls and Pitfalls

When performing any incision and drainage procedure, the location of the incision must be carefully considered to optimize the function and aesthetics once the wound heals. For bite wounds, the site of penetration will often lead directly to the location of the purulence, and the choice of incision site becomes obvious. For paronychia and eponychia, the space between the nail plate and overlying skin fold is usually the site of drainage. In a deep space infection, make the incision over the most fluctuant, inflamed area. The same is true for a felon that has pointed. Otherwise, make a mid-lateral incision deep into the volar pulp to disrupt the septa. Avoid fish-mouth incisions since these can result in a fingertip deformity after healing. For suppurative tenosynovitis, a mid-lateral incision is an excellent choice, since a Bunnel-type incision that breaks down can expose the underlying flexor tendon sheath.

Suggested Reading

1. Arons MS, Fernando L, Polayes IM. Pasteurella multocida—the major cause of hand infections following domestic animal bites. J Hand Surg 1982; 7A:47.
2. Brook I. Paronychia: A mixed infection. Microbiology and management. J Hand Surg 1993; 18:358.
3. Dellinger EP, Wertz MJ, Miller SD et al. Hand infections. Bacteriology and treatment: A prospective study. Arch Surg 1988; 123(6):745.
4. Goldstein EJ. Current concepts on animal bites: Bacteriology and therapy. Curr Clin Top Infect Dis 1999; 19:99.
5. Hausman MR, Lisser SP. Hand infections. Orthop Clin North Am 1992; 23:171.
6. Moran GJ, Talan DA. Hand infections. Emerg Med Clin North Am 1993; 11:601.

Chapter 81

Compartment Syndrome of the Upper Extremity

Zol B. Kryger and John Y.S. Kim

Introduction

Compartment syndrome of the upper extremity can occur in the arm, forearm, wrist or hand. It is defined as an increased pressure build-up within an enclosed compartment sufficient to impair muscle and nerve perfusion within that compartment. In the upper extremity, it most commonly occurs in the forearm secondary to trauma. Fractures of the humerus are the most common etiology.

Relevant Anatomy

Arm

The arm contains **two compartments**: the anterior flexor compartment (biceps and brachialis) and the posterior extensor compartment (triceps). The fascia encasing and separating these two compartments is relatively weak, and consequently fluid accumulation in one compartment will usually make its way into the other compartment and the surrounding subcutaneous tissues. Furthermore, the compartments of the arm communicate with the shoulder girdle, making compartment syndrome even more unlikely to occur.

Forearm

The forearm structures are contained within **three compartments**: an anterior flexor (volar) compartment, a posterior extensor (dorsal) compartment, and the mobile wad (superficial radial-dorsal side). These compartments do not communicate freely with one another nor with the hand distally and the arm proximally. Therefore, forearm compartment syndrome is more common than in the arm or hand.

Wrist and Hand

The wrist has **one main compartment**, the carpal tunnel. Release of the carpal tunnel should be performed at the time of forearm compartment release. The compartments of the hand are more numerous and complex than in the forearm. There are **ten compartments**: four dorsal interosseous, three volar interosseous, one thenar, one hypothenar, and one adductor pollicis compartment. In the fingers, the ligaments (Cleland's, Grayson's, and transverse retinacular) can also segregate the digit into compartments, resulting in a compartment syndrome of the neurovascular bundle.

Pathogenesis

A number of factors discussed below can lead to an increase in the pressure of a compartment, such as an accumulation of fluid within the enclosed space. As the pressure increases above 35 mm Hg, capillary perfusion becomes compromised and ischemia ensues. A vicious cycle sets in, worsening the situation: the ischemia leads to

Practical Plastic Surgery, edited by Zol B. Kryger and Mark Sisco. ©2007 Landes Bioscience.

impaired venous return, increased capillary permeability and vasodilation, all of which cause greater fluid accumulation and serve to increase the pressure even further.

The tissues most sensitive to the resulting ischemia are nerve and muscle. Greater than 6-8 hours of ischemia will cause irreversible nerve and muscle damage. The deeper compartment muscles (such as flexor digitorum profundus) are usually affected first, and the central portions of the muscle belly are more susceptible due to poor collateral circulation.

Etiology

Causes of compartment syndrome either limit the ability of the compartment to expand in size or increase its volume. In most instances, the etiology is multifactorial. For example, a displaced supracondylar humeral fracture leading to forearm edema, combined with a circumferential cast that is too tight. In the setting of trauma, multiple injuries substantially increase the risk of developing compartment syndrome. Compared to an isolated humeral shaft fracture, the combination of a humeral and radial fracture raises the risk of compartment syndrome significantly. If there is a concomitant arterial injury, the risk can approach 50%.

Factors that restrict the size of the compartment:

- Constrictive casts and dressings
- Prolonged external pressure during surgery (e.g., tourniquet)
- Prolonged external pressure during unconsciousness
- Eschar from a burn

Factors that increase the volume of the compartment:

- Fractures (usually closed) leading to edema or hemorrhage
- Spontaneous bleeding secondary to a coagulopathy
- Crush injury leading to edema
- Infection or snake bites leading to edema
- Edema from burns
- Edema from reperfusion injury, or following revascularization
- Edema from strenuous exercise
- Iatrogenic fluid infiltration

Clinical Diagnosis

The diagnosis of compartment syndrome is clinical. Measurement of compartment pressures should be used only as an adjunctive tool. The signs and symptoms are usually progressive and include the following:

- A swollen, tense compartment
- Digits in MP extension and PIP flexion
- Pain out of proportion to the underlying injury
- Pain with passive stretching of the involved muscles
- Paresthesias in the distribution of the involved nerves
- Impaired two-point discrimination test
- Motor weakness in the distribution of the involved nerves/muscles
- Intact pulses or dopplerable arterial signals

Compartment Pressure Measurements

As stated above, compartment syndrome is a clinical diagnosis and pressure measurements only aid in confirming the diagnosis. In the uncooperative or obtunded patient however, an accurate exam is not possible, and objective pressure determination can be of value if there is a high index of suspicion. The normal capillary perfusion pressure ranges from 20-25 mm Hg. Several techniques are available for measuring compartment pressure; however they differ on the threshold pressure used to make the diagnosis of compartment syndrome.

Whitesides' method describes the use of needle manometery using readily available supplies. This technique considers a pressure within 20 mm Hg of the diastolic pressure as its threshold. A modification of this method can be performed using a standard arterial line connected to an 18 gauge needle inserted into the compartment.

The **Matsen technique** involves continuous monitoring of the compartment pressure for up to 3 days. It is similar to Whitesides' technique. An absolute pressure threshold of 45 mm Hg is considered positive.

Finally, the **Murbarak method** directly measures compartment pressure using a wick (slit) catheter. There are a number of hand-held devices available that rely on this technique. Any pressure above 30 mm Hg is considered highly suggestive of compartment syndrome.

In the forearm, the needle should be inserted in the middle of each compartment, which corresponds more or less to the middle third of the forearm. In the hand, the pressure should be measured in the four dorsal and three volar interosseous compartments, as well as in the thenar and hypothenar compartments on the palmar surface. In the wrist, pressure should be measured in the carpal tunnel.

Management

Compartment syndrome is a surgical emergency. Without timely compartment release, irreversible damage will ensue. Most studies indicated that the fasciotomy should be performed within 12 hours of compartment syndrome onset. It is controversial whether to perform a fasciotomy more than 48 hours after injury. If the cause of the compartment syndrome is due to excessive external pressure, this should be immediately relieved. All dressings and cast material should be removed since this alone can significantly reduce compartment pressure and occasionally eliminate the need for surgical decompression.

Operative Treatment

The surgical release of the compartments is termed a fasciotomy. Occasionally a prophylactic fasciotomy is performed following revascularization of a limb that has been ischemic for over 4 hours. Skin incisions after fasciotomy are generally left open. They may be loosely approximated to cover any exposed nerves. The incisions can be closed several days later or skin-grafted if direct closure is not possible.

Arm

Two separate incisions are required. The flexor compartment is decompressed via an incision extending from the acromion along a line that lies just medial to the lateral bicipital sulcus. The extensor compartment is released using an incision that extends proximally from the olecranon along the lateral side of the arm over triceps. Alternatively, both compartments can be released through a single incision extending from the shoulder to the elbow crease along a line overlying the medial intermuscular septum.

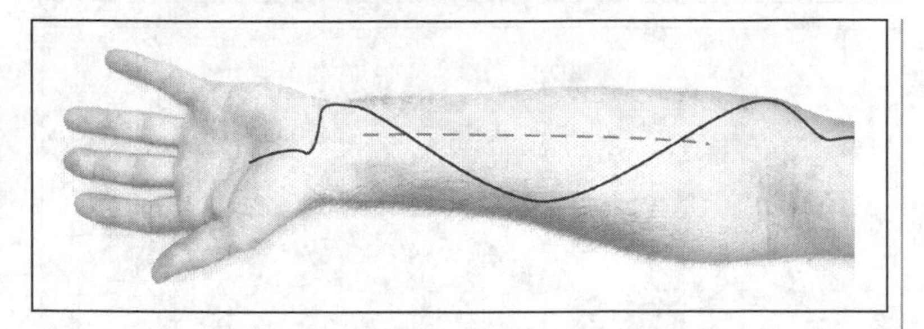

Figure 81.1. The S-shaped fasciotomy incision for release of the volar compartment and mobile wad of the forearm. The incision extends from proximal to the elbow crease and into the palm, releasing the carpal tunnel along with the forearm.

Forearm and Wrist

Release of the flexor compartment and the mobile wad should be performed first since often this will reduce the elevated pressure in the dorsal compartment. An S-shaped skin incision is made, extending from proximal to the elbow crease and into the palm (Fig. 81.1). Thus, the carpal tunnel is released along with the volar compartment. After the skin and subcutaneous tissue have been incised, the deeper fascia (including the transverse carpal ligament) is incised and the muscles are mobilized. The median nerve is identified and full release of any nerve compression is performed along its entire course. If there is suspicion of ulnar or radial nerve compromise, these nerves should also be decompressed.

If the dorsal compartment still has elevated pressures after volar release, a dorsal fasciotomy incision is made. It begins from a point 2 cm lateral and distal to the lateral epicondyle and extends about 10 cm towards the center of the wrist.

Hand

Decompression of all of the hand compartments requires multiple volar and dorsal incisions. The dorsal and palmar interosseus compartments are approached using two dorsal incisions over the second and fourth metacarpals (Fig. 81.2). The thenar and hypothenar compartments are accessed through a longitudinal incision along the radial side of the first metacarpal, and the ulnar side of the fifth metacarpal, respectively (Fig. 81.2). The carpal tunnel is released using either the standard open approach, or with an extended incision that can also allow Guyon's canal to be released (for ulnar nerve decompression). Access to the adductor compartment can be gained through the carpal tunnel release.

Postoperative Considerations

In the immediate postoperative period, urine myoglobin levels should be monitored since elevated levels can produce acute renal failure. Myoglobinuria peaks several hours after circulation is restored. Prophylaxis and treatment consist of alkalinizing the urine and keeping the patient well-hydrated. A urine output of at least 1-2 ml/kg/hr should be maintained. Serum creatinine phosphokinase (CPK) levels should also be followed since extremely high CPK levels are also a risk factor for renal failure.

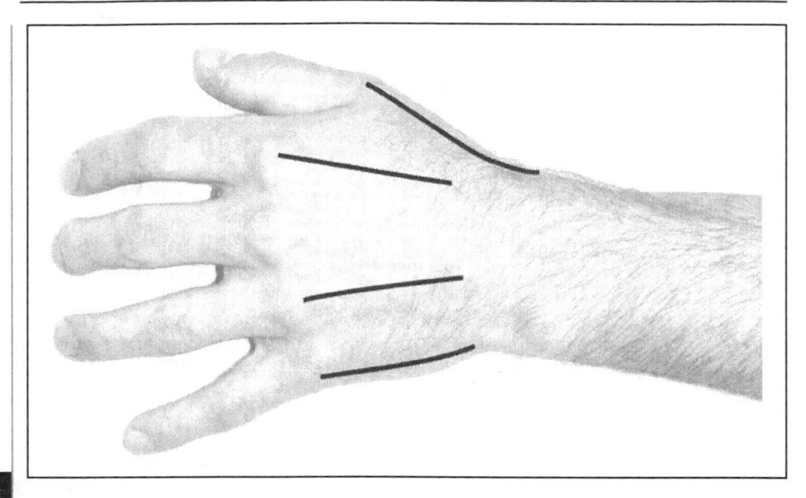

Figure 81.2. Fasciotomy incisions over the second and fourth metacarpals for releasing the dorsal and palmar interosseus compartments. Longitudinal fasciotomy incisions along the radial side of the first metacarpal and the ulnar side of the fifth metacarpal for release of the thenar and hypothenar compartments, respectively. The carpal tunnel is released via an extended incision which also allows access to Guyon's canal and the adductor compartment.

The arm is dressed in a bulky dressing and kept elevated postoperatively. The patient is brought back to the operating room after 3-5 days for wound inspection and fasciotomy closure. If direct closure is not possible, split-thickness skin grafts are used to cover the defect. This can be done immediately or within 1-2 weeks. Active and passive hand motion exercises should be initiated as soon as possible to help limit stiffness and contractures.

Complications

Unrecognized or undertreated compartment syndrome is likely to result in one or more complications. Even those cases that are recognized and treated in a timely fashion can be complicated by one of the following:

- Wound complications such as infection, dehiscence, and hypertrophic scarring
- Intrinsic hand contracture secondary to necrosis and fibrosis of intrinsic hand muscles
- Volkman's contracture of the forearm flexor muscles
- Scar contracture across the elbow or wrist leading to impaired joint motion
- Hand or arm dysfunction secondary to permanent muscle injury
- Renal failure secondary to rhabdomyolysis
- Neurologic sequelae
- Sensory nerve deficits

Pearls and Pitfalls

The diagnosis of compartment syndrome is challenging and requires a high index of suspicion. The clinical exam, underlying mechanism, and pressure measurements should be considered in making the diagnosis. Many factors can contribute to edema and swelling of the extremity without posing a risk for compartment syndrome. A tense, edematous hand or forearm without pain or nerve deficits should be watched closely but is unlikely to represent compartment syndrome. When in doubt, it is better to perform a fasciotomy, either prophylactic or therapeutic, than to wait until it is too late.

The fasciotomy should be done with extreme care to avoid injuring nerves that are otherwise not at risk from the compartment syndrome itself. The distal forearm fasciotomy incision at the wrist should be kept ulnar to the palmaris longus tendon to avoid injury to the palmar cutaneous branch of the median nerve. In the palm, the incision should be kept ulnar to the mid-axis of the ring finger.

Various "minimal" incision approaches have been described for gaining access to multiple compartments. However, the risk of incomplete compartment release or iatrogenic neurovascular injury is higher with a limited exposure. It is worth noting that in forearm compartment syndrome, release of the volar compartment will usually sufficiently decrease the pressure in the dorsal compartment sparing a dorsal incision.

Suggested Reading

1. Botte MJ, Gelberman RH. Acute compartment syndrome of the forearm. Hand Clin 1998; 14:391.
2. Gellman H, Buch K. Acute compartment syndrome of the arm. Hand Clin 1998; 14:385.
3. Ortiz Jr JA, Berger RA. Compartment syndrome of the hand and wrist. Hand Clin 1998; 14:405.
4. Serokhan AJ, Eaton RG. Volkmann's ischemia. J Hand Surg [Am] 1983; 8:806.
5. Whitesides TE, Haney TC et al. Tissue pressure measurements as a determinant for the need of fasciotomy. Clin Orthop 1975; 113:43.

Chapter 82

Replantation

Zol B. Kryger and John Y.S. Kim

Indications

General indications for replantation include amputation of multiple fingers, the thumb, the hand, or any part in a child. Loss of a single finger (excluding the thumb) is a relative indication for replantation; many surgeons will attempt replantation of distal digital injuries if at least 4 mm of intact skin proximal to the nailfold is present. Replantation of ring finger avulsions may be attempted, especially if one of the flexor tendons remains intact.

Contraindications for replantation:

- Significant systemic illness or comorbid conditions
- Concomitant life-threatening injuries
- Self-mutilation injuries
- Severe crush or avulsion injuries
- Extreme contamination
- Multiple level injuries
- Forearm or arm amputations with greater than 6 hours of warm ischemia time

Classification

Some authors classify the amputation by the flexor tendon zone (see Chapter 90). An alternative classification of amputation level has been introduced by Tamai:

Level I	Amputation at the proximal nail fold
Level II	Amputation at the DIP joint
Level III	Amputation at the middle phalanx
Level IV	Amputation at the proximal phalanx
Level V	Amputation at the superficial palmar arch

Preoperative Considerations

Ischemia Time

Amputated parts should be cooled as soon as possible since they can tolerate significantly longer "cold ischemia" than "warm ischemia" time. Ideally, the part should be wrapped in saline-soaked, cold gauze sponge and placed in a bag on ice. Digits can survive for 24-36 hours cold, compared to 8 hours warm. A hand has been successfully replanted after 54 hours of cold ischemia time. The forearm can tolerate up to 10 hours of cold ischemia and 4-6 hours of warm ischemia.

Radiographs of the amputated part and the residual extremity should be obtained to determine if there are any missing segments of bone. The patient should be consented for possible tissue grafting or free flap coverage in addition to replantation. Prior to surgery, the patient should be hydrated and warmed.

Practical Plastic Surgery, edited by Zol B. Kryger and Mark Sisco. ©2007 Landes Bioscience.

Intraoperative Considerations

Preparation of the Amputated Part

Devitalized tissue is carefully debrided, and the vessels are dissected out under the microscope. Once a vessel is identified, it is marked with a tag. Arteries are closely inspected for signs of stretching or avulsion, which is suggested by a corkscrew-like appearance termed the "ribbon" sign. Bruising along the course of the digital vessel can also be a sign of avulsion injury. Nerves and tendons are identified and tagged. The exposed bone is then minimally debrided.

Preparation of the Stump

After identification of the important vessels, nerves and tendons, the devitalized soft tissue of the stump is carefully debrided and irrigated with antibiotic solution. Large bone fragments are saved for possible grafting. The proximal bone stump is debrided. In the palm or finger, adequate exposure of the structures to be repaired may require Bruner zigzag incisions.

The Order of Repair

The order in which structures are repaired varies. The following outlines a commonly used progression:

1. **Bone shortening followed by rigid fixation**—if ischemia time is an issue, circulation should be restored prior to bony fixation
2. **Flexor tendon repair**—some authors will repair only the profundus tendon
3. **Extensor tendon repair**
4. **Arterial anastomosis**—at least one artery is usually required
5. **Venous anastomosis**—at least two veins for more proximal digital amputations. Level I amputations can be replanted without venous anastomosis
6. **Nerve repair**—this is optional, and may be delayed for a secondary procedure
7. **Soft tissue coverage**—skin grafts are sometimes required, especially in avulsion injuries. Vein grafts can be harvested with overlying skin as a composite graft

Bone Fixation

Fixation of an amputated digit can usually be achieved with crossed Kirschner wires passed retrograde through the amputated digit into the finger stump. Amputations through the proximal phalanx (Level IV) may require plate fixation for early postoperative mobilization. Transmetacarpal, transcarpal, and forearm amputations also require screw and plate fixation. This should be followed by periosteal repair, when indicated, to minimize tendon adhesion to the plates.

Vascular Repair

The arteries and veins must be trimmed back until normal intima is apparent. In order to minimize spasm, topical papavarine or concentrated lidocaine can be administered. Alternatively, a Fogarty catheter can be used to break a spasm. In crush or avulsion injuries, a vein graft may be required. Vein grafts can be taken from the dorsum of the foot, from the volar wrist, or from the dorsum of the hand. Alternatively, the lesser or greater saphenous systems can be used. The length of donor veins should be measured in situ since they shrink after harvest. Vein grafts should not be excessively long, since this can lead to kinking.

Arterial repair is usually done first. Prior to venous repair, the appearance of the replanted part should be assessed for several minutes after completion of the arterial anastomosis to confirm adequate perfusion. After the replant has warmed up, bleeding from the veins to be anastomosed should be brisk.

Level I digital amputations can be replanted without a venous anastomosis. At least two veins are required for proximal digital amputations. If necessary, digital veins can be transferred from an adjacent digit. Venous drainage can also be achieved with a proximally based cross-finger flap. If the contralateral digital artery has retrograde blood flow, it can be anastomosed to a vein to provide outflow. Finally, if no vein can be identified, the nail is removed and a heparin-soaked sponge is placed on the nail bed, or leeches are applied to the tip of the digit.

Tendon Repair

Debridement of the tendon ends should be minimized to avoid the need for tendon grafting. Most surgeons repair the extensor tendons first, and many postpone repair of the flexor tendons. In Zone II flexor tendon amputations, some surgeons will elect to repair only the FDP tendon. A core suture technique is used; this is described in detail in Chapter 90.

Nerve Repair

Nerves should be repaired primarily whenever possible with 8-0 or 9-0 suture. A description of primary nerve repair is discussed in the nerve repair chapter. When necessary, nerve grafts can be taken from a number of sites: a nonsalvageable digit, the posterior interosseus nerve, the lateral femoral cutaneous nerve, the superficial peroneal nerve, and the sural nerve. Vascularized nerve grafts have also been described. Difficult nerve repairs may be delayed after the nerve ends are identified and tagged. Sensory outcomes of digital replantation are improving. Two-point discrimination of 10 mm or less has been reported.

Soft Tissue Coverage

In order to avoid compression of the vascular anastomoses, the skin must be closed in a tension-free fashion. If this cannot be accomplished, a split-thickness skin graft should be used. In forearm or more proximal amputations, soft tissue loss may be extensive, requiring local or free flap coverage.

Special Considerations

Ring Avulsions

Outcomes for ring-avulsion replants are not as good as for amputations. Success rates range from 30-70%, depending on the extent of injury. Ring avulsions range in severity, from tearing of the soft tissue with intact circulation to complete degloving of the digit. In degloving injuries, vein and nerve grafts are required. Repair of thumb avulsions should always be attempted, while fingers avulsed through the PIP joint should be amputated.

Hand and Forearm Amputations

Amputations through the carpal region have generally good functional outcomes. Transmetacarpal amputations, however, have a much poorer long-term prognosis. This is in part due to the many small vessels that are severed in these injuries. In any case, chronic swelling of the hand is common so many surgeons will release the carpal tunnel and dorsal interosseous compartments at the time of surgery.

Of all the tissue in the upper extremity, muscle is least tolerant of ischemia. In proximal amputations, restoration of blood flow is the first priority since more muscle is involved. Prophylactic forearm fasciotomies are often performed if warm ischemia time is prolonged. Proximal amputations will usually require secondary procedures such as nerve grafts and tendon transfers.

Postoperative Care and Monitoring

At the completion of the procedure, the extremity should be placed in a well-padded splint without any circumferential dressings since swelling is inevitable. The fingertips should be exposed so that they can be monitored postoperatively.

The patient must be kept in a warm room with the extremity elevated above the heart. A continuous axillary anesthetic block can help with pain control and act as a chemical sympathectomy. Aspirin should be given since it has a potent anti-platelet effect. Although many centers use some form of anticoagulation, no randomized control trials have demonstrated a benefit to any regimen. The patient should be hydrated with intravenous fluids to avoid hypotension.

Monitoring the circulation after replantation is similar to a free-flap as discussed in the chapter on free-flap monitoring. Early detection of microvascular failure is critical for salvage of the replant. The most reliable means of monitoring a replanted digit is a clinical exam by an experienced individual combined with pulse oximetry or Doppler monitoring. SpO_2 values above 95% are normal; a saturation below 85% indicates a venous problem, and a complete lack of signal indicates and arterial problem. Arterial and venous Doppler monitoring is used in many centers; both pencil and laser Doppler can be used. Finally, removal of the nail with monitoring of nail bed bleeding provides a rudimentary means of monitoring arterial circulation.

Outcomes

The most common early complications after replantation are bleeding, infection and loss of the replant. Late complications include cold intolerance, bony non-union, nerve or tendon adhesions requiring neurolysis or tenolysis, and late necrosis of the replanted part.

Centers with a highly experienced replantation team report survival rates of 90-100% for digital replants. Level II amputations generally do better than Level I amputations. In terms of sensory recovery, two-point discrimination of 10 mm or less has been reported in a number of series. In both adults and children, persistent cold intolerance is common.

Pediatric Replantation

Many surgeons agree that unless a contraindication exists, almost all amputations in children should undergo replantation. Long-term follow-up at 15 years has demonstrated return of normal sensation, strength, and bone growth in up to 90% of digital replants in children. Distal finger tip amputations have been shown to survive with composite grafting when microvascular replantation is not possible. Even in these cases, sensory recovery can be excellent, presumably due to spontaneous neurotization. In children, bone shortening should be undertaken with caution to protect the epiphyseal growth plate so that the replanted bone will continue to grow normally.

Pearls and Pitfalls

Three of the leading replant teams, the Kleinert-Kutz group, the Buncke group, and the Tamai group, have emphasized several important points in replantation technique:

- Bone shortening to reduce tension on the vessels, tendons, and nerves
- Repair of both the artery and vein before tourniquet release
- Heparin bolus after completion of the anastomosis
- Washing out the intravascular clots from crushed digits or hands
- The use of "spare parts" from unreplantable digits
- Transposing an amputated digit onto a different stump in cases of multiple digit amputations

Suggested Reading

1. Buncke Jr HJ. Microvascular hand surgery—Transplants and replants—Over the past 25 years. J Hand Surg 2000; 25A:415.
2. Kim JYS, Brown RJ, Jones NF. Pediatric upper extremity replantation. Clin Plast Surg 2005; 32:1.
3. Kleinert HE, Jablon M, Tsai TM. An overview of replantation and results of 347 replants in 245 patients. J Trauma 1980; 29:390.
4. Lee BI, Chung HY, Kim WK et al. The effects of the number and ratio of repaired arteries and veins on the survival in digital replantation. Ann Plast Surg 2000; 44:288.
5. Pederson WC. Replantation. Plast Reconstr Surg 2001; 107:823.
6. Soucacos PN. Indications and selection for digital amputation and replantation. J Hand Surg (Br) 2001; 26B:572.
7. Tamai S, Michon J, Tupper J et al. Report of the subcommittee on replantation. J Hand Surg 1983; 8:730.
8. Weinzweig N, Sharzer LA, Startker I. Replantation and revascularization at the transmetacarpal level: Long-term functional results. J Hand Surg 1996; 21:877.

Chapter 83

Fractures of the Distal Radius and Ulna

Craig Birgfeld and Benjamin Chang

Introduction

The radius and ulna form the bony structure of the forearm. They articulate with the humerus at the elbow and the proximal carpal row at the wrist. These bones form the framework upon which the long flexor and extensor muscles of the forearm take origin. The radius lies on the thumb side of the forearm (the "radial" side) and the ulna lies on the little finger side (the "ulnar" side) of the forearm. The anatomic shape of these bones and their relationship at the joints provides for a high degree of mobility. However, the price of this mobility is propensity for injury and degenerative disorders.

Anatomy

The elbow is formed by articulations between the radius and humerus and the ulna and humerus. The ulnar-humeral joint is a stable hinge joint, allowing only flexion and extension from 0-120 degrees. The radial-humeral joint is a pivot joint, which allows the radius to rotate on the humerus. This joint provides the mobility which allows the pronation and supination of the forearm through 180 degrees.

The radius and ulna are joined by an interosseous membrane, which is flexible enough to allow rotation of the radius around the ulna in pronation and supination, but strong enough to tether the two bones in a stable relationship. Pronator and supinator muscles course between the two bones and the long flexor and extensor muscles of the wrist and fingers originate from these bones in the forearm. At the wrist, the radius and ulna support the proximal row of carpal bones. The radius articulates with the lunate and scaphoid at the lunate fossa and the scaphoid fossa. Ligaments join the scaphoid and lunate to the radius and these articulations can also be the source of arthritic pain. The distal ulna articulates with the triquetrum through the triangular fibrocartilage complex (TFCC). The radius and ulna meet at the sigmoid notch and are joined by the TFCC forming the distal-radial-ulnar-joint (DRUJ), a frequent site of injury and arthritis. The ulna styloid lies most distally and laterally and is frequently involved in fractures of the distal radius.

Radiographs

Relationships seen on normal radiographs are important to remember as these will be disrupted in fractures to the distal radius and ulna. All radiographs of the wrist should be evaluated for ulnar variance, inclination and volar tilt.

- **Ulnar variance:** on the normal AP view, the distal ulna is within ± 1-2 mm of the distal radius (Fig. 83.1).
- **Inclination:** the radius tilts toward the ulna at an angle of 22° when measured from a line perpendicular to its long axis on AP view (Fig. 83.1).
- **Volar tilt:** on lateral view, the radius tilts in a volar direction 11° when measured from a line perpendicular to its long axis (Fig. 83.2).

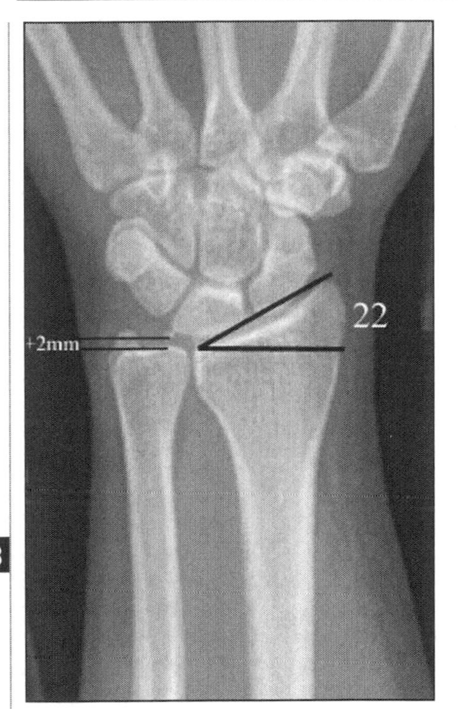

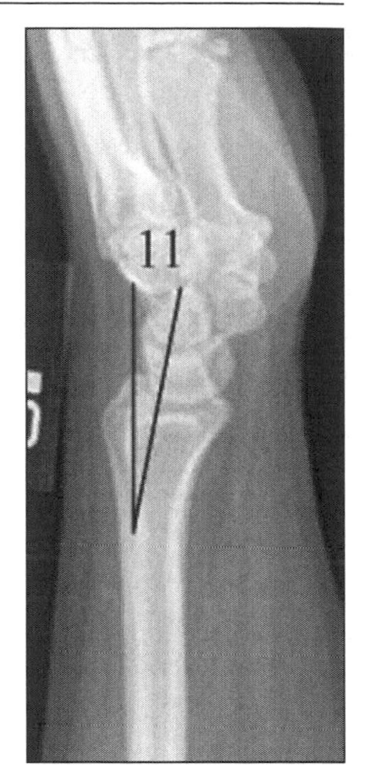

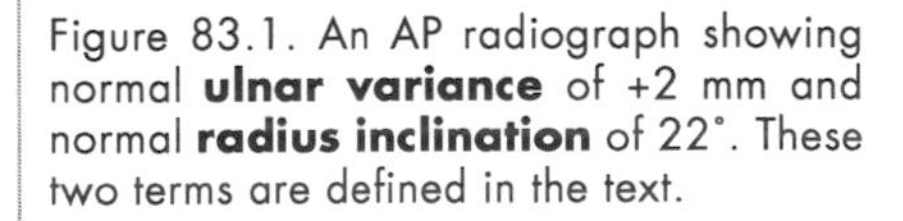
Figure 83.1. An AP radiograph showing normal **ulnar variance** of +2 mm and normal **radius inclination** of 22°. These two terms are defined in the text.

Figure 83.2. A lateral radiograph demonstrating the normal 11° **volar tilt** of the radius when measured from a line perpendicular to its long axis.

Traumatic disruptions of these relationships can destroy the normal kinematics of the wrist and lead to reduced range of motion and degenerative arthritis.

Fractures

A variety of fractures can occur in the radius and ulna after the common traumatic history of, "fall on an outstretched hand". These are amongst the most common skeletal injuries and occur more frequently in children and the elderly. Fractures of the distal radius are best described in terms of their comminution, articular involvement, displacement and angulation with particular attention paid to the measurements of ulnar variance, inclination, and volar tilt. Eponyms are frequently used for expediency, though these can lead to confusion for inexperienced evaluators. Fractures of the ulna are often multiple and can have varying degrees of clinical significance. Recall that the radius and ulna essentially form a ring between the elbow and wrist and, as is seen with the mandible, a fracture of one side of the ring will usually result in a fracture at the other.

The most common fracture, is a **dorsally** angulated, extra-articular fracture of the distal radius, otherwise known as a **Colles' fracture**. A **Smith's fracture** is a **volarly** angulated, extra-articular fracture and generally occurs from a fall onto the dorsum of the hand. A **Barton's fracture** is an intra-articular fracture and can be either volarly or dorsally angulated. A **Chauffeur's fracture** is an intra-articular fracture of the radial styloid, usually caused by forced extension and radial deviation of the wrist. The name originates from the chauffeurs who were sent to start the old Model-T Fords with the hand cranks, which would occasionally backlash onto the wrist, fracturing the radial styloid. Occasionally, a fall on an outstretched hand can occur with such force as to cause a **Monteggia fracture**, which is an unstable fracture of the proximal ulna with dislocation of the radial head. These fractures must be repaired operatively with reduction of the radial head and plate fixation of the ulnar fracture.

Evaluation

The evaluation of injuries to the radius and ulna always begins with a thorough history and physical exam of the hand. Though the injury may be immediately obvious, it is imperative that the entire hand and arm be evaluated in a careful, systematic manner so that all associated injuries are identified and treated. A thorough history of the injury is taken, including hand dominance and occupation. Ask about previous injuries to the extremity as well as relevant medical problems such as arthritis or carpal tunnel syndrome. A detailed description of the mechanism of injury can be helpful as well as associated symptoms such as numbness and tingling.

Next, inspect the arm from shoulder to fingertip and note any bruising, lacerations, and deformities. Take note of the manner in which the patient holds the arm at rest and note any pallor, cool skin or lack of sweating. Then, gently palpate the arm along its entire length to determine areas of tenderness and deformity. Evaluate the range of motion of all joints, taking into account local edema or arthritis which could limit mobility. Remember to evaluate pronation and supination of the forearm. Next, systematically check each long flexor and extensor of the wrist and fingers, as well as the intrinsic muscles of the hand. Finally, confirm that the radial, ulnar, and median nerves are intact and functioning. Entrapment or compression of these nerves can change a simple fracture of the wrist into a medical emergency requiring prompt reduction and decompression before permanent nerve damage occurs.

Example of a Correct Fracture Description

Figure 83.3 illustrates a common fracture and the correct means of describing it. An unacceptable call to one's chief resident to describe the above fracture would go something like this: "The patient has broken her right wrist". A professional call to the chief resident would be more along the lines of: "The patient is an 86 year old, right hand dominant female with a history of osteoporosis who fell on an outstretched right hand today while walking her dog. She has an obvious, painful deformity of her right wrist with no open wounds and no vascular compromise or loss of sensation. Her radiographs display an unstable, severely comminuted, intra-articular fracture of her distal radius and ulna styloid. The fracture is volarly displaced with an ulnar positive variance of 5 mm. She has an inclination of 14° and a volar tilt of -4°. There is significant articular incongruity.

Treatment

The aim of treatment is to return the fracture as nearly as possible to a normal anatomic alignment with minimal morbidity to the patient restoring adequate length (ulnar variance) and volar tilt are key. The choice of treatment depends entirely on an

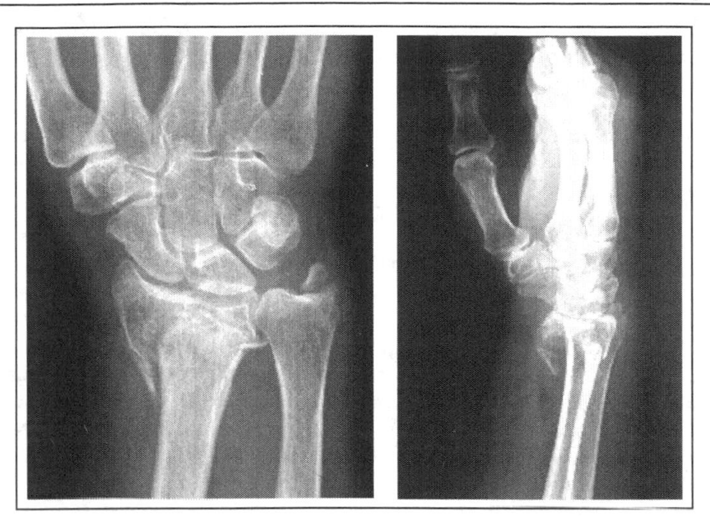

Figure 83.3. AP and lateral radiographs illustrating an example of a comminuted, intra-articular fracture of the distal radius and ulna styloid.

accurate assessment of the injury. A simple, minimally displaced fracture can be treated with cast immobilization for a period of 3-4 weeks. A displaced Colles' fracture can be treated with closed reduction, followed by application of a sugar tong splint, then cast immobilization for 4 weeks. Unstable fractures, intra-articular fractures and comminuted fractures frequently require more rigid fixation. Closed reduction with K-wire fixation is minimally invasive and adequately treats most unstable fractures that are not comminuted. However, most hand surgeons have largely shifted towards the use of a volar plate for rigid fixation. Comminuted and intra-articular fractures often need to be reduced openly in the OR, then rigidly fixed with a volar plate to hold optimal alignment. The advantage of this is an early return to activity and avoidance of joint stiffness. The downside is a trip to the OR and its inherent risks. Sometimes, the fracture is so comminuted that it won't hold a plate or the bone is too osteoporotic to hold screws and an external fixator is required. These are bulky and cumbersome but neutralize the compression force and help to maintain length.

Volar Plate Technique for Distal Radius Fractures

After exsanguination of the arm, a longitudinal incision is made over the distal portion of flexor carpi radialis (FCR). The incision is carried through the tendon sheath down to the tendon. Care must be taken to avoid injury to nearby structures such as the radial artery and palmar cutaneous branch of the median nerve. The FCR tendon is retracted radialy, and the pronator quadratus is visualized. The muscle is incised sharply, leaving a cuff of tissue on the radial side so that the muscle can be sewn back into place to cover the plate. It is then elevated in the subperiosteal plane off the distal radius with care to preserve its integrity. The periosteum of the distal radius including the fracture site is carefully elevated. The fracture is exposed, curetted, and reduced under fluoroscopy. Radial length and volar tilt must be restored. A multitude of companies produce volar plates. All are essentially locking plates with a prefabricated bend in the plate that matches the normal volar tilt. Once all the screws are placed, fluoroscopy is used to confirm the final result. The site is irrigated and the pronator reapproximated to cover as much of the plate as possible. The skin is closed with 4-0 nylon horizontal mattress sutures. A bulky dressing and volar splint are applied. The rigid fixation provided by the volar plate allows for early postoperative motion.

Pearls and Pitfalls

- X-ray all injured extremities including one joint proximal and distal to the level of injury.
- Reduction before fixation.
- When one forearm bone is fractured, look for another fracture in the other bone or a disruption of the joint at the wrist or elbow.
- Look for associated carpal injuries such as a scaphoid fracture or scpaholunate dissociation (tear of the scapholunate ligament).

Suggested Reading

1. Chang B. Principles of upper limb surgery. Chapter 65. Grabb and Smith's Plastic Surgery. 5th ed. Philadelphia: Lippincott-Raven, 1997.
2. Fernandez DL, Palmer AK. Fractures of the distal radius. Chapter 29. Green's Operative Hand Surgery. 4th ed. Philadelphia: Churchill Livingstone, 1999.
3. Leibovic SJ. Fixation for distal radius fractures. Hand Clin 1997; 13:665.
4. Trumble TE, Schmitt SR, Vedder NB. Factors affecting functional outcome of displaced intra-articular distal radius fractured. J Hand Surg 1994; 19:325.

Chapter 84

Wrist Fractures

Gil Kryger and Peter E. Hoepfner

Introduction

The wrist contains eight bones that are precisely arranged to provide maximum stability and mobility (Fig. 84.1). Although any of these bones can be fractured, the majority of injuries seen clinically and the most morbid injury is fracture of the scaphoid. Fractures of the trapezoid and lunate are extremely rare either as a fracture-dislocation injury or as an isolated injury. Fractures of the capitate are uncommon injuries and are usually in association with perilunate, trans-scaphoid, or trans-capitate fracture dislocations. The majority of this chapter will discuss scaphoid fractures and complications of nonunion.

Scaphoid Fractures

Epidemiology

The scaphoid bone (navicular bone) is the most commonly fractured carpal bone (80% of all carpal fractures). Patients, typically young, athletic, and active, fall on an outstretched arm, are injured in a motor vehicle accident, or are injured in sports. Roughly 400,000 fractures occur per year leading to significant morbidity and over 3.5 million days lost from work per year.

Anatomy

The scaphoid is a boat shaped bone with over 80% of its surface is covered by articular cartilage. It articulates with the radius and four of the remaining carpal bones. The scaphoid is the only carpal bone that spans the proximal and distal carpal rows; as such, it is involved in almost every movement of the wrist. It acts as an intercalary segment, connecting the proximal and distal carpal rows. The proximal and distal carpal rows can be thought of as two links in a chain: they are stable when pulled apart, but when compressed they will collapse. The scaphoid prevents this collapse during longitudinal compression of the wrist. However, if the scaphoid is broken (or heals with a fibrous nonunion) or if its ligamentous attachments are disrupted (scapholunate ligament tear), the carpus collapses. The blood supply is based on the superficial palmar branch (volar) and dorsal carpal branch (dorsal) of the radial artery. Both vessels enter the **distal** half of the scaphoid and supply the proximal pole via retrograde flow only. This retrograde flow is disrupted in proximal pole fractures, resulting in avascular necrosis of the proximal pole. The scapholunate ligament (dorsal), scaphocapitate ligament (palmar) and dorsal intercarpal ligaments are very important for stability. The ligaments also provide a vascular network that nourishes the scaphoid and should be preserved during surgery.

Practical Plastic Surgery, edited by Zol B. Kryger and Mark Sisco. ©2007 Landes Bioscience.

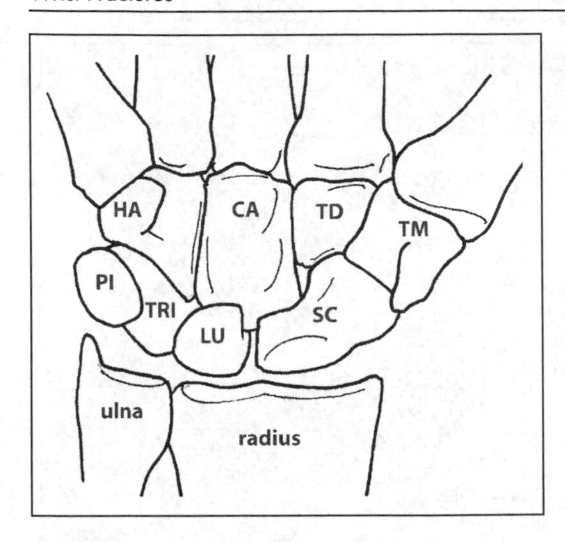

Figure 84.1. The carpal bones. HA, hamate. CA, capitate. TD, trapezoid. TM, trapezium. PI, pisiform. TRI, triquetrum. LU, lunate. SC, scaphoid.

Diagnosis

A fracture of the scaphoid should be suspected in all patients complaining of radial sided pain after a fall, in particular pain over the anatomic snuffbox (dorsally) or over the tubercle of the scaphoid volarly. Pertinent history includes the mechanism of injury, hand dominance, occupation, hobbies as well as a standard medical history.

In addition to a thorough hand exam, it is important to examine for tenderness of the snuffbox or scaphoid tubercle, pain with dorsiflexion, decreased grip strength, and dorsal swelling.

Radiographic Examination

The standard radiographs ordered include: AP, lateral, oblique, and PA of the wrist in ulnar deviation.

If the initial X-ray is negative but the exam is suspicious, an occult fracture may be present that is not evident on the plain films. A low threshold for suspicion of an occult fracture is key to its diagnosis. The wrist should be immobilized in a below-the-elbow thumb spica cast, and the patient should be seen after two weeks. If the site is still tender after 2 weeks, repeat X-rays should be obtained (95% sensitivity for repeat films). If repeat films are negative, but the patient is still tender, a bone scan is ordered. Recent data supports the use of CT and MRI in these cases as well. Patients with established nonunions of the scaphoid may present after an acute injury that aggravates the old injury. It is imperative to distinguish between an acute scaphoid fracture and a newly injured nonunion as the treatments are completely different. Nonunions typically have sclerotic edges at the fracture site whereas acute fractures will not.

Fracture Classification

Fractures can be categorized as stable or unstable.

Stable fractures may be managed conservatively or with surgery (see below). Fractures that have not healed after four months of adequate treatment may be

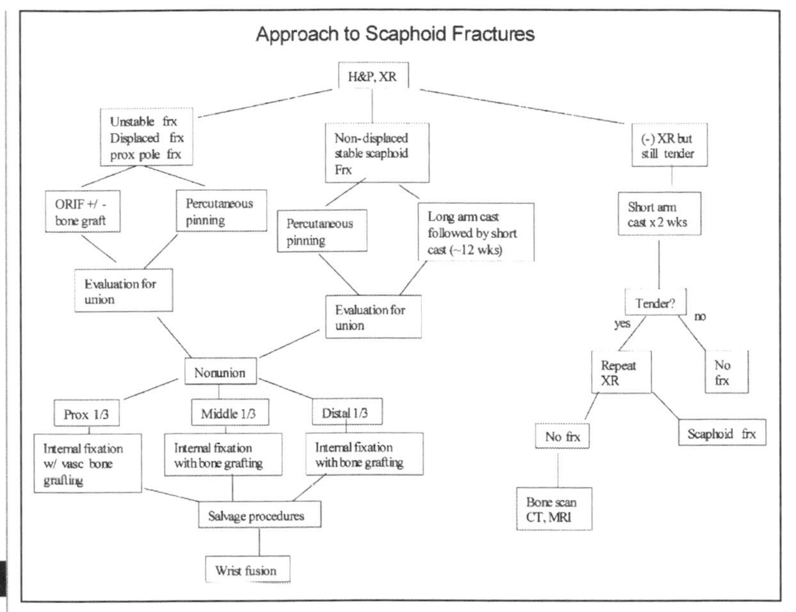

84

Figure 84.2. Algorithm for the management of scaphoid injuries.

classified as delayed unions with potential to heal if properly immobilized. Delayed unions demonstrate no sclerotic changes at the fracture site (in the presence of fracture widening and cystic changes in adjacent bone). If sclerotic changes are seen radiographically at the fracture site, the diagnosis of nonunion is made.

Unstable fractures are those that have a step-off of greater than or equal to 1 mm, any angulation, motion at the fracture site, or carpal instability. These fractures will progress to nonunion if untreated and always require surgical intervention.

Treatment

An algorithmic approach to the treatment of scaphoid fractures is shown in Figure 84.2.

Acute stable fractures: Historically, these patients were treated conservatively by immobilization for a total of 12 weeks (6 weeks in an above elbow thumb spica cast, followed by a below elbow thumb spica cast for the remaining 6 weeks). Recent studies have demonstrated that percutaneous screw fixation of acute stable scaphoid fractures should be offered to all patients that are surgical candidates. These patients return to work earlier, have earlier bony union, and have less wrist stiffness.

Displaced/unstable fractures: These are generally treated with open reduction and internal fixation. Rarely, some of these fractures may be amenable to percutaneous pinning and closed reduction to minimize soft tissue and blood supply disruption.

Proximal one-third fractures: These fractures always require surgery and should not be immobilized because of the high risk (16-42%) of nonunion and avascular necrosis.

Operative Considerations

The patient is placed supine with a well-padded tourniquet on the upper arm. Regional anesthesia (brachial plexus block) is preferred. Preoperative IV antibiotics are given. A fluoroscan and power driver are generally required.

Percutaneous screw fixation: A volar approach is used for all fractures except proximal pole fractures (because of the likelihood of missing a small proximal fragment from the volar side). A percutaneous pin is driven through the long axis of the scaphoid. This pin acts as a guide for a canulated screw that will span the fracture and provide compression and rigidity. Commonly used screws are the Accutrack, AO screw and the Herbert-Fisher cannulated screw. A second parallel pin prevents rotation while drilling the bone.

Open reduction internal fixation (ORIF): A volar approach is preferred. The surgeon should avoid stripping the bone of its ligaments as this will adversely affect blood flow and healing. Accurate reduction is obtained using a custom jig or by using pins to "joystick" the fragments into alignment. A compression screw provides rigid fixation and compression.

Bone grafting: In cases of malunion (or severe comminution), the fibrous scar is removed with curettes back to healthy bleeding bone resulting in a bone gap. Bone graft can be obtained from the iliac crest or distal radius. A screw with or without K-wire crosses the fracture and graft.

Postoperative Management

Immobilize the wrist in cast or splint (below the elbow) for 2-6 weeks (although some surgeons begin early range of motion immediately after surgery). Strengthening exercises begin roughly 3 weeks after surgery. Radiographs are obtained every 4 weeks until complete union is seen. There is new data that supports the use of adjuvant modalities to accelerate healing. External ultrasound and electromagnetic stimuli have been shown to decrease time to union and may be used in the postoperative period as well as during immobilization when surgery is not needed.

Scaphoid Nonunion

Nonunion occurs in 5-12% of all scaphoid fractures. Distal pole fractures almost always heal (nearly 100% union rate), middle one-third fractures have an 80-90% union rate, while proximal pole fractures have up to a 40% nonunion rate. Motion at the fracture prevents bony union, resulting in a fibrous pseudarthrosis at the fracture site. The highest risk of nonunion is in missed fractures, more proximal fractures, fractures associated with ligamentous injury, and displaced or comminuted fractures. Nonunion invariably results in osteoarthritis and instability of the wrist if not treated properly.

Classification

Stable nonunions are termed **Stage I**. They have a tough fibrous scar that prevents motion across the old fracture line. No degenerative changes are seen, but these will progress to unstable nonunions if not treated.

Unstable nonunions are characterized by motion across the old fracture site and associated arthritis. They are classified as Stages II-V. **Stage II** scaphoid nonunions have some degree of carpal collapse, but little or no arthritis. As the carpus collapses, the distal carpal row, namely the capitate, migrates proximally into the gap between the scaphoid fragments. The distal scaphoid fracture fragment may be displaced against the radial styloid (**Stage III**) causing radiostyloid

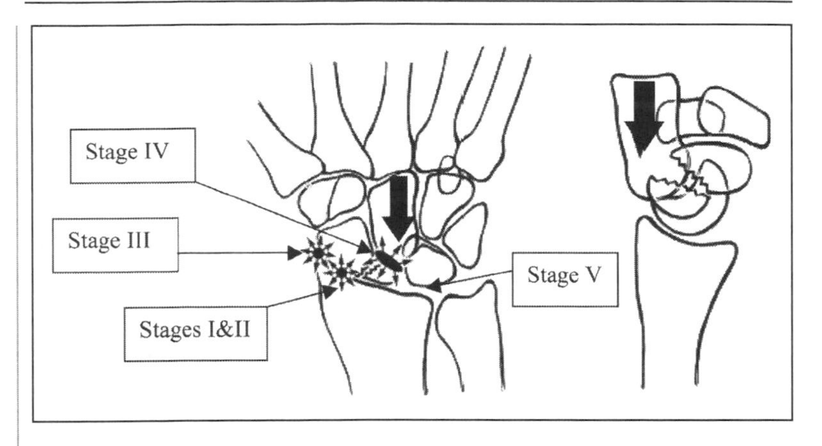

Figure 84.3. Scaphoid nonunion advanced collapse (SNAC): proximal migration (collapse) of the capitate into the gap between the scaphoid fragments displaces the fragments against adjacent bones resulting in arthritis. SLAC wrist is similar; however the capitate migrates proximally into the gap between the scaphoid and lunate.

arthritis. Additional collapse disrupts the midcarpal joint and the capitate rubs against the proximal fracture fragment and the radius resulting in midcarpal and radioscaphoid arthritis (**Stage IV**). Eventually, the distal row (mainly the capitate) pushes the lunate proximally against radius resulting in radiolunate and generalized arthritis (**Stage V**). Stages IV and V are termed scaphoid nonunion advanced collapse (**SNAC**, Fig. 84.3).

Treatment of Scaphoid Nonunion

The goal of treatment is to heal the nonunion, stabilize the carpus, and correct carpal collapse, thus relieving symptoms. After debridement of the fibrous scar, bone grafting is always required. The graft can be planned preoperatively to replace specific size and angular deformities. Wedge grafts can be placed on the volar side to correct a "humpback" deformity. Vascularized bone grafts are indicated for all proximal one-third nonunions, avascular necrosis of the scaphoid, as well as for failed nonvascularized bone graft attempts. The graft is generally based on small branches off the radial artery. Most widely used is the 1-2 intercompartmental supraretinacular artery graft (Zaidemburg Graft).

Salvage Procedures

Despite our best efforts to diagnose and fix scaphoid fractures, there will often be injuries that are missed or do not heal properly. The final common pathway of these types of injuries is degenerative arthritis of the wrist. In cases of generalized arthritis and pain, the goal is to achieve a pain-free functional wrist (with 40° of flexion and 40° of extension).

Starting with conservative approaches, the physician should counsel the patient to avoid overstraining the wrist and provide him with a wrist support. As the main patient complaint is pain, anterior and posterior neurectomies can be performed to reduce pain. Transection of the anterior and posterior interosseus nerves reduces pain and should be done when performing any other salvage procedure. Excision of the radial styloid or of any osteophyte can reduce pain by relieving impingement of the fragment on adjacent bone. The excised fragment can be used as a bone graft. Partial scaphoid excision is removal of the proximal or distal fragment without destabilizing the scapholunate ligament distally.

More complex procedures may be necessary, including scaphoid excision with four-corner fusion of the capitate, lunate, triquetrum and hamate to treat arthritis. Proximal row carpectomy involves excision of the proximal row allowing the capitate to sit in the lunate fossa of the radius. Studies have shown that the results are similar for both procedures with a proximal row carpectomy avoiding the lengthy immobilization required by four-corner fusion. The final salvage procedure for wrist arthritis is total wrist fusion.

Hamate Fractures

Fractures of the body are less common and are associated with fractures of the base of the ulnar metacarpals to form the ulnar component of axial instability patterns and should be treated as such. Nondisplaced isolated fractures can be treated with immobilization.

Hook fractures commonly occur in individuals holding golf clubs, bats or hammers. Patients complain of a dull ache when gripping. Inflammation and displacement can affect the ulnar nerve (paresthesias) or flexor tendons to the ring and small fingers (pain with flexion). Diagnosis is often delayed, but can be made with carpal tunnel views of plain films or CT and bone scans.

Acute nondisplaced hook fractures can be immobilized for 6 weeks. Displaced fractures and those that are diagnosed 1-2 months after the initial trauma are best treated with excision of the fracture fragment. Some surgeons advocate stabilization with a 1.5 mm cortical screw. After excision, mobilization is started after 1-2 weeks. With internal fixation, the hand is splinted for 6 weeks with a removable brace.

Triquetrum Fractures

These common fractures often occur in association with other wrist fractures as a result of a fall on an outstretched hand. Most commonly, a dorsal cortex chip is seen on lateral view. These fractures can usually be treated symptomatically. Less common are body fractures which are usually nondisplaced (and missed on x-ray); treatment is 4-6 weeks of immobilization. Large displaced fragments and those that involve the articulating surface require ORIF followed by 4-6 weeks of casting.

Lunate Fractures

These fractures are rare and may be a manifestation of Kienböck's disease. Treatment of nondisplaced fractures is by cast immobilization (below the elbow thumb spica) for 4-6 weeks. Displaced fractures may require ORIF.

Trapezium Fractures

These rare fractures occur as a result of compression of the trapezium between the radial styloid and thumb metacarpal. Plain films including a carpal tunnel view and oblique view will visualize the fracture; however some authors believe that CT is a better modality to define the injury and the concomitant thumb metacarpal injury. Unstable fractures, or those with 1 mm or more of depression, should be treated with ORIF (often with bone graft). Postoperative immobilization for 4-6 weeks is recommended.

Trapezoid Fractures

The least injured carpal bone, fractures of the trapezoid occur as a result of volar forces applied to the index metacarpal to cause dorsal dislocation of the bone. Acute fractures require anatomic reduction and stabilization. If closed reduction is successful, 4-6 weeks of casting (below the elbow thumb spica), or until the fracture is nontender, is needed.

Capitate Fractures

Isolated fractures are rare. Most fractures involve the proximal two-thirds of the bone as part of a perilunate injury. Nondisplaced fractures are immobilized but rotated fragments require ORIF. Diagnosis is by CT.

Pisiform Fractures

Fractures are rare and are treated with immobilization. Nonunions are treated by excision.

Wrist Instabilities

Injuries to the ligaments of the carpus are common and occur alone or in conjunction with the abovementioned fractures. It is important to understand the biomechanics of the wrist to fully understand these injuries, their natural history and treatment. A full discussion of this should be sought in a hand surgery textbook. Briefly, the wrist joint can be thought of as three joints: radiocarpal, midcarpal and carpometacarpal. These three joints flex and extend in unison to allow a full range of flexion and extension motions. Additionally, the carpal bones translate during radial and ulnar deviation. With deviation of the wrist towards the radius, the proximal bones (scaphoid, lunate and triquetrum) tend to flex volarly. With ulnar deviation, the scaphoid is extended and is better visualized on AP and PA views.

Diagnosis

As with all ligamentous injuries, the timing of diagnosis and repair is paramount. Acute injuries (<1 week) have maximum healing potential; subacute injuries (1-6 weeks) have decreased healing potential but do not demonstrate fixed deformities or arthrosis; chronic injuries (>6 weeks) usually display fixed deformities and therefore require surgical repair and reconstruction. The key is to diagnose these injuries early to maximize healing without surgery.

Classification

It is useful to categorize these injury patterns based on their radiological appearance (Fig. 84.4). On lateral view, the radius, lunate and capitate are linear while the scaphoid is flexed 45° volarly relative to the lunate. When the lunate is angled dorsally, the term dorsal intercalated segmental instability (**DISI**) is used. Because there is dissociation between the scaphoid and lunate (scapholunate ligament tear), the scaphoid is flexed beyond 80° in relation to the lunate.

Much less frequently, the lunate is angled volarly, and the term becomes volar intercalated segmental instability (**VISI**). The lunate is flexed in the same direction as the scaphoid, and the angle between them decreases to about 30°. This orientation indicated injury to the LT ligament.

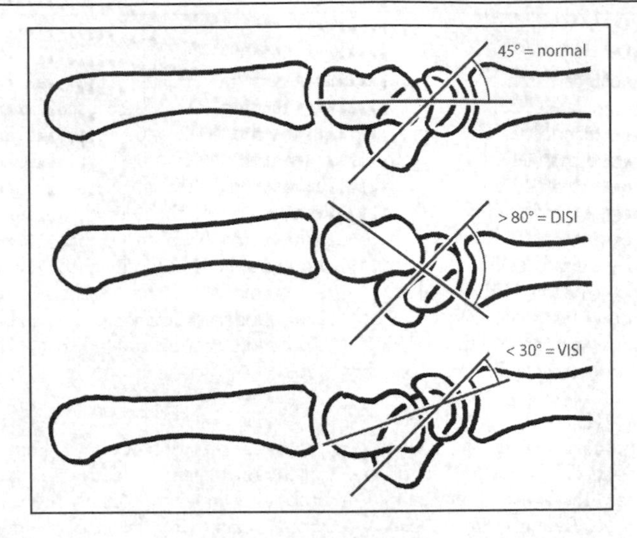

Figure 84.4. Wrist instability. Shown are the dorsal intercalated segmental instability (DISI) and the volar intercalated segmental instability (VISI) deformities.

Dorsal Intercalated Segmental Instability

The most common cause of the DISI deformity is a disruption of the scapholunate ligament (S-L). This injury should always be suspected in a radial styloid fracture. In advanced stages, the scaphoid is hyperflexed and the term rotatory subluxation of the scaphoid is also used. Mayfield and colleagues demonstrated that there is a predictable pattern of injuries as the S-L is torn, followed by a sequence of tearing of ligaments around the lunate until the carpus is dislocated; this progression correlates with the amount of energy causing the injury. The final stage of carpal dislocation is the volar dislocation of the lunate as the capitate is driven into the radiocarpal joint. Since the lunate is still attached to the radius by its volar ligaments, the lunate dislocates volarly, resting perpendicular to the radius on lateral view. This is often termed the "spilled teacup sign," and is shown in Figure 84.5.

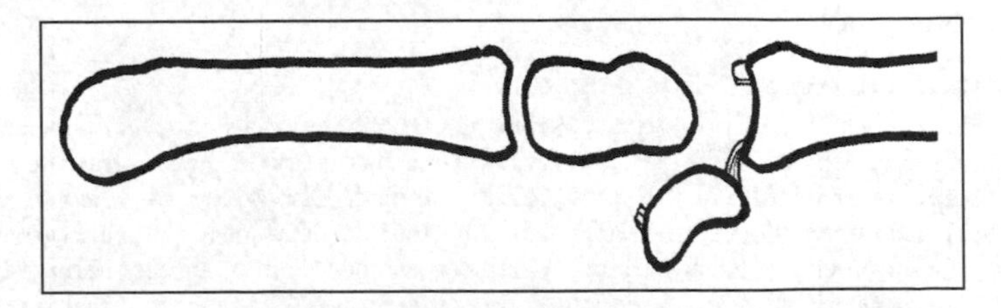

Figure 84.5. The "spilled teacup sign." Disruption of the scapholunate ligament can lead to volar dislocation of the lunate as the capitate is driven into the radiocarpal joint. Since the lunate is still attached to the radius by its volar ligaments, the lunate dislocates volarly, resting perpendicular to the radius on lateral view.

Diagnosis

Diagnosis is based on pain and tenderness in the snuffbox and over the S-L dorsally, associated with clicking sounds with motion. The **Watson scaphoid shift test** can be diagnostic. The examiner places his thumb over the scaphoid tubercle volarly, while passively deviating the wrist from ulnar to radial. The scaphoid attempts to flex volarly but is blocked by the examiner's thumb; because the S-L is disrupted, the proximal pole of the scaphoid rides up over the dorsal lip of the radius and the thumb is removed, and the scaphoid clunks back into the radioscaphoid joint and pain is elicited. The exam and radiographs should always be compared with the contralateral uninjured wrist. On scaphoid views, the scapholunate interval is >2 mm (Terry Thomas sign). As the scaphoid flexes more volarly, it can be seen on stress PA views on end (perpendicular to the radius); the distal pole overlying the proximal pole creates the "scaphoid ring sign".

Treatment

In acute and subacute settings, without rotatory subluxation of the scaphoid, treatment consists of arthroscopic debridement of the torn ligaments and stabilization of the carpus with K-wires, followed by immobilization in a below-elbow thumb spica for 4-6 weeks. Recently, direct repair of the S-L ligament with dorsal capsulodesis for augmentation has been advocated. In advanced cases termed scapholunate advanced collapse (**SLAC**), the carpus collapses in a regular pattern. Arthritis begins between the scaphoid and radial styloid and progresses to the entire radioscaphoid and capitolunate joints. As in SNAC wrist, the radiolunate joint is spared except in severe cases.

Volar Intercalated Segmental Instability

VISI deformities are much less common than DISI deformities. The most common cause of the VISI deformity is lunotriquetral (L-T) dissociation. These injuries are diagnosed by ulnar sided pain with limited range of motion in association with clunk. The provocative test is the shear test. In this test, the examiner pushes the lunate volarly and the pisiform (and triquetrum) dorsally causing a shearing force at the L-T joint. Pain is considered a positive test.

Treatment

Management of L-T injuries without frank VISI deformity is by immobilization alone. Even chronic complete tears of the L-T ligament can heal with immobilization. For more severe injuries, and those with carpal collapse, limited fusion of the wrist is indicated.

Combined Fracture-Dislocations

Extremely high energy injuries can result in carpal fracture-dislocations. The pattern of injury depends on its path across the carpus: disruption of the S-L, fracture of the scaphoid or fracture of the scaphoid and capitate. Displacement of the carpal bones can compress the neurovascular structures in the wrist resulting in acute carpal tunnel syndromes. Immediate reduction in the emergency room should be attempted, with later operative fixation. Irreducible dislocations should be taken to OR for reduction. Patterns of perilunate fracture-dislocation are shown in Figure 84.6.

Pearls and Pitfalls

- Acute, nondisplaced, nonangulated scaphoid fractures can be treated with an above-elbow thumb spica cast for at least three weeks, followed by an additional period in a below-elbow thumb spica cast with the wrist in mild radial deviation

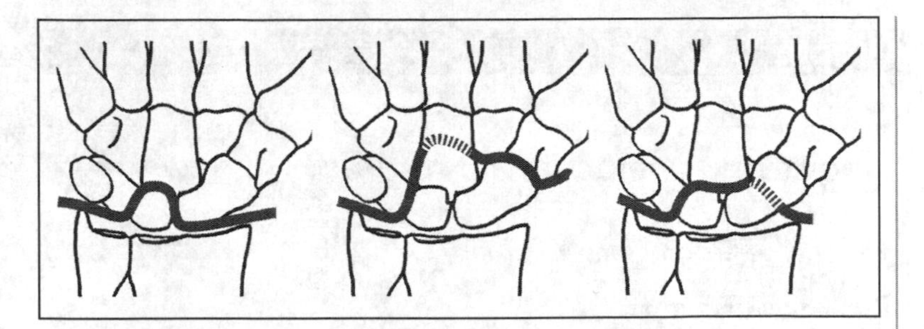

Figure 84.6. Perilunate fracture-dislocation patterns: perilunate injury with disruption of the S-L ligament (left), perilunate injury with fracture of the capitate (center), and perilunate injury with fracture of the scaphoid (right).

and palmar flexion. However, many patients, especially those who are young and active, will not agree to lengthy casting protocol. In such cases rigid fixation is required (as described above).

- Pulsed electromagnetic field therapy (PEMF) has been used to help stimulate healing in scaphoid nonunions. At the present, it is not widely used; however it should be considered in cases in which surgery is not an option since it does have a low risk profile.
- Proximal row carpectomy is an acceptable treatment for scaphoid nonunion. It can be considered even in young active patients, and long-term results have been superior to intercarpal fusion. Excision of the terminal branch of the posterior interosseous nerve during carpectomy may help reduce some of the postoperative pain.
- Hook of the hamate fractures occur in athletes gripping a club, most commonly in golfers. They can be difficult to diagnose. Pain is in the ulnar wrist and worsened by gripping and with little finger flexion against resistance with ulnar deviation of the wrist. If left untreated, they have a high risk of nonunion.
- Capitate fractures are difficult to diagnose by plain radiograph, and a CT scan is often required. The neck of the capitate can fracture in association with a scaphoid waist fracture, termed scaphocapitate syndrome. If missed, the head of the capitate can develop avascular necrosis. This injury should be suspected in a fall on an outstretched hand with the wrist in extension or from a blow to the dorsum during wrist flexion.
- Triquetrum fractures should be suspected in a fall on an outstretched hand combined with a twisting motion of the wrist. These fractures usually do not require surgery and have a low complication rate.

Suggested Reading

1. Blatt G. Scapholunate Instability. In: Lichtman DM, ed. The wrist and its disorders. Philadelphia: WB Saunders Co., 1988.
2. Bond CD, Shin AY, McBride MT et al. Percutaneous screw fixation or cast immobilization for nondisplaced scaphoid fractures. J Bone Joint Surg Am 2001; 83A(4):483.
3. Gellman H, Caputo RJ, Carter V et al. Comparison of short and long thumb-spica casts for nondisplaced fractures of the carpal scaphoid. J Bone Joint Surg Am 1989; 71(3):354.
4. Mayfield JK. Mechanism of carpal injuries. Clin Orthop Relat Res 1980; 149:45.
5. Trumble TE, ed. Principles of Hand Surgery and Therapy. Philadelphia: WB Saunders Co., 2000.
6. Zaidemberg C, Siebert JW, Angrigiani C. A new vascularized bone graft for scaphoid nonunion. J Hand Surg [Am] 1991; 16(3):474.

Chapter 85

Finger and Metacarpal Fratures

Oliver Kloeters and John Y.S. Kim

Phalangeal Fractures

Introduction

The bones of the fingers are the most commonly fractured bones in the human body, accounting for roughly 10% of all fractures. Half of all finger fractures occur in the proximal and middle phalanges, and the other half in the distal phalanx. Most phalangeal fractures can be treated nonoperatively with splinting alone. Fractures involving the articular surfaces and those with accompanying tendon or soft tissue injury are more complicated and usually require operative intervention. This chapter will focus on the commonly encountered fractures.

Relevant Anatomy

The fingers are comprised of three bones, the proximal, middle and distal phalanges. Each phalanx is further subdivided into a shaft, neck and head. The bones articulate at the proximal interphalangeal (PIP) and the distal interphalangeal (DIP) joints. The proximal phalanx articulates with the metacarpal bone at the metacarpophalangeal (MP) joint. On the palmar surface, the tendon of flexor digitorum superficialis (FDS) inserts on the volar aspect of the middle phalanx, and the flexor digitorum profundus (FDP) tendon inserts on the volar aspect of the distal phalanx. FDS flexes the PIP joint and FDP flexes primarily the DIP joint. When evaluating a finger for possible tendon injury, it is important to distinguish between these two tendons. Since FDP is the sole flexor of the distal phalanx, the PIP joint has to be immobilized in extension in order to accurately assess FDP function. Evaluation of FDS function requires immobilization of the DIP joints of the other fingers, since the tendons arise from the same muscle belly and can flex the digits simultaneously. Each extensor tendon inserts on the middle and distal phalanges, via the central and lateral bands, respectively. A more detailed discussion of flexor and extensor anatomy can be found in the chapters on flexor and extensor tendon repair.

Clinical Presentation

Presentation of a finger fracture will vary depending on the mechanism and location of the injury. Fracture signs include swelling, hematoma, ecchymosis, deformity, tenderness/pain, decreased range of motion (ROM) and accompanying soft tissue injury. In addition, the possibility of tendon avulsion, collateral ligament damage, or volar plate rupture should be determined. Assessment and documentation of any neurovascular injury must be carried out prior to any therapeutic or diagnostic manipulation, such as stress view radiographs.

Rotation deformity is an often-neglected problem and should be addressed by at the time of fracture fixation, as an untreated rotational deformity can impair ROM

Practical Plastic Surgery, edited by Zol B. Kryger and Mark Sisco. ©2007 Landes Bioscience.

and functional outcome. A useful clinical test is to ask the patient to flex his fingers while observing the position of the nails and comparing them to the other hand. As a rule, all abnormal findings in the injured hand should be compared to the other hand as individual malformation or previous injuries may lead to an inaccurate diagnosis.

Fracture Patterns

Phalangeal fractures can be classified according to the following: (1) open or closed; (2) displaced or nondisplaced; (3) the finger and phalanx involved; (4) fracture pattern: oblique, transverse, spiral, or comminuted; (5) any associated soft tissue or neurovascular injuries.

Imaging

Negative plain radiographs in three views (posterior-anterior, oblique and lateral) are ample for ruling out most phalangeal fractures. If necessary, additional stress views can be obtained in which the injured phalanx is manually stressed. Since this maneuver is usually painful, a digital block with lidocaine can be helpful. CT and MRI are not indicated for the diagnosis of a phalangeal fracture.

Transverse fractures of the proximal phalanx are more likely to result in volar displacement of the distal fragment due to traction on the proximal bone by intrinsic and extensor tendons. Middle phalanx fractures can result in distal fragment displacement either volarly or dorsally, depending on whether the fracture is proximal or distal to the FDS insertion. Fractures of the distal phalanx are usually related to crush injury involving the nail bed.

Treatment Principles

The principal goals of treatment are complete functional restoration and full ROM of the injured finger. This is achieved through precise anatomical alignment (open or closed), and early mobilization. Full immobilization (e.g., splinting) should not exceed 3 weeks as the extensor and flexor tendon systems as well as the capsules will begin to irreversibly contract and stiffen. In general, uninvolved joints should be excluded from splinting (e.g., avoid splitting the PIP and MP joints in the setting of a distal phalanx fracture). A meticulous physical examination is key to detecting rotation deformities or poor anatomical reduction. Postreduction splinting plays an important role in the treatment of almost all fractures, and various finger splints are widely available in most acute care settings. Open fractures or extensive soft tissue injury should be treated with antibiotics and tetanus prophylaxis.

Nonsurgical Treatment

Treatment of finger fractures without any form of surgical fixation consists primarily of splinting. The aim of splinting is to immobilize the fractured segment and surrounding joints without interfering with the motion of the uninvolved joints. Stack splints are used to immobilize distal phalanx fractures while allowing full ROM of both the PIP and MP joints.

Extraarticular, nondisplaced, stable phalangeal fractures are treated with closed reduction, splinting and early rehabilitation. Closed reduction is performed using axial traction with concomitant alignment of the bone. For most shaft fractures of the middle or proximal phalanges, "buddy splinting" to the adjacent finger for 1-2 weeks is sufficient. In addition, radial or ulnar gutter splints are useful as they leave the remaining digits uninvolved. Postreduction films (2 views) should be obtained after closed reduction and following one week to ensure continued fracture reduction.

Surgical Treatment

Surgical fixation and reduction are indicated for any of the following:

- Open fractures
- Unstable fractures
- Dislocated or intraarticular fractures
- Irreducible fractures

Proximal Phalanx Fractures

Different types of shaft fractures, such as transverse, oblique or spiral are in general considered to be unstable and require surgical fixation using two longitudinal K-wires. Intraarticular base fractures of the proximal phalanx are divided into three types: vertical shear fractures, compression fractures, and collateral ligament avulsions. The first two types may have impaction or rotated fragments. If this is the case, fixation is required. Stable and nondisplaced avulsion fractures are treated with splinting alone, using a buddy strap to the adjacent digit. If displacement is evident on radiographs, anatomic reduction is mandatory in order to prevent post-traumatic arthritis or persistent instability. Reduced vertical shear fractures should be stabilized using a percutaneous K-wire or an interfragmentary screw under image magnification. Fractures that have undergone bone-to-bone fixation can usually be mobilized the following day with a buddy strap to the adjacent finger. Ligament-to-bone repairs usually take longer to heal and should not be mobilized for 4 weeks. Compression fractures usually require open reduction with internal fixation (ORIF), especially if there is in an intraarticular step off of greater than 2 mm or visible angulation. Intrafragmentary screw fixation, mini-T, or minicondylar plates or used for internal fixation. The addition of a cancellous bone graft is useful to avoid over-compression and incongruity of the fragments.

Middle Phalanx Fractures

Middle phalanx fractures usually do better than proximal phalanx fractures since the DIP joint is functionally less important than the PIP or MP joints. Malunion, or post-traumatic arthrosis of the DIP joint is less severe than in the PIP or MP joints and can be treated without difficulty by arthrodesis. The vast majority of middle phalanx fractures are stable and well aligned. These can be treated with a buddy strap to the adjacent digit or gutter splints. Radiographs should be performed every 5 days for about 2 weeks to detect displacement. In the event of an unstable middle phalangeal shaft fracture, K-wire fixation is sufficient. Open fractures with extensive soft tissue damage may require ORIF.

If a shaft fracture is stable but malaligned after reduction, an extension block cast should be considered as it provides stabilization of the reduction and reduces the risk for PIP joint stiffness. The extension block should be maintained for 4 weeks followed by 2 weeks in a buddy strap. Rotational abnormalities should be ruled out by noting the alignment of the nail beds under active flexion of the fingers towards the palm and comparing them to the healthy hand. Malalignment of a fracture can occur in patients presenting weeks after the initial trauma, with a callus already present or soft tissue interposition. Often, these can no longer be reduced successfully in a closed fashion. Therefore, an ORIF is required, usually accompanied by an osteotomy along with tenolysis and capsulotomy.

Distal Phalanx Fractures

Distal phalangeal fractures are the most common finger fracture. Kaplan's classification divides them into three types: Type A, or longitudinal; Type B, or comminuted;

and Type C or transverse. Type B fractures typically result from a crush injury. This often leads to an exquisitely painful subungual hematoma. Puncture of the nail with a heated paper clip works well to release the trapped blood. Up to 2 weeks of splinting with a Stax splint is sufficient for most cases. Transverse or longitudinal shaft fractures (Type A and C) are mostly stable with minimal displacement. The finger should be splinted (usually 3 to 4 weeks) until evidence of bony union is seen on radiographs. Open and unstable shaft fractures require K-wire fixation after alignment. A detached nail should be replaced and if necessary, secured with suture fixation. The nail helps protect the underlying matrix, prevents scar formation between the nail matrix and eponychium, and guides the direction of growth of the new nail.

Fractures of the base of the distal phalanx are usually unstable and tend to angulate in a dorsal fashion due to the traction of the extensor tendons. Closed fractures are treated by splinting (e.g., using Alumafoam) of the finger in extension for at least 4 weeks. Open fractures of the base are considered to be unstable and tend to have a rotational deformity. Retrograde K-wire fixation is a straight forward method of stabilization.

Condylar Fractures

Meticulous anatomic reduction is essential in condylar fractures, as malalignment can lead to a flexion contracture. A closed reduction will generally not provide an adequate result, requiring ORIF for reconstruction. Unicondylar fractures are usually unstable and often demonstrate a rotation deformity due to traction of the collateral ligament. If closed reduction is attempted by axial traction, a reduction clamp should be used to maintain the repositioned condyle and a 0.028 inch K-wire percutaneously inserted through the uninjured condyle. Intraoperative films in all three dimensions should be obtained to insure precise alignment of the small but functionally important fragment.

Metacarpal Fractures

Introduction

A fracture of the metacarpal is often termed a "boxer's fracture" alluding to the etiology, which is usually direct or indirect trauma. Metacarpal fractures account for nearly 40% of all hand fractures. About a quarter of these are located in the neck of the 5th metacarpal bone.

The four metacarpals are tubular bones, each with an intrinsic arch. The 2nd and 3rd carpometacarpal (CMC) joints have almost no range of motion, whereas the 4th and 5th CMC joints provide 15° and 25° of relative motion, respectively.

Injuries of the metacarpal bone are clinically divided into fractures of the metacarpal base, the metacarpal shaft or neck and the metacarpal head.

Imaging

For most fractures, three-view conventional radiographs are sufficient. Severely comminuted or intraarticular fractures may also require stress views. CT scanning can be helpful for evaluation of questionable intraarticular involvement, since this can be difficult to determine on plain radiographs. The purpose of a CT scan in this setting is to assess the alignment or displacement of the fracture, and to aid in planning the operative approach.

Clinical Findings

Signs and symptoms include pain, hematoma, swelling, reduced motion and crepitus. These findings are common in metacarpal shaft fractures, while fractures of the metacarpal base and neck may have a subtle presentation.

Fractures of the shaft tend to be shortened and rotated. Rotation deformities must be addressed; if untreated they pose a risk for a weakened grip and incomplete fist closure due to overlapping fingers. Fractures of the metacarpal base may be impacted since they are often caused by an axial load, making the diagnosis easy to miss. Fractures of the metacarpal head often present with nothing more than localized swelling and pain. Crepitus can be heard but should not be elicited since it has no additional diagnostic benefit.

Treatment

Most metacarpal fractures are adequately treated with closed reduction followed by immobilization in a forearm-based splint. two-view radiographs should be taken in the follow-up period in order to confirm that the reduction has been maintained. Early mobilization of wrist and fingers is absolutely critical to achieve optimal long-term functional results. Apex and dorsal angulation are often seen in shaft fractures. There is, however, a certain degree of angulation that is acceptable (up to 10° for the second and third, up to 20° for the fourth, and up to 30° for the fifth metacarpal).

The following is a useful list of the indications for surgical treatment of a metacarpal fracture:

1. Open fractures
2. Unstable and extensively displaced fractures requiring stabilization
3. Nonreducible fractures
4. Multiple fractures
5. Old fractures that need to be reduced (>1 week post-trauma)
6. Failure of nonsurgical treatment
7. Fractures with extensive soft tissue injury

Metacarpal Base Fractures

Most base fractures are impacted and are treated by splinting alone and early mobilization. Base fractures of the 5th metacarpal often require surgical treatment if displaced greater than 2 mm. Fractures involving the CMC joint are unstable and have a strong tendency to destabilize after a successful reduction due to traction forces of the surrounding ligaments. Therefore, CMC fractures often need open reduction with pin fixation. Postoperative splinting should be maintained for 2-3 weeks followed by mobilization. The pins should be removed after 6-8 weeks.

Metacarpal Shaft Fractures

In general, metacarpal shaft fractures are treated with reduction and splinting alone. The MP and PIP joint of the affected metacarpal bone are flexed to a 90° angle. Reduction is achieved with upward directed traction on the middle phalanx and downward resistance at the level of the proximal metacarpal shaft. Either splinting (extending to the distal palmar crease) or percutaneous pins are employed to maintain the reduction result.

First Metacarpal Fractures

Introduction

Thumb fractures are usually the result of occupational or sporting-related trauma. The thumb consists of a proximal and distal phalanx and a metacarpal bone. The first CMC joint is the articulation of the trapezium and the metacarpal base and provides a wide range of motion. It is stabilized by a number of ligaments, most importantly the anterior and posterior oblique ligaments, the intermetacarpal ligaments, and the dorsal radial ligament. The ROM of the MP joint of the thumb is extremely variable. It has relatively little intrinsic stability, and the stabilizing ligaments are prone to injury. The ulnar and radial collateral ligaments stabilize the MP joint on its sides, and the volar plate stabilizes it on the palmar surface.

Fracture Patterns

Metacarpal fractures of the thumb are divided into base, shaft and head fractures. Typical signs of fracture are swelling, decreased range of motion, pain (most notably in base fractures) and occasional crepitus. Shaft fractures are rare due to the stable cortical bone structure and tend to result in apex-dorsal or apex-radial angulation. The most common fractures are of the metacarpal base, and these can be intraarticular (e.g., Bennet fracture) or extraarticular. Base fractures are usually due to a gently flexed metacarpal during an axial force. Adequate imaging of the CMC joint is essential in diagnosing a metacarpal base fracture: a true lateral view of the joint with the palm flat on the plate and wrist and hand pronated 15-35° is obtained. The imaging beam is aimed at the CMC joint from a 15° oblique angle from distal to proximal.

Extra-Articular

Extraarticular fractures are usually oblique or transverse. Closed reduction with a thumb spica cast for 4 weeks excluding the distal phalanx are usually sufficient. Reduction is obtained with axial traction, palmar directed pressure at the level of the fracture, and pronation of the distal fragment. Percutaneous pin fixation is indicated in cases of an unstable fracture postreduction (more likely in oblique fractures), open fractures, and fractures with angulation greater than 30° with shortening. External or internal fixation is required for extensively comminuted fracture types.

Intra-Articular

The Bennett fracture is an intraarticular fracture of the metacarpal base with a volar-ulnar fragment of variable size. The fragment is often kept in anatomic position through the anterior oblique ligament. Restoring full anatomic congruity is mandatory to avoid post-traumatic arthritis. If articular incongruity is less than 2 mm postreduction, the fracture can be stabilized with percutaneous pin fixation. Open or unstable fractures, irreducible fractures, or those with an incongruity greater than 2 mm postreduction are indications for ORIF.

Pearls and Pitfalls

Fractures of the metacarpals and phalanges may be associated with ligamentous injuries (collateral ligaments or carpometacarpal ligaments), and therefore a careful examination for dislocations and joint instability must also be performed. Stiffness is a prime complication of a phalangeal injury. Early mobilization is key to optimal outcome (within 3-4 weeks of injury if possible). Differences in bony healing in the adult and pediatric population should also be recognized. Since children heal faster, they can be splinted for shorter duration compared to adults (with minimally displaced metacarpal fractures for example).

Suggested Reading

1. Blair WF, ed. Techniques in Hand Surgery. Baltimore, Maryland, USA: Williams and Wilkins, 179-284.
2. Stern PJ. Fractures of the metacarpals and phalanges. In: Green DP, Hotchkiss R, Pederson WC, eds. Green's Operative Hand Surgery. 4th ed. New York: Churchill Livingstone, 1999.
3. Weiss AP, Hastings H. Distal unicondylar fractures of the proximal phalanx. J Hand Surg 1993; 18A:594.
4. Gonzales MH, Hall RF. Intramedullary fixation of metacarpal and proximal phalangeal fractures of the hand. Clin Orthop 1996; 327:47.

Chapter 86

Brachial Plexus Injuries

Mark Sisco and John Y.S. Kim

Introduction

Injuries to the brachial plexus can result in significant long-term functional impairment. Recent advances in microneurosurgery have allowed for the reconstruction of many of these injuries. This chapter will focus on specific goals and general approaches toward adult brachial plexus reconstruction.

Anatomy

The brachial plexus is formed from the branches of the C5 to T1 spinal nerves (Fig. 86.1). The spinal nerves are comprised of dorsal and ventral roots. The ventral roots carry motor fibers. The dorsal roots are made up of sensory fibers that originate in the dorsal root ganglion in or just distal to the intervertebral foramen. Each root has a root sleeve comprised of dura and arachnoid that forms the nerve sheath at the level of the intervertebral foramen.

Typically, the upper trunk derives from the C5-C6 nerve roots, the middle trunk from the C7 nerve root, and the lower trunk from C7 and T1 nerve roots. Each trunk divides into an anterior and posterior division. The anterior divisions of the upper and middle trunks comprise the lateral cord while the anterior division of the lower trunk forms the medial cord. The posterior divisions of all three roots form the posterior cord. These cords will then differentiate into terminal branches.

Epidemiology

Adult brachial plexus injuries usually result from significant trauma to the upper limb. High-speed motor-vehicle collisions, especially those involving motorcycles, are the most common mechanism of injury. Other forms of blunt trauma, including industrial accidents, falls, and pedestrian-motor-vehicle accidents, are typical causes. Penetrating trauma is a less common cause of injury and may be classified into sharp injuries and blast-effect injuries. Young men constitute the largest demographic group affected.

Mechanisms of Injury

There are five main mechanisms that lead to brachial plexus injury: (1) crush injury, caused by blunt extrinsic force; (2) traction injury, usually caused by stretching of the nerves; (3) compression injury from intrinsic force (pressure from adjacent tissues or hematoma); (4) sharp penetrating injury; and (5) blast injury (typically from gunshot wounds).

Initial Assessment

Since the mechanism of brachial plexus injury often involves significant blunt or penetrating trauma, other injuries are often present. Initial management must

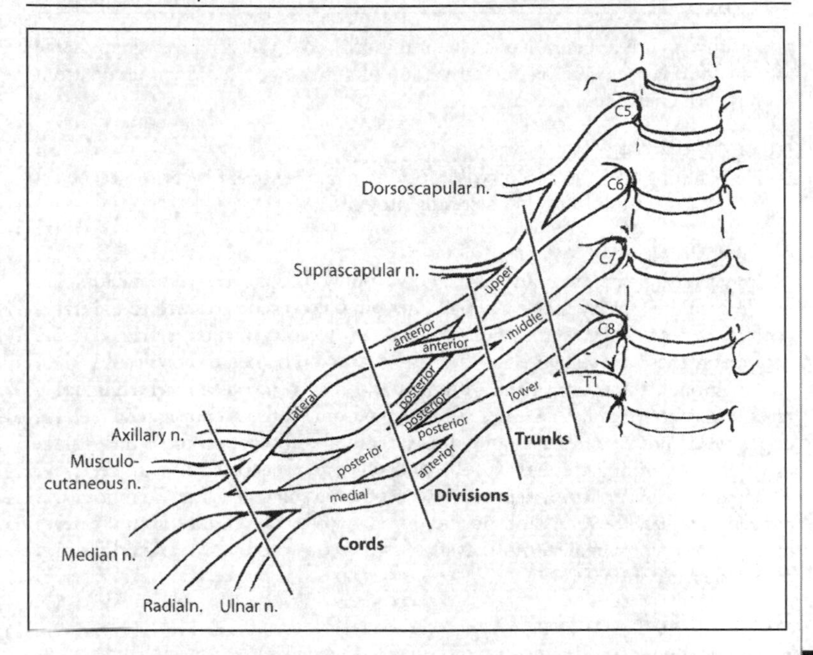

Figure 86.1. The brachial plexus.

address life-threatening injuries. As such, the diagnosis of brachial plexus injuries is often delayed until overall stabilization has occurred and mental status has improved.

History and Physical

The mechanism of injury must be accurately determined since prognosis is intimately related to etiology. Manual muscle testing should include observation of muscle atrophy and tone. The British Medical Research Council grading system for muscle function is often helpful for standardized documentation of strength. Passive range of motion should be assessed to check for contractures, followed by active range of motion testing. Median, radial, and ulnar nerve function can be assessed by testing finger and wrist motion. The musculocutaneous nerve is responsible for elbow flexion. Abduction of the shoulder is dependent on the axillary nerve. The posterior cord can be assessed by checking finger and wrist extension, elbow extension, and shoulder abduction. The medial and lateral cords can be assessed via the medial and lateral pectoral nerves, which innervate the sternal and clavicular heads of the pectoralis major, respectively. The sensory examination must be thorough, and should include testing for Tinel's sign at the supraclavicular fossa and the infraclavicular coracoid process.

Radiographic Evaluation

Plain film radiographic evaluation of the cervical spine and ipsilateral shoulder and scapula can delineate contributing bony injury or consequent bony deformity. Penetrating trauma in the setting of abnormal pulses suggests a hematoma-related plexopathy; an arteriogram may be helpful in this case. On CT myelogram, nerve root avulsion is suggested by demonstration of a meningocele, which results from tearing of the nerve root sleeve.

Injury Patterns

Determining the level of injury is critical to planning operative exposure, determining treatment modality, and assessing prognosis.

Preganglionic Injuries

Root avulsion refers to injury of the nerve root within the intervertebral foramen proximal to the sensory root ganglion. The spinal nerves are attached to the transverse processes in the lower cervical spine. As progressive traction is placed on the nerve root, these attachments are the first to break. The nerve root sleeve is then pulled through the intervertebral foramen and may tear, causing a meningocele. Finally, the nerve root itself avulses. Horner syndrome consists of ipsilateral meiosis, upper eyelid ptosis, and facial anhydrosis; it implies disruption to of the sympathetic fibers traveling through C8-T1 and suggests root avulsion at these levels. Winging of the scapula or weakness of the rhomboid suggests C5 or C6 root injury, respectively, since the long thoracic and dorsal scapular nerves arise from the nerve roots. Since no proximal stump is available for reconstruction, root avulsions may require neurotization.

Postganglionic Injuries

Postganglionic injuries, or ruptures, may have regenerative capacity and can often be grafted. Supraclavicular injuries involve the spinal nerves, trunks, or divisions. They manifest most commonly as Erb's palsy (C5-C6 injury) or total plexus injury. Erb's palsy includes loss of deltoid, supraspinatus, infraspinatus, biceps, and brachialis function and presents with internal rotation, shoulder adduction, and elbow extension. Total plexus injury presents with a flail arm. Infraclavicular injuries involve the cords and their distal branches. They often present with focal weaknesses in shoulder abduction and flexion with or without deficits in elbow, wrist, or hand strength.

Preganglionic lesions can be differentiated from postganglionic lesions by weakness of the rhomboids, serratus, and supraspinatus in combination with an electrophysiologic profile of normal sensory conduction velocities with absent somatosensory evoked potentials. It may also be possible to see meningoceles on CT myelogram or MRI. However, pre- and postganglionic injuries may coexist and accurate diagnosis may not be possible until surgical exploration.

Goals of Reconstruction

Terzis and Papakonstantinou suggest the following priorities in reconstructing target muscles:

1. Return function to the supraspinatus and deltoid muscles to stabilize the shoulder and prevent severe neck pain.
2. Restore the biceps via the musculocutaneous nerve, allowing the hand to be brought to the face. Neurotization of the ipsilateral latissimus will also assist biceps function.
3. Attempt to reestablish triceps function.
4. Reconstruct the median nerve.

Surgical Strategies

Low velocity penetrating injuries should be explored immediately and repaired primarily or with nerve graft. Injuries related to gunshot wounds typically result from blast effect rather than by direct nerve injury; as such, spontaneous recovery may be possible. Surgery for closed injuries may be performed 3-6 months after injury, after any spontaneous recovery has been assessed and before irreversible muscle atrophy has occurred.

Neurolysis

The removal of scar tissue surrounding intact nerve fibers, known as neurolysis, may help nerve regeneration.

Nerve Repair and Grafting

When nerve continuity is interrupted distal to the sensory root ganglion, nerve repair or grafting may be beneficial. Primary repair may be attempted for simple nerve lacerations. Nerve grafts, often using the sural, antebrachial cutaneous, or radial sensory nerve, are used to reconstruct traction and crush injuries.

Neurotization

Preganglionic lesions may be treated with neurotization, in which functional nerves are harvested and transferred to distal nerve root stumps. Intraplexus donors, because of their large number of axons, provide the best motor donors for neurotization. Extraplexus motor donors, such as intercostal, phrenic, accessory, and medial pectoral nerves, may be used for multiple-root avulsions. The contralateral C7 nerve root, combined with a nerve graft, may be also be used. Recently success has been reported using branches of the ulnar nerve to neurotize the musculocutaneous nerve.

Nerve Root Repair and Reimplantation

There have been several recent reports of nerve root repair in which a nerve graft is inserted into the spinal cord and attached to the avulsed distal root. Some have also attempted to reimplant the avulsed root directly into the spinal cord. It is too early to know whether these methods will prove to be successful.

Secondary Procedures

Muscle and tendon transfers may be necessary to augment function. Concomitant release of muscle contractures that develop may be helpful in restoring balanced external rotation and abduction of the upper extremity. In refractory cases, functional free muscle transfer with the gracilis may be necessary to restore distal function.

Postoperative Care

After 2-4 weeks of splinting to prevent traction on the repair, an aggressive regimen of physical therapy, occupational therapy, and electric stimulation should be initiated. Induction exercises, which stimulate the donor nerves used during neurotization should be begun when recipient motor function begins to return. Patient expectations and compliance are paramount in achieving a successful long-term outcome. Secondary reconstruction with muscle transfers and contracture releases may be necessary. Patients in whom primary nerve reconstruction is performed should be counseled on the staged nature of reconstruction.

Pearls and Pitfalls

For obvious lacerations of nerves, primary repair and reconstruction should be undertaken in the acute stage. However, for traction or closed injuries, there is a necessary period of waiting for neuropraxia to resolve prior to surgical exploration. In this latter setting, neuromas may be encountered. Nerve conduction across neuromas may be one measure of determining whether simple neurolysis will suffice or neurotization and grafting is required (poor conduction being an indication for nerve transfer or grafting). Sural nerve will give a reasonable amount of graft material with modest donor site morbidity.

Notwithstanding primary reconstruction of injured nerves, secondary deformities may still become apparent over time. For instance, internal rotation and adduction deformities may result from a relative loss of upper root function. Partial release of contracted muscle may need to be performed in conjunction with latissimus dorsi and teres muscle transfers that alter vectors of action so that better external rotation is obtained. Other standard tendon transfers in the forearm and hand can help in specific situations of persistent ulnar, median and radial nerve palsy. If it is anticipated that a functional free muscle transfer may be needed for severe plexus injury, preserving nerves to act as future donors may be important.

Often, by the time a plexus injury patient has presented for reconstruction, a significant period of time has passed. Irreversible muscle atrophy occurs by 18-24 months; a calculus of time since injury and time until reinnervation of distal targets must be considered when evaluating feasibility of primary reconstruction.

Suggested Reading

1. Chuang DC. Management of traumatic brachial plexus injuries in adults. Hand Clin 1999; 15(4):737-55.
2. Moran SL, Steinmann SP, Shin AY. Adult brachial plexus injuries: Mechanism, patterns of injury, and physical diagnosis. Hand Clin 2005; 21(1):13-24.
3. Shenaq S, Kim J. Repair and grafting of peripheral nerve. In: Mathes SJ, Hentz V, eds. Plastic Surgery. 2nd ed. (in press).
4. Shin AY, Spinner RJ. Clinically relevant surgical anatomy and exposures of the brachial plexus. Hand Clin 2005; 21(1):1-11.
5. Terzis JK, Papakonstantinou KC. The surgical treatment of brachial plexus injuries in adults. Plast Reconstr Surg 2000; 106(5):1097-1122.

Chapter 87

Nerve Injuries

Zol B. Kryger and Peter E. Hoepfner

Nerve Anatomy

Peripheral nerves are bundles (fascicles) of axons surrounded by Schwann cell sheaths. The internal architecture of peripheral nerves is very consistent and is demonstrated in Figure 87.1. The individual axons are surrounded by an **endoneurium**. This layer is not visible with surgical loupes. The axons are grouped into fascicles that are surrounded by a **perineurium**. The groups of fascicles comprise the nerve, and are collectively encircled by the **epineurium**. Both the perineurium and the epineurium can be sutured during nerve repair. There is occasional branching between fascicles, but for the most part, the fascicles travel in discreet groups throughout the length of the nerves in the arm. This is the rationale for internal neurolysis, interfascicular grafting and fascicular repair.

Classification of Nerve Injury

Table 87.1 classifies nerve injuries according to the Seddon (first column) and Sunderland (second column) classifications.

In nerves that undergo axontmesis or neurontmesis, the axons distal to the point of injury undergo **Wallerian degeneration**. As long as the endoneurium, blood supply and surrounding Schwann cells remain intact, the axon will regenerate, traveling down its original path. The rate of regeneration is estimated to average about 1-3 mm per day, but can be quite variable.

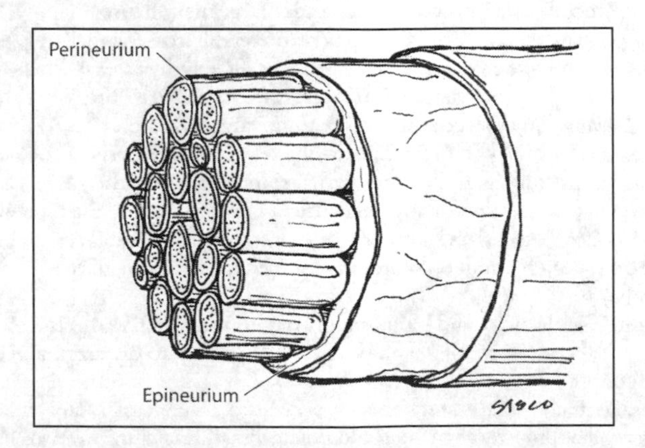

Figure 87.1. A cross section of a peripheral nerve demonstrating the perineurium and epineurium. There is an inner and an outer epineurium.

Practical Plastic Surgery, edited by Zol B. Kryger and Mark Sisco.

Table 87.1. Classification of nerve injuries

Injury Classification (Seddon)	(Sunderland)	Structure Injured	Functional Recovery
Neurapraxia	1st degree	Axon only mildly injured	Complete recovery in weeks-months
Axontmesis	2nd degree	Axon severely injured, endoneurium intact	Complete recovery in months
Neurontmesis	3rd degree	Axon and endoneurium severed	Near complete recovery
	4th degree	Fascicles and perineurium severed	Relatively good recovery
	5th degree	Complete nerve transection	Only partial recovery

Diagnosis

Any person suspected of having a nerve injury should undergo a thorough sensory and motor examination. When the nerve injury is secondary to a laceration or penetrating trauma, the point of injury is relatively easy to locate. The extent of the injury may be difficult to diagnose since a muscle can retain function even if the nerve is partially transected. Open nerve injuries should therefore be explored. Closed injuries with suspected nerve involvement pose a more challenging situation. Locating the level of nerve injury can be difficult. In these cases, EMG and nerve conduction velocity studies should be obtained about 6 weeks after the injury to locate the level of injury. Nerve exploration should therefore be deferred if only partial nerve injury is suspected and there is reasonable prospect of spontaneous recovery.

Direct Nerve Repair

Several principles are important to consider in primary neurorrhaphy:

1. **Timing of the repair**. Primary repair is preferable to secondary repair. Most authors recommend repair as early as possible. In lengthy cases, such as replants, it may not be feasible to repair injured nerves, in which case the repair should be delayed several days to weeks. There are some surgeons who prefer to always wait 2-3 weeks after injury in order to time the repair with fibrosis of the nerve (for facilitating suturing) and maximal axonal sprouting across the gap.
2. **Gap size**. After resection of any devitalized or crushed nerve, the remaining gap should not exceed 2.0-2.5 cm. If the gap exceeds this distance, it should be bridged with a nerve graft or a conduit of some sort (e.g., vein graft or polyglycolic acid tube).
3. **Tension**. The elasticity and blood supply of peripheral nerves can tolerate mobilization of about 6-8 cm of nerve. Beyond this, damage to the axons and capillaries will occur.
4. **Proper fascicular alignment**. This is essential to ensure that motor and sensory fascicles will be properly aligned. In addition to visually matching the two ends, there are several intraoperative techniques for differentiating motor and sensory fascicles.
5. **Epineurial vs. perineurial repair**. Epineurial repair alone is adequate for digital nerves using several 9-0 or 10-0 nylon sutures. Advantages include its technical ease and minimal manipulation of internal nerve structures. Its main drawback is the lack of accurate fascicular alignment. Perineurial repair properly aligns the

fascicles either individually or in groups and is used for larger nerve repair in the hand and forearm. It is more time-consuming and technically demanding. It also increases the risk of scar formation at the repair site. There is still no clear guideline as to which type of repair should be used.

Median Nerve Injury

This should be suspected in deep lacerations of the wrist or distal humerus or in penetrating injuries on the medial side of the arm that injure the brachial artery. The most consistent sign of median nerve injury is loss of skin sensibility on the palmer surface of the first three digits and loss of thenar opponens function. Patients will make a fist without the thumb and index finger folded into the palm. If the FPL, FDS and FDP of the index and middle finger are flexing normally, then the nerve injury has occurred distal to the take off of the anterior interosseous nerve. Exposure of the median nerve depends on the level of injury and is outlined in Table 87.2.

Ulnar Nerve Injury

The ulnar nerve is at risk of injury primarily from deep lacerations of the ulnar side of the wrist or from penetrating trauma to the posterior medial epicondyle. A common sensory sign of ulnar nerve injury is the loss of sensibility to the small and ulnar side of the ring fingers. Motor signs include FCU paralysis, interosseous and thumb adduction loss with a weak "key pinch," and FDP paralysis of the small and ring fingers. A long-standing injury can present with the classic "claw hand" deformity. Exposure of the ulnar nerve is outlined in Table 87.3.

Radial Nerve Injury

The radial nerve is at risk of injury primarily from fractures of the humerus or radius. Spiral, oblique humeral shaft fractures (especially open fractures) and fractures of the upper third of the radius are the ones that most commonly result in radial nerve injury. Anterior dislocation of the radial head may stretch the nerve or pin it between the radial head and the ulna. Signs of injury include loss of sensibility on the dorsum of the hand and in the first web space. Patients will be unable to extend their

Table 87.2. Median nerve exposure

Location of Injury	Key Points in Exposing the Median Nerve
Arm	• Medial incision anterior to intermuscular septum • Nerve runs adjacent to the brachial artery
Elbow	• Incision along medial border of biceps brachialis • Nerve is deep to antebrachial fascia and lacertus fibrosus
Proximal forearm	• Incision along inner border of pronator teres • Nerve passes between two heads of pronator teres • Ant. interosseous n. exits deep, proximal to pronator teres
Distal forearm	• Oblique incision between FCU and FCR tendons • Nerve lies between FDS and FDP tendons • Nerve runs deep to palmaris longus (when present)
Wrist or palm	• Incision parallel to skin crease at base of thenar eminence • Divide transverse carpal ligament • Watch for palmer cutaneous and recurrent motor branches

Table 87.3. Ulnar nerve exposure

Location of Injury	Key Points in Exposing the Ulnar Nerve
Arm	• Medial incision anterior to intermuscular septum • Nerve runs deep to the fascia of the septum
Elbow	• Incision is directly over the cubital tunnel • Incise between the olecranon and medial epicondyle
Forearm	• Incision along ulnar mid-axial line • The two head of FCU are split • Nerve runs deep to FCU on the flexor digitorum profundus
Wrist or palm	• Incision parallel to skin crease at base of thenar eminence • Incise over the pisiform • Watch for the division into sensory and motor branches

Table 87.4. Radial nerve exposure

Location of Injury	Key Points in Exposing the Radial Nerve
Proximal arm	• Posterior incision between long and medial head of triceps • Nerve runs deep to the long head of triceps • Nerve runs in the spiral groove of the humerus
Distal arm	• Nerve runs between brachialis and brachioradialis • Accompanies the terminal branch of the profunda brachia
Forearm	• Nerve runs deep to brachioradialis • Distal to brachialis tendon insertion, the nerve divides (superficial and posterior interosseous divisions)

fingers. Injury to the radial nerve in the axilla will result in paralysis of the ECRL and ECRB, triceps and brachioradialis muscles. These muscles will atrophy in long-standing nerve injury. Exposure of the radial nerve is outlined in Table 87.4.

Alternatives to Direct Repair

When direct nerve repair is not possible, several alternatives should be considered. The following chart lists common reasons precluding direct repair and some surgical options:

Reason Direct Repair not Possible	Solution
Concomitant life-threatening injuries	Delay repair until patient medically stable
Gap greater than 2.5 cm or excessive tension	Mobilize nerve, use nerve graft or conduit, shorten bone
Inability to locate the proximal nerve stump	Nerve transfer (such as intercostal n. to axillary n.)
Inability to locate the distal nerve ending	Neurotization (imbed stump into target muscle)
End organs (skin or muscle) irreversibly damaged	Neurovascular island flap or tendon transfer

Nerve Grafting

Nerve grafting is appropriate whenever there is a gap greater than 2.5 cm that persists after other measures such as nerve mobilization or transposition have been performed. Many studies have demonstrated regrowth of axons through the nerve graft and excellent return of function. The following outline covers the basic steps of nerve grafting:

- The proximal and distal nerve ends are cut back until healthy nerve is evident
- The fascicular groups are identified and isolated at different levels along the nerve
- A schematic drawing of the nerve ending cross sections is made to enable proper matching of the fascicles
- The nerve grafts are harvested and are cut slightly longer than the defect size
- At least five nerve grafts should be used for each of the three major nerves of the upper extremity (median, ulnar and radial nerves)
- The direction of the graft is **reversed** relative to its normal anatomic direction
- Grafts are used to join corresponding fascicular groups
- One to two 10-0 stitches are used in a single bite to secure the perineurium and epineurium of the graft to those of the injured nerve

Nerve Transfers

Nerve transfers should be considered when the proximal stump of the injured nerve cannot be located. The idea behind nerve transfers is to redirect part or all of a functioning motor or sensory nerve to the dennervated muscle or skin, respectively. There are many clinical applications of nerve transfers. The majority of these are for high level injuries to the nerves of the upper extremity.

Postoperative Considerations

The site of repair should be immobilized for 7-14 days. At about 3-4 weeks after surgery, range of motion exercises should be initiated. These should continue for at least 6 months. Strengthening activities should begin at about 3 months. The progress of the regenerating axons can be followed clinically with a Tinel's sign. Sensory reeducation should commence when the Tinel's sign indicates axonal regeneration has reached the target and should continue for the first year. Regenerating axons do not always make the appropriate connections with their targets. They can grow aberrantly and form a tangled mass of fibrotic nerve and scar termed a **neuroma**. Neuromas can range from subclinical and painless to exquisitely painful. There is no optimal treatment for symptomatic neuromas. Options include en bloc resection, nerve transposition, crushing, freezing, cauterization, steroid injections, and a slew of other physical and chemical treatments.

Outcomes

It may take over one year to reach maximal functional recovery after a peripheral nerve injury. All other things being equal, younger patients (especially teenagers and younger) do better. Children do the best. They usually regain superior two-point discrimination compared to older patients. Sensory recovery is usually superior to motor recovery. Another generalization is that the more proximal the injury, the worse the prognosis. The outcome is impossible to predict, but can be optimized by early repair with meticulous surgical technique and a motivated patient who will comply with the rigorous postoperative therapy that is required.

Pearls and Pitfalls

- In replants, nerve repair is occasionally delayed. In such instances, the two ends of any divided nerve should be tagged with a suture so that they can easily retrieved at the time of repair.
- Patients may have a normal motor exam despite a nerve injury, such as partial transection. Exploration is warranted if the mechanism and anatomic site of injury place the nerve at risk.
- During direct nerve repair or nerve grafting, the adjacent joints should be flexed and extended to ensure that both sides of the repair are tension free.
- In regards to nerve regeneration, one suture line is always better for axons to have to cross than two suture lines. Therefore, direct repair should be undertaken whenever possible.
- The minimal numbers of sutures should be used to achieve proper fascicular alignment.

Suggested Reading

1. Chiu DTW. Nerve repair in the upper limb. Grabb and Smith's Plastic Surgery. 5th ed. Philadelphia: Lippincott-Raven, 1997.
2. Sunderland S, ed. Nerves and Nerve Injuries. 2nd ed. Edinburgh: Churchill Livingstone, 1978:69-141.
3. Seddon HJ. Three types of nerve injury. Brain 1943; 66:237.
4. Frykman GK, Wolf A, Coyle T. An algorithm for management of peripheral nerve injuries. Orthop Clin North Am 1981; 12:239.
5. Grabb WC. Median and ulnar nerve suture. An experimental study comparing primary and secondary repair in monkeys. J Bone Joint Surg 1968; 50A:964.
6. Weber RA, Breidenbach WC, Brown RE et al. A randomized prospective study of polyglycolic acid conduits for digital nerve reconstruction in humans. Plast Reconstr Surg 2000; 106:1036.
7. Jabaley ME. Current concepts of nerve repair. Clin Plast Surg 1981; 8:33.
8. Millesi H. Fascicular nerve repair and interfascicular nerve grafting. In: Daniel, Terzis, eds. Reconstructive Microsurgery. Boston: Little Brown, 1977.
9. Nath RK, Mackinnon SE. Nerve transfers in the upper extremity. Hand Clin 2000; 16(1):131.
10. Wilgis EFS. Nerve repair and grafting. In: Green DP, ed. Operative Hand Surgery. 2nd ed. New York: Churchill Livingstone, 1988:1373-1404.
11. Mackinnon SE, Dellon AL. Algorithm for management of painful neuroma. Contemp Orthop 1986; 13:15.

Chapter 88

Vascular Trauma

Zol B. Kryger and John Y.S. Kim

Etiology

Thirty to forty percent of peripheral vascular injuries occur in the upper extremity. The most commonly injured vessel is the brachial artery, and iatrogenic trauma is the primary cause. Most iatrogenic injuries are due to diagnostic cardiac catheterizations. Ulnar and radial artery injuries account for 15-20% of peripheral vascular injury. This number has been growing due to the increasing prevalence of intravenous drugs use and the occasional accidental intra-arterial injections by users. Axillary artery injuries are infrequent. Most of the injuries to the axillary, radial and ulnar vessels are due to penetrating trauma. Gunshot wounds, followed by stab wounds, account for most of these injuries.

Vascular injury can also be due to blunt or closed trauma, usually secondary to fractures. Long bone fractures or severe dislocations account for most blunt vascular injuries. The axillary artery can be injured with a proximal humerus fracture or dislocations. The brachial artery is at risk from supracondylar humeral fractures or posterior elbow dislocations. When associated with orthopedic trauma, injury leading to arterial occlusion has a very poor prognosis.

Clinical Exam

The following signs of upper extremity vascular injury are indications for immediate surgical exploration:

- Arterial bleeding from the injury
- Diminished or absent distal pulse
- Wrist/brachial index less than 0.9
- Signs of hand ischemia
- Pulsatile or expanding hematoma
- Bruit or thrill at the site

For ruling out major vascular injury, the negative predictive value of arterial pulse indices (API) greater than 0.9 is 99%, and the sensitivity and specificity of an API less than 0.9 are 95% and 97%, respectively. Large, prospective studies have shown that the two most significant predictors of arterial injury are a pulse deficit or an abnormal API in the injured extremity. In the absence of a pulse deficit or API less than 0.9, arteriography is generally not indicated.

The following signs are equivocal and require further evaluation:

- Small, stable hematoma
- Nearby nerve injury
- Wound proximity to a major vessel
- Hemodynamic shock with no other attributable cause

Diagnostic Imaging

Arteriography is the most accurate means of detecting a vascular injury when the clinical exam is equivocal. In one large series of brachial artery injuries, only 60% of patients had an absent pulse. Patients who have a distal pulse, but other signs suggesting arterial injury, should undergo angiographic evaluation. Studies have demonstrated that it is far superior to **ultrasound** (including color flow Doppler) at diagnosing intimal injuries and other smaller arterial defects. Unless the injury is due to a high velocity projectile, emergent arteriography is not needed for penetrating trauma that is more than 5 cm from a named vessel.

Recent evidence supports the use of helical **computed tomographic angiography** (CTA) as an alternative to traditional arteriography. It is as sensitive and specific as arteriography. The main advantages of CTA are that it is less invasive and does not require the presence of an interventional radiologist. It is not, however, available at all centers. Furthermore, arteriography can be therapeutic as well: in select cases, the interventionalist can attempt endoluminal repair, such as stent placement at the site of an intimal tear.

What should be done for those patients with angiographic evidence of small intimal flaps or other insignificant vascular injuries? Nonoperative management is safe as long as the patient can be followed clinically and the following signs are absent: active hemorrhage, absent distal pulse, distal ischemia, expanding or pulsatile hematoma, bruit or thrill; 90% of patients managed nonoperatively will never require surgery.

Operative Management

As in all vascular surgical procedures, the initial goal is to gain proximal and distal control of the injured vessel. Once this is achieved and the vessel ends have been trimmed back to normal appearing artery, primary anastomosis should be attempted. Large series have shown that direct repair is possible about a third of the time. The inability to perform primary anastomosis is a result of undue tension when the two ends of the vessel are approximated, or because a segment of artery is damaged beyond repair. The second choice to direct repair is to use a piece of vein as an interpositional graft. If there is not adequate vein available, PTFE grafts should be used. Synthetic grafts are a last resort, since their long-term patency is inferior to autogenous vein. Ligation of the axillary or brachial artery should only be done as damage control in the poly-trauma patient with life-threatening injuries.

If concomitant injuries are present, arterial repair takes precedence. Fractures should be repaired following this, and the vascular anastomosis must be examined following any orthopedic procedure to ensure that the anastomotic site is intact and free from tension. Any injured vein that is not easily repairable should be ligated. Finally, any damaged nerves should be repaired using microsurgical technique. Nerve repair is discussed in detail in a separate chapter.

Axillary Artery

When the point of penetration is in the shoulder region and the pulse in the arm is absent, one may consider performing preoperative angiography in order to determine the precise point of vascular injury. This will help determine the surgical approach. When the external injury is more distal on the extremity, it is usually evident roughly where the vascular damage has occurred.

The approach for axillary artery injuries is with the arm abducted 90° and externally rotated. The incision should extend from the lower, midclavicle to the anteromedial upper arm. The axillary fascia is opened, and the pectoralis major and minor are divided near their insertion. After the segment of injured artery is identified and vascular control obtained, the other critical structures such as the axillary vein and adjacent nerves are evaluated. If the limb is ischemic from an occlusive thrombus, proximal and distal balloon angioplasty is performed. Before restoration of flow, the artery should be flushed with a heparin-containing saline solution.

Brachial Artery

The approach for brachial artery injuries is via an incision along the medial arm, between the biceps and triceps muscles. The median nerve should be identified and retracted away. If the injury is near the elbow, A "lazy S" type incision over the antecubital fossa should be made with the horizontal limbs oriented longitudinally on the arm and forearm. The brachial artery bifurcation should be exposed. If an interpositional vein graft is needed, a segment of cephalic vein can be used.

Radial and Ulnar Artery

The proximal radial artery is exposed via an incision extending from the antecubital fossa along the radial side of the forearm. The middle and distal segments are approached via an incision along the medial border of the brachioradialis muscle. The ulnar artery is approached by making an incision extending from below the medial epicondyle along the lateral border of flexor carpi ulnaris. Injuries of the radial and ulnar arteries at the wrist should be explored through longitudinal incisions.

Outcomes

Good upper limb function will be more likely the more distal in the extremity the vascular injury. Hence, isolated radial artery injuries do much better than axillary arterial trauma. The more proximal injuries have a higher likelihood of associated nerve injuries. This is especially true of blunt traumatic injuries. For instance, over 60% of brachial artery injuries have a concomitant nerve injury, usually the median nerve. Most limbs can ultimately be salvaged, with the length of ischemia prior to reperfusion being the primary determinant. Amputation rates range from 5-15%.

Isolated radial or ulnar artery injuries almost never result in hand ischemia. In fact, in the absence of an ischemic hand, the injured vessel can be safely ligated without any significant long-term complications such as hand claudication. Minor complications can occur; however the incidence is no greater than in patients who have undergone operative repair.

Complications

Early complications necessitating emergent reexploration of the site include **thrombosis, hematoma and hemorrhage**. Another serious complication requiring surgical intervention is **compartment syndrome**. It is due to the edema of reperfused ischemic muscle. Diagnosing compartment syndrome can be difficult; the hallmark is severe pain. The management of compartment syndrome is discussed in a separate chapter. A limb that has been ischemic for greater than 6-8 hours prior to reperfusion is at risk for muscle necrosis. In severe cases, amputation is often the only option.

Superficial infection, especially in grossly contaminated injuries, is the most common postoperative complication. It is usually treated with intravenous antibiotics; however debridement is sometimes necessary.

Late complications of vascular repair include **pseudoaneurysms** and **arteriovenous fistulas**. The brachial artery is a common site of postrepair pseudoaneurysm. This places the hand and fingers at risk from thromboembolic events. These usually require resection with saphenous vein graft interposition. **Nerve injuries** are another late complication that can be seen with vascular injuries at any level of the arm since the named vessels often travel in close proximity to major nerves. Nerve injury can be iatrogenic at the time of surgery or as a result of postoperative swelling or scarring at the site of repair. The most likely scenario, however, is that the nerve injury occurred at the time of the injury and was not diagnosed on the initial evaluation. Finally, minor long-term complications include **hand weakness**, **paresthesias** and **cold sensitivity**.

Pearls and Pitfalls

- Any vein that cannot be easily repaired should be clipped. There is no vein in the upper extremity that is essential. Even the axillary vein can be ligated if the repair will prolong the surgery excessively and place the patient at greater risk.
- One cannot over-emphasize the importance of a tension-free repair. Tension on the intima at the site of repair increases the risk of thrombosis and tearing of the vessels. If this cannot be achieved, a vein graft should be used. Outcomes using a vein graft are almost equivalent to direct repair.
- An arterial repair should always be reinspected following any manipulation of the limb (e.g., after fracture reduction or fixation).
- Prophylactic fasciotomy of the forearm and median nerve decompression should strongly be considered after revascularization of an arm that has been completely ischemic for more than 4 hours.

Suggested Reading

1. Aftabuddin M, Islam N, Jafar MA et al. Management of isolated radial or ulnar arteries at the forearm. J Trauma 1995; 38(1):149.
2. Busquets AR, Acosta JA, Colon E et al. Helical computed tomographic angiography for the diagnosis of traumatic arterial injuries of the extremities. J Trauma 2004; 56(3):625.
3. Dennis JW, Frykberg ER, Veldenz HC et al. Validation of nonoperative management of occult vascular injuries and accuracy of physical examination alone in penetrating extremity trauma: 5- to 10-year follow-up. J Trauma 1998; 48(2):243.
4. Fitridge RA, Raptis S, Miller JH et al. Upper extremity arterial injuries: Experience at the Royal Adelaide Hospital, 1969 to 1991. J Vasc Surg 1994; 20(6):941.
5. Frykberg ER, Crump JM, Dennis JW et al. Nonoperative observation of clinically occult arterial injuries: A prospective evaluation. Surgery 1991; 109(1):85.
6. Johansen K, Lynch K, Paun M et al. Noninvasive vascular tests reliably exclude occult arterial trauma in injured extremities. J Trauma 1991; 31(4):515.
7. Rich N, Mattox K, Hirshberg A, eds. Vascular Trauma. Philadelphia: W.B. Saunders, 2004.
8. Schwartz M, Weaver F, Yellin A et al. The utility of color flow Doppler examination in penetrating extremity arterial trauma. Am Surg 1993; 59(6):375.
9. Zellweger R, Hess F, Nico A et al. An analysis of 124 surgically managed brachial artery injuries. Am J Surg 2004; 188:240.

Chapter 89

Extensor Tendon Injuries

Zol B. Kryger and Peter E. Hoepfner

Anatomy

Extrinsic Musculature

The extrinsic muscles that extend the hand and fingers enter the wrist through six synovial-lined, dorsal compartments covered by the extensor retinaculum (Table 89.1). The extrinsic muscles of extension are innervated by the **radial nerve**. At the wrist, the tendons are surrounded by a sheath, but this sheath is not present in the hand and fingers. The four fingers are extended by EDC; however the communis tendon to the little finger is present only 50% of the time. The index and little fingers also have independent extensor muscles—EIP and EDM, respectively. These tendons usually lie ulnar and deep to the communis tendons to these two fingers.

The EDC tendons are joined proximal to the MP joints by the **juncturae tendinum**. The juncturae are almost always present between the EDC of the middle, ring and little fingers. Thus, lacerations proximal to the juncturae may not impair digit extension due to the connection to the adjacent digits. The tendons insert proximally into the MP joint volar plate (via the transverse metacarpal ligament) through attachments known as the **sagittal bands**. Distal to the MP joint, the extensor tendons divide into one central and two lateral slips. The **central slip** inserts into the middle phalanx and extends the PIP joint. The **lateral slips** reunite distally and attach to the distal phalanx, extending the DIP joint (Fig. 89.1).

The thumb is extended by three tendons: the first metacarpal by APL, the proximal phalanx by EPB, and the distal phalanx by EPL. However, the MP and IP joints of the thumb can both be extended by EPL due to the attachments of the dorsal apparatus.

Table 89.1. The dorsal compartments of the wrist. The six compartments are numbered 1-6 from radial to ulnar

1. Abductor pollicis longus (APL), extensor pollicis brevis (EPB)
2. Extensor carpi radialis brevis (ECRB), extensor carpi radialis longus (ECRL)
3. Extensor pollicis longus (EPL)
4. Extensor digitorum communis (EDC), extensor indices proprius (EIP)
5. Extensor digiti minimi (EDM)
6. Extensor carpi ulnaris (ECU)

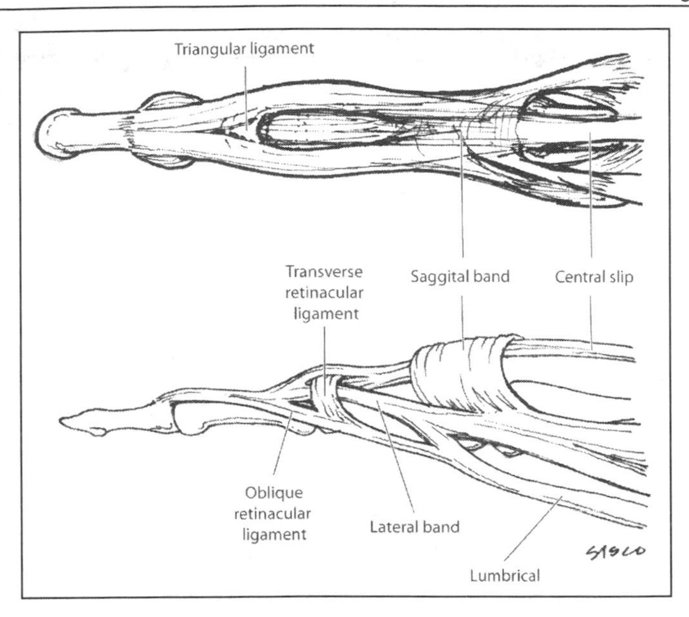

Figure 89.1. The extensor tendon mechanism.

89

Intrinsic Musculature

The intrinsic apparatus includes the dorsal interosseus muscles, the volar interossei, and the lumbricals. The four **dorsal interossei** insert onto the proximal phalanges and **abduct** the fingers and weakly flex the proximal phalanx. The three **volar interossei** do not attach to bone but rather insert onto the lateral bands that unite with the lateral slips of the extensor tendons. They **adduct** the fingers and flex the PIP joint. The **lumbricals** arise from the tendon of FDP on the palmer side and insert onto the radial lateral band of each finger. They primarily **extend the IP** joints and function as flexors of the MP joints. The intrinsic muscles are innervated primarily by the **ulnar nerve**, **except for the first and second lumbricals**, which are innervated by the **median nerve**. The intricate extensor, flexor and intrinsic systems are interconnected, aligned and stabilized with a system of ligaments found in each finger. The **triangular**, **transverse retinacular** and **oblique retinacular ligaments** perform a variety of complex actions. The pathology of these ligaments is addressed in the chapter on Dupuytren's disease. A detailed discussion of their functions is listed at the end of this chapter (see *Suggested Reading*).

Blood Supply

The extensor tendons are supplied by a number of sources. Vessels from the muscles and bony insertions travel distally and proximally, respectively, down the length of the paratenon. The dorsal aspect of the tendon is not as well vascularized as the deeper surface which receives branches from the periosteum and palmar digital arteries. Unlike the flexor tendons, the extensors have no vincular system. Synovial diffusion plays a major role in the delivery of nutrients, less so than in the flexor tendon system.

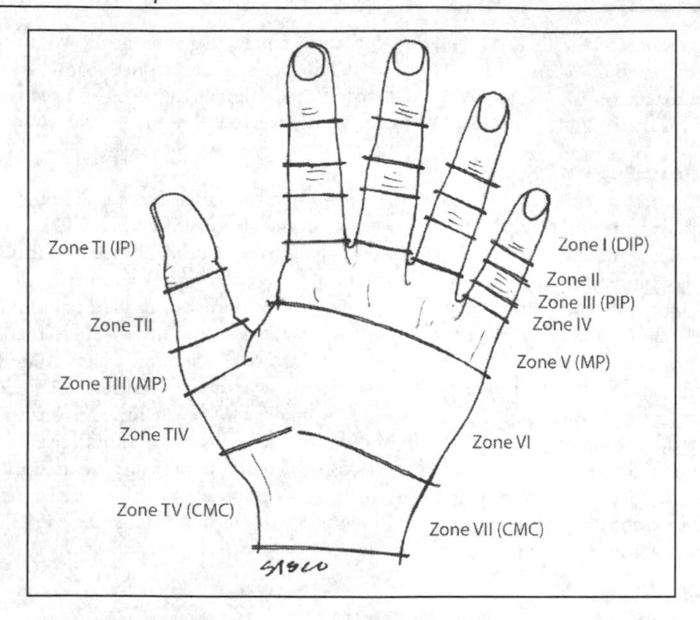

Figure 89.2. The zones of extensor tendon injury in the fingers and hand.

Zones of Injury

The dorsum of the fingers, hand and wrist can be divided into zones of injury for describing the location of tendon injury (Fig. 89.2). The odd numbered zones overlie the joints, whereas the even numbered zones overlie the bones. Injuries in zones 2, 3, and 4 of the fingers, and zone 7 in the wrist have a worse prognosis. Extensor tendon lacerations in all zones should be repaired primarily. Little to no tendon retraction occurs in injuries in zones 1-4. Proximal to the MP joints, however, tendon retraction will occur.

Zone 1 and 2 injuries (DIP and middle phalanx) can produce a **mallet finger deformity**. Zone 3 and 4 injuries (PIP and proximal phalanx) can produce the **boutonniere deformity**. These deformities and their treatment are discussed below. Zone 5 injuries (MP joint) can divide the sagittal band and displace the tendon laterally. The sagittal bands should be repaired. Injuries in this zone will usually cause the tendon stump to retract proximally. Zone 6 injuries of the dorsal hand may not result in loss of extension due to transmission of force from adjacent tendons through the juncturae. In zone 7 injuries of the wrist, the injured retinaculum should be excised, but the proximal or distal portion should be preserved in order to prevent bowstringing of the tendon.

Primary Suture Repair

Extensor tendons become very thin and flat in the hand, making their repair difficult. They can easily fray during repair. Many suture techniques have been described for primary repair of extensor tendons. Simple lacerations of the extensor tendon in zone 1 can be repaired with a figure-of-eight stitch. A number of biomechanical studies support using the **Kleinert modification of the Bunnell technique**. The modified Kessler technique is also commonly used. The chapter on flexor tendon repair illustrates and discusses these repairs in greater detail. Following primary repair of extensor tendon injuries, the finger should undergo dynamic splinting in extension for 4-6 weeks. Partial lacerations of less than 50% can usually be managed by wound care, and splinting for several weeks followed by active motion.

Tendon Rupture

Rupture of tendons in the nonrheumatoid hand usually occurs 6-8 weeks following fractures of the distal radius or carpal bones. Infection and attacks of gout also increase the risk of rupture. Tendons in the nonrheumatoid hand will rupture at the musculotendinous junction or at the bony insertion. Rarely will they rupture within the substance of the tendon itself. The most commonly ruptured tendons are the **EPL** at Lister's tubercle, as well as **EDM and EDC to the little finger**. Tendon ruptures cannot be repaired primarily in most cases, and tendon transfers are required. EPL rupture is usually treated with transfer of the EIP tendon or palmaris longus or interposition autograft. Rupture of the tendons to the little finger can be repaired by suturing the distal stump of the intact tendons of the ring finger. If the extensor muscles have not undergone contracture, tendon grafts are also an option. When performing flexor to extensor transfer, it is important to remember to transfer the tendons subcutaneously rather than below the retinaculum since wrist flexion occurs in synergy with finger extension.

Tendon Loss

Loss of segments of an extensor tendon can occur in severe burn, crush, or degloving injuries of the dorsum of the hand. Direct primary repair is generally not possible, and tendon transfers or grafts are required. Transfers can be done with both extensor and flexor tendons. A vascularized palmaris longus tendon can be transferred using a pedicled radial forearm fascial flap. A variety of free flaps containing vascularized tendons have been described. A composite dorsalis pedis free flap has been used with good results obtained. Staged tendon reconstruction with Hunter rods is also an option. Whatever technique is chosen, immediate reconstruction is preferable to a staged repair.

Postoperative Considerations

Rehabilitation

The traditional approach of static splinting without mobilization has largely been abandoned. **Dynamic splinting** with controlled motion is now standard of care. Many splints have been described. Most involve splinting the wrist in extension and allow the injured finger to extend passively and partially flex against rubber band resistance. Many studies have shown that early controlled motion decreases the incidence of postoperative adhesions and post-traumatic deformities. Proximal injuries (zones 5-7) benefit more from dynamic splinting than distal ones. Children and adults who are unable to cooperate with postoperative hand therapy should undergo static splinting.

Adhesions

Adhesions following lacerations can occur between the tendon and bone, especially if there are underlying fractures. The hallmark is **limitation of PIP extension and flexion**—due to dorsal tethering. Adhesions should be treated with 6 months of active assisted motion exercises. If this fails, **tenolysis** may be indicated. Adhesions proximal to the MP joint can also be treated with **extrinsic extensor release**, providing that the intrinsic muscles are intact. In this release, the extrinsic tendon central slip is excised just proximal to the PIP joint. Consequently, the PIP joint will be extended solely by the intrinsic muscles, and the extrinsic tendon will extend only the MP joint.

Post-Traumatic Deformities

Mallet deformity describes DIP flexion and inability to extend the joint. It results from disruption of the extensor mechanism at the distal phalanx. Injuries can be classified into discontinuity of the extensor tendon (rupture or laceration), avulsion of the tendon from its distal insertion, or fractures of the distal phalanx. Closed injuries can cause this deformity as a result of forced passive flexion of the DIP joint. If treatment is not pursued, a secondary swan-neck deformity can occur. Treatment consists of continuous immobilization of the DIP in slight hyperextension for 6-8 weeks using a splint or by percutaneous, Kirschner wire fixation. Surgical treatment should be reserved for those who fail conservative management or for fractures requiring open reduction. Surgical options include direct tendon repair, tendon grafting, arthrodesis, and a number of other reconstructive techniques.

Swan-neck deformity describes PIP hyperextension and DIP flexion. It is the progression of a mallet deformity left untreated, as a consequence of disruption of the distal extensor mechanism. The PIP joint progressively extends since all of the force of the extrinsic extensor tendon is transmitted to the PIP joint through the central slip. The DIP joint progressively flexes due to lack of extensor force combined with unopposed FDP pull on the distal phalanx. With time the volar plate becomes lax at the PIP joint. The swan-neck deformity does not respond well to conservative management, and surgical repair is usually required. Along with repair of the extensor mechanism and any avulsion fractures, the contracted intrinsic muscles and PIP joint collateral ligaments should be released. In addition, the volar plate must also be tightened. Kirschner wires are removed at 4 weeks postoperatively, and a dorsal blocking splint is generally applied at that point. Arthrodesis and arthroplasty are reserved as salvage procedures.

Boutonniere deformity describes PIP flexion and DIP/MP hyperextension. It occurs as a result of injury to the central slip over the PIP joint and volar migration of the lateral bands. This deformity is manifest at about 2-6 weeks postinjury, therefore any trauma to the PIP region should include an evaluation of the extensor tendon. Swelling of the finger can mask a developing Boutonniere deformity. In such cases, the finger should be splinted in extension and examined a few days later after the swelling decreases.

Treatment consists of splinting the PIP joint in extension for 6 weeks. The DIP joint should be mobilized during this period to aid in dorsal migration of the lateral bands. Surgical treatment is indicated for avulsions of large bony fragments from the middle phalanx, PIP joint instability and dislocation that cannot be reduced, or extensive soft tissue loss to the dorsum of the finger. The central slip can be reconstructed from the remaining extensor mechanism (e.g., centralizing portions of the lateral bands) or with a tendon graft. Postrepair splinting should be done for 2 weeks. Arthrodesis is indicated for salvage of severely injured fingers. Amputation is occasionally required for severe injuries.

Intrinsic-plus contracture describes scarring of the interosseous muscles. This occurs post-traumatically, and can be prevented by minimizing edema of the hand (elevation, ice, NSAIDs, etc), and splinting the hand in the intrinsic-plus (safe) position: MP flexion, IP extension and palmar abduction of the thumb. Intrinsic muscle necrosis and subsequent fibrosis can occur shortly after trauma to the hand.

The finding of pain on passive extension of the MP joints is an early sign of impending intrinsic muscle death. Treatment of intrinsic-plus contracture consists of conservative splinting first, followed by surgical release when necessary. The interosseous tendons can be severed at the proximal phalanx. If MP flexion contracture is present, however, the release should be performed at the musculotendinous junctions.

Post-traumatic adhesions can form between the interossei and the lumbricals distal to the transverse metacarpal ligament. Patients with this condition will present with pain upon fist-making or on forceful finger flexion. These adhesions should be dealt with surgically as soon as they are recognized. Finally, lateral band contractures after trauma can also occur. When present they can impair DIP flexion with passive PIP extension. The involved band should be excised.

Pearls and Pitfalls

- A patient who presents with a mallet finger (zone I injury), irrespective of the duration of injury, can usually be treated in a closed fashion with splinting alone. This has a high likelihood of success even in cases with a delayed presentation of weeks to months. A stiff finger may require sequential progression of the splinting of the DIP joint towards the extended position.
- A laceration in the vicinity of the DIP joint is at risk for having entered the joint space. This raises the likelihood of joint infection which can lead to failure of the repair. Delayed tendon repair should be considered in these patients.
- Zone III lacerations involving the PIP joint often involve the central slip or lateral bands, and these should be repaired. If the laceration is very close to the insertion of the central slip, there may be insufficient length of distal slip to suture. In this case, a tunnel can be created in the dorsal distal phalanx using a K-wire through which the suture ends can be passed and tied.
- Extensor tendon lacerations over the MP joints (zone V) are often due to human bites (either biting or striking an open mouth) whether the patient admits this or not. Such wounds are at very high risk of infection and the tendon should never be repaired primarily. It can be repaired a week later after the wound is no longer contaminated.
- Ruptured tendons (most commonly EPL) may present many months to years after injury. In some cases, the muscle belly has atrophied or fibrosed, and tendon transfer should be performed. Only if there is evidence of a functioning muscle should the tendon be repaired with a tendon graft.

89

Suggested Reading

1. Blair WF, Steyers CM. Extensor tendon injuries. Orthop Clin North Am 1992; 23:141.
2. Browne Jr EZ, Ribik CA. Early dynamic splinting for extensor tendon injuries. J Hand Surg 1989; 14:72.
3. Kleinert HE, Verdan C. Report of the committee on tendon injuries. J Hand Surg 1983; 8:794.
4. Landsmeer JMF. The anatomy of the dorsal aponeurosis of the human finger and its functional significance. Anat Rec 1949; 104:31.
5. Masson JA. Hand IV: Extensor tendons, Dupuytren's disease, and rheumatoid arthritis. Selected Readings Plast Surg 2003; 9(35):1-44.
6. Newport ML, Williams CD. Biomechanical characteristics of extensor tendon suture techniques. J Hand Surg 1992; 17A:1117.
7. Rockwell WB, Butler PN, Byrne BA. Extensor tendon: Anatomy, injury, and reconstruction. Plast Reconstr Surg 2000; 106:1592.
8. Verdan CE. Primary and secondary repair of flexor and extensor tendon injuries. In: Flynn JE, ed. Hand Surgery. 2nd ed. Baltimore: Williams and Wilkins, 1975.

Chapter 90

Flexor Tendon Injuries

Zol B. Kryger and Peter E. Hoepfner

Relevant Anatomy

Tendons are composed of spiral bundles of Type I collagen, ground substance, elastin, and mature fibroblasts termed tenocytes. Individual collagen bundles are covered by an endotenon. The bundles are organized into fascicles surrounded by a paratenon. The outer layer of the tendon is termed the epitenon.

In the distal forearm, the tendons of flexor digitorum superficialis (FDS) are anterior to the tendons of flexor digitorum profundus (FDP). The FDS tendons are arranged in two pairs, with the ring and middle finger tendons lying palmer to the index and little finger tendons. In the hand distal to the carpal tunnel, the tendons of FDS and FDP pair off to the digits, with the FDS lying anterior to the FDP tendon. The tendons of FDS act independently to flex the PIP joints since they arise from single muscle bundles. The tendons of FDP will often act in unison and flex several digits at the DIP joint. This is due to a common muscle origin for several of the tendons, typically the ulnar digits. At the level of the MP joint, the FDS splits into two slips, and the FDP travels through this decussation to now lay anterior.

In the digits, the tendons travel in synovial-lined tunnels called flexor tendon sheaths. The sheaths are anchored to the bones by a series of five annular pulleys, numbered A1-A5 from proximal to distal (Fig. 90.1). The odd numbered pulleys are located over the joints; the even pulleys lie over the bones. There are three thin cruciate pulleys, numbered C1-C3, that maintain tendon motion and collapse during flexion. The palmer aponeurosis lies proximal to the A1 pulley and is often referred to as the A0 pulley. It acts in unison with the first two annular pulleys. Proximal to the entrance into the digital sheath (A1 pulley), the FDS tendon lies palmer to the FDP tendon. At this point, the FDS tendon divides and becomes deep to the FDP tendon. The two portions reunite at Camper's chiasma and go on

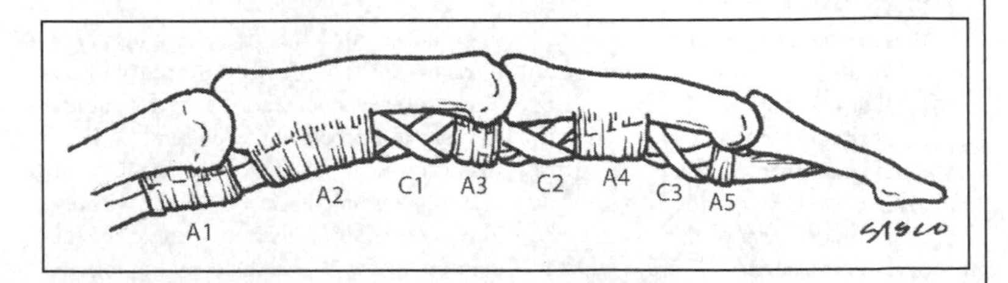

Figure 90.1. The digital flexor sheath and its pulley system. The annular pulleys are numbered A1-A5 from proximal to distal, and the cruciate pulleys are numbered C1-C3.

to attach to the middle phalanx. The FDP tendon, after passing through the FDS bifurcation, attaches to the distal phalanx.

Flexor pollicis longus (FPL) is the primary flexor of the thumb. It is the most radial structure in the carpal tunnel. It travels in its own fibrous sheath in the palm and inserts into the base of the distal phalanx of the thumb. The thumb, unlike the fingers, has two annular pulleys, A1 and A2, located over the MP and IP joints, respectively. Lying between them is an oblique pulley that is the most important of the three.

Blood Supply

The tendons have a rich vascular supply. Longitudinal vessels travel along the dorsal length of the tendons. The paired digital arteries supply segmental vessels to the sheath via the short and long vincula. Finally, the synovial fluid within the tendon sheath allows oxygen and nutrients to diffuse along its length since there are several short avascular zones over the proximal phalanx. The motion of the tendon facilitates the imbibition that delivers the nutrient-rich synovial fluid.

Biomechanics

In the neutral wrist position, only 2.5 cm of flexor tendon excursion is needed to produce digital flexion. As the wrist is flexed, the amount of tendon excursion required to flex the digits is more than tripled. Anything that causes the tendon to become flaccid and to bowstring, such as loss of an annular pulley, will result in greater excursion requirements to produce flexion. The A2 and A4 pulleys are the most important in this regard. Loss of either one will result in a substantial reduction in motion and power and a risk of flexion contracture of the digit. Rupture of the pulleys can be diagnosed with clinical exam, ultrasound or MRI.

Diagnosis

History

In obtaining the history, the posture of the hand at the time of injury should be determined. Injuries that occur with the fingers extended will result in the distal end of the tendon being located close to the wound. In contrast, injury to a flexed finger will result in the distal tendon retracting away from the wound as the finger is straightened.

Clinical Examination

Alterations in the normal resting position should be noted. Separate evaluation of both FDP and FDS function is important. Division of the FDS without injury to the FDP will not be noticeable in the resting posture. However, lacerations on the palmer surface of the fingers will usually sever the FDP tendon before the FDS tendon. FDS is evaluated by immobilizing the surrounding fingers in extension and having the patient flex the finger at the PIP joint. It is critical to isolate each joint since without DIP isolation the common muscle belly to the ulnar FDP digits may generate mock flexion at an adjacent PIP joint. FDS to the index finger is evaluated by having the patient perform a firm pulp-to-pulp pinch with the thumb. An injured FDS will cause as pseudo mallet deformity of the distal phalanx whereas an intact FDS will result in a pseudo boutonniere deformity of the distal phalanx. FDP is evaluated by immobilizing the PIP and IP joints, and evaluating flexion of the isolated DIP joint.

A complete sensory exam of the palmar surface is important since trauma to the digital nerves can occur with tendon injuries. Two-point discrimination should be performed on both the radial and ulnar aspect of each finger. The presence of nerve injury can influence the choice of incision used for exposure. Deep lacerations that disrupt the digital nerves can also sever the digital arteries. The finger can sometimes survive with intact skin even in the presence of bilateral digital artery disruption. However, at least one artery should be repaired if the tendon is also injured in order to avoid ischemia and impaired healing.

Indications for Repair

Dividing the hand into zones of injury (Zone I-V) is an internationally accepted method of classifying the location of flexor tendon injury (Fig. 90.2). As a general rule, complete flexor tendon lacerations in both the palm and the digital sheath should be repaired. Partial tendon lacerations greater than 60% should be repaired. In the past, zone II was referred to as "no man's land," since primary repair resulted in poor functional outcome. However, modern techniques have allowed the repair zone II injuries primarily.

Flexor tendon injury is not a surgical emergency; delayed primary repair (up to two weeks post-injury) can provide good long-term results. However, early repair is preferable. Although the early literature recommended against repairing FDS, most surgeons now repair both the FDP and FDS tendons. This is true for all zones of injury. Tendon repair should be attempted after bony fixation and revascularization have been achieved. Nerve repair should also be attempted when feasible.

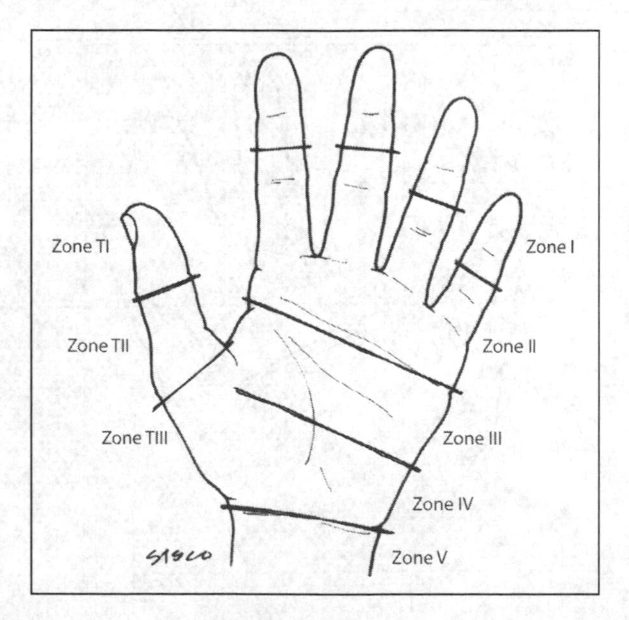

Figure 90.2. The zones of flexor tendon injury in the hand. Zone I is distal to the FDS insertion; zone II is from the A1 pulley to the FDS insertion; zone III is from the carpal tunnel to the A1 pulley; zone IV is the carpal tunnel; zone V is proximal to the carpal tunnel. The thumb is divided into three zones: TI is distal to the IP joint and T2 is distal to the MP joint.

Contraindications to primary tendon repair include: massive soft tissue injuries to the fingers or palm; inadequate skin coverage over the repair; or gross wound contamination. Some surgeons will delay primary repair if the skeletal damage is so severe that postoperative mobilization would not be possible.

Operative Technique

Since tendon transfers, tendon grafts, and arthrodesis are discussed elsewhere in this book, the focus of this chapter is primary repair.

Exposure

Rarely can the entire procedure be performed through the site of injury. There are many options for placement of the incision (Fig. 90.3). Lacerations can be extended in appropriate cases. However, if the neurovascular structures need to be explored, greater exposure is necessary. Longer incisions are usually performed in a zigzag fashion so that flexion lines in the palm and fingers are not crossed at 90° angles in order to avoid scar contracture. The surgeon must consider the position of the hand at the time of injury to determine if the tendon ends are likely to be retracted.

Tendon Retrieval

To prevent adhesions, atraumatic technique is essential. Tendon ends do not usually require debridement or shortening. It is important to keep the tendons moist throughout the procedure. In Zone I injuries involving the FDP tendon, retrieval is

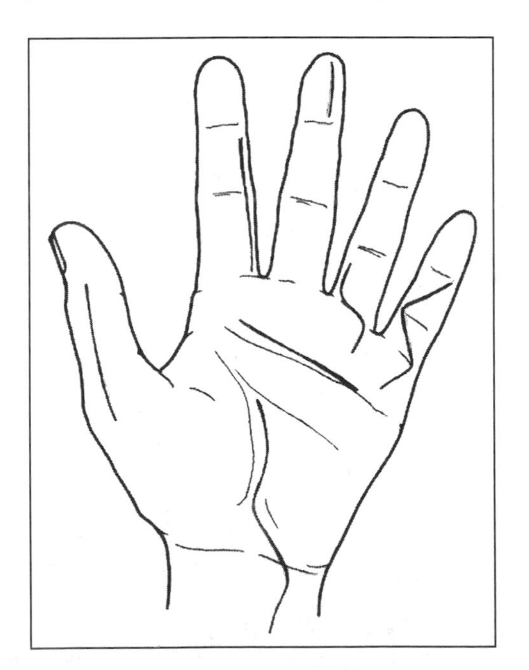

Figure 90.3. Skin incisions for exposure during flexor tendon repair. Whenever possible, the skin laceration should be incorporated into the incision. Flexion lines in the palm and fingers should not be crossed at 90°.

not usually difficult since the vincula help to anchor the tendon in place. If the DIP joint was flexed at the time of injury, the proximal stump can be retrieved by opening the cruciate pulleys (the A4 pulley should be preserved). A suture placed in the tendon can be used to pull it through the A4 pulley. Piercing the tendon transversely with a 25-guage needle can anchor it in place during repair.

In zone II injuries, the neurovascular bundles should be identified first. Either the C1 or C2 pulley is opened, whereas the A1-A4 pulleys should be preserved whenever possible. If the proximal stump is close and can be visualized, it should be carefully retrieved. A commonly described technique uses a skin hook to snare the tendon. Occasionally, "milking" the tendon towards the incision is sufficient.

If the proximal stump cannot be visualized, the Sourmelis and McGrouther technique is an excellent option. A catheter is passed through the tendon sheath from distal to proximal. The tip of the catheter is exposed via a mid palm incision proximal to the A1 pulley. The catheter is sutured to both tendons and then it is pulled distally, bringing the proximal tendon ends into contact with the distal stumps.

It is critical to ensure that the relationship of FDP and FDS at Camper's chiasma is maintained. Flexing the DIP and PIP joints will help deliver the distal stump of the FDS or FDP tendons. If insufficient distal tendon is exposed, the distal cruciate synovial sheath should be opened.

If there is insufficient remaining distal FDP stump, the proximal FDP tendon can be anchored to the base of the distal phalanx. A periosteal flap is elevated and a hole is drilled in the bone. A 3-0 suture is placed in the proximal tendon stump and passed through the bone hole to pull the tendon under the periosteal flap. The suture ends are tied over a piece of cotton and a pad-button placed over the nail. An alternative technique involves the use of bone anchors in the distal phalanx.

Suture Technique

A variety of suturing techniques have been described (Fig. 90.4). 3-0 or 4-0 braided polyester sutures are best. Locking loops are not necessary and may in fact collapse and

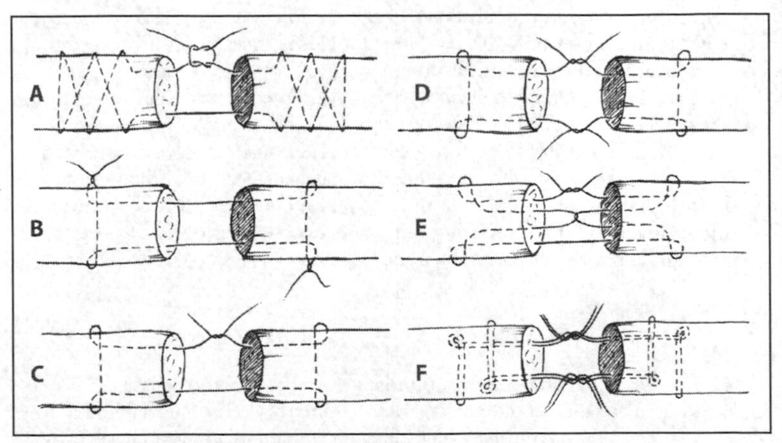

Figure 90.4. Representative techniques of end-to-end flexor tendon repair. A) Bunnell repair. B) Kessler grasping repair. C) The original Kessler repair. D) Kessler-Tajima modified repair. E) Interlock repair. F) Double loop repair.

lead to gaping. The strength of the repair is proportional to the number of sutures crossing the repair site. Studies have shown that six strands are needed in order to provide a maximally strong repair; however a six-strand repair is technically difficult and may compromise tendon nutrition and healing. As such, most surgeons use a four-strand repair. If one of the two-strand repairs shown in Figure 90.4 is used, an additional horizontal mattress stitch is placed to make it a 4-strand repair.

At the completion of the repair, an epitendinous repair using 5-0 or 6-0 monofilament suture should be run circumferentially around the repair site. This will increase the strength of the repair by up to 20% and decrease the likelihood of gaping. A number of techniques have been described, including a simple running stitch, a running lock loop, a continuous horizontal mattress, and a running-lock suture. The horizontal mattress and running-lock stitch have been shown to be the strongest.

There is ongoing debate as to the benefit of repairing the tendon sheath. Potential advantages include improved tendon nutrition, biomechanics and rate of healing. Disadvantages include the potential for narrowing, which can restrict gliding, and the technical difficulty of sheath repair.

Pulley Reconstruction

It is important to reconstruct the annular pulleys. As a minimum, the A2 and A4 pulleys should be repaired using autogenous material such as tendon, extensor retinaculum, volar plate, or fascia lata. One common technique is the Okutsu "triple loop" method which uses three loops of autogenous material encircling both the tendon and the entire bone. Most autogenous reconstructions, however, become lax with time resulting in bow-stringing of the tendon. The use of alternative alloplastic materials, such as PTFE (Gore-Tex), has shown promise in animal models in terms of biomechanical outcomes, adhesion formation, and foreign body reaction.

Tendon Rupture and Avulsions

Normal tendons usually rupture at their insertion or at the musculotendinous junction. Rupture within the substance of the tendon is unusual and is associated with conditions that weaken the tendon such as rheumatoid arthritis, infection, gout, and prior fractures. Treatment often requires tendon grafts or transfers.

Traumatic avulsions tend to occur as a result of forced extension in young males playing contact sports. The FDP tendon, especially in the ring finger, is most commonly involved. Isolated FDS rupture is rare. Severe injuries involve the avulsion of a fragment of bone along with the tendon. Treatment consists of reinserting the tendon into the base of the phalanx using a pullout suture tied over a button or recreating the tendon insertion using bone anchors. Bony avulsion requires open reduction and internal fixation of the bone fragment.

Tendon Healing

Intrinsic Healing

The first 72 hours after repair comprises the inflammatory phase, which includes neovascularization and the arrival of inflammatory cells. This is followed by a 4-week proliferative phase during which collagen and granulation tissue are formed. The remodelling phase, in which the collagen becomes organized and cross-linked, continues after this for 16 weeks. During the inflammatory phase, the strength of the repair is almost entirely due to the strength of the suture. In the early proliferative phase, the strength of the repair decreases. This loss of repair strength can be lessened by stressing the tendon by means of early mobilization with passive motion. During remodelling, the strength of the repair increases toward normal.

Extrinsic Healing

Extrinsic healing produces scar tissue and adhesions between the tendon and the surrounding soft tissues. Early mobilization with passive motion exercises can help decrease this adhesion formation while maximizing tendon excursion. Factors that can worsen adhesion formation are prolonged immobilization, tendon ischemia (e.g., vincula disruption), gaping at the repair site, discontinuity of the sheath, or trauma to the tendon or sheath.

Postoperative Care and Rehabilitation

A large, bulky, compressive dressing should be used. The hand should be immobilized and splinted in the safe position: wrist in midflexion, MP joints flexed 90°, fingers in full extension. The use of postoperative antibiotics depends on the mechanism of injury and expected risk of infection.

Numerous studies have shown that early postoperative motion stress exercises are beneficial. Repairs that are completely immobilized are half as strong at 3 weeks. This difference between the strength of mobilized and immobilized repairs continues to grow as time progresses. According to Strickland, a good rehabilitation regimen should include the following:

- The wrist and MP joints should be flexed at rest
- The PIP and DIP joints are extended at rest
- Passive flexion of the MP, PIP and DIP joints before wrist extension
- Maintenance of passive digital flexion with the wrist extended
- Frequent application of motion stress

The use of the Kleinert splint after flexor tendon repair has been popular among hand surgeons. This splint maintains the wrist in partial flexion. The tip of the finger is attached to the volar surface of the forearm with a rubber band, allowing for active extension and passive rubber band flexion. Many modifications of the original Kleinert splint are in use today. Full use of the hand should be restricted until 8-12 weeks post repair.

The main complications from primary flexor tendon repair are rupture of the repair (incidence of about 5%), adhesion formation, and joint contracture, a late complication.

Pearls and Pitfalls

- It is essential to use a delicate, atraumatic technique when handling the severed ends of a tendon. Since the entire length of tendon may be required, one cannot rely on "freshening" of the edges to remove any traumatized portions of tendon.
- In zone I injuries of the FDP tendon, anchoring the distal tendon to bone provides a stronger repair than suturing the two ends of tendon back together. However, the FDP tendon should not be advanced more than 1 cm towards the insertion on the distal phalanx since this could result in the quadriga effect.
- If a tendon graft is required, palmaris longus is the best option. If not present, plantaris or extensor tendons of the toes can be used. The needed length of graft can be determined by matching the position of the injured finger to the cascade of the uninjured fingers. The amount of tendon shortening is measured as the finger is brought into the normal cascade.

- In grossly contaminated wounds involving flexor tendon injury, repair should be delayed since the risk of infection is high. Once the wound is clean and free of necrotic tissue and purulence, delayed primary repair or reconstruction can be undertaken.
- Establishment of the correct relationship between FDS and FDP (Camper's chiasma) is essential. This can be difficult when the proximal tendon ends have retracted into the palm.
- Since fingers are often flexed during the time of injury, the tendon laceration is usually distal to the skin laceration.
- When the two ends of the tendon are approximated, the epitenon should be carefully preserved and also brought into approximation. After repair of the tendon with the internal sutures, an epitendinous repair should be performed.

Suggested Reading

1. Brunner JM. The zig-zag volar-digital incision for flexor-tendon surgery. Plast Reconstr Surg 1967; 40:571.
2. Idler RS. Anatomy and biomechanics of the digital flexor tendons. Hand Clin 1985; 1:3.
3. Leffert RD, Weiss C, Athansoulis CA. The vincula; with particular reference to their vessels and nerves. J Bone Joint Surg 1974; 56A:1191.
4. Strickland JW. Flexor tendon injuries: I. Foundations of treatment. J Am Acad Orthop Surg 1995; 3:44.
5. Strickland JW. Flexor tendon injuries: II. Operative technique. J Am Acad Orthop Surg 1995; 3:55.
6. Strickland JW. Development of flexor tendon surgery: Twenty-five years of progress. J Hand Surg 2000; 25A:214.

Chapter 91

Injuries of the Finger

Millicent Odunze and Gregory A. Dumanian

Introduction

Injuries of the fingers are extremely common. They range from minor cuts and abrasions to wounds causing serious damage to the soft tissue, nail beds, bones, tendons, or ligaments. Significant injuries may be present that are not immediately apparent. Signs and symptoms of infections, such as erythema, purulence, and fever, will often not develop for hours to days following the injury. If not treated properly, serious finger injuries can lead to permanent deformity and loss of function. Careful treatment will allow for a faster and more complete recovery. This chapter discusses finger injuries by anatomical site. Injury to the fingertip, fingernail, nail bed, phalanx, interphalangeal joints, tendons, and ligaments will all be addressed.

Causes of Finger Injuries

Injuries to the digits can occur from a variety of external forces. The forces can be categorized as **direct forces**, such as lacerations, burns, and crush, or as **indirect forces**, such as rotation (twisting), bending (stretching) and axial loading (jamming). A combination of these forces is also possible. Finger injuries are often work-related. Animal bites are another common cause. The management of finger infections is discussed in detail in the "Hand Infection" chapter.

Evaluation

Evaluation of a patient with a finger injury begins with a history of the mechanism of injury, time of injury, location of injury (e.g., home, farm, industry) and important patient-specific factors such as age, gender, handedness, occupation, avocation, systemic diseases, and prior hand problems. The patient may complain of pain to the area and deformity may be present. Physical examination often provides the diagnosis and should include assessment for skin integrity, tenderness, edema, deformity, alignment (all fingers should converge to the scaphoid area on flexion), range of motion (active and passive motion, extrinsic and intrinsic muscle function, flexor and extensor tendon function) and neurovascular status. The injury site is inspected with specific attention to the characteristics of the wound. When indicated, radiographs of the finger are obtained to assess the extent of bone injury. Intravenous antibiotics are given to patients with an open fracture, and appropriate tetanus prophylaxis is provided.

Fingertip Injuries

The fingertip is the most frequently injured part of the hand, and the middle finger is most vulnerable because it is the most distal and therefore the last to be withdrawn. Fingertip injuries are defined as those injuries occurring distal to the

Practical Plastic Surgery, edited by Zol B. Kryger and Mark Sisco. ©2007 Landes Bioscience.

insertion of the flexor and extensor tendons. Although maintenance of length, preservation of the nail, and appearance are important, the primary goal of treatment is a painless fingertip with durable and sensate skin. Considerable hand dysfunction results when a painful fingertip causes the patient to exclude the digit from use. The specific wound characteristics determine which method of treatment is optimal for a given patient. It is important to know whether there has been loss of skin or pulp and the extent of such loss. The presence of exposed bone or injury to the nail bed or perionychial tissue must be determined. In the case of amputations, it is important to establish the level and angle of injury.

Anatomy

The fingertip is the end organ for touch and is supplied with special sensory receptors that enable the hand to relay the shape, temperature, and texture of an object. The skin covering the pulp of the finger is very durable and has a thick epidermis with deep papillary ridges. The glabrous skin of the fingertip is well-suited for pinch and grasp functions. Its volar surface consists of a fatty pulp covered by highly innervated skin. The skin of the fingertip is firmly anchored to the underlying terminal phalanx by multiple fibrous septa that traverse the fatty pulp.

Classification

Allen has classified fingertip injuries based on the level of injury (Fig. 91.1):

- Type 1 injuries involve only the pulp
- Type 2 injuries involve the pulp and the nail bed
- Type 3 injuries include partial loss of the distal phalanx
- Type 4 injuries are proximal to the lunula

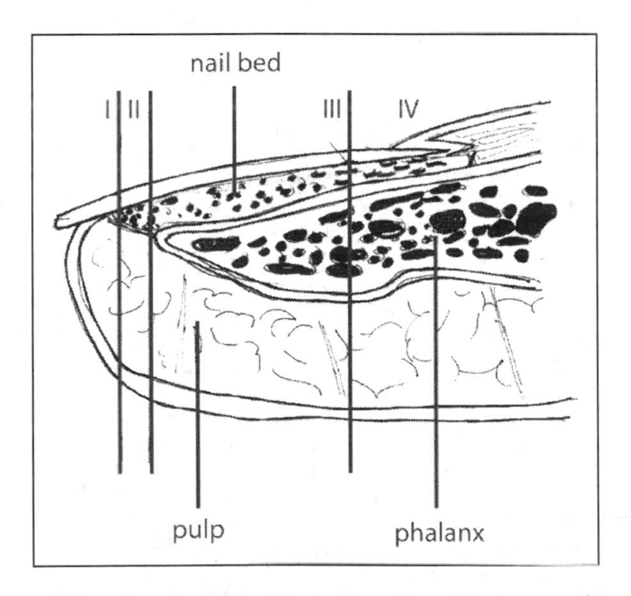

Figure 91.1. Classification of fingertip injuries. Type 1 injuries involve only the pulp; Type 2 injuries involve the pulp and the nail bed; Type 3 injuries include partial loss of the distal phalanx; Type 4 injuries are proximal to the lunula.

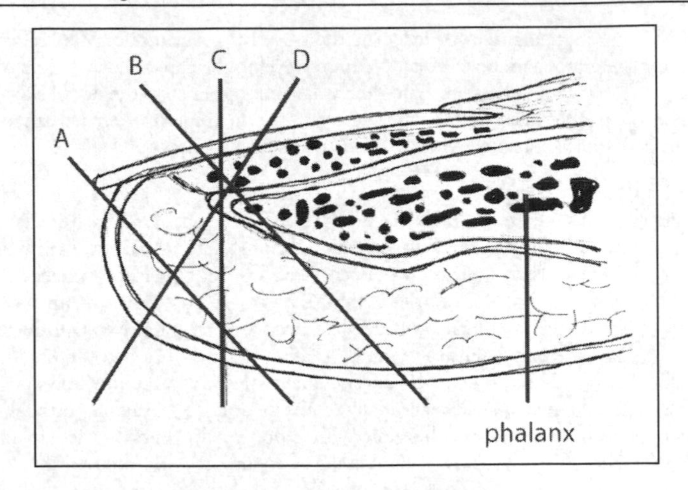

Figure 91.2. The angles of fingertip amputation. A, volar oblique without exposed bone. B, volar oblique with exposed bone. C, transverse with exposed bone. D, dorsal oblique with exposed bone.

This classification is useful to help generate a treatment plan. Additionally, tip amputations should be described in terms of the angle of injury—dorsal oblique, transverse, and volar oblique, as well as the presence of exposed bone (Fig. 91.2).

Treatment

Type 1 injuries may heal quite well by secondary intention. In contrast, Types 3 and 4 often require some type of flap coverage. Dorsal oblique and transverse injuries are more suited to local flaps. Volar oblique injuries often require a regional flap. Type 2 injuries require nail bed repair, which is discussed in the next section.

Healing by Secondary Intention

The simplest treatment of fingertip injuries is to allow the wound to heal by secondary intention. It is reserved for small defects (8 to 10 mm^2), with minimal bone exposure and minimal loss of tissue pulp. Local wound care should be performed 2 to 3 times daily with dressing changes. Healing is usually complete by 3 to 6 weeks depending on the size of the defect. In young children, this method provides good results even if larger areas of exposed bone.

Composite Grafts

If an amputated part has been recovered and it is clean and of adequate integrity, use the part for soft tissue coverage. If there is no exposed bone, de-fat the skin and suture it onto the defect. This piece will now function as a full-thickness skin graft. Minimize its thickness to enhance its chances of "taking." Even if this skin necroses it will still serve as a biologic dressing.

Revision Amputation

A simpler course of action involves shortening the digit or revision amputation. This procedure is indicated in situations in which bone is significantly degloved, and the angle of the injury is such that other options are not appropriate. Take care to limit loss of length, particularly in treating the thumb. This procedure can be performed under local anesthesia in the acute care setting. Develop the flaps to cover the tip of the digit, preferably with volar skin. Using the volar skin rather than the dorsal skin provides a more padded and durable soft tissue cover for the fingertip. Patients can return to their activities as tolerated when the soft tissues have healed.

Skin Grafts

Skin grafts can be used in fingertip injuries where there is skin loss, but adequate subcutaneous tissue is present with no exposed bone. The lack of exposed bone is paramount because skin grafts will not "take" well on bone without intact periosteum. Use this technique for injuries with skin loss of greater than 1.5 cm^2. In cases of smaller skin defects, allow the wound to heal by secondary intention. Full-thickness skin grafts provide better sensibility and durability, as well as a better cosmetic result. On the other hand, split-thickness skin grafts have a greater likelihood of "taking." Excellent hemostasis of the injury site must be obtained to avoid postoperative hematoma formation. Secure the graft with a bolster dressing that is left undisturbed for 5 to 7 days. Start physical therapy after the dressing is removed.

Local Flaps

Local flaps use adjacent local skin with its subcutaneous tissue and normal sensory end organs to cover defects. There are two common advancement flaps used for fingertip injuries—the **V-Y advancement flap** and the **homodigital triangular flap**. Both share similar principles in that a "V" incision is made adjacent to the defect. The skin and subcutaneous tissues are advanced forward, and the proximal defect is closed end-to-end. After closure, the proximal portion of the wound forms the vertical line of the "Y." The homodigital triangular flap, which is dissected more proximally, includes the digital artery within the flap. Range of motion therapy is started 7 to 10 days following local advancement flaps. These flaps are described in Chapter 92.

Regional and Distant Flaps

Regional flaps are defined as flaps taken from other parts of the hand that do not use tissue adjacent to the defect. They are well-suited for volar oblique type injuries. Owing to the postoperative immobilization required, the procedure is often discouraged in patients predisposed to finger stiffness. This includes patients older than 50 years of age, those with rheumatoid arthritis, and patients with multiple injured digits. These flaps are also not well suited for young children because of lack of compliance and the fact that simpler methods are usually adequate.

Distant flaps are defined as flaps obtained from areas of the body other than the injured limb and are primarily free-flaps. These procedures are considered in hand injuries with large soft tissue defects and provide thick, fatty coverage with little sensibility. The flaps can be developed from the chest, abdomen, groin, or opposite arm. A more extensive discussion of regional hand flaps can be found in Chapter 92.

Thumb Coverage

The thumb plays a crucial role in prehension and is involved in 50% of the function of the hand. Preservation of length of the thumb is more important functionally than in any other digit. For soft tissue defects that cannot be covered with a V-Y flap and for those measuring no more than about 2 cm in length, the **Moberg advancement flap** is the procedure of choice because it preserves length and tactile sensibility. For loss of more than two-thirds of the pulp tissue, a **cross-finger flap** from the dorsal aspect of the middle finger or a first dorsal metacarpal artery-island pedicle flap from the index finger usually provides satisfactory padding of the thumb and adequate sensibility. The **neurovascular-island flap** can help restore sensibility and padding to the ulnar side of the thumb pulp if other methods have proved unsatisfactory.

Complications

Cold intolerance after fingertip injury is common. Symptoms rarely resolve but may become more tolerable over time. Failure to resect a sufficient length of each digital nerve stump to avoid neuroma formation results in a persistently painful finger. Complications of skin grafts include hematoma, necrosis of the skin graft, and donor site complications. The resultant suture line in advancement flaps may be the cause of hypersensitivity noted by some patients. Advancement flap complications also include numbness, cold intolerance, and dysesthesias.

Fingernail and Nail Bed Injuries

The fingernail and the underlying nail bed are the most commonly injured part of the hand. The nail bed, which is the supportive tissue underneath the nail, can be damaged by laceration or crush. Development of a subungual hematoma invariably reflects nail bed injury with or without an associated fracture of the distal phalanx. Injuries of the nail bed can be divided into those that involve the germinal matrix and those that involve the sterile matrix. Germinal matrix injuries are generally more serious since nail formation originates from the germinal matrix, and an injury in this region has a higher likelihood of permanently affecting nail growth.

Anatomy

The fingernail protects the fingertip and has a major role in tactile sensation and fine motor skills. The nail complex, or perionychium, includes the nail plate, the nail bed, and the surrounding skin on the dorsum of the fingertip (paronychium). These structures are schematically shown in Figure 91.3. The fingernail is a plate of flattened cells layered together and adherent to one another. The nail bed lies immediately deep to the fingernail. The nail bed is composed of the germinal matrix, the

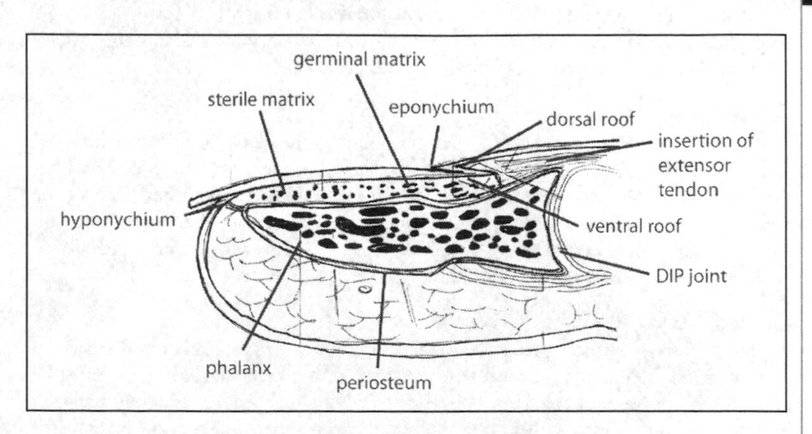

Figure 91.3. Sagittal section of the distal finger illustrating the anatomy of the nail and nail bed.

sterile matrix, and the roof of the nail fold. The germinal matrix, which produces over 90% of nail volume, extends from the proximal nail fold to the distal end of the lunula. The lunula represents the transition zone of the proximal germinal matrix and distal sterile matrix of the nail bed. The sterile matrix (ventral nail) contributes additional substance largely responsible for nail adherence. The roof of the nail fold (dorsal nail), which includes the germinal matrix, is responsible for the smooth, shiny surface of the nail plate. The hyponychium is the area immediately below the fingernail at its cut edge which serves as a barrier to subungual infection and also marks the terminal extension of bone support for the nail bed. The eponychium is the skin covering the dorsal roof of the nail fold. The paronychium is the skin at the nail margin, folded over its medial and lateral edges.

Classification

Van Beek et al have further classified acute fingernail and nail bed injuries as outlined below. This classification system provides a framework for determining the appropriate treatment.

I. Germinal Matrix Injury
- GI: Small subungual hematoma in proximal nail (25%)
- GII: Germinal matrix laceration, large subungual hematoma (50%)
- GIII: Germinal matrix laceration and fracture
- GIV: Germinal matrix fragmentation
- GV: Germinal matrix avulsion

II. Sterile Matrix Injury
- SI: Small nail hematoma (25%)
- SII: Sterile matrix laceration, large subungual hematoma (50%)
- SIII: Sterile matrix laceration with tuft fracture
- SIV: Sterile matrix fragmentation
- SV: Sterile matrix avulsion

Treatment

It is important that the nail bed be repaired with great attention to detail in order to restore its function and prevent any uncomfortable or unsightly deformities. Loupe magnification is recommended, and the use of microsurgical instruments allows easier handling of the tissue and small needle.

Grade I Injuries

Grade I injuries are treated nonoperatively unless they are painful, for which decompression or nail removal can be performed. Decompression of a subungual hematoma should be performed to relieve pain if it involves no more than 50% of the area of the nail. A hole should be placed in the nail plate over the hematoma with either a heated paper clip or an 18-gauge needle twirled like a drill. For larger subungual hematomas, the nail plate should be removed in order to repair the nail bed.

Grade II, III, and IV Injuries

The first step in treating grade II, III, and IV injuries is to carefully remove the nail by separating the nail plate from the nail matrix with a Freer elevator. Care must be taken during this step to avoid injury to the underlying nail bed. An adherent nail may indicate a grade I injury with limited nail bed involvement. The wound is then irrigated and debrided, limiting debridement to contaminated or devitalized tissues. The nail bed is then repaired under loupe magnification using 5-0 or 6-0 chromic suture.

If the proximal germinal matrix is injured, visualization may be obscured by the nail fold. Make skin incisions at 90° to the nail fold along the lateral border of the nail, then elevate the nail fold to fully evaluate the extent of the injury. If an associated phalangeal fracture is unstable, it may displace the nail bed. Unstable fractures should be pinned with a Kirschner wire, avoiding the distal interphalangeal joint if possible.

After any repair, the nail plate should be placed back into the nail fold. Replacement of the nail has several important functions: (1) it serves as a template for the new growing nail, (2) it serves as a splint for fractures, (3) it provides a biologic dressing for the nail bed, and (4) it prevents scarring of the nail fold to the nail bed. Using 4-0 or 5-0 nylon sutures, the nail is secured in place with two horizontal mattress sutures placed proximally on both sides of the nail. One or two simple sutures may be placed in the distal aspect of the nail and pulp to further secure the nail. If the nail plate is fragmented or not available, an artificial nail, a Silastic sheet, or a nail-shaped piece of nonadherent gauze is a good substitute. These can be left in place indefinitely, as they will be pushed out as the new nail regrows. Finally, the finger is bandaged to protect the digit and motion is restricted for 7 to 10 days.

Nail-Matrix Avulsions

When the proximal portion of the nail plate has been avulsed from the nail fold and lies on top of it, there is always an associated nail-bed laceration or avulsion of the germinal matrix from its proximal attachment and a fracture or epiphyseal separation. Proximal detachments or avulsions of the germinal matrix must be replaced into the nail fold. Three sutures, one at each corner and one in the middle, are sufficient. Each suture is passed from the outside to the inside of the nail fold, through the proximal edge of the germinal matrix in a horizontal-mattress fashion, and back through the base of the nail fold, exiting dorsally. After all the sutures have been passed, the germinal matrix is cinched into place by pulling proximally on the sutures, which are tied over the dorsum of the nail fold.

If the nail bed injury is associated with a loss of matrix tissue, the detached nail plate should be inspected for remnants of the nail bed that can be used for repair. If available the avulsed tissue can be carefully removed from the nail plate with a scalpel and sutured in place as a full-thickness graft directly on the distal phalanx. Defects in the sterile matrix with no tissue available for repair should be treated with a split-thickness nail bed graft. Split-thickness sterile matrix grafts are obtained by shaving the donor nail bed with a scalpel blade. The graft must be very thin (about one hundredth of an inch) to prevent a deformity from occurring at the donor site. The graft is then sutured to the surrounding nail bed. Treatment of germinal matrix avulsion with no available tissue for replacement depends primarily on the size and width of the defect. Split- or full-thickness germinal matrix grafts and local bipedicle or distally based nail bed flaps have been employed. If the entire nail bed has been completely or nearly completely avulsed and the distal phalanx remains, the best treatment is the application of a split-thickness skin graft.

Complications

Complications from nail bed repair are usually mild. Various nail deformities can occur at the injury or donor sites. Rarely, permanent loss of the fingernail can occur.

Phalangeal Injuries

Fractures of the phalanges are the most common fractures of the skeletal system, accounting for 10% of all fractures in several large series. The distal phalanx is the most commonly fractured bone of the finger. The middle phalanx is the least commonly fractured because of its high proportion of cortical to cancellous bone. Phalanges of the central digits are longer and sustain more fractures than the border digits.

Anatomy and Terminology

Each finger, except the thumb, has three phalanges: the proximal phalanx, the middle phalanx, and the distal phalanx. The thumb has only two phalanges, proximal and distal. All five digits have metacarophalangeal (MP) and interphalangeal (IP) joints. However, the thumb has only one IP joint, while the other four digits have both a distal interphalangeal (DIP) and proximal interphalangeal (PIP) joint.

Classification

Phalangeal fractures are classified as follows:

- Phalanx involved: distal, middle or proximal phalanx
- Location within the bone: base, shaft, neck, or head
- Pattern: transverse, spiral, oblique, or comminuted
- Displaced or nondisplaced
- Intraarticular or extraarticular
- Closed or open
- Stable or unstable
- Deformity: angulation, rotation, or shortening
- Associated injuries: skin, tendon, nerve, or vessel

Treatment

First and foremost, treat the patient, not the radiograph. Most stable fractures can be treated successfully by nonoperative means. These fractures are functionally stable before or after closed reduction and do well with splinting and/or buddy taping and early mobilization. The goal is restoration of normal function with the three "R's": reduction, retention, and rehabilitation. After accurate fracture reduction, the hand should be immobilized in the intrinsic plus or safe position with extremity elevation to minimize edema. Movement of the uninvolved fingers should be permitted to prevent stiffness. An exercise program should be directed toward the specific fracture with early mobilization of the injured finger. Repeat radiographs should be performed at 7-10 days to check the reduction. It is important to remember that the PIP joint is the most important joint in the fingers.

Unstable fractures cannot be reduced with a closed method or, if reduced, cannot be held in the reduced position without supplemental fixation. Closed reduction with percutaneous pinning (CRPP) or open reduction with internal fixation (ORIF) are required to provide stability and allow early mobilization. Indications for internal fixation include:

- Uncontrollable rotation, angulation, or shortening
- Multiple digit fractures that are difficult to control
- Displaced intraarticular fractures involving more than 15-20% of the articular surface
- Fracture-subluxation of the thumb and fifth finger carpometacarpal joints
- Unstable fractures: failure of closed manipulation, as in spiral fractures of the proximal phalanx or transverse metacarpal fractures
- Metacarpal head fractures
- Open fractures

For nondisplaced fractures treated in closed fashion, motion can be started within 3 weeks if the fracture is stable. Midshaft proximal phalangeal fractures require 5-7 weeks for complete bony healing. Midshaft middle phalangeal fractures require 10-14 weeks for radiographic healing of the cortical portion of the bone. Fractures requiring open reduction or severely comminuted fractures with disruption of the periosteum take twice as long to heal as simple fractures.

Complications

Loss of motion secondary to tendon adherence at the fracture site and contractures, especially at the PIP joint, are complications of phalangeal fractures. Malunion secondary to volar angulation after fractures near the base of the proximal phalanx may occur. There may be malrotation after spiral or oblique proximal and middle phalangeal fractures. Phalangeal fractures that have been percutaneously fixed may be complicated by pin tract infection. Nonunion results from bone loss, soft tissue interposition, inadequate immobilization or distraction at the fracture site.

Interphalangeal Injuries

IP injuries are **subluxations** or **dislocations** secondary to ligament or tendon injury. Joint subluxations occur with disruption of the joint's soft tissue supporting structures, but with some contact remaining between the joint surfaces. A dislocation is an injury that causes a phalanx to move out of its normal alignment with another phalanx with total loss of contact between the joint surfaces. Dislocations of the DIP joints are rare, usually dorsal, and caused by hyperextension. Dislocations of the PIP joint are most commonly dorsal but can also be volar and rotated in either direction.

Anatomy

The proximal, distal, and thumb IP joints are true hinge joints. Joint stability is provided by the three structures: radial and ulnar collateral ligaments, the volar plate, and the dorsal capsule. The dorsal capsule is very thin and provides minimal stability to the joint. The collateral ligaments and the volar plate are firm structures. The collateral ligaments limit side-to-side movement of the DIP and PIP joints. For dislocation to occur, at least two of these three structures must be disrupted.

Classification

Classification can be based on the status of the skin (closed versus open), the duration of injury (acute versus chronic), the degree of joint displacement (subluxation versus dislocation), the status of the joint surface (dislocation versus fracture-dislocation) and the ability to reduce the joint dislocation (simple versus complex). Joint dislocations and subluxations can be further subclassified based on the direction of displacement of the distal portion of the injured digit relative to the proximal portion (dorsal, volar, lateral, or medial).

Treatment

DIP Joints

Dislocations

To manage acute dorsal DIP joint dislocations, perform a closed reduction, with or without digital block anesthesia, using longitudinal traction of the distal phalanx.

Place direct pressure on the dorsal base of the distal phalanx, displacing it distally and palmarly. Postreduction radiographs should confirm congruous reduction in the joint. After joint reduction, assess joint stability and flexor and extensor tendon function. If joint instability is present after joint reduction, splint the joint for 2 to 3 weeks in 10°-20° of flexion. With acute volar DIP dislocations, avoid splinting the DIP joint in hyperextension to prevent dorsal skin wound problems. Splint the digit for 6 weeks or more to promote healing. Open dislocations of the DIP joint require irrigation, debridement, and antibiotics. When the dislocation is chronic (greater than 3 weeks) or irreducible, perform an open reduction. Irreducible DIP joint dislocations may be secondary to an interposed structure, such as the volar plate, a flexor tendon, a fracture fragment or a sesamoid bone. Remove the interposed structure to reduce the joint.

Distal Extensor and Flexor Tendon Rupture/Avulsion

Disruption of the extensor mechanism occurs if there is a sudden forceful flexion of the extended DIP joint. This results in either a disruption of the extensor tendon or an avulsion fracture with varying amounts of bone involvement. On examination, there is a characteristic flexion deformity at the DIP joint and the inability to actively extend this joint, which is known as mallet finger (Fig. 91.4). The treatment consists of continuous splinting of the DIP joint in full extension or even slight hyperextension for a minimum of 6 weeks, followed by night splinting for an additional 2 weeks. The importance of continuous splinting must be stressed because even momentary flexion of the DIP joint during the treatment period could result in resetting the treatment period back to time zero. Splinting can be effective even if initiated several weeks after the injury. Surgery is recommended for those patients with joint subluxation despite splinting, for displaced avulsion fractures with a large articular component (more than 50%). Failed splinting will also require surgery: K-wire fixation of the DIP joint in extension is usually adequate.

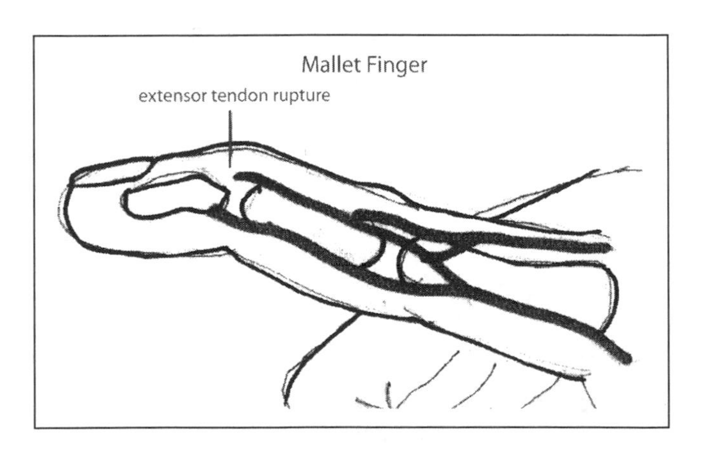

Figure 91.4. The mallet finger deformity. Disruption of the extensor tendon or an avulsion fracture causes a characteristic flexion deformity at the DIP joint and the inability to actively extend this joint.

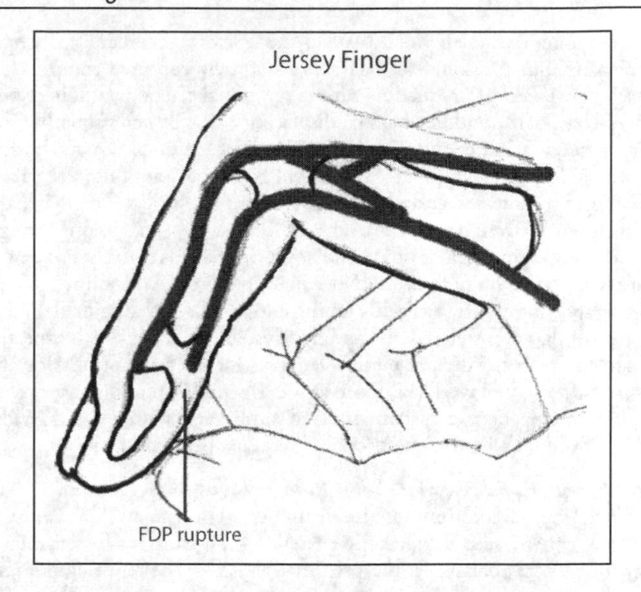

Figure 91.5. The jersey finger deformity. Forced extension of the maximally flexed DIP joint can cause avulsion of the FDP tendon, commonly a result of the ring finger catching on another player's jersey.

Avulsion of the flexor digitorum profundus tendon, "jersey finger," results from forced extension of the maximally flexed DIP joint, commonly a result of the ring finger catching on another player's jersey (Fig. 91.5). The patient presents with the loss of the normal cascade of the fingers and the inability to flex the DIP joint actively. If the tendon retracts into the palm (Type 1 injury), urgent repair is necessary to avoid progressive degeneration. If the tendon retracts to the level of the PIP joint (Type 2 injury) or is associated with a large avulsion fracture that causes the tendon to be held up at the DIP joint (Type 3 injury), early surgical intervention is preferable. However, these injuries can be repaired as late as 6 weeks after the injury. When these injuries present even later than 6 weeks, "salvage surgery" directed at the DIP joint with either DIP capsulodesis, or DIP joint fusion can be performed.

DIP Ligamentous Injuries

A ligament may be torn by a forceful stretch or blow, leaving the joint unstable and prone to further injury. Unless they accompany a dislocation, nearly all ligament injuries of the DIP joint are partial tears or sprains and thus can be treated nonoperatively. Temporary splinting for a few days for comfort should be followed by an early, vigorous active motion program.

PIP Joints

Dislocations

PIP joint dislocations are frequently associated with an injury to the volar plate, collateral ligament, extensor tendon (central slip) and/or joint articular surface. Dorsal PIP injuries without fracture can usually be reduced by closed means. A digital block is often unnecessary. The volar plate, by necessity is ruptured, usually from the middle phalanx, but the collateral ligaments are rarely ruptured completely from their attachments. Perform reduction with longitudinal traction and direct pressure on the dorsal base of the middle phalanx, displacing it distally and palmarly. Confirm congruous reduction of the joint with post-reduction radiographs. Assess both active and passive stability of the joint after reduction. For stable joints, use a resting splint with the finger compressed and almost fully extended until the digit is pain-free enough to begin early active motion. Protect against hyperextension during early motion by buddy taping the digit to the adjacent finger. Alternatively, one can use an orthoplast hand-based splint to avoid hyperextension. Protect the injured digit, especially during sporting events, by buddy taping it to an adjacent uninjured digit. Reassure the patient that persistent edema and slow resolution of stiffness is expected and continue to compress the digit at night. Decisions for the need for physical therapy can be made after several weeks of observation. Open PIP joint dislocations should be treated with irrigation, debridement, and antibiotics. Chronic and irreducible dorsal PIP dislocations are not common and require open reduction.

Proximal Extensor and Flexor Tendon Rupture/Avulsion

Palmar PIP joint dislocations involve disruption of the extensor mechanism (central slip) from either forced flexion of the extended PIP joint or a forced volar dislocation of the middle phalanx on the proximal phalanx. The resulting boutonniere deformity leads to not only a flexion deformity of the PIP joint but also a hyperextension of the DIP joint (Fig. 91.6). This characteristic deformity may take several weeks to develop. Therefore, an acutely edematous and tender PIP joint with weak active extension and dorsal tenderness should be treated even in the absence of the classic deformity. It is distinguished from the pseudoboutonniere deformity in that, although active extension is weak or not possible, full passive extension of the PIP joint is easily achieved, and hyperextension of the DIP joint does not exist. Treatment consists of an extension splint immobilizing only the PIP joint worn for a

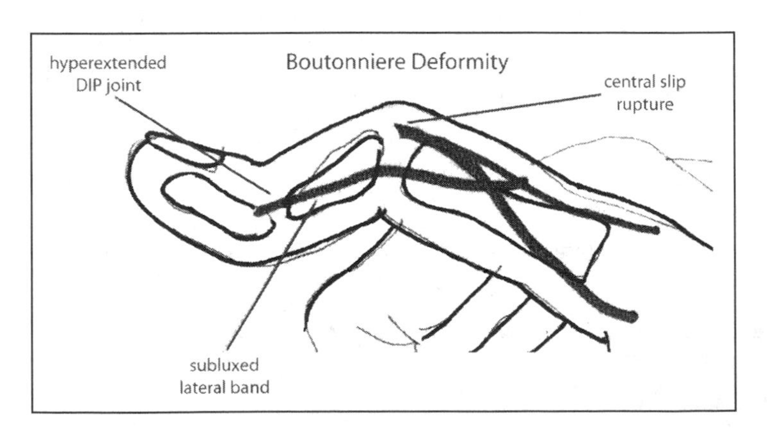

Figure 91.6. The boutonniere deformity. There is a flexion deformity of the PIP joint and a hyperextension of the DIP joint due to disruption of the extensor mechanism. This can occur as a result of forced flexion of the extended PIP joint or forced volar dislocation of the middle phalanx on the proximal phalanx.

minimum of 4 weeks, followed by progressive weaning and nighttime splinting for another 3 weeks. Similar to mallet finger immobilization, maintenance of continuous extension without flexion of the joint should be strictly enforced. Even in the setting of the chronic boutonniere deformity, splinting is the first line of treatment, since surgery often does not provide satisfactory results. Reconstruction is reserved for symptomatic, chronic deformity that fails nonoperative measures.

There are two chapters in this text devoted entirely to lacerations of extensor and flexor tendons.

PIP Ligamentous Injuries

A ligament may be torn by a forceful stretch or blow, leaving the joint unstable and prone to further injury. Forced radial deviation of the thumb results in trauma to the dorsal capsule, the ulnar collateral ligament (UCL), and the ulnar aspect of the volar plate at the MP joint. This injury, known as gamekeeper's thumb or skier's thumb, occurs most often when a skier falls on his or her pole with an open hand. Partial tears can usually be managed nonoperatively with a thumb-spica cast. This injury is managed with open repair of the ulnar collateral ligament through an ulnar incision. A Stener's lesion occurs when the UCL is trapped above the adductor aponeurosis preventing proper healing. The ligament is repaired, and bony avulsions are reduced and stabilized with Kirschner wires, a pullout wire, or screws. Postoperatively the hand is splinted from the IP joint to the elbow for 4 weeks for bony avulsions and 6 weeks for ligament repairs.

Volar Plate Ruptures

Volar plate ruptures of the PIP joint can result from a dorsal PIP joint dislocation or hyperextension injury. The volar plate usually detaches from the middle phalanx, with or without a piece of bone. This injury must be differentiated from the PIP joint fracture-dislocations in which the large size of the intraarticular fracture fragment renders the joint either unstable at full extension, or in a chronically dislocated state. Treat the mild volar plate fracture with a stable joint as any other volar plate injury, with protection against hyperextension by either a temporary dorsal-block digital splint or by strapping to an adjacent finger for 3 weeks. Encourage full flexion. Chronic volar plate ruptures can result in a swan neck deformity (Fig. 91.7). The untreated mallet finger can also lead to hyperextension of the PIP joint from unopposed and overactive pull of the extensor mechanism of the middle phalanx, also resulting in a swan-neck deformity. If the swan neck deformity is secondary to chronic volar plate rupture, nonoperative treatment includes use of an orthoplast splint or a silver, double-ring splint to help PIP joint hyperextension. Symptomatic volar plate ruptures can be helped by surgical intervention, specifically late reattachment or shortening of the volar plate, with or without some form of volar reinforcement.

Complications

Ligament and tendon injury requires joint immobilization and may require operative repair. Joint stiffness can occur and may be further worsened by intraarticular edema and resulting fibrosis. Early joint motion minimizes postinjury stiffness, but preference must be given to joint immobilization until adequate ligament stability has developed. The most common complication of gamekeeper's thumb is instability of the MCP joint due to failure of repair. It is managed with ligamentous reconstruction with a tendon graft or MCP joint arthrodesis.

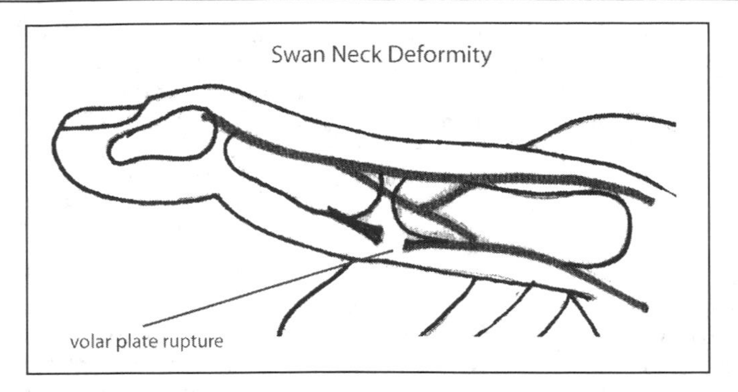

Figure 91.7. The swan neck deformity. There is a flexion deformity of the DIP joint and hyperextension of the PIP joint. Chronic volar plate ruptures can lead to this deformity. The untreated mallet finger can also lead to hyperextension of the PIP joint from unopposed and overactive pull of the extensor mechanism of the middle phalanx, also resulting in a swan-neck deformity.

Pearls and Pitfalls

- Mobilization of an injured digit by 3 weeks significantly decreases the chance of permanent loss of motion. The patient should be told to open and close their hand "against air", and to strive for motion rather than strength.
- The more unstable the reduced dislocation or fracture, the more the injury will require surgical stabilization.
- Rotational deformities do not improve with time and are a strong indication for operative repair.
- One trick that can help when using a removed fingernail as a protective dressing is to dermabond it to the nail bed. It will stay stuck for 1-3 weeks.

Suggested Reading

1. Allen M. Conservative management of fingertip injuries in adults. The Hand 1980; 12:257.
2. Craig SM. Anatomy of the joints of the fingers. Hand Clin 1992; 8:693.
3. Dray GJ, Eaton RG. In: Green DP, ed. Dislocations and Ligament Injuries in the Digits. Operative Hand Surgery 1978; 3:149.
4. Fassler PR. Fingertip injuries: Evaluation and treatment. J Am Acad Orthop Surg 1996; 4:84.
5. Hart RG, Kleinert HE. Fingertip and nailbed injuries. Emerg Med Clin North Am 1993; 11:755.
6. Rosenthal EA. Treatment of fingertip and nail bed injuries. Orthop Clin North Am 1983; 14:675.
7. Russell RC, Cases L. Management of fingertip injuries. Upper Extremity Trauma and Reconstruction 1989; 94:1298.
8. Van Beek AL, Kassan MA, Adson MH et al. Management of acute fingernail injuries. Hand Clin 1990; 6:23.
9. Zook EG. Anatomy and physiology of the perionychium. Hand Clin 1990; 6:1.

Chapter 92

Soft Tissue Coverage

Hongshik Han and Zol B. Kryger

Introduction

The hand is the most frequently injured part of the body. The majority of hand wounds are punctures or lacerations that can be closed primarily in the acute care setting. More severe injuries, however, often require the recruitment of additional tissue in order to obtain adequate coverage. This chapter presents the commonly used local flaps for soft tissue coverage of the hand, including the fingers and thumb. Although a comprehensive array of flaps will be presented, the list is not exhaustive. In addition, extensive defects may necessitate the use of a free flap. The various free flaps used for soft tissue coverage of the hand are discussed elsewhere in this book.

Flaps for Finger Coverage

The **volar V-Y advancement flap** (Fig. 92.1), also called the Atasoy or Kleinert flap, is most commonly used for the dorsal-oblique finger tip amputation with exposed bone. The inverted-triangle, V-shaped flap is elevated on the volar pad, and the distal advancement of the wound is closed in Y-fashion. With complete division of the fibrous septae and flap mobilization, 1 cm of advancement is routinely obtained.

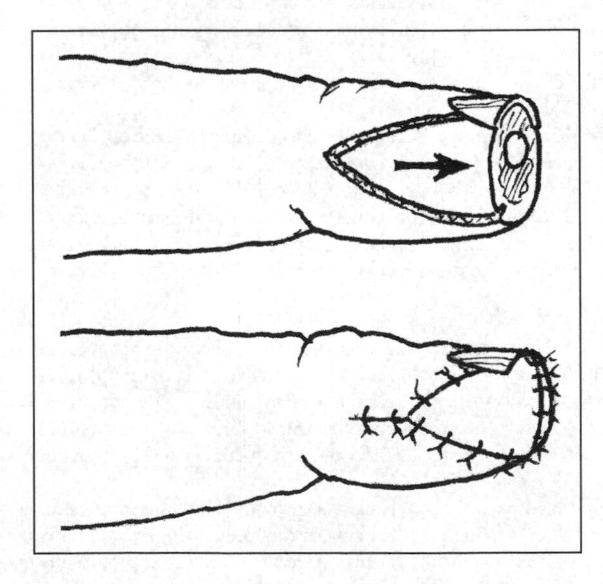

Figure 92.1. V-Y advancement flap for fingertip amputations. The flap can also be volar-based within the fingertip pulp.

The **bilateral triangular advancement flap**, also called the Kutler flap, was classically used for the transverse, or volar-oblique finger tip amputation with exposed bone. It is now rarely employed because (1) only about 3-4 mm advancement is obtained, (2) often it creates an insensate fingertip, and (3) it can create a sensitive sagittal scar at the finger tip.

The **oblique triangular flap** is used for the volar or oblique finger tip amputation with exposed bone. It combines the advantages of the volar V-Y advancement flap and bilateral triangular advancement flap. It can easily be converted to a neurovascular island pedicle flap if more elevation is required, in which case a skin graft is used to cover the donor site.

The **cross-finger flap**, also termed the trans-digital flap, is most often used for the volar pad injury that requires more tissue coverage than is possible with advancement-type flaps. Often there will be exposed bone, tendon, or distal interphalangeal (DIP) joint. There are multiple donor digits for this flap. Any adjacent finger can serve as a donor. The flap should be elevated superficial to the paratenon and slightly larger than the defect. The donor site is closed with a full-thickness skin graft (FTSG), and the two digits are held together with a splint or K-wires for 2 weeks. After this period, the flap is separated from the donor digit.

The **reverse cross-finger flap** is a variation of the traditional cross-finger flap. The papillary and reticular dermis are divided, and the reticular dermis is used to cover the dorsal defect on the adjacent finger. The papillary dermis is sewn back into the donor site to heal as a random flap.

The **Hueston and Souquet flaps** are both lateral palmar rotation-advancement flaps used to cover tip amputations with exposed bone. The Hueston flap includes only one neurovascular bundle at the base of the flap, whereas the Souquet flap includes both. In both flaps, a back cut is created that requires skin grafting.

The **thenar flap** is most often used to cover the index and long finger tip amputations. The donor site is found by placing the injured finger tip(s) over the thenar eminence, and an H-shaped incision is made to bury the stump in the thenar pad. The flap is separated after about 2 weeks.

The **dorsal middle finger flap** is a potentially sensate, **neurovascular island flap** that can be used to cover defects in all the fingers, except the thumb. The flap is similar to a cross-finger flap but extends more proximally to include a vascular bundle. The dissection often needs to progress as proximal as the common digital artery. A dorsal sensory branch that can be anastomosed with a recipient digital nerve is elevated with the flap. The donor site must be closed with a FTSG.

Flaps for Thumb Coverage

The **Moberg flap** is an advancement flap designed to preserve the amputated thumb's length. It is a robust and sensate flap mobilizing the volar, proximal thumb tissue with both neurovascular bundles. The Moberg flap can easily be converted into a bipedicled flap if more advancement is needed. However, it has the potential of leaving the thumb metacarpophalangeal (MP) and interphalangeal (IP) joints in a flexed position.

The **neurovascular island pedicle flap** is used to provide padded, sensate skin to the thumb. The donor flap is from the ulnar side of the middle finger. The flap's neurovascular bundle with its surrounding fat may need to be dissected proximally to the level of the superficial arch. A FTSG is used to close the donor site. A contraindication to the use of this flap is a middle finger that cannot be adequately perfused with only the digital artery on the radial side. This can be tested in a manner similar to the Allen test at the wrist.

The **racquet flap**, also called the Holevich flap, is used to provide sensate skin to the thumb, especially for chronic median nerve lesions. It is based on the second dorsal metacarpal artery. Dorsal sensory branches of the superficial radial nerve are included to provide sensation to the volar thumb. The racquet flap can also be used to provide pliable tissues to the first web space.

The **kite flap**, also termed the Foucher flap, is an extension of the racquet flap and is used to provide sensate skin to a scarred and denervated thumb pad. This flap includes the skin over the index MP joint. It is elevated from distal to proximal superficial to the paratenon, along with the first dorsal metacarpal artery. A skin tunnel is created, and the flap is passed through the tunnel to reach the recipient site. A FTSG is used to close the donor site.

The **homodigital island flap**, also called the annular flap or Goumain flap, is used to preserve thumb length and to provide sensate coverage to the amputated thumb tip, particularly at the proximal phalanx level. A circular incision is made around the thumb 2 cm proximal to the defect, and once the neurovascular bundles are freed, the circumferential flap advances distally to cover the defect. A FTSG is used to close the donor site.

Flaps for Coverage of the Hand and Proximal Fingers

The **dorsal island digital flap** is an axial-type, island flap based on the dorsal digital artery. Its size can be up to 3 x 3 cm over the dorsal proximal phalanx. It is used to cover the ipsilateral or adjacent finger defect in the PIP region. The preservation of the dorsal digital veins and artery is crucial, and the defect is covered with a FTSG.

The **fillet flap** is used to cover a proximal digital amputation using the salvaged soft tissues from the amputated digit. The bone, tendon, and pulp tissues are filleted off the skin, and the resulting skin flap is used to cover the dorsal or palmar wound. Any attachments of the skin to the hand should be preserved.

The **retrograde radial forearm flap** is based on the radial artery with the intact palmar arch. It is a robust flap and has undergone several variations. A positive Allen test is a contraindication for the use of this flap. Radial artery reconstruction after the transfer is usually not required. The pedicle length is sufficient to cover almost any hand defect. The dissection begins at distal volar forearm radial to flexor carpi ulnaris (FCR) to expose the radial artery and its venae commitantes. The dissection continues between the FCR and brachioradialis, and when the distal flap edge is reached, the flap is elevated from its ulnar border until the septal perforators are reached. The pedicle and the septal perforators are preserved and kept contiguous with the flap. The dissection continues from the flap's radial border until the septal perforators are reached. The branches off to the FCR and the proximal radial artery-venae commitante are ligated to elevate the flap. The donor site is closed with a split-thickness skin graft (STSG).

If the skin and subcutaneous tissues are not needed, it can be raised as an adipofascial flap. When harvested as an adipofascial flap, the skin is incised in a zigzag fashion, and the antebrachial fascia is exposed. The flap is designed on the fascia. The dissection is the same except the skin is spared, leaving the antebrachial fascia intact underneath. The donor site is closed primarily, and the flap is inset into the defect with a STSG covering the flap at the recipient site.

The **retrograde radial forearm fascial flap** is a distally-based, random volar antebrachial fascia turnover flap. It differs from the retrograde radial forearm flap in that the radial artery is left in situ, and the vascular supply comes from the distal perforating branches of the radial artery. It can be used to cover both volar and dorsal hand defect. The dissection begins by making an S-shaped skin incision just deep to the hair follicles over the volar forearm. The branches of the radial antebrachial nerves are protected. The flap is designed over the fascia at least 3-4 cm wide and dissected in a proximal to distal direction until a point 5 cm proximal to the radial styloid. The flap is turned over, and a STSG is used to cover the flap.

The **posterior interosseous forearm flap** is another retrograde fasciocutaneous flap based on the posterior interosseous artery (PIA). It is used to cover defects in the following areas: first webspace-thumb, dorsal hand-dorsal PIP, and anterior wrist-palm. The flap is centered in the axis between the lateral epicondyle and ulnar styloid with elbow in full flexion. The PIA originates at the proximal and middle third junction of this axis. There are 7-14 fasciocutaneous perforators distal to this point along the axis, and the center of the flap should be distal to this point. The PIA terminates 2 cm proximal to the ulnar styloid by way of anastomoses with the dorsal wrist arcade. The dissection begins at this point in a distal to proximal direction, and the PIA is identified. The posterior interosseous nerve is located radial to the artery and must be protected throughout the dissection. The flap dissection continues proximally to the main perforator with ligation of all muscular branches. The flap is elevated in an ulnar to radial direction with the pedicle being ligated proximal to the flap with and release of the septum from its ulnar shaft attachment.

The **dorsal ulnar artery flap** is a fourth retrograde fasciocutaneous flap of the forearm. It is used to cover defects of the ulnar-volar or dorsal hand. It is based on the dorsal branches of the ulnar artery that originates 2-5 cm proximal to the pisiform bone. The dissection begins 2 cm proximal to the pisiform, and with ulnar retraction of flexor carpi ulnaris, the dorsal ulnar branch is seen arising from the ulnar artery. The flap is centered along the ulnar axis with palmaris longus forming the volar border and the fourth extensor digitorum communis tendon forming the dorsal border. The proximal and middle third junction of the ulnar forearm forms the distal flap border.

Rotation flaps are useful for coverage of the dorsal fingers and the dorsum of the hand, where the skin is more lax. They are designed to redistribute the tension over the larger radian of the flap edges. The use of a back cut, also called Burow's triangle, further helps to reduce the tension at the tip of the flap. Transposition flaps, such as the Limberg or rhomboid flap, are also useful on the dorsal hand. The design of these flaps is discussed in greater detail in Chapter 11.

Pearls and Pitfalls

There are numerous options available for soft tissue coverage of the hand. Whenever possible the simplest technique, namely direct closure, should be used. The next more complex option is the local random flap, followed by the pedicle flap. At the top of the reconstructive ladder is the microvascular free tissue transfer to the hand. Split-thickness skin grafts and the use of tissue expanders are also good options for coverage when there is an adequate tissue bed without exposed tendon, artery, nerve or bone, and the adjacent skin is healthy. The choice of reconstructive technique depends on the wound geometry, the amount of soft tissue required, and the local wound conditions, such as the immediate necessity to cover exposed bone or tendon. Conditions that limit joint motion, (e.g., arthritis and Dupuytren's disease), or impair circulation, (e.g., Raynaud's or heavy smoking), should always be taken into account when planning the reconstruction.

Suggested Reading

1. Atasoy E, O'Neill E. Local flap coverage about the hand. Atlas Hand Clin 1998; 3(2):179.
2. Chase RA. Historical review of skin and soft tissue coverage of the upper extremity. Hand Clin 1985; 1:599.
3. Germann G. Principles of flap design for surgery of the hand. Atlas Hand Clin 1998; 3(2):33.
4. Gilbert A. Pedicle flaps of the upper limb. Philadelphia: Lippincott, 1992.
5. Lister G. The theory of the transposition flap and its practical application in the hand. Clin Plast Surg 1981; 8:115.
6. Martin D, Bakhach J, Casoli V et al. Reconstruction of the hand with forearm island flaps. Br J Plast Surg 1990; 43:290.
7. Weinzweig N, Chen L, Chen ZW. The distally based radial forearm fasciocutaneous flap with preservation of the radial artery: An anatomic and clinical approach. Plast Reconstr Surg 1994; 94:675.
8. Zancolli EA, Angrigiani C. Posterior interosseous island forearm flap. J Hand Surg {Br} 1988; 13B:130.

Chapter 93

Carpal Tunnel Syndrome

David S. Rosenberg and Gregory A. Dumanian

Introduction

Carpal tunnel syndrome (CTS) is the most common compressive neuropathy, developing due to compression of the median nerve at the wrist. The syndrome affects roughly three percent of the adult American population and is three times more common in women than in men. The condition is bilateral in half of all patients. In the United States, roughly 500,000 operations are performed each year to decompress the carpal tunnel at an annual economic cost which exceeds $2 billion. The prevalence of this condition has been reported to be higher in persons who perform repetitive motion activities, but the significance of this finding has been challenged in the literature.

Clinical Presentation

Patients with CTS typically report intermittent pain and paresthesias in the median nerve distribution of the hand, comprising the thumb, index finger, middle finger, and radial half of the ring finger. Sensation in the thenar area of the palm tends to be unaffected, due to the fact that this area of the hand is innervated by the palmar cutaneous branch of the median nerve which "branches" off of the median nerve proximal to the carpal tunnel. Symptoms generally progress gradually over a period of months to years, and they are typically worse at night. These nighttime paresthesias are a classic symptom of a patient with CTS, which is attributable to the tendency for the wrist to flex during sleep. Symptoms may also be aggravated during the day by activities for which the wrist is flexed or extended for prolonged periods, or when the hands are vibrated, such as when holding a steering wheel.

Chronic median nerve compression can lead to weakness of the thenar intrinsic muscles, producing decreased dexterity. Patients often report that their grasp is weak and that they have difficulty holding objects. Muscle atrophy develops, producing a visible depression in the thenar eminence. In selected patients, the median nerve compression affects only the motor fascicles and sensory finding are absent. Patients with this condition are often unaware of the problem until there is obvious muscle atrophy.

Etiologies

The primary pathophysiology leading to the development of CTS is an **increase in interstitial pressure** in the carpal tunnel (CT). This increased pressure has numerous causes that can be classified into four main categories: **idiopathic or spontaneous**, **intrinsic** factors, **extrinsic** factors, and **exertion/overuse** conditions. A second pathophysiologic mechanism exists apart from increased interstitial pressure, and that is **neuropathic factors**. A breakdown of some of these factors follows.

Practical Plastic Surgery, edited by Zol B. Kryger and Mark Sisco. ©2007 Landes Bioscience.

I. Idiopathic/spontaneous

IIa. Intrinsic factors outside the nerve that increase CT volume

A. Conditions altering fluid balance
1. Pregnancy
2. Renal failure
3. Chronic hemodialysis
4. Myxedema
5. Acromegaly
6. Menopause
7. Oral Contraceptives
8. Congestive heart failure
9. Thyroid disease

B. Inflammatory conditions
1. Rheumatoid arthritis
2. Gout
3. Amyloidosis
4. Pseudogout
5. Lupus erythematosis
6. Scleroderma
7. Nonspecific tenosynovitis
8. Dematomyositis
9. Infection:
   - Pyogenic (bacterial)
   - Mycobacterium
   - Fungal
   - Lyme disease
   - Viral
   - Parasitic (guinea worm)

C. Tumor and tumor-like masses
1. Ganglion
2. Lipoma
3. Fibroma
4. Pigmented villonodular tenosynovitis

D. Anatomical abnormalities
1. Vascular malformations
2. Anomalous muscles:
   - Proximal origin of lumbrical muscles
   - Distal muscle of flexor digitorum superficialis muscle
   - Abnormal insertion of palmaris longus
   - Anomalous slip of flexor pollicis longus
   - Palmaris profundus

E. Hemorrhagic disorders-hemorrhage within the carpal tunnel
1. Hemophilia
2. von Willebrand disease
3. Acute leukemia
4. Anticoagulants
5. Rupture of aneurysm of a persistent median artery

F. Traumatic injuries
1. Posttraumatic scarring within the tunnel-traction neuropathy
2. Trauma causing hemorrhage within the CT

IIb. Intrinsic factors inside the nerve that increase CT volume

A. Neurilemmoma
B. Lipofibroma
C. Neurofibroma
D. Neuroma

III. Extrinsic factors that alter tunnel contour

A. Acute fractures
B. Acute carpal dislocations/subluxations
1. Lunate dislocation
2. Rotatory subluxation of scaphoid

C. Intercarpal arthritis

IV. Exertional/overuse conditions

V. Neuropathic factors

A. Diabetes mellitus
B. Multiple myeloma
C. Alcoholism
D. Vitamin toxicity
E. Nutritional deficiency
F. Exposure to industrial solvents
G. Hand-arm vibration syndrome
H. Medication-lithium, beta blockers, ergot overdose

Anatomy

The carpal tunnel is a space defined by the concave arch of the carpus enclosed by the transverse carpal ligament (TCL). The scaphoid, trapezium, and sheath of the flexor carpi radialis (FCR) make up its radial margin. The ulnar boundary consists of the triquetrum, the hook of the hamate, and pisiform. Ten structures course through the carpal tunnel. These include the median nerve, four flexor digitorum superficialis (FDS) and four flexor digitorum profundus (FDP) tendons (Fig. 93.1), all ensheathed by the ulnar bursa, and the flexor pollicis longus, found at the radial side of the canal and surrounded by the radial bursa. The flexor carpi radialis tendon (FCR) travels through its own separate osteofibrous tunnel. The TCL itself attaches medially to the pisiform and laterally to the tuberosity of the scaphoid bone. It is thickest (0.6 to 2 mm) at the junction of the mid and distal thirds and thin at its proximal and distal ends. In individuals with CTS the ligament is much thicker and can be up to 6 mm wide at its thickest portion.

The median nerve is the most superficial structure within the carpal tunnel and is covered by a layer of cellulo-adipose tissue. The nerve lies directly under the TCL, in the radiopalmar portion of the canal. The superficialis and profundus tendons of the index finger lie immediately dorsal to the median nerve. The nerve divides distally into five sensory branches and the recurrent motor branch. In 80% of cases, the motor branch arises from the radiopalmar region of the median nerve. The remainder of patients has the origin in the central location, and a small percentage take-off from the ulnar aspect. Ten percent of patients have multiple motor branches. The palmar sensory cutaneous branches (PSCB) of the median nerve branches off 5-7 cm proximal to the wrist crease, often found between the palmaris longus and FCR tendons. The PSCB terminate in the subcutaneous tissues overlying the thenar musculature.

A sensory communicating branch between the median and ulnar nerves in the palm is present in 80% of patients, with roughly half lying within millimeters of the TCL. Injuries to this nerve have been reported during carpal tunnel release, leading to paresthesias or dysesthesias in the ulnar digits.

Physical Examination

Clinical history and physical examination including provocative testing are more easily performed than electrodiagnostic evaluation, and they are more appropriate tools for the ambulatory setting. Because patients suffering from CTS frequently report symptoms of numbness and tingling in the radial three digits, a thorough **sensory exam** should be done on initial assessment. The most common tests for sensibility utilize static or moving two-point discrimination, Semmes-Weinstein monofilament testing, and vibrometry. Sensibility testing is conceptually divided into innervation density testing and threshold testing. Static and moving two-point discrimination are density tests. Static testing measures slowly adapting neuron receptors (those mediating the sensation of constant touch or pressure) and moving two-point discrimination measures quickly adapting fibers (which mediate transient touch or movement). Innervation density measures several overlapping peripheral receptor fields. Testing for innervation density is a significant part of the exam because it may be perceived as normal in the presence of mild or even moderate compression of a peripheral nerve due to the presence of multiple overlapping fields.

Threshold tests such as vibrometry or Semmes-Weinstein monofilaments evaluate single nerve fibers innervating a single receptor or group of receptor cells. Threshold testing is more likely to show a gradual and progressive change in value as a

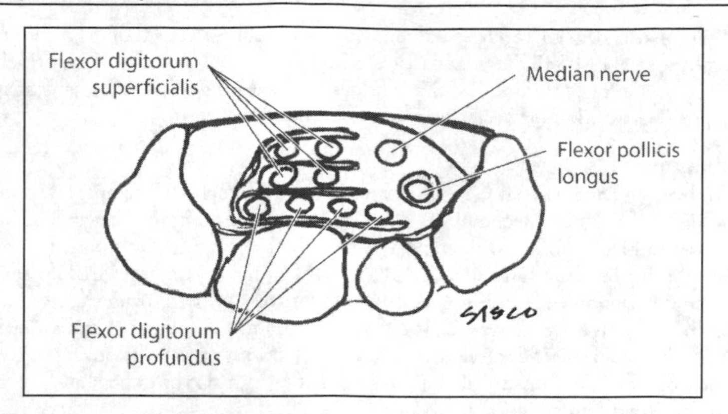

Figure 93.1. A schematic cross-section of the carpal tunnel.

greater and greater number of nerve fibers are affected, as seen in nerve compression. For CTS, threshold testing is the preferred method for evaluating hand sensibility. Although vibrometry is more sensitive, the nylon Semmes-Weinstein monofilaments are more user-friendly, and therefore more often used.

Provocative testing also assists in the diagnosis of CTS. These tests are based on the principle that stressing an already injured median nerve will exacerbate the symptoms of pain, paresthesias and numbness. Three tests are most common: **Tinel's test** (median nerve percussion), Phalen's test (wrist-flexion) and the **median nerve compression test**. **Phalen's test** is performed with the forearms upright, allowing the elbows to rest on a hard surface. The wrists are allowed to drop into flexion for 30-60 seconds, causing reproduction of symptoms. Eliciting Tinel's sign involves tapping gently along the course of the median nerve at the wrist from proximal to distal. A positive response is recorded if the patient perceives tingling in the median nerve distribution. The median nerve compression test is performed with the examiner gently applying sustained pressure with their thumb over the patient's carpal canal. Paresthesias in the distribution of the median nerve, which resolve when the pressure is released, within 30 seconds, are indicative of a positive test.

Motor testing of the thenar eminence can be a useful adjunct to the exam, especially in patients with long-standing symptoms. Compare the profiles of the thenar eminences of both hands, and check the strength of the abductor pollicis brevis by testing resistance to palmar abduction, while using a simultaneous comparison to the opposite hand. This test is less sensitive in patients with pain from basilar joint arthritis.

Many hand surgeons feel that electrodiagnostic testing should be obtained in any patient with symptoms of CTS. There is argument over this issue, but valid electrodiagnostic testing is regarded as the "gold standard" for CTS. In order to make the diagnosis of CTS, electrodiagnostic testing must show slowing of median sensory or motor nerve conduction velocity at the wrist, prolonged distal motor or sensory latency, or denervation of the abductor pollicis brevis muscle. Electromyography (EMG) may also demonstrate that a patient is suffering from other conditions such as pronator teres syndrome or cervical radiculopathy. But is EMG absolutely necessary to make the diagnosis? Most articles conclude that in a classical CTS presentation there is no need for electrodiagnostic testing, but in atypical symptoms

EMG is mandatory. There have been reports of successful treatment of CTS despite initially normal EMG findings, and in a U.S. survey only 33% of surgeons routinely use electrophysiolgic studies. EMG itself does not have prognostic value, and there is no good correlation between electrophysiology and symptoms.

Management

In treating patients with CTS, the objectives are to ameliorate symptoms, optimize physical performance, and improve function. Nonsurgical management includes minimizing work and leisure exposures which aggravate symptoms, neutral wrist splinting in the evening, nonsteroidal anti-inflammatory drugs (NSAIDS) and/or steroid injections. Steroid injection into the carpal tunnel provides temporary relief in about 40-80% of patients, lasting several months. Steroid injections which provide favorable relief after treatment are associated with patients having a positive surgical outcome. Occupational and physical therapy can also be of benefit. Indications for surgical intervention include:

- Lack of response to conservative management over a 3-6 month period
- Moderate/severe symptoms associated with significantly prolonged distal sensory and/or motor latencies
- Slowed sensory conduction velocities
- Denervation of the abductor pollicus brevis muscle on EMG

Surgical Release of the Carpal Tunnel

Local anesthesia with sedation is often used with carpal tunnel release, but the optimal method of anesthesia must be decided upon between the surgeon and the anesthesiologist. The patient is placed supine on the operating table and the forearm is supinated on a hand table. A well padded pneumatic tourniquet is placed on the upper arm and antibiotic prophylaxis is given. The tourniquet is inflated to 250 mm Hg (except in patients with hemodialysis shunts).

Prior to administration of the local anesthetic, surface markings should be made to assist in designing an incision that avoids injury to the deep and superficial arches, recurrent motor branch, palmar cutaneous branch and ulnar neurovascular bundle (Fig. 93.2).

The incision is started at Kaplan's cardinal line (a line parallel with the palmar wrist crease, beginning at the apex of the thumb-index web and ending slightly distal to the hook of the hamate. Aim toward the ulnar border of the palmaris longus and continue the incision until reaching the proximal wrist crease. The incision may be continued 2-3 cm into the distal forearm in an ulnar/oblique direction to aid in localization of the median nerve and ulnar neurovascular bundle. The dissection is continued bluntly through subcutaneous tissue allowing identification of the palmar fascia. This fascia is incised longitudinally in a direction parallel to the ulnar border of the palmaris longus. The borders of the transverse carpal ligament are visualized, including the perivascular tissue protecting the superficial arch distally. The ulnar neurovascular bundle is identified and protected. Proximally, the skin and soft tissues are retracted volarly to expose the antebrachial fascia. Blunt dissection proceeds above and below the fascia to protect the palmar cutaneous and median nerves. Division of the TCL is continued into the antebrachial fascia for 1-2 cm. The distal median nerve and recurrent motor branch are then inspected. Residual distal fibers of the flexor retinaculum and any fascia constricting the motor branch are carefully divided. The tourniquet is deflated, careful hemostasis is obtained, the wound is irrigated, and the

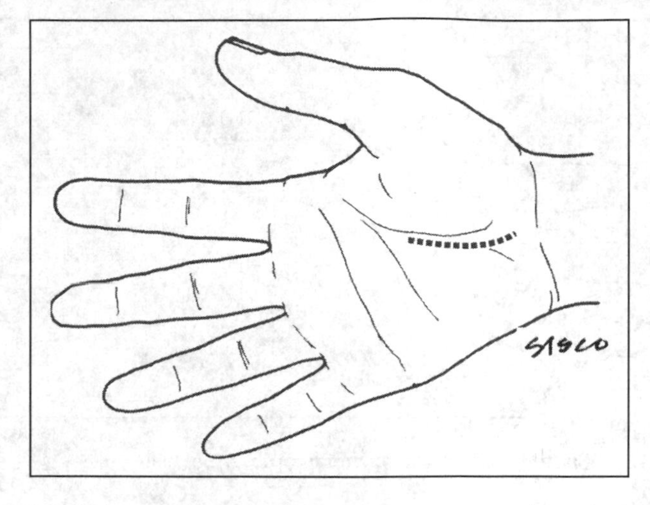

Figure 93.2. Surface markings for the incisions in open carpal tunnel release.

skin closed with 4-0 nylon. A bulky, compressive hand dressing is applied with a short-arm volar splint to maintain the wrist in slight extension.

Endoscopic Carpal Tunnel Release

The endoscopic carpal tunnel release (ECTR) was first described by Okutsu in 1987. Since then, several modifications to the technique have been made. The majority of the literature in ECTR deals with the Chow and Agee technique. The indications for ECTR and traditional CT release are similar. There are several situations where ECTR is not the optimal choice. These include cases where the patient has undergone a previous CT release, patients with prior wrist fusion, cases where the surgeon suspects proliferative tenosynovitis, and any suspicion of a mass within the CT.

The dual portal ECTR, described by Chow, involves making a small incision just 1 cm proximal to the wrist flexion crease in the midline of the forearm (Fig. 93.3). A tourniquet is applied after positioning the patient similar to the manner of a traditional CTR. The volar antebrachial fascia is entered and a curved dissector is passed into the CT, hugging the hook of the hamate. A "washboard" sensation is felt as the tip of the dissector passes over the ridges in the dorsal surface of the TCL. If this "washboard" sensation is not felt, the dissector is either palmar to the TCL or in Guyon's canal. The distal edge of the TCL is identified and marked on the overlying skin. A Ragnell retractor is used to retract the antebrachial fascia and open the proximal port. The slotted cannula/obturator is placed into the CT after removing the dissector. The cannula is safely placed, the wrist and fingers are extended and the hand is placed in the hand-holder. A distal port is established by making a 1 cm transverse incision over the tip of the cannula. The cannula system is passed from the distal port. A 30°, 4 mm endoscope is passed into the proximal end of the cannula system.

Once the surgeon is confident there are no intervening structures in the operative field, the probe blade is inserted through the proximal port. It is advanced until it engages the distal end of the TCL and then pushed proximally, cutting the distal

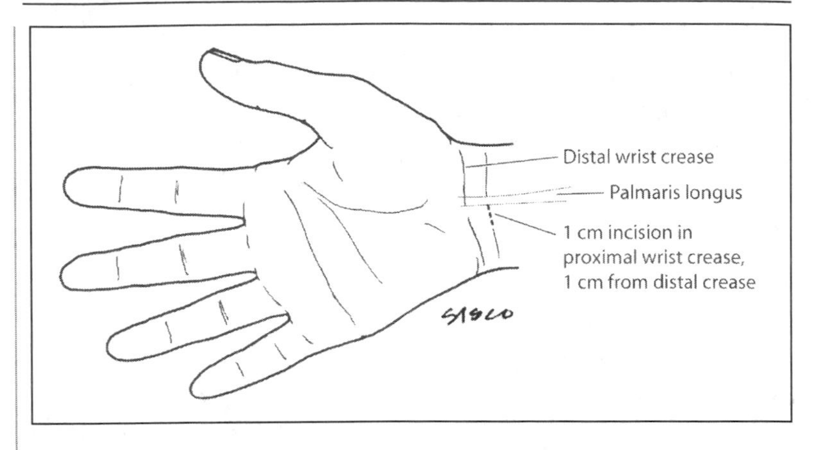

Figure 93.3. Surface markings in the endoscopic carpal tunnel release.

1.5 cm of the TCL. The blade is removed and the triangular blade is inserted through the distal portal to a point midway between the proximal and distal edge of the TCL. An incision is made large enough to accept the retrograde blade and deep enough to penetrate the full thickness of the TCL. The triangular blade is replaced by the retrograde blade, which is then used to cut the rest of the distal edge of the TCL from proximal to distal. The retrograde blade is then placed in the proximal port and engages the proximal end of the incision made in the TCL. The retrograde blade is pulled through the proximal TCL from distal to proximal. This last cut completes the release of the TCL. The obturator is replaced and the cannula system is removed. Wound closure is done with a 4-0 polypropylene subcuticular suture at the wrist and a single vertical mattress stitch at the palm. A bulky compression dressing is applied. The tourniquet is released and no splint is needed.

Complications

Complications from carpal tunnel release are quite rare and include the following:

- Immediate Complications
  - Infection—very rare
  - Palmar cutaneous branch injury—most frequent nerve injury
  - Nerve injury—more likely in endoscopic release
    - Median nerve
    - Motor branch of median nerve
    - Ulnar nerve
  - Vascular injury
  - Tendon laceration
- Delayed Complications
  - Recurrence of symptoms—in 10-20% of patients
  - Incomplete release of the TCL
  - Weakened grip
  - Reflex sympathetic dystrophy
  - Tendon bowstringing

Postoperative Considerations

After surgery, the dressing is exchanged for a light bandage and removable splint at three to seven days. Patients are encouraged to at least wear the splint when sleeping for 2 weeks after surgery as this aids in preventing tendon bowstringing. Finger motion is encouraged and early wrist motion initiated to prevent adhesions between the median nerve and surrounding tendons. Most patients return to work within 4-6 weeks after open release and within 2-3 weeks after endoscopic release. Worker's compensation patients usually require significantly longer time (2-3 months) until they return to work.

Pearls and Pitfalls

- CTS is the most common compressive neuropathy.
- Symptoms, most often pain and paresthesias in the median nerve distribution, are often worse at night and after repetitive motion activities.
- Electrodiagnostic testing is not required in order to correctly diagnose CTS.
- Resolution of symptoms with steroid injection is associated with positive outcome with surgical treatment.
- Tell patients that pain will be relieved within days after surgery; however, motor and sensory improvement take much longer (up to 12 months).

Suggested Reading

1. Atroshi I, Gummesson C, Johnsson R et al. Prevalence of carpal tunnel syndrome in a general population. JAMA 1999; 282:153.
2. Katz JN, Simmons BP. Carpal tunnel syndrome. N Engl J Med 2002; 346:1807.
3. Michelsen H, Posner M. Medical history of carpal tunnel syndrome. Hand Clin 2002; 18:257.
4. Nagle DJ. Endoscopic carpal tunnel release. Hand Clin 2002; 18:307.
5. Steinberg DR. Surgical release of the carpal tunnel. Hand Clin 2002; 18:291.

Chapter 94

Cubital Tunnel Syndrome

David Rosenberg and Gregory A. Dumanian

Introduction

Ulnar nerve compression at the elbow, or cubital tunnel syndrome, is the second most common nerve entrapment in the upper limb after carpal tunnel syndrome. It classically presents with arm and hand pain, and associated motor and sensory deficits. Pain, however, is the predominant complaint. The compression is often transient, with symptoms relieved by the patient simply changing arm position. Systemic diseases such as diabetes, alcoholism, renal failure and malnutrition can predispose patients to compressive neuropathies. Compression of the ulnar nerve at the elbow is frequently idiopathic; there is often a component of compression which is initially dynamic, but as fixed structural changes occur over time, the compression becomes static.

Relevant Anatomy

The ulnar nerve is typically compressed at one of several sites which can extend from 10 cm proximal to the elbow to 5 cm distal to the joint. The first proximal point of compression is at the **arcade of Struthers**, a musculofascial band nearly 2 cm in width. Anatomically, this point of compression is found an average of 8 cm proximal to the medial epicondyle. The arcade of Struthers should not be confused with the far less common ligament of Struthers, which is associated with compression of the median nerve. Next, there is potential for compression by the **medial intermuscular septum**. This structure can impinge the nerve in patients who have undergone previous cubital tunnel surgery. Next, the **medial head of the triceps** muscle can compress the ulnar nerve in the region proximal to the medial epicondyle. Hypertrophy of the triceps may compress the nerve (sometimes seen in weight lifters), or the nerve can snap over the medial muscle fibers causing friction neuritis. Compression of the ulnar nerve can also occur at the level of the medial epicondyle. Valgus deformity of the elbow from a previous epiphyseal injury or a malunited supracondylar fracture both are described as causing ulnar nerve compression at this level.

The **cubital tunnel** is a fibro-osseous groove bounded anteriorly by the medial epicondyle and laterally by the olecranon and the ulnohumeral ligament. The medial boundary of the groove is covered by a fibroaponeurotic band which in the normal state is loose and permits longitudinal gliding of the nerve with elbow flexion and extension. Compression at this site can be due to several factors grouped into three categories: conditions within the groove (fracture fragments, arthritic spurs, hypertrophic bone, and ganglia), factors outside the groove (external compression, or presence of an anomalous anconeus epitrochlearis muscle), and conditions which predispose the nerve to displacement outside the groove during elbow flexion (congenital laxity of the fibroaponeurotic covering over the epicondylar groove, or a traumatic tear in the covering).

The most common site of ulnar nerve compression is located just distal to the cubital tunnel proper, where the nerve traverses a tunnel between the ulnar and humeral heads of the flexor carpi ulnaris muscle. The roof of the tunnel is referred to as **Osbourne's ligament**. A small bony tubercle is palpable in the groove of the ulna just under Osborne's ligament. Clinically, patients tend to display a sensitive Tinel's sign at this spot. At the time of nerve decompression, the nerve tends to be "softest" at this spot, and a color change in the nerve (yellow proximal to this spot and white distal) can occur there. The fifth site of compression is point at which the ulnar nerve leaves the flexor carpi ulnaris in the proximal third of the forearm. As it exits the muscle, the nerve penetrates a fascial layer to lie between the flexor digitorum superficialis and flexor digitorum profundus muscles. The nerve can be constricted by this fascia as it traverses it producing a neuropathy.

Clinical Assessment

The symptoms of ulnar nerve compression can vary from numbness and paresthesias in the ring and little fingers to severe pain on the medial aspect of the elbow and dysesthesias radiating distally into the hand. Patients with early stages of neuropathy may not complain of any weakness although they may comment on some deterioration of hand function. Physical exam should begin at the neck. Cervical disk disease must be ruled out, as well as brachial plexus compression due to thoracic outlet syndrome. There are a number of provocative tests to evaluate for compression at these two sites. In the absence of tenderness or pain in the neck or plexus, the likelihood of significant nerve compression is low.

The elbow is inspected for deformity, and the range of motion is evaluated. The ulnar nerve is palpated along its course and in the epicondylar groove during elbow flexion for any nerve subluxation or dislocation. Local tenderness anywhere along the course of the nerve identifies a likely site of compression. A provocative test for ulnar nerve compression is the elbow flexion test, in which the elbow is placed in full flexion with the wrist in full extension for 1-3 minutes. The test is considered positive if paresthesias or numbness occur in the ulnar nerve distribution. The area of the sensory deficit can assist in distinguishing nerve compression at the elbow from one at the wrist. Ulnar nerve compression at the wrist in Guyon's canal tends to spare dorsal sensibility because the area is innervated by the dorsal sensory branch of the ulnar nerve which branches 5-6 cm proximal to the ulnar styloid.

Sensibility testing for light touch and vibration are helpful as they reflect innervation density which is only compromised after the presence of axonal degeneration. Muscle weakness tends to occur later than numbness although occasionally inability to adduct the fifth finger (positive Wartenberg sign) is found in early presentations. Weakness affects the intrinsic muscles of the hand more often than the extrinsic muscles in the forearm. Comparing the strength of the ulnar nerve-innervated first dorsal interosseous muscle to the median nerve-innervated abductor pollicis brevis muscle is important.

Electrodiagnostic studies are often obtained when nerve compression is suspected but are not essential when the diagnosis is obvious on clinical exam. These tests are valuable when clinical symptoms and findings are equivocal, when the site of nerve compression is unclear or thought to be at multiple levels, or when polyneuropathy or motor neuron disease is suspected. Radiographic examination of the elbow is necessary in patients with traumatic and arthritic conditions of the elbow. The required views include the routine anteroposterior, oblique and lateral views, and also a view imaging the epicondylar groove. Osteophytes or bone fragments from the medial trochlear lip are often seen in these patients. Magnetic resonance imaging is not an essential part of the diagnosis.

Differential Diagnosis

The differential diagnosis of ulnar nerve compression includes nerve compression which affects the origins of the ulnar nerve in the cervical spine (C8-T1 nerve roots) or the brachial plexus (medial cord). Cervical disk disease is the major cause of these conditions, followed by spinal tumors and syringomyelia. The medial cord of the brachial plexus may be compressed by thoracic outlet syndrome or a Pancoast tumor. Rarely the ulnar nerve is compressed at more than one site, a phenomenon known as "double crush" syndrome. When neural function is compromised at one level, it increases the susceptibility for nerve compression at another level, likely due to impaired axoplasmic flow. Other conditions to exclude include diabetes mellitus, vitamin deficiencies, alcoholism, malignant neoplasms and hypothyroidism.

Operative vs. Nonoperative Management

Ulnar nerve compression is classified as acute, subacute and chronic.

Acute and Subacute Nerve Compression

Acute conditions are frequently due to a single event, such as blunt trauma to the medial aspect of the elbow or an acute fracture. Subacute compression takes longer to develop. It is seen in those who continually rest on their elbows and in patients confined to bed due to illness or surgery. In these individuals, nerve compression usually improves if the nerve irritation is reversed. Nonoperative management includes patient education to avoid elbow flexion and prolonged pressure at the elbow, and splinting. Splints are worn for three to four weeks and removed only for bathing. The course of splinting is followed by active range-of-motion exercises. NSAIDs may be prescribed, but steroid injections are avoided. Patients with no muscle weakness who opt for nonoperative care are reassessed regularly for muscle strength. Progressive weakness is an indication for surgery whether or not there is a change in symptoms.

Surgical intervention becomes necessary if conservative care fails to relieve local tenderness, numbness or paresthesias. In the absence of muscle weakness, however, there is no urgency to surgical intervention. If activities of daily living, work, or leisure-time activities are compromised-in the absence of any meaningful improvement with conservative treatment, then surgery is recommended. Patients with mild muscle weakness that persists for three to four months have an indication for surgery as well.

Chronic Nerve Compression

Those patients presenting with chronic neuropathy associated with weakness frequently fail nonoperative management and require surgery. Factors influencing successful surgical outcome include patient age, the duration of nerve compression, and the severity of muscle weakness and numbness. The worst prognosis is seen in those with muscle atrophy, severe weakness, and patients with sensibility deficits in innervation density and two-point discrimination. Such patients may have minimal or no improvement in sensation or strength following surgery if the neuropathy is that far advanced.

Operative Treatment

Decompression of the ulnar nerve can be performed with or without nerve transposition. **Decompression without transposition** typically refers to a localized surgery where the ulnar nerve passes through the cubital tunnel, namely, between the two heads of the flexor carpi ulnaris. The surgery involves sectioning Osborne's ligament. The decompression should be limited distally to reduce the risk of nerve subluxation. The limit of dissection is the line drawn from the medial epicondyle to the tip of the olecranon. The entire surgery can be carried out under local anesthesia and involves limited dissection. The ideal patient for this surgery is one who presents with recurrent symptoms of ulnar neuropathy due to swelling of the flexor carpi ulnaris muscle with repetitive activities (e.g., violionists), as well as findings localized to the cubital tunnel. Decompression in situ proximal to the epicondylar groove is not commonly performed. It is indicated in cases of hypertrophy of the medial head of the triceps (typically seen in bodybuilders) and if snapping of this medial head with elbow flexion leads to symptoms. Decompression in situ is contraindicated for severe cases of ulnar nerve compression, particularly in posttraumatic scarring, when the nerve should be transferred to an unscarred area.

In the mid 1950s, the treatment of ulnar neuropathy by decompression accompanied by **medial epicondylectomy** was popularized. This was advocated because it was felt that a transposed nerve could still be irritated by slipping onto the apex of the medial epicondyle. The disadvantages of this procedure are that it fails to release the most distal potential sites of compression and it does not relieve traction forces on the nerve as effectively as transposition. Over-excision of the epicondyle can destabilize the elbow by damaging the medial collateral ligament which can lead to a postoperative valgus deformity. Despite these potential disadvantages, medial epicondylectomy is effective in mild cases of cubital tunnel. It is a relatively quick and simple surgery that avoids direct trauma or manipulation of the nerve.

Nerve decompression with transposition is the most popular method of treatment of cubital tunnel, and it has several theoretic advantages. First, the nerve is removed from its irritated location and repositioned to a new site. Second, the nerve is effectively lengthened several centimeters by transposing it into a new volar pathway. There are three locations for ulnar nerve transposition: subcutaneous, intramuscular, and submuscular.

Subcutaneous transposition is the most commonly used method because of its ease and high success rate. This method is ideal for the elderly, the obese and in patients with arthritic joints. The disadvantages of this method are that it fails to decompress the nerve at the most distal site and the nerve remains vulnerable to repeated trauma.

Intramuscular transposition is the most controversial of the three approaches. Proponents claim it has less scarring than submuscular transposition while others have found scarring to be a common complication. Advocates also feel that the intramuscular position offers more protection than the subcutaneous location. **Submuscular transposition** has several advantages. It insures that all five potential sites for nerve compression have been explored and allows repositioning of the nerve in an unscarred location where there is no influence from traction or external compressive forces. This method of transposition requires greater dissection and can potentially cause more nerve ischemia. Another drawback to submuscular transposition is the risk of an elbow flexion contracture while waiting for the flexor-pronator muscle group to heal. Submuscular transposition is contraindicated when there is scarring of the joint capsule or a joint distorted by arthritis or malunion.

Surgical Technique

Whichever technique is chosen, there are some basic steps that tend to be followed consistently. The operation is performed under regional or general anesthetic. Local anesthesia can be used only for decompression in situ without nerve transposition. Additionally, the surgery should be performed under tourniquet. For the three types of transposition (described below), the incision begins 8 to 10 cm proximal to the medial epicondyle. The medial intermuscular septum is palpated, and the incision is made directly over it. The incision continues along the epicondylar groove and ends 5 to 7 cm distal to the epicondyle over the course of the ulnar nerve. When dissecting through the subcutaneous tissue, one must be sure to protect the posterior branches of the medial antebrachial cutaneous nerve. It can be found within 6 cm of the epicondyle, either proximal or distal to it. Along the course of the ulnar nerve, the fascia immediately posterior to the medial intermuscular septum is incised. In succession, the fibroaponeurotic covering of the epicondylar groove is released, followed by Osborne's ligament at the cubital tunnel, and the fascia of the flexor carpi ulnaris. The fibrous edge of the medial intermuscular septum is excised especially near the epicondyle where the septum is thicker and wider. Care is taken when mobilizing the nerve to minimize traction on it. The ulnar nerve is dissected free along a distance of 15 cm, from the arcade of Struthers to the deep flexor pronator aponeurosis. Epineurolysis may be performed in areas where the epineurium is thickened.

Subcutaneous transposition of the ulnar nerve requires that it be secured in its new position in a stable fashion. This prevents the nerve from slipping back into the epicondylar groove. Creating a fasciodermal sling is the preferred method. A flap based near the medial epicondyle is created and reflected medially. The flap is passed posterior to the nerve and anchored to the subcutaneous tissue with a suture. Alternatively, the flap may be based and reflected laterally. The joint is immobilized postoperatively with the forearm in pronation and the elbow bent at 90° (some surgeons advocated splinting at 45°) for two weeks, followed by hand therapy.

Intramuscular transposition involves creating a trough in the flexor-pronator muscle mass with division of the fibrous bands to provide a suitable bed for the ulnar nerve. Submuscular transposition requires the entire flexor-pronator muscle mass to be detached from its origin. There are several means of performing this, including dividing the tissue in a Z-plasty approach, leaving a 1 cm cuff of tissue on the bone to ease reattachment, and removing the muscle sharply from bone. The muscle can even be removed along with a portion of the medial epicondyle. Following detachment, the flexor carpi ulnaris is released for a short distance from the ulna distal to the insertion of the collateral ligament. The ulnar nerve is transferred onto the bed of the brachialis muscle proximal to the joint line and on the capsule of the joint distally. The nerve branch to the flexor carpi ulnaris is often dissected away from the main body of the ulnar nerve to prevent its tethering or kinking. Finally, the flexor-pronator muscle mass is reattached to the medial epicondyle. The tourniquet is deflated, hemostasis obtained, and the subcutaneous tissue closed. A posterior splint is applied with the wrist in neutral position, the forearm in neutral rotation, and the elbow flexed at 45°. The patient is immobilized for one week, followed by active range of motion exercises. Most patients resume full activity by 10 weeks post surgery although for sports that require throwing up to 6 months of therapy may be needed.

Pearls and Pitfalls

- The initial treatment of acute and subacute neuropathy is nonoperative; for chronic neuropathy associated with muscle weakness or other symptoms that do not improve with conservative treatment, surgery is often required.
- Treatment is most successful in those with mild neuropathy.
- Subcutaneous nerve transposition is the least complicated technique. It is the preferred approach in the elderly and in neuropathy due to an acute fracture, elbow arthroplasty, or for secondary neurorrhaphy.
- Medial epicondylectomy for ulnar nerve decompression tends to have a high recurrence rate, though the surgery is quick and effective in mild cases.
- Intramuscular transposition can be associated with severe postoperative perineural scarring.
- Submuscular transposition is the preferred method for most chronic neuropathies that need surgery. It also the preferred approach when prior surgery has not been successful.

Suggested Reading

1. Kleinman WB, Bishop AT. Anterior intramuscular transposition of the ulnar nerve. J Hand Surg [Am] 1989.; 14(6):972-9.
2. Matev B. Cubital tunnel syndrome. Hand Surg 2003; 8(1):127-31.
3. Mowlavi A, Andrews K, Lille S et al. The management of cubital tunnel syndrome: A meta-analysis of clinical studies. Plast Reconstr Surg 2000; 106(2): 327-34.
4. Posner MA. Compressive ulnar neuropathies at the elbow: I. Etiology and diagnosis. J Am Acad Orthop Surg 1998; 6(5):282-8.
5. Tetro AM and Pichora DR. Cubital tunnel syndrome and the painful upper extremity. Hand Clinics 1996; 12(4):665-77.

Chapter 95

Trigger Finger Release

Hakim Said and Gregory A. Dumanian

Overview

In normal physiology the tendons of the hand glide smoothly through the slippery double synovium of the flexor sheath, which envelops them as they course into the digits. A series of fibrous pulleys retain the tendon with sufficient tension to allow gliding but prevent bowstringing. Imbalance in this delicate relationship results in wear and tear, inflammation, excessive friction, and impairment in function. The ensuing inability to smoothly flex or extend the digits is termed triggering, and the condition of trigger finger is one of the most commonly seen problems in hand surgery. Fortunately, it is also one of the most treatable. The term **stenosing tenosynovitis** of the flexor sheath is also used to describe triggering of a finger.

Anatomy

The anatomy of the flexor mechanism of the fingers involves a series of pulleys whose purpose it is to prevent bowstringing and to keep the tendons taut throughout the flexion arc. Pulleys are described as annular or cruciate and numbered from proximal to distal (Fig. 95.1). Even numbered pulleys lie over the phalangeal shafts, while the odd pulleys overlie joints. Mechanical studies reveal that at least half the A2 pulley should be preserved for efficient transfer of flexor traction into the fingers, while in the thumb the oblique pulley provides the critical functional role in flexion.

Physiology

In heavy, forceful use of the hands, there is significant mechanical loading of the distal edge of the **A1 pulley** over the MCP joint. Unless a significant injury has occurred elsewhere on the tendon, this leading edge is overwhelmingly the site of triggering in almost every case, even if other areas of tendon nodularity are appreciable. Wear injury to the tendon at this point initiates an inflammatory cycle, and the increased swelling exacerbates the size mismatch between the tendon and its pulleys. This cycle is perpetuated by some combination of fraying, inflammation, microtrauma and ischemia although the dominant factor may vary from patient to patient. The natural history of this condition eventually leaves the tendon unable to glide and effectively locks the digit in place, most commonly in flexion.

Presentation

Patients typically describe either a painful area on the volar surface over a proximal phalanx, physical locking of one or more digits, or both. Occasionally, patients report having a "broken finger." More commonly, however, clicking, popping or even burning with finger motion is the cardinal complaint. Some patients report a traumatic incident preceding the symptoms although most do not. For reasons that are not yet clear, diabetic patients have an increased likelihood of developing triggering.

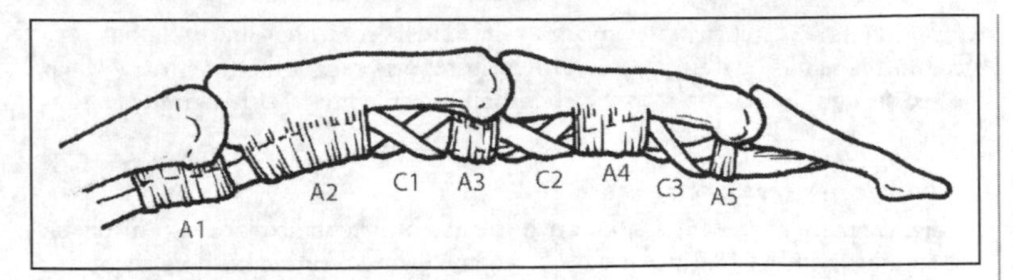

Figure 95.1. The annular and cruciate pulley system of the digital flexor sheath. The pulleys are numbered from proximal to distal. Note that most triggering occurs at the A1 pulley.

Grading

The degree of pathology occurs along a continuum, classified into four grades primarily based on the existing impairment of excursion:

Grade 1—Pretriggering

The mildest presentation involves only nodularity of the tendon sheath without any difficulty in tendon gliding. An inflamed area of tendon is usually palpable, tender to exam, and may be crepitant. Patients often report initially bumping something during grasp and still feeling persistent soreness along the tendon for weeks after.

Grade 2—Active Triggering

The next grade includes clicking or "catching", which is nonlocking (actively correctible). The name "trigger finger" derives from the nonlinear effort involved with this kind of flexion, similar to the trigger mechanism of a firearm.

Grade 3—Passive Triggering

This grade involves consistent locking of the digit during flexion that is only correctable by passive means (usually the opposite hand is used). Note that even in this grade, passive flexion causes minimal tendon excursion and will **not** cause locking.

Grade 4—Incarceration

When the nodularity of the tendon is greater than the diameter of the flexor sheath and/or pulleys, a digit is described to be incarcerated, or locked. Failure to promptly release this stage (i.e., within 24 hours) can result in permanent impairment of motion if the digit is permitted to scar in place. Often these can be reduced with slow, even, sustained passive traction and then managed as a lower grade lesion.

Treatment

Management of trigger finger, or stenosing tenosynovitis, can be divided into **closed**, **percutaneous** and **open treatment**.

Closed Treatment

Corticosteroid injection is the mainstay of treatment for triggering, as it improves the size discrepancy between the tendon and its enclosure, primarily by reducing the swelling of the tendon. The effectiveness of one injection has been reported in the 70–80% range for resolving triggering. A second injection raises the cure rate to 85% for first time sufferers with isolated early grade triggering. Steroid effect depends on its persistence in the injection site, and its continued anti-inflammatory effect on the tendon within the canal. Recent reports suggest better efficacy when injected around the tendon sheath rather than directly into it, although the reasons for this are not well established.

Percutaneous Treatment

Higher grade or persistent lesions can be treated by percutaneous release. Involving local anesthesia and an 18 gauge needle, percutaneous intervention has been advocated by some authors for moderate (grade 2 or 3) lesions. The needle tip is inserted, cutting edge oriented longitudinally, to avoid nerve injury, and used to divide the fibers of the A1 pulley. A history of consistent triggering is key to this approach as it allows for immediate identification of complete release, since the pulley cannot be visualized by this technique. Because the digital nerves cross the path of the tendon sheath for the thumb and border digits, many surgeons restrict this treatment to long and ring digit release, in order to minimize the risk of nerve injury. Others have reported its use safely in all digits. After release, steroid injection is often given as an adjunct treatment. Detractors of this technique suggest that the blind movement of the needle during attempts to cut the pulley can be more traumatic than a carefully controlled open release.

Open Release

For lesions in border digits with refractory triggering, an open release should be considered. These can be approached by a variety of volar incisions over the MCP joints: transverse, longitudinal or oblique. Also feasible under local anesthesia, open release involves blunt dissection to the level of the sheath and sharp division of the A1 pulley under direct visualization. Usually traction is applied to deliver both flexor tendons through the wound, ensuring that they are no longer constrained by the pulley.

Pearls and Pitfalls

Although recommended injection amounts are variable, the senior author favors a reduced steroid load—0.2 ml of 10 mg Kenalog, which has not changed its efficacy in his hands. This decreases the adverse effects of steroids including fat atrophy, discoloration and hyperglycemia, which can be seen with traditional doses.

As a rule, axial incisions should be made in an inter-nervous plane whenever possible. In the case of the thumb, where the digital nerves run as close as 1 mm beneath the skin of the proximal digital crease, this may provide a valuable safety net for the surgeon especially in the context of resident teaching. Triggering in the thumb is also noteworthy for its peculiar presentation of pain which can radiate up the forearm. Moreover, the principal pulley to preserve for normal thumb flexion is the oblique pulley, which bears careful scrutiny given its proximity to the release site in the thumb.

While most triggering is manageable using these techniques, triggering in the context of rheumatoid arthritis usually requires tendon debulking to restore a more normal ratio of tendon to sheath, rather than pulley releases. Occasionally, even a slip of flexor digitorum sublimis must be excised to restore more normal functional relations in these patients.

Suggested Reading

1. Carlson Jr CS, Curtis RM. Steroid injection for flexor tenosynovitis. J Hand Surg [Am] 1984; 9:286.
2. Newport ML, Lane LB, Stuchin SA. Treatment of trigger finger by steroid injection. J Hand Surg [Am] 1990; 15:748.

Chapter 96

Ganglion Cysts

Hakim Said and Thomas Wiedrich

Overview

Ganglion cysts at any location in the body are believed to be formed by overgrowth or herniation of the synovial capsule, producing a pedunculated bulge associated with a nearby joint. These are most commonly seen in the wrist and hand, but can also be seen at other joints (e.g., Baker's cyst in the knee).

Etiology

There is some suggestion that trauma to the joint capsule may play a role in their development. Degenerative joint disease and post-traumatic changes to a joint have been linked to the formation of these cysts. Some forms of ganglion cyst require removal of a degenerative spur off the nearby joint to prevent repetitive capsular trauma and recurrence. Repeated wear and loading are thought to be factors, and individuals who are more active (i.e., gymnasts) are more prone to their development. Women are also more prone than men, which may be related to increased connective tissue laxity. Nevertheless, healthy joints can also develop ganglion cysts for reasons that are unclear.

Physiology

Ganglion cysts are usually smooth, round, discrete masses, formed by a collection of joint fluid encapsulated by a whitish synovial sac. There is normally a small stalk leading to the joint which may produce a one-way valve effect, trapping synovial fluid under pressure in the associated cyst. Technically, these have been identified as pseudocysts since they represent diverticulae off the joint capsule. The fluid contents are similar to normal joint fluid but can be concentrated to the point of being thick and gelatinous. The main components of these contents are globulin, albumin, hyaluronic acid, glucosamine. Characteristically, these contents can be easily trans-illuminated as a diagnostic aid.

Presentation

Ganglion cysts represent 50-70% of all soft tissue masses of the hand and wrist. Peak incidence occurs in 20-40 year old patients, but can occur at any age. There is a threefold increased incidence in women relative to men. Typical presentation involves a 1-3 cm firm or rubbery transilluminating mass, which is nonfixed, but does not slide with tendon movement. Early manifestation is noted as a mild discrete bulge, but increasing size can result in discomfort or pain, especially at extreme hyperflexion or extension. A large size also poses a cosmetic concern to many patients. X-rays confirm only soft tissue involvement of the mass but may also demonstrate associated degenerative changes in the adjacent joint. Patients who are

Practical Plastic Surgery, edited by Zol B. Kryger and Mark Sisco.

symptomatic without an appreciable mass may have an occult cyst, which can be diagnosed by MRI.

Common Sites of Occurrence

The most common site of occurrence is in the wrist, where up to 80% of cysts will occur. In order of frequency, ganglion cysts present on the dorsal wrist, volar wrist, volar hand and distal digit.

Dorsal Wrist (80%)

The common form of the most frequent mass off the wrist, these lesions tend to be pedunculated off a stalk extending from the distal edge of the scapholunate ligament.

Volar Wrist (20%)

Second most frequent, these can be radiocarpal (65%), scaphotrapezial (34%) or radial artery adherent (54%) originating from between the radial artery and flexor carpi radialis.

Volar Retinaculum (5%)

Significantly rarer, these are found at the proximal digital crease or the volar side of the MP joint. They can be mistaken for tendon sheath tumors; however they mostly (80%) originate from the A1 or A2 pulley and are not mobile with tendon sliding.

Distal Digit (5%)

Also known as a **mucous cyst**, these masses typically extend from a degenerative spur off an arthritic DIP joint in an older patient. They cause a slow growing, nontender nodule over the dorsal distal phalanx, eventually producing a characteristic grooving of the nail bed. Occasionally, these become infected through a tract from the nail distally and then present as a paronychia, even though chronic nail grooving hints at the chronicity of the true underlying diagnosis.

Treatment

Management of ganglion cysts can be divided into **closed**, **percutaneous**, **open** and **endoscopic treatment**.

Closed Treatment

Presentation is usually characteristic enough that observation is an option. Since activity has been thought to exacerbate this condition, splinting and reduction in active use may result in improvement in some cases. Treatment can usually be delayed until the patient's symptoms are aggravating enough to warrant intervention. Given the waxing and waning in the size of these cysts, the patient may elect to wait until it is clear that things will not resolve or improve. Unfortunately, observation may reveal improvement or spontaneous resolution, but this is frequently followed by recurrence at a later date. Closed treatment by traumatic rupture (so-called "Bible therapy") is mentioned largely for historical purposes.

Percutaneous Treatment

Many authors advocate aspiration, which involves percutaneous needle drainage, usually with a large enough bore needle to permit withdrawal of the viscous contents (20 gauge or greater). This has the side effect of leaving a sufficient drainage tract for remaining contents to be directly expressed by compression. Some surgeons include compression as a significant component of this approach to prevent reaccumulation of the cyst contents. Aspiration alone has a high reported recurrence rate (50-85%).

Open Treatment

For ganglion cysts of any significant size and for lesions refractory to conservative interventions, many surgeons advocate open removal. This approach allows for removal of the synovial stalk and is associated with a dramatically lower recurrence rate in a number of large series. This modality results in some postoperative joint stiffness and a short but significant scar. In the case of mucous cysts, excision and skin elevation for osteophyte debridement can require local rotational flaps of the fingertip for coverage.

Endoscopic Approach

The most recent development in treatment options involves endoscopic approach to wrist ganglion cysts. This method allows for visualization and resection of the synovial sac and the stalk leading to the wrist with minimal disruption of the overlying soft tissues. This approach has been credited with the least stiffness postoperatively, although, all-told, the small port scars comprise roughly the same total length as the open approach scar. Recurrence using this technique is similar to that of open approach in most recent series.

Pearls and Pitfalls

Understanding the underlying problems is key in successful management of ganglion cysts. Mucous cysts, for example, are now understood to be related to degenerative spurs at the DIP joint level. Failure to address this at the time of excision will result in recurrence. Similarly, the stalk of a dorsal wrist ganglion has been well described as emanating from under the distal edge of the scapholunate ligament in essentially 100% of cases. It is incumbent, then, to follow the stalk to this location and to remove any channels from the joint in a 5 mm radius at the time of resection in order to prevent recurrence. This consideration should be kept in mind even though the actual cyst may seem far removed from its most likely origin. Incisions placed directly over cysts, but distant from their origins, will make it difficult to follow the stalk fully back and thus risks recurrence.

Suggested Reading

1. Angelides AC, Wallace PF. The dorsal ganglion of the wrist: Its pathogenesis, gross and microscopic anatomy, and surgical treatment. J Hand Surg 1976; 1(3):228.
2. Minotti, Taras. J Am Soc Surg Hand May 2002; 2(2).
3. Nahra ME, Bucchieri JS. Ganglion cysts and other tumor related conditions of the hand and wrist. Hand Clin 2004; 20(3):249.

Chapter 97

Stenosing Tenosynovitis

Zol B. Kryger and John Y.S. Kim

Introduction

Tenosynovitis, or inflammation of the tendon sheath, is a misnomer since not all cases demonstrate classic findings of inflammation. It can be loosely classified into two broad categories: infectious or overuse/stenosis. This chapter will focus on stenosing tenosynovitis of the dorsal compartments. Stenosing flexor tenosynovitis of the digits (i.e., trigger finger) is discussed in a separate chapter in this book.

Stenosing Tenosynovitis

Since most cases are due to repetitive trauma and overuse, the dominant hand is most commonly involved. The incidence is much higher in women (10:1), supporting the notion that there is an underlying autoimmune role in this condition. Associated conditions include rheumatoid arthritis, gout, renal disease, various endocrine disorders, osteoarthritis and pregnancy.

Flexor Tenosynovitis

The most common form of flexor tenosynovitis is the trigger finger, or stenosis at the A1 pulley. This condition is discussed in a separate chapter. Flexor carpi radialis is another commonly involved tendon.

Extensor Tenosynovitis

Tenosynovitis of the extensor tendons involves thickening of the extensor retinaculum and narrowing of the fibroosseous tendon canal. It commonly occurs in the first dorsal compartment where the abductor pollicis longus (APL) and extensor pollicis brevis (EPB) tendons travel (Table 97.1). This condition is known as de Quervain's disease.

Table 97.1. The dorsal compartments of the wrist. The six compartments are numbered 1-6 from radial to ulnar

1. Abductor pollicis longus (APL), extensor pollicis brevis (EPB)
2. Extensor carpi radialis brevis (ECRB), extensor carpi radialis longus (ECRL)
3. Extensor pollicis longus (EPL)
4. Extensor digitorum communis (EDC), extensor indices proprius (EIP)
5. Extensor digiti minimi (EDM)
6. Extensor carpi ulnaris (ECU)

Practical Plastic Surgery, edited by Zol B. Kryger and Mark Sisco. ©2007 Landes Bioscience.

Tenosynovitis of the second dorsal compartment can also occur. This is termed intersection syndrome, and the pain is located where the EPB and APL tendons cross over the second compartment tendons (ECRL and ECRB). The pain is more proximal than in de Quervain's disease. Tenosynovitis of the third compartment presents with pain and swelling over Lister's tubercle. Untreated it can lead to rupture of the extensor pollicis longus (EPL) tendon. Other extensor tendons that may also demonstrate tenosynovitis include extensor carpi ulnaris (ECU), extensor indices proprius (EIP) and extensor pollicis longus (EPL).

De Quervain's Tenosynovitis

Clinical Presentation

The onset of de Quervain's disease is usually gradual. There is a history of a single traumatic episode in less than 25% of cases. Patients will often describe a repetitive or prolonged activity that involves overexertion of the thumb. In other cases, the thumb will be in a static position with postural deviation of the wrist, exerting stress on EPB and APL. The dominant hand is usually involved, and the symptoms are most often intense. Some patients will awaken from sleep due to pain.

Differential Diagnosis

De Quervain's is often confused with arthritis of the first carpometacarpal (CMC) joint. Although they can coexist, the diagnosis of de Quervain's is supported by tenderness over the first dorsal compartment at the radial styloid and a positive Finkelstein's test—the patient experiences pain in the first dorsal compartment during ulnar deviation of the hand while making a fist (with the thumb tucked below the fingers). Furthermore, plain radiographs of the wrist will usually be normal. CMC arthritis will cause pain over the joint when grinding the extended thumb in circles (positive grind test). Radiographs will also demonstrate arthritic changes in the CMC joint.

Treatment

The initial approach to stenosing tenosynovitis should be conservative management: an injection of a corticosteroid mixed with lidocaine. Over 80% of cases will be cured by corticosteroid injection. The addition of splint immobilization and rest do not improve the outcome. If conservative treatment fails, surgical release of the affected compartment can be tried. The first dorsal compartment can contain either one or two tunnels, depending on whether EPB and APL travel together or separate. There can even be a third tunnel containing an anomalous tendon. Inadequate release of all the tunnels can result in persistent symptoms.

Pearls and Pitfalls

- The first dorsal compartment usually has more than just two tendon slips (i.e., APL and EPB) and may even have two or more distinct fibroosseus tunnels. Each of the two tendons can have multiple slips. During surgery, each separate compartment should be identified and the intervening septae divided. The various tendon slips should be inspected one at a time to ensure proper gliding free of excess synovium.
- Stenosing tenosynovitis is a common finding in patients with rheumatoid arthritis, and tendon rupture is one of the worst complications that can result. Treatment if these patients should focus on prevention of rupture. NSAIDs and

infrequent steroid injections are the mainstay of medical therapy, and surgical tenosynovectomy is reserved for patients who fail conservative management.
- The diagnosis of gout should always be ruled out (by joint aspiration and microscopic fluid examination) in any patient presenting with tenosynovitis of multiple sites in the hand or in whom a history of trauma or repetitive motion is not present.

Suggested Reading

1. Anderson BC, Manthey R, Brouns MC. Treatment of DeQuervain's tenosynovitis with corticosteroids: A prospective study of the response to local injection. Arthritis Rheum 1991; 34:793.
2. Froimson AF. Tenosynovitis and tennis elbow. In: Green, ed. Operative Hand Surgery. New York: Churchill Livingstone, 1993:1989-2000.
3. Moore SJ. De Quervain's tenosynovitis: Stenosing tenosynovitis of the first dorsal compartment. J Occup Environ Med 1997; 39(10):990.
4. Richie III CA, Briner Jr WW. Corticosteroid injection for treatment of de Quervain's tenosynovitis: A pooled quantitative literature evaluation. J Am Board Fam Pract 2003; 16:102.
5. Weiss AP, Akelman E, Tabatabai M. Treatment of de Quervain's disease. J Hand Surg 1994; 19(4):595.

Chapter 98

Radial Artery Harvest

Zol B. Kryger and Gregory A. Dumanian

Preoperative Considerations

Harvest of the radial artery is most commonly performed for obtaining conduit for coronary artery bypass graft surgery. Although many cardiothoracic surgeons perform the harvest themselves, some still use the expertise of the hand and plastic surgeon. If the vascular exam of both arms is symmetric, then the nondominant hand is traditionally used. Safe removal of the radial artery is contingent on the ability of the ulnar artery to adequately perfuse the hand. There are two noninvasive tests used to determine if the ulnar artery has adequate blood flow. Aside from the status of the ulnar artery, it is not clear what other factors are contraindications to radial artery harvest. Active infection of the forearm, a history of Raynaud's phenomenon, prior arterial trauma to the forearm, and an absent palpable radial artery pulse are some commonly cited contraindications.

Allen Test

This is the first screening test that should be performed on all potential patients. The surgeon uses his thumb and fingers to compress the radial and ulnar arteries at the wrist. The patient exsanguinates the hand by making a fist several times and then opens the hand so that the fingers are in a relaxed and gently extended position. The examiner then releases pressure from over the ulnar artery. Capillary refill time in the hand is noted. A normal Allen test is refill in less than 5 seconds, and greater than 5 seconds indicates an abnormal Allen test. About 85-90% of patients will have a normal Allen test. If the test is normal, radial artery harvest can be performed without further testing.

Duplex Ultrasound

If the Allen test is abnormal, bilateral duplex ultrasonography or pulse volume recordings should be performed of hands and fingers, with and without radial artery compression. Over 90% of these patients will have a normal noninvasive exam and can proceed to radial artery harvest safely. In most cases, one of the two hands will demonstrate preserved flow pattern with radial artery compression, and consequently, safe harvesting of the radial artery. In the rare case that both hands demonstrate abnormal arterial flow to the hands, radial artery harvest is contraindicated.

Intraoperative Considerations

Relevant Anatomy

The radial artery enters the forearm between the biceps tendon (lateral) and bicipital aponeurosis (medial). The dissection should be on the radial side of the

Practical Plastic Surgery, edited by Zol B. Kryger and Mark Sisco. ©2007 Landes Bioscience.

bicipital aponeurosis to avoid injury to the brachial and ulnar arteries and median nerve. The artery continues down the length of the forearm, running deep to the brachioradialis muscle. Exposure of the artery is best achieved with lateral retraction of this muscle. As it travels more distally, the artery is sandwiched by the brachioradialis laterally and the superficial flexor compartment medially. It travels over the pronator teres, flexor digitorum superficialis and flexor pollicus longus (FPL)—in that order. In the mid forearm, care must be taken to avoid injury to the superficial radial nerve by minimizing lateral traction of the brachioradialis. The lateral antebrachial cutaneous nerve must also be avoided in the mid forearm by retracting it laterally along with the brachioradialis and the short and long extensor carpi muscles during exposure of the artery. In the distal forearm, the radial artery becomes very superficial. It is bounded by the tendon of brachioradialis and radius bone laterally, the flexor carpi radialis (FCR) tendon medially, and FPL dorsally. Most of the perforating branches of the radial artery emerge in the distal forearm and are at risk for avulsion if care is not taken.

Operative Technique

The arm is positioned extended and supinated. The incision extends from a fingerbreadth lateral to the biceps tendon to a point just medial to the radial styloid at the wrist crease. The fascia between brachioradialis and FCR is divided. Some surgeons will administer a loading dose of a calcium-channel blocker followed by continuous infusion in order to prevent radial artery spasm. Initial dissection should begin in the mid forearm where the artery is best visualized. The artery should be retracted upward lightly with a vessel loop. All branches should be clipped and divided. From this point, the dissection should progress both proximally and distally. The distal artery is ligated using suture ligation at the level of the radial styloid. The proximal artery is both suture ligated and tied off using a 2-0 silk tie. After the artery is divided and removed from the patient, it is placed in papaverine-soaked gauze. Closure of the wound should be done in two layers. Some surgeons will leave a flat Jackson-Pratt drain in for 1-2 days. For dressings, only gauze and gentle ACE wrap compression is required.

Endoscopic Technique

In the past few years, endoscopic harvest of the radial artery has been performed at a number of centers. The artery is harvested through a 3 cm wrist incision with the aid of a Harmonic Scalpel. Some surgeons perform a counter incision at the elbow to divide and ligate the artery, although this can be done endoscopically. The complication rate from this procedure is somewhat higher than in the traditional open approach. Most of the complications are related to bleeding, such as conversion to an open procedure and postoperative hematoma. Patient satisfaction with the endoscopic technique is high (80-90%), and it is likely that this technique will gain popularity in the future.

Postoperative Considerations

Removal of the radial artery is extremely well-tolerated. Surgical site infection rate is low (4%) compared to the saphenous vein harvest site (18%). Radial sensory neuropathy is reported in 10-20% of patients, manifested by dorsal hand numbness. Noninvasive vascular tests demonstrate no difference between the operated and nonoperated hands after radial artery harvest. Cold intolerance, hand claudication, neuropathies, grip strength, and sensory discrimination are also no different between the two hands after surgery. Nevertheless, some patients report more disability, pain and physical limitations of the operated hand; however these findings are not supported by objective measurements. The lack of postoperative complications can be explained by the fact that the caliber and flow rate of the ulnar artery increases after radial artery harvest. Blood flow in the brachial artery remains constant. In summary, removal of the radial artery does not decrease blood flow to the hand, and does not result in any clinically evident detrimental changes.

Pearls and Pitfalls

In the well selected patient with preserved blood flow to the hand with radial artery compression, the major source of morbidity is injury to small nerves at the distal aspect of the incision. A tourniquet helps to visualize these small antebrachial cutaneous nerves for preservation. The arm with the best vascularity should be used, rather than the nondominant limb.

Suggested Reading

1. Abu-Omar Y, Mussa S, Anastasiadis K et al. Duplex ultrasonography predicts safety of radial artery harvest in the presence of an abnormal Allen test. Ann Thorac Surg 2004; 77:116.
2. Allen RH, Szabo RM, Chen JL. Outcome assessment of hand function after radial artery harvesting for coronary artery bypass. J Hand Surg 2004; 29A:628.
3. Casselman FP, La Meir M, Cammu G et al. Initial experience with an endoscopic radial artery harvesting technique. J Thorac Cardiovasc Surg 2004; 128(3):463.
4. Dumanian GA, Segalman K, Mispireta LA et al. Radial artery use in bypass grafting does not change digital blood flow or hand function. Ann Thorac Surg 1998; 65:1284.
5. Dumanian GA, Segalman K, Buehner JW et al. Analysis of digital pulse-volume recordings with radial and ulnar artery compression. Plast Reconstr Surg 1998; 102(6):1993.
6. Reyes AT, Frame R, Brodman RF. Technique for harvesting the radial artery as a coronary artery bypass graft. Ann Thorac Surg 1995; 59:118.
7. Royse AG, Royse CF, Maleskar A et al. Harvest of the radial artery for coronary artery surgery preserves maximal blood flow of the forearm. Ann Thorac Surg 2004; 78(2):539.

Chapter 99

Common Anomalies of the Hand and Digits

Zol B. Kryger and John Y.S. Kim

Epidemiology and Terminology

Polydactyly (duplicated digits) is the most common congenital anomaly of the hand. The incidence ranges from 1 in 300 in African Americans to 1 in 3000 in Asians and Caucasians. **Syndactyly** (webbed digits) is the second most common anomaly. In Caucasians, the incidence is 7 in 10,000. Roughly a third of cases are hereditary. These inherited cases are typically bilateral and occur more frequently in males.

Other less common anomalies include **radial club hand**, in which the radius is hypoplastic or absent, **symphalangism** (fused phalanges), **clinodactyly** (deviated digit) and **camptodactyly** (flexed digit). **Apert** and **Poland syndrome** are two congenital syndromes associated with hand defects. **Constriction ring syndrome** (CRS) is a noninherited, congenital anomaly that can affect the hand and other parts of the body. There are a number of other less common congenital hand anomalies that are not covered in this chapter.

Polydactyly

Preoperative Considerations

Congenital hand duplications are classified as preaxial (radial side), central or postaxial (ulnar side). Duplications in Blacks and Native Americans are usually postaxial, whereas in Caucasians they are usually preaxial at the thumb. If the opposite pattern is found in a child, referral to a geneticist and evaluation for a genetic syndrome is warranted (Tables 99.1, 99.2).

Intraoperative Considerations

Treatment of thumb duplications involves creating a single, mobile, well-functioning thumb. The ulnar one is usually the better of the two thumbs, however in some cases, a new thumb is created by combining the best available parts. Other principles of this repair include reattachment of the thenar intrinsic muscles, reconstruction of the ulnar collateral ligament, and release of the thumb

Table 99.1. Classification of postaxial (fifth ray) duplications

Classification	Description
Type I	A soft tissue nubbin without bony attachment (common)
Type II	A duplicated digit with skeletal attachments
Type III	Complete ray duplications along with the metacarpal bone

Practical Plastic Surgery, edited by Zol B. Kryger and Mark Sisco. ©2007 Landes Bioscience.

Table 99.2. Preaxial (thumb) duplications are classified from distal to proximal

Classification	Site of Duplication
Type I	Distal phalanx
Type II	Interphalangeal joint (common)
Type III	Proximal phalanx
Type IV	Metacarpophalangeal joint (common)
Type V	First metacarpal
Type VI	Carpometacarpal joint

index space with a Z-plasty. In addition, of the thumb's three mobile joints, at least two should be preserved. Long-term outcomes are usually less than optimal, especially in the more proximal duplications. Strength, mobility and function are almost never as good as in the unaffected side.

Syndactyly

Preoperative Considerations

Syndactyly, or webbing of the digits, is classified based on the extent of webbing and the nature of the interconnected tissue (Table 99.3).

Intraopertive Considerations

Simple, incomplete syndactyly can often be treated with skin flaps only, sparing the need for skin grafts. Correction of syndactylized digits is usually performed as a staged procedure since simultaneous release of both the radial and ulnar side can compromise vascular supply to the digit. This is a procedure that relies on meticulous, atraumatic technique. Zigzag incisions should be used on the palmar surface, and the flaps should be divided equally between the two digits. The volar flap base should be kept wider than the base of the dorsal flap. Full-thickness skin grafts are commonly required, especially for commissure reconstruction. The digits are closed from distal to proximal, preferably using absorbable sutures in young children. In cases of complex syndactyly the bone should be covered with soft tissue and burred down until it is smooth.

Postoperative Considerations

The web spaces must be packed with gauze dressings so that there is no undesirable healing of raw surfaces to one another. In some centers, dressing

Table 99.3. Classification of syndactyly

Classification	Description
Simple, complete	Soft tissue connection only, webbing extends to fingertips
Simple, incomplete	Soft tissue connection only, webbing terminates more proximally
Complex	Either bony or cartilaginous connections
Complicated	Duplicated skeletal parts located in the interdigit space

changes are performed under general anesthesia. Postoperative immobilization is essential. Inadequate protection of the surgical site can lead to infection, skin graft loss and dehiscence. Many surgeons will cast the child's entire arm with the elbow flexed at 90°. Straight arm or shorter casts are no match for the active, determined child.

Insufficient commissure release or closure under tension at the time of the original surgery can lead to scar contracture at the web space or "recurrence" of the webbing. This complication will often require surgical revision, especially for the first web space, which is undoubtedly the most important one. Additional full-thickness skin grafts or recruitment of local soft tissue is required for release of the web space. Overall, the revision rate for syndactyly repair is about 10%.

Radial Clubhand

Preoperative Considerations

The incidence of this anomaly is 1 in 50,000 births. In can occur as an isolated congenital defect, or as part of a syndromic condition such as VATER syndrome (vertebral, anorectal, tracheoesophageal, renal, and radial anomalies). Radial club hand ranges from mild radial hypoplasia to complete aplasia of the radius (Table 99.4). Syndromic cases are more severe. In most cases, the ulna is subsequently deformed, demonstrating bowing and thickening. The wrist is usually deviated to the radial side, and the muscles of the forearm and wrist are often fused and shortened. The thumb is often hypoplastic or absent. Elbow flexion is minimal but improves with time.

Treatment

Treatment begins with limb stretching by the parents. Nighttime splinting can help maintain the limb in the stretched position. Skeletal traction using an external fixating device is also an option. Once the wrist is in a neutral position, it is stabilized and centralized. The treatment should begin around the age of 6 months. This process greatly relies on the use of tendon transfers. Once the wrist is centralized, K-wire fixation maintains its position for at least three months. If the radial muscles are inadequate for tendon transfers, wrist arthrodesis can be considered. Syndromic radial club hands have a much poorer outcome, since the deformity is usually more severe. Once wrist alignment is achieved, pollicization is performed in a subsequent procedure.

Table 99.4. Classification of radial clubhand

Classification	Description
Type I	Proximal or distal radial deficiency, very little deviation or bowing
Type II	Shortened radius, deviated hand, bowed and thickened ulna
Type III	Partial radius, hand is deviated and ulna is bowed (most common)
Type IV	Complete radius aplasia, sublaxed and deviated hand, bowed ulna

Symphalangism

Defined as fused phalanges at the interphalangeal joints, this condition is seen in syndromic patients such as those with Apert syndrome or Poland syndrome. Both syndromes involve fusion of phalanges primarily in the central three rays. The metacarpophalangeal joints are spared. In addition to syndromic cases, true symphalangism has a genetic basis and presents with PIP involvement, fused digits and long, slender fingers.

Treatment of symphalangism should focus on correcting digital angulation and rotation rather than finger stiffness. PIP joint fusion should be done at 10°, 30°, 40° and 50° for the index, middle, ring and small finger, respectively. The IP joint can be reconstructed using a number of techniques, ranging from silicone caps and spacers to vascularized second-toe joint transfer. Most techniques provide less than optimal results. When possible, autogenous material is preferred in children.

Camptodactyly

Defined as a flexion deformity of the digit or thumb, this condition often involves the PIP joint of the fifth finger. It develops as a result of an imbalance between the forces of flexion and extension at the involved joint. Abnormal insertions of the lumbricals or interossei can result in excessive flexion forces and consequent flexion deformity of the joint. Involved digits will be stiff and have decreased range of motion. On radiographic evaluation, the phalanx will have a flat condyle (instead of the normal rounded shape) and flat articulating surface.

Contractures of less than 50° have good outcomes with conservative treatment alone. Stretching and splinting are usually adequate. Contractures greater than 70° usually indicate a long-standing deformity, and joint fusion is required. Any surgical release of a contracture must be followed by intensive ranging exercises and splinting at night.

Constriction Ring Syndrome

Other terms used to describe this condition include annular band syndrome, amniotic bands, and congenital amputations. There is no positive inheritance pattern in CRS. It occurs sporadically in utero. The mechanism is believed to be the detachment of strands of the chorionic sac that wrap around a body part of the fetus. Examples include fingers, toes and even entire limbs. The anatomy proximal to the point of constriction is entirely normal. Distal, however, the tissue is deprived of oxygen and develops abnormally. The constricted site has a characteristic ring appearance. This band can be superficial or extend down to the periosteum. It can completely encircle the digit, or only partially span its circumference.

Treatment of CRS consists of excising the scarred, constricting ring, and filling the resulting contour deformity with advancement or rotational flaps. Z-plasties alone will not correct the problem. Toe-to-thumb transfer works well in these patients, since the anatomy proximal to the constriction site is normal and the tendons, nerves and blood vessels are adequate for reconstruction.

Trigger Thumb

This condition is essentially a stenosing tenosynovitis of the flexor tendon sheath. Many cases are bilateral. The differential diagnosis includes congenital absence of the extensor pollicus longus which results in unopposed flexion forces on the thumb.

The children present at a young age with thumbs that lock and trigger upon extension. As the volar plate is stretched, a compensatory hyperextension of the MP joint will occur. A-1 pulley release should be performed by age 2. Prior to that, there is still a possibility of spontaneous resolution of the problem. The treatment of trigger fingers is discussed in greater detail elsewhere in this book.

Pearls and Pitfalls

- Release of the first web space and creation of a mobile, opposable thumb is perhaps the most functional operation a pediatric hand surgeon can perform for syndactyly.
- Adequate release of the first web space can be achieved with a Z-plasty. Although a single large Z-plasty may seem adequate, it is often better to use the four-flap Z-plasty for optimal length and contour within the depth of the web space.
- When releasing syndactylized digits, the skin flaps should be equally distributed between the two digits. It is risky to allot all of the flap tissue to one digit and skin graft the other.
- In thumb polydactyly, the ulnar thumb is usually better. If both thumbs are similar in size, appearance, and function, the one with the better ulnar collateral ligament at the MP joint should be preserved. Occasionally, a new thumb must be created using the best parts from each partner.
- In distal phalanx duplications, the nail should be narrowed and the paronychial fold recreated. In proximal phalanx duplications, the collateral ligaments must be recreated but not using the extensor mechanism. Adequate gliding of an extensor tendon requires a smooth layer beneath it.
- The optimal age for correction of constriction bands and separation of digits is controversial. Many surgeons advocate operating on the child towards the end of the first year of life. Hand and foot procedures should be done before the child begins to explore the environment with his hands and to walk.

Suggested Reading

1. Brown PM. Syndactyly—a review and long term results. Hand 1977; 9(1):16-27.
2. Cohen MS. Thumb duplication. Hand Clin 1998; 14(1):17-27.
3. Graham TJ, Ress AM. Finger polydactyly. Hand Clin 1998; 14(1):49-64.
4. Miura T. Congenital constriction band syndrome. J Hand Surg [Am] 1984; 9A(1):82-8.
5. Steenwerckx A, De Smet L, Fabry G. Congenital trigger digit. J Hand Surg [Am] 1996; 21(5):909-11.
6. Upton J. Congenital anomalies of the hand and forearm. In: McCarthy JG, May J, Littler JW, eds. Plastic Surgery. Vol 8. Philadelphia: W.B. Saunders, 1990.
7. Van Heest AE. Congenital disorders of the hand and upper extremity. Pediatr Clin North Am 1996; 43(5):1113-33.

Chapter 100

Dupuytren's Disease

Oliver Kloeters and John Y.S. Kim

Definition

Dupuytren's contracture is an abnormal thickening of the subcutaneous palmar tissue with potential involvement of the digital fascial structures. Symptoms can progress from benign nodules and cords to significant contractures and impairment of function.

Etiology

The precise etiology of this condition has yet to be elucidated; however, histopathologic changes such as increased deposition of extracellular matrix, excessive tissue proliferation and the alterations of myofibroblasts have been observed. Patients with diabetes mellitus, HIV infection, previous myocardial infarction, tobacco abuse, and epilepsy have a higher risk for Dupuytren's disease. A history of trauma may also be a predisposing factor as Dupuytren's disease can be seen after nonspecific hand injuries and Colles' fractures. The term Dupuytren's diathesis refers to the presence of a spectrum of proliferative fibroplastic diseases, such as Lederhose disease, a contracture of the medial plantar fascia of one or both feet, and Peyronie's disease, a fibroplastic contracture of the penis.

Epidemiology

Heredity is a clear predisposing factor as the contracture is most common in patients with Northern European heritage (Scandinavian, Celtic) and rare in African-Americans and Asians. Dupuytren's affects men 5-10 times more frequently than women.

Pathophysiology

A number of different theories exist addressing the pathophysiology of Dupuytren's disease:

1. **Murrell's theory** is that Dupuytren's contracture is initiated by multifactorial vascular compromise of the palmar vessels. The ischemia causes liberation of endothelial xanthine oxidase-derived free radicals. The ensuing cycle of cell damage, fibroblast proliferation and collagen deposition ultimately leads to a further narrowing of the vessels with more ischemia (and therefore explains the chronic and progressive nature of the disease).
2. **McFarlane's theory** advocates an "intrinsic theory," stating that the characteristic fibrotic changes derive from pathologically altered normal fascial tissue. The disease therefore occurs along anatomic routes (diseased fascia) with predictable progression to cords.

Practical Plastic Surgery, edited by Zol B. Kryger and Mark Sisco.

3. **Hueston** posits an "extrinsic theory" that claims that Dupuytren's disease occurs via nodules that are generated by de novo metaplastic transformation of fibrofatty tissue which will then become cords.
4. **Gosset** distinguishes cords from nodules. He believes that nodules arise de novo and that cords arise from palmar fascia. Therefore, nodules and cords represent different forms rather than different stages of Dupuytren's disease (synthesis theory).

Relevant Anatomy

During the development of Dupuytren's disease, fascia and ligaments become thickened cords, resulting in MCP, PIP and sometimes DIP joint contractures in the digits. Nodules and cords will form in the palm. MCP joint contracture occurs when the pretendinous bands—the longitudinal extension of the palmar aponeurosis—become fibroblastic. PIP joint contracture is common in Dupuytren's disease and related to the formation of cords: the central cord and the spiral cord. These develop out of four anatomical structures: the pretendinous band, the spiral band, the lateral digital sheet and Grayson's ligament (Fig. 100.1). With advanced disease, the neurovascular bundle will be displaced superficially and toward the palmar midline. Less common, PIP joint contracture can originate from a single lateral cord derived from the lateral digital sheet; however, this attenuated form does not usually lead to a midline shift of the neurovascular bundle. Natatory ligaments may also be involved, leading to an adduction deformity at the webspaces.

Clinical Presentation

The typical patient with early Dupuytren's disease is a middle-aged, male Caucasian of northern heritage presenting with palmar nodules (most often at the base of the ulnar digits). The palmar skin becomes more and more atrophic and adherent to the nodules. In later stages MP and PIP flexion contractures, or compensatory DIP joint hyperextensions can be observed due to progressive cord formation proximal to the nodules. Sometimes in more aggressive forms of Dupuytren's disease, knuckle pad formation on the dorsum of the PIP joint can also be observed.

Nonsurgical Treatment

Many nonsurgical therapies have been evaluated over the last decades, including vitamin E, DMSO, allopurinol, colchicines, α-interferon and calcium-channel

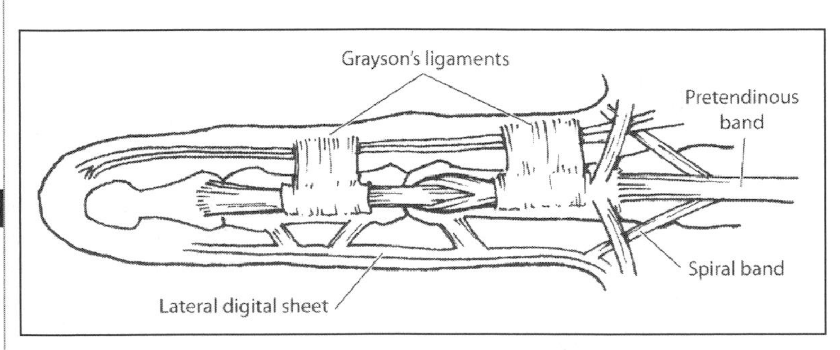

Figure 100.1. The primary structures that become pathologic cords in Dupuytren's disease. (Note that Cleland's ligament is not involved.)

blockers. These treatments have not demonstrated long-term reliability. Inconsistent data exists concerning treatment with ultrasonic therapy and corticosteroid injections. Static or dynamic splinting, either pre- or postoperatively, is the oldest technique in the treatment of Dupuytren's but is generally used as an adjunctive modality and not as the principal method of treatment.

The most promising nonsurgical approach under investigation is the ultrasonographic guided percutaneous fasciotomy with collagenase, an enzyme deriving from *Clostridium histolyticum*. In a recent study, direct injection of collagenase in the contracted cord in select clinical situations showed excellent results in 90% of involved MP joints and 66% of PIP joints. Further studies are needed to confirm the efficiacy and safety of this novel treatment and refine inclusion criteria.

Surgical Treatment

Indications

In and of itself, the existence of the disease does not mean surgery is required. The decision for surgery should be based on the degree of contracture, the extent of involved joints, and the patient's age and specific situation. It should also be noted that while surgery is the mainstay of treatment for advanced disease, recurrence rates range from 26-80%.

Widely accepted indications for surgery are:

a. Flexion contracture of the MP joint greater than 30°
b. Flexion contracture of the PIP joint of any degree
c. Concomitant web space contracture, impaired neurovascular status and loss of articular cartilage

Contraindications include:

a. Infected or macerated skin, especially when skin grafts will be required
b. Noncompliant patients
c. Advanced concomitant arthritis

Surgical Principles and Strategies

Surgical treatment varies from simple fasciotomy to regional fasciectomy or subtotal removal of the fascia. Generally, fasciectomy is advocated. On rare occasions, dermatofasciectomy may be required for advanced, recalcitrant cases with extensive involvement of the overlying skin. Incisions are either performed in a longitudinal or transverse fashion in the palm. Oblique Bruner's incisions can be used for digital incisions. Alternatively, a series of short transverse skin incisions in the palm and phalanges can limit the number and length of scars. This later technique can be particularly useful when the palm and multiple fingers are involved. Less common now is the McCash technique in which an open palm incision is made, and the transverse incision heals secondarily or is covered by a full-thickness skin graft.

For severe contractures in the digits, extreme care must be taken with even the skin incision since the contracture has displaced the neurovascular structures superficially so that the nerve and vessels may literally be up against the dermis. Since the tourniquet is inflated, the vessels may be barely visible. For such severe contractures, it is useful to identify both ulnar and radial neurovascular bundles proximally in the palm and trace them distally into the cord. Separating the scarred structures from the vital digital nerve and vessels will be necessarily tedious. Additionally loss of domain can occur with severe contractures, and it may be necessary to perform grafts or rarely flaps to cover the exposed structures in the digit. Loss of domain in the palm can be allowed to heal secondarily via the McCash technique.

Complications

Early Complications

Early complications include infection, hematoma, skin loss and nerve or vascular injury. Risk of hematoma can be reduced if meticulous cauterization of small bleedings is performed after tourniquet release. The most dreaded complication is irreversible neurovascular injury. The tourniquet should be released to ensure appropriate digital perfusion after contracture release. If ischemia persists, then revascularization procedures with vein grafts may be necessary.

Late Complications

Late complications include recurrence, reflex sympathetic dystrophy (RSD) and loss of flexion. Young patients with a family history are the most prone to recurrence. Recurrence should be differentiated between **local** recurrence of Dupuytren's tissue at the former site of resection or as **distant** recurrence elsewhere in the hand resulting in flexion or extension deformity. Surgical treatment of Dupuytren's disease poses an increased incidence for development of RSD, with a 3-5 times higher incidence in female patients. Ensuring that the patient is compliant with the postsurgical rehabilitation decreases the risk of subsequent RSD. Overall, 10-20% of all Dupuytren's patients who undergo surgery experience one or more complication.

Postsurgical Therapy

Postoperative physical therapy with both active and passive range of motion is mandatory. Static splinting may also be instituted to ameliorate persistent contracture.

Pearls and Pitfalls

The use of limited transverse incisions across the palm and digits is an effective way to perform fasciectomies without extensive scarring. Incisions should be placed within preexisting creases and in the case of adjacent digits may be extended in the palm. The putative advantage of limiting incisions is that postoperative rehabilitation is facilitated. If there are sites of severe contracture, delicate extension should be performed after their release. Gentle manipulation and rewarming over time should return perfusion to the digit. Only rarely will persistent ischemia occur. In such cases, the patient should be kept in the recovery area until the surgeon has determined whether revascularization is necessary.

Suggested Reading

1. Saar JD, Grothaus PC. Dupuytren's disease: An overview. Plast Reconstr Surg 2000; 106:125.
2. McFarlane RM, Jamieson WG. Dupuytren's contracture: The management of one hundred patients. J Bone Joint Surg [Am] 1966; 48:1095.
3. Hueston JT. Dupuytren's contracture. Lancet 1986; 22(2):1226
4. Lubahn JD. Dupuytren's disease. In: Trumble TE, ed. Hand Surgery Update 3. Hand, Elbow and Shoulder. Illinois: American Society for Surgery of the Hand, 2003:393-401.

Chapter 101

Reflex Sympathetic Dystrophy

Zol B. Kryger and Gregory A. Dumanian

Introduction and Terminology

Reflex sympathetic dystrophy (RSD) is an ill-defined constellation of symptoms including pain and dysfunction, typically associated with direct or indirect injury to one or more nerves. Many hand surgeons and pain specialists believe that patients are often incorrectly diagnosed with RSD. Both underdiagnosis and overdiagnosis are common. Some authorities prefer to use the terms **complex regional pain syndrome** (CRPS) or the more specific definition, **sympathetically maintained pain syndrome** (SMPS). This is in contradistinction to those patients with complex regional pain that is sympathetically independent (SIPS). Nevertheless, the term RSD is still widely used and this chapter will refer to this term.

All patients with RSD will report some inciting injury. This can be surgical or traumatic. The injury can be open, such as a laceration, peripheral nerve surgery, or carpal tunnel release. Conversely, the injury can be closed, as in the case of a Colles' fracture. Other less common causes include insults remote from the site of pain such as myocardial infarction, stroke, gastric ulcers and spinal cord injury. Conditions that can produce signs and symptoms similar to RSD and must be excluded include herpes zoster, phantom pain, various neuralgias, and metabolic or compressive neuropathies.

Pathophysiology

The pathophysiology of RSD is not well understood. Both the central and peripheral nervous systems most likely play a role in the development of RSD. What is clear is that the sympathetic system plays a role in the early stages of RSD. One theory is that persistent excitation of certain spinal cord neurons result in a sympathetic vasomotor response and consequently, pain and altered blood flow. Another hypothesis is that some sensory nerves in the spinal cord become hypersensitive after an inciting trauma, resulting in normal stimuli being perceived as painful. A variety of causes related to alterations in the peripheral nerves have been proposed. These include alterations in vasomotor tone or sympathetic nerves, abnormal stimulation of sensory nerves after partial injury, ectopic peripheral pacemakers, and a host of other abnormalities. Finally, some researchers have proposed that RSD has a purely inflammatory pathogenesis.

Diagnosis

Clinical Exam

Most patients will present with the triad of **dysfunction**, an **abnormal sympathetic response** and **pain**. The other cardinal signs and symptoms of RSD are edema, stiffness and discoloration. Secondary signs are trophic changes, vasomotor instability, temperature changes, pseudomotor changes, demineralization of bones,

Practical Plastic Surgery, edited by Zol B. Kryger and Mark Sisco. ©2007 Landes Bioscience.

and palmar fibromatosis. Sympathetic blockade should eliminate some or all of these symptoms; however this is not always the case.

The hallmark of RSD is pain. The pain is out of proportion to the exam and is often described as severe burning that lasts long after the stimulus is removed. Pain with light touch is common as well. Extreme sensitivity to cold is a frequent complaint. This can be tested by applying a drop of acetone or ethyl chloride spray to the sensitive area. Nevertheless, RSD patients have used the gamut of terms to describe their pain, including sharp, dull, crushing, cramping and aching. Edema of the affected extremity is common. Joint stiffness to both active and passive motion has been described. The stiffness can be profound. Motion produces pain, resulting in decreased movement of the joint and subsequent decrease range of motion. Discoloration can range from an intense red to a gray, ashen color, and the color often changes during a single examination. The dorsal surface is more characteristically erythematous in appearance, whereas the palmar surface is more likely to be cyanotic or bluish in color.

Trophic changes most commonly manifest with a glossy, shiny appearance to the skin. The subcutaneous tissue can atrophy as well. Decreased capillary refill can occur as a result of vasomotor instability. Conversely, rapid capillary refill can be seen. The affected side should always be compared to the normal one. Furthermore, the involved side is often colder or warmer than the contralateral side. Hair growth is often dramatic. Hyperhydrosis is common early in the course of RSD, and dryness in the later stages. Osteoporosis of any of the bones in the arm or hand can be seen. Plain films of the hand are an important diagnostic test. Finally, nodules and hardening of the palmar surface can be seen, similar to that of Dupuytren's disease.

There is controversy regarding the existence of an RSD-prone personality. Many of these patients will have some sort of personality disorder or psychiatric disturbance, such as hypochondria, depression or anxiety.

Triphasic Bone-Scans

This is probably the best diagnostic test for RSD. The three phases of this scan are the arterial phase, the venous or blood pool phase, and the metabolic phase. A diagnosis of RSD is supported by increased perfusion in the first phase, increased venous pooling in the MCP and digits during the second phase, and asymmetric uptake in the joints of the affected side in the metabolic phase. There is debate as to which phase of the scan is most important in the diagnosis. Increased tracer uptake in all the joints of the affected hand during the third phase has yielded the highest sensitivity and specificity. In summary, the triphasic bone-scan is a good study for confirming late, severe cases of RSD. However, it has not proven to be sufficiently accurate in diagnosing the very early stages of RSD.

Sympathetic Blockade

The response to sympathetic blockade is both diagnostic and therapeutic. A stellate ganglion block is given with success of the block confirmed by warming of the extremity and a Horner's syndrome. If the patient reports symptomatic improvement in response to the block, the diagnosis of RSD is confirmed. Lack of a response, however, does not exclude RSD. In summary, individuals who do not meet the criteria for RSD or SMPS should be approached with the notion that they suffer from some form of complex regional pain that is not sympathetically maintained and cannot be treated by sympathectomy.

Intravenous alpha-Adrenergic Blockade

Bier block administration of intravenous phentolamine, an alpha-adrenergic receptor antagonist, should produce sympathetic blockade. If the signs and symptoms are sympathetically-mediated as in the case of RSD, they should diminish in response to this infusion. It is important to administer saline in a blinded fashion to eliminate the placebo effect.

Classification

- Lankford classified RSD into five categories:
  - **Minor causalgia**: A mild form of RSD seen after injury to a sensory nerve in the forearm, hand or fingers.
  - **Major causalgia**: The more severe form of causalgia in which pain and dysfunction are prominent. It occurs as a result of injury to mixed motor and sensory nerve.
  - **Minor traumatic dystrophy**: Mild RSD with an inciting trauma, but no known nerve injury.
  - **Major traumatic dystrophy**: The form most commonly thought of when the term RSD is used. Seen after trauma or fracture of the upper extremity, without specific nerve involvement.
  - **Shoulder and hand syndrome**: RSD due to remote injury such as an MI or cervical spine injury. Symptoms begin in the shoulder and spread to the hand, sparing the elbow.
- SMPS can be classified into Type I and II:
  - **Type I**: What is thought of when the term RSD is used. Pain follows and inciting event and is out of proportion to the exam. The other findings typically associated with RSD are usually present.
  - **Type II**: This type of SMPS describes causalgia, similar to the definition given in the Lankford classification.

Staging

- RSD can also be thought of in terms of its stage: early, established or late.
  - **Early RSD**: Defined as the first three months of symptoms. Pain is often burning and can be caused be even light touch. Discoloration, hyperhydrosis, and increased temperature are often present.
  - **Established RSD**: Defined as the period between three and twelve months of symptoms. Pain is still the dominant feature. Skin dryness, joint stiffness, contractures and osteoporosis are common. The temperature of the hand gradually goes from warm seen in early RSD to cold, as compared to the other side.
  - **Late RSD**: Defined as the final stage of RSD, twelve months or longer after onset of symptoms. The pain may become less severe during this stage, however flare-ups can occur. Stiffness and joint contracture are the most prominent features of late RSD. The skin can become thickened and nodular, and severe osteoporosis is not uncommon.

Treatment

The overriding goal of treatment for RSD is elimination of persistent sources of pain. Simple measures such as relieving pressure points or elevation of the extremity can be very helpful. Local and regional nerve blocks help neutralize sensory nerves as well as providing a chemical sympathectomy.

The **stellate ganglion block** is the most effective regional nerve block. It has been demonstrated to provide some degree of relief; however results are variable. Numerous studies have been published with good results ranging from zero to 100%. However, little long-term data is available, and few studies are randomized. A satisfactory block is indicated by warming of the upper extremity and a Horner's sign (unilateral pupillary constriction, ptosis, anhydrosis and facial flushing). Conventional stellate blocks are done with lidocaine or bupivicaine. Good results have been obtained with narcotic blocks (e.g., fentanyl) in refractory cases. Usually repeated biweekly blocks are required. For patients unable to tolerate weekly treatments, a continuous stellate block for 3 to7 days has been used successfully.

Although not widely used in the U.S., sympathetic inhibition can also be achieved using an **intravenous regional block** with anti-adrenergic agents such as bretylium, guanethidine or reserpine. These agents are infused intravenously into an extremity using the Bier block technique to isolate the upper extremity. Other drugs such as steroids and NSAIDs have been used as well. Good long-term pain relief has been demonstrated with this technique.

A variety of **oral medications** have been used to treat RSD. Several drug regimens, such as a short course of oral corticosteroids, nightly amitryptyline, and select calcium channel blockers have met with good success. Oral phenoxybenzamine and other anti-adrenergic drugs have been used with mixed results. Calcitionin and phenytoin have been used to relieve symptoms of RSD; however their use has met with mixed results.

Physical therapy should consist of active range of motion of all joints from the shoulder to the DIP joints. Hand therapy should not be done while the patient is actively in pain. It can be performed immediately following sympathetic blocks when substantial pain relief has been achieved. Progressive **stress loading** without joint motion is also recommended. It involves the use of active traction and compression exercises. Static splints can be used to keep the hand in the intrinsic plus position.

Adjunctive treatments can be helpful in dealing with RSD that does not respond to traditional sympathetic blocks and hand therapy. Biofeedback, psychotherapy, smoking cessation, and transcutaneous electrical nerve stimulation have all been attempted.

Surgical sympathectomy should be reserved for severe, prolonged cases, and those that are refractory to other treatment modalities. The procedure consists of transection of the upper thoracic sympathetic chain via an extrapleural, axillary approach. The T2 and T3 sympathetic nerves must be completely transected. Success rates up to 90% have been reported. More recently, sympathectomies have been performed under video-assisted thoracoscopic surgery (VATS).

Long-Term Outcomes

Very few studies have addressed the sequelae of patients successfully treated for RSD. Overall, long-term results have been disappointing. At one year post-treatment, roughly half of patients have cold intolerance or pain with cold weather. Trophic changes persist in about a third of patients. Joint swelling and stiffness, as well as decreased grip strength are also common complaints. In summary, RSD and SMPS are still poorly understood. The diagnosis of these conditions can be challenging, and their treatment even more so. Active and future research will undoubtedly shed greater light on these syndromes and offer promise for those who suffer from them.

Pearls and Pitfalls

1. It is important for the treating physician to realize that almost any injury can be the inciting cause for RSD. The earlier the inciting injury is recognized, the more likely treatment is to be successful.
2. Pain free movement is probably the best therapeutic modality against RSD. Nerve blocks and oral analgesics combined with physical therapy is a treatment goal.
3. The extremity surgeon treating this condition is a coach or motivator for the patient, more so for this disease process than almost any other. The patient needs frequent counseling about the disease process and the expected length of treatment.

Suggested Reading

1. Dzwierzynski WW, Sanger JR. Reflex sympathetic dystrophy. Hand Clin 1994; 10:29.
2. Lankford LL. Reflex sympathetic dystrophy. In: Hunter JM et al, eds. Rehabilitation of the Hand-Surgery and Therapy. 3rd ed. St. Louis: CV Mosby, 1990.
3. Nath RK, Mackinnon SE, Stelnicki E. Reflex sympathetic dystrophy. The controversy continues. Clin Plast Surg 1996; 23:435.
4. Zyluk A. The sequelae of reflex sympathetic dystrophy. J Hand Surg (Br) 2001; 26:151.

Appendix I

Part A: Important Flaps and Their Harvest

Zol B. Kryger and Mark Sisco

Groin Flap

The groin flap is a **fasciocutaneous Type A** flap based on the **superficial circumflex iliac** system. The skin is innervated from the **T12 lateral cutaneous nerve**. This flap is used primarily as a **rotational flap** for coverage in the abdominal wall and perineum; it can also be used as a free flap for distant coverage.

The skin of the lateral groin is elevated along an axis parallel and 3 cm inferior to the inguinal ligament (Fig. AI.1). The skin flap can measure up to 25 x 10 cm. The pedicle originates from the femoral artery roughly in the femoral canal. The main vein drains into the saphenous vein just distal to the fossa ovalis.

The skin is incised down to the fascia. The flap is elevated distal to proximal in the plane superficial to tensor fascia lata, which serves as the distal extent of the flap. It is dissected free from the anterior superior iliac spine (ASIS), inguinal ligament and external oblique fascia. The deep fascia that envelops the sartorius is included in the flap. The pedicle is dissected to the medial edge of the sartorius muscle—the proximal limit of the flap.

If the skin is elevated as an island flap, the pedicle should be identified in the femoral triangle through a transverse incision, before the skin island incision is made. The flap is elevated from distal to proximal as described above, until the dissection meets the pedicle.

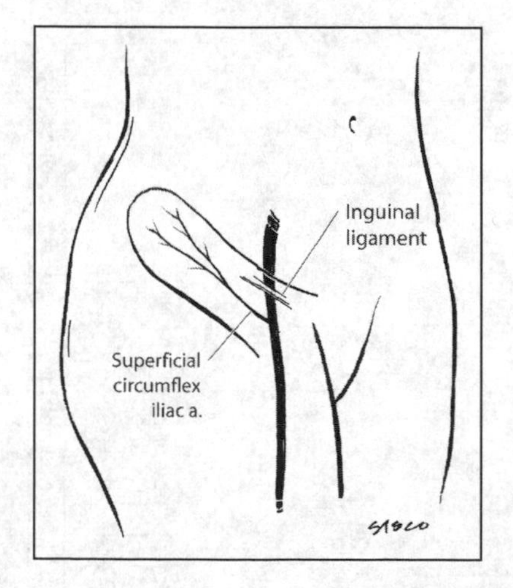

Figure AI.1. The groin flap and HS blood supply.

Table AI.1. Fasciocutaneous and perforator flaps

Flap	Type*	Size (cm)	Blood Supply (D=dominant, M=minor)	Common Uses
Forehead	A	22 x 7	D: Superficial temporal a. M: Supratrochlear a. M: Supraorbital a	Middle and inferior face, oral cavity, nose, frontal and maxillary sinus
Nasolabial	C	2 x 5	D: Angular a. M: Alar branches of superior labial a.	Nose, lips, floor of mouth
Temporoparietal fascia	A	12 x 9	Superficial temporal a.	Scalp, upper and middle face, ear, free flap
Deltopectoral	C	10 x 20	D: Perforators 1-3 of IMA M: Perforators 4-5 of IMA	Middle and inferior face, neck, oral cavity, esophagus
Gluteal thigh	A	12 x 30	D: Superior gluteal a. D: Inferior gluteal a. M: Branches of lateral circumflex femoral a. M: 1st perforator of profunda femoris	Sacrum, ischium, perineum
Scapula	B	20 x 7	D: Circumflex scapular a.	Neck, axilla, shoulder, back, free flap (mandible)
Radial forearm	B	10 x 40	D: Radial a. M: Musculocutaneous branches of radial recurrent a. and inferior cubital a.	Forearm, elbow, wrist, hand, free flap
Groin	A	25 x 10	D: Superficial circumflex iliac	Upper extremity, abdominal wall, perineum, free flap
Lateral thigh	B	7 x 20	D: 1st-3rd perforators of profunda femoris	Ischium, greater trochanter, free flap
Medial plantar	B	12 x 6	D: Medial plantar a. M: Musculocutaneous perforators from medial plantar a.	Plantar foot, ankle, free flap (contralateral foot)
Anterolateral thigh (ALT)	B, C	12 x 20	D: Septocutaneous branches of descending branch of lateral circumflex a. M: Musculocutaneous branches of descending branch of lateral circumflex a.	Trunk, abdomen, groin, thigh, free flap
Deep Inferior epigastric perforator (DIEP)			D: Perforator from deep inferior epigastric a.	Free flap (breast)
Thoracodorsal perforator (TAP)			D: Perforator from thoracodorsal a.	Free flap (breast, extremity)
Superior gluteal perforator (SGAP)			D: Perforator from superior gluteal a.	Sacrum, ischium, lower back, free flap

*Mathes-Nahai classification of fasciocutaneous flaps. Type A, direct cutaneous pedicle. Type B, septocutaneous pedicle. Type C, musculocutaneous pedicle

Table AI.2. Muscle and musculocutaneous flaps

Flap	Type*	Size (cm)	Blood Supply (D = dominant, M = minor)	Common Uses
Temporalis	III	10 x 20	D: Anterior deep temporal a. D: Posterior deep temporal a. M: Branches of middle temporal a.	Facial reanimation, orbit, palate, mandible
Pectoralis major	V	15 x 25	D: Pectoral branch of thoracoacromial a. M: Pectoral branch of lateral thoracic a. S: Internal mammary (1-6) and intercostal (5-7) perforators	Face, chest, neck, shoulder, axilla, sternum, upper extremity
Gluteus maximus	III	24 x 24	D: Superior gluteal a. D: Inferior gluteal a. M: Branches of lateral circumflex femoral M: 1st perforator of profunda femoral	Sacrum, ischium, trochanter, vagina, free flap
Latissimus dorsi	V	25 x 35	D: Thoracodorsal S: Posterior intercostal perforators and lumbar perforators	Scalp, neck, trunk, breast, abdomen, upper extremity, free flap
Rectus abdominis	III	25 x 6	D: Superior epigastric a. D: Deep inferior epigastric a. M: Subcostal and intercostal a.	Breast, trunk, abdomen, groin, perineum, pelvic floor, free flap
Serratus	III	15 x 20	D: Lateral thoracic a. D: Branch of thoracodorsal a.	Head, thorax, axilla, posterior trunk, intrathoracic, free flap
Gracilis	II	6 x 24	D: Ascending branch of medial circumflex femoral a. M: 1-2 branches of SFA	Perineum, groin, penis, vagina, free flap
Hamstring	II	15 x 45	D: 1st-3rd perforators of profunda femoris M: Branch of inferior gluteal a. M: Superior lateral genicular a.	Ischium
Tensor fascia lata	I	5 x 15	D: Ascending branch of lateral circumflex femoral a.	Abdomen, groin, perineum, trochanter, ischium, free flap
Vastus lateralis	I	10 x 25	D: Descending branch of lateral circumflex femoral a. M: Transverse branch of lateral circumflex femoral a. M: Posterior branch from profunda femoris M: Superficial branch of lateral superior genicular a.	Ischium, trochanter, groin, perineum, knee, abdomen
Gastrocnemius	I	20 x 8	D: Medial sural a. D: Lateral sural a. M: Anastomotic vessels from sural a.	Knee, lower thigh, upper leg
Soleus	II	8 x 28	D: Muscular branches of popliteal D: First 2 branches of posterior tibial a. D: First 2 branches of peroneal a. M: Segmental branches of posterior tibial a.	Middle and lower leg
Fibula	V	3 x 40	D: Nutrient branches of peroneal a. M: Periosteal and muscular branches of peroneal a.	Tibia, free flap (mandible)

*Mathes-Nahai classification of muscle and musculocutaneous flaps. Type I, one pedicle. Type II, dominant and minor pedicles. Type III, two dominant pedicles. Type IV, segmental pedicles. Type V, one dominant pedicle and secondary segmental pedicles.

Table AI.3. Visceral flaps

Flap	Type*	Size (cm)	Blood Supply (D = dominant, M = minor)	Common Uses
Omentum	III	40 x 60	D: Right gastroepiploic a. D: Left gastroepiploic a.	Head and neck, trunk, intrathoracic, abdomen, groin, perineum, free flap
Jejunum	I	7-25	D: Jejunal a. (from SMA)	Free flap (esophagus)

*See classification in Table AI.2.

Rectus Abdominis Flap

The rectus abdominis flap can be harvested as either a **muscle or musculocutaneous, Type III flap**. The two dominant pedicles are **the superior and deep inferior epigastric** arteries. Minor pedicles include the intercostals and subcostal arteries, with the T8 subcostal artery usually being the largest. The muscle and overlying skin are innervated by **segmental motor and cutaneous intercostal (7-12) nerves**, respectively. It is an extremely useful flap used in breast, perineal and vaginal reconstruction, and as coverage in the thorax, abdomen, posterior trunk and groin. For these purposes, it is primarily used as a **rotational flap or island pedicle flap**. It is also an extremely versatile **free flap** based on the deep inferior epigastric vessels.

Contraindications for use of the rectus abdominus flap include:

- Unilateral subcostal incision (Kocher incision) for an ipsilateral flap based on the superior pedicle
- Bilateral subcostal incisons (Chevron incision) for any flap based on the superior pedicle
- Low transverse incison (Pfannenstiel incision) for any muscle flap based on the inferior pedicle (exception: a deep inferior epigastric perforator flap)
- Any portion of the skin island that is lateral to a prior skin incision should not be used
- Prior use of the internal mammary artery is a relative contraindication for a superiorly based flap
- History of major external iliac vascular surgery is a relative contraindication for an inferiorly based flap, unless angiography confirms otherwise.

For harvesting a muscle flap, either a longitudinal paramedian or low transverse skin incision is used. For the musculocutaneous flap, the skin island can be marked in multiple horizontal or vertical patterns. A transverse (horizontal) skin island can be up to 21 x 8 to 21 x 14 cm in size. This skin can be divided into zones: zone 1 is over the ipsilateral rectus; zone 2 is over the contralateral rectus; zone 3 is lateral to the ipsilateral rectus; zone 4 (least reliable skin) is lateral to the contralateral rectus.

After the skin paddle is marked, the inferior border is incised down to the anterior rectus sheath. The skin and subcutaneous fat are elevated from lateral to medial off the fascia. The dissection is slowed several centimeters lateral to the midline where the musculocutaneous perforators are encountered. The superior border of the skin island should be incised only after confirming that the donor skin will close without excessive tension. Alternatively, the superior incision can be made first, followed by the inferior incision once it is clear that the abdominal skin will come together without undue tension.

The anterior rectus sheath is opened sharply in a longitudinal direction exposing the rectus muscle (Fig. AI.2). The muscle is dissected free from its sheath, with care taken not to violate the posterior sheath. For the inferiorly-based flap, the muscle is divided at or near the costal margin. The superior epigastric artery and vein are divided at the medial border of the muscle. For the superiorly-based flap, the muscle is divided at the level of the pubis symphysis. The deep inferior epigastric artery and vein can be dissected for several centimeters prior to division. This can serve as an alternative pedicle for microvascular anastomosis if the superior pedicle is insufficient. Care must always be taken to avoid injuring the musculocutaneous perforators feeding the skin paddle.

AI

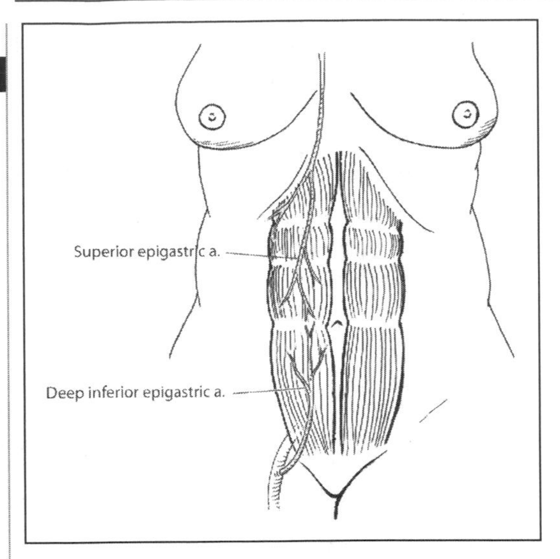

Figure AI.2. The rectus abdominus musculocutaneous flap and its dual blood supply.

If only a muscle flap is required, the abdominal skin and subcutaneous fat are elevated off the anterior abdominal wall. The rectus sheath can then be opened as described above without concern for the musculocutaneous perforators.

Once the rectus muscle is harvested, the donor site is closed. The anterior rectus sheath can be closed primarily using a running or interrupted permanent suture. If the fascial edges are frayed or primary closure will create under undue tension, a synthetic mesh can be used to replace the missing segment. If necessary, small tears in the fascial edges during primary fascial closure can be reinforced with an overlying piece of mesh.

Fibula Composite Flap

The fibula free flap can be harvested as either an **osseous or osseofasciocutaneous (composite), Type V flap**. In addition, cuffs of muscle are usually incorporated in order to protect the blood supply. The dominant pedicle is the **nutrient branch** of the peroneal artery. The minor pedicles are the **periosteal and muscular branches** of the peroneal artery. The sensory nerve supply is from the **superficial peroneal nerve**. This flap is used primarily as a **free flap** for mandibular reconstruction or for reconstruction of the ipsilateral tibia and femur.

If an osseous flap is needed, a longitudinal incision is made along the posterior border of the fibula from the head of the fibula to the lateral malleolus. If an osseofasciocutaneous flap is used, the skin territory should be marked as a vertical ellipse over the middle third of the fibula. The skin island can span from 6 cm below the fibular head to 8 cm above the distal fibula, and it can measure up to 5 x 15 cm. The width of the skin island can be extended; however closure will require a skin graft.

For harvesting the osseous flap, the lateral compartment is opened, and the peroneus longus and brevis are detached from the fibula leaving a small cuff of muscle attached. The common and superficial peroneal nerves are identified and preserved.

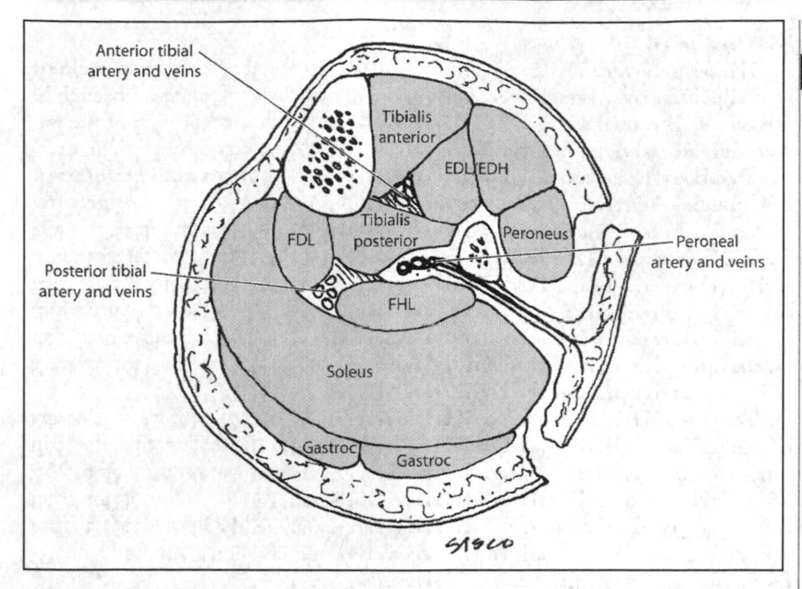

Figure AI.3. A cross section of the leg showing the compartments of the leg and the anterior and posterior approaches to the deep posterior compartment.

The proximal and distal osteotomies are performed (4 cm inferior to fibular head and 6 cm superior to lateral maleolus). As long as at least 6 cm of distal fibula is left intact, ankle stability will be preserved.

A cross section of the leg showing the compartments in relationship to the fibula are shown in Figure AI.3. In the anterior approach, the anterior compartment muscles are released followed by division of the interosseous membrane and entry into the superficial and deep posterior compartment musculature. The peroneal artery and vein are divided upon entry into the deep posterior compartment, followed by tibialis posterior and flexor hallucis longus. In the posterior approach, the posterior compartments are released first followed by entry into the anterior compartment and release of its musculature. Cuffs of muscle should be included with the flap.

The dominant pedicle usually enters the middle third of the fibula on the medial side via the nutrient foramen 14-19 cm (average of 17 cm) below the styloid process. The peroneal artery and vein can be dissected proximally as needed to achieve adequate length. Shortening of the bone should be done in the subperiosteal plane in order to avoid injuring the blood supply.

For harvesting the osseofasciocutaneous flap, the skin island is incised circumferentially down through the deep fascia. Anteriorly the skin island is elevated subfascially off the anterior and lateral compartment musculature. The posterior skin island is elevated subfascially off the lateral gastrocnemius and soleus muscles. The remainder of the dissection proceeds as described above. Special care is taken to preserve the septocutaneous and musculocutaneous perforators. A cuff of flexor hallucis longus and soleus should be preserved.

Pectoralis Major Flap

The pectoralis major flap (Fig. AI.4) can be harvested as either a **muscle or musculocutaneous, Type V flap**. The dominant pedicle is **the pectoral branch of the thoracoacromial artery**. The minor pedicle is the **pectoral branch of the lateral thoracic artery**. Minor segmental pedicles include the first through sixth **internal mammary perforators**, and the fifth through seventh **intercostal perforators**. The muscle is innervated by the **lateral** (superior) and **medial** (inferior) **pectoral nerves**. The sensory innervation is from the **intercostal (2-7) nerves**. It is a versatile **pedicle flap** for coverage of defects of the face, chest, neck, shoulder, axilla, sternum and upper extremity flap. It can also be used in the intrathoracic cavity. By dividing either its origin or insertion, it can serve as a **rotational flap or island pedicle flap** for many sites. The pectoralis major muscle flap can also serve as a **functional flap** for the upper extremity (elbow flexion). Rarely, it is used as a **free flap** for coverage in the head and neck or perineum.

The flap can be elevated with the entire skin paddle covering the muscle, or any part of it. The borders of the flap are the clavicle, the anterior axillary line, the sixth intercostal space and the parasternal line. The skin territory can measure up to 20 x 28 cm. For head and neck reconstruction, a smaller skin paddle is used: an inframammary skin island in women and a parasternal paddle in men. Most small skin defects can be closed primarily. Larger skin paddles will leave a donor site requiring skin grafting or secondary flap closure.

Harvesting the pectoralis major muscle is relatively simple. If only muscle is required, a horizontal incision is made below the clavicle, vertically along the axillary line or midsternal. The skin and subcutaneous fat are dissected free of the muscle. As an island flap based on the thoracoacromial pedicle, the muscle fibers are divided from their origin and from the clavicle, and the muscle is dissected from medial to

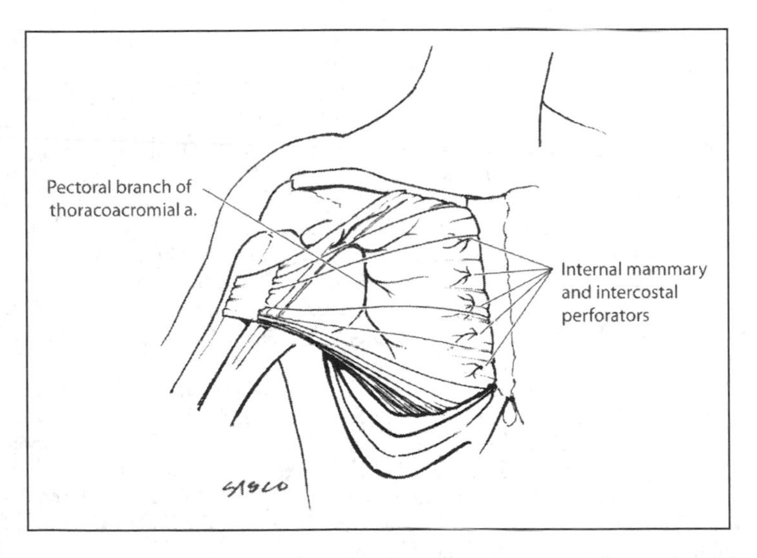

Figure AI.4. The pectoralis major muscle flap and its blood supply.

lateral. The muscle is divided just lateral to the pedicle, which enters the muscle from the deep side around the junction of the middle and lateral thirds of the clavicle.

As a turnover flap (reverse flap), the muscle is divided laterally at the level of the lateral border of pectoralis minor, preserving the lateral one-third of the muscle. The muscle is dissected from lateral to medial, preserving the vascular and nerve supply to the remaining portion. The dissection continues to within 2-3 cm from the sternal border until the internal mammary perforators are visualized.

For head and neck reconstruction, the horizontal, infraclavicular skin incision is used. A skin paddle is often needed as described above. The skin island is incised and the muscle divided distally. Mobilization occurs in a superior direction towards the clavicle. If the entire muscle is not needed, a wide central strip of muscle is often sufficient to vascularize the skin pedicle. The musculocutaneous flap can be pulled through the clavicular incision.

If functional muscle transfer for the upper extremity is required, an anterior axillary line incision is used. The portion of muscle required is outlined. The muscle is dissected free from the subcutaneous tissue, and the origin at the ribs and sternum is divided. The muscle is mobilized from medial to lateral towards the humerus. Care is taken to preserve the blood supply and motor nerves. The muscle can be tunneled through the axilla onto the arm. Elbow flexion can be achieved by suturing the pectoralis to the biceps tendon.

Latissimus Dorsi Flap

The latissimus dorsi flap (Fig. AI.5) can be harvested as either a **muscle or musculocutaneous, Type V flap**. The dominant pedicle is the **thoracodorsal artery**. Minor segmental pedicles include a medial and lateral row of **posterior intercostals and lumbar perforators**. The muscle is innervated by the **thoracodorsal nerve** which travels with the vascular pedicle. The sensory innervation is from the **intercostal nerves**. It is a versatile **pedicle flap** for coverage of defects of the neck, trunk, breast, abdomen and upper extremity. The latissimus muscle flap can also serve as a **functional flap** for the upper extremity (elbow extension or flexion). It is also used as a **free flap** for coverage in the scalp, lower and upper extremity-especially when a thin flap is required.

The skin island, when required, is marked. Options include oblique skin islands—with the superior end towards the midline or the axilla, superior posterior skin island, superior transverse, inferior transverse, lateral or vertical orientation. The dominant pedicle is marked in the posterior axilla entering the lateral deep surface of the muscle about 15 cm below the humeral insertion. The skin is incised around the island beveling away from the skin island. The muscle fibers of the latissimus are identified: superiorly, with the scapula and trapezius muscle. The fibers are divided, separating the muscle from the scapula. The attachments to the vertebral column are divided. The lumbrosacral fascia is divided to the level of the posterior axillary line. The minor pedicles are divided as they emerge from the posterior intercostals and lumbar vessels. The direction of dissection progresses towards the axilla where the pedicle is located. The dissection should allow sufficient flap mobility without compromising the pedicle. If needed, the insertion to the humerus can be divided to gain mobility.

If a muscle flap alone is required, the location of the skin incision is variable. Once the skin is incised and the superficial side of the muscle is exposed, the skin flaps are elevated off the muscle exposing its entire dimensions. The muscle is freed

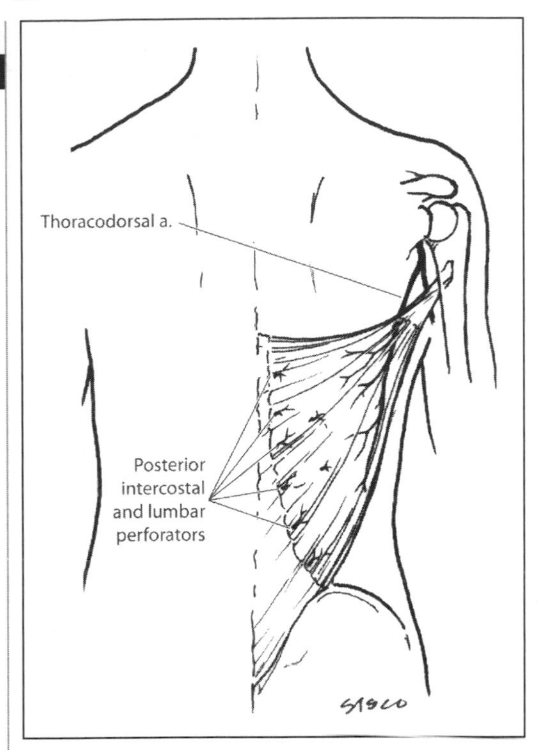

Figure AI.5. The latissimus dorsi muscle flap and its blood supply.

from its origin at the midline and superiorly as described above. Dissection proceeds towards the axilla and the pedicle. If the muscle is to be used as a free flap, adequate pedicle length can easily be achieved. Once the pedicle is clearly identified, the crossing branches to the serratus are divided. The circumflex scapular artery and branches to teres major are also divided. The insertion can be divided once the pedicle is completely dissected.

Serratus Flap

The serratus flap (Fig. AI.6) can be harvested as either a **muscle or musculocutaneous, Type III flap**. The dominant pedicles are the **lateral thoracic artery and branches of the thoracodorsal artery**. The muscle is innervated by the **long thoracic nerve**. The sensory innervation is from the **intercostal (2-4) nerves**. It is a versatile **pedicle flap** for coverage of defects of the head, thorax, axilla and posterior trunk. It can also be used in the intrathoracic cavity. By dividing either its origin or insertion, it can serve as a **rotational flap or island pedicle flap**. It is used as a **free flap** for coverage in the head and neck or limbs. It can also be used as a **functional flap** for facial reanimation. It can be harvested with the latissimus muscle as a combined flap. It can also be elevated as an **osseomusculocutaneous flap** by harvesting a portion of a rib along with the flap since the ribs are vascularized through the attachments of the serratus to the periosteum.

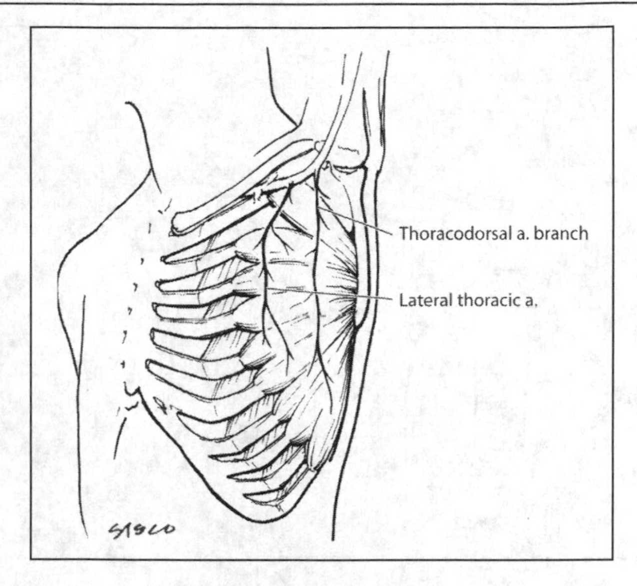

Figure AI.6. The serratus flap and its blood supply.

The serratus flap is often harvested without a skin paddle. The skin is incised diagonally across the axilla. The muscle slips are identified. The upper three slips are vascularized by the lateral thoracic artery. The remaining lower slips receive their blood supply from the thoracodorsal branches. Preservation of the upper slips will decrease the chances of scapular winging. Therefore, only the lower three or four slips should be harvested based on the branches from the thoracodorsal artery. The slips are divided from the ribs anteriorly and dissected posteriorly towards the scapula. The muscle is divided and elevated. The long thoracic nerve runs on the superficial surface of the muscle. This nerve should be preserved during the dissection. The thoracodorsal pedicle can be lengthened by dividing the branches to the latissimus.

If an osseomusculocutaneous flap is needed, the serratus is harvested with a portion of the 5th or 6th rib. The muscle slips to the desired rib are preserved, and the rib is dissected in the extrapleural plane.

Omental Flap

The omental flap (Fig. AI.7) is a **Type III visceral flap**. It has two dominant pedicles: the **right or left gastroepiploic arteries.** It can be used as a **pedicle flap** for coverage in the head and neck, trunk, intrathoracic region, abdomen, groin and perineum. It is useful as a **free flap** for head and neck reconstruction or as coverage in the extremities.

The greater omentum lies between the greater curvature of the stomach and the transverse colon. After exposure of the peritoneal cavity, the omentum is released from its attachments to the colon along the antimesenteric border. The vascular branches from the gastroepiploic arch to the greater curvature are divided. The desired pedicle is chosen, and the other pedicle is ligated. When the right pedicle is chosen, the omentum is mobilized to within 3 cm of the pylorus. If it is to be harvested as a free

AI

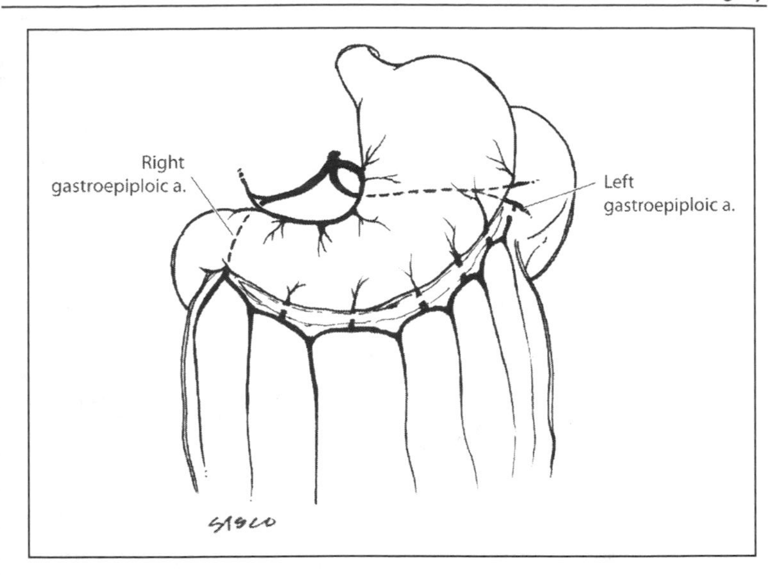

Figure AI.7. The omental flap and its blood supply.

flap, the gastroepiploic artery and vein can be dissected further for increased pedicle length. If the left pedicle is chosen, the right pedicle is divided and the omental dissection continues to within 7 cm of the gastrosplenic ligament.

Gracilis Flap

The gracilis flap (Fig. AI.8) can be harvested as either a **muscle or musculocutaneous, Type II flap**. The dominant pedicle is **the ascending branch of the medial circumflex femoral artery**. The minor pedicles are the first and second **branches of the superficial femoral artery**. The muscle is innervated by the **anterior branch of the obturator nerve** which enters it on its deep surface, superior to the vascular pedicle. The sensory innervation is from the **intercostal nerves**. It is a versatile **pedicle flap** for coverage of defects of the abdomen, pelvis, perineum, groin, penis and vagina. The gracilis muscle flap can also serve as a **functional flap** for facial reanimation. It is also used as a **free flap** for coverage in the head and neck and extremities.

For muscle flap elevation, a linear incision is made 2-3 cm posterior to a line connecting the pubis and the medial condyle. The gracilis is posterior to the adductor longus. The musculotendinous insertion of the gracilis lies posterior to the sartorius and saphenous vein. The tendon is isolated with a penrose drain and then divided. As the muscle is mobilized proximally, the minor pedicles will be encountered as they enter the medial muscle belly. If the dominant pedicle is chosen, these minor pedicles are divided. The dominant pedicle can be exposed by medial retraction of the adductor longus. It passes over adductor magnus as it enters the gracilis on its deep surface about 10 cm inferior to the pubic tubercle. If additional pedicle length is required, as is the case for a free flap, the pedicle can be dissected proximally after

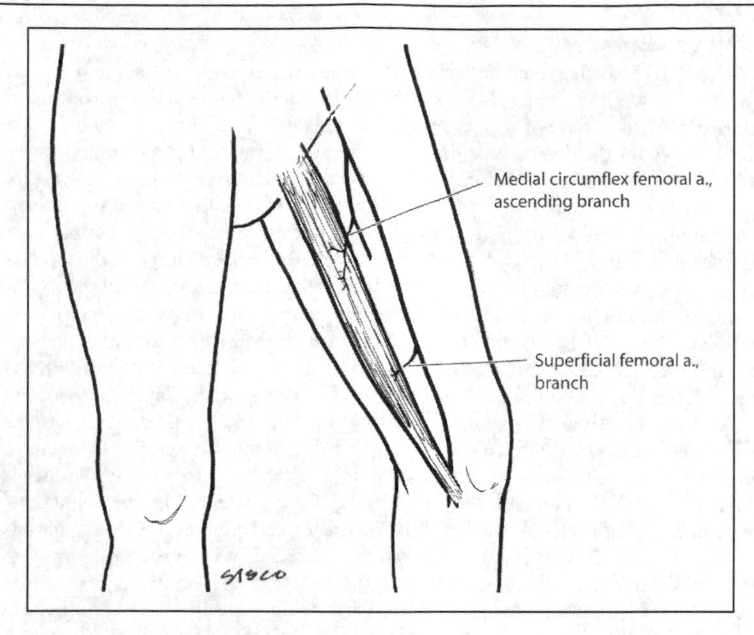

Figure AI.8. The gracilis muscle flap and its blood supply.

division of the branches to the adductor magnus and longus muscles. The anterior branch of the obturator nerve should be identified as it enters the muscle superior to the point of entry of the pedicle.

The skin island, if required, should be a vertical or horizontal ellipse overlying the proximal or mid portions of the muscle. Once the distal gracilis muscle is exposed, the relationship of the skin island to the underlying muscle should be confirmed prior to incising its entire border. If it has been drawn too distally, it should be redrawn in a more proximal position. The skin island should be incised from distal to proximal down to the level of the fascia. The deep surface of the skin island can be sutured to the muscle in order to avoid traction injury to the musculocutaneous perforators. The gracilis musculocutaneous flap does not have a robust and reliable skin paddle. Surgical delay should be considered if a large skin paddle is required.

Radial Forearm Flap

The radial forearm flap can be harvested as an **osseofasciocutaneous** or **fasciocutaneous Type B flap.** Its dominant pedicle is the **radial artery** and minor pedicles are **musculocutaneous branches of the radial recurrent artery** and the **inferior cubital artery.** The skin paddle can measure up to 10 x 40 cm, requiring skin grafting of the donor site in most cases. Primary donor site closure can be achieved if a very small skin paddle is used. This flap can be used locally in the arm, forearm or hand for coverage. It is a versatile free flap, often used in head and neck reconstruction.

The harvesting of this flap and the special considerations involved in its use are discussed in Part B of this appendix.

AI

Gluteus Flap

This flap can be harvested as a **musculocutaneous, muscle** or **fasciocutaneous flap** (see below). It is a **Type III flap** with two dominant pedicles, the **superior** and **inferior gluteal arteries** (Fig. AI.9). Minor pedicles not commonly used as sole blood supply for the flap include the first perforator of the profunda femoral artery and branches of the lateral circumflex femoral artery. The skin paddle can measure up to 24x24 cm. This flap is used locally for coverage of pressure sores of the sacrum and ischium, as well as in the trochanteric region if other flaps (e.g., TFL flap) are not available. Bilateral flaps can be advanced medially to close a large midline sacral defect. It can also be used for reconstruction of pelvic and vaginal defects.

The gluteal **fasciocutaneous perforator flap** is a **Type A flap** that can be used locally for coverage of spinal defects and pressure sores. It has also been uses successfully as a free flap, most notably for breast reconstruction. Most commonly the flap is based on the superior gluteal artery perforator, hence the common name for this flap is the **SGAP**.

For fasciocutaneous flap harvest, the skin paddle is marked: either a transposition or V-Y advancement pattern is used, keeping in mind the ability to close the donor site directly. The skin is incised through the superficial fascia, down to the gluteal muscles. The flap is elevated along with the muscle fascia. Care is taken to locate the perforators. The proximal perforators to be saved are skeletonized to allow greater flap mobility. Those that hinder flap mobility are ligated and divided. When used locally, the flap is transposed or advanced into the defect and sewn into place with two layers. For the free SGAP, the perforators are traced back to the superior gluteal artery, and the pedicle is dissected proximally to gain sufficient length.

For musculocutaneous flap harvest, either a superior or inferior skin paddle can be utilized. The superior skin paddle is based on the superior gluteal artery, and the inferior paddles on the inferior gluteal artery (The entire muscle and buttock skin can be

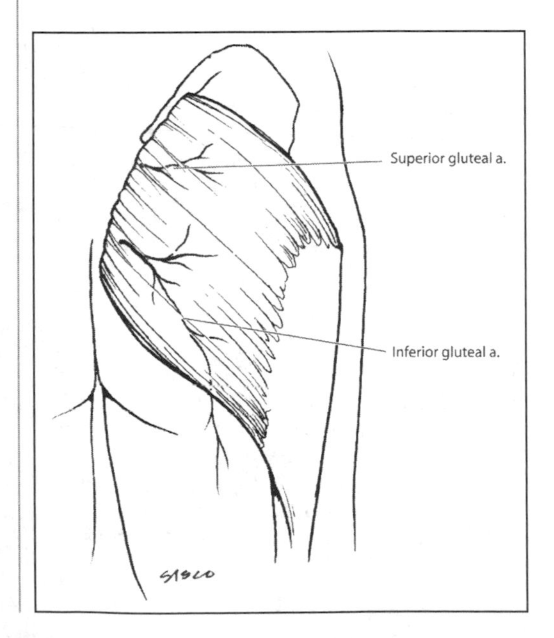

Figure AI.9. The gluteus muscle flap and its blood supply.

based on the inferior artery). The muscle flap can be advanced or rotated. The skin and subcutaneous tissue are divided. For rotational flaps, the muscle insertion (greater trochanter and IT band) is also divided. The inferior and lateral borders of the muscle are divided. The muscle is detached from its origin. For ambulatory patients, the inferior portion of the muscle with its insertion and origin should be preserved. The pedicle and the sciatic nerve are located using the piriformis muscle as a landmark (the sciatic nerve emerges from beneath this muscle). The flap is inset into the defect, and the muscle fascia is sewn to the contralateral gluteus maximus fascia. The subcutaneous tissue is closed in a second layer followed by skin. The donor site is closed directly over a suction drain, and additional drains placed under and over the flap. If direct closure is not possible, skin grafting the donor site is an option.

Anterolateral Thigh (ALT) Flap

The **ALT flap** is a **Type B** or **C fasciocutaneous flap** is an extremely versatile free flap used most commonly for head and neck reconstruction. It can also be raised as a purely **cutaneous free flap** without the underlying fascia. It is based on either **septocutaneous** or **musculocutaneous branches** (much more common) of the descending branch of the **lateral circumflex artery**. A skin paddle up to 12 x 20 cm in size can be harvested; however any flap wider than 10 cm is difficult to close primarily.

The skin island is marked by drawing a line down the lateral thigh: from the ASIS to the superolateral corner of the patella. The midpoint of this line represents the site at which the greatest concentration of perforators can be found. A 6 cm diameter circle (centered at the midpoint of the line) will capture the main perforators and should be included within the flap. The medial border incision is made first. The dissection can be either subfascial or suprafascial. The subfascial dissection is safer and will create a fasciocutaneous flap; however the underlying muscles will bulge out and make the closure more difficult. In addition, there is greater risk of injury to the sensory and motor nerves that travel within and just below the fascia. The suprafascial dissection will yield a purely cutaneous ALT flap, sparing the nerves and muscular fascia. This dissection is challenging due to the smaller size of the perforators above the fascia. The perforators are traced back to the lateral circumflex artery pedicle. In only about 10-15 % of cases is there an adequate septocutaneous vessel. In the majority of cases, the perforators are musculocutaneous and must be followed to the pedicle through the vastus lateralis and rectus femoris muscles.

Once sufficient pedicle length has been obtained, the lateral skin incision can be made. The advantage of not incising the entire skin border early on is most notable in head and neck reconstruction. The harvest of the flap can be performed simultaneously with the resection and recipient site preparation. Prior to making the final skin incision, the size of the required skin island will be known based on the size of the defect and any size adjustments can be made. After the lateral aspect of the flap is raised and joined with the medial dissection, the pedicle is ligated and the flap transferred. A skin island less than 10 cm in width can usually be closed primarily over suction drainage. If this is not possible, the majority of the donor site is closed, and a skin graft is used to cover the remaining central portion.

Suggested Reading

1. Mathes SJ, Nahai F. Reconstructive Surgery: Principles, Anatomy, and Technique. New York: Churchill Livingstone, 1997.
2. Strausch B, Vasconez LO, Findley-Hall EJ. Grabb's Encyclopedia of Flaps, 2nd edition. Boston: Little Brown, 1998.

Part B: Radial Forearm Free Flap

Peter Kim and John Y.S. Kim

Indications

The radial forearm free fasciocutaneous flap has become one of the mainstays of head and neck reconstruction. The radial forearm flap provides thin, pliable tissue from a reliable donor site based on a long vascular pedicle. It is described for coverage of soft tissue defects in the head and neck, the posterior trunk, and upper and lower extremities. Additionally, the radial forearm free flap has been used for reconstruction of defects of the esophagus and penis.

Numerous modifications have broadened the application of this flap. A portion of the radial cortex (no greater than one-third the circumference of the radius) can be harvested along with the skin paddle in reconstructing composite defects such as those seen in marginal mandibulectomies and palatal resections. The tendon of the palmaris longus can be included in the flap to be used as a sling in lip reconstruction. The lateral and/or medial antebrachial sensory nerve can be included to create a neurosensory flap, as is used in neophallus reconstruction.

Preoperative Considerations

Allen Test

This is the first screening test that should be performed on all potential patients. The surgeon uses his thumb and fingers to compress the radial and ulnar arteries at the wrist. The patient exsanguinates the hand by making a fist several times, and then opens the hand so that the fingers are in a relaxed and gently extended position. The examiner then releases pressure from over the ulnar artery. Capillary refill time in the hand is noted. A normal Allen test is refill in less than 5 seconds, and greater than 5 seconds indicates an abnormal Allen test. About 85-90% of patients will have a normal Allen test. If the test is normal, surgery can proceed without further testing.

If the Allen test is abnormal, bilateral duplex ultrasonography or pulse volume recordings should be performed of hands and fingers, with and without radial artery compression. Over 90% of these patients will have a normal noninvasive exam and can proceed to radial forearm flap harvest safely. In most cases, one of the two hands will demonstrate preserved flow pattern with radial artery compression, and consequently, safe harvesting of the flap. In the rare case that both hands demonstrate abnormal arterial flow to the hands, use of this flap is contraindicated.

The donor site scar must be addressed during preoperative counseling. The scar can be particularly unsightly in obese patients due to the high "step-off" between the muscle bed and the surrounding skin. Other flaps may need to be considered if the patient is particularly concerned about the appearance of the donor site scar. Another relative contraindication is the use an osseocutaneous flap in postmenopausal women. Osteoporosis places these patients at increased risk of developing a postoperative fracture.

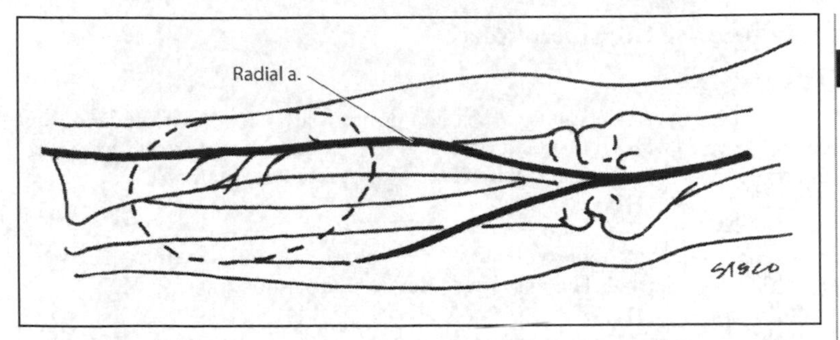

Figure AI.10. The course of the radial artery. The skin island can be designed anywhere along its axis (dashed area).

Flap Elevation

Harvest of the flap can usually be performed rapidly in a bloodless field using a tourniquet. The radial forearm flap is a Type II fasciocutaneous flap. Its dominant pedicle is the radial artery. Deep venous drainage is via the venae comitantes. Superficial venous drainage is through the cephalic vein which is routinely included with the flap. One large (up to 10 x 40 cm) or multiple smaller skin islands can be harvested anywhere in the volar forearm from the antecubital fossa to the wrist along the axis of the radial artery (Fig. AI.10).

As it bifurcates off the brachial artery, the radial artery courses between the brachioradialis and the pronator teres muscle bellies. As it progresses distally, it travels between the tendons of the brachioradialis and the flexor carpi radialis. Accordingly, the distal incision is made first, identifying the flexor carpi radialis and brachioradialis tendons, as well as the underlying artery. The cephalic vein lies radial to the brachioradialis tendon. It can be ligated and divided at this stage.

The proximal incision is made, and the brachioradialis and the flexor carpi radialis are identified. The radial artery can be found deep to the brachioradialis muscle and superficial to the pronator teres. An incision is carried out along the ulnar border of the skin island. The flap is then elevated from this ulnar border, working toward its arterial axis. The flap is elevated off the flexor digitorum superficialis, palmaris longus and the flexor carpi radialis. Particular care is taken to preserve the peritenon of the tendons. Just radial to the flexor carpi radialis lays the intermuscular septum carrying the fasciocutaneous perforators.

Lastly, the radial skin incision is made, and the deep fascia is dissected off the brachioradialis muscle ulnarward. The perforators are identified at the ulnar border of the brachioradialis along the intermuscular septum. The brachioradialis is retracted radially to reveal the radial artery. The radial artery is cross-clamped distally and, if adequate arterial filling of the hand is demonstrated, the radial artery is ligated and transected. The flap is elevated with dissection of the radial artery and its venae comitantes. The cephalic vein can be dissected further proximally to provide an additional long venous pedicle. The radial forearm flap is then ready for transfer to the intended defect.

Postoperative Considerations

Donor Site Closure

Part of the donor site can be closed in a primary fashion. The remainder of the defect is covered with a split-thickness skin graft. The donor site is dressed according to preference and the hand is splinted for 5-7 days to ensure graft take.

Complications

Complications are not very frequent at the donor site, and harvest of this flap is generally well-tolerated. The following complications can occur:

- Infection
- Partial or complete skin graft loss
- Cold intolerance
- Hand claudication and other neurosensory changes
- Poor cosmetic appearance of the arm
- Radius fractures (osteocutaneous flap)

Pearls and Pitfalls

- Peritenon covering the flexor carpi radialis, brachioradialis, FDP and FDS tendons should be preserved to minimize skin graft loss and tendon desiccation.
- If the hand is found to be ischemic during cross-clamping of the radial artery, a venous interposition graft may rarely be needed to maintain adequate perfusion.
- Care should be taken to preserve the sensory branch of the radial nerve during dissection of the radial artery.
- It is important to include the intermuscular septum to preserve the septocutaneous perforators. The more distal the skin island, the longer the vascular pedicle.

Suggested Reading

1. Abu-Omar Y, Mussa S, Anastasiadis K et al. Duplex ultrasonography predicts safety of radial artery harvest in the presence of an abnormal Allen test. Ann Thorac Surg 2004; 77:116.
2. Bardsley AF, Soutar DS, Elliot D et al. Reducing morbidity in the radial forearm flap donor site. Plast Reconstr Surg 1990; 86(2):287.
3. Evans HB. The radial forearm flap. In: Buncke HJ, ed. Microsurgery: Tranplantation, Replantation: An Atlas Text. 4th ed. Philadelphia: Lea and Febiger, 1991: Chapter 14.
4. Evans GRD et al. The radial forearm free flap for head and neck reconstruction: A review. Am J Surg 1994; 168:446.
5. Villaret DB, Futran NA. The indications and outcomes in the use of osteocutaneous radial forearm free flap. Head Neck 2003; 25(6):475.
6. Yang G, Chen B, Gad Y. Forearm free skin flap transplantation. Nat Med J China 1981; 61:139.

Appendix II

Surgical Instruments

Zol B. Kryger and Mark Sisco

Introduction

The purpose of this Appendix is to provide illustrations of the basic surgical instruments used in plastic surgery. It is aimed towards the medical students, junior residents, and non-plastic surgeons who are largely unfamiliar with the names of the instruments and the many subtle differences. These images were reprinted with generous permission from Teleflex Medical.

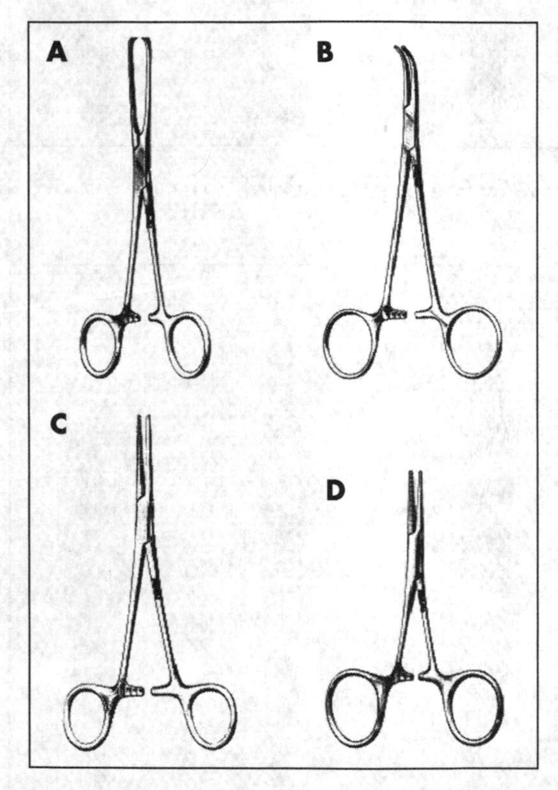

Figure AII.1. Allis tissue forceps (A), Mixter forceps (B), Kelly forceps (C), Hartman mosquito forceps (D).

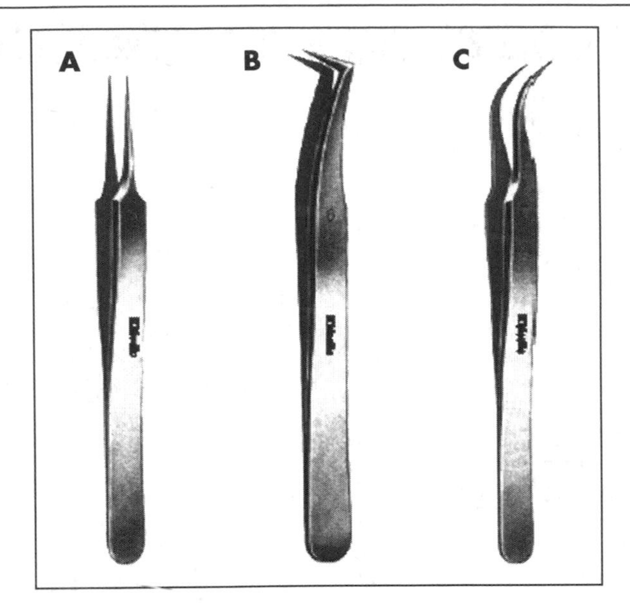

Figure AII.2. Jeweler forceps (straight, angled, curved).

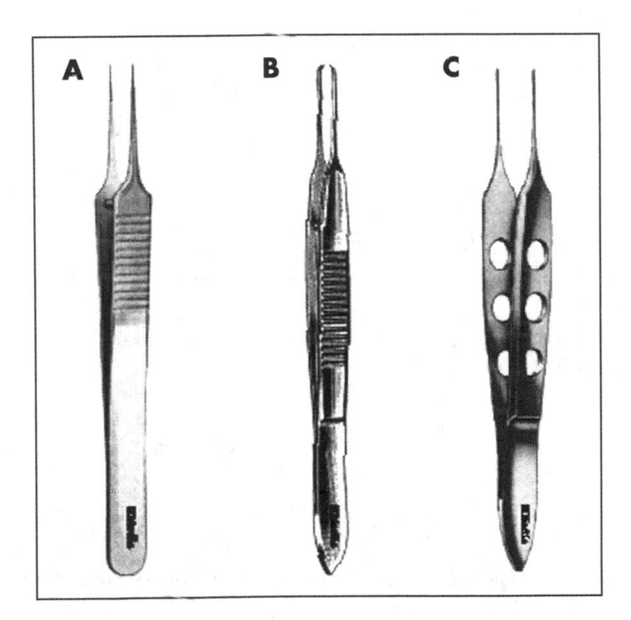

Figure AII.3. Micro tying forceps (A), Castroviejo micro forceps (B), Bishop Harmon forceps (C).

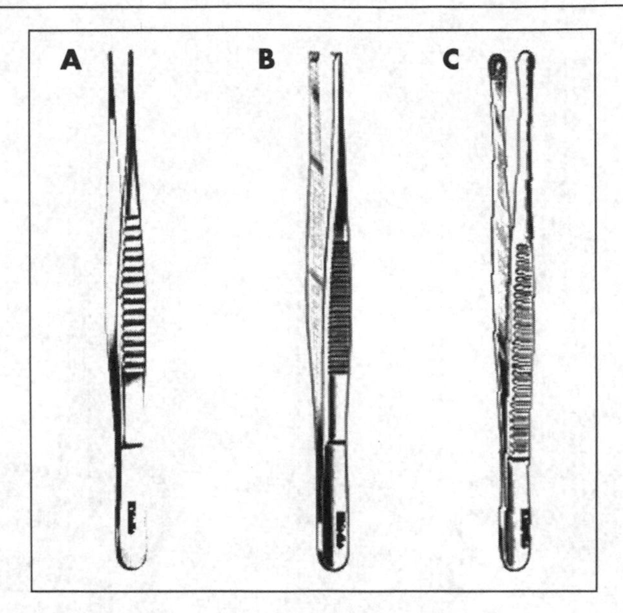

Figure AII.4. Debakey atraumatic forceps (A), Tissue forceps (B), Russian forceps (C).

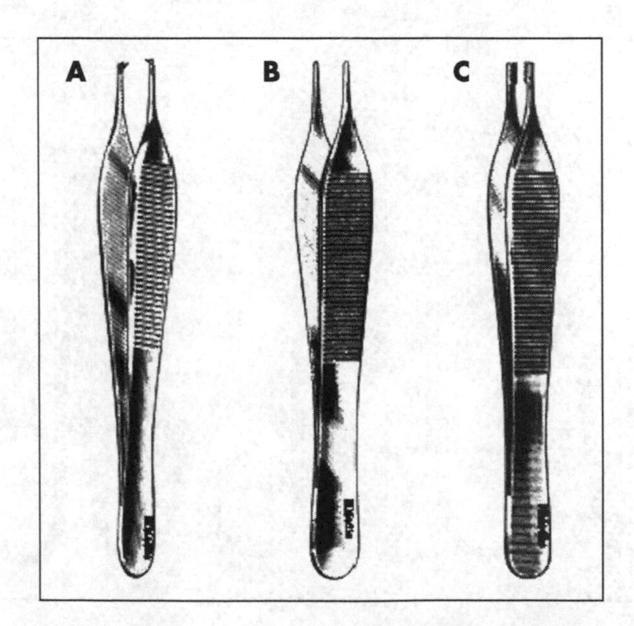

Figure AII.5. Adson tissue forceps or "Adson with teeth" (A), Adson dressing forceps or "smooth Adson forceps" (B), Adson Brown forceps (C).

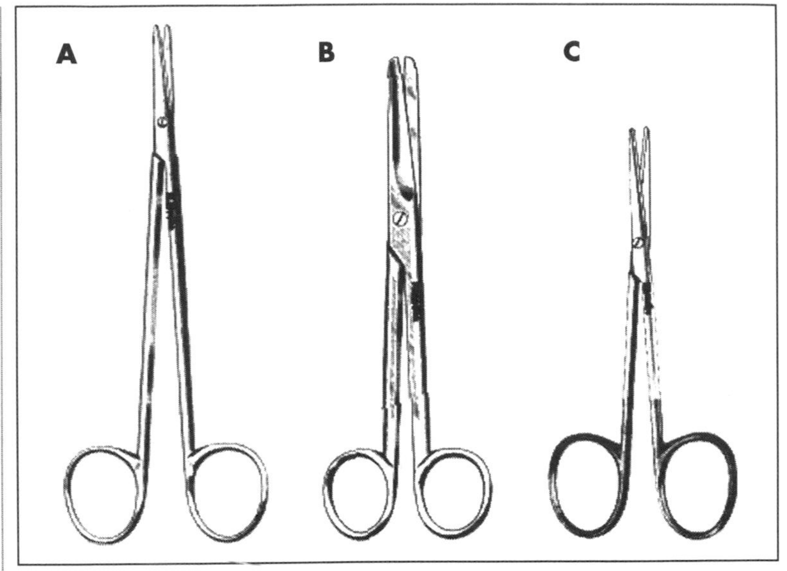

Figure AII.6. Metzenbaum scissors (A), Mayo scissors (B), Strabismus scissors (C).

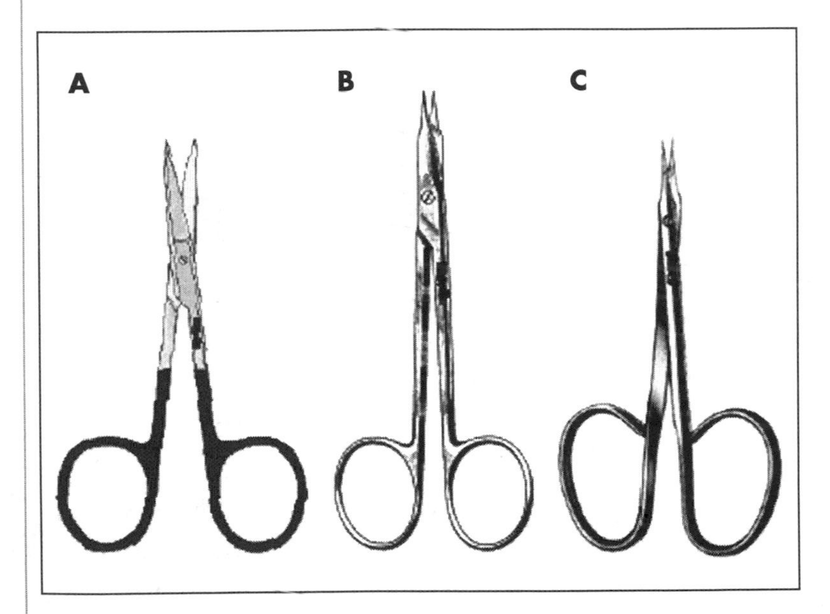

Figure AII.7. Iris scissors (A), Stevens tenotomy scissors (B), Gradle scissors (C).

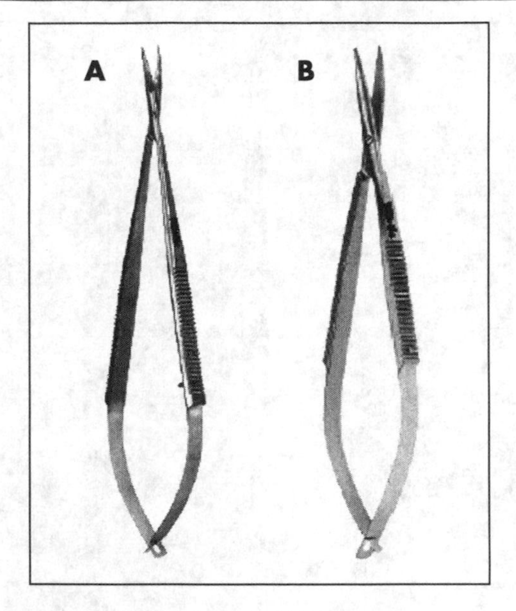

Figure AII.8. Micro scissors (A), tenotomy scissors (B).

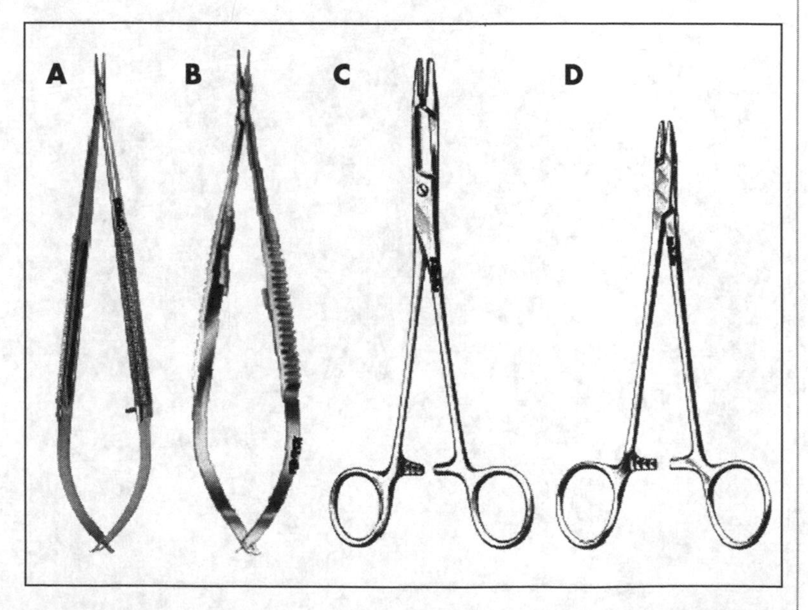

Figure AII.9. Barraquer needle holders (A), Castroviejo micro needle holder (B), Olsen-Hager needle holders (C), Webster/Brown needle holders (D).

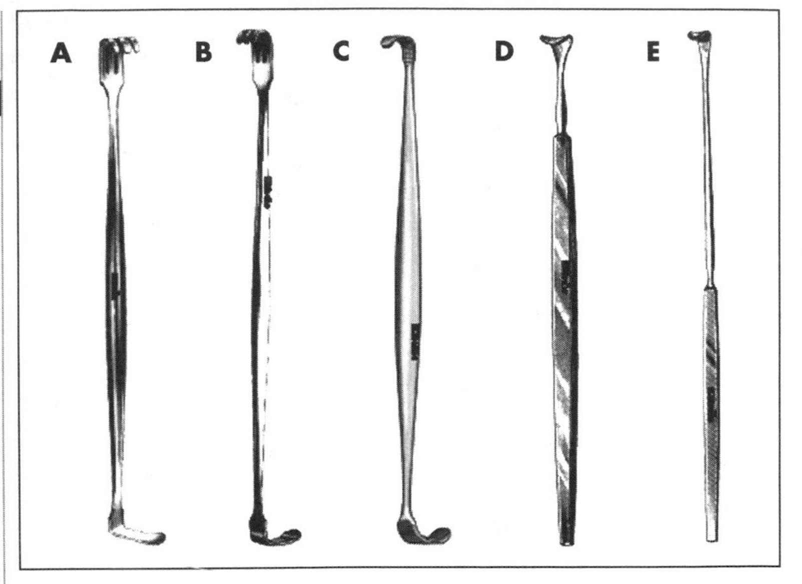

Figure AII.10. Mathieu retractor (A), Senn retractor (B), Ragnell-Linde retractor (C), Cushing vein retractor (D), Love nerve retractor (E).

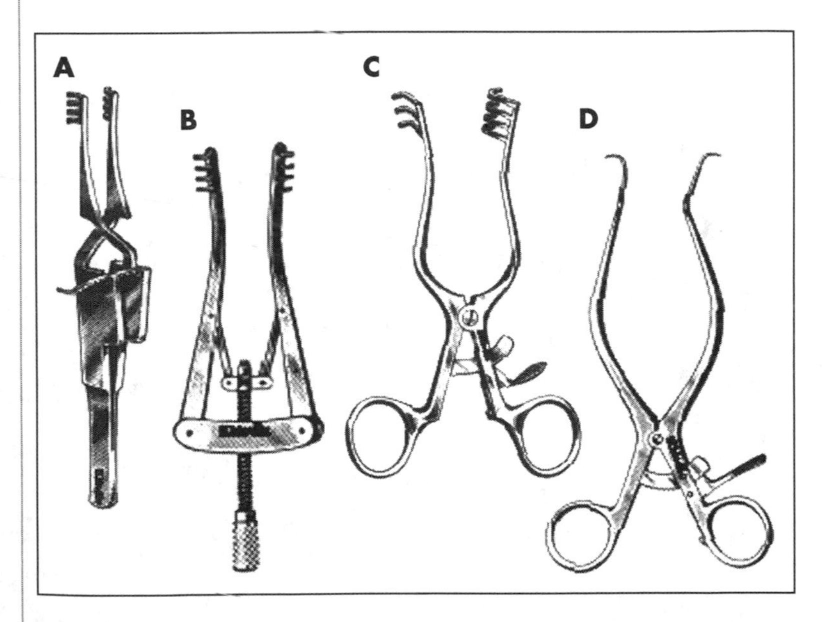

Figure AII.11. Self-retaining retractor (A), Alm retractor (B), Weitlaner retractor (C), Gelpi retractor (D).

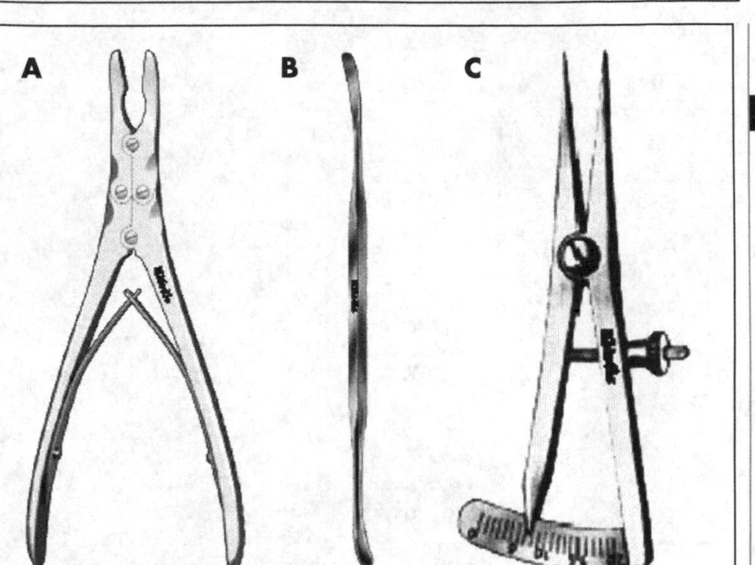

Figure AII.12. Ruskin/Beyer rongeur (A), Periosteal elevator (B), Castroviejo calipers (C).

Index

A

Abscess, collar-button 464
Abdomen 25, 63, 134, 150, 260, 264, 266, 311, 312, 317, 423, 424, 426, 550, 608-611, 615, 617, 618
  open 25
Abdominal compartment syndrome (ACS) 159
Abdominal wall defect 97, 98, 267, 309
Abdominoplasty 38, 262, 269, 423-429
  mini-abdominoplasty 423
Abductor pollicis longus (APL) 465, 466, 479, 533, 586, 587
Acanthosis 122
Acellular dermal matrix 311
Acne 119, 336, 375, 436, 440, 445, 446, 455, 456
Actinic keratosis 121, 126, 436, 440
Actinic lentigine 368, *see also* Solar lentigine
Adaptec™ 15, 17
Adenoma, sebaceous 124, 436
Adenoma, benign pleomorphic 166
Adhesion 6, 26, 297, 311
Aging 46, 368, 377, 386, 387, 404, 413, 417, 446, 447
Alar cartilage 351, 353, 354, 394, 398
Alginate 16, 18, 19, 84
Allen test 478, 548, 563, 589, 622
Allevyn™ 18
Alloderm 59, 98, 151, 153, 221, 262, 311, 342
Allograft 25, 146, 151, 212, 213
Alloplastic framework 345, 346
Alpha-hydroxy acid 441, 443, 448
Alveolar closure 350
Amputation 43, 80, 82-84, 111, 123, 136, 289, 318, 319, 322, 485, 492, 493-496, 531, 537, 548, 549, 561-563, 595
  revision 549
Anastomosis 27, 57-60, 85, 97, 98, 109, 111, 268-272, 297, 464, 466, 494, 530, 611
  technique
    end-to-end 57, 58, 543, 550
    end-to-side 57, 58
Anesthesia 29, 31-36, 38-41, 74, 86, 128, 210, 245, 246, 279, 280, 284, 320, 346, 352, 360, 378, 380, 381, 383, 385, 387, 392, 396, 403, 404, 426, 431, 434, 436, 438, 443, 447, 451, 454, 455, 457, 459, 505, 550, 555, 570, 577, 578, 582, 594
  agent 29, 32, 33
  axillary block 33, 495
  local 26, 27, 29-33, 38, 39, 41, 42, 128, 133, 210, 245, 284, 378, 381, 383, 387, 390, 392, 396, 403, 404, 426, 431, 436, 447, 454, 455, 460, 480, 550, 570, 577, 578, 582
    eutectic mixture of local anesthetics (EMLA) 31
  nerve block 33-36, 383, 436, 443, 455, 603-605
    infraorbital 34, 443
    median 35
    mental 34, 443
    radial 35
    digital (ring) 36
    supraorbital 34, 378, 443
  regional block 33, 443, 604
    intravenous 38, 604
  ring block 459, 460
  stellate ganglion block 602, 604
  tumescent 39
Angiofibroma 123
Angiogenesis 2, 93
Angiosome 51
Ankle-brachial index (ABI) 82, 88, 90, 529
Annular band syndrome 595
Annular pulley *see* Tendon, pulley

Annulus of Zinn 231
Anterior horizontal mandibular osteotomy (AHMO) 400-402, 405
Anterior lamella 178, 388
Anterior nasal spine (ANS) 243, 329, 330, 351, 353
Anterior open bite 252, 256, 325, 332
Antia-Buch procedure 170
Antibiotic 4, 8, 11, 20, 23, 24, 28, 65, 66, 70, 73-79, 84, 92, 93, 97, 101, 109, 111, 112, 114-116, 122, 143, 147, 159, 169, 191, 226-229, 246, 250, 255, 280, 284, 286, 293, 295, 301, 305, 308, 355, 373, 381, 385, 410, 411, 426, 428, 429, 431, 436, 455
  prophylaxis 77-79, 226
  spacer 116
Antiemetic 39, 41, 372, 375, 455
Antihelix 373, 382-385
Antitragus 382
Apert syndrome (acrocephalosyndactyly) 336-338, 592
Apligraf 92, 93, 151, 152
Apocrine cystadenoma 124
Apocrine hair follicle 2, 123, 124, 145, 150
Aponeurosis 178, 180, 231, 263, 296, 312, 360, 377, 387, 465, 539, 559, 578, 589, 590, 598
Aquaphor™ 15, 17, 103, 456
Arcuate expanse 388
Arcuate line 263, 265
Arcus marginalis 379, 380, 388
Arm 55, 75, 113, 124, 157, 185, 275, 279, 288, 323, 355, 366, 409, 426, 466, 473, 486, 488, 490, 492, 499, 500, 502, 505, 520, 523, 525, 526, 530-532, 550, 567, 570, 571, 574, 589-591, 594, 602, 615, 619, 624
Arteriography 82
Arteriovenous malformation (AVM) 139, 143, 144
Arteries
  anterior tibial 318
  brachial 623
  dorsal nasal 396
  inferior epigastric 268, 272, 310, 311, 317, 424, 611
  insufficiency 13, 28, 60, 87, 88, 93, 147
  intercostal 424
  internal mammary 294, 297, 406, 611
  labial 194, 203, 206-208, 396
  occlusive disease 89, 90
  ophthalmic 225, 387
  popliteal 80, 318
  posterior tibial 318, 609
  radial 113, 214, 464, 465, 500, 502, 506, 529, 531, 563, 564, 584, 589-591, 619, 622-624
    harvest 589-591
  superficial inferior epigastric 272, 310
  superficial temporal 63, 347, 377, 378, 382
  superior epigastric 263, 264, 267, 269, 270, 296, 311, 424, 611
  superior labial 194, 206, 396
  supratrochlear 191, 377
  thoracodorsal 275, 276, 297, 609, 615-617
Arterial pulse indices 529
Arthritis 389
Articulare 330
ASA class 76, 430
Audiologist 351, 360
Augmentation 38, 51, 66-69, 260, 271, 341, 398-400, 403, 404, 406-408, 411, 417-419, 421, 422, 431, 446-450, 510
Auricle 69, 168, 169, 190, 234, 243, 340, 344-347, 373, 382-385
  cartilage graft 190, 234
  deformity 168, 169
  framework 69, 243, 344-347
  prosthesis 346
Avulsion 29, 242, 256, 472, 474, 483, 492-494, 512, 514, 520, 521, 537, 544, 552, 553, 556-590
Axon 29, 220, 221, 224, 521, 523, 524, 527, 528
Axontmesis 523, 524

B

Bacitracin 20, 21, 160
Baker's classification 411
Basal cell carcinoma (BCC) *see* Cancer
Basal cell nevus syndrome 128
Basion 329
Beclapermin 23
Bell's palsy 217
Benelli 419, 420, 422
Berkow formula 156, 157
Bernard-Burow's technique 199, 202, 208, 209
Betadine 21, 23, 458
Betel leaf 165
Bicarbonate 30, 33, 39, 437, 444
Bicoronal synostosis 336, 337
Bilateral sagittal split 332, 341, 400
Biobrane 149, 151, 152
Biofilm 73
Biomechanics 226, 508, 540, 544
BioPlastique 66-68
Biopsy 74, 83, 90, 92, 101, 110, 121, 125, 130, 132-134, 143, 164, 259, 410, 414, 450
Bites
  human 8, 169, 481-484, 538
  cat 482, 483
  dog 482, 483
B-K mole syndrome 122
Blepharoplasty 38, 44, 374, 337, 381, 386-392, 436, *see also* Eyelid
  lower eyelid 386, 388-393
  negative vector 390
  pseudocoloboma 337
  subciliary 390
  transconjunctival 391, 392
  upper eyelid 386-390, 392, 393, 520
Bone 322, 323, 335, 336, 338, 341, 343, 346, 350, 355, 358, 359, 362, 363, 366, 367, 373, 377, 378, 380, 381, 394, 395, 397, 399, 400, 402-404, 448, 462-467, 469, 470, 472, 473, 479, 482, 492, 493, 495-497, 500-508, 510, 512-517, 526, 529, 534-536, 539, 543-545, 547-550, 552, 554-556, 559, 561-564, 568, 574, 576, 578, 590, 592, 593, 601, 602, 613
  crib 213
  fixation 493, 514
    nonvascularized 212, 213, 506
  palatine 231, 359
  pisiform 462-464, 478, 503, 508, 510, 526, 564, 568
  scan 83, 92, 295, 503, 507, 602
    triphasic 602
Botulinum toxin 451-453
Bowen's disease 126, 314, 436
Brachial plexus 300, 505, 518-522, 575, 576
Brachioradialis 477, 526, 531, 563, 590, 623, 624
Brachydactyly 337
Branchial arch 340, 348
Breast 7, 38, 39, 43, 47, 61, 63, 66, 67, 79, 258-263, 266, 267, 269-272, 274-280, 283-285, 288-293, 300, 406-422, 608, 609, 611, 615, 620
  anatomy 263, 274, 275, 284, 406
  asymmetry 421
  augmentation 38, 67, 260, 406-411, 417, 418, 421
    periareolar incision 260, 409, 416, 419
    subpectoral pocket 279, 420
  cancer 258-260, 262, 272, 278, 406, 407, 414, 415, 417
    ductal carcinoma in situ (DCIS) 258, 259, 261
    lobular carcinoma in situ (LCIS) 258
    liposuction-assisted mastectomy 415
    lumpectomy 259, 261
    male 415
    radiation 258-261, 278, 281, 406
    staging 258
    treatment 258, 259, 262
  capsule 261, 262, 410-412
  capsulectomy 66, 262, 280, 408, 411
  capsulorraphy 280
  feeding 288, 354, 357, 408, 421
  glandular ptosis 418
  implants 66, 280, 406-408, 412, 417

Breast 7, 38, 39, 43, 47, 61, 63, 66, 67, 79, 258-263, 266, 267, 269-272, 274-280, 283-285, 288-293, 300, 406-422, 608, 609, 611, 615, 620
saline 39, 66, 67, 278, 280, 281, 407, 411, 412
silicone 66-68, 278, 281, 412
inframammary crease 275, 276, 407, 418, 420
mammaplasty 288-293, 406-412, 422
mastopexy 38, 260, 271, 406, 409, 412, 417-422
vertical 419
nipple asymetry 421
nipple-areolar complex (NAC) 63, 260, 283, 288-293, 406, 409, 412, 415, 417-422
nipple-areola reconstruction (NAR) 271, 283-287
nipple-areolar necrosis 286, 288, 421
periareolar incision 260, 409, 416, 419
pseudo-ptosis 419
ptosis 260, 278, 279, 289, 290, 406, 410, 417-422
grade 418-420
reconstruction 47, 61, 66, 259-261, 263, 266, 267, 269, 271, 272, 274, 275, 278, 283, 284, 300, 620
delayed 260-262, 269, 275, 278
drains 270, 271, 276, 277, 280, 281, 292, 293, 301, 415, 416
immediate 260, 261, 275
implant exchange 275, 280, 281
seroma 64, 271, 276, 277, 297, 410, 411, 415, 416, 421
reduction 38, 271, 287, 288-293, 413-416
circumareolar 289, 418-420
free nipple graft See Grafts, free nipple
inferior pedicle 288-291, 293, 415, 421, 611
medial pedicle 288-290, 292, 293, 421
patterns 288-291, 293
vertical scar 288-290, 292, 418-420
Wise pattern 288-291, 293, 419, 420
Browlift 229, 377-381
coronal 378-380
direct 379
endoscopic 229, 380
Buddy splinting 513
Buddy taping 554, 558
Buerger's disease (thromboangiitis obliterans) 89
Bulla 91, 122
Bunnell technique 535, 543
Bupivicaine 29-33, 133, 437, 604
Burn 15-17, 20, 22, 25, 43, 44, 61, 64, 87, 100, 117, 118, 129, 145, 147, 151-162, 168, 169, 171, 174, 198, 283, 431, 441, 442, 445, 487, 536, 547
excision 61
rule of nine 156, 157
superficial 155
total body surface area (TBSA) 154-157
Burow's triangle 52, 208, 209, 564

C

Camper's chiasma 465, 539, 543, 546
Camptodactyly 592, 595
Cancer 13, 16, 95, 96, 121, 126, 127, 130, 131, 134, 135, 138, 163-165, 167, 168, 188, 198, 214, 258-262, 270-272, 278, 283, 314, 346, 404, 406, 407, 414, 415, 417, 418, 481
breast *see* Breast, cancer
carcinoma 121, 126-130, 164-167, 180, 188, 198, 258, 414, 415, 436, 483
basal cell carcinoma (BCC) 126-129, 180, 188, 198, 436
nodular 127
squamous cell (SCC) 121, 126-130, 164, 180, 436
endophytic 164
superficial 436

skin 96, 126, 127, 130, 131, 168, 188, 314
staging 126, 127, 134, 135, 164, 258
treatment 258, 259, 283, 417
Canthopexy, lateral 391-393
Canthus 182, 186, 235, 372, 377, 381, 386, 391
cantholysis 179, 182, 185, 186
canthotomy 179, 182, 185, 186, 234, 236, 240, 392
lateral 178, 231, 235, 372, 377, 381, 386, 389, 391
ligament 182
tendon 178, 186, 231, 232, 242, 245, 389
Capillary malformation 140, 143
lymphatic 138, 143
simple (port wine stain) 139, 140, 143, 434
venous 139, 140, 143
Capsulodesis 510, 557
Capsulotomy 280, 408, 411, 514
Carcinoma *see* Cancer
Carpal tunnel
release 464, 489, 568, 570, 571, 572, 601
endoscopic (ECTR) 571, 572
palmar cutaneous branch 35, 36, 466, 491, 500, 566, 570, 572
palmar sensory cutaneous branch (PSCB) 464, 568
syndrome (CTS) 464, 479, 499, 510, 566-573
Carpenter syndrome 337, 338
Carpometacarpal (CMC) joint 462, 463, 470, 479, 515-517, 554, 587, 593
Cartilage 7, 44, 97, 121, 127, 128, 147, 160, 163, 169-175, 177, 183-185, 188, 190-192, 198, 214, 216, 234, 243, 246, 302, 341, 344-346, 348, 350-355, 382-385, 394-399, 502, 599
costal 175, 190, 234, 263, 264, 296, 302, 344-346
framework 344-346
Cast, thumb spica 503, 504, 508, 510, 517
Cellulitis 70, 72, 83, 143, 294, 410, 482, 483
Central slip 466, 533, 537, 538, 557, 558
Centric relation 328, 333, 401
Cephalogram 328-330, 332, 340, 367
Cephalometric analysis 327-329, 332, 333, 400, 401
Cervical spine injury 245, 603
Charcot foot 81, 91
Chemical peel 435, 436, 440-445
Baker-Gordon 441, 444
deep 440-445
glycolic acid 440, 441, 443-445
medium-depth peel 440, 443-445
phenol peel 441, 444, 445
salicylic acid peel 440-445
superficial 440, 442-445
trichloroacetic acid (TCA) 441-444
Chemotherapy 13, 135, 165, 259, 261, 281, 317, 481
Chest wall defect 299-303
full-thickness 300
Chin augmentation 403
Chondritis 169
Chondroid syringoma 124
Chromophore 433-435
Clark's level 131, 134
Cleft lip 43, 44, 128, 340, 343, 348-355, 356, 357, 362, 404
Millard 43, 44, 47, 188, 348, 352, 354
unilateral 349, 352
Cleft palate 337, 338, 348, 349, 352, 356-367, 404
fistula 366
lengthening 362
V-Y pushback 362
plane 329, 330
primary palate 350, 356, 359
repair 352, 360, 361, 366, 367, *see also* Flaps, mucoperichondrial
Clonidine 39, 41, 372
Closed reduction 241, 245, 251-254, 256

Closure 3, 4, 6, 8, 9, 11, 12, 14, 16, 22, 24, 46, 52, 59, 60, 62, 63, 70-72, 75, 78, 84-86, 92, 97, 104, 107, 114-116, 129, 133, 144, 147, 150-153, 167, 169, 170, 176, 178, 179, 181, 182, 185-187, 190, 196, 199, 201, 203, 206, 208-210, 257, 263, 269, 270, 275, 280, 284-286, 292, 293, 295-300, 303-305, 307, 309-311
Cocaine 29, 31, 32, 245
Collagen 2, 3, 16, 20, 22, 62, 84, 96, 117, 119, 145, 151-153, 220, 368, 389, 412, 421, 435, 437, 438, 444, 445, 448-450, 456, 539, 544, 597
  type I 539
Collagenase (Novuxol®) 22, 103, 599
Collateral ligament 463, 474, 478, 512, 514, 515, 517, 537, 555, 557-559, 577, 578, 592, 596
Colonized wound 72
Columella 128, 188, 189, 350-354, 394, 396, 397
Comminution 245, 246, 471, 505
Commissure reconstruction 198, 593
Comorbid condition 9, 74, 76, 102, 140, 299, 492
Compartment, dorsal 465, 479, 489, 491, 533, 586, 587
Compartment pressure 319
Compartment syndrome 158, 159, 318-320, 463, 486-491, 531
  abdominal (ACS) 159
Complex regional pain syndrome (CRPS) 601
Complication 38, 45, 47, 56, 60, 61, 63, 64, 68-70, 76, 79-81, 83, 97, 101, 104, 109, 113, 114, 118, 134, 142, 149, 166, 169, 187, 203, 206, 209, 215, 225, 227, 228, 236, 238, 242, 253, 255, 260-262, 267, 269, 271, 272, 274, 277, 286, 287, 294, 297, 299, 304, 309, 312, 313, 336, 346, 366, 372, 375, 376, 380, 381, 385, 391, 392, 396, 398, 399, 409-411, 415, 416, 421, 423-425, 429-432, 436, 440, 444, 445, 448, 455, 456, 461, 485, 490, 495, 502, 511, 517, 531, 532, 545, 551, 553, 555, 559, 572, 577, 587, 590, 591, 594, 600, 624
Component separation 310, 312
Composite 49, 51, 69, 170, 174-176, 181, 183, 190, 191, 198, 212, 213, 255, 283, 301, 374, 391, 398, 493, 495, 536, 549, 612, 622
Computed tomography (CT) 83, 121, 134, 140, 218, 226, 229, 233, 237, 238, 240, 241, 245, 249, 250, 256, 344, 404, 503, 507, 508, 511, 513, 515, 520, 566, 567, 571
  angiography (CTA) 530
Concha cavum 382
Concha cymba 382
Condyle 107, 158, 211, 221, 248, 250, 252, 253, 256, 328, 330, 342, 515, 595, 618
Condylion 330
Congestion 27, 28, 59, 60, 87
Conjunctiva 178, 182-185, 187, 232, 235, 240, 246, 387, 388, 391
Constriction ring syndrome (CRS) 592, 595
Contracture 9, 10, 43, 75, 150, 155, 181, 185, 189, 191, 196, 198, 261, 262, 266, 277, 355, 358, 361, 407, 409-412, 417, 421, 476, 490, 515, 519, 521, 536, 537, 538, 540, 542, 545, 555, 577, 594, 595, 597-600, 603
Contracture release 521, 600
Contraindications 17, 25, 30, 259, 278, 283, 295, 360, 406, 417, 436, 440-442, 455, 492, 495, 542, 562, 563, 589, 599, 611, 622
Corticosteroid 9, 77, 118, 119, 142, 437, 581, 587, 599, 604
  injection 118, 119, 581, 587, 599
CosmoDerm 449
CosmoPlast 449
Craniofacial syndrome 324, 334, 336-338
Craniosynostosis 334, 335, 337, 338

Creatinine phosphokinase (CPK) 489
Crossbite 256, 325
Crouzon syndrome (craniofacial dysostosis) 337, 338
Cruciate pulley 465, 539, 543, 581
Cryotherapy 118, 119, 124, 128, 129
Cubital tunnel 526, 574, 575, 577, 578
Cup ear 383
Cupid's bow 193, 350-354
Curasol™ 15, 17
Curettage 126, 128, 129, 142
Cutinova Hydro™ 18
Cutis marmorata telangiectatic congenita malformation 139
Cutting 7, 42, 149, 215, 352, 384, 571, 582
Cylindroma 124
Cyst 121, 122, 166, 414, 447, 461, 504, 583-585

D

Dacrocystorhinostomy 231
Dacron® 7, 66-68
de Quervain's disease 465, 479, 586, 587
Debridement 3, 4, 13-19, 22, 23, 25, 74, 75, 83-85, 89, 90, 92, 93, 97, 98, 103-105, 111, 114-116, 147, 151, 155, 159, 162, 169, 294, 295, 298-300, 303, 305, 308, 320, 322, 429, 461, 482, 483, 494, 506, 510, 532, 542, 552, 556, 558, 585
Debriding agent 23, 74, 84, 103, 104
Deep temporal 216, 217
Deep vein thrombosis (DVT) 38, 271, 425, 429, 432
Deformational plagiocephaly 334, 335, 338
Deformities 43, 81, 83, 91, 141, 148, 168, 169, 172, 173, 176, 189, 190, 196, 198, 227, 228, 232, 233, 237, 238, 242, 243, 245, 251, 278, 296, 300, 327-330, 332, 336, 342, 348, 350-354, 357, 358, 361, 373, 383-386, 397-399, 401, 414, 419, 420, 422, 430, 447
- auricle 168, 169
- boutonniere 477, 535, 537, 540, 558, 559
- buffalo hump 430
- claw hand 467, 477, 525
- ear 168, 383
  - constricted 383
  - microtia 340-347
    - classification 343
  - prominent ear 382-384
- inverted-V 397, 399
- mallet 477, 535, 537, 538, 540, 556, 559, 560
- nasal *see* Nasal, deformities
- open roof 399
- parrot beak 398
- pseudo mallet 477, 540
- swan-neck 475, 537, 559, 560

Dental injuries 97, 214, 242, 253, 255, 256
Dentoalveolar 255, 256
Dermabrasion 123, 436, 440, 454-457
Dermagraft 151, 152
Dermalon 7
Dermaplaning 454
Dermasanding 454
Dermatofibromas 123
Dermatofibrosarcoma protuberan 123, 130
Dermatoheliosis 435
Dermatome 140, 148, 454
Dermis 2, 6-8, 61, 71, 75, 96, 102, 122, 123, 134, 145-148, 150, 153, 155, 216, 222, 270, 285, 287, 291, 351, 368, 371, 378, 416, 418, 434, 435, 437, 438, 441, 444, 448-450, 454, 456, 457, 464, 562, 599
Dermoid cyst 121, 414
Dexon 7, 66, 311
Diabetes 1, 13, 20, 23, 39, 75, 77, 80, 81-90, 93, 102, 115, 151, 255, 318, 372, 389, 421, 567, 574, 576, 597
- Doppler 82, 88, 90, 105
- foot ulcer 20, 23, 75, 80, 81, 83-86, 89, 93, 151

Diagnostic imaging 530
Diagnostic procedure 351, 360, 415

Diazepam 32, 39, 40, 357, 414, 443, 455
Differential diagnosis 141, 143, 481, 576, 587, 595
Dihydrotestosterone (DHT) 458
Diplopia 236, 238, 392
Dislocation 252, 463, 473, 479, 499, 502, 508-511, 517, 525, 529, 537, 555-560, 567, 575
Distraction osteogenesis (DO) 214, 338, 341
Dog ear 9, 191, 285
Doppler 60, 82, 88, 90, 97, 105, 112, 158, 270, 271, 277, 460, 495, 530
  ultrasonography 60
  implantable internal 60
Dorsal carpal arch 466
Dorsal intercalated segmental instability (DISI) *see* Fracture, wrist
Dorsal interossei 463, 478, 534
Drains 77-79, 101, 106-109, 111, 112, 136, 149, 196, 228, 270, 271, 276, 277, 280, 281, 292, 293, 297, 301, 307, 312, 375, 384, 415, 416, 428, 429, 483, 590, 607, 621
Dressings 3, 8, 9, 11-19, 22-26, 31, 47, 74, 75, 84, 90, 92, 93, 103, 104, 110, 111, 118, 133, 142, 146, 147, 149, 151-153, 162, 180, 250, 286, 287, 292-294, 310, 311, 346, 384, 385, 403, 416, 421, 435, 437, 456, 461, 487, 488, 490, 495, 501, 545, 549, 550, 553, 560, 571-573, 590, 593, 627
  alginate 16, 18, 19, 84
  compression 9, 90, 92, 572
  nonadherent 15, 17, 26, 384
  vaseline-impregnated gauze 17
  wet-to-dry 13, 14, 22, 74, 79, 84, 103, 111, 147
  wet-to-moist 13, 14, 103
Ductal carcinoma in situ (DCIS) 258, 259, 261
DuoDERM™ 15, 18, 149
Duplex ultrasound 60, 589
Dupuytren's contracture 597, 600
Dupuytren's disease 463, 464, 476, 534, 565, 597-600, 602
Dynamic splinting 536, 599

E

Ear 9, 27, 35, 62, 63, 67, 69, 128, 129, 136, 141, 160, 168-175, 177, 181, 183, 185, 191, 219, 251, 285
  anatomy 382
  otohematoma 169
  pinna 382, 383
Earlobe cleft 176
Earlobe reconstruction 170, 176, 177
Eccrine poroma 124
Ectropion 181, 183, 187, 235, 236, 386, 390-393, 444, 453
Elastogel™ 15, 17
Elbow flexion test 575
Electrodessication 126, 128
Electrodiagnostic study 575
Electrodiagnostic testing 569, 573
Electromyography (EMG) 524, 569, 570
Embolic phenomena 88
Embolization 142, 144
Embryology 348, 356
Endoneurium 523, 524
Endoscopic evaluation 227
Endoscopic approach 229, 235, 253, 585
Endoscopy-assisted ORIF 253
Endotenon 539
Endothelial cell 2, 12, 57, 72, 138, 140, 142
Enophthalmos 232-234, 237, 238, 242
Entrapment 232, 233, 238, 479, 499, 574
Entropion 236
Enzymatic debriding agent 23, 103
Epicel 152, 153
Epidermal appendage 123, 124, 145, 147
Epidermodysplasia verruciformis 126
Epidermoid cyst 121, 122

Epinephrine 26, 29-33, 36, 39, 133, 148, 246, 353, 366, 381, 396, 403, 431, 432, 436, 437, 456, 459
Epinephrine toxicity 31, 432
Epineurium 219, 523, 527, 578
Epistaxis 226, 396
Epitenon 539, 546
Epithelial cyst 121
Epithelialization 2-4, 8, 12, 15, 20, 22, 23, 149, 152, 155, 160, 285, 436, 437, 442, 456
Eponychium 466, 515, 552
Erb's palsy 520
Erythroplakia 164
Erythroplasia of Queyrat 126, 436
Escharotomy 158, 159
Ethibond 7
Ethilon 7
Ethmoid bone 243, 244
Exchange 409
Excision 148, 335, 378, 384, 390, 392, 398, 415, 419, 420, 423, 425, 483, 506-508, 511, 577, 585
Excisional biopsy 121, 125, 130, 132, 133
Exfoliation 443
Exophthalmos 233, 236, 327
Extensor
  carpi radialis brevis (ECRB) 465, 477, 526, 533, 586, 587
  carpi radialis longus (ECRL) 465, 477, 526, 533, 586, 587
  carpi ulnaris (ECU) 465, 533, 586, 587
  digiti minimi (EDM) 465, 466, 533, 536, 586
  digitorum communis (EDC) 465, 466, 477, 533, 536, 564, 586
  indices proprius (EIP) 465, 466, 533, 536, 586, 587
  pollicis brevis (EPB) 465, 466, 477, 479, 533, 586, 587
  pollicis longus (EPL) 465, 466, 477, 479, 533, 536, 538, 586, 587
  retinaculum 465, 533, 544, 586
Extrinsic healing 545
Extrinsic musculature 533
Eyelid 7, 34, 46, 63, 128, 129, 136, 145, 146, 178-187, 221, 231, 235, 238, 245, 327, 337, 368, 386, 387-392, 404, 444, 447, 448, 452, 453, 520, *see also* Blepharoplasty
  anatomy 178, 180, 387, 388
  Asian 389
  lesion 136, 180
  lower 34, 128, 178-181, 187, 182-186, 221, 231, 235, 238, 246, 327, 372, 386, 388-393, 404, 447, 448, 450
    reconstruction 181, 183
  upper eyelid 34, 178-183, 185-187, 386-390, 392, 393, 520
    reconstruction 183, 185
  posterior lamella 178, 183, 388
  preseptal suborgicularis oculi fat (SOOF) 386, 388
  ptosis 452, 520
  reconstruction 178-181, 183, 185-187
  tumor 178, 180

F

Facelift *see* Rhytidectomy
Facial
  aging 46, 368, 377, 404, 446, 447
  anatomy 368, 377, 378, 400
    muscles 195, 218
  asymmetry 400
  expression 8, 193, 217-219, 221, 351, 368, 369, 377
  growth abnormalities 358
  nerves *see* Nerves, facial
  paralysis 217, 224
    reanimation 217
  rejuvenation 435, 438, 440-445
  relationship 327
Fascia 49-51, 62, 69, 83, 98, 105, 107-109, 115, 123, 133, 147, 148, 153, 155, 165, 216, 217, 221, 231, 237, 238, 261-263,

269, 270, 272, 274-276, 279, 280, 291-293, 297, 301, 307, 310-312, 318-322, 345, 347, 368, 369, 371, 373, 374, 376, 378-380, 384, 388, 406, 415, 418, 423, 424, 426-430, 464, 486, 489, 525, 526, 531, 536, 544, 563, 564, 570, 571, 575, 578, 590, 597-599, 607-609, 611-613, 615, 619-621, 623
Fascial compartment 319
Fasciectomy 599, 600
Fasciotomy 4, 135, 158, 159, 319, 488-491, 495, 532, 599
Fat pad 368, 370-373, 378, 386-388, 390, 391
Feeding difficulty 357
Felon 465, 482, 484, 485
Femoral head excision 105
Fentanyl 40-42, 455, 604
FGF receptor 334, 338
Fibroblast 2, 3, 12, 21, 23, 61, 72, 88, 93, 96, 117, 145, 152, 438, 449, 450, 539, 597
Fibrosis 416, 439, 447, 490, 524, 537, 559
Filling port 62, 64
Filling rate 62
Finger 27, 33-36, 60, 88, 136, 158, 276, 337, 366, 462-467, 470, 475-480, 482, 484-486, 491-497, 499, 507, 511-516, 519, 525, 526, 531-542, 544-548, 550, 551, 553, 554, 556-564, 566, 568, 571, 573, 575, 580, 581, 586, 587, 589, 595, 596, 599, 603, 622
  anatomy 462, 512, 533, 539, 548, 551, 580
  coverage 542, 548-550, 561-564, 585
  injury 462-464, 466, 467, 476-478, 480, 491-494, 497-499, 507, 511-519, 525, 526, 533, 535-560
  malrotation 555
  metacarpophalangeal (MP) joint 462, 463, 466, 474-478, 512-514, 516, 517, 533-535, 537-539, 541, 545, 559, 562, 563, 584, 593, 596, 599
  nail bed 132, 466, 483, 494, 495, 513, 514, 547-549, 551-553, 560, 584
    anatomy 551
    injury 547, 551-553
  oblique pulley 465, 540, 580, 582
  tip amputation 495, 549, 561, 562
Fingernail 466, 547, 551-553, 560
Fingertip 465, 466, 476, 485, 495, 499, 547-551, 561, 562, 585, 593
  injury 547-551
    nail-matrix avulsion 553
    sterile matrix 466, 551-553
Finkelstein's test 479, 587
Firbrinolysin (Fibrolan®) 22
Fissure 182, 231, 329, 337, 371, 386, 389, 391, 392
Fitzpatrick skin type 128, 442
Flaps 4, 8-11, 24, 27, 28, 44, 46, 47, 49-64, 75, 85, 86, 92, 93, 98, 100-102, 104-116, 129, 130, 133, 136, 144, 146, 147, 165, 167, 169-179, 181-186, 188-192, 196-210, 212-215, 217, 223, 228, 230, 235, 246, 260-264, 266-272, 274-277, 279-281, 283-287, 291, 292, 294-298, 300-312, 314-318, 320-323, 341, 342, 345, 347, 348, 353-355, 362-363, 365, 366, 372-375, 379, 384, 390, 420, 421, 423-425, 427-429, 458-461, 464, 492, 494, 495, 526, 530, 536, 543, 549-551, 553, 561-564, 578, 585, 593, 595, 596, 600, 607-624
  Abbe 199, 201-204, 206-209
  adipofascial 563
  advancement 44, 52, 53, 105, 107, 108, 170, 175, 179, 182-184, 190, 191, 196, 197, 200, 206, 208-210, 295, 296, 310, 353, 354, 460, 550, 551, 561, 562, 578, 595, 607, 620, *see also* Rotational flap
  annular 563

anterolateral thigh (ALT) 167, 317, 608, 621
atasoy 561
axial 49-51, 206, 563
banner 171
bell 286
bilobed 53, 190
bipedicle 173, 190, 362, 363, 553, 562
bipedicled tripier 181, 184
cheek advancement (Mustarde) 179, 181-185
clavicular tubed 54
composite 49, 51, 170, 174-176, 183, 190, 191, 212, 213, 374, 536, 612
conchal chondrocutaneous transposition 173
cross-finger 494, 550, 562
Cutler-Beard 179, 185, 186
C-V 286
deep inferior epigastric perforator (DIEP) 47, 272, 342, 608, 611
delay 51, 269, 424
deltopectoral 167, 608
distant 52, 54, 55, 202, 210, 301, 309-312, 314, 549, 550
double opposing semicircular 53, 55, 285
dorsal island digital 563
dorsal middle finger 562
dorsal ulnar artery 564
dorsalis pedis fasciocutaneous 321, 322
Estlander 199, 201, 202, 207
facial artery musculomucosal (FAMM) 191, 197
fasciocutaneous 49-51, 56, 57, 105, 107, 109, 113, 115, 310, 315, 320-323, 564, 607, 608, 612, 613, 619-623
fibula composite 612
fibula free 50, 167, 213, 214, 609, 612
fillet 563
harvest 57, 167, 263, 272, 275, 620, 622
forearm 92
forehead 44, 54, 63, 181, 185, 190-192, 608
Foucher 563
free bone-muscle 215
free fibula 214
free fibula osteocutaneous 315
free 47, 50, 52, 54, 56, 60, 61, 85, 86, 92, 115, 144, 165, 167, 190, 191, 199, 202, 210, 212-214, 223, 271, 272, 296, 297, 301, 315, 320, 322, 323, 464, 492, 494, 536, 561, 607-612, 614-622
  free flap monitoring 271
gastrocnemius 92, 320, 321, 609
gluteal perforator 105, 106, 608
gluteus myocutaneous 106, 620
goumain 563
gracilis myocutaneous 51, 56, 112, 210, 312, 314-317, 320-323, 609, 618, 619
groin 54, 310, 314, 323, 607, 608
Holevich 563
homodigital island 563
homodigital triangular 550
Hueston 562, 598
iliac crest free 213, 323
iliolumbar 310
island 54, 105, 107, 108, 276, 526, 551, 562, 563, 607, 614, 616
jejunal interposition free 167, 609
juri 460
Karapandzic 199, 201, 202, 205, 206
kite 563
Kleinert 561
Kutler 562
labial mucosal advancement 196, 197
lateral arm 113, 323
lateral palmar rotation-advancement 562
latissimus dorsi 51, 56, 63, 274-277, 296, 297, 301-303, 306, 307, 312, 322, 323, 609, 615-617
Limberg 53, 54, 564
Littler neurovascular island 54

local 52, 62, 86, 113, 179, 181, 182, 185, 190, 210, 283, 284, 297, 300, 302, 306, 310-312, 314, 316-318, 320, 322, 366, 549, 550, 561
locoregional 190, 191, 310, 311, 314
loss 56, 59, 192, 203, 271, 274
microvascular 75
Moberg advancement 550, 562
modified double opposing tab 285
modified Hughes 179, 182, 183, 185
modified star 285, 286
mucoperichondrial 190, 191, 246
muscle 51, 56, 57, 85, 86, 105, 106, 111, 114-116, 147, 181, 183, 184, 215, 235, 264, 274, 279, 295-298, 300-308, 311, 317, 320, 341, 390, 611, 612, 614-616, 618-621
musculocutaneous 51, 105, 107, 203, 274, 275, 294, 300, 301, 609, 611, 612, 615-617, 619, 620
  type II 618
  type III 611, 616, 620
  type V 614, 615
myocutaneous 50, 51, 54, 61, 63, 92, 98, 105-109, 167, 213, 261, 263, 264, 266-272, 274, 295, 296, 311, 314, 315-317
nasolabial 181, 190, 191, 198-200, 608
neurovascular island 54, 526, 551, 562
neurovascular island pedicle 562
oblique rectus abdominis myocutaneous (ORAM) 98, 315-317
oblique triangular 402, 562
omental 306, 315, 609, 610, 617, 618
omental pedicle 297, 307
osseocutaneous 50, 622
paraspinous muscle 306-308
pectoralis major 167, 295, 296, 302, 609, 614
pedicle 51, 52, 54, 167, 173, 263, 264, 266, 269-272, 314, 317, 320, 321, 362,536, 551, 562, 564, 611, 614-618
pedicled groin 314
perforator 105, 106, 266, 272, 307, 317, 608, 609, 611, 620
pericranial 228, 229
posterior thigh 105, 108, 315-317, 320
racquet 563
radial artery island fasciocutaneous 113
radial forearm 92, 536, 563, 564, 608, 619
radial forearm free 167, 191, 210, 214, 315, 622-624
radial forearm-palmaris longus tendon free 210
random 49, 50, 85, 109, 113, 562, 564
reconstruction 86, 111, 115, 130, 165, 173, 190, 261, 274, 298, 310, 311, 314, 315
rectangular 52, 53
rectus abdominis 51, 54, 98, 112, 115, 263-272, 274, 301, 302, 311, 312, 317, 320-322, 609, 611
rectus abdominis myocutaneous 98, 317
regional 61, 296, 314, 549, 550
retroauricular 175
retrograde radial forearm fascial 564
retrograde radial forearm 563, 564
reverse cross-finger 562
rhomboid 53, 54, 190, 198, 314, 564
rotation 44, 52, 53, 62, 106, 108, 112, 191, 192, 200, 203, 295, 296, 301, 312, 348, 353-355, 458, 459, 562, 564, 595
S 285
semicircular 53, 55, 179, 182, 183, 185, 186, 285
serratus 302, 323, 609, 616, 617
skate 285, 287

skin 49, 50, 55, 92, 171, 173, 175-177, 181, 235, 260, 261, 269, 270, 275, 279, 280, 285, 304, 307, 310, 314, 317, 348, 372, 373, 375, 460, 563, 593, 596, 607, 615
soleus muscle 320, 321, 609
souquet 562
superficial inferior epigastric artery (SIEA) 272, 310
supramalleolar 322
sural artery 322
tarsoconjunctival 179, 182, 183, 185
temporoparietal fascial 345, 347, 608
Tenzel 179, 182, 185, 186, *see also* Flaps, semicircular
TFL fasciocutaneous 107, 311, 620
TFL myocutaneous 98, 105, 107, 108, 115, 311, 312, 321, 322, 609, 620
thenar 54, 562
thoracoepigastric 310, 608
tongue 167, 197, 198, 366
trans-digital 562
transposition 52, 53, 62, 173, 200, 296, 564
TRAM 47, 51, 54, 63, 263, 264, 266-272, 274
supercharged 269, 272
muscle-sparing 272
trapezius muscle 276, 302, 306, 307, 615
triangular advancement 562
triangular 9, 53, 348, 353-355, 550, 562
tripier 181, 184
tubed 54, 171
turnover 96, 190, 295, 296, 322, 564, 615
selection 105
V-Y advancement 44, 52, 53, 105, 107, 108, 170, 175, 179, 182-185, 190, 191, 196, 197, 200, 206, 208-210, 295, 296, 310, 320, 353, 354, 384, 460, 550, 551, 561, 562, 578, 595, 607, 620
Flexor carpi radialis (FCR) 36, 464, 500, 525, 568, 584, 586, 563, 568, 590, 623, 624
Flexor carpi ulnaris (FCU) 36, 113, 467, 477, 481, 525, 526, 531, 563, 564, 575, 577, 578
Flexor digitorum profundus (FDP) 463-467, 476, 477, 487, 494, 512, 525, 526, 534, 537, 539-546, 557, 568, 575, 624
avulsion 557
Flexor digitorum superficialis (FDS) 464-466, 476, 477, 512, 513, 525, 539-541, 543, 544, 546, 567, 568, 575, 590, 623, 624
Flexor pollicis longus (FPL) 464, 465, 476, 525, 540, 567, 568, 590
Floor of mouth 165, 608
Fluence 433, 434
Flumazenil 40
Fluorescein 60, 279
Foot ulcer 23, 75, 80, 81, 83-86, 93, 151
Forearm 92, 113, 157, 167, 191, 210, 214, 315, 462, 466, 467, 478, 486-495, 497, 499, 501, 516, 522, 525, 526, 531, 532, 536, 539, 545, 563, 564, 569-571, 575, 578, 582, 589, 590, 594, 603, 608, 619, 622, 623
amputation 493, 494
Forehead 34, 44, 54, 63, 67, 124, 129, 167, 181, 185, 190-192, 220, 226, 227, 328, 334, 335, 368, 377-379, 381, 444, 452, 453, 608
Fractures 57, 68, 87, 100, 188, 225-229, 231-243, 245, 246, 248-257, 319, 320, 326, 362, 399, 462, 465, 470-474, 476, 477, 479, 481, 486, 487, 497-517, 525, 529, 530, 532, 536, 537, 544, 547, 551-557, 559, 560, 567, 574, 576, 579, 597, 601, 603, 622, 624
anterior table 226-228
Barton's 499
base 514, 516, 517
Bennett 471, 517
Boxer's 472, 515

capitate 502, 503, 505-511
Colles' 498, 500, 597, 601
Chauffeur's 471, 499
classification 226, 239, 241, 248, 250, 251, 503, 505, 508, 514, 554
dislocation 502, 510, 511, 555, 559
distal phalanx 251, 472, 513, 514
distal radius 462, 472, 497-500, 512, 515-517, 536, 537, 551, 553
extra-articular 498, 499, 513, 517, 554
facial 225, 237, 239-243, 255
  complex 241
  complications 242
finger 512-514
first metacarpal 471, 516
frontal sinus 225-230, 242
  biomechanics 226
  cerebrospinal fluid (CSF) leak 226-229
  cranialization 228, 229
  displaced anterior table 227
  displaced posterior table 228
  anterior and posterior table 225, 226, 228
hamate 507, 511
humeral 477, 487, 525, 529
  supracondylar 487, 529
Le Fort I 239, 240, 241
lunate 479, 502, 507, 510, 511
mallet 472
mandible 240, 241, 248-257, 507, 515
  angle 248, 249, 253, 254
  adentulous case 240
  bilateral subcondylar 252, 256
  body 248, 253, 507
    favorable 248, 253, 254
  compression plate 254
  children 241, 253, 255, 256
    pediatric dentoalveolar trauma 255, 256
  condylar 248, 250-254, 256, 515
    favorable 251
    high condylar 252
  open bite 251, 252, 256
  subcondylar 250, 252, 253, 256
  symphyseal and parasymphyseal 211, 214, 248, 250, 254, 256
  unfavorable or displaced condylar 248, 251
maxillary 239-242
metacarpal 471, 472, 507, 515-517, 554
  base 471, 515-517
  shaft 515, 516
Monteggia 499
nonunion 242, 254, 257, 470, 502-506, 508, 511, 555
nasoorbitoethmoid (NOE) 188, 243, 245, 246
pattern 231, 239, 248, 249, 513, 517
palate 241
percutaneous screw fixation 235, 251, 254, 504, 505, 514
phalanx 472, 512-517, 537, 551, 553-555,
  middle 514, 516, 537, 555
    midshaft 555
proximal 248, 251, 252, 254, 505, 506
  midshaft 555
  phalanx 513, 514, 554, 555
pisiform 508
scaphoid 465, 470, 479, 501-506, 510, 511
shaft 477, 487, 513-517, 525
Smith's 498
trapezium 507
trapezoid 502, 507, 508
triquetrum 473, 507, 511
unstable 255, 499, 500, 503-505, 507, 514-517, 553, 554, 559, 560
  unstable nonunion 503, 505
wrist 243, 473, 478, 497-499, 501, 502-511
  dorsal intercalated segmental instability (DISI) 473, 478, 508-510

zygomatic 231, 237, 238, 241, 242
arch 231, 238, 242
ulna 474, 497-500
Freckles 123
Free tissue transfer 47, 56, 59, 92, 188, 191, 210, 296, 297, 301, 307, 311, 317, 318, 320, 322, 342, 564
Frey's syndrome 166
Frontonasal duct 226-228
Fronto-orbital advancement 335, 336
Frosting 442-444

G

Gamekeeper's thumb (Skier's thumb) 474, 478, 559
Gardner's syndrome 122
Gauze 13-15, 17, 19, 27, 74, 84, 103, 149, 162, 286, 287
Genioplasty 341, 400-405
Genitalia 155, 157
defects 314, 315
Giant nevi 61
Gillies' approach 43, 44, 238, 242, 379
Gingiva 165, 255, 327
Gingivoperiosteoplasty 350
Girdlestone arthroplasty 105
Glabellar complex 452, 453
Glabellar line/region 69, 191, 368, 378, 381, 447, 451, 452
Gland 2, 71, 75, 81, 96, 97, 123, 124, 130, 145, 150, 163, 165, 166, 178, 216, 217, 284, 368-371, 387, 390, 406, 409, 413, 416, 418, 440, 456
Glomus tumor 123
Glossectomy 164
Glossoptosis 338
GLUT-1 140, 144
Gnathion 330
Goltz-Gorlin syndrome 128
Gonial angle 330, 332
Gonion 330
Gore-Tex® 66, 67, 69, 311
Gout 481, 536, 544, 567, 586, 588
Graft 15-17, 24, 25, 44, 47, 49, 50, 52-54, 57, 59, 61, 63, 65, 69, 75, 85, 92, 93, 110-113, 115, 129, 133, 136, 146-150, 152, 153, 159, 167, 171-176, 178, 179, 181-185, 187-192, 197, 198, 212-215, 220, 221, 223, 224, 228, 233-235, 240, 246, 283, 285, 288, 291, 292, 294, 300, 301, 303, 307, 309-312, 314, 315, 317, 320, 335, 341, 342, 344, 346, 347, 350, 366, 392, 397-399, 416, 446, 447, 454, 460, 461, 490, 493-495, 505, 506, 507, 514, 521, 522, 524, 526, 527, 530-532, 536-538, 542, 544, 545, 549-551, 553, 559, 562-564, 589, 593, 594, 596, 599, 600, 612, 621, 624
bone 190, 191, 212-214, 228, 234, 235, 240, 242, 301, 335, 338, 341, 350, 355, 362, 399, 505-507, 514
buccal mucosal 183, 184
cartilage 171-175, 183, 185, 190, 234, 398, 399
chondrocutaneous composite 175, 183
composite 175, 176, 181, 183, 190, 198, 212, 213, 283, 301, 398, 493, 495, 549
conchal cartilage graft procedures, contralateral 172, 173
costochondral 21
cross-facial nerve 220, 223, 224
dialysis 112
fascial 310, 311
fat grafting 342, 446-448, 450
free nipple 263, 288, 291, 292, 415
hard palate mucosal 184
infection 111
mucosal 179, 182-184
nerve 220, 221, 223, 224, 320, 494, 495, 521, 524, 526-528
PTFE 69, 530
prosthetic 25, 111, 112
rib 212, 214, 301, 341
septal chondromucosal 183, 184
sheet 149

skin 4, 15-17, 24, 25, 47, 50, 52, 53, 59, 61, 63, 75, 85, 90, 92, 107, 110, 115, 129, 133, 145-150, 152, 153, 155, 159, 167, 171-176, 182, 183, 185, 188-192, 198, 200, 214, 284, 285, 300, 301, 303, 305, 307, 310-312, 314, 315, 323, 344, 347, 392, 454, 490, 493, 494, 549-551, 553, 562-564, 593, 594, 596, 599, 612, 614, 619, 621, 624
full-thickness (FTSG) 61, 63, 145, 146, 148, 150, 152, 153, 155, 175, 179, 181, 190, 291, 292, 347, 549, 550, 553, 562, 563, 593, 594, 599
partial-thickness 145, 152, 153
split-thickness (STSG) 17, 47, 50, 75, 92, 119, 146-150, 152, 153, 167, 172, 176, 181, 314, 315, 347, 454, 490, 494, 550, 553, 563, 564, 624
"taking" 549, 550
Grind test 479, 587
Groin 54, 75, 79, 111, 112, 113, 263, 310, 312, 314, 316, 323, 550, 607-611, 617, 618
Growth abnormality 256
Growth factor (GF) 1, 2, 9, 20, 23, 72, 73, 84, 93, 96, 102, 117, 119, 141, 334, 338, 152
Guyon's canal 464, 467, 489, 490, 571, 575
Gynecomastia 413-416, 430
pseudogynecomastia 413-415

H

Hair 2, 26, 62, 63, 78, 82, 88, 90, 91, 121, 123, 124, 127, 131, 145, 150, 168, 172, 173, 199, 200, 210, 214, 346, 347, 372, 373, 375, 378, 434, 445, 458-461, 564, 602
follicle 2, 123, 124, 145, 150, 459
loss 82, 88, 91, 375, 458, 459
restoration 458-461
Hamate 462-464, 472, 477-479, 503, 507, 511, 568, 570, 571
Hand 6, 7, 23, 33, 35, 36, 43, 44, 56, 66, 123, 126, 143, 145, 146, 155, 157, 158, 214, 240, 250, 252, 289, 307, 328, 337, 338, 404, 409, 425, 431, 434, 436, 440, 454, 456, 460, 462-470, 476-484, 486, 488-496, 498-500, 503, 507, 508, 511, 513-515, 517, 520, 522, 525, 529, 531-533, 535-537, 539-542, 545, 547, 548, 550, 551, 554, 558-561, 563, 564, 566, 567, 569-571, 574, 575, 578, 580, 581, 583, 584, 586-592, 594-597, 600-604, 608, 619, 622-624
amputation 492-496
anatomy 462-467, 533
compartments 463, 486
exam 476-480, 499, 503
infection 481, 482, 547
intrinsic hand muscle 463, 467, 478, 480
lumbricals 463, 467, 478, 534, 538, 567
soft tissue coverage 65, 493, 494, 549, 561-564
Hard-tissue-replacement (HTR) polymer 66-68, 215, *see also* Dacron® and Mersilene®
Harvest 50, 56, 57, 61, 65, 98, 110-113, 146-148, 150, 153, 167, 175, 176, 181, 183, 187, 189, 191, 203, 212-214, 224, 229, 230, 263, 264, 267-270, 272, 275, 276, 292, 296-298, 312, 316, 447, 450, 493, 589-591, 607, 620-624
Healing, secondary intention 4, 72, 75, 100, 104, 111, 128, 129, 196, 198, 200, 300, 305, 375, 483, 549, 550
Hearing 218, 343, 344, 350, 351, 357-361, 510
Helical advancement 170
Helical crus 174, 382, 383, 460

Helical rim 168, 171, 345, 382, 383
Helix 2, 35, 169, 170, 172, 173, 175, 177, 345, 346, 382, 383, 385, 435, 449
Hemangioma 138-144
  congenital 138-140
  infantile 138-140, 142
Hemangiopericytoma 139
Hematoma 13, 27, 29, 60, 64, 109, 146, 149, 166, 232, 235, 236, 245, 246, 271, 276, 277, 297, 320, 336, 346, 372, 375, 380, 384, 385, 391, 396, 409-411, 414-416, 421, 426, 429-431, 447, 452, 512, 515, 518, 519, 529-531, 550-552, 590, 600
Hemifacial microsomia 340, 343
Hendrickson classification 241
Hernia 79, 112, 271, 272, 297, 307, 310-312, 424, 425
  incisional 309, 424
Hidradenitis suppurativa 75
Hirudo medicinalis 27
Hook of hamate 478
Horner syndrome 388, 520, 602, 604
House-Brackmann grading system 218
Human papilloma virus 165
Hutchinson's sign 132
Hyaluronic acid 352, 448, 449, 450, 583
Hydrocolloid 15, 18
Hydrogel 13, 15, 17, 19, 84, 92, 97, 103, 111
Hydroquinone 442, 456
Hydrotherapy 104, 159
Hydroxyapatite 69, 229, 230
Hylaform 449
Hyperkeratosis 122
Hyperpigmentation 91, 96, 119, 132, 436-438, 456
Hypertelorism 336, 337, 357
Hyponychium 466, 552
Hypopharynx 163, 165, 167
Hypopigmentation 96, 436, 439, 442, 456
Hypoplasia 256, 336, 337, 340-343, 393, 401-403, 446, 594
H-zone 128, 130

I

Imbibition 146, 149, 175, 540
Implant 62-66, 68-70, 114-116, 212, 228, 234, 236, 261, 262, 266, 274-281, 284, 342, 345, 346, 374, 399, 400, 403, 404, 406-412, 417-422
Incision 4, 6, 8, 13, 20, 44, 52, 55, 59, 62-64, 70, 72, 76-79, 112-114, 117, 118, 122, 125, 130, 132, 133, 158, 159, 169, 170, 173, 175, 177, 182, 183, 185, 186, 191, 196, 198, 199, 203, 206, 208, 210, 214, 227, 229, 234, 235, 238, 240, 241, 246, 253, 257, 260, 269, 270, 275, 276, 279-281, 288-294, 297, 309, 312, 317, 319, 348, 353, 354, 362-366, 373-380, 384, 390, 391, 397, 403, 404, 407-410, 415-420, 422, 424-429, 447, 460, 482-485, 488-491, 493, 500, 525, 526, 531, 541-543, 550, 553, 559, 562-564, 570-572, 578, 582, 585, 590, 591, 593, 599, 600, 607, 611, 612, 614, 615, 618, 621, 623
Incisional biopsy 132, 133
Incompetence 366, 453
Infection 1, 8, 11, 13, 14, 18-20, 23, 25, 50, 57, 63, 64, 68-85, 100-102, 104, 109, 111, 113-119, 126, 129, 139, 142, 146, 147, 149, 150, 152, 153, 162, 165, 169, 174, 187, 196, 212, 213, 217, 226, 228, 234, 236, 242, 253, 260-262, 271, 277, 287, 294-299, 301, 304, 305, 308-312, 322, 336, 345, 346, 349, 352, 357, 358, 360, 376, 381, 385, 389, 403, 404, 409-411, 415, 421, 425, 429, 431, 436, 441, 447, 455, 456, 461, 465, 466, 481-485, 487, 490, 495, 532, 536, 538, 544-547, 552, 555, 567, 572, 589, 590, 594, 597, 600, 624

Inferior epigastric vessel (IEA) 51, 263, 264, 266-270, 611
Inferior orbital fissure (IOF) 231
Infraorbital groove 232
Infraorbital rim 231, 233-235, 238, 240, 242, 243, 330, 351, 389
Inhalation injury 155, 160, 161
Inheritance, pattern 357, 595
Injury 1, 2, 4, 11, 13, 15, 29, 33, 37, 38, 57, 63, 69, 76, 80, 87, 88, 95-97, 100, 102, 105, 108, 109, 117, 142, 145, 147, 154-156, 158, 160, 161, 166, 169, 203, 205, 214, 217, 226-228, 231, 233, 236, 238-242, 245, 248, 249, 253, 255-257, 261, 270-272, 276, 277, 296, 297, 299, 301, 303, 304, 310, 314, 318-320, 322, 326, 336, 368-370, 373-376, 380, 381, 396, 403, 408, 410, 425, 426, 431, 433-435, 440, 444-446, 462-464, 467, 468, 476-478, 480, 481, 487, 488, 490-494, 497-513, 515-533, 535-557, 559-562, 567, 568, 570, 572, 574, 580, 582, 590, 591, 597, 600, 601, 603, 605, 619, 621
Innominate fascia 369
Inosculation 146, 149, 191
Instability 72, 81, 297
Insufficiency 9, 13, 28, 30, 60, 87, 88, 93, 147, 151, 214, 364, 445
Integra 152, 153
Intercanthal width 327, 328
Intercostal artery 424
Intercostal vessel 263, 424
Interferon-α (IFNα) 135, 141, 142
Interleukin-2 (IL-2) 135
Internal mammary artery (IMA) 294, 297, 298, 302, 406, 608, 611
Internal mammary vessel 264, 268, 270
Internal nasal valve 188, 350, 396, 398
Interossei, volar see Volar, interossei
Interosseous compartment 158, 488, 495
Interosseous membrane 462, 497, 613
Interphalangeal (IP) joint 136, 146, 462, 463, 465, 466, 470, 475, 477, 478, 512, 533, 534, 537, 540, 541, 547, 553-555, 559, 562, 593, 595
Interpolation 54
Interpupillary width 328
Intersection syndrome 587
Intracranial pressure (ICP) 335, 336
Intraocular pressure 233, 236, 391, 392
Intravelar veloplasty 362, 363
Intrinsic plus position 482, 554, 537, 604
Intrinsic-plus contracture 537, 538
Iodoflex® 21
Iodoform gauze™ 18
Iodosorb® 21
Iritis 233
Irradiated tissue 95, 96, 146, 213, 299
Irrigation 58, 78, 97, 115, 169, 298, 411, 483, 484, 556, 558
Ischemia time 322, 492, 493, 495
Isolagen 450
Isotretinoin 436, 441, 455

J

J point analysis 330, 332
Jersey finger 557
Joints 3, 66, 68, 69, 81, 83, 85, 90, 105, 114, 115, 126, 127, 134-136, 143, 146, 155, 211, 214, 250, 252, 300, 328, 462, 463, 465, 466, 469, 470, 474-479, 482-485, 490, 492, 494, 497, 499-501, 506, 508-510, 512-517, 528, 533-535, 537-541, 543, 545, 547, 553-560, 562, 563, 565, 569, 574, 577, 578, 580, 582-585, 587, 588, 593, 595, 596, 598, 599, 602-604
  distal interphalangeal 553
  distal-radial-ulnar (DRUJ) 462, 497
  stiffness 500, 559, 585, 602, 603

K

Kanavel's sign 484
Kaplan's cardinal line 514, 570
Kaposiform hemangioendothelioma (KHE) 139, 140, 142
Kasabach-Merritt syndrome 139, 140, 142
Keloid 9, 117-119, 122, 176, 375, 384, 436, 441, 455
Kenalog 456
Keratinocyte 12, 21, 72, 74, 96, 122, 132, 151, 152
Keratitis 185
Keratoacanthoma 121
Keratosis 121, 126
Kernahan classification 348, 349, 356
Key pinch 467, 525
Kleeblattschadel (Clover leaf skull) 334, 336
Kleinert modification 535
Kleinert splint 545
Klinefelter's syndrome 413-415
Klippel-Trenaunay syndrome 140
Knot security 6, 7
Kunt-Simonowsky procedure 391, 393
K-wire 238, 500, 505, 510, 514, 515, 538, 556, 562, 594
    fixation 238, 500, 514, 515, 556, 594

L

Labiomandibular fold 372
Labiomental angle 401, 402
Labiomental fold 193, 194, 202, 208, 371, 401
Lacrimal canaliculi 178
Lacrimal gland 178, 387, 390
Lacrimal sac 178, 231, 245, 386, 389
Lactation 408, 417, 421
Lagophthalmos 187, 245, 386
Langerhans cell 71
Larynx 163, 165, 348
Laser 119, 122, 124, 141-143, 214, 347, 380, 433-440, 444, 495, 511
    ablative 435, 437, 438
    Er:YAG 435-437
    Erbium:glass 435, 437, 438
    intense pulsed light (IPL) 435, 437
    Nd:YAG 434, 435, 437
    nonablative laser 435, 437-439
    peel 435
    pulsed dye laser (PDL) 434, 435, 437, 438
    pulsed electromagnetic field (PEMF) therapy 511
    rejuvenation 435
    resurfacing 433-439, 440, 444
    surgery 433, 435
    therapy 122, 142, 143
Laxity 187, 200, 203, 210, 281
Le Fort I osteotomy 241, 332, 341, 401, 404
Leeches 27, 28, 59, 494
    suppliers 28
    therapy 27, 28, 59
Leukoplakia 126, 164, 196
Levator aponeurosis 178, 180, 231, 387
Lidocaine 26, 29-34, 36, 39, 133, 246, 353, 403, 431, 432, 436, 437, 443, 448, 449, 456, 459, 493, 513, 587, 604
    toxicity 31, 432
Ligament 111, 182, 231, 256, 312, 368, 371-373, 376, 381, 387, 462-464, 471, 473, 474, 478, 479, 486, 489, 501, 502, 505, 507-512, 514-517, 525, 533, 534, 537, 538, 547, 555, 557-559, 568, 570, 574, 575, 577, 578, 584, 585, 592, 596, 598, 607, 618
Ligament of Struthers 574
Ligamentous injury 505, 508, 557, 559
Limb 9, 10, 11, 63, 80, 82-85, 87, 88, 90, 93, 111, 113, 135, 142, 158, 182, 183, 186, 285, 289-292, 318, 319, 384, 420, 421, 481, 488, 518, 531, 532, 550, 574, 591, 594, 595, 616
    perfusion 135
    salvage 85, 93, 318
Limberg 53, 54, 564
Line of election 10

Linea alba 263
Lip 34, 43, 44, 69, 123, 128, 129, 163, 191, 193, 194, 196-210, 219-222, 248, 255, 328, 340, 343, 348-357, 359, 361, 362, 364, 368-371, 397, 400, 401, 404, 447, 453, 510, 576, 608, 622
- anatomy 193, 194, 350, 351
- lower lip 34, 193, 194, 196-210, 219-222, 248, 328, 370, 400, 453
  - length 196, 203, 328
  - reconstruction 198-201, 203, 205-208
- reconstruction 193-210
- upper lip 34, 193, 194, 196, 198-201, 203, 204, 206-209, 219-221, 328, 348-351, 353, 355, 397, 401, 453
  - length 328
  - reconstruction 198, 201, 203, 206-208

Lipodystrophy 368, 430, 438
Lipomas 123, 414, 430, 567
Liposuction 38, 39, 41, 289, 415, 416, 423-425, 430-432, 446
- anatomy 430
- -assisted mastectomy 415
- complications
  - in morbidly obese patient 271
  - from the tumescent solution 432
- power-assisted (PAL) 431
- suction-assisted lipectomy (SAL) 430, 431
- tumescent technique 431
- ultrasound-assisted 416, 430, 431

Lobular carcinoma in situ (LCIS) 258
Local muscle transposition 221
Local wound care 20, 84, 85, 93, 96, 104, 109, 111, 115, 298, 299, 549
Lumbosacral region 308
Lymph node dissection 133, 134, 136, 259
- axillary dissection 259
- elective dissection (ELND) 133, 134

Lymphocyte 1, 2
Lymphoscintigraphy 134
Lyofoam™ 16, 18

M

Macrogenia 401, 402
Macrophage 1, 2, 22, 68, 71, 72, 96, 102
Macrostomia 337, 341, 342
Macule 122
Mafenide 160, 169
Magnetic resonance angiography (MRA) 82, 90, 102, 111
Magnetic resonance imaging (MRI) 83, 92, 101, 123, 134, 140, 143, 166, 412, 414, 503, 513, 520, 540, 576, 584
Major causalgia 603
Major traumatic dystrophy 603
Malar and buccal fat pad 371, 372
Malar augmentation 66, 68, 69, 400, 404
Malar fold 371
Male pattern baldness 62, 458
Malformation 138-144, 333, 357, 513, 567
Malnutrition 13, 77, 304, 574
Malocclusion 240, 242, 248, 251, 252, 255, 256, 326, 327, 333, 361, 400, 402, 404
Malunion 242, 254, 505, 514, 555, 577
Mandible 34, 69, 166, 211-213, 217, 234, 240-242, 248, 249, 252, 254-257, 324-330, 332, 333, 340-342, 348, 357, 358, 361, 368, 370-373, 376, 400-403, 405, 498, 608, 609, 612
- anatomy 211, 212
- edentulous patient 252, 255
- prominence 348
- pseudomacrogenia 401
- pseudomicrogenia 401, 402
- ramus hypoplasia 341
- reconstruction 211-215, 612
- retaining ligament 371-373
- vertical ramus osteotomy (VRO) 332

Mandibular plane 329, 330, 405

Mangled extremity severity score (MESS) 318, 319
Marionette line 368, 372, 438, 448
Marjolin's ulcer 109, 129
Marlex 301, 311
Mastectomy 47, 61, 258-263, 270, 274-279, 281, 283, 285, 412, 415-417
subcutaneous 412, 415
Mastication 211, 221, 255, 340, 341, 350
Mastopexy 38, 260, 271, 406, 409, 412, 417-422
Mathes-Nahai classification 51, 264, 309, 608, 609
Mature 73, 117, 127, 138, 375, 539
Maxilla 34, 69, 231, 234, 237, 239-244, 255, 257, 324, 325, 327-330, 332, 333, 336-338, 341, 348-351, 355, 356, 358, 359, 361-363, 366, 371, 374, 393, 394, 401, 402, 404, 608
prominence 348, 349, 402
sagittal hyperplasia 332, 402
sagittal hypoplasia 332, 401, 402
transverse hypo/hyperplasia 332
vertical deficiency 401
vertical excess 401
Maxillomandibular fixation (MMF) 240-242, 250-254, 256, 257
Maxon 7
McCash technique 599, 600
McGregor's patch 371
McGrouther technique 543
Medial canthal tendon (MCT) 178, 231, 242, 245, 389
Medial canthus 386
Mediastinitis 294-298
Medpor® 66, 67, 69, 403
Meibomian gland 178, 387
Melanin 96, 123, 132, 433, 434, 438
Melanocyte 96, 122, 123, 132
Melanoma 121-123, 126, 130-136, 180, 483
acral lentiginous 132
Breslow thickness 131
juvenile (Spitz nevus) 122
Lentigo maligna 133, 180
nodular 132
superficial spreading 132
subungal 132, 136
Mental foramen 34-36, 257, 403, 404
Menton 328-330, 400
Mepivicaine 29, 31
Mersilene® 7, 66, 68, 311
Mesh 66-69, 74, 76, 149, 152, 230, 234, 267, 270, 272, 301, 303, 310-313, 315, 612
absorbable 310-312
polyamide 67, 69, *see also* Supramid® and Nylamid®
Metastatic potential 126, 127, 129
Methicillin-resistant *S. aureus* (MRSA) 21, 77, 79
Methylmethacrylate 66-68, 234, 301, 303, 333
Methylprednisolone 246
Microdermabrasion 454
Microgenia 400-402
Micrognathia 337, 341, 357, 360, 401
Microneurovascular transplant 223
Microsomia 340-343
craniofacial 340-342
Microstomia 206, 209
Microsurgery 27, 51, 56-60, 213, 272, 296, 300, 301
Microtia *see* Deformities, ear
Microvascular disease 80, 422
Midazolam 32, 40, 41, 455, 459
Middle crura 396
Middle ear disease 352, 358
Mineral oil 148, 191, 444
Miniplates 68, 246, 336, 402
Minor causalgia 603
Minor traumatic dystrophy 603
Mirault 348
Mitek anchor 403
Model surgery 328, 332, 333
Modiolus 193, 198, 207, 351
Mohs micrographic surgery (MMS) 128-130
Moist desquamation 96, 97
Molding (orthotic cranioplasty) 335, 352, 410
Moles *see* Nevus
Monitoring 17, 19, 26, 40, 41, 56, 59, 60, 156, 271

Mucocele 226, 228, 229
Mucoepidermoid carcinoma 166
Mucous cyst 584, 585
Munro and Lauritzen classification 341, 342
Muscles 8, 16, 38, 41, 46, 49-51, 56, 57, 61-63, 80, 85-87, 98, 100, 105-108, 111-116, 127, 133, 147, 148, 153, 155, 158, 165, 167, 178, 180-184, 191, 193-196, 198, 203, 206, 209, 210, 213, 215-224, 228, 230-235, 237, 238, 242, 245, 248, 251, 253, 254, 257, 261-267, 269-272, 274-277, 279-281, 294-298, 300-308, 311, 312, 316-323, 335, 340, 341, 348, 350, 351, 353, 354, 359, 360, 362-365, 368-374, 376-379, 381-383, 386-392, 402, 403, 405, 406, 409, 410, 412, 424, 429, 447, 451-453, 462, 463, 465-467, 477, 478, 480, 486, 487, 489, 490, 495, 497, 499, 500, 512, 519-522, 524, 526, 527, 531, 533, 534, 536-540, 547, 566, 567, 569, 570, 574-579, 590, 592, 594, 607, 609, 611-623
- corrugator 369
- external oblique 296, 312
- facial expression 8, 217-219, 221, 351, 369, 377
- frontalis 63, 217, 369, 377-379, 381, 452, 453
- gluteal 105
- gracilis 51, 56, 112, 210, 224, 312, 314-317, 320-323, 521, 609, 618, 619
- inferior tarsal 183, 234, 388
- internal oblique 213, 263, 312
- latissimus dorsi 51, 56, 63, 215, 274-277, 296, 297, 301-303, 306, 307, 312, 322, 323, 520, 522, 609, 615-617
- levator labii superioris 193, 219, 351, 371
- levator palpebrae superioris 178, 180
- levator superioris alaeque 351
- levator veli palatini 359, 360, 363
- mastication 221, 340, 341
- masseter 165, 216, 221-223, 237, 251, 253, 254, 371, 372
- Müller's 178, 180, 183, 387, 388
- orbicularis oculi 178, 219, 235, 245, 369, 371, 374, 377-379, 386, 388, 390, 392, 447, 452, 453
- orbicularis oris 193, 194, 203, 206, 207, 210, 219, 222, 348, 351, 350, 370, 453
- palatoglossus 359, 360
- palatopharyngeus 359, 360
- pectoralis major 167, 262, 279, 280, 281, 295, 296, 302, 406, 409, 415, 519, 531, 609, 614
- pectoralis 275, 276, 279, 281, 296-298
- platysma 63, 216, 217, 219, 221, 369-371, 373, 376, 453
- power 477
- proccrus 191, 219, 369, 377-381, 452
- rectus abdominis 51, 54, 98, 112, 115, 263-272, 274, 301, 302, 311, 312, 317, 320-322, 609, 611
- rectus femoris 112, 311, 312, 320, 621
- sartorius 51, 111-113, 607, 618
- serratus 275, 276, 280, 281, 302, 323, 406, 520, 609, 616, 617
- temporalis 221, 222, 228, 230, 238, 242, 251, 378, 379, 384, 609
- vastus lateralis 51, 108, 115, 312, 320, 321, 609, 621
- zygomaticus major 193, 219, 224, 369, 370, 371, 373, 374, 376

Myelomeningocele 304, 308
Myofibroblast 3, 12, 96, 597
Myoglobinuria 489

N

Nail 81, 83, 86, 132, 133, 466, 483, 485, 492, 494, 495, 513-515, 543, 547-549, 551-553, 560, 584, 596
  lunula 466, 548, 552
  paronychia 482-485, 584, 596
  paronychium 466, 551, 552
  perionychium 466, 551
Naloxone 40
Narcotics 41, 250, 280, 293, 301, 380, 410, 604
Nasion 329, 330
Nasojugal crease/fold 368, 371, 372, 447
Nasolabial fold 10, 63, 128, 191, 193, 194, 209, 219, 221, 368, 371-374, 438, 447-450
Nasopharynx 163, 165, 167
Neck 3, 6, 44, 58, 60, 61, 63, 100, 124, 125, 128, 132, 134, 136, 139, 140, 157, 163-167, 210, 216, 218, 248, 250, 288, 300, 348, 357, 368, 369, 373, 376, 436, 440, 462, 475, 511, 512, 515, 520, 537, 554, 559, 560, 575, 608-610, 614-619, 621, 622
  dissection 136, 164-167, 373, 374, 376
Necrosis 4, 6, 10, 22, 28, 46, 55, 64, 72-74, 84, 88-90, 96, 97, 111, 117, 153, 158, 234, 235, 246, 252, 253, 271, 274, 277, 286, 288
Necrotizing fasciitis 83
Needle 4, 7, 31, 33-36, 58, 101, 134, 166, 167, 169, 277, 279, 288, 414, 448-451, 460, 488, 543, 552, 582, 584, 629
  reverse-cutting 7
  tapered 7
Negative pressure wound therapy 24-26, 104, 115, 310, 322, *see also* Vacuum-assisted closure
Negative vector *see* Blepharoplasty
Neomycin/Polymyxin (Neosporin®) 20, 21
Neovascularization 72, 79, 89, 144, 146, 544
Nerves 29, 33-36, 38, 39, 51, 63, 80, 81, 107, 108, 112, 122, 145, 147, 149, 150, 158, 165-167, 178, 195, 196, 206, 210, 214, 216-221, 223-225, 231, 232, 236-239, 242, 245, 249, 253, 257, 266, 270, 275, 276, 300, 315, 316, 318-320, 324, 336, 340, 341, 343, 368-383, 396, 403-406, 408-410, 426, 436, 443, 451, 455, 462-464, 466, 467, 476, 477, 479, 480, 486-491, 493-496, 499, 500, 506, 507, 511, 518-534, 540, 541, 551, 554, 562-564, 566-570, 572-579, 582, 590, 591, 595, 599-601, 603-605, 607, 611, 612, 614-619, 621, 622, 624, 630
  anatomy 33, 217, 370, 369, 523
  antebrachial cutaneous 210, 214
  anterior interosseous 245, 476, 525
  auriculotemporal 35, 377, 382
  cutaneous 315, 396, 406, 426, 494, 519-521, 578, 579, 590, 591, 607
  digital 36, 467, 524, 540, 541, 551, 562, 582, 599
  ethmoidal 245, 396
  glossopharyngeal 220
  facial 63, 165, 166, 195, 196, 216-220, 223, 224, 253, 340, 341, 343, 368-376, 378-380, 383
    cross-facial 220, 223, 224
    frontal branch 216, 217, 369, 374
    temporal branch 216, 217, 378-380
    zygomatic branch 216, 217, 219, 368-371, 374, 376
  fascicles 219, 523-525, 527, 539, 566
  graft *see* Graft, nerve
  great auricular 373, 376, 382
  grind test 479, 587

hypoglossal 220, 371
hypokalemia 432
infraorbital 34, 196, 232, 236, 237, 242, 245, 324, 376, 396, 404, 443
injury 63, 166, 242, 249, 253, 318, 320, 336, 374, 375, 381, 408, 410, 467, 476, 477, 520, 521, 523-529, 531, 532, 541, 582, 603
  classification 523, 524, 603
  facial 63, 166, 253, 375
  median 467, 476, 525
  postganglionic 520
  preganglionic 520
  radial 477, 525
  ulnar 467, 476, 477, 525
marginal mandibular 340, 370, 371, 373, 376
medial antebrachial cutaneous 578
median 35, 463, 464, 466, 467, 476, 479, 480, 489, 491, 499, 500, 519, 520, 522, 525, 527, 531, 532, 534, 563, 566-570, 572-575, 590
  compression 479, 566, 569
  exposure 525
motor 29, 196, 216, 217, 220, 266, 270, 381, 451, 476, 569, 615, 621
  function 476
oculomotor 178
perineurium 147, 219, 523, 524, 527
radial 35, 466, 467, 476, 477, 480, 489, 522, 525-527, 533, 563, 590, 624
  exposure 526
nerve repair 219, 224, 320, 493, 494, 521, 523-526, 528, 530, 541
  direct 219, 524, 526, 528
  primary neurorrhaphy 524
supraorbital 34, 377, 378, 396, 443
supratrochlear 245, 377, 378
sural 220, 221, 224, 494, 522
transfer 220, 522, 526, 527
transposition 527, 577-579
trigeminal 33, 122, 196, 225, 231, 238, 324, 377
ulnar 35, 36, 158, 300, 463, 464, 466, 467, 476, 477, 489, 507, 519, 521, 525, 526, 534, 568, 572, 574-579
  compression 574-577
  exposure 526
Neurapraxia 213, 524
Neurolysis 495, 521-523, 578
Neuroma 219, 220, 522, 527, 551, 567
Neurontmesis 523, 524
Neuropathy 80, 81, 89-91, 233, 236, 238, 288, 318, 479, 566, 567, 573-577, 579, 590, 591, 601
  compressive 566, 573, 574, 601
Neurosurgical consultation 229
Neurotization 495, 520-522, 526
Nevus 122-125, 128, 132, 133, 140
  blue 123
  blue-rubber bleb 140
  compound 122
  intradermal 122, 123
  junctional 122
  nevocellular 122
  of Ota 122
Nonablative radiofrequency resurfacing 438
Noninvoluting congenital hemangioma (NICH) 139, 140, 142
Norwood classification 458
Nonunion, unstable 503, 505, *see also* Fractures and Scaphoid
Nose 31, 34, 35, 63, 66-69, 78, 123, 128, 129, 141, 142, 160, 161, 163, 183, 184, 188-193, 199, 226, 228, 234, 239-246, 248, 328, 329, 335, 338, 348-358, 360, 362-369, 377, 378, 383, 394-399, 444-455, 459, 608
  anatomy 188, 189, 243, 244, 350, 351, 394, 395
  anterior nasal spine (ANS) 243, 329, 330, 351, 353
  bone 188, 239, 241, 243-245, 378, 394, 395, 397, 399

cartilage 234
lateral 188, 243, 350, 354, 355, 394-397, 399
lower 188, 243, 350, 355, 394-396
upper 188, 243, 354, 394, 396, 397, 399
deformities 189, 190, 228, 242, 243, 245, 350, 397-399
boxy tip 398
dorsal hump 397-399
saddle-nose 243, 399
wide nasal bridge 398
widened alar base 350, 398
lateral crura 188, 396-398
medial crura 188, 351, 394-398
nasalis 219, 351
posterior nasal spine (PNS) 329, 330, 332
reconstruction 63, 188-192, 348
septum 163, 188, 225, 228, 243-246, 350, 394, 396, 397, 399
spine 243, 246, 329, 330, 332, 350, 351, 353, 354
valve 188, 243, 350, 396, 398
Novuxol® 22
NSAID 452, 537, 570, 576, 587, 604
Nurolon 7
Nutrition 1, 3, 9, 20, 78, 84, 85, 97, 100-102, 109, 146, 161, 253, 305, 357, 441, 544, 567
optimization 101
Nylamid™ 69

O

Obstructive sleep apnea (OSA) 342, 400, 405
Occlusal plane 325, 330, 332, 333
Occlusion 8, 11, 59, 82, 88, 90, 112, 118, 119, 211, 229, 240-242, 248, 251, 252, 255-257, 319, 324, 325, 327, 328, 333, 340-342, 358, 361, 385, 400-402, 404, 444, 529
centric 328, 333
distal 327
overbite 325, 327, 330
Oculocardiac reflex 233, 238
OMENS classification 341
Omentum 297, 306, 307, 315, 610, 617
Ondansetron 39, 40
Op-site™ 14, 17, 103, 149
Opthalmoplegia 392
Optic neuropathy 233, 236, 238
Oral cavity 33, 163, 165, 167, 324, 340, 371, 608
Oral commissure 35, 63, 220-222, 224, 340, 370, 373
Orbital fissure 231
Orbital hematoma 245
Orbital rim 69, 226, 231-235, 238, 240-243, 330, 337, 351, 372, 374, 377-381, 386-389, 391, 453, 459
Orbital septum 234, 235, 387-391
Orbital wall 231, 235-237, 243, 342
Orbitale 329, 330
Oronasal fistula 362, 366
Oropharynx 75, 163, 165, 167
Orthodontics 341, 350, 352, 361
Orthognathic surgery 328, 332, 361, 401
Orthopedic shoe 85, 93
Orthoplast splint 559
Osbourne's ligament 575
Osteoarthritis 463, 470, 475, 505, 586
Osteomyelitis 16, 20, 23, 25, 74, 83-85, 92, 101, 103, 104, 109, 110, 116, 294, 295, 297, 298, 308, 318, 481, 485
Osteosynthetic plate 213
Osteotomy 241, 332, 335, 341, 398, 400-405, 514, 613
Otohematoma 169
Otoplasty 382-384
Overjet 325, 327

P

p53 tumor-suppressor gene 129
Palatal closure 361-364, 366
timing 361, 366
Palatal plane 329, 330

Palate 128, 163, 183, 184, 197, 239, 241, 337, 338, 340, 343, 348-350, 352, 356-364, 366, 367, 369, 404, 609
anatomy 359
Palmar cutaneous branch 35, 36, 466, 491, 500, 566, 570, 572
Palmar fascia 464, 570, 598
Palmar sensory cutaneous branch (PSCB) 464, 568
Palmer aponeurosis 465, 539
Palpebral fissure 182, 337, 371, 386, 389, 391, 392
Panafil® 22, 103
Panniculectomy 423
Panorex® 249, 251, 256, 328, 340
Papain-urea (Accuzyme®) 22, 75, 103
Papillary dermis 145, 434, 437, 441, 444, 456, 562
Papule 121-124, 127, 448
Parakeratosis 122
Paralysis 370, 452, 467, 477, 525, 526
Paralyzed face, reanimation of 217
Parasymphyseal region 447, 448
Parkes Weber syndrome 140
Parkland formula 156
Paronychia, chronic 483
Parotid duct (Stenson's duct) 165, 371, 372
Parotid gland 165, 216, 217, 369-371
Parotidectomy 166, 216
Pars marginalis 351
Pediatric dentoalveolar injury 255, 256
Pelvic defect 98, 314-317
Pelvic floor 314-317, 609
Pelvic floor and perineum, reconstruction 315
Penile skin, defects 314
Penis, defects 314, 315
Perfusion 9, 13, 24, 49, 58, 62, 81, 82, 85, 135, 154, 156, 158, 159, 271, 486, 488, 494, 600, 602, 624
Perineal defect 314-317
Perineurium 147, 219, 523, 524, 527
Peripheral neuropathy 81, 89, 90
Peripheral vascular disease (PVD) 13, 77, 88, 102, 113, 115, 318
Periprosthetic infection 114
Perko (Zurich) approach 364
Pfeiffer syndrome 337, 338
PHACES 140
Phalen's test 479, 569
Pharynx 163, 247, 348, 369
Philtral column 193, 194, 348, 350-353, 355
Philtrum 193, 194, 196, 198, 201, 203, 348, 350, 353, 371
Photoaging 368, 436, 438
Photothermolysis 433
Physical therapy 119, 154, 224, 521, 550, 558, 570, 600, 604, 605
Pierre Robin sequence 357, 338, 357
Pilar cyst 122
Pilomatricoma 124
Pin tract infection 555
Pinna 382, 383
Pisiform bone 564, 568
Pitanguy's line 217, 369, 378
Plagiocephaly 334, 335, 338
Plaque 88, 121, 122, 124, 126, 128
Platelet-derived growth factor (PDGF) 1, 23, 84, 93, 96
Plummer Vinson syndrome 164, 165
Pogonion 330
Poland syndrome 592, 595, *see also* Apert syndrome
Polydactyly 337, 592, 596
Polymethylmethacrylate (PMMA) 68
Polymorphonuclear leukocyte (PMN) 1, 2
Polymyxin 21
Polytetrafluoroethylene (PTFE) 66, 69, 311, 403, 530, 544, *see also* Teflon®, Gore-Tex®, Proplast®
Porion 329
Porous polyethylene 69, 234, 345, *see also* Medpor®
Port wine stain *see* Capillary malformation
Position of function 476
Postaxial (fifth ray) duplications, classification 592
Posterior nasal spine (PNS) 329, 330, 332

Postoperative nausea and vomiting (PONV) 38, 41, 247, 372
Post-thrombotic syndrome 87
Post-traumatic 234, 238
Povidone-iodine 21, 23
Preaxial (thumb) duplication 593
Premalignant lesion 121, 126
Premaxilla 240, 329, 350, 351, 359, 361
Pressure relieving strategies 103
Pressure sore 25, 100-106, 109, 110, 129, 620
  treatment 102
Pressure Sore Staging System 102
Pressure therapy 322, 104, 118
Procollagen 2
Prominence 81, 100, 101, 237, 348, 349, 374, 384, 385, 397, 402, 404, 478, 479
Prophylaxis 77, 455
Proplast® 66, 69
Propofol 40, 455
Prosthesis 46, 76, 114-116, 167, 213, 262, 275, 276
Provocative test 479, 480, 510, 568, 569, 575
Proximal interphalangeal joint (PIP) 81, 83, 462, 463, 465, 466, 476, 477, 487, 494, 512-514, 516, 533-540, 543, 545, 554, 555, 557-560, 563, 564, 595, 598, 599
  PIP ligamentous injury 559
Pruzansky classification 340, 341
Pseudoaneurysm 532
Pseudogynecomastia 413-415
Pseudomonas aeruginosa 73, 160, 294
Pseudoptosis 393, 418
Pterygoid 251, 253, 254
  lateral 251, 340
Pterygomaxillary fissure 329
Ptosis 260, 278, 279, 289, 290, 368, 369, 372, 378, 379, 381, 386, 392, 393, 403, 406, 410, 417-422, 452, 520, 604
Pulse oximetry 40, 445, 495
Punctae 178
Pyogenic granuloma 139, 140, 142, 481

R

Radial artery *see* Artery, radial
Radial clubhand 594, 595
Radial forearm 92, 167, 191, 210, 214, 315, 536, 563, 564, 608, 619, 622, 623
Radial *see* Nerve, radial
Radial styloid 471, 479, 499, 505, 507, 509, 510, 564, 587, 590
Radial-humeral joint 497
Radiated pelvis 98
Radiation 1, 13, 15, 65, 74, 95-98, 115, 118, 119, 128, 129, 140, 146, 147, 164-166, 176, 188, 258-261, 266, 271, 272, 278, 281, 283, 299-301, 314, 317, 318, 368, 406, 433, 441
Radiation and reconstruction 261
Radiation enteric fistula 97
Radiation injury 15, 95, 96, 97, 301
Radiation therapy 115, 118, 119, 128, 147, 166, 259, 261, 278, 281, 314, 317
Radiesse 448, 450
Radius 214, 462, 473, 479, 497-500, 502, 505-510, 525, 536, 585, 590, 592, 594, 595, 622, 624
Ramus condyle unit (RCU) 252, 328
Rapidly involuting congenital hemangioma (RICH) 139, 140
Raynaud's disease 89, 565
Raynaud's phenomenon 89, 589
Reactive fibrosis 96
Reactivity 6, 7
Reconstructive ladder 47, 48, 114, 133, 318, 564
Reconstructive surgery 43-48, 50, 65, 66, 80, 85, 130, 133, 141, 168, 178, 188, 193, 198, 259, 262, 263, 274-278, 295, 299, 304, 314, 318, 324, 341, 342, 403
Rectus sheath 263, 265, 268-270, 272, 297, 312, 611, 612
Recurrence 75, 85, 93, 98, 101, 102, 109, 117-119, 122, 127-130, 135, 180, 259, 261, 262, 298, 310-312, 366, 418, 422, 572, 579, 583-585, 594, 599, 600

Reduction 31, 38, 196, 211, 228, 233, 238, 240-242, 245, 246, 250-254, 256-258, 260, 271, 287-289, 291, 357, 384, 399, 402, 413, 415, 417, 440, 444, 458-461, 463, 499-501, 504, 505, 508, 510, 513-517, 532, 537, 540, 544, 554-556, 558, 584
Reduction mammaplasty 288
Reflex sympathetic dystrophy (RSD) 572, 600-605, *see also* Complex regional pain syndrome
Regnault classification 418
Regranex® 23, 84
Relationship 116, 143, 213, 216, 248, 251
Relaxed skin tension line (RSTL) 8, 125, 133, 210
Repair, delayed primary 541, 546
Replantation 27, 59, 60, 291, 322, 326, 492-496
  pediatric 495
  upper extremity 495
Resorbable plate 229, 230, 234, 240
Resorcin 441, 443
Restoration 43, 85, 92, 212, 240, 242, 255, 300, 309, 458, 459, 461, 495, 513, 531, 554
Restylane 448, 449
Resuscitation 39, 41, 154, 156, 158, 159, 162, 432
Retaining ligament 368, 371-373, 381
Reticular dermis 145, 437, 441, 444, 456, 562
Retinacular ligament 463, 534
Retracted ala 398
Retrobulbar hematoma 236, 391
Retrogenia 400-403
Retrognathia 338, 400-403
Retroorbicularis oculi fat (ROOF) 386, 389
Rhabdomyolysis 490
Rheumatoid arthritis 389, 474, 475, 544, 550, 567, 582, 586, 587
Rhinophyma 436, 455
Rhinoplasty 31, 38, 394-398
  closed 397
  columellar strut 397
  open 397
  open roof deformity 399
  tip projection 397, 398
Rhytidectomy 30, 31, 38, 44, 216, 217, 368-376, 430, 456
  anatomy 368
  composite 374
  deep plane 374, 456
  skin/subcutaneous 373
  SMAS/muscle 368, 369, 373, 374
  subperiosteal lift 374
  sub-SMAS dissection 372, 374, 376
Rhytides 377-379, 435, 436, 438, 439, 440
Rigid fixation 234, 241, 242, 251-254, 298, 493, 500, 501, 505, 511
Ring avulsion 494
Rotation 44, 52, 53, 61, 62, 108, 112, 191, 192, 200, 203, 211, 252, 295, 296, 301, 312, 344, 348, 353-355, 400, 401, 458-462, 478, 497, 505, 512, 513, 515, 516, 520-522, 547, 554, 562, 564, 578, 595

S

Saethre Chotzen syndrome (craniocephalosyndactyly) 337
Safe position 545, 554
Sagittal band 466, 474, 533, 535
Sagittal synostosis 334, 335
Salivary gland, minor 166
Salivary gland tumor 165, 166
Salvage procedure 261, 312, 347, 506, 507, 537
Scalp 4, 6, 8, 32, 33, 46, 62, 122-124, 129, 136, 172, 175, 190, 216, 219, 230, 369, 378, 379, 383, 458-461, 608, 609, 615
Scaphocapitate ligament 502, 511
Scaphocapitate syndrome 511
Scaphocephaly 334
Scaphoid 382, 462-465, 468, 470, 476, 479, 497, 501-511, 547, 567, 568

bone (navicular bone) 464, 465, 470, 502, 568
fossa 184, 382, 462, 497
nonunion 505, 506, 511
nonunion advanced collapse (SNAC) 506, 510
nonunion, treatment 506
ring sign 510
shift 479, 510
Scapholunate (SL) joint 479
Scapholunate advanced collapse (SLAC) 506, 510
Scapholunate interval 510
Scapholunate ligament 473, 501, 502, 507-509, 584, 585
Scar 2, 3, 6-11, 45, 46, 65, 67, 75, 76, 97, 100, 117-119, 122, 129, 140-142, 176, 181, 196, 188, 189, 193, 201, 206, 210, 220, 224, 235, 260, 267, 277, 278, 280, 285, 287-290, 292, 293, 315, 345, 346, 348, 354, 355, 358, 373, 375, 378-380, 384, 397, 398, 404, 409, 411, 415-420, 422, 424, 429, 430, 435, 436, 438, 441, 449, 454-456, 459-461, 490, 505, 506, 515, 521, 525, 527, 542, 545, 562, 581, 585, 594, 599, 622
atrophic 119, 435, 438
contracture 9, 10, 196, 355, 358, 490, 542, 594
hypertrophic 9, 45, 64, 117-119, 122, 375, 381, 384, 409, 411, 422, 429, 436, 455, 456, 490
immature 117, 119
linear hypertrophic 117-119
mature 117, 375
lengthening 11
reorientation 10
vertical 288-290, 292, 418-420, 429
Schirmer's tests 390, 393
Scleral show 235, 327, 386, 392, 393
Sclerotherapy 143
Screwdriver test 480
Scrotum, defects 314
Sculptra™ 342, 448
Sebaceous adenoma *see* Adenoma
Sebaceous (holocrine) gland 71, 96, 97, 123, 124, 130, 145, 150, 456
Sebaceous hyperplasia 124, 436
Seborrheic keratosis 121, 436, 455
Secondary intention 4, 72, 75, 100, 104, 111, 128, 129, 196, 198, 200, 298, 300, 305, 375, 461, 483, 549, 550
Secondary palate 350, 356, 357, 359, 367
Secondary procedure 203, 408, 421, 430, 447, 493, 495, 521
Sedation 38-42, 245, 284, 366, 378, 383, 403, 426, 431, 442, 443, 447, 455, 457, 459, 460, 570
conscious 38-41, 284, 378, 383, 455
intravenous 38-40, 455
Selection 10, 19, 39, 86, 105, 214, 248, 266, 386, 403, 406, 418, 423, 445
Sella 329, 330, 332
Sella-Nasion-A point (SNA) 329, 330, 332, 506, 510
Sella-Nasion-B point 329, 330
Semilunar line 312
Semmes-Weinstein monofilament 90, 568, 569
Sentinel node biopsy (SLNBx) 132, 133, 134, 259, 330, 332
Separation of parts 98, 312
Septal hematoma 246, 396
Septum 163, 188, 225, 228, 234, 235, 243-246, 319, 350, 387-391, 394, 396, 397, 399, 488, 525, 526, 564, 574, 578, 623, 624
Seroma 13, 64, 109, 146, 149, 271, 276, 277, 297, 307, 308, 410, 411, 415, 416, 421, 429, 431
Shea's modification 102
Shoulder and hand syndrome 603
Sialocele 166
Silicone 66-68, 117-119, 152, 153, 234, 278, 281
Silicone gel sheeting 118, 119

Silvadene® 20, 21, 97
Silver nitrate 160
Silver sulfadiazine 20, 21, 160, *see also* Silvadene®
Skier's thumb *see* Gamekeeper's thumb
Skin 2, 4, 6-11, 13-19, 23-28, 34-36, 46, 47, 49-53, 55, 59-64, 69-72, 75-79, 81, 83, 85, 86, 88-93, 95, 96, 98, 100, 102, 105-108, 110, 112-115, 117-133, 135, 136, 143, 145-153, 155, 156, 159, 162, 163, 167-178, 180-185, 187-192, 196, 198, 200, 201, 203, 206, 208, 210, 213, 214, 226, 227, 235, 260-264, 266-271, 274-281, 283-285, 288-293, 296, 297, 300, 301, 303-305, 307, 309-317, 320-323, 336, 340, 341, 344-348, 351, 353, 354, 368, 369, 371-377, 379, 381, 384, 386, 388-390, 392, 393, 396, 398, 404, 412, 414-416, 418-427, 429-431, 433-445, 447-452, 454-457, 460, 461, 464-467, 476, 480, 485, 488-490, 492-494, 499, 501, 525-527, 541-543, 546-556, 562-564, 570, 571, 582, 585, 593, 594, 596, 598-600, 602, 603, 607, 611-615, 617, 619-624
  cancer *see* Cancer, skin
  cooling 434
  flora 71, 72
  graft *see* Graft, skin
  lesion 121-125, 130, 390
    benign 121-125
  substitute 93, 145, 146, 150, 151, 153
  tag 123, 340, 341
Smoking 1, 9, 13, 75, 78, 86, 88, 272, 286, 349, 368, 375, 421, 425, 455, 565, 604
Snuffbox 35, 465, 479, 503, 510
Soft tissue coverage 65, 80, 214, 299, 301, 345, 347, 493, 494, 549, 561-564
Solar elastosis 368
Solar keratosis 121, 126, *see also* Actinic keratosis
Solar lentigine 436
Solar lentigo 440
Sourmelis and McGrouther technique 543
Spina bifida 304
Spine 111, 239, 243, 245, 246, 255, 256, 269, 300, 304, 307, 308, 329, 350, 351, 353, 354, 426, 519, 520, 576, 603, 607
Spinal cord 100, 102, 105, 108, 109, 304, 521, 601
  cord injury 100, 102, 105, 108, 109, 601
  wound 25, 300, 304-308
Squamous cell carcinoma (SCC) *see* Cancer, carcinoma
Staging 102, 126, 127, 134, 135, 164, 258, 603
Stahl ear 383
Staphylococcus aureus 21, 73, 76-78, 294, 410, 481-484
Staphylococcus epidermidis 76, 410
Staples 6, 149, 280, 292, 380, 420, 460
Static suspension procedure 221
Stem cell 144, 215
Stenosing tenosynovitis 465, 580, 581, 586-588, 595
Steri-strip® 8, 118, 119, 286, 292, 410
Stomodeum 348, 349
Stratum corneum 71, 122, 132, 440, 442, 443, 464
Strickland 545
Strip craniectomies 335
Striped Y classification 356
Sturge-Weber syndrome 140
Stylomastoid foramen 216, 369
Subcilliary approach 234, 235, 240, 390
Subcutaneous tissue 50, 61, 75, 76, 83, 102, 105-108, 123, 128, 129, 145, 148, 200, 206, 208, 263, 264, 266-270, 276, 280, 281, 283, 286, 287, 292, 309, 341, 353, 368, 369, 377, 379, 386, 424, 430, 446, 464, 486, 489, 550, 563, 568, 570, 578, 597, 602, 615, 621

Sublingual gland 166
Submandibular duct 165, 166, 371
Submandibular node 165, 196
Submandibular salivary gland 166, 371
Substitute 20, 25, 30, 46, 68, 93, 103, 145, 146, 150, 151, 153, 187
Subtarsal approach 235, 240
Subungual hematoma 515, 551, 552
Sulfamylon 160, *see also* Mafenide
Sunscreen 436, 456
Superficial musculoaponeurotic system (SMAS) 165, 216, 217, 368-370, 372-376, 378
Superficial palmar arch 464, 492
Superficial temporal fascia 369, 378-380
Superior orbital fissure (SOF) 231
Superior spinal wound 306, 307
Supramid® 66, 69
Supraorbital notch 34, 380
Supratarsal fold 387, 389
Supratrochlear vessel 230
Surgical instruments 552, 625-631
Surgilene 7
Suspensory ligament of Lockwood 231
Sutures 4-9, 11, 27, 57-59, 68, 69, 74, 78, 98, 109, 111, 118, 149, 182, 183, 185, 189, 191, 196, 203, 206, 208, 210, 219-222, 224, 231, 235-238, 246, 257, 270, 272, 275-277, 279, 280, 282, 285, 286, 288, 292, 293, 301, 303, 311, 315, 329, 334, 335, 337, 345, 354, 355, 374, 375, 379-381, 384, 385, 390-392, 397, 398, 403, 404, 420, 421, 426, 428, 429, 447, 459, 460, 494, 501, 515, 523, 524, 528, 535, 536, 538, 543-546, 549, 551-553, 572, 578, 590, 593, 612, 615, 619
  absorbable 6-8, 182, 185, 196, 210, 280, 292, 354, 380, 391, 593
  braided 7, 543
  catgut 6, 7
  gut 6, 7, 285, 447
  epitendinous 544, 546
  monocryl 7, 270, 292
  monofilament 6, 7, 90, 282, 544, 569
  multifilament 6, 7
  Mustarde mattress 384, 385
  nonabsorbable 6, 7, 68, 182, 185, 196, 270, 310, 404, 420, 426
  nylon 4, 6, 7, 149, 191, 286, 345, 354, 501, 524, 553, 569, 571
  PDS 6, 7, 282
  poliglecaprone 7
  polydiaxanone 7
  polyester 6, 7, 543
  polyethylene (PE) 66, 68, 69
  polyethylene terephthalate 68
  polyglactin 6, 7
  polyglycolic acid (PGA) 6, 7, 67, 68, 524
  polyglyconate 7
  poly(L-lactide) (PLLA) 67, 68
  polypropylene 6, 7, 459, 460, 572
  polytetrafluoroethylene (PTFE) 69
  prolene 6, 7, 66, 67, 270, 272, 292
  removal 6, 7, 191, 392
  silk 6, 7, 219, 590
  technique 4, 5, 11, 57, 494, 535, 543
    buried stitches 5, 6, 270, 426
    horizontal mattress 5, 6, 98, 403, 501, 544, 553
    intradermal 6, 118, 270
    running 4, 5, 6, 257, 270, 280, 292, 403, 426, 428, 459, 544, 612
    simple interrupted 4, 5
    subcuticular 5, 6, 280, 292, 428, 572
    vertical mattress 5, 6, 572
  vicryl 270, 292
Sweat gland 2, 75, 81, 96, 97, 145, 150, 166
Sympathetic blockade 602, 603
Sympathetically maintained pain syndrome (SMPS) 601-604
Symphalangism 592, 595
Syndactyly 43, 337, 592-594, 596
Syringoma 124, 436

Index

T

Tamai 492, 496
Tamoxifen 258, 259
Tangential excision 148
Tarsal plate 178, 180, 182, 234, 387-389, 391
Tattooing 279, 283, 286-288, 434, 455
Tear trough 371
Teeth 34, 212, 240, 241, 248, 249, 253-257, 324-328, 332, 340, 350, 361, 362, 401, 403, 404, 483, 627
  anatomy 324
  deciduous (primary) 324-326
  permanent 241, 324-326
Teflon® 66, 67, 69
Tegaderm™ 14, 15, 103, 149, 292
Telangiectasias 142, 368, 438
Telecanthus 232, 242, 245
Temporal fascia 216, 217, 237
Temporomandibular joint (TMJ) 69, 211, 214, 250, 252, 253, 256, 340, 341
  reconstruction 214
Temporoparietal fascia 216, 238, 345, 347, 369, 378, 379, 608
Tendon 16, 21, 35, 36, 66, 67, 80, 84, 85, 97, 112, 115, 146, 147, 178, 186, 210, 214, 231, 232, 242, 245, 320, 322, 360, 378, 380, 389, 463-466, 472, 476, 477, 480, 482-485, 491-496, 500, 507, 512, 513, 515, 521, 522, 525, 526, 533-546, 547, 548, 554-559, 562-565, 568, 572, 573, 580-584, 586, 587, 589, 590, 594-596, 615, 618, 622, 623, 624
  canthus 178, 186, 231, 232, 242, 245, 389
  distal extensor and flexor 556
  extensor 35, 463, 466, 472, 476, 477, 480, 483, 493, 494, 512, 513, 515, 533-538, 544, 545, 547, 548, 556-559, 564, 586, 587, 595, 596
    evaluation 477
    injury 476, 533-538, 548, 556-559
    juncturae tendinum 466, 477, 533
  flexor 464, 465, 476, 477, 480, 484, 485, 492-494, 507, 513, 534, 536, 539-543, 545, 546, 556, 558, 559, 582, 595
    injury 539-546, 548, 556-559
    "no man's land" 541
    repair
      four-strand repair 544
      Kleinert modification 535
      Okutsu "triple loop" method 544
    sheath 484, 539, 580, 581
  injury 476-478, 480, 512, 533-548, 554-559, 562, 572, 580
    post-traumatic adhesion 538
    zones 535, 541
  intrinsic healing 544
  lateral canthal 178, 231, 389
  medial canthal (MCT) 178, 231, 242, 245, 386, 389
  repair 67, 320, 493, 494, 512, 535-538, 541, 542, 544, 546, 542, 543, 545
    six-strand repair 544
  paratenon 146, 147, 534, 539, 562, 563
  pulley 465, 484, 539-541, 543, 544, 580-582, 584, 586, 596
    annular 465, 484, 539-541, 543, 544, 580-582, 586
    reconstruction 544
  retrieval 542
  rupture/avulsion 476, 477, 485, 512, 536-538, 544, 545, 556-559, 587
  palmaris longus 210, 464, 491, 525, 536, 545, 564, 568, 622, 623
Tennison repair 348
Tenolysis 495, 514, 536
Tenon's capsule 231
Tenosynovitis 465, 479, 482, 484, 485, 567, 571, 580, 581, 586-588, 595
  infectious 484

Tensor fascia lata (TFL) 51, 98, 105, 107, 108, 115, 297, 301, 311, 312, 320-322, 607, 609, 620
Terry Thomas sign 510
Tetracaine 29, 31
Tevdek 7
Thermal relaxation time (TRT) 433, 434
Thromboangiitis obliterans (Buerger's disease) 89
Thumb 34, 35, 136, 158, 337, 414, 456, 462, 463, 465-467, 470, 474, 476-479, 492, 494, 497, 503, 504, 507, 508, 510, 516, 517, 525, 533, 537, 540, 541, 550, 551, 554, 555, 559, 561-564, 566, 569, 570, 580, 582, 587, 589, 592-596, 622
  coverage 550, 561-563
  duplication 592
  laxity 478
Tibialis anterior or posterior 88, 115
Tinel's sign 220, 224, 479, 519, 527, 569, 575
Tissue
  engineering 215
  expander 61-63, 66, 261, 262, 275, 278-281, 283, 564
  expansion 61-64, 67, 175, 274, 275, 278, 280, 286, 300, 301, 347
    intraoperative expansion 62
TNM classification/staging system 126, 127, 134, 135, 164, 258
Tongue 163-165, 167, 197, 198, 220, 357, 366, 369
Tonsils 163, 165
Topical therapy 97
Torticollis 335
Total hip arthroplasty (THA) 114, 115
Total knee arthroplasty (TKA) 114, 115
Toxicity 21, 29-32, 66, 85, 89, 95, 432, 444, 567
Tragal pointer 216
Tragus 34, 217, 343-345, 371, 373, 374, 378, 382
TransCyte 151, 152
Transforming growth factor-β (TGF-β) 1, 96, 102, 117, 119, 334
Transparent film 14, 17
Transplant 46, 129, 203, 212, 215, 219, 223, 460, 461, 481
Transposition 52, 53, 61, 62, 105, 173, 191, 200, 219, 221, 270, 294, 296, 301, 305, 344, 420, 423, 527, 564, 577-579, 620
Transversalis fascia 263
Transverse carpal ligament (TCL) 464, 489, 525, 568, 570-572
Trauma 4, 6, 7, 13, 43, 46, 60, 61, 66, 76, 81, 86-88, 97, 138, 141, 155, 156, 161, 168, 169, 196, 198, 211, 217, 231, 236, 240, 241, 243, 250, 252, 254, 255, 260, 283, 299, 303, 309, 310
Traumatic wound 25, 299
Treacher Collins (mandibulofacial dysostosis) 337, 404
Tretinoin 436, 441, 442
Triamcinolone 118
Triangular Burow's 209
Triangular fibrocartilage complex (TFCC) 462, 480, 497
Trichiasis 187, 389
Trichilemmal cyst 122
Trichoepithelioma 124, 436
Trichofolliculoma 124
Trigger finger 580-582, 586, 596
Trigger thumb 595
Trigonocephaly 334
Triple-V incision 415
Trophic change 88, 601, 602, 604
Tuberous sclerosis 123
Tufted angioma 139
Tumor 73, 117, 123, 125-127, 129-136, 138-140, 142, 143, 163-166, 178, 180, 196, 211, 212, 217, 223, 258, 259, 261, 299, 303, 309, 393, 455, 567, 576, 584
  vascular 138-140, 142, 143
Turribrachycephaly 334
Two-flap palatoplasty 364, 365
Two-point discrimination 476, 487, 494, 495, 527, 540, 568, 576

U

Ulnar 35, 36, 158, 214, 300, 462-467, 474-480, 489-491, 497-499, 503, 507, 508, 510, 511, 513, 517, 519, 521, 522, 525-527, 529, 531, 533, 534, 539-541, 551, 555, 559, 562-564, 568, 570, 572, 574-579, 586, 587, 589-593, 596, 598, 599, 622, 623
- collateral ligament (UCL) 463, 474, 478, 555, 559, 592, 596
- nerve *see* Nerve, ulnar
- styloid 474, 564, 575
- variance 497-499
- humeral joint 497

Ultrasound 60, 143, 410, 414, 416, 430, 431, 505, 530, 540, 589
Ultraviolet radiation (UV) 128, 188, 368, 434
Unit 328, 346, 436, 441, 444, 448, 451, 462
Unna boot 75, 90
Upper extremity 69, 88, 158, 288, 318, 320, 467, 486, 495, 521, 527, 529, 532, 603, 604, 608, 609, 614, 615
Uvula 356, 359-362
- bifid 356, 361, 362

V

Vacuum-assisted closure (VAC) 16, 22, 24, 25, 84, 92, 104, 147, 153, 296, 298, 299, 305, *see also* Negative pressure wound therapy
Vagina 314-316, 609, 611, 618, 620
- defects 314, 317, 620, *see also* Vulvo-perineal surface

Valium 39
Vascular
- disease 77, 80, 87-89, 102, 113, 115, 267, 318, 389, 421, 422
- exam 82, 478, 589
- graft infection 111
- malformation 138-144, 567
- trauma 529-532

Vasculitide 89
VASER system 431
VATER syndrome 594
Velar closure 364
Velocardiofacial syndrome 338
Velopharyngeal 358-360, 362-364, 366, 367
- function 359, 362-364
- incompetence 366
- insufficiency (VPI) 364, 366
- mechanism 358

Venous 9, 13, 16, 18, 20, 27, 28, 41, 47, 59, 60, 68, 75, 87, 88, 90-93, 104, 123, 129, 138-140, 143, 147, 151, 156, 159, 203, 225, 268, 269, 336, 377, 406, 409, 487, 493-495, 602, 623, 624
- congestion 27, 28, 59, 60
- insufficiency 13, 87, 151

Vermilion border 163, 203, 206, 210, 351, 353
Versed 40, 42, 80
Vesicle 122, 143
Vessel 1, 25, 39, 49, 51, 52, 54-61, 80, 82, 85, 88-90, 93, 102, 105, 108, 111, 112, 140, 143-145, 148-150, 155, 194, 206, 214, 230, 261, 263, 264, 266-270, 272, 274-276, 297, 300, 307, 319, 362, 365, 376, 377, 387, 391, 424-426, 429, 434, 452, 609, 611, 615, 621
Vibrometry 568, 569
Vicryl 6, 7, 66, 67, 270, 292, 311
Vincula 534, 540, 543, 545
Virchow's law 334
Volar 462-466, 478, 479, 484-486, 488, 489, 491, 493, 497-502, 505, 506, 508-510, 512, 513, 517, 533, 534, 537, 544, 545, 548-550, 555-564, 571, 577, 580, 582, 584, 593, 596, 623
- intercalated segmental instability (VISI) 473, 478, 508-510
- interossei 463, 478, 534
- plate 463, 466, 500, 501, 512, 517, 533, 537, 544, 555-560, 596
  - rupture 512, 559, 560
- tilt 497-501

Volkman's contracture 490
Vomer bone 188, 243
von Langenbeck procedure 362, 363, 366
Vulvo-perineal surface defects 314

W

Wallerian degeneration 523
Wartenberg sign 575
Warthin's tumor 166
Watson scaphoid shift test 479, 510
Webster incision 415
Wedge and star-wedge excisions 169, 170, 201, 202, 287
Wegener's disease 89
Wegener's granulomatosis 389
Wetting solution 39
Wharton's duct 166, 371
White roll 193, 194, 196, 199, 201, 210, 351, 353, 354
Whitnall's ligament 387
Whitnall's tubercle 231, 391
Witches chin 403
Wounds 1-4, 6-9, 11-28, 30, 43, 52, 60, 63, 65, 67, 70-76, 78-93, 95-98, 100, 102-104, 106, 109-118, 122, 128, 129, 133, 144-155, 159, 160, 162, 170, 180, 182, 189, 190, 196, 199, 211, 225, 229, 241, 252, 261, 262, 276, 283, 286, 294-301, 303-313, 318-323, 346, 366, 375, 381, 384, 385, 392, 398, 410, 422, 425, 435, 438, 441, 460, 481-485, 490, 499, 518, 521, 529, 536, 538, 540, 542, 546-550, 552, 556, 561, 563, 564, 570, 572, 582, 590
  care 1-3, 20-23, 84, 85, 93, 96, 100, 103, 104, 109-111, 115, 147, 152, 159, 160, 162, 298, 299, 536, 549
  classification 102, 294
  closure 3, 4, 6, 8, 9, 11, 12, 14, 24, 71, 78, 97, 116, 133, 144, 150, 152, 153, 182, 276, 298-300, 303-305, 307, 309-311, 318, 320, 366, 384, 572, 561, 590
    emergent 310
    improper wound edge alignment 8
    primary closure 4, 8, 9, 11, 78, 133, 182, 310, 311
      delayed 4, 78
    vacuum-assisted closure (VAC) 16, 22, 24, 25, 84, 92, 104, 147, 153, 296, 298, 299, 305
  chronic 12, 16, 20, 23, 24, 71, 73, 92, 93, 96, 109, 129, 147
  contaminated 3, 8, 72, 78, 104, 311, 313, 546
  diabetic 80-86, 91, 93
  healing 1-3, 6-9, 12, 13, 18-21, 24, 72, 75, 78, 80-82, 85, 86, 93, 96, 102, 109, 113, 114, 147, 262, 283, 301, 441
    delayed 12
    inflammatory phase 1-3, 12, 72, 544
    remodeling phase 1, 2, 12, 544
    proliferative phase 1, 2, 12, 139, 544
  tensile strength 2, 6, 25, 96, 310, 311
  spinal 300, 304-307, 308
  sternal 294-298, 300
  therapy 24-26, 84, 104, 115, 310, 322
Wrist 36, 243, 462, 464-467, 473, 474, 476-480, 486, 488-491, 493, 497-499, 501-511, 516, 517, 519, 520, 525, 526, 529, 531, 533, 535, 536, 540, 545, 563, 564, 566, 568-573, 575, 578, 583-587, 589, 590, 594, 608, 622, 623
  anatomy 462-465, 486, 502, 503, 533, 568
  fractures *see* Fractures, wrist
  injuries 535
    rotatory subluxation 509, 510, 567
  instability 508, 509
  proximal row carpectomy 507, 511
  wrist/brachial index 529

X

Xanthelasma 436
Xenograft 146
Xeroform™ 15, 17, 103, 149, 160, 282

Z

Z-plasty 9-11, 53, 54, 118, 170, 177, 179, 182, 183, 185, 186, 196, 363, 364, 460, 578, 593, 596
  double-opposing 363, 364
Zofran 39, 372, 455
Zyderm 449
Zygomaticotemporal branch 377
Zyplast 449